Global Trends in Health, Technology and Management II

Anil Kumar Saxena • Aaruni Saxena
Editors

Global Trends in Health, Technology and Management II

Proceedings of the 2nd International Symposium GTHTM-2025

Springer

Editors
Anil Kumar Saxena
Global Institute of Pharmaceutical
Education and Research
Kashipur, India

Aaruni Saxena
Nottingham City Hospital, University
of Nottingham
Nottingham, UK

International Symposium on Global Trends in Health, Technology and Management, gthtm, gthtm2025, GTHTM, Uttarakhand, India, 20250215, 20250217, https://conference.ghtmf.com

ISBN 978-3-032-12319-0 ISBN 978-3-032-12320-6 (eBook)
https://doi.org/10.1007/978-3-032-12320-6

This Springer imprint is published by the registered company Springer Nature Switzerland AG
The registered company address is: Gewerbestrasse 11, 6330 Cham, Switzerland

Preface

The International Symposium on Global Trends in Health, Technology, and Management (GTHTM-2025) was organized by the Global Health Techno Management Forum (GHTMF) and the Global Institute of Pharmaceutical Education and Research (GIPER), Kashipur, from February 15 to 17, 2025, at Tiger Camp Resort, Jim Corbett. This event was in continuation of previous international symposia, including GTHTM-2024, Drug Design and Development Research (DDDR-2021), Current Trends in Pharmaceutical and Medical Sciences (CTPMS-2020), Computational Methods in Toxicology and Pharmacology Integrating Internet Resources (CMTPI-2017), and the International Seminar on Pharmaceutical Education and Research (ISPER-2010).

The GTHTM-2025 symposium, conducted in a hybrid (online and offline) mode, commenced with the welcoming of the Chief Guest, Prof. (Dr.) Y. K. Gupta, and Guest of honor, Prof. (Dr.) Diwan S. Rawat (Kumaun University, Nainital), along with participants from 10 different countries. The inaugural session featured keynote addresses by the Chief Guest and Prof. (Dr.) Soniya Nityanand, Vice-Chancellor of KGMU, Lucknow. A highlight of the session was the release of the proceedings of the first GTHTM-2024 symposium, published by Springer.

The symposium featured 21 invited/plenary lectures and several oral and poster presentations, covering a diverse range of topics related to healthcare technologies, including researches on drug design and discovery related to cancer, Alzheimer's disease, microbial and tropical diseases, lifestyle disorders, as well as emerging areas like artificial intelligence, machine learning, environmental protection, green agriculture, and the study of phytochemicals and natural products.

The GTHTM-2025 proceedings comprise 30 chapters as the main content contributed by researchers from various domains in the symposium. The first 14 chapters deal with the medicinal chemistry covering new approaches/strategies involved in drug design and discovery, synthesis SARs, QSARs, and docking studies of molecules for different diseases. The rest 16 chapters describe the phytochemicals, natural products, and general topics related to healthcare technologies. In addition, the messages and details of the organizing and advisory committees are included in front matter, while abstracts of eight invited lectures, one best and other poster presentations uncovered in 30 articles in the back matter of the proceedings.

It has been a great privilege to edit and compile this volume along with my co-editor, Dr. Aaruni Saxena. I sincerely thank him, as well as all the contributing authors, for their dedication and efforts in preparing their chapters. My gratitude also goes to the reviewers for their timely and constructive feedback. I am especially thankful to the series editors, particularly Prof. (Dr.) Gunda Georg, for their continuous support and cooperation.

Lastly, I extend my heartfelt thanks to Sofia Costa, Hemavathy Manivannan, and the entire Springer team for their invaluable assistance. I hope this volume serves as a meaningful resource for researchers working in the field of health and techno-management.

Kashipur, Uttarakhand, India

Anil Kumar Saxena

Patron

Prof. (Dr.) Anil Kumar Saxena, Hon'ble Chairman
Global Institute of Pharmaceutical Education and Research, Kashipur, India

Organizing Committee

Chairman

Prof. (Dr.) Deepak Teotia

Organizing Secretary

Prof. (Dr.) Sisir Nandi

Co-organizing Secretary

Sarfaraz Ahmed

Members

Arun Kumar, Rajan Kaushik, Muneer Alam, Swati Tamta, Anjali Negi, Shahjad, Divyanshu Kumar, Vineeta Pandey, Harsh Sharda, Raja, Adil, Sarfaraj, Kuldeep Singh, Rahul Kumar, Fareed, Puneet Joshi

International Advisory Committee

Prof. Alain Carpy (CNRS, Bordeaux, France)
Prof. Athina Geronikaki (Aristotle University, Greece)
Prof. Emil Reisinger (University Medical Center, Rostock, Germany)
Prof. Esin AKI-YALÇIN (Ankara University, Faculty of Pharmacy, Ankara) Dr. Imran Ahmad (Jina Pharmaceuticals Inc, USA)
Prof. Ismail YALÇIN (Ankara University, Faculty of Pharmacy, Ankara) Dr. James Devillers (CTIS, France)
Prof. Johann Gasteiger (Friedrich-Alexander-Universität, Germany) Prof. Krishnan Balasubramanian (Arizona State University Tempe, AZ) Prof. Lakshmi P. Kotra (University of Toronto)

Ludy Suryantoro (Team lead, Multisectoral engagement, WHO, Switzerland) Prof. Marjan Vracko (National Institute of Chemistry, Slovenia)
Prof. Om Prakash (Kansas State University, KS)
Dr. Omer Saka (EY Parthenon Health Sciences and Wellness, Switzerland) Prof. Subhash C. Basak (University of Minnesota, USA)
Prof. Dr. med. Thomas Hohlfeld (University of Düsseldorf) Dr. Ved Srivastava (Chemistry Intarcia Therapeutics Inc, USA)
Prof. (Dr.) Vladimir Poroikov (Institute of Biomedical Chemistry, Russia)

National Advisory Committee

Dr. Alok Saxena (Advisor, GTC, Vorro Technology, United States) Prof. B. Jayaram (IIT, New Delhi)
Dr. Dinesh Patel (MD and CEO, Themis Medicare, Mumbai) Prof. D.S. Rawat (Vice Chancellor, Kumaun University, Nainital) Prof. Indira Ghosh (Ex- Professor, SCIS JNU, New Delhi)
Prof. Kunal Roy (Professor and Ex-Head Pharmaceutical Technology, Jadavpur University, Kolkata)
Dr. Madhu Dikshit (Ex-Director, CSIR-CDRI, Lucknow)
Dr. O.N. Tripathi (President, The Academy of Life Sciences, Lucknow)
Prof. Poonam Tandon (Vice Chancellor, Deen Dayal Upadhyaya Gorakhpur University, Gorakhpur)
Prof. P.V. Bharatam (Professor and Head, NIPER, Mohali)
Prof. Dr. Rashmi Kumar (Ex- Professor and Head, Department of Pediatrics, KGMU, Lucknow)
Prof. Dr. Sandeep Kumar (Professor & Founder Director AIIMS Bhopal) Dr. Sanjay Batra (Chief Scientist, CSIR-CDRI, Lucknow)
Prof. Shailendra K. Saraf (Dean (Pharmacy), Dr. APJAKTU & Director (Pharmacy), BBDNIIT, Lucknow)
Prof. Shubhini A. Saraf (Director, NIPER, Raebareli)
Prof. Soniya Nityanand (Vice Chancellor, KGMU, Lucknow) Prof. T.K. Chakraborty (IISC, Bangalore)
Dr. Vineet Kumar (Scientist-G, Chief Scientist and Professor, Forest Research Institute, Dehradun)
Dr. Yatendra Kumar (Ex-President, Wockhardt)
Prof. Y. K. Gupta (Ex- Head, Dept. of Pharmacology, AIIMS, New Delhi)

वीर माधो सिंह भण्डारी
उत्तराखण्ड प्रौद्योगिकी विश्वविद्यालय
(उत्तराखण्ड सरकार द्वारा अधिनियम 415/2005 द्वारा स्थापित पूर्ववर्ती उत्तराखण्ड तकनीकी विश्वविद्यालय)
VEER MADHO SINGH BHANDARI
UTTARAKHAND TECHNICAL UNIVERSITY
(Formerly Uttarakhand Technical University Established by Act no. 415/2005 by Uttarakhand Government)

Prof. (Dr.) Onkar Singh
Vice Chancellor
प्रो० (डॉ०) ओंकार सिंह
कुलपति

Office Address :
Chakrata Road, P.O. Chandanwari, Premnagar, Suddhowala
Dehradun - 248007, Uttarakhand (India)
Phone No. : 0135-2774068,
Email : vc@uktech.ac.in, Website : www.uktech.ac.in

Ref. No. 223/VC/UTU/25 Date. 04-02-2025

Message

I am delighted to know that Global Institute of Pharmaceutical Education and Research (GIPER), Kashipur is organizing an international symposium on "Global Trends in Health, Technology and Management" (GTHTM-2025) during February 15-17, 2025. This conference will provide a platform for brain storming among the leading scientists, young researchers, academicians and students globally. The discussions will pave the way for the advancements in interdisciplinary fields of drug design, discovery and development, innovations and future directions in pharmaceutical sciences and help in the sustainable development.

I congratulate the organizers and extend my best wishes for the successful organization of this symposia.

04/02/25
(Prof. Onkar Singh)
Vice Chancellor

Dr. Dhan Singh Rawat

Minister

Medical Health and Medical Education, Co-operative, Higher Education, Sanskrit Education, School Education

Govt. of Uttarakhand

Vidhan Sabha Bhavan
Dehradun

Room No. 18

Ph. 0135-2666410 (Off.)

0135-2666411 (Fax)

Mob. 9412370255

MESSAGE

It is my great pleasure to note **that Global Institute of Pharmaceutical Education and Research (GIPER), Kashipur** is organizing an international symposium on **"Global Trends in Health, Technology and Management" (GTHTM-2025) from 15 to 17 February, 2025.**

It makes an excellent international platform of brain storming scientific sessions for the leading scientists, young researchers, academicians and students globally to discuss the interdisciplinary fields of drug design, discovery and development research. Innovations and Future Directions in Pharmaceutical Sciences to attain the sustainable development goals for the human benefit.

I thank the organizing committee and warm wishes for the success of this symposia.

(Dr. Dhan Singh Rawat)

GLOBAL INSTITUTE OF PHARMACEUTICAL EDUCATION & RESEARCH (GIPER)
D. PHARM, B.PHARM, M. PHARM
Approved by : AICTE & PCI, New Delhi
Affiliated to : UBTE Roorkee and Uttarakhand Technical University, Dehradun

Ref. No. .... Date 12/02/2025

Message

It is with great pleasure that we announce that the Global Institute of Pharmaceutical Education and Research (GIPER), Kashipur along with Global Health Techno Management Forum (GHTMF) is organizing the 2nd "International Symposium on Global Trends in Health, Technology, and Management" (GTHTM-2025) in the scenic and serene surroundings of Jim Corbett National Park at The Tiger Camp, Ram Nagar, Uttarakhand during February 15-17, 2025.Building on the tremendous success of GTHTM-2024 where scientists from across the globe including WHO delegates from Switzerland actively participated, the upcoming symposium aims to further extend the initiated valuable discussions. It will also address a wide range of topics, spanning drug design and discovery in the fields of pharmaceutical, medical and biomedical sciences, along with newer technologies involving artificial intelligence in stem cell therapy, microbial resistance, digital and public health.

We are pleased to announce the successful publication of the proceedings from the "International Symposium on Global Trends in Health, Technology, and Management" (GTHTM-2024) by Springer Nature. This achievement reflects the high-quality research and discussions from the symposium, where experts addressed global challenges in health, technology, and management. The proceedings highlight innovative ideas and reinforce the symposium's global impact, serving as a valuable resource for scholars and professionals in these fields. We are excited to continue this collaboration with Springer Nature for GTHTM-2025. This continued collaboration with a prestigious publishing house underscores the significance of the event and its contributions to advancing knowledge in these critical areas.

This International Symposium serves as a venue to unite stakeholders from diverse disciplines to share ideas and work together for societal progress. I hope that the discussions here will focus on translating the advancements in these areas into tangible benefits for the general public, improving health and the environment, with GIPER and GHTMF playing a key role in this effort.

I warmly welcome all participants and look forward for their active involvement in ensuring the success of this event. I also wish to express my gratitude to the distinguished scientists, educators, business leaders, and professionals for honouring us with their presence. My sincere appreciation goes to the organizing committee, staff, and students of GIPER, Kashipur, for their dedicated efforts. I am confident that this event will be a great success.

Anil Kumar Saxena

Dr. Anil Kumar Saxena
Chairman

Contents

Innovations in Drug Design: Case Studies from Discovery to Development

Anil Kumar Saxena[1](✉)

[1] Global Institute of Pharmaceutical Education and Research, Kashipur, India
anilsak@gmail.com

Abstract. Drug design is a critical aspect of drug discovery and development, a complex, costly, and time-consuming process aimed at identifying promising drug candidates for human therapeutics. The process consists of two main stages: first, the identification of both the target and the lead molecules, and second, the more challenging step of optimizing the lead molecule to develop a candidate drug. Approaches to drug design are essential in both discovering new leads and optimizing existing ones. Traditionally, lead identification relied on natural products, understanding disease mechanisms, random screening, and human patron recognition. However, the advent of Computer-Aided Drug Design (CADD) techniques has significantly eased the challenges in drug discovery. These methods, based on ligand- and structure-based approaches, involve developing predictive models using tools like QSAR (Quantitative Structure-Activity Relationship), pharmacophores, and docking studies with CAMM (Computer-Aided Molecular Modeling Techniques). These models enable efficient virtual screening, facilitating the identification of lead molecules from vast molecular databases, including focused libraries. CADD also shows promise in repositioning existing generic drugs, which could be particularly beneficial for the Indian pharmaceutical industry. Once suitable leads are identified, they can be further optimized using CADD to address challenges encountered during preclinical and clinical trials, ultimately providing effective drug candidates and new leads. The approaches significantly shorten the timeline from the initial identification of the lead molecules to the selection of the candidate drugs and have been successfully applied in the development of neuroleptics, antihypertensive, anti- Alzheimer's, and antitubercular agents.

Keywords: SBDD · QSAR · QSIAR · Molecular hybridization · Virtual screening

1 Introduction

The pursuit of discovering new chemical entities (NCEs) as drugs necessitates comprehensive study. It demands significant time and financial resources and requires the amalgamation of chemistry and biology. The latest estimates indicate that the duration for a new chemical to enter clinical trials is around 8.3 years, with a median expenditure of $985 million [1], and in terms of time to reach market it may take a maximum of

A. K. Saxena, A. Saxena (Eds.): GTHTM 2025, pp. 1–18, 2026.
https://doi.org/10.1007/978-3-032-12320-6_1

12 years [2, 3] and a total cost of 2.8 billion dollars [4]. The early phase of the drug discovery during the preclinical development incorporates identification and validation of the target, hit discovery, and optimization, ultimately resulting in the selection of a candidate drug for further drug development including both short- and long-term toxicity assessments and phase 1, 2, and 3 clinical trials, before its marketing approval [5]. The drug discovery phase, a critical component of innovation, historically relied on serendipitous discoveries or traditional trial-and-error methods, often necessitating decades and substantial resources with minimal certainty of success [6]. Both time and financial resources require conservation, prompting the research and development divisions of pharmaceutical companies to explore solutions to these constraints through developments in computational technology and molecular biology in the late 20th century.

The most crucial step in the design of the molecule is the innovation of chemical scaffold/sub-structure(s) followed by their annelation and incorporation leading to the new prototypes which act as lead molecules and the process is termed as lead generation. The sub-structural concept and pattern recognition in the identification of essential structure requirements (sub-structural/ pharmacophore) and their incorporation into the new novel chemical entities has always been behind the mind of medicinal chemist in designing the novel molecules. This is followed by the optimization of observed activity of the lead molecules through structure-activity relationship (SAR) and quantitative structure-activity relationship (QSAR) studies. The incorporation of different sub-structures for the same activity have led to molecules with increased activity and is termed (sub-structural analysis). Drug discovery has significantly advanced through substructural analysis and pattern recognition, which aid in identifying essential structural requirements (sub-structure/pharmacophore). These insights are then integrated into designed molecules to enhance their activity, utilizing ever evolving contemporary methods and resources, including computer-aided molecular modeling (CAMM) and quantitative structure-activity relationships (QSAR). While, incorporation of sub-structures with different biological activities into one molecule have led to molecules with the profile of two or more activities depending upon the activities of the incorporated sub-structures and is termed as medicinal chemical hybridization (MCH) also known as molecular hybridization [7, 8]. These approaches gained momentum with the advancements in molecular biology including genomics, proteomics and computer aided drug design (CADD) techniques. The CADD includes the state-of-art techniques viz. ligand and structure-based approaches coupled with virtual screening tools. Their integration in drug design has revolutionized the drug discovery. By using the state of the art computational algorithms and software, scientists can simulate and predict interactions between drug candidates and biological targets, making the drug design process significantly more efficient and cost- effective. The emergence of computational tools in drug design has resulted in significant enhancements in the speed and precision of novel drug development. These methodologies not only reduce the time from discovery to market but also improve the overall understanding of complex biological processes (Fig. 1). The rational design of drugs using computational modelling has set new standards in targeting specific and personalized medicine. One of the most prominent applications of CADD is virtual screening which uses computational power to screen large chemical libraries for quick identification of the compounds most likely to bind effectively to a target protein. The

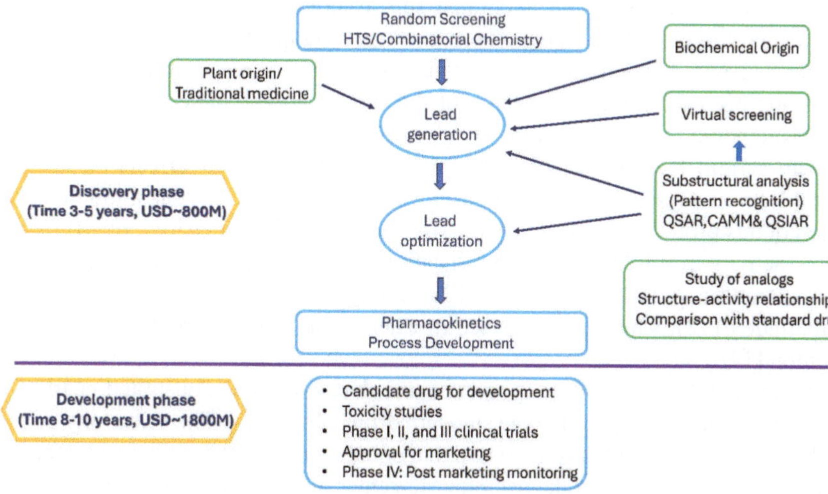

Fig. 1. The components of CADD used in drug discovery phase.

automation of screening techniques has revolutionized the identification of active compounds by allowing researchers to evaluate thousands of potential drug candidates rapidly. Recent years have witnessed an increase in machine computing capacity, improvements in machine learning, and the accrual of omics data [9].

Recently, the advent of artificial intelligence (AI) has revolutionized [10, 11] the identification of hit and lead compounds, as well as expedited the validation of drug targets, hence enhancing optimization in drug design and discovery [12, 13]. These advances have collectively transitioned the field from traditional empirical methods to a modern, technology-driven paradigm that emphasizes precision and efficiency in drug design.

2 Case Studies

The above discussed concepts have been applied in the design and optimization of biological activity and some of such case studies from our own work are as follows.

2.1 Design and Synthesis of Dopamine D2 Antagonists as Neuroleptic Agents

The concept of substructural analysis and pattern recognition is crucial for identifying essential structural requirements, such as sub-structures and pharmacophores. Incorporation of these elements into designed molecules followed by optimization of their activity through CADD, including quantitative structure-activity relationships (QSAR) and computer-aided molecular modelling (CAMM), has been beneficial in drug discovery. The primary method for structural modifications aimed at generating new prototype molecules involves the creation of rigid analogs which stabilize specific preferred conformations of the integrated sub-structures. Such molecules are sometimes new prototypes which may be used in integrating the active pharmacophores for the desired activity keeping in view the essential physicochemical properties.

This approach facilitated the design of the new prototype molecule octahydropyrazinopyridoindole **(1)** by integrating two significant scaffolds, tryptamine and piperazine, into a rigid framework (Fig. 2) which are linked to their effects on the CNS (Central Nervous System) and CVS (Cardiovascular System). The substitution at the 2 position of this prototype by an active pharmacophore (p-fluorobenzoylpropyl) for neuroleptic activity resulted in the identification of a new neuroleptic drug, 2-γ-(4-fluorobenzoylpropyl)-1,2,3,4,6,7,12,12a-octahydro(2′,1′:6,1)pyrido(3,4-b)indole (Centbutindole) **(2),** which was licensed to Chemosyn, Mumbai in 1987 for commercialization [14].

The compound centbutindole (generic name: biriperone) demonstrated nearly five-fold greater potency compared to the standard drug haloperidol. The selectivity of centbutindole for 5HT2a/D2 (1.24) surpassed that of haloperidol (0.61) and was comparable to risperidone (1.61). Due to the asymmetric center at C12a, the two enantiomers (−) C-12aS and (+)C-12aR were synthesized and assessed for their neuroleptic activity and dopamine (D2) receptor binding. The (−)C-12aS isomer exhibited a selectivity that was 1200 times greater than that of the corresponding (+)C-12aR isomer [15–17].

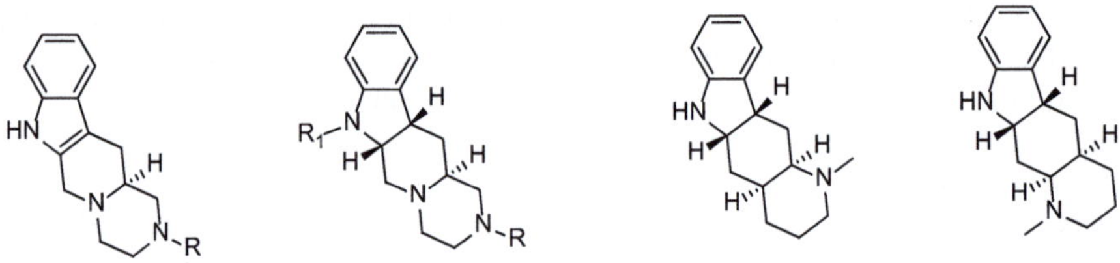

1: R=H
2: R=$(CH_2)_3COC_6H_4$-4-F

3: R=R1=$(CH_2)_3COC_6H_4$-4-F

4

5

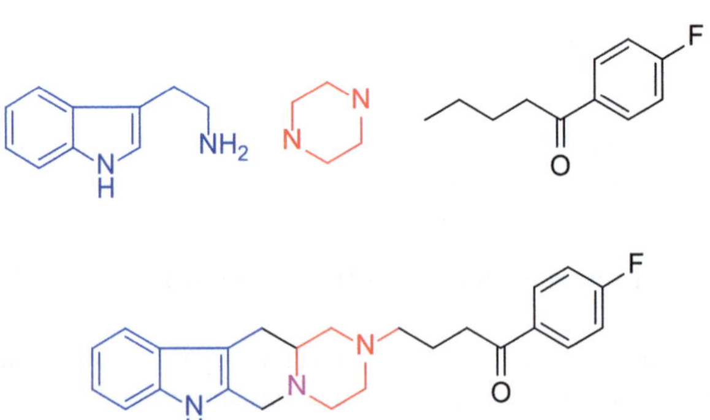

Fig. 2. Integration of different cores in a single molecule.

The analysis of electrostatic and molecular lipophilicity potential maps via conformational analysis using CAMM, in comparison with other dopamine (D2)-receptor antagonists, elucidated the distinctions of centbutindole from other neuroleptics and facilitated the identification of potential additional benefits of centbutindole relative to existing neuroleptics [18]. Further the QSAR studies, in the centbutindole analogs (Eq. 1) indicated the variation in activity was described by the parabolic relationship with lipophilic parameter and the positive contribution of an indicator variable In1 describing the importance of pharmacophoric feature where aromatic ring is separated from the neuroleptic N by 4 carbon atoms [19].

$$\log 1/C = -1.592(\pm 0.313) + 0.317(\pm 0.430)\pi_R - 0.115(\pm 0.105)(\pi_R)^2 + 1.270(\pm 0.389)\,\mathrm{In1} \quad (1)$$

$$N = 18, R = 0.742\,EV = R^2 = 0.550, S = 0.579, F = 5.73$$

These studies resulted in the design and synthesis of an additional equipotent neuroleptic molecule, 2,7-di-γ-[4-fluorobenzoylpropyl]. -1, 2, 3, 4, 6, 6a, 7, 11b, 12, 12a-decahydropyrazino[2′,1′:6,1] -pyrido(3,4-)indole **(3)** which demonstrated a superior profile compared to centibutindole regarding 5HT2a/D2 (1.41) selectivity. The comparison of the parameters defining the nitrogen coordinate relative to the catechol ring in (−) 6aR–apomorphine, a dopamine agonist, and (−)12aS-centbutindole revealed similar orientations of the attached hydrogens which suggested that substituting the catechol portion of apomorphine with the indole moiety could result in the development of novel dopamine agonists. The hypothesis was validated by synthesizing analogs **4** and **5** which exhibited significant dopaminergic activity in the reserpine-induced rigidity test (ED_{50} = 2.35 and 3.11 mg/kg) and the catatonia test (ED_{50} = 6.78 and 8.05 mg/kg), respectively [20–25].

2.2 Design and Synthesis of Anti-Hypertensive Agents

The approach of 'freezing' the preferred conformation of established active compounds and/or integrating active sub-structures into new prototypes (medicinal chemical hybridization-MCH) has been utilized in various processes which has resulted in the design, synthesis, and analysis of numerous molecules from various heterocyclic systems. These synthesized molecules were examined for their structure-activity relationships which included both 2D and 3D QSARs. This approach has led to the identification of various new chemical entities (NCEs) currently at different phases of drug development, including anti-hypertensives such as 1-(3-methylphenyl)-4-(2-β-quinolylethyl) piperazine **(6)** (Centhaquin) and 1-[(3-methylphenyl)piperazin-1-yl]-3-[thio(4-acetamido)phenyl]propane **(7)** [26–28]. Centhaquine marketed as (Lyfaquin®) has recently been used as resuscitative agent in the treatment of hypovolemic shock [29]. Both compounds **6** and **7** functioned as centrally acting hypotensive agents, resulting in a 57% and 51% reduction in blood pressure in anesthetized cats following both intravenous and intraduodenal administration; however, compound **6** exhibited a longer duration of action compared to compound **7**.

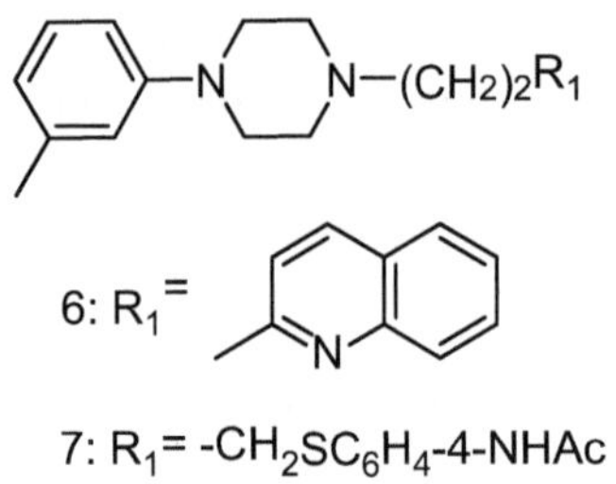

Hypotension was associated with an enhanced response to intravenous epinephrine in the majority of experiments, while the nictitating membrane response remained unchanged. The responses to intravenous acetylcholine, histamine, and isoprenaline remained largely unchanged. Given the observation that **6** did not exhibit a significant effect when administered to spinal transected cats, yet demonstrated activity in decerebrated cats, it suggested that **6** primarily functions through a central site of action and was more potent than **7**.

The formulation of QSAR in the context of biological activity data derived from whole animal experiments is complex and includes both pharmacodynamics and pharmacokinetics. The in vivo effects of the compounds were ascertained through complex and non-linear relationships that were time and dose-dependent. The precise nature of these relationships is typically ascertainable only under specific experimental conditions and is generally considered to be unknown. The incorporation of cross product terms in regression analysis enhances the complexity of data analysis and interpretation. Thus, the back propagation – Artificial Neural Network (ANN) methodology was utilized for the first time to elucidate the time-dependent hypotensive effects observed in cats, using physicochemical parameters and time as independent descriptors. The hydrophobicity and electronic parameters of the substitutes account for the observed variance. The back propagation methodology of artificial neural networks was employed for the first time to clarify the time-dependent hypotensive effects observed in felines, utilizing physicochemical parameters and time as independent variables. The hydrophobicity and electronic properties of the substitutes explained the observed variance [30].

2.3 Case Studies on AChE

Alzheimer's disease (AD) is defined by the deterioration of memory and cognitive functions in elderly individuals, impacting a significant global population [31, 32]. Acetylcholinesterase (AChE, E.C.3.1.1.7) is crucial for the hydrolysis of the neurotransmitter acetylcholine following its release [33]. The predominant therapeutically accessible pharmacological agents for Alzheimer's disease (AD) treatment are acetylcholinesterase inhibitors (AChEIs). These medications offer only symptomatic relief and experience a decline in therapeutic efficacy over time [34]. Consequently, there is an urgent requirement for novel cholinesterase inhibitors that possess a broader therapeutic window for the treatment of Alzheimer's disease. Recent researches have primarily concentrated on strategies that target the AChE enzyme, as well as additional targets such as butyrylcholinesterase (BChE), amyloid-β (Aβ), β-secretase-1 (BACE), metal, antioxidant properties, and free radical scavenging capacity [35]. An innovative approach to the production of new chemical entities (NCEs) entails the creation of hybrid molecules that incorporate sub-structures with recognized pharmacological characteristics into a single scaffold [36]. The primary sub-structures utilized in the development of these molecules originate from various chemical classes, including carbamates, acridine, huperzine, quinoline and other heterocyclic analogs. Subsequent optimization of activity has been achieved through structural modifications of prototype molecules to develop the structure-activity relationship (SAR). This methodology has resulted in the creation of new molecules. Substituted 1,2,3,4-tetrahydroquinolin-6(or-7)-yl carbamates are examples of molecules developed through the methodology with AChE inhibitory activity along with additional beneficial pharmacological characteristics. These compounds **(8–11)** have been reported as anti-Alzheimer's agents [37–40] and were developed by systematic virtual screening (VS) experiments, which included the development of 3D-pharmacophore models, the screening of a virtual library, synthesis, and pharmacological evaluation. The initial lead compound 2-chlorophenylcarbamic acid 1-benzyl-1,2,3,4-tetrahydroquinolin-6-yl **(8)** [41] showed IC_{50} value 3.31 μM [38]. The modifications at targeted site-1 and site-2 of lead compound **8** due to the substantial aromatic characteristic at the active site gorge and the primary hydrophobic interactions of AChE with its inhibitors. The propargyl and 2,4-dichlorobenzyl groups, which exhibited higher hydrophobicity, were identified as the most suitable groups for site-1 as R_1. The significant modifications at site-2 of the lead compound **8** involved the incorporation of both substituted aromatic and long-chain (C6–C8) alkyl groups, resulting in the development of a series of novel compounds **9** and **10**. Among these three molecules the most promising compound, **9** (AChE, $IC_{50} = 0.70$ μM) showed almost comparable AChE inhibitory activity with rivastigmine (*m*AChE $IC_{50} = 1.11$ μM) during in vitro studies in Swiss albino mice brain (*m*AChE) [39]. Interestingly, the docking studies in the substituted 1,2,3,4-tetrahydroquinolin-7-yl carbamates analogs revealed that analog **10** was more potent than **9** [40]. The molecule **10** (*m*AChE $IC_{50} = 0.10$ μM), which contains a benzyl group on the carbamate end, demonstrated an 11-fold enhancement in AChE inhibitory activity relative to the established drug rivastigmine, and exhibited comparable or slightly superior AChE inhibition when compared to tacrine. Compound **10** demonstrated potential for enhancing learning in mice, as indicated by behavioral studies that show results comparable to those of the drugs rivastigmine and tacrine in vivo. The molecules **9–11** were assessed using the aldicarb-sensitivity assay and in vivo passive avoidance test, which resulted in the identification of novel orally active AChE inhibitors that enhanced cognition impaired by scopolamine in Swiss male mice. The percentage of learning improvement for molecule **10** was similar to that of donepezil, yet weaker than that of tacrine.

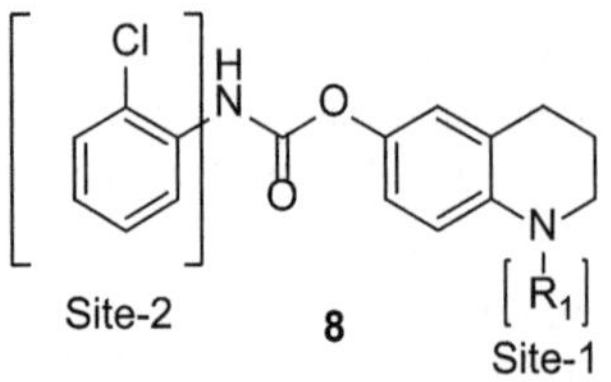

8

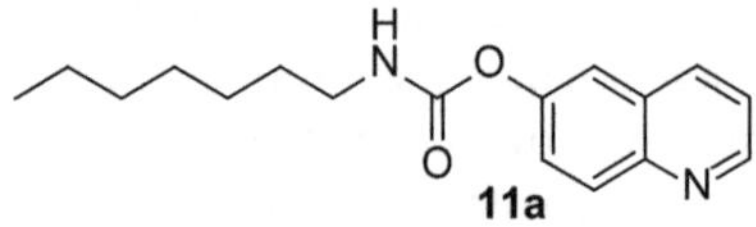

11a

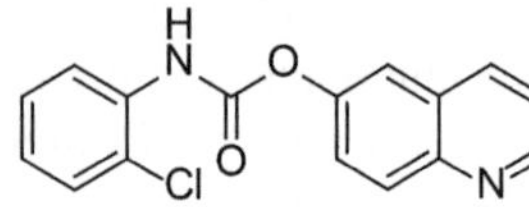

11b

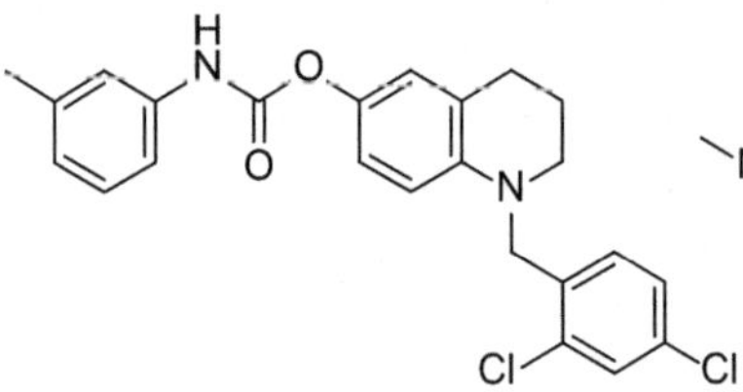

9

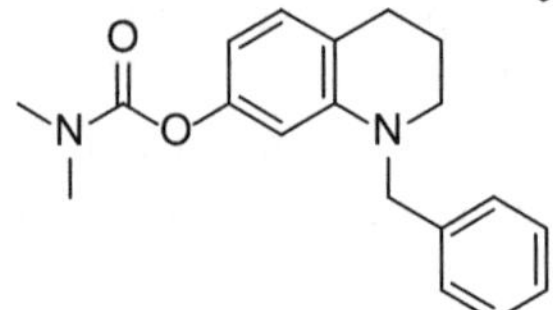

10

2.4 Case Studies in Tuberculosis Treatment

Tuberculosis (TB), caused by *Mycobacterium tuberculosis*, remains a leading cause of mortality due to infectious diseases globally. The World Health Organization's Global Tuberculosis Report 2024 indicates that tuberculosis resulted in around 1.3 million fatalities in 2023, along with 10.6 million new cases [42]. Although TB is preventable and treatable, but still, it continues to pose a considerable public health challenge, particularly in low- and middle-income nations. Tuberculosis (TB) treatment depends on multi-drug regimens (for example the WHO approved DOTS therapy); nevertheless, the emergence of multidrug-resistant (MDR) and extensively drug-resistant (XDR) TB highlights the critical necessity for novel therapeutic targets. MDR-TB can be treated by second-line drugs such as 6-fluoroquinolones and one injectable (kanamycin/capreomycin/amikacin) that were used the past few decades [43]. Targets for tuberculosis (TB) drugs aim to disrupt essential pathways in *Mycobacterium tuberculosis* to impede bacterial growth and survival. Primary targets encompass enzymes implicated in cell wall biosynthesis, specifically InhA and DprE1, which are crucial for sustaining the distinctive mycolic acid-rich outer layer. Additional significant targets include DNA gyrase, essential for DNA replication, and RNA polymerase, which is inhibited by rifampicin. Inhibitors of protein synthesis, particularly those targeting ribosomal function, are crucial in this context. Furthermore, research is being conducted on metabolic enzymes, including alanine racemase and isocitrate lyase [44]. Targeting these pathways offers various strategies for addressing both drug-sensitive and drug-resistant tuberculosis [45]. Recent literature suggest the quest for new anti-tubercular drugs [46]. A possible approach involves targeting the bacterial energy metabolism, namely the F_1F_0 ATP synthase complex, which is crucial for ATP synthesis in *Mycobacterium tuberculosis*. This enzyme is not present in human mitochondria in the same structural configuration, rendering it a selective and efficacious therapeutic target [47]. ATP synthase is considered essential for optimal growth in *M. tuberculosis* [48] and in the non-pathogenic model strain *Mycobacterium smegmatis* [49]. There have been various diarylquinolines (DARQ) which have been studied against tuberculosis. TMC207, one of the (DARQ), has displayed highest potency with minimum inhibitory concentration (MIC) of 0.06 µg/ml. TMC207 is now recognized as Bedaquiline [50]. Bedaquiline [51] has significantly transformed treatment approaches. In 2012, the US Food and Drug Administration (FDA) approved the first new class of drugs for tuberculosis in four decades [52]. The approval coincided with the development of additional new drugs, namely pretomanid and delamanid, alongside linezolid and clofazimine. This shift altered the tuberculosis treatment paradigm, resulting in the initial multidrug combination that markedly diverged from the conventional regimen and possessed the capacity to address pre-existing resistance to first-line therapies [53]. Several DARQs and related analogs have been investigated for their Mtb ATP synthase inhibition and antitubercular activities. These molecules were designed using the state of art approaches of CADD for identifying the essential structural features for Mtb ATP synthase inhibition and their incorporation. The most potent and recently developed ATP synthase inhibitors from these studies have been given in Fig. 3.

The SAR and docking studies on the synthesized compounds showed that compound **12** (*N*-(7-chloro-2-methylquinolin-4-yl)-N-(3-((diethylamino)methyl)-4-hydroxyphenyl)-2,3-dichlorobenzenesulfonamide) displayed comparable activity in hypoxic environment against non- replicating bacteria. Compound **12** was designed by altering the 2,3 dichlorophenyl group of the parent molecule to inhibit ATP synthase enzyme and showed MIC and IC_{50} values of 3.12 µg/mL and 0.51 ± 0.030 µM respectively. Its CC_{50} value was greater than 300 µg/mL showing its non-toxic nature to Vero

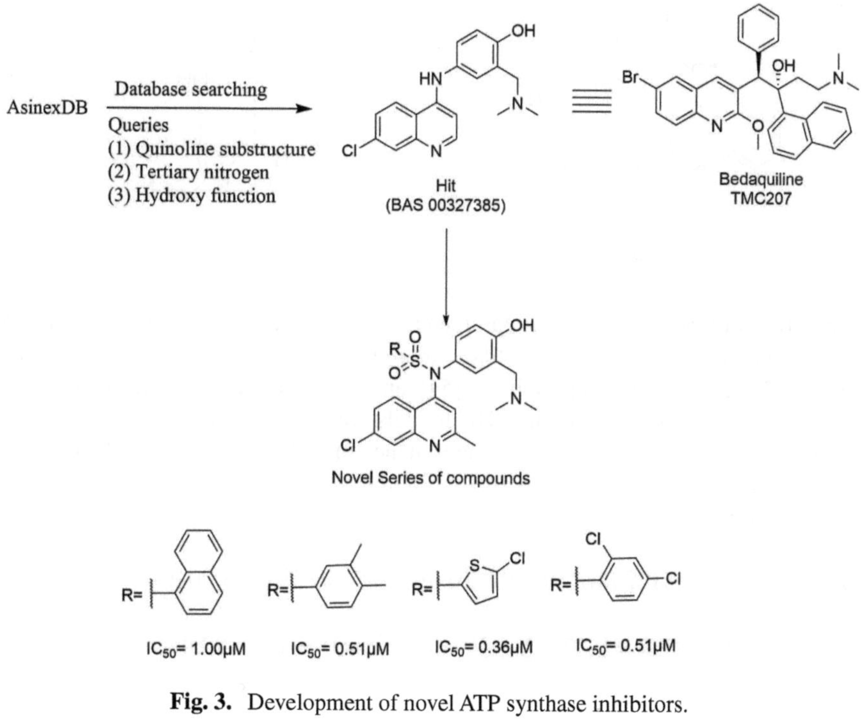

Fig. 3. Development of novel ATP synthase inhibitors.

cell line. The mode of binding of these compounds have also been discussed by the authors [54].

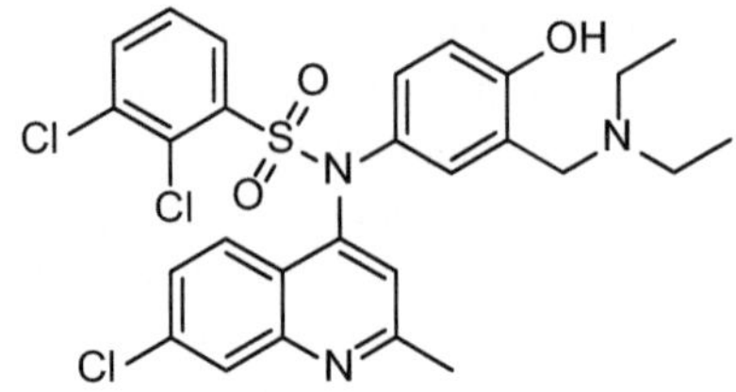

12

These studies were followed by the design and synthesis of many bisquinoline compounds where the phenyl group is replaced with a quinoline moiety of the parent compound (TMC207) and among them 4-(dimethylamino)-1,1-bis(2-methoxyquinolin-3-yl)-2-phenylbutan-2-ol (**13**) and 4-(dimethylamino)-1,1-bis(2- methoxyquinolin-3-yl)-2-(2-napthyl) phenylbutan-2-ol (**14**), lowered intracellular colony forming units (CFUs) of *M. tuberculosis* to about 90% and 91% respectively which indicated the high potency of the compounds. The respective ATP synthase IC_{50} value of the two compounds were 0.07 μM and 0.03 μM and both of them fit into the binding pocket of the enzyme as proved by molecular docking studies [55].

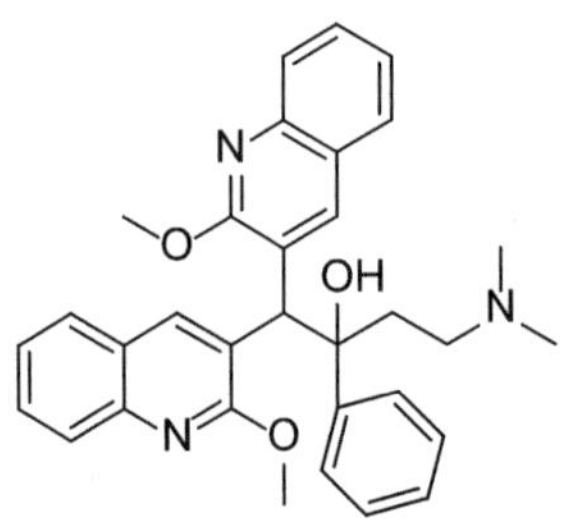

13

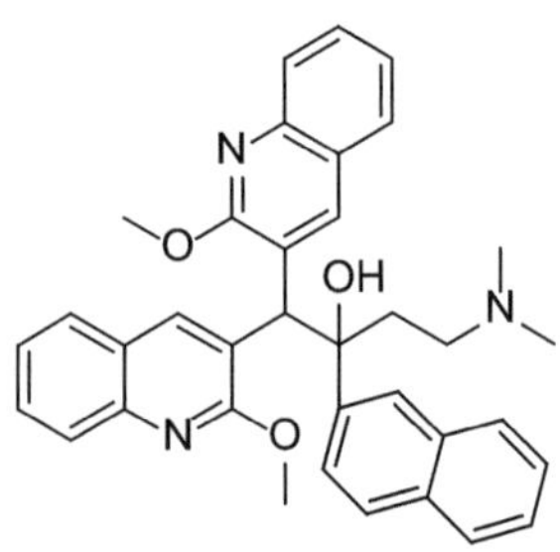

14

The effects of physicochemical parameters, including electronic, steric and hydrophobicity factors, on the ATP synthase inhibitory activity of 4-substituted amino sulphonyl-2-methyl-7-chloroquinolines [54] and bis-quinoline core [55] have been investigated through QSAR studies. The models are presented as Eq1, Eq2, and Eq3. Since the molecules contained stereocenter, hence models for both R and S configuration were generated.

For R configuration

$$pIC50 = 0.923(\pm 0.122)MR - 0.209(\pm 0.086)PC - 0.032(\pm 0.006)CMA + 3.191 \quad (2)$$

$$\mathrm{N = 25, R = 0.871 EV = R^2 = 0.76, S = 0.394, F = 22.44}$$

For S configuration

$$pIC50 = 0.955(\pm 0.123)MR - 0.174(\pm 0.088)PC - 0.032(\pm 0.006)CMA + 2.627 \quad (3)$$

$$N = 25, R = 0.877\,EV = R^2 = 0.77, S = 0.384, F = 23.97$$

For RS configuration (mean)

$$pIC50 = 0.954(\pm 0.122)MR - 0.185(\pm 0.086)PC - 0.033(\pm 0.006)CMA + 3.129 \quad (4)$$

$$N = 25, R = 0.883\,EV = R^2 = 0.78, S = 0.382, F = 24.26$$

The Molar Refractivity (MR), which contributes positively, along with the negatively contributing Partition Coefficient (PC) and Connolly Molecular Area parameters (CMA), has demonstrated high predictive capability, as evidenced by external data sets. The studies concluded that the mean values of the computed 3D parameters of enantiomers can be utilized in QSAR analysis for racemic compounds. The QSAR model established for the antitubercular activity in whole *M. Tuberculosis* H37Rv demonstrated a strong correlation and exhibited a similar reliance on the same descriptors, specifically MR and CMA (Eq. 5).

$$pMIC = 0.735(\pm 0.108)MR - 0.031(\pm 0.005)CMA + 3.167 \quad (5)$$

$R = 0.922$ EV $= R^2 = 0.85$, $N = 11$, $S = 0.222$ $F = 23.14$

The above results indicated that ATP Synthase is the target enzyme for the antitubercular activity of these molecules [56].

The data set of these molecules [54, 55] has recently been utilized to develop quantitative predictive models through a novel approach known as QSIAR. The approach takes into account the drawback in the docking algorithm of the docking software where the docking scores usually do not correlate with the biological activity. The QSIAR combines Structure-Based Drug Design (SBDD) with Quantitative Structure-Activity Relationship (QSAR) to produce effective models. The QSIAR considers the specific interactions between a ligand and the amino acid residues at the active site as independent variables, while biological activity is treated as a dependent parameter, quantified to elucidate the observed activity.

$$BA = Const. + a1AR1 + a2AR2 + \dots\dots\dots anARn$$

where, BA = binding energy/affinity, biological activity/docking scores;

AR = interaction with amino acid residue

The QSIAR methodology involves the utilization of, number of total interactions between the ligand and the amino acid residues, in developing QSAR models where the total number of interactions (combined weightage) is taken as independent parameter and the biological activity is taken as dependent parameter. It is important to mention here that during the development of the QSIAR methodology, a single weightage protocol was also utilized wherein number of interactions between any single amino acid

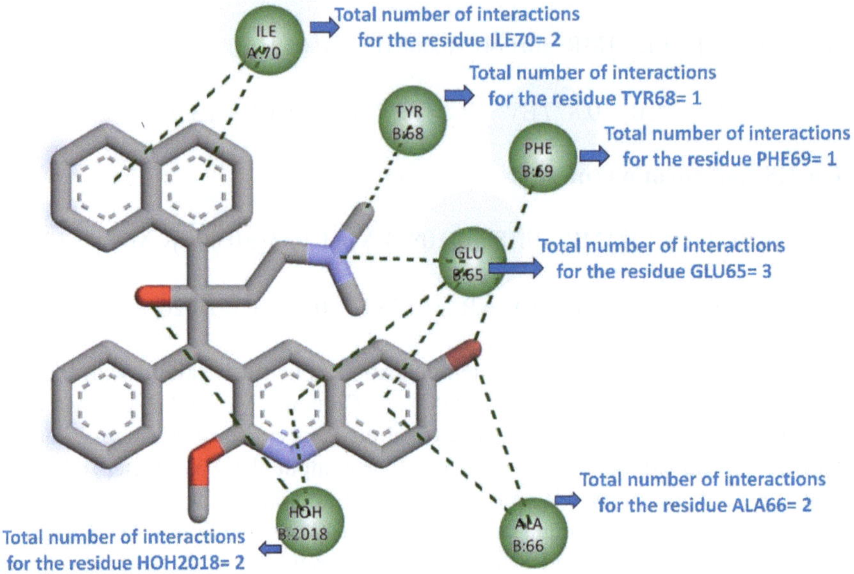

Fig. 4. 2D interaction diagram of Bedaquiline in complex with 4V1F (ATP synthase) explaining the QSIAR methodology.

residue and ligand was taken as '1' irrespective of the fact that more than 1 interaction may be present in between that amino acid residue and the ligand molecule. The results obtained from the single weightage were inferior to combined weightage protocol which may be attributed to the subtle contributions of each interaction present. The resulting dataset from the methodology is shown in Fig. 4.

The application of the QSIAR methodology on the above dataset [54, 55] resulted in a preliminary equation which was reduced systematically to reach to equation with just one parameter Eq. (6).

$$pIC_{50} = -0.428 + 0.161(\pm 0.01582)\,\mathrm{Glu}\,65b \tag{6}$$

$$N = 24; R^2 = 82.5\%; r - Sq(adj) = 81.7\%; S = 0.301; r - Sq(\mathrm{pred}) = 77.71\%$$

The Eq. 6 was validated through molecules of the external dataset and showed a good correlation (R = 0.8518) between the pIC_{50} and pIC_{90} values which was indicative of its robustness [57]. The same methodology was also applied to different molecules of diverse class. This class of compounds belonged to the class of imidazo[1,2-a] pyridine ethers, and squaramides [58]. The application of QSIAR to this dataset again yielded a one parameter equation (Eq. 7).

$$pIC50 = -0.845 + 1.094(\pm 0.137)Glu65\text{b} \tag{7}$$

$$N = 28, R^2 = 71.00\%, r - Sq(adj) = 69.88\%, F = 63.65, s = 0.615, r - Sq(pred) = 67.75\%$$

The parameter (Glu65B) was the same which was obtained through the datasets of substituted chloroquinolines [54] and bisquinolines [55] (Eq. 6) which indicates the importance of the amino acid residue. Both the results prompted for combining all the molecules which expanded the dataset to 52 molecules. The molecules were divided into training and test dataset of 27 and 25 molecules each. The developed and validated quantitative model, represented by Eq. 8, was further validated using an additional test set comprising of four molecules [59] and showed good correlation (R = 0.76) between the predicted pIC_{50} values and the pIC_{90} [60].

$$pIC_{50} = 0.259 + 0.3802(\pm 0.0655)Glu65\text{b} - 0.793(\pm 0.191)\text{HOHb} - 0.353(\pm 0.125)Phe69\text{b} + 0.287(\pm 0.176)Ala66\text{b} - 0.870(\pm 0.348)\text{I} \tag{8}$$

$$N = 27, R^2 = 67.74\%, r - Sq(adj) = 60.06\%, F = 8.82, S = 0.647, r - Sq(pred) = 48.60\%$$

The results obtained in Eqs. 6 and 7 along with the docking results in terms of important interacting residues and their corresponding ligand were used in designing a focussed library. Three molecules of the focussed library were synthesized and were subsequently predicted by the best model (Eq. 7). The structure and the predicted activity of the molecules are given in Fig. 5.

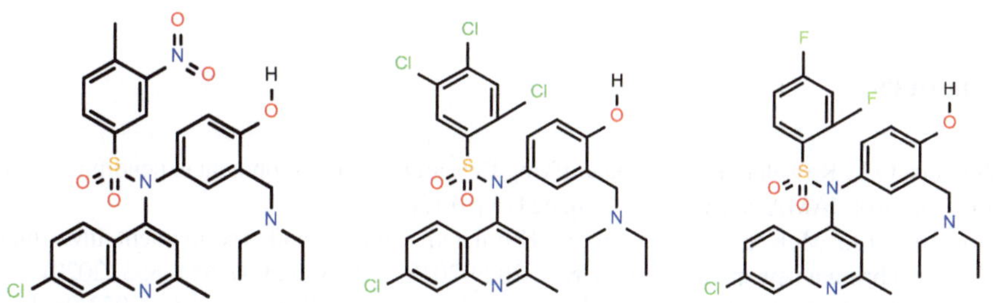

Predicted pIC_{50}=0.1168μM   Predicted pIC_{50}=0.1494μM   Predicted pIC_{50}=0.5296μM

Fig. 5. The structure and the predicted activity of the 3 molecules from the generated focussed library.

3 Conclusions

The computer-aided drug design (CADD) includes QSAR and CAMM and contributes in both crucial steps the lead generation and lead optimization steps of drug discovery process by using the LBDD and SBDD approaches. The discussed case studies outline the significant impact of computer-aided drug design (CADD) in developing novel chemical entities and drug candidates across various therapeutic classes. Neuroleptic agents were designed using substructural analysis, pattern recognition, and medicinal chemical hybridization (MCH), facilitating the identification of effective molecular frameworks and led to the invention of a neuroleptic drug (Centbutindole) approved for marketing. For antihypertensives, strategies such as "freezing" the preferred bioactive conformations of known drugs and integrating active sub-structures into new prototypes using MCH/molecular hybridization resulted in new antihypertensive agents. In the search for anti-Alzheimer's agents, systematic virtual screening (VS) was used, involving the development of 3D-pharmacophore models and screening of virtual compound libraries to identify potential candidates. Antitubercular compounds were discovered through a novel and innovative approach, Quantitative structure-interaction Activity Relationship (QSIAR), which correlates the number of interacting amino acid residues with biological activity. These examples underscore the growing importance of CADD in modern drug discovery, offering a rational and efficient approach to identify and optimize leads. CADD reduces time, cost, and experimental effort by predicting drug-likeness and activity in silico before synthesis. As computational tools advance and integrate with AI and machine learning, CADD is poised to play an even more transformative role in the future of personalized and precision medicine, accelerating the development of safer and effective therapeutics.

Funding None.

References

1. Wouters, O.J., Kesselheim, A.S.: Quantifying Research and Development expenditures in the drug industry. JAMA Netw. Open. **7**(6), e2415407 (2024)
2. Wouters, O.J., McKee, M., Luyten, J.: Estimated research and development investment needed to bring a new medicine to market, 2009-2018. JAMA. **323**(9), 844–853 (2020)
3. Stock, B.: Drug development – the four phases. https://biostock.se/en/2023/01/drug-development-the-four-phases/ (2023)
4. Möhrle, J.J.: How long does it take to develop a new drug? Lancet Regional Health Europe. **43**, 100998 (2024). https://doi.org/10.1016/j.lanepe.2024.100998
5. Hughes, J.P., Rees, S., Kalindjian, S.B., Philpott, K.L.: Principles of early drug discovery. Br. J. Pharmacol. **162**(6), 1239–1249 (2011). https://doi.org/10.1111/j.1476-5381.2010.01127.x
6. Pina, A.S., Hussain, A., Roque, A.C.: An historical overview of drug discovery. Methods Mol. Biol. (Clifton, NJ). **572**, 3–12 (2009). https://doi.org/10.1007/978-1-60761-244-5_1

7. Viegas-Junior, C., Danuello, A., da Silva, B.V., Barreiro, E.J., Fraga, C.A.: Molecular hybridization: a useful tool in the design of new drug prototypes. Curr. Med. Chem. **14**(17), 1829–1852 (2007). https://doi.org/10.2174/092986707781058805
8. Marchesi, E., Perrone, D., Navacchia, M.L.: Molecular hybridization as a strategy for developing Artemisinin-derived anticancer candidates. Pharmaceutics. **15**(9), 2185 (2023). https://doi.org/10.3390/pharmaceutics15092185
9. Jia, P., et al.: The roles of computer-aided drug synthesis in drug development. Green Synth. Catal. **3**(1), 11–24 (2022). https://doi.org/10.1016/j.gresc.2021.11.007
10. Struble, T.J., et al.: Current and future roles of artificial intelligence in medicinal chemistry synthesis. J. Med. Chem. **63**(16), 8667–8682 (2020). https://doi.org/10.1021/acs.jmedchem.9b02120
11. Paul, D., Sanap, G., Shenoy, S., Kalyane, D., Kalia, K., Tekade, R.K.: Artificial intelligence in drug discovery and development. Drug Discov. Today. **26**(1), 80–93 (2021). https://doi.org/10.1016/j.drudis.2020.10.010
12. Mak, K.K., Pichika, M.R.: Artificial intelligence in drug development: present status and future prospects. Drug Discov. Today. **24**(3), 773–780 (2019). https://doi.org/10.1016/j.drudis.2018.11.014
13. Sellwood, M.A., Ahmed, M., Segler, M.H., Brown, N.: Artificial intelligence in drug discovery. Future Med. Chem. **10**(17), 2025–2028 (2018). https://doi.org/10.4155/fmc-2018-0212
14. Ghosh, P.: Novel active pharmaceutical ingredients from India: the actors-part-I. MGM J. Med. Sci. **8**, 73 (2021). https://doi.org/10.4103/mgmj.mgmj_12_21
15. Saxena, A.K., Jain, P.C., Anand, N., Dua, P.R.: Agents acting on the central nervous system. 15. 2-substituted 1,2,3,4,6,7,12,12a-octahydropyrazino [2′,1′:6,1]pyrido[3,4-b]indoles. New class of central nervous system depressants. J. Med. Chem. **16**(5), 560–564 (1973). https://doi.org/10.1021/jm00263a036
16. Rusig, I., Léger, J.M., Laguerre, M., Saxena, A.K., Carpy, A.: Structure and molecular properties of (1)-Centbutindole. Comparison with haloperidol. J. Chem. Crystallogr. **25**(8), 443–451 (1995). https://doi.org/10.1007/BF01665699
17. Saxena, A.K, et al.: 2-Substituted-1,2,3,4,6,7,12,12A-octahydropyrazino(2{40 ,1{40 :6,1) pyrido(3,4-B)indoles. United States Patent US3917599A (1973)
18. Chakrabarty, R., et al.: Rational design, synthesis and evaluation of (6aR∗,11bS∗)-1-(4-fluorophenyl)-4-{7-[4-(4-fluorophenyl)-4-oxobutyl]1,2,3,4,6,6a,7,11b,12,12a (RS)-decahydropyrazino[2′,1′:6,1]pyrido[3,4-b]indol-2-yl}-butan-1-one as a potential neuroleptic agent. Bioorg. Med. Chem. **15**(23), 7361–7367 (2007). https://doi.org/10.1016/j.bmc.2007.07.018
19. Saxena, A.K., et al.: QSAR studies in substituted 1,2,3,4,6,7,12,12a-octahydropyrazino[2′,1′:6,1]pyrido[3,4-b]indoles—a potent class of neuroleptics. Bioorg. Med. Chem. **11**(9), 2085–2090 (2003). https://doi.org/10.1016/s0968-0896(02)00652-1
20. Kumar, Y., Mehta, P., Saxena, A.K., Gulati, A., Singh, H.K., Anand, N.: Synthesis of cis & trans 1-substituted 1,2,3,4,4a,5,11,11a-octahydro-6H-pyrido(3,2-b) carbazoles, 4-substituted-1,2,3,4,4a,5,6,11c-octahydro-7H-pyrido(2,3-c)-carbazoles, cis-4-methyl-1,2,3,4,4a,5,6,12b-octahydro-7H-pyrido(2,3-c)-acridine & cis-1-methyl-1,2,3,4,-4a,5,12,12a-octahydro-6H-pyrido(3,2-b)acridine - a new class of potential antiparkinsonian agents. Ind. J. Chem. **30B**, 213–221 (1991)
21. Saxena, A.K., Singh, H.K., Dhawan, B.N., Anand, N.: A process for the synthesis of Cis-1-Alkyl Substituted- 1,2,3,4,4a,5,11 C-Otahydro-6h-Pyrido (2,3-C). India Patent IN167493 (1991)
22. Saxena, A.K., Singh, H.K., Dhawan, B.N., Anand, N.; A process for the synthesis of novel cis 1,2,3,4,4a,5, 1, 11a-octahydro-6h-pyrido (3,2-b) carbazole. India Patent IN165919 (1990)

23. Saxena, A.K., Singh, H.K., Dhawan, B.N., Anand, N.: A process for the synthesis of cis-4-methyl 1,2,3,4,4a,5,6,11 c-otahydro-6h-pyrido (2,3-c) carbazole. India Patent IN167494 (1991)
24. Saxena, A.K., Singh, H.K., Dhawan, B.N., Anand, N.: A process for the synthesis of cis-1-alkyl substituted- 1,2,3,4,4a,5,11,11 a-otahydro-6h-pyrido (3,2-b) carbazole. India Patent IN167492 (1991)
25. Saxena, A.K., Singh, H.K., Dhawan, B.N., Anand, N.: A process for the synthesis of novel cis 1,2,3,4,4a,5,5,11 c-octabydro- 7h-pyrido (2,3-c) carbazole. India Patent IN167491 (1991)
26. Rusig, I., Laguerre, M., Carpy, A., Saxena, A.: ChemInform abstract: comparison of molecular properties of D2Receptor antagonists from different chemical families of neuroleptics. ChemInform. **27** (2010). https://doi.org/10.1002/chin.199621274
27. Srimal, R., Dhawan, B., Rao, J., Saxena, A.: A preliminary report on the hypotensive activity of 1-[(3-tolyl-piperazin-1-yl)-3-thio-acetamidophenyl]propanes. Eur. J. Pharmacol. **186**, 1831 (1990)
28. Saxena, A., Rao, J., Srimal, R., Audry, E., Carpy, A.: Synthesis and QSAR studies on hypotensive N-[3-aryl(thio/-sulfino)propyl]piperazines/-piperidines. Ind. J. Chem. **32B**, 1249–1257 (1993)
29. Gulati, A., et al.: Resuscitative effect of Centhaquine (Lyfaquin(®)) in hypovolemic shock patients: a randomized, multicentric, controlled trial. Adv. Ther. **38**(6), 3223–3265 (2021). https://doi.org/10.1007/s12325-021-01760-4
30. Saxena, A.K., Schaper, K.-J.: QSAR analysis of the time dependent hypotensive activity of 1-[3-(4-substituted phenylthio)propyl]-4-(3,4-substituted phenyl)piperazines using artificial. Neural Netw. **22**(9–10), 965–983 (2003). https://doi.org/10.1002/qsar.200330850
31. Mount, C., Downton, C.: Alzheimer disease: progress or profit? Nat. Med. **12**(7), 780–784 (2006). https://doi.org/10.1038/nm0706-780
32. Cummings, J.L., Doody, R., Clark, C.: Disease-modifying therapies for Alzheimer disease: challenges to early intervention. Neurology. **69**(16), 1622–1634 (2007). https://doi.org/10.1212/01.wnl.0000295996.54210.69
33. Morrison, H.: Chapter 1 - acetylcholinesterase. In: Morrison, H. (ed.) Enzyme Active Sites and their Reaction Mechanisms, pp. 1–4. Academic Press (2021). https://doi.org/10.1016/B978-0-12-821067-3.00001-5
34. Ismail, Y.A., Haitham, Y., Walid, M., Mohamed, H., El-Satar, Y.M.A.: Efficacy of acetylcholinesterase inhibitors on reducing hippocampal atrophy rate: a systematic review and meta-analysis. BMC Neurol. **25**(1), 60 (2025). https://doi.org/10.1186/s12883-024-03933-4
35. Grill, J.D., Cummings, J.L.: Current therapeutic targets for the treatment of Alzheimer's disease. Expert. Rev. Neurother. **10**(5), 711–728 (2010). https://doi.org/10.1586/ern.10.29
36. Gontijo, V.S., et al.: Molecular hybridization as a tool in the design of multi-target directed drug candidates for neurodegenerative diseases. Curr. Neuropharmacol. **18**(5), 348–407 (2020). https://doi.org/10.2174/1385272823666191021124443
37. Shakya, N., Fatima, Z., Nath, C., Saxena, A.K.: Substituted carbamic acid quinolin-6-yl esters useful as acetylcholinesterase inhibitors. WO2006070394 (2006)
38. Chaudhaery, S.S., et al.: Novel carbamates as orally active acetylcholinesterase inhibitors found to improve scopolamine-induced cognition impairment: pharmacophore-based virtual screening, synthesis, and pharmacology. J. Med. Chem. **53**(17), 6490–6505 (2010). https://doi.org/10.1021/jm100573q
39. Roy, K.K., Tota, S., Tripathi, T., Chander, S., Nath, C., Saxena, A.K.: Lead optimization studies towards the discovery of novel carbamates as potent AChE inhibitors for the potential treatment of Alzheimer's disease. Bioorg. Med. Chem. **20**(21), 6313–6320 (2012). https://doi.org/10.1016/j.bmc.2012.09.005

40. Roy, K.K., Tota, S., Nath, C., Shukla, R., Saxena, A.K.: Substituted 1, 2, 3, 4-tetrahydroquinolin-7-yl carbamates as acetylcholinesterase inhibitors for treatment of alzheimer's disease. US20130324573 (2013)
41. Shakya, N., Fatima, Z., Nath, C., Saxena, A.K. Substituted carbamic acid quinolin-6-yl esters useful as acetylcholinesterase inhibitors. WO2006070394 (2006)
42. (GTB) GPoTaLH.: Global tuberculosis report 2024. https://www.who.int/publications/i/item/9789240101531 (2024)
43. Nuermberger, E.L., Spigelman, M.K., Yew, W.W.: Current development and future prospects in chemotherapy of tuberculosis. Respirology (Carlton, Vic). **15**(5), 764–778 (2010). https://doi.org/10.1111/j.1440-1843.2010.01775.x
44. Sanchez, S., Demain, A.L.: Metabolic regulation and overproduction of primary metabolites. Microb. Biotechnol. **1**(4), 283–319 (2008). https://doi.org/10.1111/j.1751-7915.2007.00015.x
45. Saxena, A.K., Singh, A.: Mycobacterial tuberculosis enzyme targets and their inhibitors. Curr. Top. Med. Chem. **19**(5), 337–355 (2019). https://doi.org/10.2174/1568026619666190219105722
46. Ahmed, S., Sisir, N., Saxena, A.K.: An updated patent review on drugs for the treatment of tuberculosis (2018-present). Expert Opin. Ther. Pat. **32**(3), 243–260 (2022). https://doi.org/10.1080/13543776.2022.2012151
47. Novoa-Aponte, L., Soto Ospina, C.Y.: *Mycobacterium tuberculosis* P-type ATPases: possible targets for drug or vaccine development. Biomed. Res. Int. **2014**, 296986 (2014). https://doi.org/10.1155/2014/296986
48. Sassetti, C.M., Boyd, D.H., Rubin, E.J.: Genes required for mycobacterial growth defined by high density mutagenesis. Mol. Microbiol. **48**(1), 77–84 (2003). https://doi.org/10.1046/j.1365-2958.2003.03425.x
49. Tran, S.L., Cook, G.M.: The F1Fo-ATP synthase of mycobacterium smegmatis is essential for growth. J. Bacteriol. **187**(14), 5023–5028 (2005). https://doi.org/10.1128/jb.187.14.5023-5028.2005
50. Hongmanee, P., Rukseree, K., Buabut, B., Somsri, B., Palittapongarnpim, P.: In vitro activities of cloxyquin (5-chloroquinolin-8-ol) against *Mycobacterium tuberculosis*. Antimicrob. Agents Chemother. **51**(3), 1105–1106 (2007). https://doi.org/10.1128/aac.01310-06
51. Khoshnood, S., et al.: Bedaquiline: current status and future perspectives. J. Global Antimicrobial Resistance. **25**, 48–59 (2021). https://doi.org/10.1016/j.jgar.2021.02.017
52. Mahajan, R.: Bedaquiline: first FDA-approved tuberculosis drug in 40 years. Int. J. Appl. Basic Med. Res. **3**(1), 1–2 (2013). https://doi.org/10.4103/2229-516x.112228
53. Nasto, B.: Not-for-profit to launch antibiotic against drug-resistant tuberculosis. Nat. Biotechnol. (2019). https://doi.org/10.1038/d41587-019-00034-3
54. Singh, S., et al.: Novel, potent, orally bioavailable and selective mycobacterial ATP synthase inhibitors that demonstrated activity against both replicating and non-replicating *M. tuberculosis*. Bioorg. Med. Chem. **23**(4), 742–752 (2015). https://doi.org/10.1016/j.bmc.2014.12.060
55. Kalia, D., et al.: Synthesis and anti-tubercular activity of conformationally-constrained and bisquinoline analogs of TMC207. MedChemComm. **6**(8), 1554–1563 (2015). https://doi.org/10.1039/C5MD00131E
56. Saxena, A.K., Alam, M.: ATP synthase inhibitors as anti-tubercular agents: QSAR studies in novel substituted Quinolines. Curr. Top. Med. Chem. **20**(29), 2723–2734 (2020). https://doi.org/10.2174/1568026620666200903163515
57. Ahmed, S., Prabahar, A.E., Saxena, A.K.: Molecular docking-based interactions in QSAR studies on *Mycobacterium tuberculosis* ATP synthase inhibitors. SAR QSAR Environ. Res. **33**(4), 289–305 (2022). https://doi.org/10.1080/1062936X.2022.2066175

58. Tantry, S.J., et al.: Discovery of Imidazo[1,2-a]pyridine ethers and Squaramides as selective and potent inhibitors of mycobacterial adenosine triphosphate (ATP) synthesis. J. Med. Chem. **60**(4), 1379–1399 (2017). https://doi.org/10.1021/acs.jmedchem.6b01358
59. Guillemont, J., Meyer, C., Poncelet, A., Bourdrez, X., Andries, K.: Diarylquinolines, synthesis pathways and quantitative structure-activity relationship studies leading to the discovery of TMC207. Future Med. Chem. **3**(11), 1345–1360 (2011). https://doi.org/10.4155/fmc.11.79
60. Ahmed, S., Prabahar, A.E., Saxena, A.K.: Molecular docking-based interaction studies on imidazo[1,2-a] pyridine ethers and squaramides as anti-tubercular agents. SAR QSAR Environ. Res. **34**(6), 435–457 (2023). https://doi.org/10.1080/1062936X.2023.2225872

Development of Allosteric Inhibitors Against Cyclin-Dependent Kinase 2 (CDK2)

Kelsey A. Holdaway[1(✉)] and Gunda I. Georg[1]

[1] Department of Medicinal Chemistry and Institute for Therapeutics Discovery and Development, College of Pharmacy, University of Minnesota, Minneapolis, MN, USA
skluz032@umn.edu; georg@umn.edu

Abstract. Cyclin-dependent kinase 2 (CDK2) is a well-established target for anti-cancer therapies and, more recently, non-hormonal male contraception, given its pivotal role in cell cycle regulation and meiosis. However, the development of effective, FDA-approved CDK2 inhibitors has been hindered by the challenge of achieving selectivity, particularly over the closely related kinase CDK1. Allosteric inhibitors offer a promising alternative by targeting non-catalytic binding sites, thus providing the potential for greater selectivity and fewer off-target effects. Recent crystallographic studies have identified four distinct allosteric pockets on CDK2, providing new avenues for drug development. Notable allosteric ligands, such as **EF-3-006**, demonstrate both potent CDK2 binding (K_D of 18 nM) and impressive selectivity over CDK1, along with functional effects like meiosis arrest in mouse testicular explants, highlighting their potential for non-hormonal contraception. Despite these promising findings, allosteric kinase inhibitors remain an underexplored area of research. Targeting these newly discovered allosteric pockets may enable the development of selective, efficacious therapies for cancer and contraception, overcoming the challenges faced by traditional ATP site inhibitors.

Keywords: Cyclin-dependent kinase 2 · Anti-cancer · Non-hormonal male contraceptive · Allosteric inhibition · Negative cooperativity

1 Introduction

1.1 Cyclin-Dependent Kinases

Cyclin-dependent kinases (CDKs) are a family of serine/threonine kinases responsible for regulating the cell cycle upon activation by binding partner cyclins. CDKs 1, 2, 4, and 6 regulate the cell cycle of the twenty proteins comprising the CDK family [1]. Specifically, CDK1 coordinates the cell cycle progression through mitosis, CDK2 facilitates the G_1/S transition, and CDK4/6 initiates cell cycle entry in response to growth factor signaling [2]. Knockout mouse studies have demonstrated the redundancy inherent amongst cell cycle CDKs, with only CDK1 determined to be essential [3, 4].

A. K. Saxena, A. Saxena (Eds.): GTHTM 2025, pp. 19–41, 2026.
https://doi.org/10.1007/978-3-032-12320-6_2

Monomeric CDKs are inactive and require two distinct steps for activation, which include phosphorylation of the activation loop (A-loop) and binding of the cognate cyclin [5]. As CDK expression levels remain relatively stable throughout the cell cycle, activation is predominantly driven by the association of the cyclin-binding partner (Fig. 1). Approximately thirty cyclins are known to partner CDKs in humans [6]. As such, cognate cyclin partners of cell cycle CDKs demonstrate a characteristic oscillating expression pattern that ensures appropriate coordination of each cell cycle stage [7].

Several structural features of cell cycle CDKs are essential for activation, which include the A-loop, the αC helix (also known as the PSTAIRE helix), and the DFG motif (Fig. 2). Importantly, phosphorylation on a specific residue (Thr160 in CDK2) within the A-loop by CDK-activating kinase and cyclin binding induces a conformational change, resulting in activation. Specifically, monomeric CDK2 has an A-loop-in

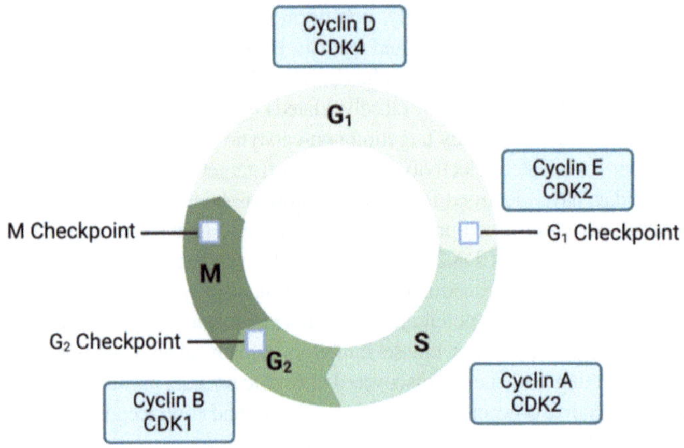

Fig. 1. Important CDK cell cycle regulators with respective cyclin binding partners.

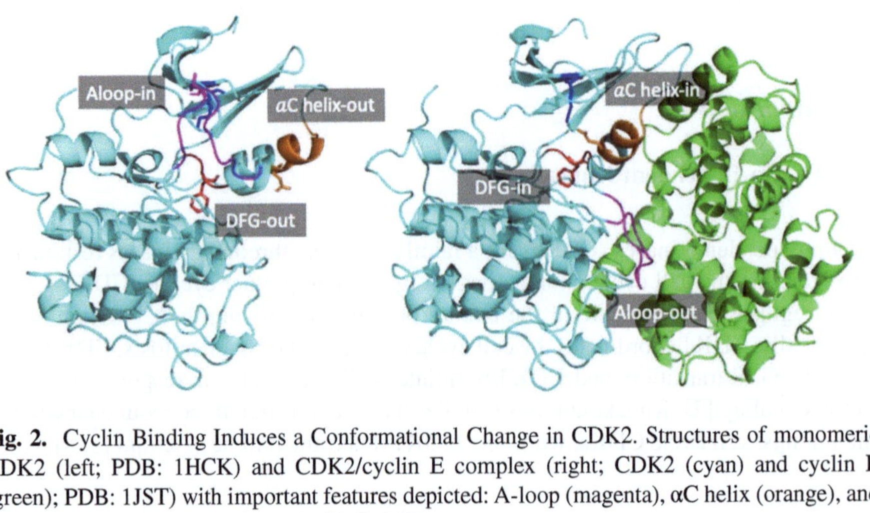

Fig. 2. Cyclin Binding Induces a Conformational Change in CDK2. Structures of monomeric CDK2 (left; PDB: 1HCK) and CDK2/cyclin E complex (right; CDK2 (cyan) and cyclin E (green); PDB: 1JST) with important features depicted: A-loop (magenta), αC helix (orange), and DFG motif (red). Image generated with PyMOL.

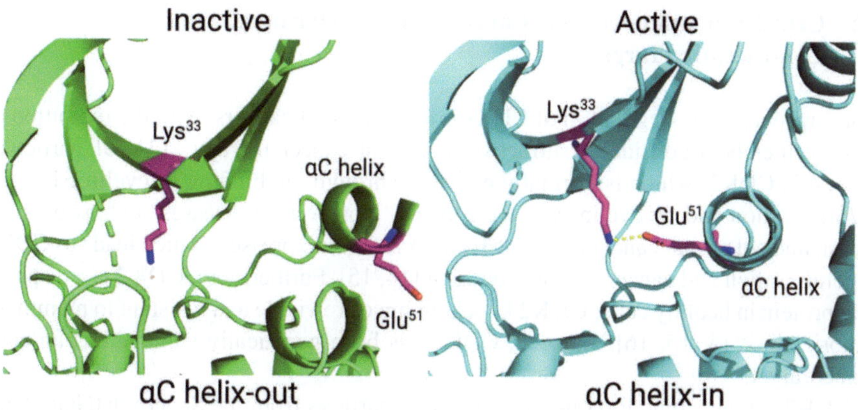

Fig. 3. αC Helix of CDK2. Inactive (green) and active (cyan) conformations of CDK2. The αC helix is swung outwards in the inactive conformation (left; PDB: 1HCL) and moves inward in the active conformation (right; PDB: 1JST), which allows for a Lys-Glu salt-bridge. Image generated with PyMOL.

and DFG-out orientation, which blocks the αC helix from moving toward the active site and, thus, creates an auto-inhibited conformation [8]. In the active state, the αC helix-in conformation allows for forming a Lys-Glu salt bridge, and the A-loop-out orientation recognizes peptide substrate after phosphorylation on Thr160 (Fig. 3) [9].

Unlike CDK1/2, CDK4 remains in an inactive state even after cyclin binding [10, 11]. Despite phosphorylation of the A-loop, the αC helix-out conformation persists, similar to the observed conformation for inactive, monomeric CDK2. As such, it has been proposed that substrate engagement might be an essential element toward CDK4 activation [7]. The structurally diverse activation states for cell cycle CDKs may be something that can be exploited to generate CDK-specific allosteric inhibitors in the future.

Allostery also plays a key role in kinase signaling, which requires careful regulation. The activity of CDK complexes is regulated by post-translational modifications, as well as protein-protein and protein-ligand interactions, leading to specific conformational changes that modulate CDK function [7]. These diverse CDK sub-populations may harbor unique druggable pockets, presenting opportunities to identify allosteric inhibitors with enhanced selectivity and alternate mechanisms of action. For example, the transition states of CDK2 have been characterized, revealing that the activation of CDK2 is regulated by conformational sampling between the A-loop-in and three distinct A-loop-out states [12]. Among these, the A-loop-out$_3$ conformation is the only state associated with appreciable kinase activity [12]. This suggests that the alternative states that monomeric CDKs can adopt could provide novel structural pockets for small molecule targeting, enabling the development of highly selective inhibitors.

1.2 CDK2 as a Validated Anti-Cancer and Non-Hormonal Male Contraceptive Target

Not surprisingly, dysregulation of CDKs is a hallmark of cancers, with the inhibition of specific members considered attractive targets in cancer therapy [13]. Of particular interest is CDK2, which is normally activated through the binding of cyclin E1 at the G_1/S transition and by cyclin A in the S phase. However, in diseases such as ovarian, breast, and colorectal cancers, cyclins E and A are overexpressed, which leads to CDK2 hyperactivation and cancer cell proliferation [14, 15]. Furthermore, CDK2 is a dispensable protein in healthy cells; CDK2 knockout mice are viable and resistant to mammary tumorigenesis [4, 15, 16]. As such, CDK2 has been genetically validated as an anti-cancer target.

CDK2 can also bind to non-classical cyclin partners from the Speedy/RINGO family of proteins, such as Spy1, encoded by the *Spyda* gene, which can alter the conformation of the kinase (Fig. 4) [17]. The CDK2/Spy1 complex is activated irrespective of the post-translational modifications required to activate the CDK2/cyclin complex completely [18]. The overexpression of Spy1 overrides cell cycle checkpoints, resulting in increased susceptibility to tumor formation [19–21]. Elevated Spy1 levels are also observed in drug-resistant estrogen receptor-positive and triple-negative breast cancers (TNBC) [21]. In a Spy1$^-$ knockout model of TNBC, a reduction of proliferation, increased response to chemotherapy, and delayed onset of tumor formation were observed [19, 22]. Thus, targeting Spy1 may reinstate checkpoints and increase sensitivity to therapy in patients who do not respond to chemotherapy, including immunotherapy. Additionally, a crystal structure of the CDK2/Spy1 complex suggests that the binding pocket differs from that of the CDK2/cyclin complex, indicating that the binding affinity of traditional ATP site CDK2 inhibitors could be altered [18, 23]. The difference between the activated CDK2/Spy1 complex and that of the CDK2/cyclin complex might contribute to the clinical failure observed for CDK2 inhibitors in patients with cancer types expressing elevated Spy1 levels such as breast, prostate,

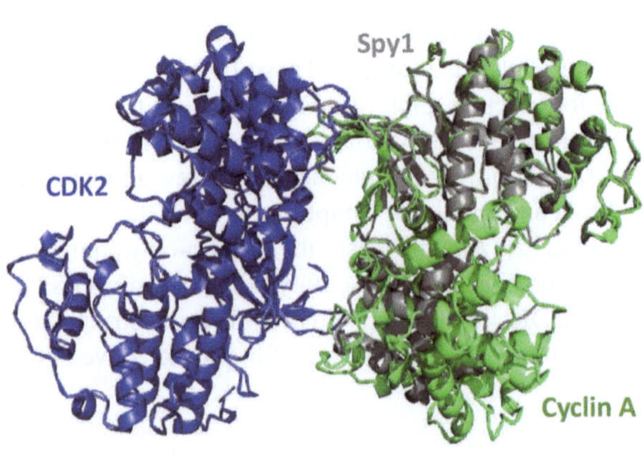

Fig. 4. Overlay of CDK2 (blue) in complex with Spy1 (grey, PDB: 5UQ2) and cyclin A (green, PDB: 1FIN).

ovarian, liver, brain, and blood cancers [24–28]. Therefore, allosteric inhibitors of CDK2 can potentially disrupt both mechanisms of CDK2 activation.

CDK2 has also been genetically validated as a non-hormonal male contraceptive target with knockout mice viable but sterile [4, 29–32]. The mice lack spermatocytes, spermatids, and sperm but retain spermatogonia [4, 31]. The CDK2 binding partner Spy1 is expressed in all tissues but has a 20+ fold higher expression in the testis [33]. As such, *Spyda*$^{-/-}$ knockout mice showed the same viable but sterile phenotype as *Cdk2*$^{-/-}$ mice with no additional abnormalities, demonstrating spermatocyte arrest in the same stage as *Cdk2*$^{-/-}$ mice [30, 31, 34].

When comparing the anatomy, histology, and mating behavior of *Cdk2*$^{-/-}$ mice to wild-type mice, it was found that they were identical except for their gonads [4, 31]. Primary mouse embryonic fibroblasts (MEFs) isolated from the knockout mice proliferated normally, though with a delayed S phase, suggesting compensatory mechanism(s) during mitosis [4, 31, 35]. Silencing CDK1 resulted in reduced proliferation in MEFs [36]. Furthermore, complete genetic replacement of the CDK1 locus with CDK2 led to a non-viable phenotype, indicating that while CDK1 can compensate for the loss of CDK2 during mitosis, importantly, it cannot replace the role of CDK2 in meiosis [4, 31, 35]. Later studies demonstrated that mouse embryos lacking CDK2, CDK3, CDK4, and CDK6 developed normally, with CDK1 binding to all cyclins and driving the mammalian cell cycle [37, 38]. Therefore, a selective CDK2 inhibitor is anticipated to primarily impact spermatogenesis, with minimal harmful effects on somatic cells.

To date, no exquisitely selective CDK2 inhibitor has been FDA-approved [39]. However, several selective inhibitors are currently undergoing clinical trials, such as **tagtociclib** (PF-07104091), **BLU-222**, **AZD8421**, **INX-315**, and **INCB123667** (Table 1) [40]. **Tagtociclib** demonstrated 115-fold selectivity for CDK2/cyclin E1 over CDK1/cyclin A2 and more than 200-fold selectivity over CDK4/cyclin D1 and CDK6/cyclin D3 in cellular assays [41]. It is currently being investigated in Phase II clinical trials for small cell lung cancer, ovarian cancer, breast cancer (NCT04553133), and breast cancer and solid tumors (NCT05262400). Likewise, **BLU-222** demonstrated 25-fold selectivity for CDK2/cyclin E1 over CDK1/cyclin B1 and more than 150-fold selectivity over CDK4/cyclin D1 and CDK6/cyclin D3 in cellular assays [42]. **BLU-**

Table 1. Cellular efficacy and selectivity for CDK2 inhibitors currently undergoing clinical trials.

	NanoBRET Cellular IC_{50}, nM				
	CDK2/cyclin E1	CDK1/cyclin A2	CDK4/cyclin D1	CDK6/cyclin D3	CDK9/cyclin T1
Tagtociclib[a]	20	2300	>4100	>5000	>5000
BLU-222[a]	17.7	452.3[a]	5104.6	2621.7	2697.7
AZD-8421[c]	9	–	–	–	–
INX-315[a]	2.3	374[b]	–	–	2950
INCB123667[c]	53	692	873	1582	5273

[a]CDK1/cyclin B1
[b]HEK-293 cells
[c]Cell line not disclosed.

222 is undergoing a Phase 1/2, open-label, first-in-human clinical trial (NCT05252416) to evaluate safety, tolerability, PK/PD, and efficacy as a single agent and in combination therapy for patients with advanced solid tumors. **AZD-8421** demonstrated selectivity for CDK2 over CDK1, CDK4, and CDK6 (data not reported) and is currently in a Phase 1/2, first-in-human clinical trial (NCT06188520) for patients with advanced breast cancer and metastatic high-grade ovarian cancer [43]. In cellular assays, **INX-315** demonstrated 163-fold selectivity for CDK2/cyclin E1 over CDK1/cyclin B1 [44]. The first-in-human Phase 1/2 clinical trial (NCT05735080) with **INX-315** is currently recruiting patients with advanced/metastatic cancers, including breast cancer patients who progressed on a prior CDK 4/6 inhibitor regime and *CCNE1*-amplified solid tumors who progressed on standard of care treatment. In addition, **INCB123667** demonstrated 13-fold selectivity for CDK2 over CDK1 and is currently undergoing an open-label, Phase 1 clinical trial (NCT05238922) for patients with advanced solid tumors [45].

Although several promising CDK2 compounds are currently undergoing clinical trials, selective targeting of CDK2 remains challenging. The structural homology amongst kinases complicates the development of specific inhibitors, as most kinase inhibitors historically target the well-conserved ATP binding site. This lack of selectivity has led to undesirable off-target effects due to the inhibition of multiple kinases in clinical settings [40]. To improve selectivity, targeting less-conserved allosteric pockets presents a promising alternative [46].

1.3 Allosteric Approach

Allosteric inhibition is a promising strategy in drug development, particularly for proteins like CDK2, which currently lack an FDA-approved selective inhibitor [39, 40]. Allosteric inhibitors represent a class of small molecules that interact with a kinase in a manner distinct from the traditional ATP binding site, inducing conformational changes that regulate the kinase activity. Type III inhibitors bind at allosteric sites near the ATP binding site and typically stabilize the inactive or "closed" conformation of the kinase, preventing kinase activation by ATP, and consequently blocking downstream signaling. Allosteric Type IV inhibitors bind at sites far removed from the ATP binding site. The binding of Type IV inhibitors typically significantly alters the kinase's function, such as promoting an inactive conformation or disrupting its interaction with other proteins in the signaling pathway.

This allosteric approach offers several advantages, especially in challenging targets like CDK2, which is crucial in cell cycle regulation but shares structural similarities with other CDKs, complicating the development of selective inhibitors [47]. By targeting allosteric sites, inhibitors can achieve greater selectivity for CDK2, reducing the risk of off-target effects often seen with ATP-competitive inhibitors. This selectivity, particularly over CDK1, is important to avoid harming normal cells that rely on other CDKs for essential processes. Furthermore, allosteric inhibitors can reduce toxicity by modulating the activity of CDK2 without fully inhibiting its function, offering a more controlled way to interfere with cell cycle progression. This could be of particular importance when considering the role of CDK2/Spy1 in meiosis but not mitosis.

Allosteric inhibition may also help overcome resistance mechanisms. Traditional ATP-competitive inhibitors are vulnerable to mutations in the ATP binding site, a problem an allosteric inhibitor could bypass. By binding to non-ATP sites, allosteric modulators might provide a more durable and effective treatment, especially in cancers where resistance to conventional CDK inhibitors is a concern. This approach becomes even more significant in the context of breast cancer, where CDK4/6 inhibitors are the standard of care. Despite their effectiveness, nearly all patients either exhibit de novo resistance (10% of cases) or develop acquired resistance within 24–28 months, with a substantial proportion of resistance driven by the upregulation of the binding partner of CDK2 [19, 48, 49]. As such, combinations of CDK4/6 with CDK2 inhibitors are currently being explored in several clinical studies, as it has been found that CDK2 and CDK4/6 compensate for each other to maintain the hyperphosphorylation of the retinoblastoma protein, which promotes tumor progression [41, 50, 51]. Additionally, CDK2 allosteric inhibitors are anticipated to be effective against cancers resistant to the ATP-competitive CDK2 inhibitors currently under development. As such, allosteric modulators could circumvent this mechanism of resistance, offering a more sustainable therapeutic strategy and improving outcomes for patients who otherwise would relapse.

The FDA-approval of allosteric inhibitors for other targets highlights the growing success of this approach in drug development [52]. **Trametinib, cobimetinib, binimetinib** and **selumetinib** are Type III FDA-approved MEK1/2 inhibitors for cancer therapy [53]. It has been suggested that fewer than 10% of all kinases possess a Type III allosteric site [54]. The Type IV kinase inhibitor of BCR-ABL1 used to treat chronic myeloid leukemia, **asciminib** (Scemblix), allosterically binds to the myristoyl–binding pocket and is effective against most ABL1 kinase domain mutations that confer resistance to ATP-competitive tyrosine kinase inhibitors [55]. Ultimately, these therapies offer more precise targeting with fewer side effects, making allosteric inhibition a compelling strategy for selectively modulating kinases.

Allosteric inhibitors of kinases (Type III and Type IV) offer an alternative strategy to the traditional ATP-competitive inhibitors. They offer advantages in selectivity, overcoming resistance due to mutations, and the ability to modulate kinase function in a more nuanced manner. These inhibitors have already led to FDA-approved therapies, with the potential for further innovations in precision medicine. While no selective CDK2 inhibitors are currently approved, the allosteric approach holds significant potential for creating safer, more effective treatments.

2 CDK2 Allosteric Inhibitors

Developing Type III and IV kinase inhibitors that target less conserved allosteric pockets adjacent to or distal from the ATP binding site could enhance selectivity and reduce therapeutic resistance commonly seen with Type I kinase inhibitors. While most CDK2 inhibitors are known to interact with the ATP binding site, new approaches have emerged to exploit recently identified allosteric pockets. Four allosteric pockets in CDK2 have been crystallographically characterized, and two additional pockets have

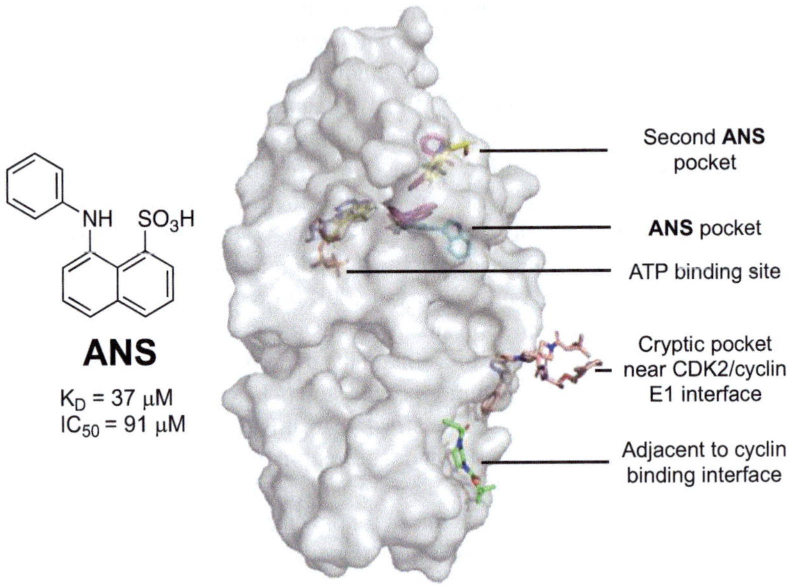

Fig. 5. Allosteric Pockets of CDK2. Four allosteric pockets have been crystallographically verified on CDK2 (grey, surface rendered 20% transparent), shown with representative ligands bound (PDB: 7RWF, 5OSJ, 5FP6, 8VQ4, 3PXF). Another two pockets have been putatively discovered via computational methods along the cyclin binding interface. Image generated with PyMOL.

been suggested based on computational methods (Fig. 5). Furthermore, adducts with a covalent ligand were identified at multiple modification sites through LC-MS/MS analysis.

2.1 ANS Pocket Allosteric Inhibitors

To identify potential allosteric binding pockets on CDK2, the extrinsic fluorophore **8-anilino-1-naphthalene sulfonate** (**ANS**) was used to probe human CDK2 and cyclin A using fluorescence spectroscopy and protein crystallography [56]. From this, it was demonstrated that **ANS** interacts with free CDK2 in a saturation-dependent manner with an apparent K_D of 37 μM. Co-crystallography revealed that **ANS** binds near the ATP site and the αC helix of CDK2, forming key interactions with the catalytic Lys33 and the DFG motif (Fig. 6; PDB: 3PXZ, 3PY1, 3PXF, 3PXQ, 4EZ77, 7UG1). Furthermore, a second **ANS** molecule binds near the first, interacting with residues in the αC helix [56]. It was ultimately found that **ANS** binding induced significant structural changes in CDK2, leading to an αC helix conformation that prevents cyclin A binding. These results suggested that the **ANS** pocket could be a novel target site for allosteric inhibitors that disrupt CDK-cyclin interactions.

The **ANS** pocket is the most extensively studied, with over 20 ligand-bound co-crystal structures deposited in the PDB. To develop a high-throughput assay to discover additional chemical scaffolds that bind within the allosteric pocket, the **ANS** analog ***p*-**

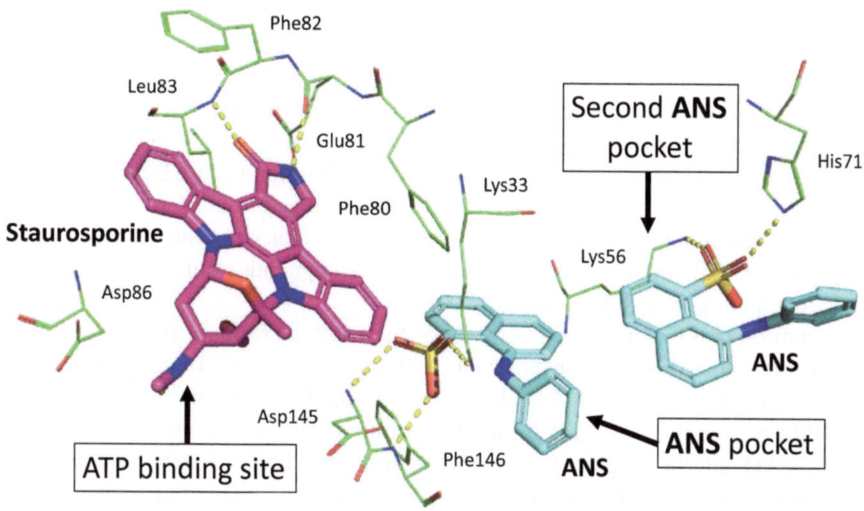

Fig. 6. Co-Crystal Structure of CDK2 Bound to Staurosporine and **8-Anilino-1-Naphthalene Sulfonate (ANS)**. Two molecules of **ANS** (cyan) bind in an allosteric pocket distinct from the ATP site binder **staurosporine** (pink; PDB: 4EZ7). Within the **ANS** pocket, **ANS** forms key interactions with the catalytic Lys33 residue and the DFG motif. A second **ANS** molecule binds near the first, interacting with residues in the αC helix. Image generated in PyMOL.

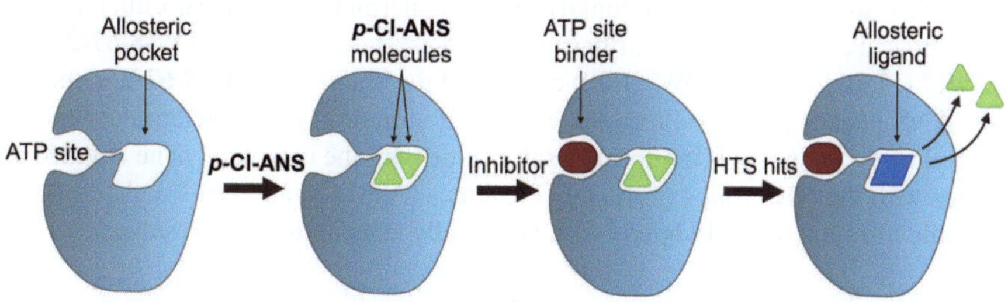

Fig. 7. ***p*-Cl-ANS** Displacement Assay. The allosteric pocket in CDK2 accommodates two fluorescent ***p*-Cl-ANS** molecules, which are displaced upon binding of allosteric ligands. Adapted from *ACS Chem. Biol.* **2011**, *6* (5), 492–501.

Cl-ANS was exploited due to its improved fluorescent nature. In this assay, the addition of allosterically binding compounds leads to the displacement of ***p*-Cl-ANS** from the hydrophobic interior of the allosteric pocket to the aqueous solvent, resulting in a decrease in fluorescence, which can be quantified (Fig. 7). Utilizing this assay, ~20 k compounds from our in-house collection were selected for a high-throughput screen (HTS) based on their similarity to **ANS**, especially compounds containing a negatively charged moiety, such as a carboxylic or sulfonic acid. Compounds were tested at a 30 μM screening concentration, with hits causing >50% loss of ***p*-Cl-ANS** fluorescence selected for advancement. These hits were then tested in a dose-response, both in the presence and absence of the ATP site binder **staurosporine**, to differentiate between inhibitors targeting the ATP and allosteric sites. Hits that produced IC_{50} values <50 μM with minimal change between the presence and absence of **staurosporine** ("**staurosporine** shift" < two-fold) were chosen for repurchase, ultimately resulting in 10 hits capable of decreasing the fluorescent signal of ***p*-Cl-ANS** in a concentration-dependent manner in the presence and absence of **staurosporine**.

In parallel, a virtual HTS (vHTS) was conducted with ~720 k structures from our in-house collection using the crystal structure of **ANS** bound to CDK2. Of the 260 vHTS hits identified, 230 were commercially available and tested for inhibition using the ***p*-Cl-ANS** displacement assay at a 30 μM screening concentration. Of the seven hits that produced >25% inhibition at 30 μM, only three compounds exhibited concentration-dependent inhibition with IC_{50} values <100 μM.

To confirm that the hits were not false positives and the result of fluorescence interference, orthogonal nonfluorescent HSQC NMR was conducted. Using known ligands that bind at either the ATP site or **ANS** site, an NMR "fingerprint" was established. From this, we could discern that of the 11 hit compounds tested, three were assumed to be either ATP site binding ligands or nonbinding, five showed indeterminate binding due to protein aggregation in the assay, and three demonstrated binding to the allosteric site. Of the three allosteric binders, two were ortho-substituted benzoic acids. This led to the discovery of **TW-8-67-2** as the primary scaffold for further development (Fig. 8) [57].

Verification of allosteric binding of **TW-8-67-2** was performed using the ***p*-Cl-ANS** displacement assay in the absence (IC_{50} = 4.6 μM) and presence (IC_{50} = 5.7 μM) of **staurosporine**, which demonstrated an essentially unchanged IC_{50} value. Binding to CDK2 was also assessed orthogonally with isothermal titration calorimetry (ITC), which showed a K_D of 5.2 μM. Furthermore, a co-crystal structure of **TW-8-67-2** bound to CDK2 (PDB: 7RWF, 8FOW, 8FP0) confirmed the binding of **TW-8-67-2** in the allosteric **ANS** pocket (Fig. 9). From the co-crystal structure, it was determined that the carboxylate on the phenyl ring forms a salt bridge with the catalytic lysine (Lys33) and engages with the NH groups of the DFG residues (Asp145 and Phe146). Additionally, the indole nitrogen forms H-bonds with potentially critical αC helix residues.

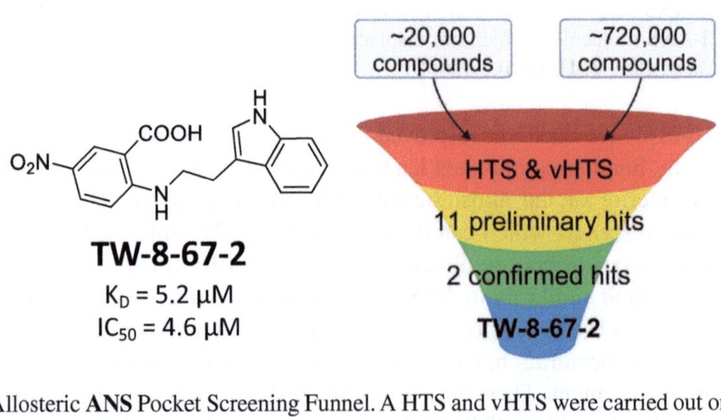

Fig. 8. Allosteric **ANS** Pocket Screening Funnel. A HTS and vHTS were carried out on the **ANS** pocket of CDK2. This led to the identification of **TW-8-67-2** as the lead scaffold for optimization.

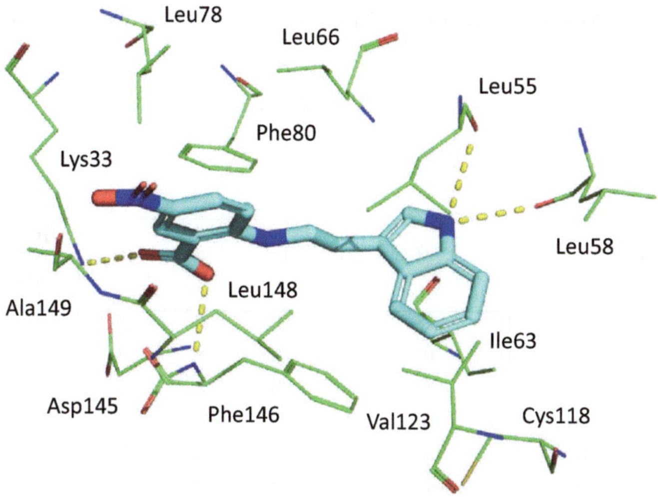

Fig. 9. Crystal Structure of CDK2 in Complex with **TW-8-67-2.** The binding of **TW-8-67-2** (cyan) in the allosteric pocket (green) was confirmed with crystallography. Important H-bonding interactions with the DFG motif, catalytic lysine, and the αC helix residues are shown (PDB: 7RWF). Image generated in PyMOL.

Subsequent structure-activity relationship (SAR) studies began by exploring substituents at different positions on the phenyl ring (Table 2). A compound without substituents at R^1, R^2, or R^3 (compound **1**) resulted in a complete activity loss in the ***p*-Cl-ANS** assay. Nitro substituents at R^1 (compound **2**) or R^2 (compound **5**) showed reduced binding affinity or equipotent affinity, respectively. In general, the presence of a substituent at the R^3 position is the driver of binding affinity. As such, other electron-withdrawing groups at R^3 were explored. Replacing the nitro group with a trifluoromethyl group (compound **19**) improved the binding affinity by four-fold, and the larger pentafluorosulfanyl group (compound **23**) showed a 10 + -fold improvement.

Likewise, the indole ring was explored by the introduction of chloro groups at R^4 (compound **28**), R^5 (compound **30**), R^6 (compound **34**), and R^7 (compound **32**). Adding chloro groups at R^4 and R^5 resulted in equipotent binding affinities compared to the parent compound **TW-8-67-2**. The introduction of a chloro group at R^7 worsened binding substantially. However, substituents at the R^6 position (compounds **34–36**) improved the binding affinity for CDK2 by 10 + -fold. Overall, halogens at the R^6 position on the indole were crystallographically confirmed to fill a hydrophobic pocket at this position that was previously unoccupied by the parent compound **TW-8-67-2**, which accounts for the improved affinity.

Upon identifying potency elements on the phenyl and indole rings, the essential elements were combined to yield compounds with higher affinity for CDK2 (Table 3). The replacement of the nitro group with a trifluoromethyl group and the addition of various halogen (**EF-2-275** and **EF-3-006**) or nitrile (**EF-4-084**) groups at the indole C6 position resulted in 29- to 50-fold improvement over the parent compound **TW-8-67-2** in the ***p*-Cl-ANS** assay and K_D values <50 nM. Combining the pentafluorosulfanyl group

Table 2. SAR of **TW-8-67-2** exploring the effects of substituents on the phenyl and indole rings.

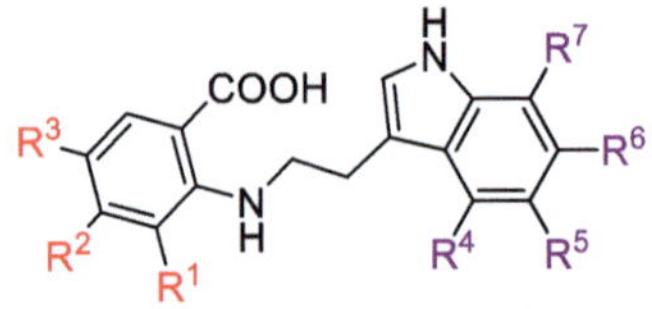

Compound	R¹	R²	R³	R⁴	R⁵	R⁶	R⁷	IC₅₀, *p*-Cl-ANS (μM)
TW-8-67-2	H	H	NO_2	H	H	H	H	4.6
1	H	H	H	H	H	H	H	>300
2	NO_2	H	H	H	H	H	H	38
5	H	NO_2	H	H	H	H	H	6.1
19	H	H	CF_3	H	H	H	H	1.1
23	H	H	SF_5	H	H	H	H	0.30
28	H	H	NO_2	Cl	H	H	H	3.9
30	H	H	NO_2	H	Cl	H	H	5.8
34	H	H	NO_2	H	H	Cl	H	0.44
32	H	H	NO_2	H	H	H	Cl	15
35	H	H	NO_2	H	H	F	H	0.37
36	H	H	NO_2	H	H	Br	H	0.25

Table 3. Combining potency features from phenyl and indole ring SAR to yield high affinity analogs.

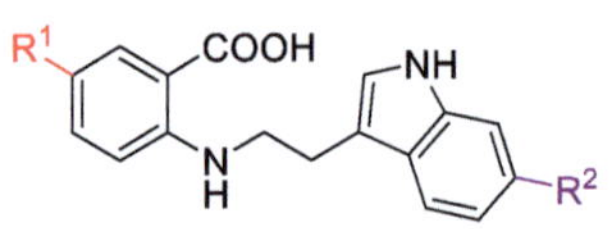

Compound	R¹	R²	IC₅₀, *p*-Cl-ANS (nM)	K_D, ITC (nM)	K_D, SPR (nM)
TW-8-67-2	NO_2	H	4600	5200	2600
19	CF_3	H	1100	2300	--
EF-2-275	CF_3	Cl	92	31	23 ± 2
EF-3-006	CF_3	Br	160	18	16 ± 2
EF-4-084	CF_3	CN	100	23	6.1 ± 1.2
23	SF_5	H	300	260	--
EF-4-177	SF_5	CN	87	7.4	--

on the phenyl ring with the nitrile substituent at the C6 position on the indole ring (**EF-4-177**) resulted in the highest affinity compound of the series with a K_D of 7.4 nM. Ultimately, this SAR revealed the first-in-class crystallographically confirmed allosteric compounds (PDB: 7RWE, 7RXO, 7S85, 7RWF, 7S4T) demonstrating low nanomolar affinities for CDK2, as measured by ITC and surface plasmon resonance (SPR) [57].

Additionally, data from cellular thermal shift assays and SPR indicated that **EF-3-006** exhibited strong selectivity for CDK2 over CDK1, and less than 80% inhibition was observed across a panel of 371 kinases at a 10 μM concentration (Fig. 10) [58]. Further investigation into the interaction of this compound series with cyclin binding showed that, despite their strong affinity for CDK2, the compounds poorly inhibited CDK/cyclin complexes in commercial kinase activity assays [58]. A Förster resonance energy transfer (FRET) assay was used to monitor the kinase's conformational states, which revealed a negatively cooperative relationship with cyclin, preventing the compounds from binding effectively to the **ANS** pocket [58]. In addition to FRET, CDK2 ADP-Glo assays were used to demonstrate the negative cooperativity. For compound **EF-3-006**, it was found that at a 1:1 p-CDK2:cyclin molar ratio, the IC_{50} was ~eight-fold more potent than at a 1:10 M ratio and > 100-fold more potent than at a 1:100 M

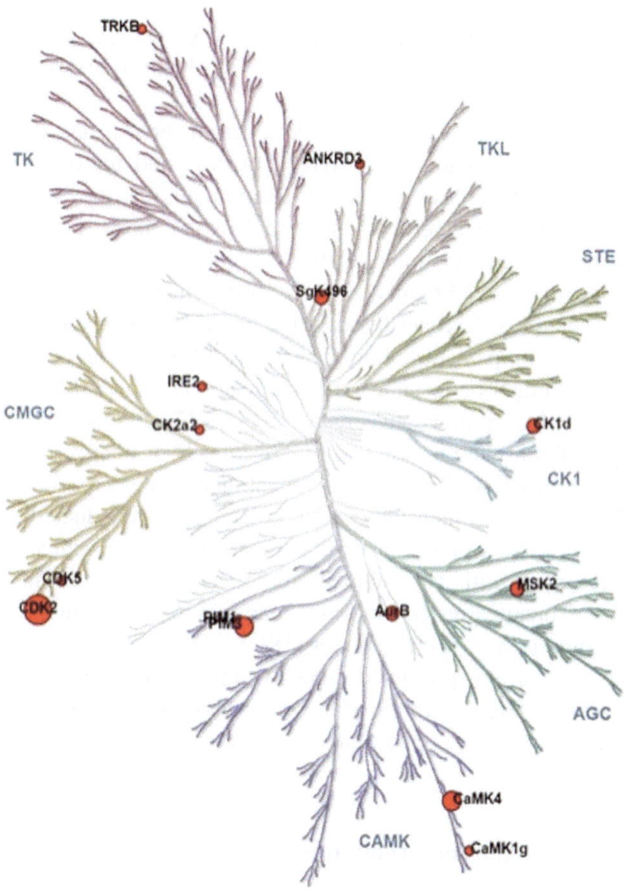

Fig. 10. Kinome Selectivity of EF-3-006. **EF-3-006** did not show >80% inhibition of any kinase at a 10 μM concentration. Only 4% of the 371 kinases tested showed >50% inhibition [58].

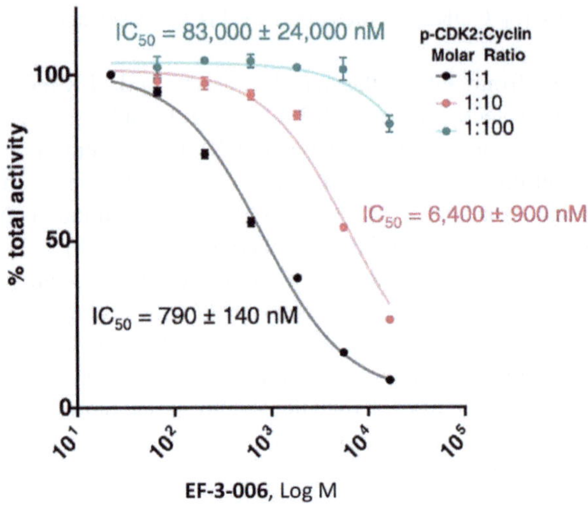

Fig. 11. EF-3-006 Exhibits Negative Cooperativity with Cyclin A. As cyclin concentration increases in an activity assay, the inhibitory potential of **EF-3-006** decreases [58].

ratio (Fig. 11). Consequently, in cancers like ovarian, breast, and colorectal, where cyclin overexpression leads to CDK2 hyperactivation and tumor cell proliferation, further optimization of this series will be necessary to address the negative cooperativity with cyclin and enhance the efficacy in these disease contexts [14, 15].

In the testis, the meiotic function of CDK2 is regulated by non-classical cyclin partners from the Speedy/RINGO family, such as Spy1, which can alter the conformation of the kinase. Given the differences between the mitotic CDK2/cyclin complex and the meiotic CDK2/Spy1 complex, developing inhibitors that selectively disrupt one mechanism over the other might be feasible. As such, lead compounds **EF-3-006** and **EF-4-177** showed potential as contraceptive agents. **EF-4-177** demonstrated a long half-life in human ($T_{1/2}$ = 68 min) and mouse ($T_{1/2}$ = 278 min) liver microsomes. Additionally, oral administration of **EF-4-177** in mice demonstrated a long exposure in both plasma and testes. A 28-day longitudinal study with CD-1 mice was conducted with **EF-4-177**, which revealed a 45% reduction in sperm count compared to the vehicle treatment [57]. Furthermore, treatment with **EF-3-006** in spermatocyte chromosome spreads from mouse testicular explants arrested meiosis, recapitulating a phenotype similar to that seen in *Cdk2*$^{-/-}$ and *Spdya*$^{-/-}$ spermatocytes [58]. While the mechanism of meiotic disruption remains unclear, it is presumed to involve interference with Spy1 binding. The potential to exploit differences in the affinities of various binding partners to achieve selectivity presents an exciting opportunity to investigate other therapeutically relevant kinases.

2.2 Other Reported Allosteric Inhibitors of CDK2

A virtual HTS was once again employed to probe the **ANS** pocket, leading to the testing of 35 compounds using the same ***p*-Cl-ANS** displacement assay as in the previous study on **ANS** pocket binders [59, 60]. Among these compounds, two exhibited single-digit

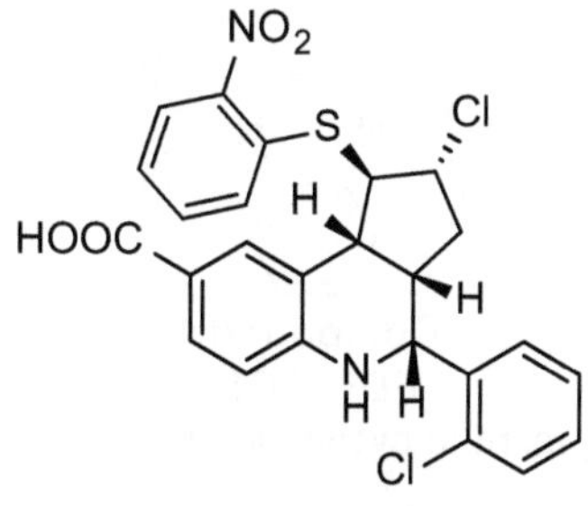

Compound 3f
ANS pocket
K_D = 43 μM
IC_{50} = 11 μM

Fig. 12. Chemical structure of **Compound 3f**, binding pocket, and biological activity data.

micromolar EC_{50} values. Competition assays were then conducted in the presence of the potent ATP site inhibitor **staurosporine**, which showed no significant changes in the **ANS** displacement curves, indicating that these compounds do not bind to the ATP site. However, when tested for inhibition of CDK2/cyclin A kinase activity at concentrations up to 100 μM, none of the compounds demonstrated significant inhibition. Preincubating a compound at 100 μM with CDK2 before adding cyclin A resulted in 30% inhibition of kinase activity, suggesting the need for more potent analogs. In human breast cancer cell lines (MDA-MB-231 and ZR-75-1), the most effective compound exhibited an IC_{50} of 4 μM, while the others showed IC_{50} values greater than 20 μM [59]. Notably, there was no correlation between the **ANS** displacement assay results and the antiproliferative cellular assay, prompting further SAR investigations. These studies led to the identification of **Compound 3f**, which showed an IC_{50} of 11 μM in the **ANS** displacement assay and a K_D of 43 μM for GST-CDK2 binding, as determined by microscale thermophoresis (Fig. 12) [61]. To further explore the binding site of **Compound 3f**, site-directed mutagenesis was performed on residues in the allosteric pocket [62]. The results revealed a significant decrease in **Compound 3f** binding (with a K_D of 361 μM) in the triple mutant K33A/K56A/H71A, suggesting that **Compound 3f** binds to the same allosteric site as **ANS**, although crystallographic confirmation is still pending.

Fragment screening approaches have also been employed to identify small molecules capable of binding allosterically to CDK2, with promising results. Through this strategy, the acrylamide fragment **Compound 1** was discovered (Fig. 13) [63]. Crystallographic analysis confirmed that **Compound 1** binds specifically to the Cys177 residue, located near the cyclin binding interface of CDK2. Despite its binding affinity, **Compound 1** was found to be a weak inhibitor of the CDK2/cyclin A2 complex, achieving only 83% inhibition at a concentration of 0.5 mM.

In addition, another fragment, **Compound 2**, was identified through a similar screening process (Fig. 14) [64]. Crystallographic data showed that **Compound 2** binds at the ATP site of CDK2 and at a highly conserved secondary site. This secondary site corresponds to the second **ANS** pocket, which had been previously characterized and is

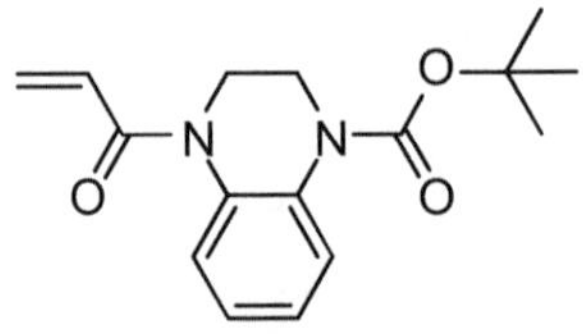

Compound 1
PDB: 5OSJ
Adjacent to cyclin binding interface

Fig. 13. Chemical structure of **Compound 1** and binding pocket.

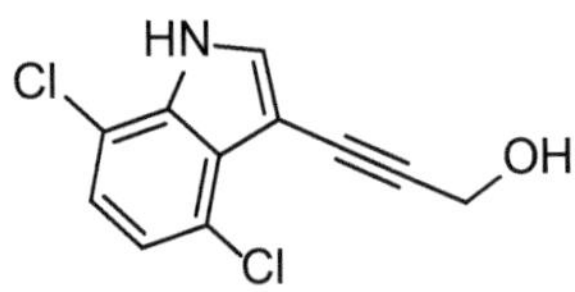

Compound 2
PDB: 5FP6
Second **ANS** pocket

Fig. 14. Chemical structure of **Compound 2** and binding pocket.

thought to play a role in disrupting the CDK2/cyclin A interface [56]. Identifying these two fragments (**Compound 1** and **Compound 2**) suggests that both could serve as starting points for developing novel allosteric inhibitors of CDK2. With further optimization, these compounds may lead to more potent inhibitors capable of modulating CDK2 activity in a highly selective manner, offering potential therapeutic applications.

An activity-based protein profiling strategy combined with X-ray crystallography was employed to uncover a cryptic allosteric pocket on CDK2 at the interface of the CDK2/cyclin E1 binding site [65]. The compound **I-125A**, described in the patent literature [66], was employed to inhibit the CDK2/cyclin E1 complex, and partial inhibition of CDK2/cyclin E1 activity was observed in functional assays (Fig. 15). The co-crystallography of CDK2/cyclin E1 and **I-125A** revealed a previously unrecognized binding pocket adjacent to the kinase C-lobe and A-loop but distal to the ATP binding site. Comparing the co-crystal structure with the apo-structure highlighted the significant structural rearrangements needed to accommodate **I-125A** binding. This suggested that the movement of the A-loop disrupts the phosphorylation of Thr160, a critical step for the full activation of CDK2. As a result, the 2,6-diazaspiro[3.4]octane class of inhibitors could serve as a scaffold for future optimizations in developing novel allosteric inhibitors targeting the CDK2/cyclin E1 complex.

Computational methods were employed to identify potential peptides that could bind to a non-catalytic pocket near the CDK2/cyclin E1 interface [67]. In particular, molecular dynamics simulations and dynamic network analysis were integrated to uncover correlations not apparent in static structural models. These insights guided the selection of candidate peptides for further investigation. Lead compounds were then

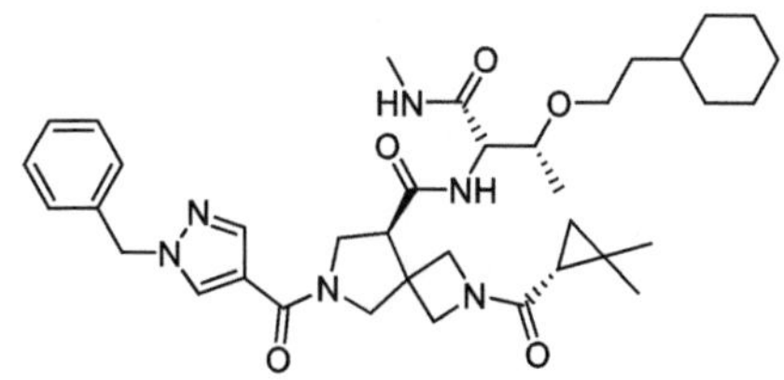

I-125A
PDB: 8VQ4
Cryptic pocket near CDK2/cyclin E1 interface
IC_{50} = 0.13 μM

Fig. 15. Chemical structure of **Compound I-125A**, binding pocket, and biological activity data.

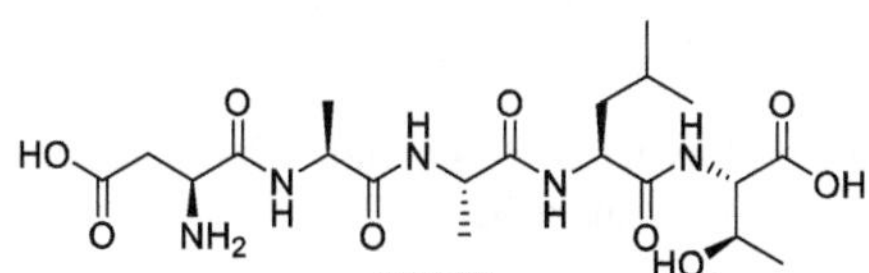

DAALT
Interface of CDK2/cyclin E1
K_D = 0.47 μM

Fig. 16. Chemical structure of **DAALT**, binding pocket, and binding affinity data.

evaluated for their ability to disrupt the CDK2/cyclin E1 complex through immunoprecipitation, followed by a kinase assay. Among the peptides tested, **DAALT** demonstrated the most significant reduction in CDK2 kinase activity, indicating it had the strongest effect on complex dissociation (Fig. 16). Further binding assays revealed that **DAALT** binds to CDK2 with a K_D of 0.47 μM. However, in the presence of 60 μM ATP, the binding affinity decreased nearly 100-fold. This observation led to the hypothesis that **DAALT** may compete with ATP either by directly occupying the orthosteric ATP binding site or indirectly by binding in a way that prevents ATP from stabilizing the conformation required for efficient enzyme activity. Additional studies will be needed to determine the binding site definitively and confirm the precise mechanism of action of **DAALT**.

Virtual HTS was used to identify potential CDK2/cyclin A3 complex inhibitors [68]. From this screening, **B2** emerged as the lead in subsequent biochemical assays, showing an IC_{50} of 52 μM against the CDK2/cyclin A3 complex (Fig. 17). Competition assays conducted in the presence of ATP revealed no significant change in the IC_{50} values, suggesting that **B2** inhibits the complex through an allosteric mechanism. Additionally, GST pull-down assays demonstrated that **B2** effectively disrupts the CDK2/cyclin A3 complex in a concentration-dependent manner. Based on these findings, the authors hypothesized that **B2** binds in a non-catalytic pocket near the interface of the CDK2/cyclin A3 complex. However, further analysis will be required to validate this hypothesis. Compound **B2** contains a phenyl imine substructure, making it a potentially problematic screening hit that could be a pan-assay interference compound (PAIN) [69].

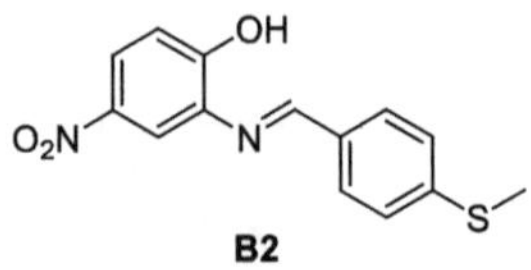

B2
Near interface of CDK2/cyclin A3
IC_{50} = 52 μM

Fig. 17. Chemical structure of **B2**, binding pocket, and biological activity data.

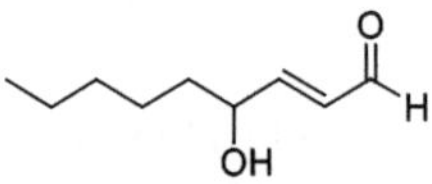

4-hydroxy-2-nonenal (HNE)
Sites of modification:
His60, His71, Lys129, His161,
Cys177, His268, His283, His295

Fig. 18. Chemical structure of **4-hydroxy-2-nonenal** (**HNE**) and sites of covalent modification identified through mass spectrometry analysis.

A proteomic approach combined with mapping to the CDK2 crystal structure revealed that the well-known electrophile **4-hydroxy-2-nonenal** (**HNE**) covalently modifies CDK2 at multiple solvent-exposed histidine residues and a lysine residue (Fig. 18) [70]. Notably, **HNE** modifies His71, which is located at the interface of the cyclin E/A binding site, and His161, which is immediately adjacent to Thr160, a critical phosphorylation site for CDK2 kinase activity. In vitro studies using a colorectal cancer cell line (RKO) also showed a modification of Cys177. **HNE** treatment reduced kinase activity in CDK2/cyclin A complexes, which correlated with a delay in the progression of the cell cycle into the S phase. No significant changes were observed in **HNE**-treated CDK2/cyclin E complexes. Overall, covalent inhibition of CDK2, in the context of either cancer treatment or contraception, could be a viable strategy if selectivity is achieved.

3 Future Directions

Future research on allosteric inhibitors of CDK2 should prioritize the development of more potent compounds with higher affinity and slower off-rates. Current allosteric inhibitors, such as **EF-3-006**, exhibit fast association and dissociation rates, whereas cyclin has a significantly slower dissociation rate—more than ten-fold lower [58, 71]. This off-rate disparity allows cyclin to outcompete allosteric inhibitors, particularly at higher cyclin concentrations. Designing inhibitors with slower off-rates, closer to those of cyclin, could enhance their efficacy and prolong their duration of action.

One strategy to achieve reduced off-rates is the development of covalent inhibitors. **Compound 1** and **HNE** have demonstrated potential as covalent inhibitors, along with the co-crystallized orthosteric inhibitor **NU6300**. As the first described irreversible inhibitor of CDK2, **NU6300** forms a covalent bond with Lys89 near the ATP binding site [72]. This discovery provides valuable insights into covalent inhibition strategies and highlights the feasibility of using this approach to create potent and selective CDK2 inhibitors. Future efforts could focus on designing novel covalent allosteric inhibitors that combine the selectivity advantages of allosteric binding with the potency and irreversibility of covalent inhibition.

Another promising direction is the exploration of bivalent inhibitors. Developing bivalent compounds that simultaneously target allosteric and orthosteric sites could yield more potent and selective CDK2 inhibitors. This approach can potentially integrate the benefits of both inhibition mechanisms, resulting in compounds with improved pharmacological profiles while addressing some limitations of current inhibitors [73, 74].

4 Summary

Allosteric inhibitors of CDK2 represent a promising and underexplored area of drug discovery, offering the potential for more selective and targeted therapies than traditional ATP site inhibitors. These allosteric compounds bind to non-catalytic pockets on CDK2, avoiding the highly conserved ATP binding site shared across CDKs, particularly CDK1, which often leads to unwanted off-target effects. Recent crystallographic studies have confirmed the presence of four distinct allosteric pockets on CDK2, providing valuable structural insights that could aid in the design of more selective inhibitors. A notable inhibitor of the most well-characterized allosteric pocket is **EF-3-006**, which has been reported as an allosteric inhibitor with an impressive 18 nM K_D for CDK2 and remarkable selectivity over CDK1 [57]. **EF-3-006** also demonstrated the ability to arrest meiosis in mouse testicular explants, showcasing its potential for contraceptive applications. This selective inhibition is key in therapeutic settings, such as cancer treatment and contraception, where precise modulation of CDK2 is desired without compromising the function of CDK1. Developing these allosteric inhibitors could pave the way for more targeted therapies, offering greater specificity in modulating CDK2 activity while reducing the risk of side effects associated with broader kinase inhibition.

References

1. Wood, D.J., Endicott, J.A.: Structural insights into the functional diversity of the CDK–cyclin family. Open Biol. **8**, 180112 (2018)
2. Morgan, D.O.: Cyclin-dependent kinases: engines, clocks, and microprocessors. Annu. Rev. Cell Dev. Biol. **13**, 261–291 (1997)
3. Diril, M.K., et al.: Cyclin-dependent kinase 1 (Cdk1) is essential for cell division and suppression of DNA re-replication but not for liver regeneration. PNAS. **109**, 3826–3831 (2012)

4. Berthet, C., Aleem, E., Coppola, V., Tessarollo, L., Kaldis, P.: Cdk2 knockout mice are viable. Curr. Biol. **13**, 1775–1785 (2003)
5. Morris, M.C., Gondeau, C., Tainer, J.A., Divita, G.: Kinetic mechanism of activation of the Cdk2/cyclin a complex: key role of the C-lobe of the Cdk. J. Biol. Chem. **277**, 23847–23853 (2002)
6. Malumbres, M.: Cyclin-dependent kinases. Genome Biol. **15**, 1–10 (2014)
7. Hope, I., Endicott, J.A., Watt, J.E.: Emerging approaches to CDK inhibitor development, a structural perspective. RSC Chem. Biol. **4**, 146 (2023)
8. De Bondt, H.L., Rosenblatt, J., Jancarik, J., Jones, H.D., Morgan, D.O., Kim, S.-H.: Crystal structure of cyclin-dependent kinase 2. Nature. **363**, 595–602 (1993)
9. Jeffrey, P.D., et al.: Mechanism of CDK activation revealed by the structure of a cyclinA-CDK2 complex. Nature. **376**, 313–320 (1995)
10. Takaki, T., Echalier, A., Brown, N., Hunt, T., Endicott, J., Noble, M.: The structure of CDK4/cyclin D3 has implications for models of CDK activation. PNAS. **106**, 4171–4176 (2009)
11. Zhang, J., Yang, P.L., Gray, N.S.: Targeting cancer with small molecule kinase inhibitors. Nat. Rev. Cancer. **9**, 28–39 (2009)
12. Majumdar, A., et al.: Allostery governs Cdk2 activation and differential recognition of CDK inhibitors. Nat. Chem. Biol. **17**, 456–464 (2021)
13. Roskoski Jr., R.: Cyclin-dependent protein serine/threonine kinase inhibitors as anticancer drugs. Pharmacol. Res. **139**, 471–488 (2019)
14. Au-Yeung, G., et al.: Selective targeting of cyclin E1-amplified high-grade serous ovarian cancer by cyclin-dependent kinase 2 and AKT inhibition. Clin. Cancer Res. **23**, 1862–1874 (2017)
15. Akli, S., Van Pelt, C.S., Bui, T., Meijer, L., Keyomarsi, K.: Cdk2 is required for breast cancer mediated by the low-molecular-weight isoform of cyclin E. Cancer Res. **71**, 3377–3386 (2011)
16. Ray, D., Terao, Y., Christov, K., Kaldis, P., Kiyokawa, H.: Cdk2-null mice are resistant to ErbB-2-induced mammary tumorigenesis. Neoplasia. **13**, 439–444 (2011)
17. Karaiskou, A., Perez, L.H., Ferby, I., Ozon, R., Jessus, C., Nebreda, A.R.: Differential regulation of Cdc2 and Cdk2 by RINGO and cyclins. J. Biol. Chem. **276**, 36028–36034 (2001)
18. McGrath, D.A., Fifield, B.A., Marceau, A.H., Tripathi, S., Porter, L.A., Rubin, S.M.: Structural basis of divergent cyclin-dependent kinase activation by Spy1/RINGO proteins. EMBO J. **36**, 2251–2262 (2017)
19. Fifield, B.A., Vusich, J., Haberfellner, E., Andrechek, E.R., Porter, L.A.: Atypical cell cycle regulation promotes mammary stem cell expansion during mammary development and tumourigenesis. Breast Cancer Res. **26**, 106 (2024)
20. McAndrew, C.W., Gastwirt, R.F., Donoghue, D.J.: The atypical CDK activator Spy1 regulates the intrinsic DNA damage response and is dependent upon p53 to inhibit apoptosis. Cell Cycle. **8**, 66–75 (2009)
21. Ferraiuolo, R.M., Fifield, B.A., Hamm, C., Porter, L.A.: Stabilization of c-Myc by the atypical cell cycle regulator, Spy1, decreases efficacy of breast cancer treatments. Breast Cancer Res. Treat. **196**, 17–30 (2022)
22. Al Sorkhy, M., et al.: The cyclin-like protein Spy1/RINGO promotes mammary transformation and is elevated in human breast cancer. BMC Cancer. **12**, 45 (2012)
23. Meister, D., Hayward, J., Fifield, B.-A., Porter, L., Trant, J.: Spy1-activated cyclin dependent kinase 2 shows reduced affinity for current synthetic small molecule inhibitors of the CDK2-cyclin complex. ChemRxiv, September 12 (2023)
24. Fifield, B.-A., Qemo, I., Kirou, E., Cardiff, R.D., Porter, L.A.: The atypical cyclin-like protein Spy1 overrides p53-mediated tumour suppression and promotes susceptibility to breast tumourigenesis. Breast Cancer Res. **21**, 1–18 (2019)

25. Lu, S., et al.: Spy1 participates in the proliferation and apoptosis of epithelial ovarian cancer. J. Mol. Histol. **47**, 47–57 (2016)
26. Zhang, L., Shen, A., Ke, Q., Zhao, W., Yan, M., Cheng, C.: Spy1 is frequently overexpressed in malignant gliomas and critically regulates the proliferation of glioma cells. J. Mol. Neurosci. **47**, 485–494 (2012)
27. Ke, Q., et al.: Expression and prognostic role of Spy1 as a novel cell cycle protein in hepatocellular carcinoma. Exp. Mol. Pathol. **87**, 167–172 (2009)
28. Hang, Q., et al.: Expression of Spy1 protein in human non-Hodgkin's lymphomas is correlated with phosphorylation of p27 Kip1 on Thr187 and cell proliferation. Med. Oncol. **29**, 3504–3514 (2012)
29. Rhee, K., Wolgemuth, D.J.: Cdk family genes are expressed not only in dividing but also in terminally differentiated mouse germ cells, suggesting their possible function during both cell division and differentiation. Dev. Dyn. **204**, 406–420 (1995)
30. Viera, A., Rufas, J.S., Martinez, I., Barbero, J.L., Ortega, S., Suja, J.A.: CDK2 is required for proper homologous pairing, recombination and sex-body formation during male mouse meiosis. J. Cell Sci. **122**, 2149–2159 (2009)
31. Ortega, S., et al.: Cyclin-dependent kinase 2 is essential for meiosis but not for mitotic cell division in mice. Nat. Genet. **35**, 25–31 (2003)
32. Faber, E.B., Wang, N., Georg, G.I.: Review of rationale and progress toward targeting cyclin-dependent kinase 2 (CDK2) for male contraception. Biol. Reprod. **103**, 357–367 (2020)
33. Fagerberg, L., et al.: Analysis of the human tissue-specific expression by genome-wide integration of transcriptomics and antibody-based proteomics. Mol. Cell. Proteomics. **13**, 397–406 (2014)
34. Mikolcevic, P., et al.: Essential role of the Cdk2 activator RingoA in meiotic telomere tethering to the nuclear envelope. Nat. Commun. **7**, 11084 (2016)
35. Satyanarayana, A., Berthet, C., Lopez-Molina, J., Coppola, V., Tessarollo, L., Kaldis, P.: Genetic substitution of Cdk1 by Cdk2 leads to embryonic lethality and loss of meiotic function of Cdk2. Development. **135**, 3389–3400 (2008)
36. Aleem, E., Kiyokawa, H., Kaldis, P.: Cdc2-cyclin E complexes regulate the G1/S phase transition. Nat. Cell Biol. **7**, 831–836 (2005)
37. Santamaria, D., et al.: Cdk1 is sufficient to drive the mammalian cell cycle. Nature. **448**, 811–815 (2007)
38. Lau, H.W., et al.: Quantitative differences between cyclin-dependent kinases underlie the unique functions of CDK1 in human cells. Cell Rep. **37**, 109808 (2021)
39. Xie, Z., et al.: Lessons learned from past cyclin-dependent kinase drug discovery efforts. J. Med. Chem. **65**, 6356–6389 (2022)
40. Giarolla, J., Holdaway, K., Nazari, M., Aiad, L., Sarkar, B., Georg, G.: Targeting cyclin-dependent kinase 2 (CDK2) interactions with cyclins and speedy 1 (Spy1) for cancer and male contraception. Future Med. Chem. **17**, 607 (2025)
41. Shen, C., et al.: PF-07104091, a first-in-class CDK2-selective inhibitor for the treatment of HR+/HER2-breast cancer and CCNE1 high ovarian cancer. Cancer Res. **84**, 5709–5709 (2024)
42. Brown, V., et al.: BLU-222, an investigational, potent, and selective CDK2 inhibitor, demonstrated robust antitumor activity in CCNE1-amplified ovarian cancer models. In: Proceedings of the American Association for Cancer Research Annual Meeting, vol. 2022, pp. 8–13 (2022)
43. Denz, C.R., et al.: Targeting the cell cycle with AZD8421, a potent and highly selective CDK2 inhibitor. Cancer Res. **84**, 5730–5730 (2024)
44. Dietrich, C., et al.: INX-315, a selective CDK2 inhibitor, induces cell cycle arrest and senescence in solid tumors. Cancer Discov. **14**, 446–467 (2024)
45. Chand, S., et al.: Development of a CDK2-selective small molecule inhibitor INCB123667 for the treatment of CCNE1hi breast cancers. Cancer Res. **83**, 10.1158 (2023)

46. Wu, P., Clausen, M.H., Nielsen, T.E.: Allosteric small-molecule kinase inhibitors. Pharmacol. Ther. **156**, 59–68 (2015)
47. Kalra, S., Joshi, G., Munshi, A., Kumar, R.: Structural insights of cyclin dependent kinases: implications in design of selective inhibitors. Eur. J. Med. Chem. **142**, 424–458 (2017)
48. Papadimitriou, M.C., Pazaiti, A., Iliakopoulos, K., Markouli, M., Michalaki, V., Papadimitriou, C.A.: Resistance to CDK4/6 inhibition: mechanisms and strategies to overcome a therapeutic problem in the treatment of hormone receptor-positive metastatic breast cancer. BBA-Mol. Cell. Res. **1869**, 119346 (2022)
49. Rampioni Vinciguerra, G.L., et al.: CDK4/6 inhibitors in combination therapies: better in company than alone: a mini review. Front. Oncol. **12**, 891580 (2022)
50. Freeman-Cook, K.D., Hoffman, R.L.: Recent advances in the discovery of cylclin-dependent kinase 2 (CDK2) selective inhibitors. MCR, 283–311 (2023)
51. Arora, M., et al.: Rapid adaptation to CDK2 inhibition exposes intrinsic cell-cycle plasticity. Cell. **186**, 2628–2643 (2023)
52. Roskoski Jr., R.: Properties of FDA-approved small molecule protein kinase inhibitors: a 2024 update. Pharmacol. Res. **200**, 107059 (2024)
53. Laufer, S., Bajorath, J.: New horizons in drug discovery-understanding and advancing different types of kinase inhibitors: seven years in kinase inhibitor research with impressive achievements and new future prospects. J. Med. Chem. **65**, 891–892 (2021)
54. Pan, Y., Mader, M.M.: Principles of kinase allosteric inhibition and pocket validation. J. Med. Chem. **65**, 5288–5299 (2022)
55. Schoepfer, J., et al.: Discovery of asciminib (ABL001), an allosteric inhibitor of the tyrosine kinase activity of BCR-ABL1. J. Med. Chem. **61**, 8120–8135 (2018)
56. Betzi, S., et al.: Discovery of a potential allosteric ligand binding site in CDK2. ACS Chem. Biol. **6**, 492–501 (2011)
57. Faber, E.B., et al.: Screening through lead optimization of high affinity, allosteric cyclin-dependent kinase 2 (CDK2) inhibitors as male contraceptives that reduce sperm counts in mice. J. Med. Chem. **66**, 1928–1940 (2023)
58. Faber, E.B., et al.: Development of allosteric and selective CDK2 inhibitors for contraception with negative cooperativity to cyclin binding. Nat. Commun. **14**, 3213 (2023)
59. Rastelli, G., Anighoro, A., Chripkova, M., Carrassa, L., Broggini, M.: Structure-based discovery of the first allosteric inhibitors of cyclin-dependent kinase 2. Cell Cycle. **13**, 2296–2305 (2014)
60. Martin, M.P., Alam, R., Betzi, S., Ingles, D.J., Zhu, J.Y., Schonbrunn, E.: A novel approach to the discovery of small-molecule ligands of CDK2. Chembiochem. **13**, 2128–2136 (2012)
61. Christodoulou, M.S., et al.: Probing an allosteric pocket of CDK2 with small molecules. ChemMedChem. **12**, 33–41 (2017)
62. Carlino, L., et al.: Structure-activity relationships of hexahydrocyclopenta[c]quinoline derivatives as allosteric inhibitors of CDK2 and EGFR. ChemMedChem. **13**, 2627–2634 (2018)
63. Craven, G.B., et al.: High-throughput kinetic analysis for target-directed covalent ligand discovery. Angew. Chem. Int. Ed. Engl. **57**, 5257–5261 (2018)
64. Ludlow, R.F., Verdonk, M.L., Saini, H.K., Tickle, I.J., Jhoti, H.: Detection of secondary binding sites in proteins using fragment screening. PNAS. **112**, 15910–15915 (2015)
65. Zhang, Y., et al.: Expanding the ligandable proteome by paralog hopping with covalent probes. bioRxiv (2024)
66. Kirman, L.C., et al.: CDK2 inhibitors and methods of using the same. WO2022165513A2022165511 (2022)
67. Chen, H., et al.: Break CDK2/cyclin E1 interface allosterically with small peptides. PLoS One. **9**, e109154 (2014)

68. Hu, Y., Li, S., Liu, F., Geng, L., Shu, X., Zhang, J.: Discovery of novel nonpeptide allosteric inhibitors interrupting the interaction of CDK2/cyclin A3 by virtual screening and bioassays. Bioorg. Med. Chem. Lett. **25**, 4069–4073 (2015)
69. Baell, J.B., Nissink, J.W.M.: Seven year itch: Pan-assay interference compounds (PAINS) in 2017 - utility and limitations. ACS Chem. Biol. **13**, 36–44 (2018)
70. Camarillo, J.M., Rose, K.L., Galligan, J.J., Xu, S., Marnett, L.J.: Covalent modification of CDK2 by 4-hydroxynonenal as a mechanism of inhibition of cell cycle progression. Chem. Res. Toxicol. **29**, 323–332 (2016)
71. Putta, S., Villegas, C.A., Rubin, S.M.: Differences in binding affinity among cell-cycle CDK and cyclin pairs. J. Mol. Biol. **437**, 168952 (2025)
72. Anscombe, E., et al.: Identification and characterization of an irreversible inhibitor of CDK2. Chem. Biol. **22**, 1159–1164 (2015)
73. Lee, S., Kim, J., Jo, J., Chang, J.W., Sim, J., Yun, H.: Recent advances in development of hetero-bivalent kinase inhibitors. Eur. J. Med. Chem. **216**, 113318 (2021)
74. Wittlinger, F., et al.: Linking ATP and allosteric sites to achieve superadditive binding with bivalent EGFR kinase inhibitors. Commun. Chem. **7**, 38 (2024)

Evaluation of 2-phenylthiazolidin-4-one Analogues as Inhibitors of the SARS-CoV-2 Main Protease

Anthi Petrou[1(✉)], Athina Geronikaki[1(✉)], Hathaichanok Chuntakaruk[2,3], Leentje Persoons[4(✉)], Thanyada Rungrotmongkol[2,3], Dominique Schols[4(✉)], Aliki Papadimitriou-Tsantarliotou[4(✉)], Ioannis S. Vizirianakis[5,6 (✉)], Steven De Jonghe[4(✉)], and Fliur Z. Macaev[7(✉)]

[1] School of Health, Department of Pharmacy, Aristotle University of Thessaloniki, Thessaloniki, Greece
anthi.petrou.thessaloniki1@gmail.com; geronik@pharm.auth.gr
[2] Program in Bioinformatics and Computational Biology, Graduate School, Chulalongkorn University, Bangkok, Thailand
[3] Center of Excellence in Structural and Computational Biology, Department of Biochemistry, Faculty of Science, Chulalongkorn University, Bangkok, Thailand
[4] Laboratory of Virology and Chemotherapy, Rega Institute for Medical Research, Department of Microbiology, Immunology and Transplantation, Herestraat Leuven, Belgium
leentje.persoons@kuleuven.be; dominique.schols@kuleuven.be; alikipapadi@pharm.auth.gr; steven.dejonghe@kuleuven.be
[5] Laboratory of Pharmacology, School of Pharmacy, Aristotle University of Thessaloniki, Thessaloniki, Greece
[6] Department of Health Sciences, School of life and Health Sciences, University of Nicosia, Nicosia, Cyprus
ivizir@pharm.auth.gr
[7] Laboratory Organic Synthesis, Institute of Chemistry, Moldova State University, Chişinău, Moldova
fliur.macaev@sti.usm.md

Abstract. The continuous emergence of SARS-CoV-2 mutations has prolonged the COVID-19 pandemic and raised concerns about the effectiveness of current vaccines. In this study, we present the evaluation of several thiazole/benzothiazole-based thiazolidinone derivatives, selected from a pool of 85 designed compounds through molecular docking, as potential inhibitors of the SARS-CoV-2 main protease. The experimental data revealed that among the fifteen compounds chosen, five compounds showed inhibitory activity with IC_{50} within the range of 0.19–13.15 μM. By assessing the cellular effects of these molecules, we observed that they didn't have the capacity to affect the cellular viability of human normal MRC-5 cells considering them safe. Collectively, these results support the potential antiviral properties of thiazole/benzothiazole-based thiazolidin-4-one derivatives as promising candidates for the development of COVID-19 therapeutics.

A. K. Saxena, A. Saxena (Eds.): GTHTM 2025, pp. 43–56, 2026.
https://doi.org/10.1007/978-3-032-12320-6_3

Keywords: First SARS-CoV2 main protease · Virtual screening · Synthesis · Molecular docking · Molecular dynamics simulations

1 Introduction

The first report of an outbreak by novel corona virus in Wuhan, China was first reported in December 2019 and despite any attempts to contain it there, this virus spread worldwide causing a pandemic. This virus belongs to Coronaviridae family. It is one of the RNA viruses, responsible for human respiratory distress ranging from common cold, dry cough, high fever, lungs infection to severe pneumonia. This novel coronavirus caused pandemic which is responsible to kill more than 7.09 million deaths worldwide [1, 2]. On January 10, 2020, Zhang et al. [3] uploaded the novel virus's first genome to GenBank (MN908947) and other databases.

The respiratory tract cells are most affected by this virus's high pathogenicity, which leads to respiratory illnesses that can progress into serious, perhaps, fatal pathologies. The SARS-CoV-2 genome codes for accessory, structural, and non-structural proteins that are vital to the viral replication cycle [4–6]. The primary potential targets for antiviral treatments are spike protein (S), 3C-like protease (3CLpro) or ***main protease*** (M^{pro}), papain-like cysteine protease (PLpro), and RNA-dependent RNA polymerase (RdRp) [7].

The SARS-CoV-2 main protease (Mpro, also known as 3CLpro) plays an indispensable role in viral replication. After infection, the viral RNA genome is translated into two large polyproteins, pp1a and pp1ab. These polyproteins must be cleaved at more than 11 distinct sites to release the functional non-structural proteins (nsps) that form the replication–transcription complex (RTC). Mpro is responsible for catalyzing most of these cleavage events, thereby enabling the production of essential viral enzymes such as the RNA-dependent RNA polymerase (RdRp) and the helicase. Without this proteolytic processing, the virus cannot replicate efficiently. Importantly, because Mpro recognizes specific cleavage sequences that differ from those targeted by human proteases, selective inhibition of Mpro can effectively block viral replication with minimal off-target toxicity. This makes Mpro a validated and highly attractive drug target for the development of anti-SARS-CoV-2 and pan-coronavirus therapeutics [8, 9]. This enzyme is a prime target for prospective anti-COVID medications due to its critical function in the replication cycle of the virus and involvement in the cleavage of polyproteins. But the most significant is that this enzyme has a high degree of conservation among *Coronaviridae* family members, this is very important because it can be used for the design and discovery of new pan-CoV (Pan-Coronavirus) Direct-Acting Antivirals. Moreover, the binding site of this enzyme is highly different from that of known human cysteine proteases, allowing specific inhibitions with reduced toxicity.

6 LU7 is the first main protease structure submitted to the Protein Data Bank [10, 11]. This structure opened the door to in silico research for developing SARS-CoV-2 main protease inhibitors. A potent method for the quick discovery of lead molecules for clinical usage is molecular docking studies [12–14]. It gives us the possibility to simu-

late the interactions between a small molecule ligand and a macromolecular protein and describe how it behaves in the binding site.

The current study is a continuation of our previous work on finding suitable inhibitors for the main protease in SARS-CoV2 using computational methodologies.

2 Materials and Methods

2.1 Molecular Docking

The SARS-CoV-2 main protease (Mpro) structure in complex with the inhibitor N3 (PDB ID: 6 LU7) was selected from the Protein Data Bank (https://www.rcsb.org) [10, 15]. Protein preparation was performed using SPDBV [16] to fix missing residues, followed by the removal of water molecules and the co-crystallized ligand. Polar hydrogens and charges were added using AutoDock Tools, and the final structure was saved in pdbqt format.

Ligands were sketched in ChemDraw 12.0 [18], optimized using MMFF94 force fields in LigandScout [19], and saved as mol2 files. Partial charges were computed followed by the generation of conformers, and the best conformer was selected. The structures were then converted to pdbqt format using AutoDock Tools by adding polar hydrogens, Gasteiger and Kollman charges, and defining torsions.

Molecular docking was conducted using AutoDock 4.2 [20]. Conformational sampling was performed with the Lamarckian Genetic Algorithm (LGA), which combines a genetic algorithm with local search optimization to explore ligand flexibility. The scoring function applied was the AutoDock semi-empirical free energy force field, which estimates binding affinity by accounting for van der Waals, hydrogen bonding, electrostatics, desolvation, and torsional entropy terms. The docking grid was centered at $x = -10.46$, $y = 4.1037$, $z = 73.009$, with dimensions of $30 \times 40 \times 25$ Å to encompass the active site. To ensure specific binding, docking was repeated with an expanded grid covering the entire enzyme. Docking validation was performed by re-docking N3, yielding an RMSD of 0.91 Å. The best binding poses and interactions were analyzed using Biovia Discovery Studio Visualizer and LigandScout [19].

2.2 ADMET Properties

The ADMET properties were predicted using the pkCSM server [21], as previously reported in [22].

2.3 Chemistry

The melting points were measured using the MEL-TEMP II device (LAB Devices, Holliston, MA, USA). ^{1}H NMR nuclear magnetic resonance spectra were acquired in DMSO-d_6 or $CDCl_3$ using an Agilent spectrometer operating at 500 MHz. Chemical shifts are given in parts per million (ppm), using tetramethylsilane (TMS, $\delta = 0$ ppm) as the internal reference. The progression of reactions was tracked using thin-layer chromatography (TLC) on F254 silica gel plates (Merck, Darmstadt, Germany). All solvents and reagents were obtained from Aldrich Chemie (Steinheim, Germany) and were of analytical grade. The synthesis of compounds 1–15 followed previously described procedures [22].

2.3.1 Synthesis of Compounds A1 and A2. Equal amounts of suitable aromatic amine and benzaldehyde (1.0 mmol) are stirred and heated under reflux in dry toluene, followed by the addition of 5% L-proline, which acts as a catalyst. The stirring and heating continue for 8 h until the complete reaction of the aromatic amine. The progress of the reaction is monitored by thin-layer chromatography using a solvent mixture of toluene:ethanol (8:2).

(Z)-4-((2,6-difluorobenzylidene)amino)-N-(thiazol-2-yl)benzenesulfonamide (A**1**).

Yield: 46%. m.p. 223–224 °C. R_f: 0.62 (Toluene/Ethanol, 8/2). ^{1}H NMR (500 MHz, DMSO-d_6) δ 8.03–7.97 (m, 2H), 7.76 (dd, J = 1.9, 8.6 Hz, 2H), 7.25 (s, 1H), 7.15 (s, 3H), 3.14 (s, 1H), 3.11 (s, 1H), 3.11 (s, 1H). ^{13}C NMR (126 MHz, DMSO-d_6) δ 168.92, 148.67, 139.20, 134.57, 128.17, 126.46, 105.25.

(Z)-N-(4-fluorobenzylidene)-4-(4-fluorophenyl)thiazol-2-amine (A**2**).

Yield: 83%. m.p. 251–252 °C. R_f: 0.74 (Toluene/Ethanol, 8/2). ^{1}H NMR (500 MHz, DMSO-d_6) δ 9.05 (s, 1H), 8.08 (t, J = 6.7 Hz, 1H), 7.75 (s, 4H), 7.44 (d, J = 9.0 Hz, 1H), 7.36 (dq, J = 7.2, 7.8, 23.6 Hz, 2H), 7.06 (dt, J = 8.3, 21.7 Hz, 1H).

2.4 Inhibition of SARS-CoV-2 3CLpro Enzymatic Activity by Synthesized Compounds

The main protease (3CLpro) enzymatic assay was carried out using the SARS-CoV-2 specific 3CLpro assay kit purchased from BPS Biosciences (Catalog #79955-1, San Diego, CA, USA), in accordance with the manufacturer's protocol as previously outlined [16]. IC_{50} values of the tested compounds were determined using the GraphPad Prism 9 software with non-linear regression (curve fitting). The reference inhibitor GC376, supplied by BPS Biosciences, exhibited an IC_{50} value of 0.439 μM, aligning with the manufacturer's protocol value and confirming the validity of the in vitro assay.

2.5 Evaluation of Cellular Viability

The cytotoxicity was evaluated in MRC-5 cells under standard conditions, as previously reported in [16, 22].

3 Results and Discussion

3.1 Rational Design of Compounds

The SARS-CoV-2 main protease (Mpro) cleaves its substrate at specific sites, specifically before serine (Ser), alanine (Ala), or glycine (Gly) residues and immediately after glutamine (Gln), which follows leucine (Leu). This cleavage site is denoted as Leu-Gln

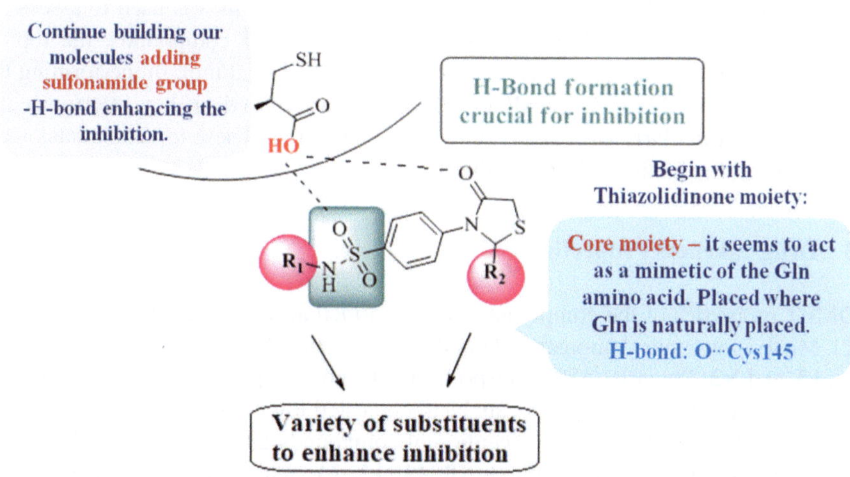

Fig. 1. The design of novel Mpro inhibitors.

↓ Ser/Ala/Gly ↓ (Fig. 1). The Leu residue is positioned near key residues such as His41, Met49, Tyr54, and Met165, while Gln is situated close to Cys145 and surrounded by Asn142, Glu166, His163, and His172. The catalytic function of Mpro relies on the dyad formed by Cys145 and His41, where His41 acts as a proton acceptor and Cys145 serves as a nucleophile [23, 24]. Consequently, for a molecule to exhibit strong binding affinity, it must effectively interact with these critical catalytic residues.

Previous studies [16] have shown that the thiazolidinone scaffold serves as a bioisostere of the Gln residue found in the enzyme's natural substrate. This moiety is positioned within the active site between Glu166 and the catalytic Cys145, where Gln would naturally be located. Additionally, the carbonyl oxygen of thiazolidinone forms hydrogen bonds with Cys145, reinforcing its interaction within the binding site. Moreover, fluorine substituents were found to enhance inhibitory activity [16]. Considering the findings from our prior research [16, 18], we preserved the fluorine and thiazolidinone functionalities while introducing a sulfonamide linker to mimic the native amide bond of the substrate, potentially enabling interactions with Cys145 (Fig. 1).

In the end, to evaluate whether the presence of the sulfonamide bond and thiazolidinone moiety is crucial for inhibition, we synthesized two additional compounds that lack these functionalities. By comparing the inhibitory activity of these modified compounds with the original ones containing the sulfonamide and thiazolidinone, we aim to assess the importance of these structural properties for the binding affinity and overall inhibitory potency against the novel coronaviral main protease.

3.2 Molecular Docking Studies

Molecular docking simulation was carried out to evaluate the binding affinity of designed compounds for SARS-CoV-2 Mpro. Using the crystal structure (PDB ID: 6 LU7), a series of thiazolidinone derivatives were docked as racemates, with the best conformers selected based on RMSD values and binding free energy. Additionally, the

PASS (Prediction of Activity Spectra for Substances) program was used to assess the biological activity potential of the compounds. Among 85 compounds, the top 15 showed binding free energies between −10.54 and − 7.4 kcal/mol, outperforming the reference inhibitor N3 (−7.7 kcal/mol). Selection was based on key intermolecular interactions, particularly with His41, Cys145, and Glu166. These top candidates were chosen for further analysis (Fig. 2, Table 1).

3.3 Drug-Likeness Properties

ADMET properties of the compounds was performed according our previous findings [22]. Most compounds demonstrated good Caco-2 permeability, except for compounds **2–6**, **13**, and **N3**. Potential ABC transporter-mediated transport was observed for compounds **1, 5, 10, 15**, and **N3**. VDss analysis suggested low tissue distribution for compounds **1, 7, 8,** and **14**. With the exception of compounds **10** and **15**, all displayed low BBB permeability, while compounds **10, 11, 12, 14**, and **15** showed potential CNS penetration. Metabolic predictions indicated that most compounds act as CYP2D6 and CYP3A4 substrates or inhibitors. Hepatotoxicity was predicted for compounds **1–8, 13,** and **N3**, whereas the remaining compounds were classified as non-toxic.

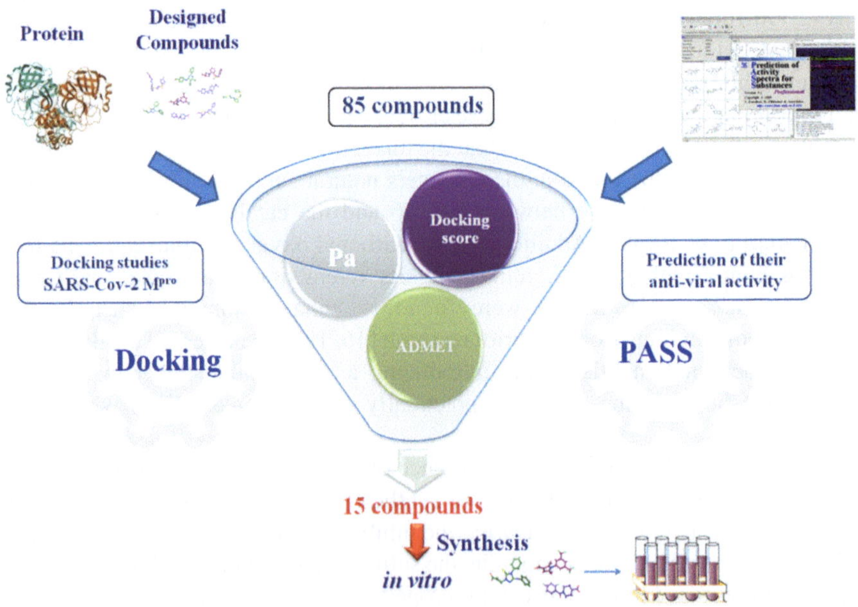

Fig. 2. Workflow diagram.

Table 1. Predicted binding free energies of selected compounds to SARS-CoV-2 Mpro.

No	Structure	Est. free energy of binding (kcal/mol)	Residues involved in hydrogen bond formation	Estimated Inhibition Constant, Ki (μM)
1	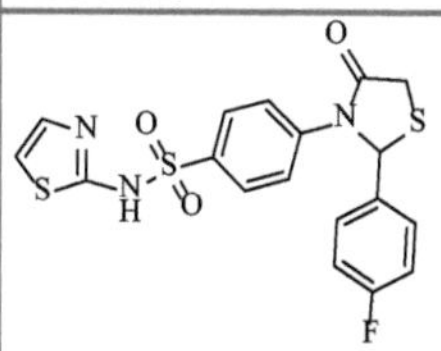	−8.93	Gln189, Gln192	0.92
2	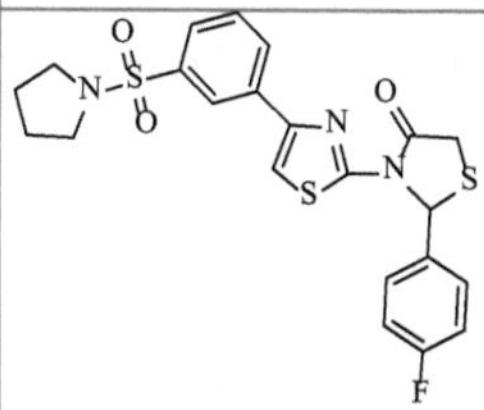	−8.65	Glu166, Cys145	1.56
3	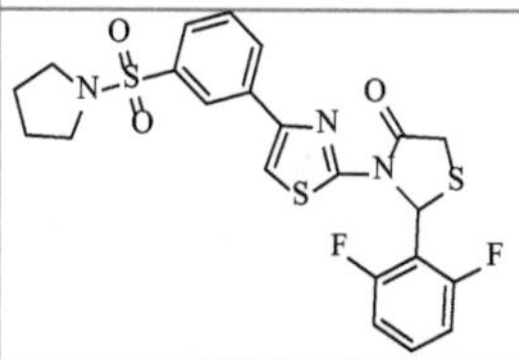	−8.18	His164, Glu166	1.93
4	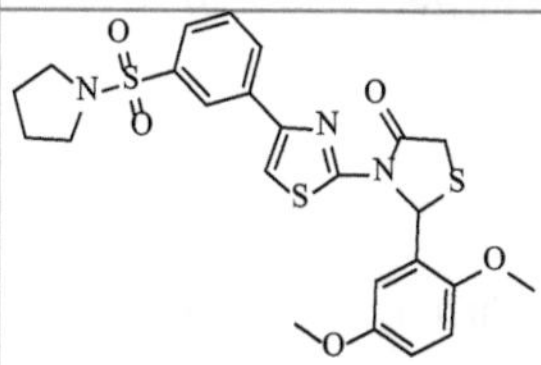	−8.16	His164, Glu166	2.13
5	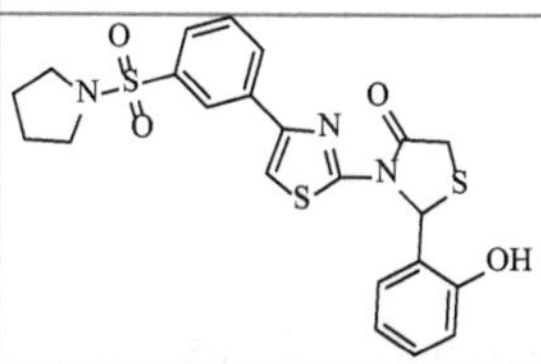	−7.43	Gly143, Cys145	2.49
6	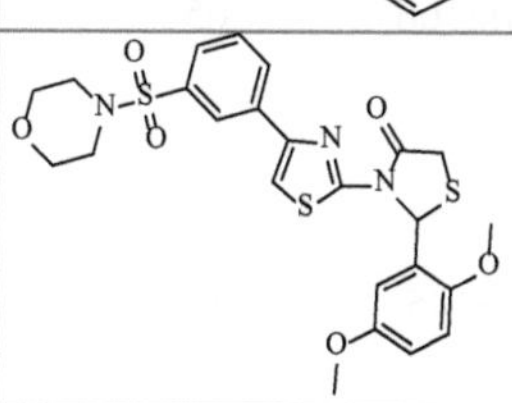	−7.41	Asn142, Cys145	4.16
7	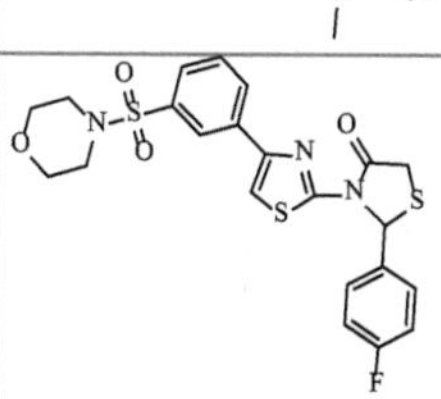	−8.06	His41, Asn142	2.02

(continued)

Table 1. (continued)

No	Structure	Est. free energy of binding (kcal/mol)	Residues involved in hydrogen bond formation	Estimated Inhibition Constant, Ki (μM)
8	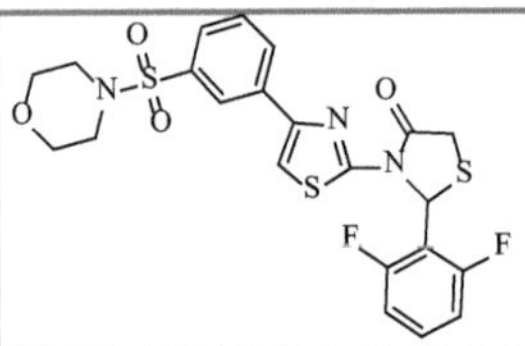	−7.52	Met165, Glu166	3.19
9	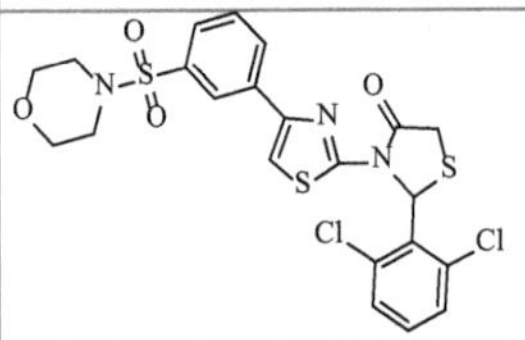	−8.26	Cys145, Glu166	2.57
10	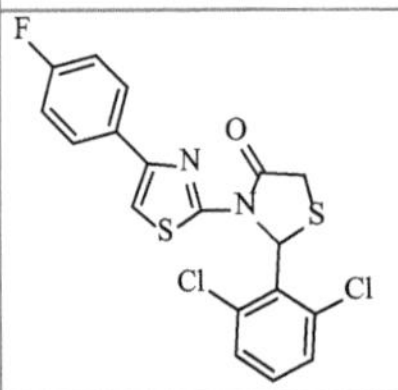	−8.01	Met165, Glu166	1.87
11	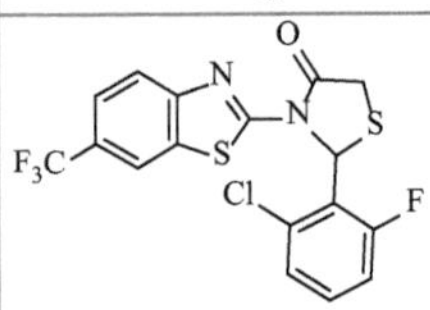	−8.82	Cys145, Glu166	1.89
12	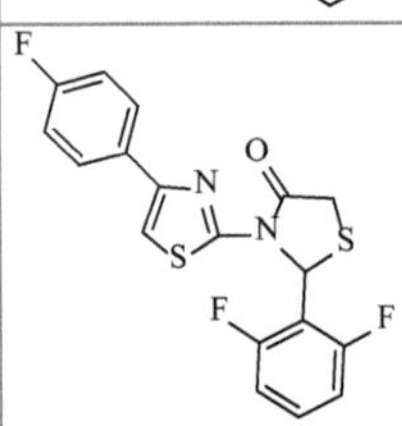	−9.21	Cys145, Glu166	1.89
13	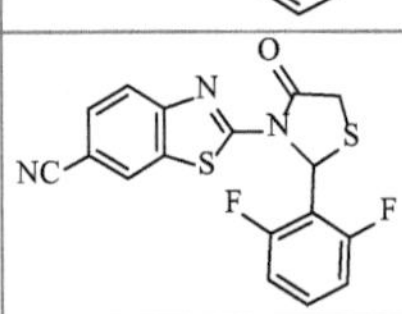	−10.54	Cys145, Glu166	1.65
14	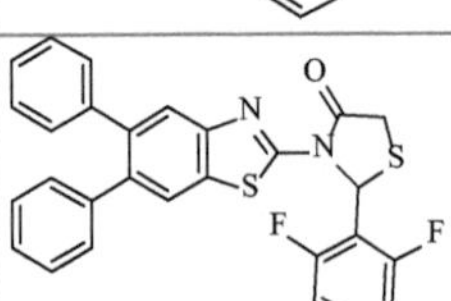	−8.13	Ser144, Cys145	3.19
15	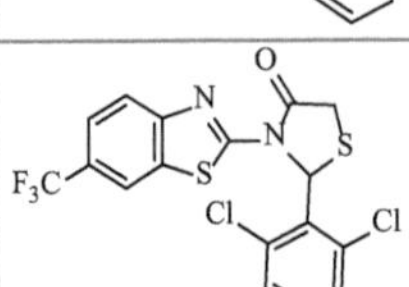	−7.56	Cys145	4.92

(continued)

Table 1. (continued)

No	Structure	Est. free energy of binding (kcal/mol)	Residues involved in hydrogen bond formation	Estimated Inhibition Constant, Ki (μM)
A1	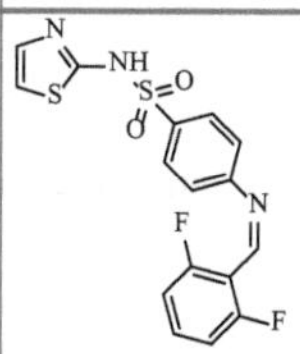	−7.20	Cys145, Asn142	5.98
A2	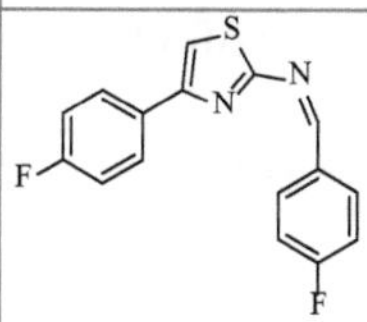	−5.94	Asn142	40.78
N3	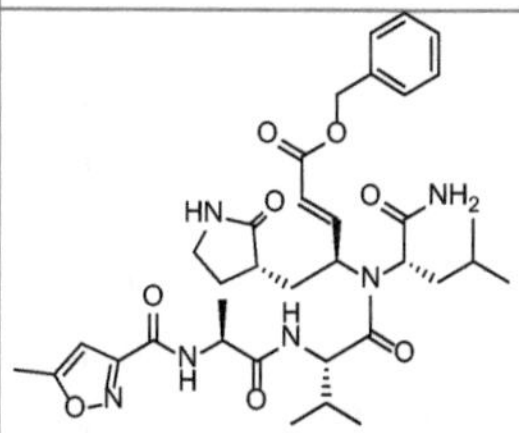	−7.70	Glu166, Gln189	3.58

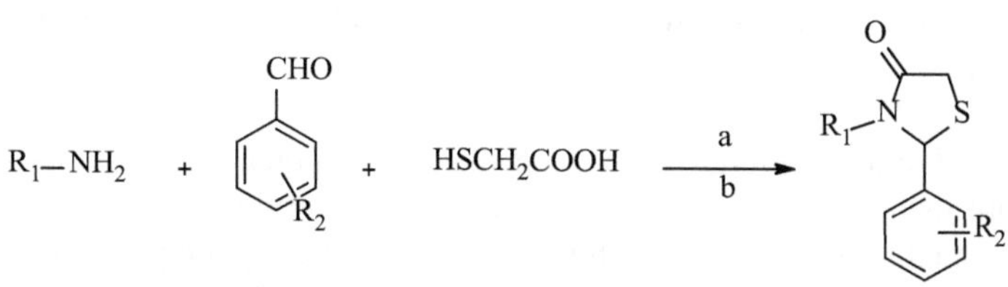

Scheme 1. General synthetic procedure. Reagents and conditions: (a) Conventional method: Dry toluene, reflux for 18–32 h, (b) microwave-assisted technique: 100 °C, power 100 W, 30 min.

3.4 Chemistry

Compounds **1-15** were synthesized using a one-pot method following Scheme 1, as detailed in our previous publication [22]. For certain compounds, a microwave-assisted one-pot procedure was utilized (Scheme 1). All compounds were acquired as racemates and synthesized compounds were characterized by various spectroscopic methods (^{1}H NMR, ^{13}C-NMR).

Compounds **A1** and **A2** were synthesized using one-pot method following Scheme 2.

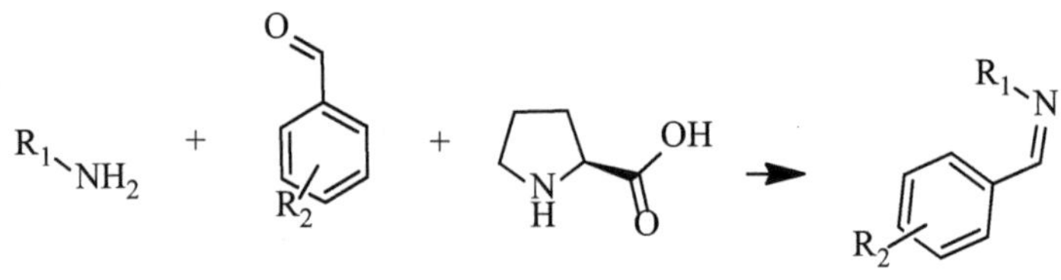

Scheme 2. General synthetic procedure of compounds A1 and A2. Reagents and conditions: Dry toluene, reflux for 8 h.

Table 2. %Inhibition at 50 μM and IC_{50} values of compounds **1–15** for SARS-CoV-2 3CL protease.

No	% Inhibition at 50 μM	IC_{50} (μM)	No	% Inhibition at 50 μM	IC_{50} (μM)
1	89.23	3.571	**10**	9.17	>50
2	5.14	>50	**11**	91.37	13.25
3	0	>50	**12**	100	0.192
4	88.12	4.291	**13**	0	>50
5	0	>50	**14**	0	>50
6	2.82	>50	**15**	10.63	>50
7	100	0.234	**A1**	83.29	5.15
8	44.37	>50	**A2**	0	>50
9	11.97	>50	**GC376**	100 at 100 μM	0.439

3.5 Biological Evaluation- Inhibition of SARS-CoV-2 3CLpro Enzymatic Activity

Fifteen thiazolidinone derivatives were evaluated as potential inhibitors of SARS-CoV-2 Mpro using a biochemical assay, with GC376 serving as the positive control. The assay aimed to determine the inhibitory activity of each compound by measuring their ability to suppress Mpro enzymatic function. The results, summarized in Table 2, provide insights into the potency of these derivatives compared to GC376. Key parameters such as IC_{50} values and percentage inhibition were analyzed to assess their effectiveness (Fig. 3).

The results demonstrated that several compounds exhibited inhibitory activity against SARS-CoV-2 Mpro, with IC_{50} values ranging from 0.19 to 13.25 μM. Notably, six out of the fifteen tested derivatives showed potent inhibition, with IC_{50} values between 0.192 and 5.14 μM. Among them, compound **12** displayed the strongest activity, achieving an IC_{50} of 0.192 μM, followed by compound **7** (0.234 μM), compound **1** (3.571 μM), and compound **4** (4.291 μM). Importantly, compounds **12** and **7** outperformed the reference inhibitor GC376, indicating their potential as lead candidates for further development.

The structure-activity relationship analysis revealed that the presence of fluoro and methoxy groups on the phenyl ring enhances Mpro inhibition. The most favorable structural feature for activity was identified as a **4-(morpholinosulfonyl)phenyl-thiazolyl moiety combined with a 2,6-difluorophenyl ring,** as seen in compound **12**. In contrast, replacing both fluorine atoms with chlorine led to compound **10**, which

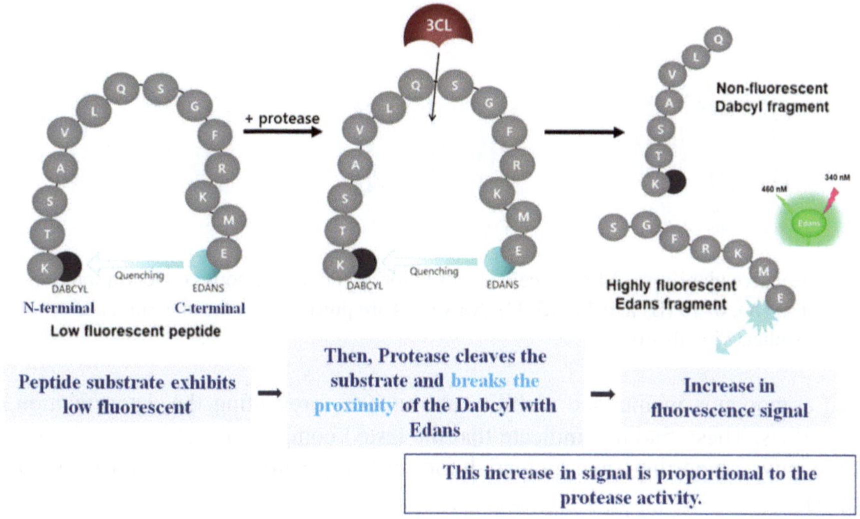

Fig. 3. SARS-CoV-2 3CLpro enzymatic assay.

exhibited no inhibitory activity. Similarly, introducing a cyano group at position 6 of the benzothiazole ring (compound **13**) or incorporating a 5,6-diphenyl substitution on the benzothiazole core (compound **14**) had a detrimental effect on Mpro inhibition.

Moreover, the crucial role of the sulfonamide bond in the inhibition of SARS-CoV-2 main protease can be clearly observed when comparing the inhibitory effects of compounds **A1** and **A2**. Compound **A2**, which lacks the sulfonamide bond, exhibited no significant inhibition, demonstrating that the presence of this functional group is vital for the molecule's activity. In contrast, compound **A1**, which retains the sulfonamide bond, showed a measurable inhibitory effect with an IC_{50} value of 5.15 μM. This stark difference in inhibitory potency highlights the essential contribution of the sulfonamide group to the overall binding affinity and inhibitory efficacy against the protease. The data strongly suggests that the sulfonamide bond is a critical component for the effective inhibition of the enzyme, underscoring its importance in the design of potent inhibitors for novel corona virus main protease.

These findings highlight the critical role of specific functional groups in optimizing inhibitory potency and provide valuable insights for further structural modifications aimed at improving activity. Future studies will focus on refining these derivatives to enhance their pharmacokinetic properties and therapeutic potential.

3.6 Assessment of Cellular Viability

The cytotoxicity of compounds **1–15** was evaluated in MRC-5 cells at concentrations of 0.1 μM, 1 μM, and 10 μM, following the methodology previously described in [22]. No significant toxicity was observed at the lower concentrations, suggesting a favorable safety profile. However, at 10 μM, compounds **13** and **15** exhibited the highest cytotoxicity, with cell viability reduced to 55.0% and 57.9%, respectively. Despite this, all

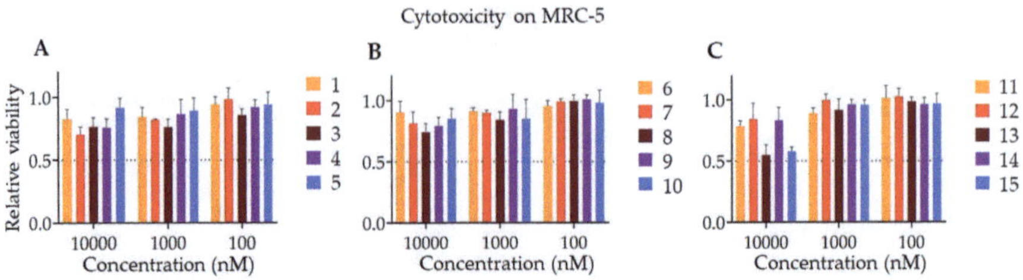

Fig. 4. Relative viability of MRC-5 cells exposed for 24 h under various concentrations of compounds **1–5** (A), **6–10** (B) and **11–15** (C). All values are presented as mean ± standard deviation (SD) of triplicate incubations.

tested compounds maintained viability above 50%, preventing the determination of CC_{50} values. These findings indicate that the tested compounds generally exhibit low cytotoxicity, supporting their potential for further development as Mpro inhibitors (Fig. 4).

4 Conclusion

In this study, we screened 85 potential SARS-CoV-2 Mpro inhibitors. Based on molecular docking results, the 15 analogues with the lowest binding energy were selected for biochemical evaluation. Among these, the most potent compounds demonstrated IC_{50} values below 1 μM in enzymatic assays, indicating strong inhibitory activity.

In silico pharmacokinetic analysis suggested that these compounds possess favorable drug-like properties for oral administration. Notably, compounds **12**, **7,** and **1** displayed strong interactions with key Mpro residues involved in substrate stabilization. The thiazolidinone core effectively mimicked the Gln residue of the enzyme's natural substrate, contributing to high binding affinity. These compounds also exhibited low predicted free binding energy, consistent with their potent in vitro antiviral activity.

Moreover, the crucial role of the sulfonamide bond in the inhibition of SARS-CoV-2 Mpro was clearly demonstrated by comparing compounds **A1** and **A2**. Compound **A2**, which lacks the sulfonamide bond, showed no inhibition, while compound **A1**, which retains the sulfonamide bond, exhibited a notable IC50 value of 5.15 μM. This stark contrast in inhibitory potency highlights the importance of the sulfonamide group in the molecule's ability to effectively inhibit the enzyme.

Overall, these results emphasize the potential of the selected compounds as promising Mpro inhibitors and lay the groundwork for further optimization and development of antiviral therapeutics targeting COVID-19.

Acknowledgement. Macaev Fliur thanks the Ministry of Education and Research of the Republic of Moldova for supporting the subprogram MetNatVal code 010601.

References

1. WHO.: Novel Coronavirus e China. Available online: https://www.who.int/csr/don/12-january-2020-novel-coronavirus-china/en/. Accessed on 15 Dec 2022 (2020)
2. Zhu, N., et al.: A novel coronavirus from patients with pneumonia in China, 2019. N. Engl. J. Med. **382**(8), 727–733 (2020)
3. Huang, C., Wang, Y., Li, X.: Clinical features of patients infected with 2019 novel coronavirus in Wuhan, China. Lancet. **395**, 497–506 (2020)
4. Sun, J., et al.: COVID-19: epidemiology, evolution, and cross-disciplinary perspectives. Trends Mol. Med. **26**, 483–495 (2020)
5. Du, L., He, Y., Zhou, Y., Liu, S., Zheng, B., Jiang, S.: The spike protein of SARS-CoV—A target for vaccine and therapeutic development. Nat. Rev. Microbiol. **7**, 226–236 (2009)
6. Lu, R., Zhao, X., Li, J., Niu, P., Huang, B., Zhu, N.: Genomic characterisation and epidemiology of 2019 novel coronavirus: implications for virus origins and receptor binding. Lancet. **395**, 565–574 (2020)
7. Ghanbari, R., et al.: Existing antiviral options against SARS-CoV-2 replication in COVID-19 patients. Future Microbiol. **15**, 1747–1758 (2020)
8. Amin, S.A., Banerjee, S., Singh, S., Qureshi, I.A., Gayen, S., Jha, T.: First structure–activity relationship analysis of SARS-CoV-2 virus main protease (Mpro) inhibitors: an endeavor on COVID-19 drug discovery. Mol. Divers. **25**, 1827–1838 (2021)
9. Fu, L., et al.: Both Boceprevir and GC376 efficaciously inhibit SARS-CoV-2 by targeting its main protease. Nat. Commun. **11**(1), 4417 (2020)
10. Jin, Z., Du, X., Xu, Y., Deng, Y., Liu, M., Zhao, Y.: Structure of Mpro from SARS-CoV-2 and discovery of its inhibitors. Nature. **582**, 289–293 (2020)
11. Yin, S., Huang, M., Li, D., Tang, N.: Difference of coagulation features between severe pneumonia induced by SARS-CoV2 and non-SARS-CoV2. J. Thromb. Thrombolysis. **51**, 1107–1110 (2021)
12. Gimeno, A., et al.: The light and dark sides of virtual screening: what is there to know? Int. J. Mol. Sci. **20**(6), 1375 (2019)
13. Alzain, A.A.: Insights from computational studies on the potential of natural compounds as inhibitors against SARS-CoV-2 spike omicron variant. SAR QSAR Environ. Res. **33**, 953–968 (2022)
14. Arwansyah, A., et al.: Molecular modelling on multiepitope-based vaccine against SARS-CoV-2 using immunoinformatics, molecular docking, and molecular dynamics simulation. SAR QSAR Environ. Res. **33**, 649–675 (2022)
15. https://www.rcsb.org/structure/6lu7. Accessed 13 Oct 2022
16. Guex, N., Peitsch, M.C.: SWISS-MODEL and the Swiss-PdbViewer: an environment for comparative protein modeling. Electrophoresis. **18**, 2714–2723 (1997)
17. Petrou, A., et al.: Thiazole/Thiadiazole/Benzothiazole based Thiazolidin-4-one derivatives as potential inhibitors of Main protease of SARS-CoV-2. Molecules. **27**, 2180 (2022). https://doi.org/10.3390/molecules27072180
18. Cousins, K.R.: Computer review of ChemDraw ultra 12.0. J. Am. Chem. Soc. **133**, 8388 (2011)
19. Wolber, G., Langer, T.: LigandScout: 3-D pharmacophores derived from protein-bound ligands and their use as virtual screening filters. J. Chem. Inf. Model. **45**, 160–169 (2005)
20. Morris, G.M., et al.: AutoDock4 and AutoDockTools4: automated docking with selective receptor flexibility. J. Comput. Chem. **30**, 2785–2791 (2009)
21. https://biosig.lab.uq.edu.au/pkcsm/prediction, Melbourne, Australia. Accessed 13 Oct 2022

22. Tratrat, C., Petrou, A., Fesatidou, M., Haroun, M., Chohan, M., Geronikaki, A.: Rational design, synthesis, and biological evaluation of novel Thiazole/Thiazolidinones multitarget anti-human immunodeficiency virus molecules. Pharmaceuticals. **18**, 298 (2025)
23. Ho, B.L., Cheng, S.C., Shi, L., Wang, T., Ho, K., Chou, C.: Critical assessment of the important residues involved in the dimerization and catalysis of MERS coronavirus Main protease. PLoS One. **10**, e0144865 (2015)
24. Marzi, M., Vakil, M., Bahmanyar, M., Zarenezhad, E.: Paxlovid: mechanism of action, synthesis, and in Silico study. Biomed. Res. Int. **2022**, 7341493 (2022)

Engineered Antibiotics: A Strategy to Combat Multi-Drug-Resistant Bacteria: A Review on Antibiotics Modification

Bangmayee Dash[1], Susmita Chakrabarty[1], Nimay Chandra Giri[2], and Bhabani Shankar Nayak[3](✉)

[1] DRIEMS University, Tangi, Cuttack, Odisha, India
[2] Centurion University of Technology and Management, Jatni, Bhubaneswar, Odisha, India
[3] KIIT School of Pharmacy, KIIT Deemed to Be University, Bhubaneswar, Odisha, India
bhabanishankar.nayak@kims.ac.in

Abstract. Drugs, which are known as antibiotics, treat bacterial infections in both humans and animals. Needless to say, a vast array of microorganisms, including fungi and bacteria, develop antibiotics. Contaminations brought on by drug-resistant bacteria have become a significant problem for public healthcare systems in recent years. In actuality, the introduction of antibiotics has contributed to the rise of bacteria resistant to multiple drugs. In this review, we have tried to analyze detailed aspects of the antimicrobial resistance (AMR) mechanism conferred by both gram-negative and gram-positive bacteria by taking ESKAPE, i.e., *E. coli, Staphylococcus aureus, Klebsiella pneumoniae, Acinetobacter baumanii, Pseudomonas aeruginosa, Enterobacter* spp. etc. respectively. Herein, we discussed novel redesigning methods for conventional antibiotics and provided a viewpoint on the subject. There is an acute need to discover and utilize unconventional methods for antibacterial drug development to block and tackle these global threats. It is essential to reduce the excessive use of antibiotics and reduce the possibility of unknown MDR bacteria. Furthermore, it is important to discover new classes of modified antibiotics to enhance their lifespan and deal with MDR bacteria.

Keywords: Antibiotic · Antimicrobial resistance · Multi-Drug-Resistant Bacteria · Biofilm · Drug Modification

1 Introduction

Resistance to Antibacterial agents threatens the effective avoidance and therapy of a steadily increasing range of infections caused by various pathogens such as bacteria, parasites, viruses, and fungi. Antibiotic resistance occurs when bacteria alter over time and no longer react to drugs, forming infections that are difficult to treat, enhancing the

A. K. Saxena, A. Saxena (Eds.): GTHTM 2025, pp. 57–84, 2026.
https://doi.org/10.1007/978-3-032-12320-6_4

possibilities of disease spread, and enhancing the death rate, which is the primary reason for modifications of antibiotics and the development of novel antibiotics to treat the superbugs.

1.1 Antibiotics

American microbiologist Selman Waksman first coined the term antibiotic in the year of 1942 to describe the chemical substances released by certain microorganisms to inhibit the growth of other microorganisms in their vicinity. Antibiotics are drugs that combat bacterial infections in people and animals. In Webster's dictionary, it is defined as a chemical or substance produced by microorganisms that inhibits the growth of another microorganism. The term "ANTIBIOTIC" emanates from "ANTIBIOSE", which means in opposition to the existence of microorganisms like bacteria. According to the World Health Organization, AMR is the primary public health problem. Based on observation, it can be estimated that in 2019, nearly 1.27 million people died due to AMR. It affects the treatment processes like surgery, cesarean sections, and cancer treatment [1]. Antibiotics can also be expounded as the secondary metabolites of some microorganisms. Innumerable microorganisms like bacteria and fungi produce antibiotics that are efficacious inimical to bacteria. These act as wonder drugs for humanity. The first antibiotic, which is used in human infections, is pyocyanase. Selman Waksman introduced the term antibiotics. From 1940 to 1960, it was a period referred to as the golden age of antibiotics because infectious diseases were under control. In 1899, Emmerich and Low discovered the therapeutic potential of *Pseudomonas aeruginosa.* The first modern antibiotics began in 1854 by Antoine Bechamp, are aniline and atoxyl. In 1907, Salvarsan, the first antibiotic against *Syphills,* became the most prescribed drug. Chloramphenicols isolated from the *Actinomycete Streptomyces* venezuelae 1947. In 1949, Macrolides, which are the most useful antibiotics produced by *Saccharopolyspora erythrea.* Polymyxin E does *Paenibacillus polymyxa* produce a lipopeptide antibiotic. Alexander Fleming, the Scottish physician, discovered penicillin in 1928. He invented it fortuitously while performing his experiment on the bacteria species of *Staphylococcus*. There was contamination in unwrapped Petri dishes with the growth of mold and the appearance of the clear zone- Alexander Flemming, his famous saying was "When I woke up just after dawn on September 28, 1928, I certainly did not plan to revolutionize all medicine by discovering the world's first antibiotic, or bacteria killer. I assume that was precisely what I did [2].

Antibiotics can be grouped into seven categories, like the mode of activity, depending as the source, spectrum, chemistry, function, and ayurvedic. We can divide antibiotics into bactericidal (which destroys the bacteria), examples are Gentamycin, Tobramycin, Amikacin, Aminoglycosides, Vancomycin, Penicillin V, Penicillin G, and Fluoroquinolones. The bacteriostatic antibiotics (which prevent the growth of the bacteria) such as Trimethoprim, Erythromycin, Macrolides, and Chloramphenicol. Broad-spectrum antibiotics such as Aminoglycosides (Tobramycin, Amikacin, Neomycin E), Chloramphenicol, and Rifamycins. These antibiotics are effective in opposition to both gram-negative and gram-positive bacteria. Narrow-spectrum antibiotics include Cephalosporins, Vancomycin, Cloxacillin, and Penicillin. Depending on the source, these are categorized as synthetic, natural, and semi-synthetic. Antibiotics classification that is based on chemistry is an Aminoglycoside, β –β-β-lactams, Fluoroquinolones, Macrolides, and Nitroimidazoles. These are also classified according to their function, like inhibitors of nucleic acid, inhibitors of the cell wall, inhibitors of metabolic pathways, inhibitors of cell membranes [3]. Interestingly, some Indian Ayurvedic medicine shows antibacterial activity, such as Dashmula churna, Triphala churna, and Haritaki churna [4].

1.2 Impact of Antibiotics on Bacteria

Antibiotics kill bacteria by inhibiting DNA replication, protein synthesis, transcription, cell wall synthesis, and cell membranes.

1.2.1 DNA Replication. Quinolones are derived from Nalidixic acid, which was a result of the synthesis of nalidixic acid. Oxolinic acid is the first generation of quinolone, ciprofloxacin is the second generation of quinolones, and Levofloxacin and Gemifloxacin are the third generation and, fourth generation respectively. It obstructs the conservation of chromosomes, targets topoisomerase 2 and topoisomerase 4, and inhibits the rejoining of DNA strands. In the case of gram-positive bacteria, topoisomerase-4 is the target, and in the case of gram-negative bacteria, topoisomerase 2 is the primary target. Quinolone - topoisomerase-DNA complex is formed when antibiotics bind to the enzyme and break the double-stranded DNA. In this way, the DNA replication fork is blocked and the process is inhibited.

1.2.2 Cell Wall. The major function of an antibiotic is to inhibit the cell wall of bacteria. Peptidoglycan is a major element and it contains N-acetylglucosamine (NAG) and N-acetylmuramic acid (NAM) with crosslinked by pentapeptides (L-Ala-D-Glu-L-Lys-D-Ala-D-Ala). There is a difference between the pentapeptides of gram-positive and gram-negative bacteria. In the case of gram-positive bacteria, it is L-Ala-D-Glu-L-Lys-D-Ala-D-Ala, whereas Gram-negative bacteria contain diaminopimelic acid, ligase, and alanine groups. These enzymes are inactivated by seromycin. β-Lactam drugs bind to the Penicillin-binding proteins. PBPS is present in the bacterial cytoplasm. Some examples of β – β-β-β-lactam antibiotics are cephalosporins, penicillin, carbapenem, and monocyclic monobactam. They all have four-membered lactam rings. β- β-lactam ring imitates the D–Ala–D–Ala group of the pentapeptides. The carbonyl group of β-lactam rings reacts with the serine (Ser) residue of penicillin-binding proteins. Glycopeptide antibiotics like avoparcin, ristocatin, vancomycin, and teicoplanin. These antibiotics react with the C-terminal D-Ala-D-Ala of murein precursors. Vancomycin is the antibiotic of the last resort.

1.2.3 Transcription and Translation. RNA polymerase (RNAP) of bacteria is 350 kDa and it consists of core enzymes α_2 ββ'ω, β & β' are large subunits. Rifamycins (RIF) bind to the β subunit, and Sorangcins also bind to the active site of the core enzyme [5]. Bacterial ribosome is of 70 s type, i.e., 30 s + 50 s. It contains three major steps: initiation, elongation, and termination. Antibiotics like aminoglycosides and tetracyclines

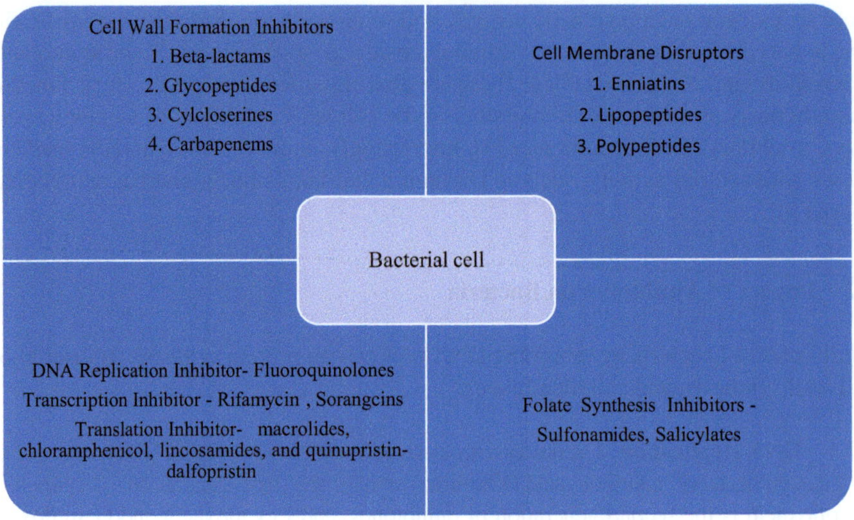

Fig. 1. Impact of Antibiotic on bacteria Cell [6].

target conserved sequences of 16S rRNA of the 30S subunit. Other antibiotics like macrolides, chloramphenicol, lincosamides, and quinupristin-dalfopristin target 23 rRNA of the 50 s subunit. The impact of antibiotics on the bacterial cell is shown in Fig. 1 [6].

2 Antibiotic Resistance

Antibiotic resistance is an important threat to all, including animals and humans worldwide, and indirectly, it will be a major issue for medical officers. Genes that are in opposition to antibiotics are detected in all habitats, like nature, modified, and clinical. The major cause of antibiotic resistance is the overuse of it worldwide. This review article aims to draw attention to antibiotics, which are bullet guns for bacteria, but resistant to them due to biofilm is a major problem faced by researchers to develop modified antibiotics by combining old antibiotics with chelating ligands, herbal plant extracts, and proteomics to enhance the lifespan of conventional drugs. Resistance to antibiotics was described to come about when the antibiotics lose their potential to hamper the growth of bacteria. The bacteria become resistant and continue to grow in the presence of the same antibiotics. Normally, antibiotics are effective against bacteria, but the resistant bacteria require elevated concentrations of the same drugs. The appearance of resistance to antibiotics was noticed soon after the introduction of new antibiotic compounds. Antibiotic resistance is a global problem due to the use of antibiotics. It is of two types: Natural resistance and Acquired resistance. Natural resistance is universally shared with microorganisms. It is independent of exposure to previous antibiotics. Acquired resistance includes horizontal gene transfer methods like transformation, transduction, conjugation, and transposition [7]. Antibiotic resistance is a major warning to the global antibiotic arsenal and the power to cure nosocomial disease.

2.1 Biochemical Mechanism

Five major mechanisms are involved in resistance to antibiotics: limiting uptake of the drug, formation of biofilm, active drug efflux, inactivation of the drug, and modification of the drug target.

2.2 Limiting Intake of Drugs

There is a natural difference in the microorganisms to limit the ingestion of antibiotics. In gram-positive bacteria, the lipid polysaccharide layer provides a barrier to antibiotics. For example, the High lipid content of mycobacteria provides a barrier to hydrophilic drugs, and the formation of a thick cell wall in *Staphylococcus aureus* shows resistance to vancomycin. Antibiotics enter the bacterial cell via porin channels. In gram-negative bacteria, porin channels allow entry to hydrophilic molecules. A decrease in the number of porins present and mutations in porin channels are the two important ways in which it can limit drug uptake.

2.3 Efflux Pump

Proteins are involved in this transport in the extrusion of antibiotics from the bacterial cell to the external environment. These are common in both gram-positive and gram-negative bacteria. Based on structure and energy source, it is of five types: −1. The ATP binding cassette (ABC) family, the multi-drug and toxic compound extrusion (MATE) family, the small multidrug resistance (SMR) family, the major facilitator superfamily (MFS), and the resistance nodulation superfamily (RND). It uses the proton motive force to throw out the antibiotics from cells.

2.4 Inactivation of Drugs

In this resistance mechanism, degradation of antibiotics and transfer of any chemical group to antibiotics are two main ways in which bacteria inactivate the drugs. The transfer of acetyl, phosphoryl, and adenyl groups inactivates antibiotics. Such as β-lactamases, which are hydrolysis enzymes. β-Lactam antibiotics are slowed by β-β-lactamases (cephalosporinase and penicillinase). Based on their structure, it is of four types (A, B, C, D), and the basis of substrate specificity is of three types: cephalosporinases, serine β-lactamase, and metallo (zinc-dependent) β-lactamases. NDM-1 (New Delhi metallo-beta-lactamase −1) is identified in *E. coli*. It makes the bacteria resistant to a broad range of β-lactam antibiotics. In gram-negative bacteria, it is the most common resistance. IMP-1 (imipenem resistance) and VIM-1 (Verona integron-encoded Metallo beta-lactamase) are the most widely distributed carbapenem-resistant Enterobacteriaceae. The tetX gene inactivates tetracycline by hydrolysis. There are three types of aminoglycosides modifying enzymes like aminoglycoside acetyltransferase, aminoglycoside phosphotransferase, and adenyl transferase, and chloramphenicol is inactivated by chloramphenicol acetyltransferase.

2.5 Target Modification

In the bacterial cell, many components target the antibiotics. It is a prominent mechanism of the molecular target and can arise due to point mutations in selected genes. Small changes in amino acid sequence can change the protein structure. In the case of gram-positive bacteria, resistance to β-lactams drugs is due to changes in the number and structure of penicillin-binding proteins (PBPs). In the case of vancomycin-resistant Enterobacteriaceae, there are five phenotypes of vancomycin resistance, like van A, van B, van C, van D, and van E. Among them, van C and van E replace the terminal D-Ala with D-ser. The detailed mechanism of antimicrobial resistance is shown in Fig. 2 [8].

2.6 Biofilm

Bacterial communities form biofilms by colonization. It shields the bacteria from the action of the host immune system and gives protection from antibiotics by forming a slimy layer. The matrix of biofilm contains proteins, DNA, and polysaccharides, which makes it tough for antibiotics to reach the bacteria. Biofilm is indirectly involved in antibiotic resistance [9, 10]. The primary supposition is the prospect of imperfect penetration of the drug into the slimy layer. There is the absence of a genetic blockade to the spreading of solutes the size of drugs through the matrix. The second one is based on the modification of the microenvironment within the biofilm. The third one is the subpopulation of bacteria inside the film. Biofilm helps in antimicrobial resistance, which is shown in Fig. 3 [11]. One of the major mechanisms which is associated with antibiotic resistance is Quorum Sensing or Density Sensing. It can be defined as the process by which bacteria share their population density. Major biological activities depend on this mechanism and are based on gene regulation. On the enhancement of population density, bacterial cells start to form various signaling molecules that impact various biochemical processes, such as antibiotic resistance, formation of biofilm, and pathogenicity factors. Quorum sensing is responsible for the regulation of genes of bacteria that are associated with drug resistance. An example of signaling molecules for Gram-positive bacteria is auto-inducing peptides, and in the case of Gram-negative bacteria is acyl-homoserine lactones. The major mechanism of Quorum sensing is the

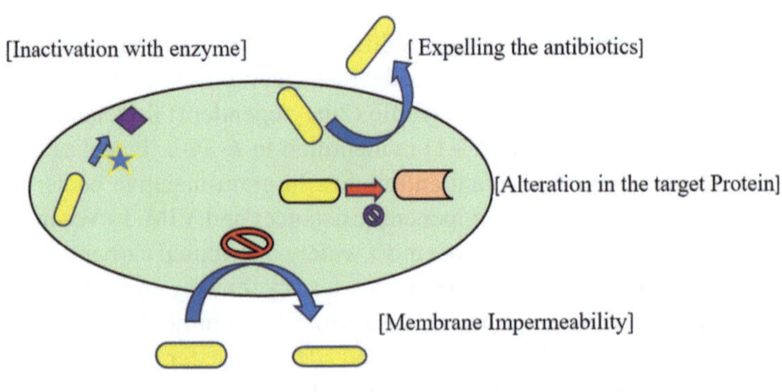

Fig. 2. Antibiotic Resistance Mechanism [8].

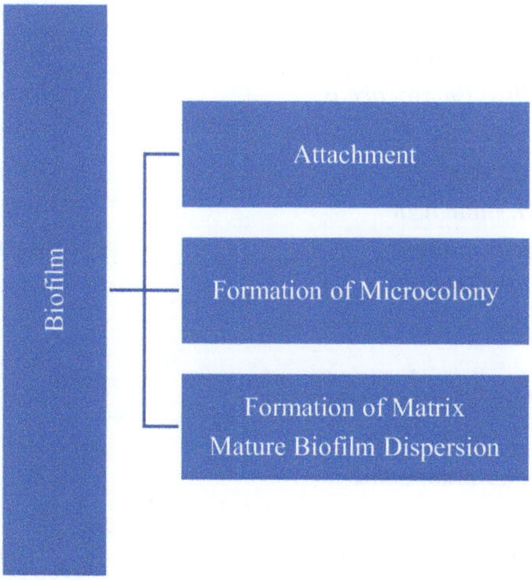

Fig. 3. Antimicrobial Resistance due to Biofilm [11].

formation of a signaling molecule, followed by the next step, which includes identifying the signaling molecules by the receptor molecules of the bacterial cell and activation of the synchronized property of the bacterial population. It is directly proportional to the antibiotic resistance mechanism by forming microbial biofilms increases the pathogenicity and reduces the impact of drugs [12].

3 Multi-Drug Resistant Bacteria: the Superbugs

Multi-drug resistant Bacteria can also be termed superbugs for the world. These superbugs are responsible for increasing the infection and death rates rapidly. The World Health Organization has noticed that the post-antibiotic era will result in their increase in infections, and small infections on the wound may result in death if we are unable to act against resistance. In the USA, more than 63000 patients die each year due to hospital-acquired infections caused by bacteria, and multidrug-resistant bacteria affect more than 25000 people. A researcher reviewed a comparative scenario of antibiotic resistance in India, where carbapenem drugs lost their activity due to metallo-β-lactamase. MBL fabricating *P. aeruginosa* is appearing as a principal conviction for healthcare-associated infections. In India widespread presence of MBL makes bacterial assortments from 7% to 65%. In Sikkim, Methicillin-resistant *S. aureus* was detected in 111 out of 291 clinical samples (38.21%). Clinical samples appeared as vancomycin-intermediate and vancomycin-resistant *S. aureus*. Antibiotic resistance rates of various organisms in India are shown in Table 1 [13]. These multidrug-resistant organisms (MDROs) indirectly increased hospital costs. The European Centre for Disease Prevention and Control (ECDC) and the Centers for Disease Control and Prevention (CDC) were categorized as different patterns of resistance related to the public health and medical sector. Multidrug resistance (MDR), Extensive drug resistance (XDR), and

Table 1. List of MDR Bacteria [15].

Sl. No.	Multi-Drug-Resistant Bacteria	Antibiotics
1	*Lactobacillus, Enterococcus*	Vancomycin
2	*E. coli, V. cholerae, P. mirabilis*	Ampicillin
3	*E. coli, S. aureus*	Rifampicin
4	*Shigella, H. influenzae*	Tetracycline
5	*Streptococcus mutans*	Bactriacin
6	*E. coli, V. cholerae*	Ciprofloxacin
7	*S. aureus, Streptococcus pyogenes*	Erythromycin
8	*E. coli*	Neomycin
9	*Dictyostelium discoideum*	Nystain
10	*S. aureus, S. epidermidis*	Gentamycin
11	*Acinetobacter baumannii*	Polymyxin B
12	*S. aureus*	Kanamycin
13	*Mycobacterium tuberculosis*	Isoniazid
14	*S. aureus*	Methicillin
15	*E. coli, V. cholerae*	Norfloxacin

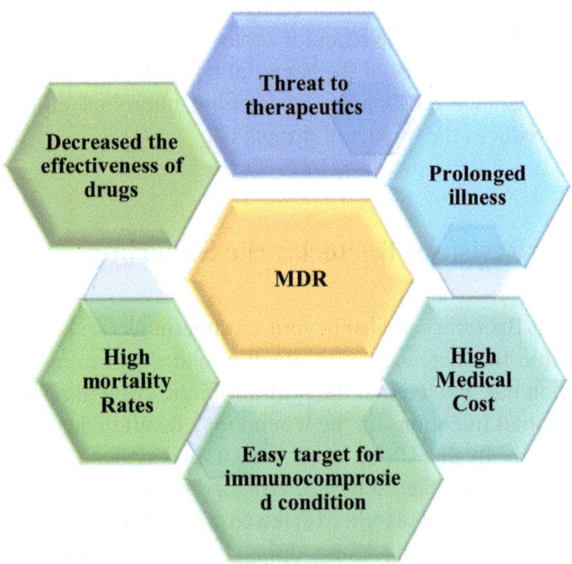

Fig. 4. Problem associated with MDR [14].

Pan Drug resistance (PDR) are different types of resistance for bacteria. MDR can be defined as when an organism is resistant to more than one antibiotic in vitro, which retards their antibiotic treatment. XDR can be explained as when bacteria are resistant to multiple drugs. PDR can be defined as when bacteria are resistant to all available antibiotics. Clinical Laboratory Standards Institute (CLSI) categorized it by antibiotic susceptibility testing. The problem associated with MDR is shown in Fig. 4 [14].

According to the World Health Organization (WHO, 2017), 12 families of bacteria pose a significant warning to Mankind. The list of MDR bacteria is of three types like critical, high, and medium. Critical MDR bacteria present a particular threat in medical settings, nursing homes, blood catheters, and ventilators. The list of MRD is shown in Table 1 [15].

These bacteria are associated with infections of the bloodstream and pneumonia. Unfortunately, these are resistant to cephalosporin and carbapenems. High mortality rates and high medical costs are associated with multidrug resistance and influence the efficacy of antibiotics. Resistance to antibiotics gives rise to a barrier in disease control by increasing the possibility of spreading a resistant pathogen, lowering the effectiveness of treatment, and thus, the consequence is an extended time of infection in patients. The more expensive therapies also have treatment costs. Cancer, chemotherapy, and organ transplantation have contributed to the development of MDR. Thus, it is a complex problem requiring concerted efforts of microbiologists, ecologists, health care specialists, educationalists, policymakers, and pharmaceutical industry workers. There is an urgent need for new antibiotics due to the unavoidable development of resistance that follows the introduction of antibiotics to the clinic. The main master plan of the pharmaceutical industry for the development of new therapeutics has been a modification of existing antibiotics. However, it has turned out to be increasingly difficult to launch new antibiotics to meet the needs of the community. Combinatorial chemistry, adjuvant therapy, and engineered antibiotics have a significant impact on the discovery of novel antibiotics. Many attempts have been made to develop safe, more effective treatment protocols to overcome the challenges of antibiotic resistance. The antibiotic resistance scenario in India and worldwide is shown in Figs. 5 and 6 below [16].

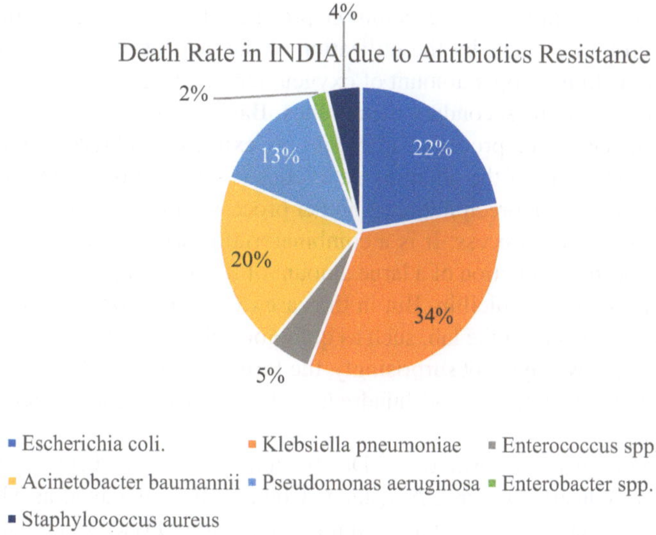

Fig. 5. Antibiotic resistance in India [16].

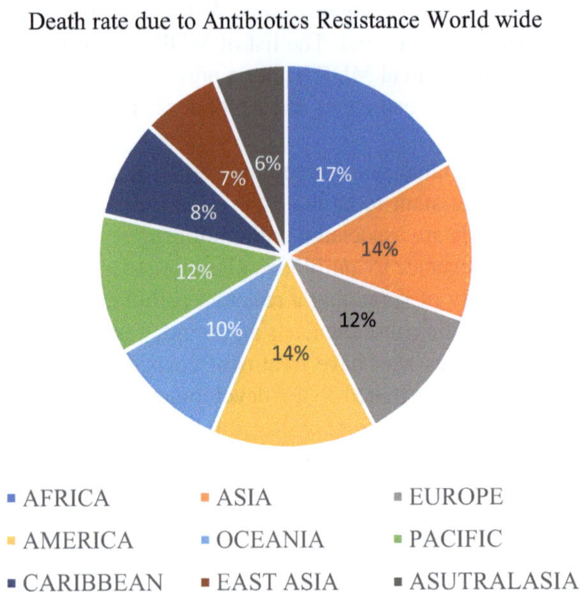

Fig. 6. Antibiotic Resistance all over the world [16].

4 Formulation of Modified Antibiotics

The major processes of formation of antibiotics are fermentation, semi-synthetic, and biosynthesis. Fermentation is the common process of forming antibiotics used by industrial Microbiology labs by the utilization of the area of interest microorganism, liquid growth medium, proper amount of oxygen, temperature, pH, and media. We can consider antibiotics to be secondary metabolites. Based on the observation of population size and the cell death process. Antibiotics are extracted and formed into a crystalline Product. The purity of the obtained antibiotics is obtained by using ion exchange, adsorption, and chemical precipitation. Another process for the formation of antibiotics is semi- semi-synthetic process. It is a combinatorial process of natural fermentation and lab work for the production of a large amount of it. An example of a semi-synthetic penicillin is ampicillin, Penicillin. But in the case of synthetic drugs, which are made completely synthetically in the lab, such as quinolone class, nalidixic acid, and chloroquine. However, nowadays, not surprisingly, the formulation of antibiotic production is a significant focus of interest, and hundreds of biosynthetic gene clusters have been analyzed [17].

Nafithromycin was discovered by Dr. Jitendra Singh on 20.11.2024 from the Biotechnology Industry Research Assistance Council. It is marketed as Miqnaf. This novel antibiotic's major role is to target community-acquired Bacterial Pneumonia, and it is more effective than azithromycin. The major significance of the drug is to boast superior safety and no side effects [18].

4.1 Vancomycin

It is one of the major glycopeptide drugs that is effective in opposition to MDR bacteria such as MRSA (methicillin-resistant *Staphylococcus aureus*) infections. Vancomycin ($C_{66}H_{75}C_{12}N_9O_{24}$) is produced by *Streptomyces orientalis.*

Zhao et al. had described a master plan to design nanoprobes in opposition to methicillin-resistant *Staphylococcus aureus.* This phenomenon is operated by the receptive-detachment of SiO_2 –Cypate-Vancomycin. Cationic PAH, VAN, and PAA-PEG molecules were combined to form the desired nanoprobes [19]. Chen et al. had altered vancomycin with nanomaterials against both gram-positive and gram-negative bacteria. The adaptation of glycopeptide antibiotics on the ostensible of the nanomaterials. Engrossingly, the designed VAN enhances the diameter. VAN –engineered with silica-bound iron oxide nanoparticles [20]. Okano et al. had changed VAN pocket alterations outlined to furnish both D-Ala-D-Lac and D-Ala-D-Ala, holding together which provoc vancomycin resistance. Surface modification enhanced the antibacterial. Here, an alteration in the c-end of VAN with quaternary ammonium salt. The C-end of van altered with 4-chlorobiphenyl methyl (CBP) gives more vigorous antibiotics [21]. Xie et al. had described [Ψ[C[=NH}NH]TPG^4] VAN, analogous modified to reveal the affinity for bacterial cell wall [22]. Ahmed et al. had reported that transportan 10 [TP10], it is an amphipathic cell-pungent peptide with effective translocation property, was mixed with vancomycin. Four conjugates were designed: van-PEG3-TP10 and combined with N- or C-end or lys 7 of the TP10 molecules [23]. Nigam et al. had proposed a method to construct novel Vancomycin mixed with cathelicidin peptides. Van destroys the formation of the peptidoglycan layer. Cathelicidins are derived from fish, mammals, and reptiles and have the unique property of self-promoting uptake [24]. Cochrane et al. had reported that congeners of tridecaptin A_i [Octyltridecaptin A_i [Oct-TriA$_1$] were effective in opposition to carbapenem-resistant *Klebsiella pneumonia* [25]. Binda et al. had reviewed that the growth of semi-synthetic glycopeptides drugs centered on made use of vancomycin analogs led to a preference for modifications like Teicoplanin [dalbavanin], Oritavacin is obtained from chloroeremomycin [26]. Xi et al. had reported that periodortitis is a customary disease caused by biofilms of teeth. It was cured by designing a multifunctional corona vesicle with intrinsic bacteriocidal properties and increased drug delivery capabilities. It is composed of poly[€-caprolactone]-block-poly[Lysine-stat-phenylalanine][PCL-b-p[lys-stat-phe] and poly[ethylene oxide] block-poly-[€-caprolatone [27]. Wu et al. had modified vancomycin to have a trimethylammonium salt. Here, engineered vancomycin enhanced the bactericidal effect as compared to vancomycin against vancomycin-resistant *Enterococci* (VRE) [28]. Campbell et al. had synthesized modified vancomycin as an anti-biofilm compound against *Staphylococcus aureus.*Ethyl-4-ethoxybenzoic acid and methyl gallate are combined with vancomycin [29]. Fang et al had reported that modification of vancomycin drugs with extracellular cancellous bone linked with electrostatic forces and bonds [30]. Ghaffar et al had reported a combination of vancomycin with metal-organic framework and chitosan. This modification increased the bactericidal effect of drugs against opposition to *aureus,* and it is a novel antibiotic treatment against multidrug-resistant bacteria [31]. Zhou et al. had modified vancomycin with

L-homocarnosine[peptide]. The antibacterial study of modified drugs was conducted against *S. aureus*. It is successfully used to treat osteomyelitis and bone infection [32].

4.2 Tetracycline

Tetracycline is synthesized from the Streptomyces genus of actinobacteria. It is formed by using various polyketide antibiotics and a tertiary alpha-hydroxy ketone, as shown in Fig. 7 [33].

Sum pe had designed a reasoned strategy by modifying the carboxamide group of tetracyclines and combining it with chelocardins. It was synthesized by *Amycolatopsis sulphurea* [34]. Salman had proposed a novel method for the modification of tetracycline antibiotics with the help of a synthetic polymer that is, polyarylonitrile fabric1-Ethyl-3-(3-dimethylamino propyl)cto (EDC) is used to disable drugs, and it enhances the bacteriocidal property [35]. Shi et al. had reported a novel method to modify the conventional tetracycline with the help of ovalbumin. Here, polyethyleneimine is combined with ovalbumin and gold nanoparticles for the detection of drugs and modification [36]. Dey et al. had approached a unique method of eradication of bacterial biofilm by mixing naringin and drugs such as Tetracycline and ciprofloxacin. The murine macrophage cell line was used to study the cytotoxicity and bactericidal effect of Naringin. This modified tetracycline successfully removed the premature biofilm on the surface of a urinary catheter. The antibacterial activity of modified drugs was observed against *P. aeruginosa* [37].

4.3 Neomycin

It is an aminoglycoside antibiotic having the molecular formula $C_{23}H_{46}N_6O_{13}.3H_2SO_4$. A major application of this enzyme is to treat various nosocomial bacterial infections. The structure of this antibiotic is shown in Fig. 8 [38].

Bera et al. had proposed a unique way to combat MDR bacteria by combining amino-glycosidic antibiotics neomycin and lipid molecules [cationic detergents]. Modification of neomycin with lipid molecules showed cationic head moieties with the polycationic lipid molecules [39]. Wang et al. had reported the expansion of the introductory gadolinium-constructed pathogen determined MRI inhibitory agent. Here,

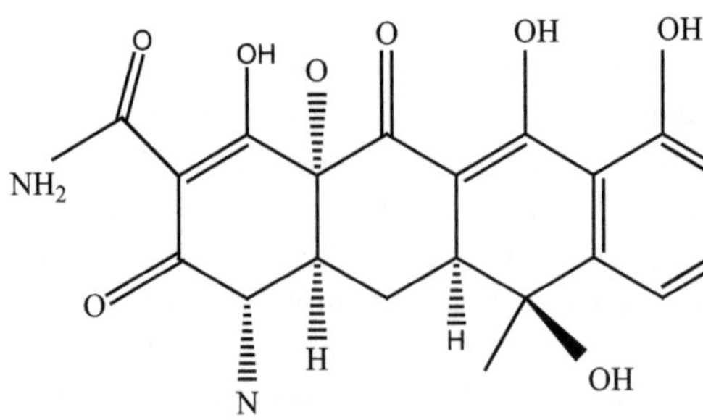

Fig. 7. The chemical structure of Tetracycline [33].

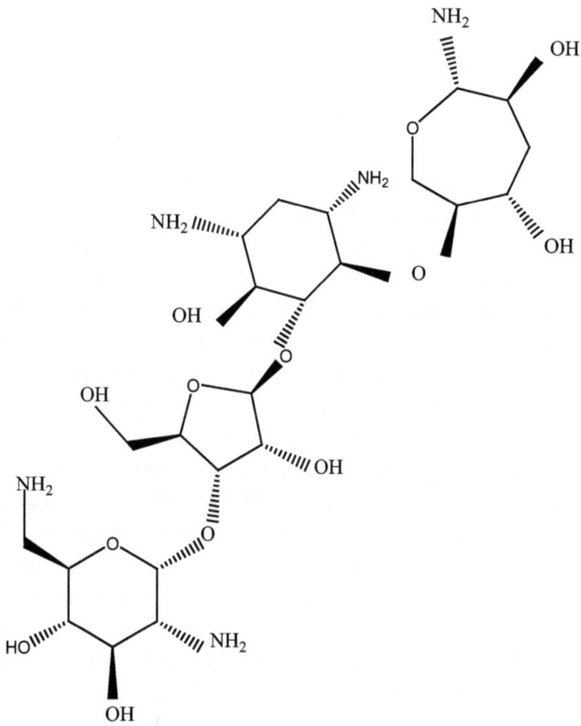

Fig. 8. The chemical structure of Neomycin [38].

neomycin is combined with Dotarem [FDA approved]. Neomycin acts on the activity that is idiosyncratic to bacteria. Cy5 Fluorophore combined through lipid linker to neomycin. The combinatorial effect of neomycin-Gadolinium modified drugs was effective against both Gram-positive and Gram-negative bacteria [40]. Mathivanan et al. had proposed a novel research work by combining poly(styrene sulfonic acid-co-maleic acid) along with polyvinyl alcohol and neomycin to form nanofiber mats. This engineered neomycin's effect was observed for wound healing in Wistar rats, and both gram-negative and gram-positive bacteria were killed by it. For wound dressing materials, it can be successfully used [41]. Zhang Y, et al. had reported that a combination of aminoglycoside drugs and inorganic Cu_2O showed antibacterial activity. But its effect was increased when neomycin was used against *E.coli.* They hypothesized that Cu_2O helps to damage the cell wall of bacteria and proposed a novel method for the modification of neomycin B, Kanamycin A, Amikacin, and Neamine with cationic amphiphiles and oligocationic amphiphiles. Here, they reported that polyol-altered neomycin antibiotics showed an excellent bactericidal effect [256 fold] [42]. Degtyareva, et al. Had approached a method by combining neomycin with 12 anthraquinones. The engineered neomycin inhibits the prokaryotic translation, and the bactericidal effect was studied under *Escherichia coli, Enterococcus faecalis, Citrobacter freundii, Shigella flexneri, Serratia marcescens, Proteus mirabilis, Enterobacter cloacae, and Staphylococcus epidermis* [43].

4.4 Ampicillin

Ampicillin is a beta-lactam antibiotic used to treat various bacterial infections caused by susceptible, usually gram-positive, organisms. The chemical structure of this enzyme is shown in Fig. 9 [44].

Ashley Brown1, et al. Had proposed the combination of silver nanoparticles and gold nanoparticles with ampicillin. It was effective against both gram-positive and gram-negative pathogens. In addition to this, thiol interchange with sodium citrate crowned with AuNPs formed a novel way for drug delivery. And utilizing Fe-absorption expressions by combining beta-lactam drugs and Fe chelators like catechol and hydroxamic acid. The antimicrobial property of modified AMP drugs increased when combined with clavulanic acid. Tetrameric acid is used to form bonds between metals and drugs. These novel beta-lactam drugs are effective against gram-positive bacteria [45]. Khatoon, et al. Had reported a one-step formation of AgNPs by utilizing Ampicillin. Due to the reduction that occurred in the metallic silver to form NPs in the presence of AMP. The antimicrobial effect of AMP-AGNPS was observed on multidrug-resistant bacteria [46]. Grau, et al. Had approached the modification of Ampicillin with silicon dioxide. Due to the hydrogen bonding and amino moieties of ampicillin, it is stable in nature. Maximum absorption at basic pH [47]. Topsakal, et al. Had modified Ampicillin with gold nanoparticles along with polyvinyl alcohol for tissue development. These chemical scaffolds enhanced the bactericidal effect of Ampicillin [48]. Magadla, et al. had combined ampicillin with 1-[2-Methoxyethyl]piperidine with Zn phthalocyanic compound. These modified drugs are combined with silica nanoparticles and 1,3-propane sultone. The antibacterial property was studied against *Staphylococcus aureus* [49]. H.Al-Noor, et al. had Approached a novel method to engineer Ampicillin with the help of chromium, Iron, Cobalt, Cadmium, and Mercury. Schiff base as primary ligand and nicotinamide as next. Ampicillin is mixed with Chlorobenzophenone. The bacteriocidal property of novel ampicillin was studied in opposition to *P. pseudomonas, E. coli, S. aureus*, and *B. subtilis* microbes [50].

4.5 Beta-Lactam Antibiotics

These are bactericidal that react with the cell wall of the bacteria by using penicillin-binding proteins. The drugs are effective in opposition to both Gram-negative and Gram-positive bacteria. The chemical structure of this drug is shown in Fig. 10 [51].

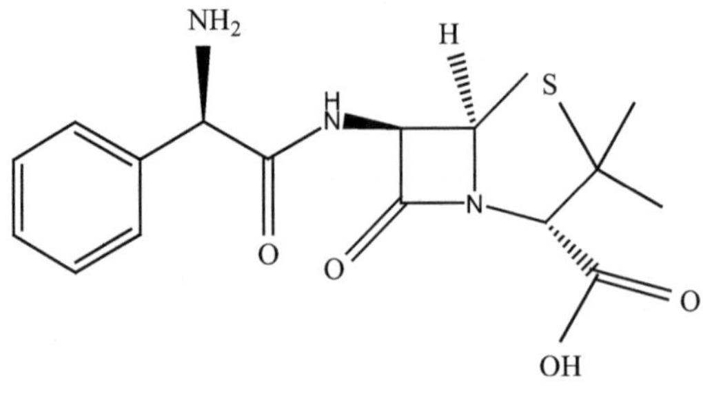

Fig. 9. The chemical structure of Ampicillin [44].

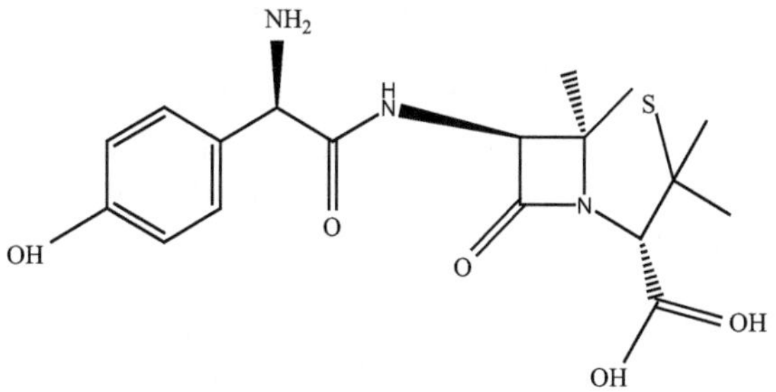

Fig. 10. The chemical structure of Beta-lactam drug [51].

R.V. Thappeta et al. Had combined old antibiotics (Vancomycin, Oxacillin, Meropenem) with chitosan-based oligolysine antibacterial peptide. It alone decreased biofilm. Interestingly eradicated preformed biofilm when mixed with drugs and enhanced the antibacterial property of antibiotics against different MDR strains. CSM5-K5 treatment reduced bacterial burdens by 1–3 orders of magnitude, and it acted synergistically with oxacillin, vancomycin, and streptomycin to kill MRSA, VRE, and MDR *E. coli*. Low-level resistance to CSM5-K5 that did arise for MRSA increased susceptibility to the β-lactam antibiotic oxacillin. This report showed the feasibility & benefits of using this synthetic cationic peptide as an alternative to, or in combination with, traditional antibiotics to treat MDR bacteria [52]. Tauri et al. Had approached the N-sulfonyloxy-2-azetidinones with beta-lactam drugs like AMP, OXA, and enhanced the activity of the old antibiotics. C_4 groups of combined antibiotics, particularly the hydrophobic group of the beta-lactamase enzyme, due to its electrophilic property. N-Hydroxy-4-bromomethyl is prepared to form N-sulfonyloxy beta-lactam drugs. It is a novel way to engineer beta-lactam drugs [53]. Tauran et al. had proposed a method for enhancing the drugs in opposition to MDR bacteria. Salmochelins are a c-glycosylated form of enterobactin. These are essential for Fe assimilation in bacteria. Combination of β-lactam drugs [Ampicillin/Amoxicillin] with glycosylated enterobactin. It increased the antimicrobial property of conventional drugs against *E. coli* [54]. Miramoth et al. had combined β-lactam drugs, penicillin G, with squalene. PNG formed pH- pH-sensitive and pH-insensitive chemical bonds, had showed different properties due to the interaction between the hydrophobic chain and the polar head of PNG. These novel bioconjugates showed magnificent antibiotic loading. These engineered beta-lactam drugs were effective against *S. aureus* [55]. Yenn et al. It has been reported that Beta-lactam antibiotics can be combined with adjuvants to kill the superbugs. It is another method to increase the lifespan of magical bullets. Stigmasterol is a major component of the plant. It is also used as an anticancer agent and food additive. When it is combined with ampicillin, it enhances the antibacterial activity of β-lactams antibiotics. This was effective against both gram-positive and gram-negative bacteria [56]. Turos et al. It has been reported that modifying penicillin antibiotics and polyacrylate nanoparticles against methicillin-resistant *S. aureus* and methicillin-susceptible *S. aureus*. Polyacrylate nanoparticles showed antibacterial activity, and butyl acrylate, styrene were used in the conversion of water-insoluble antibiotics into acrylate form. During emulsion polymerization, β β-lactam antibiotics could be incorporated into the nanoparticles. They used five acrylate forms of penicillin with antibiotics in aqueous media. Penicillin acrylamide was prepared by them using penicillin G potassium salt. 6-aminopenicillanic acid used in NP 2 and NP 3. Carbohydrate acylated with 6-aminopenicillanic acid is used in the preparation of NP4. Triethylamine with penicillin G monomer was used to prepare NP5. NP 6 and NP7 without acrylate side chain. From the minimum inhibitory concentration, they observed that Np1, Np4 and Np 6, and Np 3 showed antibacterial activity against both MRSA and MSSA. Penicillin G nanoparticles (Np1 and Np 4) are more effective than Np2 and Np 3 [57]. Ketikids et al. Had combined Silver ions with Penicillin G and synthesized a novel antibiotic as $[Ag\{Pen\}\{CH_3OH\}]_{2.}Ag$ combined with the drug's carboxylic moieties and amide group of the Nitrogen atom. The bactericidal effect of these modified beta-lactam drugs was examined against *Staphylococcus aureus, Staphylococcus epidermidis, and Pseudomonas aeruginosa*. It was effective in opposition to *Pseudomonas aerugionas* as compared to PenNa [58].

4.6 Tobramycin

Tobramycin is synthesized by *Streptomyces tenebrarius*. It is an aminoglycoside antibiotic. This drug is used to treat various bacterial infections that are associated with Gram-negative infections, Fig. 11 [59]

Boffoli, et al. Had reported the modification of tobramycin with synthetic glycopolymers. Manno pyranose and galacto pyranose are combined with sodium 2-acrylamide-2-methyl-1-propane sulfonate to form a carbohydrate-containing polymer. The formed polymer was mixed with tobramycin. This modified tobramycin has successfully eradicated the biofilm formed by the *Pseudomonas aeruginosa.* Here, polyacrylamide polymer is used to link the sugar groups and to study the antibacterial effect of modified drugs [60]. Khan, et al Had approached a unique method for the modification of conventional tobramycin. Here, tobramycin is combined with phyto molecules-coated zinc oxide nanoparticles, gallic acid. The bactericidal property of modified drugs was studied in opposition to *Staphylococcus aureus, Pseudomonas aeruginosa, and Escherichia coli. Coli* [61]. Ham, et al. Had reported a combination of tobramycin with ginger and its different analogs. The antibacterial and antibiofilm property of ginger and tobramycin was studied against *Pseudomonas aeruginosa* [62]. Kashyap, et al. Had reported a combination of Tobramycin with colistin or Ciprofloxacin. The bacteriocidal property of engineered tobramycin was studied against *Acinetobacter baumannii* [63]. R.Tait, et al. It has been reported that the combinatorial therapy of tobramycin with ceftazidime for the treatment of cystic fibrosis caused by Hypermutable *Pseudomonas aeruginosa.* The antibacterial property of tobramycin increased in combination with Ceftazidime [64]. Puvvadi et al. Reported a combination of tobramycin with Tris-buffered chelating source CaEDTA in the presence of iron. The antibacterial property of the synthesized tobramycin was studied against *P. aeruginosa* [65].

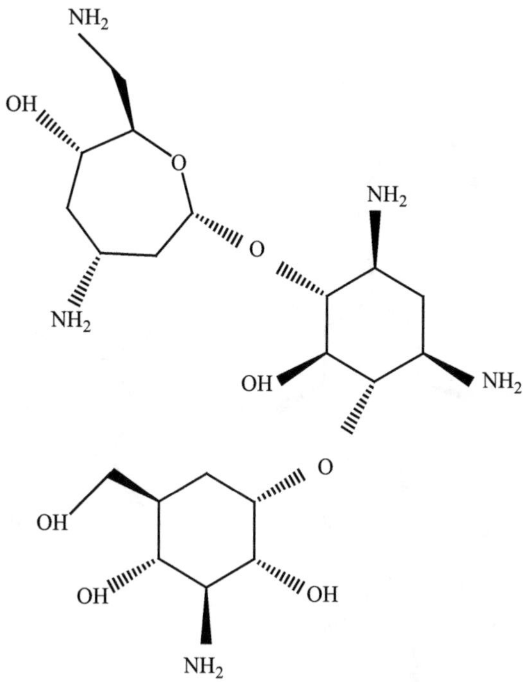

Fig. 11. The chemical structure of Tobramycin [59].

4.7 Amikacin

Amikacin is used to treat serious bacterial infections in various parts of the body. This antibiotic belongs to the aminoglycoside antibiotics. Chemical structure is shown in Fig. 12 [66].

Farhan, et al. Had reported that modification of amikacin with imipenem. The bactericidal property of amikacin was studied against *Pseudomonas aeruginosa*. A cooperative mechanism was observed between amikacin and imipenem [67]. Magallon et al. Had reported a combination of amikacin with zinc ion and pyrithione. The bacteriocidal property of engineered amikacin was studied against *Acinetobacter baumannii* and *Klebsiella pneumoniae* bacteria [68]. A.Darlow, et al. Had reported the modification of amikacin with Fosfomycin. The bacteriocidal property of amikacin was studied against *Enterobacterales and* successfully used in neonatal sepsis [69]. Wang, et al. had reported a combination of amikacin with ceftazidime/avibactam. The bactericidal and bacteriostatic properties of synthesized amikacin were studied against *Klebsiella pneumoniae* resistance to carbapenem [70].

4.8 Ciprofloxacin

It is a fluoroquinolone antibiotic that is used to treat various types of bacterial infections. The chemical structure of this antibiotic is shown in Fig. 13 [71].

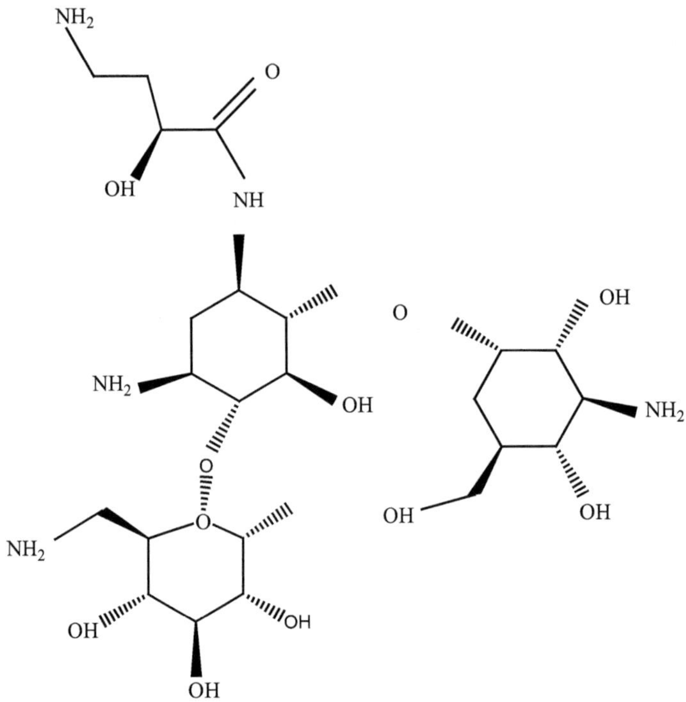

Fig. 12. The chemical structure of Amikacin [66]

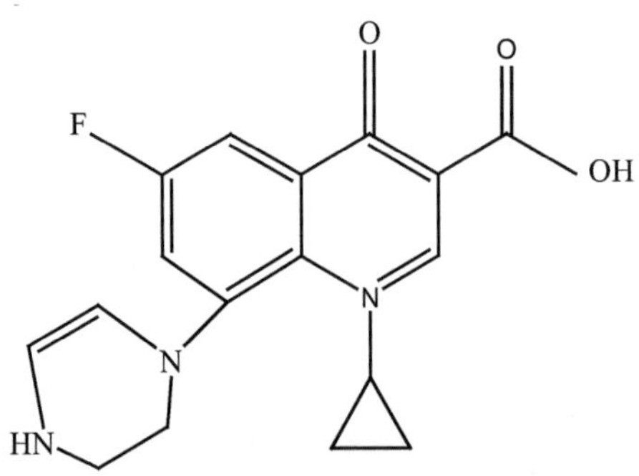

Fig. 13. The chemical structure of Ciprofloxacin [71].

Soukarieh et al. Had reported the modification of ciprofloxacin with polyurethane β-cyclodextrin. The antimicrobial property of modified ciprofloxacin was examined against *S. aureus, Sarcinalutei, Bacillus cereus, and Bacillus subtilis* [72]. Chrysouli et al Had been reported that Fluorescent probes were blended using a CuI-catalysed azide-alkyne cycloaddition to connect a Ciprofloxacin azide imitative with alkyne-substituted green and blue fluorophore derivatives, keeping antibacterial properties in opposition to Gram-negative and Gram-positive bacteria. The use of confocal fluorescent microscopy showed intracellular penetration, which was substantially increased in the presence of carbonyl cyanide 3-chlorophenylhydrazone as an efflux pump inhibitor in *Escherichia coli* [73]. Bykowska A, et al Had modified Ciprofloxacin hydrochloride with lanthanum. By using the sol-gel method, carrageenan and sodium alginate aerogels were combined with lanthanum. The antibacterial effect and absorption property of modified drugs increased [74]. M.Ibramin et al. It has been reported that alteration of the nitrogen atom of the C7 of the ciprofloxacin enhances the absorption. Ciprofloxacin combined with sulfonamide. These novel antibiotics showed more bactericidal effects against DNA gyrase and DNA topoisomerase 4 of *Staphylococcus aureus*. It was effective against both gram-positive and gram-negative bacteria [75]. Alsughayer et al. Had combined ciprofloxacin with ethyl cyanoacetate and ethyl acetoacetate to synthesize cyanoacetyl piprazinyl dihydroquinoline derivative 3, and oxobutanoyl piprazinyl dihydroquinoline derivative 5. *N*-alkylated derivatives 8–10 were synthesized by combining ciprofloxacin with chloroacetonitrile, chloroacetyl acetone, chloroacetone, and carbonate salts. These modified drugs showed an effective bactericidal effect against ciprofloxacin-resistant pathogens [76].

4.9 Gentamicin

Gentamicin is an essential antibiotic for the treatment of serious urinary tract infections. This drug shows bactericidal effects against bacteria, which are responsible for meningitis and infection of the blood. The chemical structure of gentamicin is shown in Fig. 14 [77].

Monti, et al. Had proposed an effective strategy for synthesizing biocompatible delivery of antibiotics related to nano-biomaterials by forming hybrid molecules. The modification of gentamicin was based on gold nanoparticles, citrate, and chitosan. These hybrid molecules were used in the removal of bacterial biofilm. $AuCl_3$ and Trisodium citrate were required to prepare gentamicin-chitosan-gold nanoparticles [78]. Xiao, et al. A novel method to enhance the bactericidal effect of conventional gentamicin drugs by combining them with a sponge containing nano-cellulose. The novel sponge was prepared by cellulose nanofibers, cellulose acetoacetate, and 3-aminopropyl(triethoxy)silane by cross-linking. FT-IR, XPS, EDX, SEM, Nitrogen adsorption-desorption measurement, and compressive test were used to observe the covalent link between gentamicin and nanocellulose sponge. Gentamicin, along with CNF sponges, exhibits high-level bacteriocidal properties in opposition to *E. coli* and *S. aureus* [79]. Liu Y, et al. Had proposed a unique method to engineer gentamicin with chitosan film. The surface modification of conventional drugs with chitosan was formed by the following methods-First a chitosan film was formed chitosan solution. The chitosan film surface was mixed with citric acid. At last, gentamicin was combined. The modified antibiotics showed higher antibacterial properties [80]. Yuanhao Qiu et al. Had approached a method for the modification of gentamicin with phosphatidylcholine and chitosan. The zeta potential of the novel drug was observed with the help of dynamic light scattering, and a scanning electron microscope was used to observe the surface structure. Interestingly, the novel gentamicin showed biofilm-penetrating properties. Increased the bactericidal effect of old Gentamicin [81]. Akhtar, et al. Had approached a novel strategy to engineer Gentamicin with poly lactic-co-glycolic acid and chitosan. The antibacterial property of new drugs was studied against *Bacillus subtilis* [82].

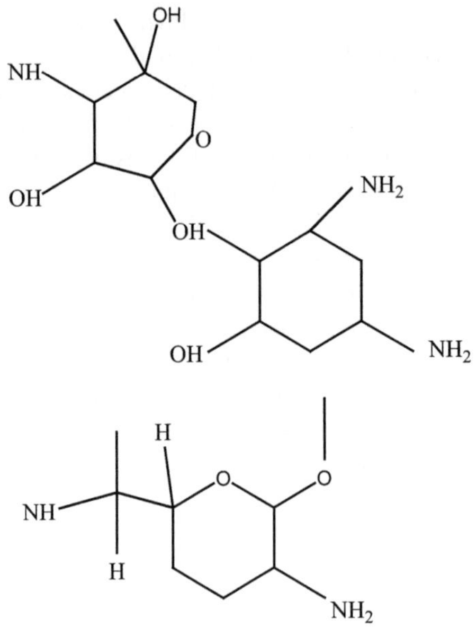

Fig. 14. The chemical structure of Gentamicin [77].

4.10 Colistin

Colistin is synthesized by *Bacillus colistinus,* which is effective against various gram-negative bacteria. It is a polymyxin drug made up of polymyxins E1 and E2.

Luther et al. Had designed their research work in opposition to gram-negative pathogens, which are resistant to carbapenem and cephalosporin (third generation). Natural products can also be used to design synthetic antibiotics. Chimerical drugs contain one β-hairpin peptide macrocycle attached to polymyxin and colistin. These novel drugs showed high antibacterial activity against the ESKAPE pathogen. First OMP- the inhibiting drug is Murepavadin [83]. J.J.Augilera-Correa et al. Had approached a novel method to engineer colistin with mesoporous silica nanoparticles. Here, colistin is combined with moxifloxacin and Arabic gum [84]. Nguyen et al. Had reported the modification of colistin with NCL 195. It increased the bactericidal effect of colistin against *gram-negative bacteria* [85]. Dubashynsaya et al. Had modified colistin with hyaluronic acid. Carboxyl moieties of hyaluronic acid and amino moieties of colistin showed polymeric interaction. These engineered drugs enhanced the bactericidal effect in opposition to the ESKAPE pathogens. Among these, it is effective against *Pseudomonas aeruginosa* [86].

4.11 Cefotaxime

It is a compound of cephalosporin composed of acetoxymethyl and [2-(2-amino-1,3-thiazol-4-yl)-2-(methoxy imino)acetyl]amino side groups. The chemical structure of cefotaxime is shown in Fig. 15 [87].

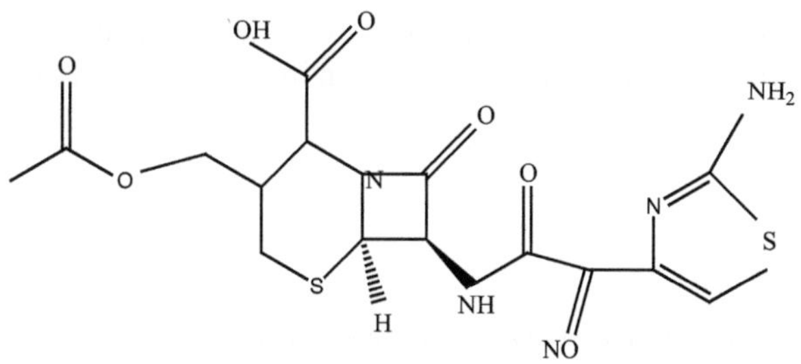

Fig. 15. The chemical structure of Cefotaxime [87].

Halawani, et al. Had experimented on the conjugation of cefotaxime & silver nanomaterials. The antibacterial effect of modified drugs was examined against *E.coli and* MRSA. The combinatorial effect of cefotaxime–silver nanoparticles increased the high antibacterial activity as contrast to Cefotaxime and AgNps. It is a novel method to enhance the life span of old drugs [88]. Mainradi, et al. Had modified cefotaxime with Amoxicillin. The bacteriocidal effect of engineered cefotaxime was studied in opposition to *Enterococcus faecalis*. From this report, synergistic co-relation was observed between Cefotaxime and Amoxicillin with the help of the penicillin-binding protein competition assay [89]. *Ku¨hn,* et al. Had approached a novel method for the modification of cefotaxime. In this research work, cefotaxime combined with levofloxacin was used to enhance the bactericidal property of conventional Cefotaxime against penicillin-resistant pneumococcal bacteria with the help of a checkerboard assay [90].

4.12 Erythromycin

This antibiotic is synthesized from *Saccharopolyspora erythraea,* and it is effective in opposition to both gram-positive and gram-negative bacteria.

Rastogi, et al. Had observed the potent antibacterial properties of erythromycin with 4′-carbamates of 6,11-di-O-methylerythromycin A. This novel antibiotic was effective in opposition to gram-positive bacteria such as MRSA and *Streptococcus Pneumoniae,* which are resistant to erythromycin. These are the compounds used to modify the conventional drugs such as 6,11-Di-O-methyl-2¢-O-acetyl erythromycin A, 6,11-Di-O-acetyl-400-O-acylimidazolylerythromycinA,6,11-Di-O-methyl-400-O-(N-N-N-propyl carbamoyl)erythromycin A, and 6,11-Di-O-methyl-400-O-(N-butyl carbamoyl) erythromycin A. The antibacterial property of the modified drug was studied against MRSA, along with proteomics data [91]. Fallah, et al. Had reported a unique method to engineer erythromycin with nanomaterials. Here, the conventional drugs are combined with nano dendrimers G1[PEG] and G2[DCC + PEG]. The antibacterial study of novel drugs was conducted against opposition to *Staphylococcus aureus, Staphylococcus epidermidis, Staphylococcus saprophyticus, and Pseudomonas aeruginosa.* As compared to erythromycin, the modified drugs with G2 nano dendrimer were efficacious against both Gram-negative and Gram-positive pathogens [92]. Xu, et al. Had approached a method for the modification of old erythromycin by the combined treatment with propanol. The antimicrobial property was studied against *Saccharopolyspora erythraea*. C 13 was used to observe the variation in the cell activity [93].

5 Discussion

The purpose and scope of this review cover the modern antibiotic modification methods to combat superbugs by enhancing the life period of antibiotics. Modification of drugs includes adjuvant therapy, by using chelating ligands, and Ayurvedic herbs. It is essential to discover new drugs, two-drug delivery simultaneously. Many serious and devastating bacterial diseases were eliminated with the help of no less than seventeen non-identical antibiotics that have been discovered until now. Due to the global misapply of presently available antibiotics, there is a quickening rate of appearance of new drug-resistant strains, which can bring out critical diseases. After fifty years of clinical antibiotics utilization, opposition to specific multidrug resistance has appeared as a forbidding issue. This matter will need the cooperation of many controls along with the alternation of novel antibiotics. The redesigning of conventional antibiotics and surface modification of drugs discussed above have the potential to provocation antibiotic resistance in the forthcoming years. Hence, future approaches also incorporate the combination of currently marketed antibiotics not only with metal salts but also with their distinct physical property like oxidation. We can also develop novel antibiotics with organic particles and colloidal forms. This is another way to synthesize drugs that are more efficacious. For this, it is essential to improve the molecular tools. It is important to create new conduits for antibiotics, which are much more rapid and uncomplicated, to combat the emerging multidrug-resistant bacteria. Furthermore, we can apply these engineered antibiotics for the healing of chronic wounds, the eradication of biofilm, and the extension of the life span of old antibiotics. Limitations of antibiotic modifications deal with the minimum inhibitory concentration to eradicate biofilm or Chronic wound healing. Engineered Antibiotics should be effective in low concentrations and enhance their life span against the superbugs.

6 Conclusion

The expansion of engineered antibiotics is a worldwide healthcare concern, for which it is essential to combat the multidrug-resistant bacteria by eradicating their biofilm. According to WHO, among 32 types of antibacterial agents, only 6 drugs are effective in opposition to the superbugs and considered novel antibiotics. Nowadays, it is a major topic of concern for the researcher. The formation of new and advanced technologies has been implemented in the synthesis of novel agents. Antibiotic resistance remains a major health issue worldwide, and it requires urgent intervention. By utilizing combinatorial technology, Postbiotics, novel antibiotics, can be formed. Furthermore, the major aim of this review article is to form new drugs that have more antibacterial properties with a minimum concentration and increase the immune power of individuals.

Creating public awareness regarding antibiotic resistance requires a multi-faced approach, such as public awareness campaigns, utilization of media channels, and education and training should be integrated into the school curriculum, and healthcare professional training. Foster the partnership with the policymakers and public to address antimicrobial resistance, along with the proper use of antibiotics.

Acknowledgement. The author wishes to thank Biju Patnaik University of Technology, Rourkela, for providing facilities to complete this review study. The authors also wish to thank the GTHTM-2025 Organizing Team for allowing me to participate in the conference and presentation as well.

Conflict of Interest The authors declare no conflict of interest, financial or otherwise.

Funding None.

References

1. Magiorakos, A.P., et al.: Multidrug-resistant, extensively drug-resistant and pan drug-resistant bacteria: an international expert proposal for interim standard definitions for acquired resistance. Clin. Microbiol. Infect. **18**, 268–281 (2012)
2. Tan, S.Y., Tatsumura, Y.: Alexander Fleming (1881-1955): discoverer of penicillin. Singapore Med. J. **56**, 366–367 (2015)
3. Flemming, H.C., Wingender, J., Szewzyk, U., Steinberg, P., Rice, S.A., Kjelleberg, S.: Biofilms: an emergent form of bacterial life. Nat. Rev. Microbiol. **14**, 563–575 (2016)
4. Tambekar, D.H., Dahikar, S.B.: Antibacterial activity of some Indian Ayurvedic preparations against enteric bacterial pathogens. J. Adv. Pharm. Technol. Res. **2**, 24 (2011)
5. Five bacteria types claimed 6.8 lakh lives in India in 2019: Lancet. (2022)
6. Ahmad, M., Aduru, S.V., Smith, R.P., Zhao, Z., Lopatkin, A.J.: The role of bacterial metabolism in antimicrobial resistance. Nat. Rev. Microbiol., 1–16 (2025)
7. Ahmed, W., Zhai, Z., Gao: Adaptive antibacterial biomaterial surfaces and their applications. Mat. Bio. **2** (2019)
8. Chakrabarty, S.: Quantitative phytochemical investigation, antibacterial potency, and drugability assessment of three Indian medicinal plants leaf extracts using bioinformatics tools. Int. J. Exp. Res. Rev 42 (2024); Blair, J.M., Richmond, G.E., Piddock, L.J.: Multidrug efflux pumps in Gram-negative bacteria and their role in antibiotic resistance. Fut. Microbiol. **10** 1165–77 (2014)
9. Chakrabarty, S., Mishra, M.P., Bhattacharya, D.: Targeting microbial bio-film: an update on MDR gram-negative bio-film producers causing catheter-associated urinary tract infections. Appl. Biochem. Biotechnol. **6**, 2796–2830 (2022)
10. Campbell, E.A., et al.: Structural mechanism for rifampicin inhibition of bacterial RNA polymerase. Cell. **104**(6), 901–912 (2001)
11. Nandi, S.: Recent advances in ligand and structure-based screening of potent quorum sensing inhibitors against Antibiotic resistance induced bacterial virulence. Recent Pat. Biotechnol. **10**(2), 195–216 (2016)

12. Sahoo, K.C., Tamhankar, A.J., Johansson, E., Lundborg, C.S.: Antibiotic use, resistance development, and environmental factors: a qualitative study among healthcare professionals in Orissa, India. BMC Public Health. **10** (2010)
13. Wang, Y., et al.: Comparison of solithromycin with erythromycin in enterococcus faecalis and enterococcus faecium from China: antibacterial activity, clonality, resistance mechanism, and inhibition of biofilm formation. Jour. Anti. **74** (2022)
14. Aminov, R.I.: A brief history of the antibiotic era: lessons learned and challenges for the future. Front. Microbiol. **1**, 134 (2010)
15. Kumar, S.G., Adithan, C., Harish, B.N., Sujatha, S., Roy, G., Malini, A.: Antimicrobial resistance in India: a review. J. Nat. Sci. Biol. Med. **2**, 286–291 (2013)
16. Mogasale, V.V., Saldanha, P., Pai, V., Rekha, P.D.: A descriptive analysis of antimicrobial resistance patterns of WHO priority pathogens isolated in children from a tertiary care hospital in India. Sci. Rep. **11**, 5116 (2021)
17. Painter, C., Limmathurotsakul, D., Roberts, T., van Doorn, H.R., Mayxay, M., Ashley, E.A.: Sustainable antimicrobial resistance surveillance: time for a global funding mechanism. Lancet Infect. Dis. **25**(2), e99–e103 (2025)
18. Veeraraghavan, B., et al.: Activity of novel lactone ketolide nafithromycin against multicentric invasive and non-invasive pneumococcal isolates collected in India. JAC Antimicrob. Resist. **3**(2), dlab066 (2021)
19. Zhao, Z., Yan, R., Yi, X., Li, J.: Bacteria-activated Theranostic Nanoprobes against methicillin-resistant Staphylococcus aureus infection. ACS Nano. **5**, 4428–4438 (2017)
20. Chen, Y.H., et al.: Vancomycin-loaded nanoparticles enhance sporicidal and antibacterial efficacy for *Clostridium difficile* infection. Front. Microbiol. **10** (2019)
21. Okano, A., Isley, N.A., Boger, D.L.: Peripheral modifications of [Ψ[CH_2 NH]Tpg ^{4}]vancomycin with added synergistic mechanisms of action provide durable and potent antibiotics. Proc. Natl. Acad. Sci. **114**, 26 (2017)
22. Xie, J., Pierce, J.G., James, R.C., Okano, A., Boger, D.L.: A redesigned vancomycin engineered for dual -ala-ala and -ala- -lac binding exhibits potent antimicrobial activity against vancomycin-resistant bacteria. J. Am. Chem. Soc. **35** (2011)
23. Nigam, A., Almabruk, K.H., Saxena, A., Yang, J., Mukherjee, U., Kaur: Modification of rifamycin polyketide backbone leads to improved drug activity against rifampicin-resistant mycobacterium tuberculosis. J. Biochem. **30** (2014)
24. Cochrane, S.A., Li, X., He, S., Yu, M., Wu, M., Vederas, J.: Synthesis of Tridecaptin-Antibiotic conjugates with in vivo activity against gram-negative bacteria. J. Med. Chem. **58**, 9779–9785 (2015)
25. Binda, E., Marinelli, F., Marcone, G.L.: Old and new Glycopeptide antibiotics: action and resistance. J. Ant. **4**, 572–594 (2014)
26. Xi, Y., Wang, Y., Gao, J., Xiao, Y., Du, J.: Dual Corona vesicles with intrinsic antibacterial and enhanced Antibiotic delivery capabilities for effective treatment of biofilm-induced periodontitis. ACS Nano. **12**, 13645–13657 (2019)
27. Wu, Z.C., Boger, D.L.: Exploration of the site-specific nature and generalizability of a trimethylammonium salt modification on vancomycin: a-ring derivatives. Tet. **24**, 3160–3165 (2019)
28. Campbell, M., Cho, C.Y., Ho, A., Huang, J.Y., Martin, B.: 4-Ethoxybenzoic acid inhibits Staphylococcus aureus biofilm formation and potentiates biofilm sensitivity to vancomycin. Int. J. Antimicrob. Agents. **56** (2020)
29. Fang, B., et al.: Extracellular matrix scaffold crosslinked with vancomycin for multifunctional antibacterial bone infection therapy. Biomaterials. **268** (2021)
30. Ghaffar, I., Imran, M., Perveen, S., Kanwal, T., Saifullah, S., Bertino, M.F.: Synthesis of chitosan-coated metal organic frameworks (MOFs) for increasing vancomycin bactericidal potentials against resistant S. aureus strain. Mater. Sci. Eng. C. **105** (2019)

31. Zou, Z., et al.: Vancomycin modified copper sulfide nanoparticles for photokilling of vancomycin-resistant enterococci bacteria. Colloids Surf. B Biointerfaces. **189** (2020)
32. Sum, P.E.: Recent developments in tetracycline antibiotics. Curr. Pharm. **18**, 119–132 (1998)
33. NCBI.: https://pubchem.ncbi.nlm.nih.gov/compound/Tetracycline. 03 Jan 2025
34. Salman, M.Y., Taşkan, E., Hasar, H.: Comparative potentials of H2- and O2-MBfRs in removing multiple tetracycline antibiotics. Process Saf. Environ. Prot. **167**, 184–191 (2022)
35. Shi, H., Ou, M.Y., Cao, J.P., Chen, G.F.: Synthesis of ovalbumin-stabilized highly fluorescent gold nanoclusters and their application as Hg2+ sensor. RSC Adv. **5** (2015)
36. Dey, P., Parai, D., Banerjee, M., Hossain, S.T., Mukherjee, S.: Naringin sensitizes the antibiofilm effect of ciprofloxacin and tetracycline against Pseudomonas aeruginosa biofilm. IIJMM. **3** (2020)
37. Wang, N., Luo, J., Deng, F., Huang, Y., Zhou, H.: Antibiotic combination therapy: a strategy to overcome bacterial resistance to aminoglycoside antibiotics. Front. Pharmacol. **13** (2022)
38. NCBI.: https://pubchem.ncbi.nlm.nih.gov/compound/Neomycin-C. 03 Jan 2025
39. Mathivanan, J., Bai, Z., Chen, A., Sheng, J.: Design, synthesis, and characterization of a novel 2′-5′-linked amikacin-binding aptamer: an experimental and MD simulation study. ACS Chem. Biol. **18** (2022)
40. Nitanan, T., Rojanarata, T., Ngawhirunpat, T., Opanasopit, P.: Neomycin-loaded poly(styrene sulfonic acid-co-maleic acid) (PSSA-MA)/polyvinyl alcohol (PVA) ion exchange nanofibers for wound dressing materials. Int. J. Pharm. **71** (2013)
41. Khalil, M.A., Maghraby, G.M., Sonbol, F.I., Allam, N.G., Ateya, P.S.: Enhanced efficacy of some antibiotics in the presence of silver nanoparticles against multidrug-resistant *Pseudomonas aeruginosa* recovered from burn wound infections. Front. Microbiol. **27** (2021)
42. Degtyareva, N.N., et al.: Antimicrobial activity, AME resistance, and A-site binding studies of Anthraquinone-neomycin conjugates. ACS Infect. Dis. **3**, 206–215 (2017)
43. Brown, A.N., Smith, K., Samuels, T.A., Lu, J., Obare, S.O., Scott, M.E.: Nanoparticles functionalized with ampicillin destroy multiple-Antibiotic-resistant isolates of Pseudomonas aeruginosa and Enterobacter aerogenes and methicillin-resistant Staphylococcus aureus. Appl. Environ. Microbiol. **78**, 2768–2774 (2012)
44. NCBI.: https://pubchem.ncbi.nlm.nih.gov/compound/Neomycin-C. 10 Jan 2025
45. Khatoon, N., Alam, H., Khan, A., Raza, K., Sardar, M.: Ampicillin silver nano formulations against multidrug resistant bacteria. Sci. Rep. **29** (2019)
46. Noseda, G., Román, G., Juan, J.: Advance on adsorption of amino-functionalized silica nanocarrier for the delivery of therapeutic ampicillin as a drug model. Inorg. Chem. Commun. **123** (2021)
47. Topsakal, A., Midha, S., Yuca, E., Tukay, A., Sasmazel, H.T., Kalaskar, D.M.: Study on the cytocompatibility, mechanical and antimicrobial properties of 3D printed composite scaffolds based on PVA/gold nanoparticles (AuNP)/ampicillin (AMP) for bone tissue engineering. Mater. Today Commun. **28** (2021)
48. Magadla, A., Openda, Y.I., Mpeta, L.S., Nyokong, T.: Evaluation of the antibacterial activity of gallic acid anchored phthalocyanine-doped silica nanoparticles towards Escherichia coli and Staphylococcus aureus biofilms and planktonic cells. Photodiagn. Photodyn. Ther. **42** (2023)
49. Al-Noor, T.H., Mohapatra, R.K., Azam, M., Karem, L.K.A., Mohapatra, P.K., Ibrahim, A.A.: Mixed-ligand complexes of ampicillin-derived Schiff base ligand and nicotinamide: synthesis, physicochemical studies, DFT calculation, antibacterial study, and molecular docking analysis. J. Mol. Struct. **1229** (2021)
50. Thappeta, K.R., Vikhe, V., Yong, Y.S., H., A.M.: combined efficacy of an antimicrobial cationic peptide polymer with conventional antibiotics to combat multidrug-resistant pathogens. ACS Infect. Dis. **6** (2020)

51. NCBI.: https://pubchem.ncbi.nlm.nih.gov/compound/Amoxicillin. 10 Jan 2025
52. Tarui, A., Tanaka, A., Ueno, M., Sato, K.: Synthesis of a fluorinated ezetimibe analog using radical allylation of α-bromo-α-fluoro-β-lactam. J. Chem. **39** (2015)
53. Turan, B., Şendil, K., Şengül, E., Gültekin, M.S., Taslimi, P., Gulçin, İ.: The synthesis of some β-lactams and investigation of their metal-chelating activity, carbonic anhydrase and acetylcholinesterase inhibition profiles. J. Enzyme Inhib. Med. Chem. **31** (2016)
54. Sémiramoth, N., Di Meo, C., Zouhiri, F., Valetti, S., Gorges, R.: Self-assembled Squalenoylated penicillin bioconjugates: an original approach for the treatment of intracellular infections. ACS Nano. **6**, 3820 (2012)
55. Yenn, T.W., Khan, M., Amiera, N., Chean, L., Ibrahim, D.: Stigmasterol: an adjuvant for beta-lactam antibiotics against beta-lactamase positive clinical isolates. Ster. **128** (2017)
56. Turos, E., et al.: Penicillin-bound polyacrylate nanoparticles: restoring the activity of beta-lactam antibiotics against MRSA. Bioorg. Med. Chem. Lett. **12** (2007)
57. Ketikidis, I., Banti, C.N., Kourkoumelis, N., Tsiafoulis, C.G., Hadjikakou, S.K.: Conjugation of penicillin-G with silver(I) ions expands its antimicrobial activity against gram-negative bacteria. Ant. **25** (2020)
58. Boffoli, D., Bellato, F., Avancini, G., Gurnani, P., Yilmaz, G., Romero: tobramycin-loaded complexes to prevent and disrupt Pseudomonas aeruginosa biofilms. Dr U. Del.. Tra. Res. **28** (2022)
59. NCBI.: https://pubchem.ncbi.nlm.nih.gov/compound/Tobramycin 10 Jan 2025
60. Khan, S.A., Shahid, S., Mahmood, T., Lee, C.S.: Contact lenses coated with hybrid multifunctional ternary nanocoatings (Phytomolecule-coated ZnO nanoparticles: Gallic acid: tobramycin) for the treatment of bacterial and fungal keratitis. Acta Biomater. **128**, 262 (2021)
61. Ham, S.Y., Kim, H.S., Lee, J.H., Byun, Y., Park, H.D.: Combined treatment of 6-Gingerol analog and tobramycin for inhibiting Pseudomonas aeruginosa infections. Microbiol. Spectr. **9** (2021)
62. Kashyap, S., Kaur, S., Sharma, P., Capalash, N.: Combination of colistin and tobramycin inhibits persistence of Acinetobacter baumannii by membrane hyperpolarization and down-regulation of efflux pumps. Microbes Infect. **24** (2021)
63. Tait, J.R., et al.: Pharmacodynamics of ceftazidime plus tobramycin combination dosage regimens against hypermutable Pseudomonas aeruginosa isolates at simulated epithelial lining fluid concentrations in a dynamic in vitro infection model. J. Glob. Antimicrob. Resist. **55** (2021)
64. Puvvadi, R., Mikkelsen, H., McCahon, L., Grogan, S., Ditcham, W., Reid, D.W.: Role of Tris-CaEDTA as an adjuvant with nebulized tobramycin in cystic fibrosis patients with Pseudomonas aeruginosa lung infections: a randomized controlled trial. J. Cyst. Fibros. **2** (2021)
65. Farhan, S.M., Raafat, M., Abourehab, M.A.S.: Effect of imipenem and amikacin combination against multi-drug resistant *Pseudomonas aeruginosa*. J. Antibiot. **10** (2021)
66. NCBI.: https://pubchem.ncbi.nlm.nih.gov/compound/Amikacin. 10 Jan 2025
67. Magallon, J., et al.: Amikacin potentiator activity of zinc complexed to a pyrithione derivative with enhanced solubility. Sci. Rep. **12** (2022)
68. Darlow, C.A.: Amikacin combined with Fosfomycin for treatment of neonatal sepsis in the setting of highly prevalent antimicrobial resistance. Antimicrob. Agents Chemother. (2021)
69. Wang, F.: Evaluation of ceftazidime/avibactam alone and in combination with amikacin, colistin, and tigecycline against Klebsiella pneumoniae carbapenemase-producing K. Pneumoniae by in vitro time-kill experiment. PLoS One. **16** (2021)
70. Soukarieh, F., Gurnani, P., Romero, M., Halliday, N.: Design of quorum sensing inhibitor-polymer conjugates to penetrate *Pseudomonas aeruginosa* biofilms. ACS Macro Lett. **12**, 314 (2023)

71. NCBI.: https://pubchem.ncbi.nlm.nih.gov/compound/Ciprofloxacin 10 Jan 2025
72. Chrysouli, M.P., et al.: Ciprofloxacin conjugated to diphenyltin: a novel formulation with enhanced antimicrobial activity. Dalton Trans. **49** (2020)
73. Bykowska, A., et al.: Phosphine derivatives of ciprofloxacin and norfloxacin, a new class of potential therapeutic agents. J. Chem. **38** (2014)
74. Ibrahim, N.M., Fahim, S.H., Hassan, M., Farag, A.E.: Design and synthesis of ciprofloxacin-sulfonamide hybrids to manipulate ciprofloxacin pharmacological qualities: potency and side effects. Eur. J. Med. Chem. **228** (2022)
75. Alsughayer, A., Sagheer, F.: Antibiotic resistance and drug modification: synthesis, characterization and bioactivity of newly modified potent ciprofloxacin derivatives. Bioorg. Chem. **108** (2021)
76. Monti, S., Jose, J., Sahajan, A., Kalarikkal, N., Thomas, S.: Structure and dynamics of gold nanoparticles decorated with chitosan–gentamicin conjugates: ReaxFF molecular dynamics simulations to disclose drug delivery. Phys. Chem. **24** (2019)
77. NCBI.: https://pubchem.ncbi.nlm.nih.gov/compound/Gentamicin. 10 Jan 2025
78. Xiao, Y., Rong, L., Wang, B., Mao, Z.: Light-weight and high-efficacy antibacterial nanocellulose-based sponge via covalent immobilization of gentamicin. Carbohydr. Polym. **200** (2018)
79. Liu, Y., et al.: Gentamicin modified chitosan film with improved antibacterial property and cell biocompatibility. Int. Mac. **58** (2017)
80. Qiu, Y., Xu, D., Sui, G., Wang, D., Wu, M., Han, L.: Gentamicin decorated phosphatidylcholine-chitosan nanoparticles against biofilms and intracellular bacteria. Int. J. Biol. Macromol. **156**, 640 (2020)
81. Akhtar, B., Muhammad, F., Aslam, B., Saleemi, M.K., Sharif, A.: Pharmacokinetic profile of chitosan modified poly lactic co-glycolic acid biodegradable nanoparticles following oral delivery of gentamicin in rabbits. Int. J. Biol. Macromol. **164**, 1493 (2020)
82. Luther, A., et al.: Chimeric peptidomimetic antibiotics against gram-negative bacteria. Nature. **576**, 452 (2019)
83. Correa, J.J., Gisbert, M., Mediero, A., Carias, R.A., Jiménez, C., Esteban, J.: Arabic gum plus colistin coated moxifloxacin-loaded nanoparticles for the treatment of bone infection caused by Escherichia coli. Acta Biomater. **137** (2022)
84. Nguyen, H.T., Venter, H., Veltman, T., Williams, R., O'Donovan, L.A., Russell, C.C.: In vitro synergistic activity of NCL195 in combination with colistin against gram-negative bacterial pathogens. Int J Anti. **57** (2021)
85. Dubashynskaya, N.V., Raik, S.V., Dubrovskii, Y.A., Shcherbakova, E.S., Demyanova, E.V., Shasherina, A.Y.: Hyaluronan/colistin polyelectrolyte complexes: promising antiinfective drug delivery systems. Int. J. Biol. Macromol. **187**, 157 (2021)
86. Halawani, E.M., Hassan, A.M., Gad El-Rab, S.M.: Nanoformulation of biogenic cefotaxime-conjugated-silver nanoparticles for enhanced antibacterial efficacy against multidrug-resistant bacteria and anticancer studies. Int. J. Nanomedicine. **15**, 889–1901 (2020)
87. NCBI.: https://pubchem.ncbi.nlm.nih.gov/compound/Cefotaxime. 14 Jan 2025
88. Mainardi, J.L., Gutmann, L., Acar, J.F., Goldstein, F.W.: Synergistic effect of amoxicillin and cefotaxime against enterococcus faecalis. Antimicrob. Agents Chemother. **39**(9), 1984–1987 (1995)
89. Kühn, F., Cottagnoud, M., Acosta, F., Flatz, L., Entenza, J., Cottagnoud, P.: Cefotaxime acts synergistically with levofloxacin in experimental meningitis due to penicillin-resistant pneumococci and prevents selection of levofloxacin-resistant mutants in vitro. Anti. Micro. **47** (2007)

90. Rastogi, N., Goh, K.S., Berchel, M., Bryskier, A.: In vitro activities of the ketolides telithromycin (HMR 3647) and HMR 3004 compared to those of clarithromycin against slowly growing mycobacteria at pHs 6.8 and 7.4. Antimicrob. Agents Chemother. **10** (2000)
91. Fallah, F., Zargar, M., Yousefi, M., Alam, A.N.: Synthesis of the erythromycin-conjugated nanodendrimer and its antibacterial activity. Eur. J. Pharm. Sci. **123** (2018)
92. Xu, F., et al.: Exploring the metabolic fate of propanol in industrial erythromycin-producing strain via ^{13}C labeling experiments and enhancement of erythromycin production by rational metabolic engineering of Saccharopolyspora erythraea. Bio. Commun. **542** (2021)
93. Ning, Q., et al.: Combined effects of nanosized polystyrene and erythromycin on bacterial growth and resistance mutations in Escherichia coli. J. Hazard. Mater. **422** (2022)

In Silico Studies of 1,2,4-Triazole Derivatives by Targeting DHFR Againsts *staphylococcus aureus*: Pharmacophore Generation, Atom-Based 3D-QSAR, Molecular Docking and ADMET

Saket[1], Ankush[1], Arun Kumar[1], and Neelam Vashisth[1(✉)]

[1] Department of Pharmaceutical Sciences, Gurugram University, Gurugram, Haryana, India
neelam.vashisth@gurugramuniversity.ac.in

Abstract. Antibiotics, which disrupt essential processes for bacterial growth and proliferation, are crucial in combating bacterial infections. However, their misuse and overuse have resulted in increasing antimicrobial resistance, highlighting the urgent need for new antibiotics. Compounds with a 1,2,4-triazole ring exhibit diverse biological activities, with extensive research demonstrating their significant antimicrobial properties. In silico approaches are used on previously synthesized ninety-five 1,2,4-triazole derivatives against *staphylococcus aureus* for protein Dihydrofolate reductase (DHFR). The protein Wild-type *staphylococcus aureus* DHFR in complex with NADPH and Trimethoprim (PDB ID 2W9G), were used for the model development against the *S. aureus*. Alongside, the four features Ligand based pharmacophore hypothesis and Atom based 3D-QSAR model was generated with statistical parameters include correlation coefficient for the training set ($Q^2 = 0.7995$), regression coefficient for test set ($R^2 = 0.9808$) which proves the reliability of model. Contour visualization was done for the best docked and least docked compound. Furthermore, the best docked molecule was taken and through CheMBL database ligand based virtual screening was performed. Later, ADME/T studies were carried out on top five compounds from the virtual screened database. Among the five compounds, the compounds CheMBL 208844 and CheMBL 275068 were found to be most potent antimicrobial agents against *S. aureus*. These compounds may be considered as lead for further development of antimicrobial agents.

Keywords: 1,2,4-triazoles · Antimicrobial activity · Atom based 3D-QSAR · Dihydrofolate Reductase · ADME/T

1 Introduction

Antimicrobial agents are those agents which can kill microorganisms or abort their microbial growth. These microorganisms are minute microscopic living things present in our surroundings capable of inducing infectious aliments such as influenza, measles, skin infection and blood infection [1]. There are number of pathogen bacteria which

A. K. Saxena, A. Saxena (Eds.): GTHTM 2025, pp. 85–116, 2026.
https://doi.org/10.1007/978-3-032-12320-6_5

developed the AMR due to various reason like Fluroquinolone-resistant *Escherichia coli,* Carbapenem-resistant *klebstella pneumonia,* Vancomycin-resistant *enterococci* and Methicillin-resistant *Staphylococcus aureus*, Vancomycin-resistant *E.coli* [2]. Every year, approximately 700,000 people die due to antimicrobial resistance about 2.8 million antimicrobial resistant infections occur every year and if this continues then the mortality rate will become 10 million deaths globally by 2050 has been estimated [3].

Heterocyclic compounds are organic cyclic compounds that include at least one hetero atom. Nitrogen, Oxygen, Sulphur are most frequent heteroatoms, although heterocyclic rings containing other heteroatoms are also commonly recognised [4]. Triazole belong to five membered heterocycles which has accumulated a growing and scientific interest from the scientific community due to their broad application in pharmaceuticals, agricultural, material sciences fields [5].

The presence of 1,2,4-triazole moiety within various heterocycles significantly impacts the lipophilicity, polarity and hydrogen bonding capacity of molecule. This influence often leads to enhancements in the pharmacological, pharmacokinetic, toxicological and physicochemical properties of the compounds [6]. 1,2,4-triazole have a wide range of pharmacological properties which are Antimicrobial [6], Anticonvulsant [7], Antidepressant [8], Anti-inflammatory [9], Analgesic [10], Antifungal [11], Anticancer [12], Anti-parasitic [13], Anti-urease [14], 5α reductase inhibitory [15].

In light of these alarming issues of AMR, it is imperative to explore innovative approaches to combat AMR, and one promising avenue is to target the DHFR protein [16]. This protein has emerged as a pivotal point of intervention due to its essential role in microbial replication, and its gain resistance to drugs like trimethoprim has highlighted the urgency for novel treatment strategies [17].

Drug design involves shaping and charge of molecules in such a way that they will bind to a biomolecular target by interacting with them. The key approach for this is computer modeling which is synonymous to computer-aided drug design or in silico drug design. Ligand-based drug design (or indirect drug design) is dependent on understanding other molecules that have high binding affinities to a target biomolecule. It includes Pharmacophore model and QSAR [18]. The pharmacophore is the molecular scaffold that defines a molecule's vital features which are responsible for its biological activity. Atom-based QSAR is the correlation of 3D properties of the atoms with their biological activity by using distinct statistical parameters. This method uses atom types and their occupancy of a grid of cubes as independent variables for fitting and predicting properties. Through QSAR visualization interactions of an atom with their features are interpreted includes: Negative ionic charges (N), positive ionic charges (P), a hydrogen bond donor (D), hydrophobic (H), electron withdrawing (W) and miscellaneous (X) [19].

2 Materials and Methods

2.1 Dataset

The dataset of all the previously synthesized 95 1,2,4-triazole derivatives with diverse MIC ranging from 1.206 μg/ml to 3.486 μg/ml against *S. aureus* were fetched and compiled from the published work dated above 2011(Table 2). Furthermore, for the development of pharmacophore hypothesis and 3D-QSAR model MIC value (μg/ml) was converted into pMIC [log mol. weight/conc$_n$ (μg/ml)] [20].

2.2 Application Software and Web-Based Software

Computational studies including pharmacophore hypothesis generation, generation of atom-based 3D-QSAR model, virtual screening and molecular docking, all are performed in Schrodinger's suite (Maestro version 13.6), LLC, New york, 2022. The studies conducted for non-profit free accessible web tools like ChemBl and ADMET Lab2.0 tool was performed online.

2.3 The Generation of 3D Wild-Type *staphylococcus aureus* DHFR in Complex with NADPH and Trimethoprim (PDB ID 2W9G)

The 3D structure of Wild-type *staphylococcus aureus* DHFR in complex with NADPH and trimethoprim (PDB ID 2W9G, resolution **1.95 Å)** with no mutation was retrieved from protein data bank (PDB) (https://www.rcsb.org/structure/2W9G) [21]. It comprises of ligand Trimethoprim with a subunit B. For the visualization of dihedral angles ψ against φ of amino acid residues in the protein structure. Ramachandran plot was used. It shows the empirical distribution of data points in single structure. The distribution of red and yellow regions in the plot characterized the promisingly favoured and thoughtfully authorized regions (Fig. 1).

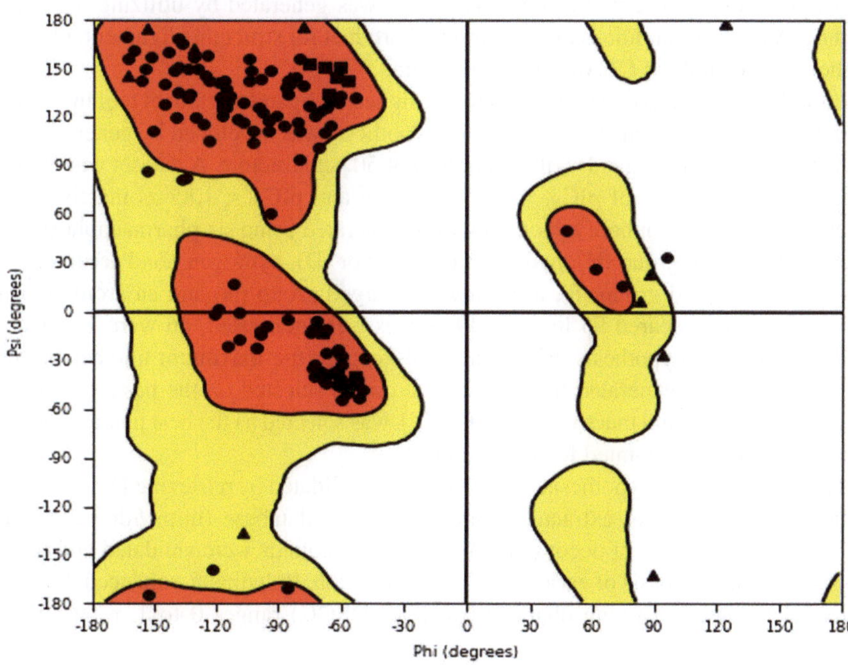

Fig. 1. Graphical representation of Ramachandran plot.

2.4 Protein Preparation and a Grid Generation for Selected Target Dihydrofolate Reductase Complexed with Folate and NADP+

The protein preparation wizard was used to get the PDB ID and by reviewing the structure, unnecessary atoms and chains, unbound ligands and water molecules were removed from the protein. After the addition of missing residues and hydrogen bonds, grid generation was allowed by using Glide module. OPLS-2005 force field was used for the shape refinement and energy minimization of a protein [22].

2.5 Preparation of Ligands

All the collected 95 1,2,4-triazole derivatives were sketched in 2D form using ChemDraw 16.0 version and converted into their 3D form by the LigPrep module (2022, Schrodinger, LLC, NY). All the ligands were prepared by selecting them from a Project table of LigPrep module. The compounds were neutralized with ionization states adjusted at a pH 7.4 ± 0.0 with maximum ligand size of 500 atoms and no conformers were generated for the compounds. OPLS-2005 force field was used for the energy minimization.

2.6 A Pharmacophore Model Generation

A ligand-based pharmacophore model is used to identify and extract the possible interactions between ligand-receptor complexes as it is an ensemble of common steric and electronic features. The pharmacophore model was generated by utilizing a PHASE module. A PHASE module is multi-computed product for structure alignment, pharmacophore model and 3D-QSAR model generation.

Initially, all the prepared 95 ligands were aligned by ligand alignment tool in PHASE module. After the alignment of the ligand, hypothesis was developed by generating the conformers for all the ligands with maximum of 50 per structure. A dataset was created with threshold activity of $pIC_{50} \geq 1.85$ as active and $pIC_{50} < 1.85$ as inactive compounds. The pharmacophore hypotheses were generated using six pharmacophoric features: hydrophobic group (H), hydrogen bond donor (D), hydrogen bond acceptor (A), a positively charged group (P), a negatively charged group (N) and an aromatic ring (R). Among the prepared 95 ligands, 49 actives were generated, 46 were inactive. A total of 7 distinct hypothesis were generated keeping the maximum number of sites four. Among the 7 generated hypothesis best one is selected on the basis of survival score of the actives and inactive [23]. HHRR_1 was selected as the best pharmacophore hypothesis with four pointed features (Table 1).

The best generated hypothesis was selected and validated by retrieving Dihydrofolate reductase decoy dataset extracted from the DUD-e database (https://dude.docking.org/). The thorough set of decoys and 49 active compounds were validated and model was assessed on the basis of enrichment factor (EF) 1%, Boltzmann-enhanced discrimination of receiver operating characteristics (BEDROC) ranges 0 to 1, robust initial

Table 1. All 7 generated hypothesis by PHASE module and their parametric scores.

Hypothesis	Survival Score	Site Score	Vector Score	Volume Score	BEDROC
HHRR_1	5.102	0.3458	0.7539	0.3363	0.6372
AHRR_1	4.9759	0.295	0.742	0.3271	0.511
AHRR_2	4.9711	0.2997	0.7279	0.3275	0.4994
AAHR_1	4.6626	0.2585	0.6263	0.2867	0.4749
AAHR_2	4.6486	0.2515	0.616	0.293	0.4781
AARR_1	4.4972	0.2196	0.6363	0.2958	0.257
AHRR_3	4.468	0.1848	0.5705	0.2028	0.3485

enhancement (RIE), receiver operation characteristics (ROC) ranges 0 to 1 and area under the accumulation curve (AUAC) 0 (bad) to 1 (excellent) [24].Laterally, the validated hypothesis was further utilized for the 3D-QSAR analysis.

2.7 Atom Based 3D-QSAR Model

The 3D-QSAR is an attempt of correlation of 3D properties of the ligands with the biological potency by using distinct mathematical expressions. The Atom based 3D-QSAR model was developed by the PHASE module of Schrodinger (Maestro version 13.6) [25]. The best HHRR_1 generated hypothesis comprises of ligands were selected for the development of QSAR model by separating the dataset of compounds into training set (64 compounds) and test set (31compounds). Dataset was randomly divided into 70% of the compounds as training set and remaining 30% as a test set (Table 2).

Atom based QSAR tool of the PHASE module with maximum Partial least square (PLS) factor of 5 and maintained 1 Å grid spacing for the model. Leave-one-out (LOO) regression analysis of internal validation method was used with statistical parameters and six features pointed are categorised as; Negative ionic charges (N), positive ionic charges (P), a hydrogen bond donor (D), hydrophobic (H), electron withdrawing (W) and miscellaneous (X). For the better perception of structural optimization, QSAR visualization was performed for the most active compound and least active compound [26] (Fig. 1).

2.8 Molecular Docking and Virtual Screening

Virtual screening is a data processing technique used to discover the drug database library of small or large molecules recognizing the structures which are most probably befitting the specific target protein. CheMBL is a freely accessible chemical database, utilized to recognize the potent lead compounds. The compound 1,2,4-triazole was screened for drug similarity search from the CheMBL database and total 620 hit leads were retrieved. Further, their 2D similar structures were downloaded and subjected to molecular docking with the protein Dihydrofolate reductase (PDB ID: 2W9G) by the use of a Glide module. The 'Extra precision mode' from the ligand docking of a glide

Table 2. Dataset of Atom based 3D-QSAR model built by HHRR_1 hypothesis with their observed and predicted activity (pMIC).

Compound no.	Chemical Structure	QSAR set	Observed activity	Predicted activity
SC1	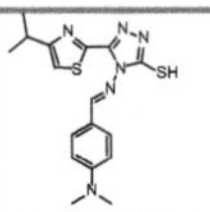	Test	1.367	1.745
SC2	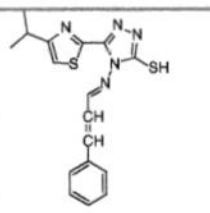	Training	1.648	1.648
SC3	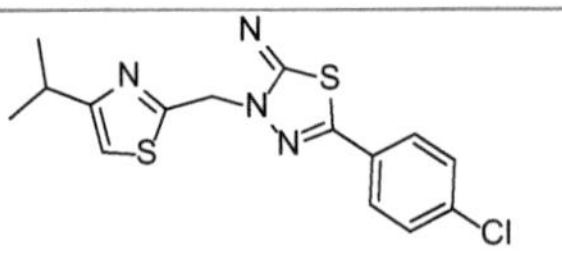	Test	1.655	1.680
SC4	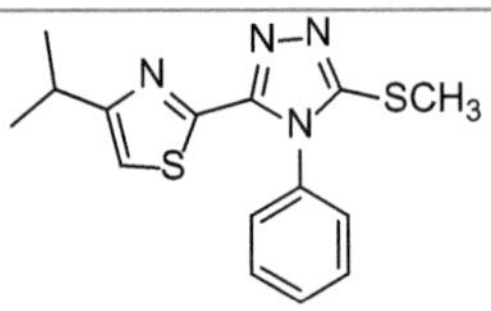	Training	1.296	1.311
SC5	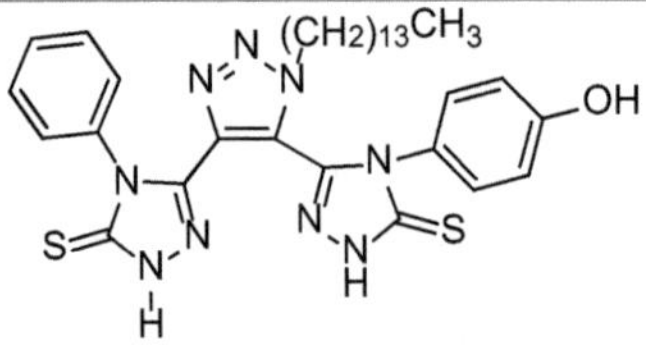	Training	1.898	1.882
SC6	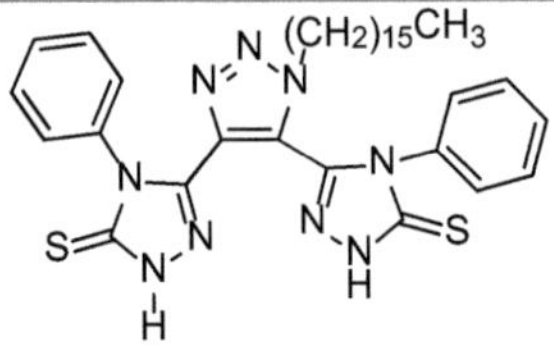	Test	2.207	1.822
SC7	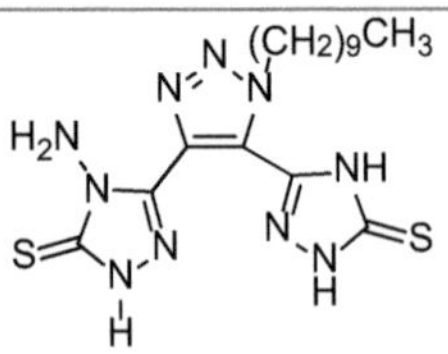	Training	1.422	1.386
SC8	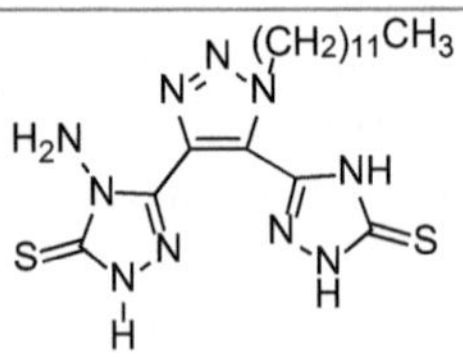	Training	1.751	1.732

(continued)

Table 2. (continued)

Compound no.	Chemical Structure	QSAR set	Observed activity	Predicted activity
SC9	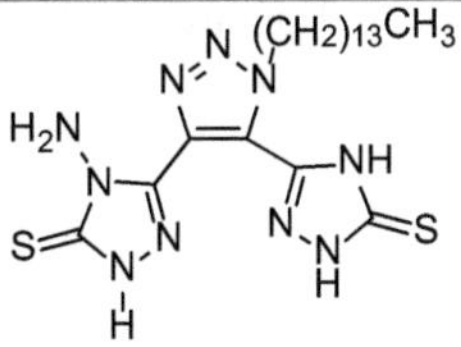	Test	2.078	1.719
SC10	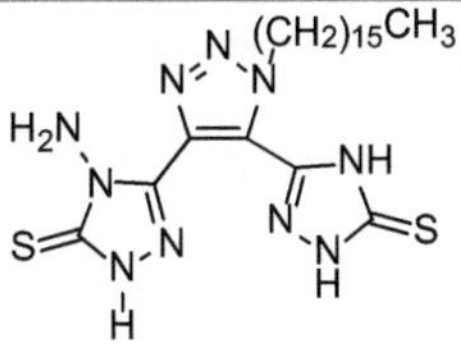	Training	2.705	2.743
SC11	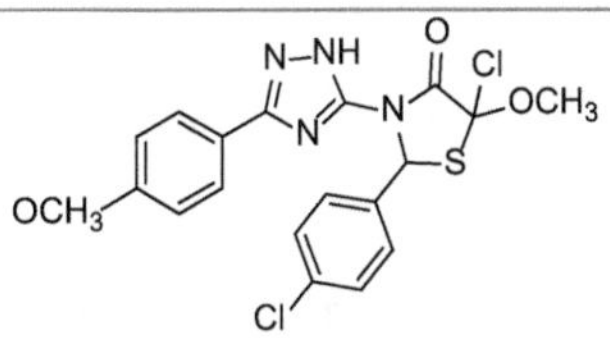	Test	1.751	1.560
SC12	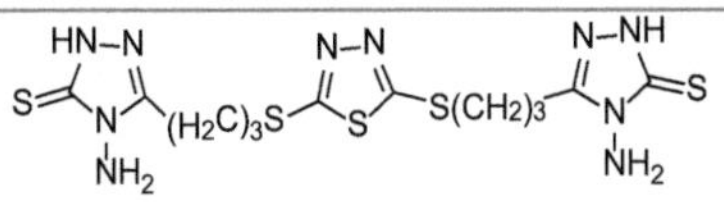	Training	1.762	1.756
SC13	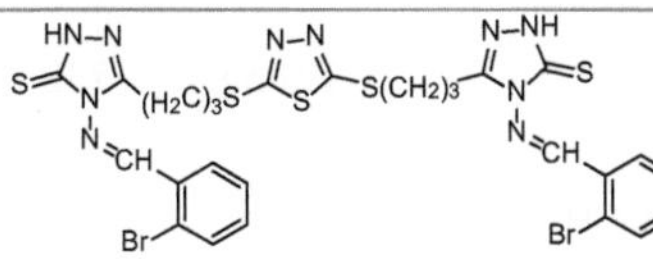	Training	1.998	2.004
SC14	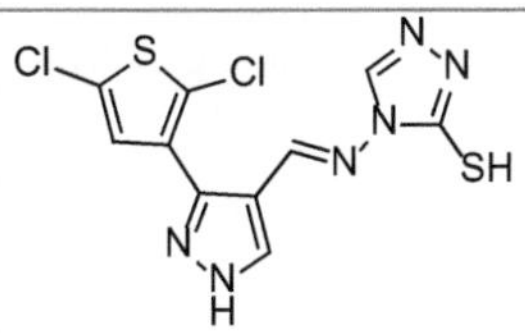	Training	2.155	2.129
SC15	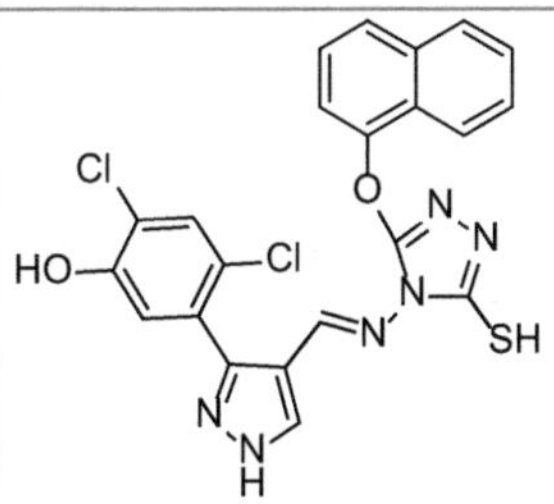	Test	1.899	2.116

(continued)

Table 2. (continued)

Compound no.	Chemical Structure	QSAR set	Observed activity	Predicted activity
SC16	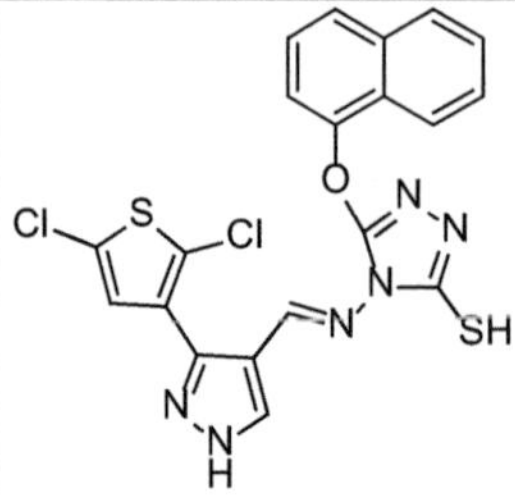	Training	2.492	2.526
SC17	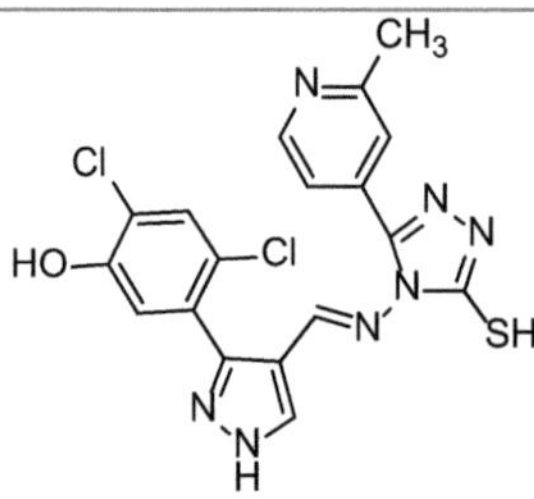	Training	2.139	2.216
SC18	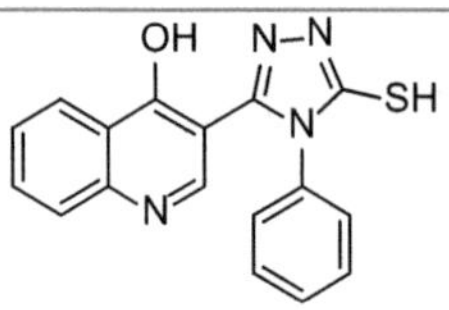	Training	1.793	1.804
SC19	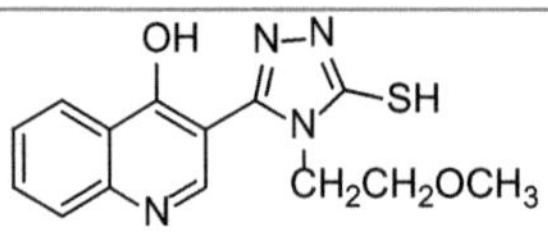	Training	1.772	1.796
SC20	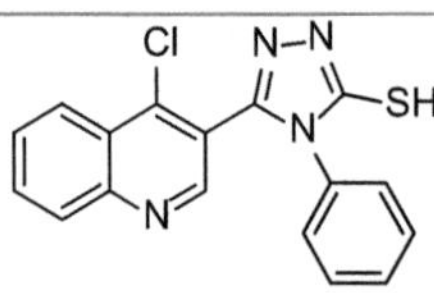	Training	1.813	1.782
SC21	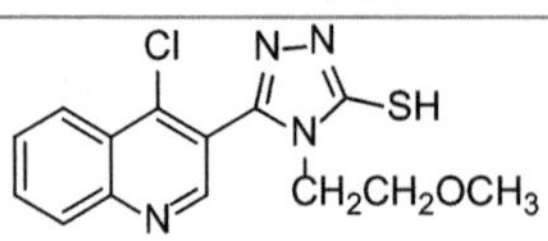	Test	1.492	1.856
SC22	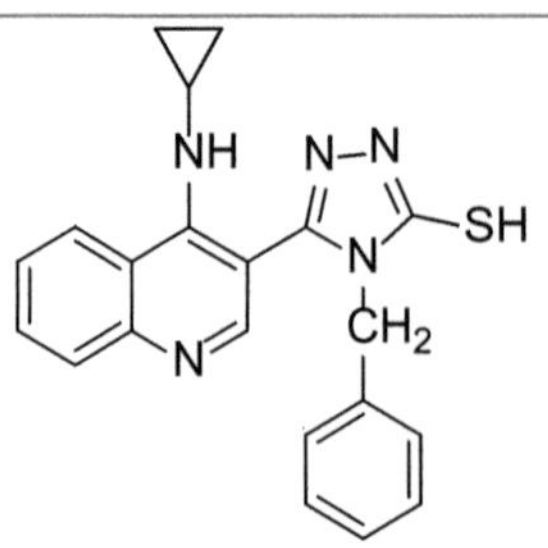	Test	1.849	1.834

(continued)

Table 2. (continued)

Compound no.	Chemical Structure	QSAR set	Observed activity	Predicted activity
SC23	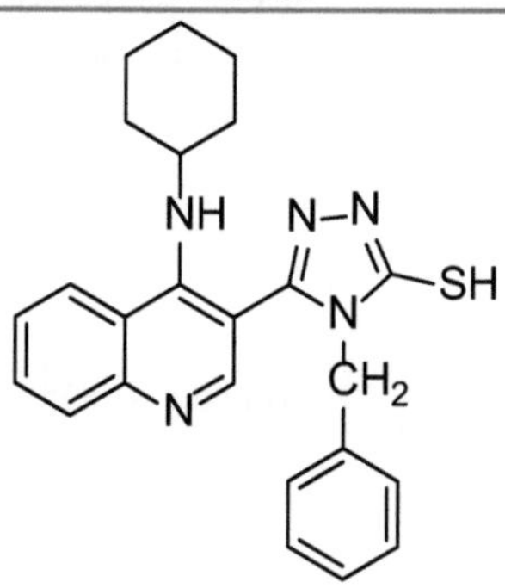	Test	1.596	1.821
SC24	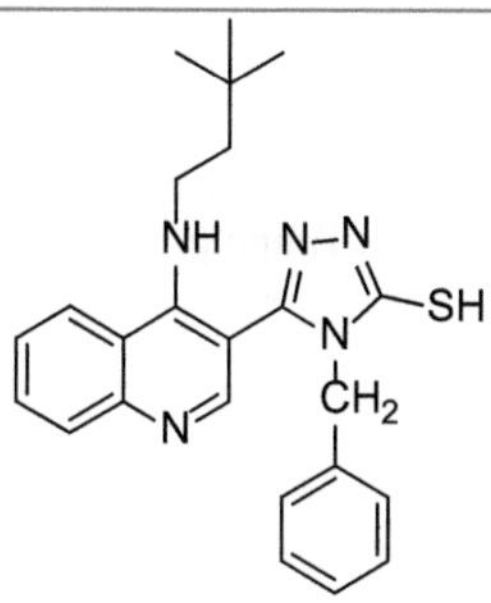	Training	1.598	1.604
SC25	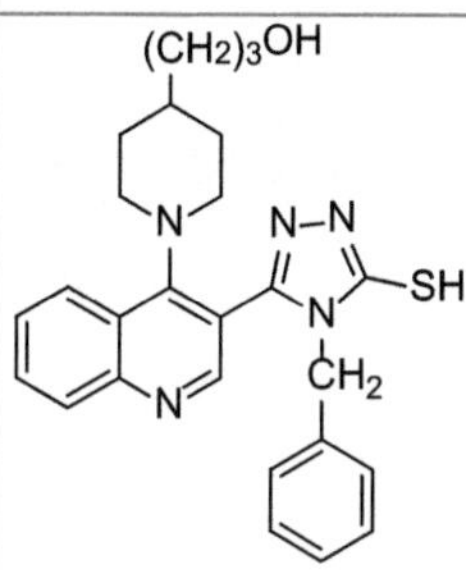	Training	1.634	1.681
SC26	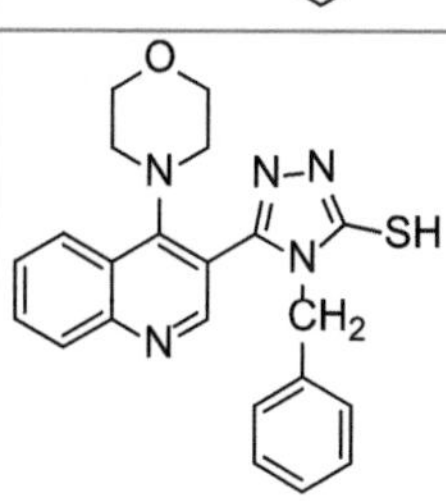	Training	1.877	1.936

(continued)

Table 2. (continued)

Compound no.	Chemical Structure	QSAR set	Observed activity	Predicted activity
SC27	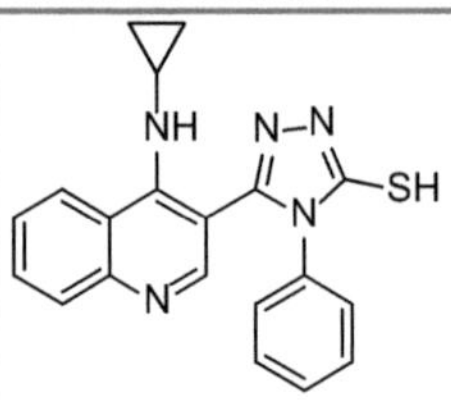	Training	1.835	1.825
SC28	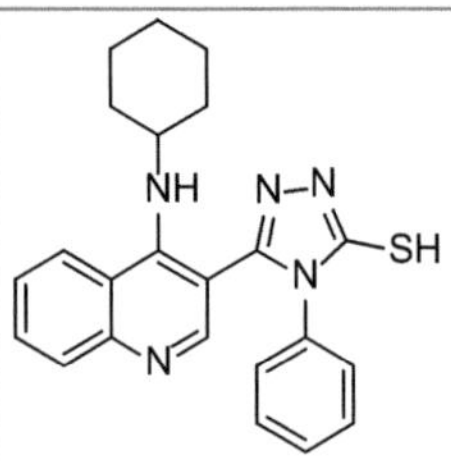	Training	1.875	1.871
SC29	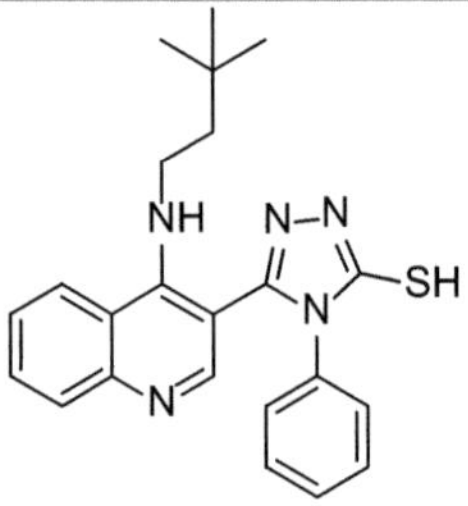	Training	1.877	1.869
SC30	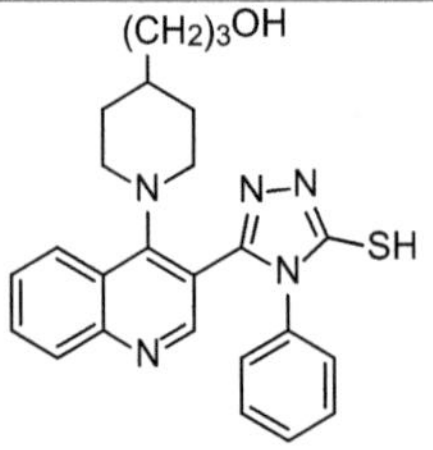	Training	1.914	1.917
SC31	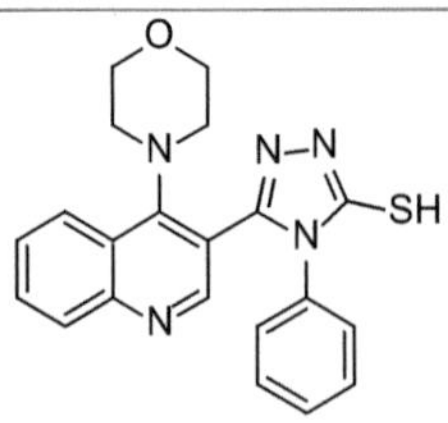	Test	1.864	1.960

(continued)

Table 2. (continued)

Compound no.	Chemical Structure	QSAR set	Observed activity	Predicted activity
SC32	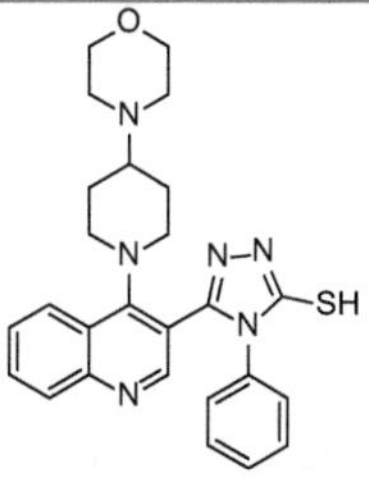	Training	1.937	1.923
SC33	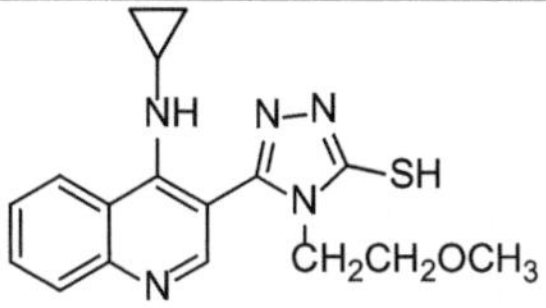	Training	1.816	1.778
SC34	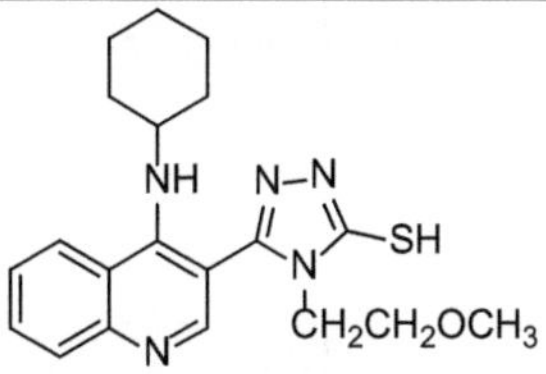	Training	1.858	1.824
SC35	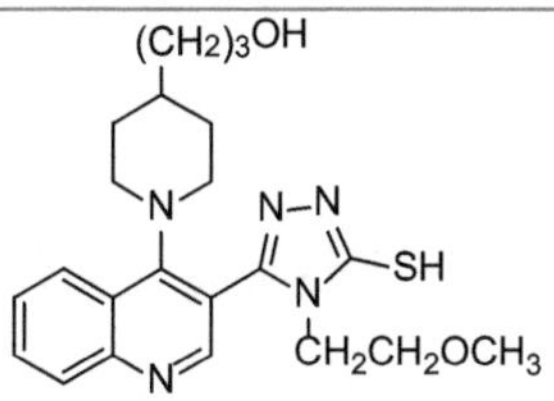	Test	1.899	1.819
SC36	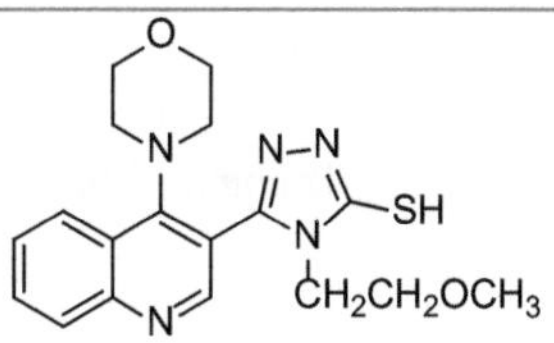	Test	1.847	1.813
SC37	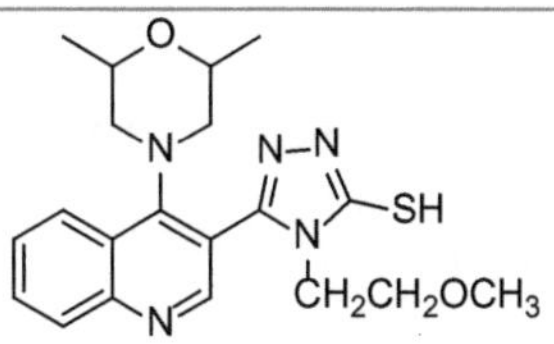	Training	1.873	1.806

(continued)

Table 2. (continued)

Compound no.	Chemical Structure	QSAR set	Observed activity	Predicted activity
SC38	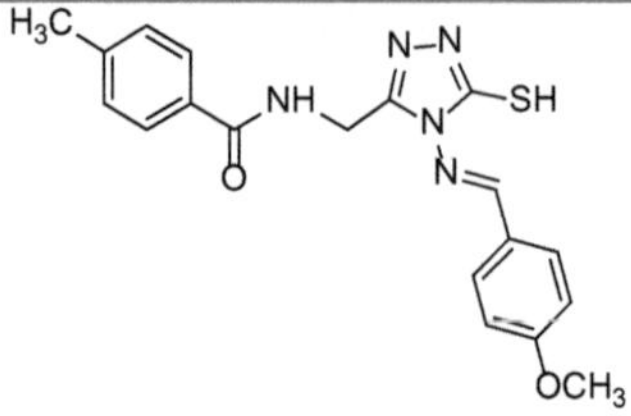	Training	2.087	2.056
SC39	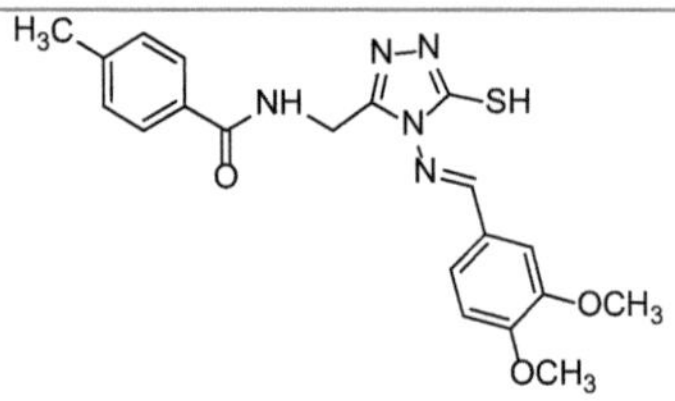	Test	2.119	1.994
SC40	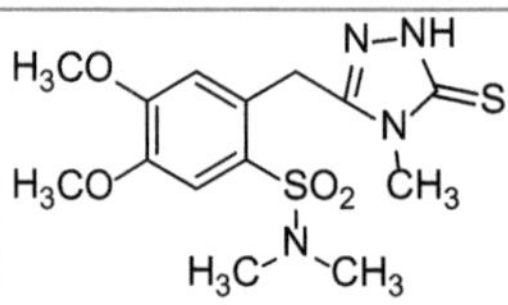	Training	3.140	3.109
SC41	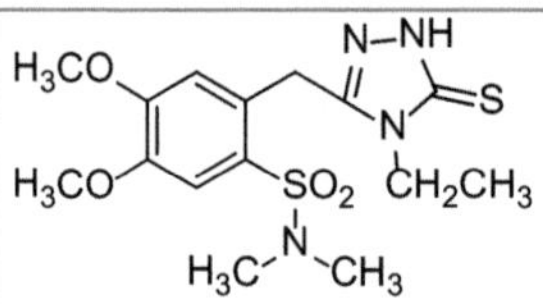	Test	3.172	3.089
SC42	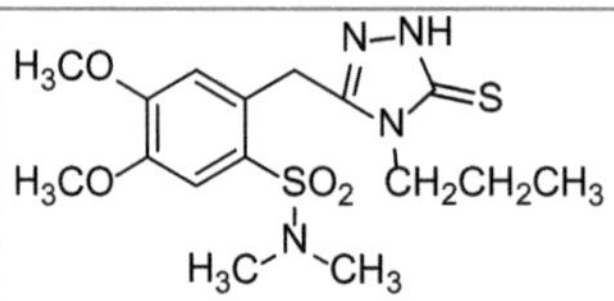	Training	2.904	2.869
SC43	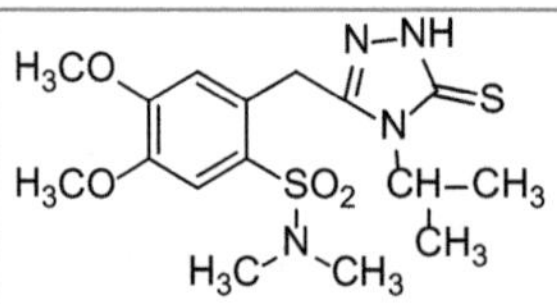	Training	3.023	3.049
SC44	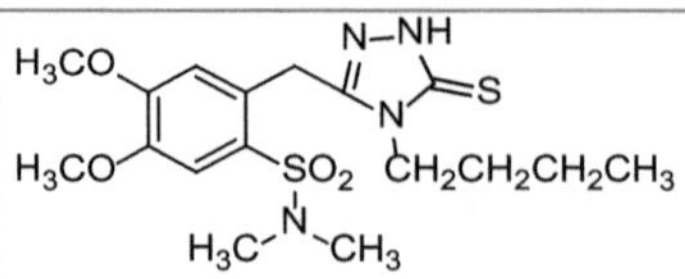	Test	3.237	3.165

(continued)

Table 2. (continued)

Compound no.	Chemical Structure	QSAR set	Observed activity	Predicted activity
SC45	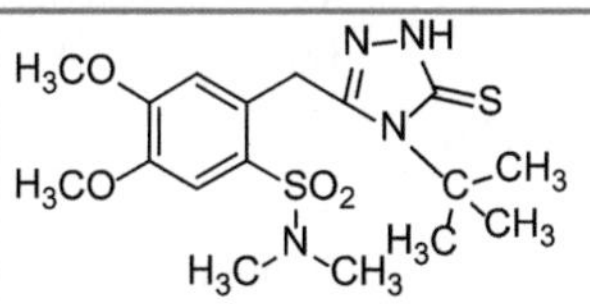	Test	3.237	3.001
SC46	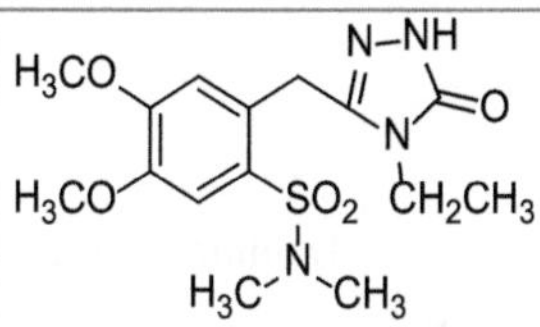	Test	3.137	2.861
SC47	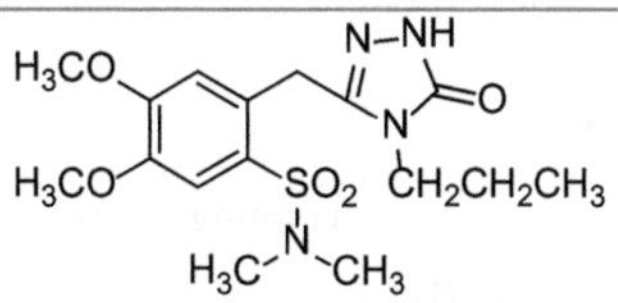	Training	3.170	3.178
SC48	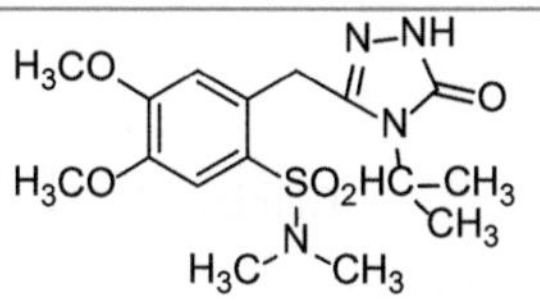	Training	2.994	3.033
SC49	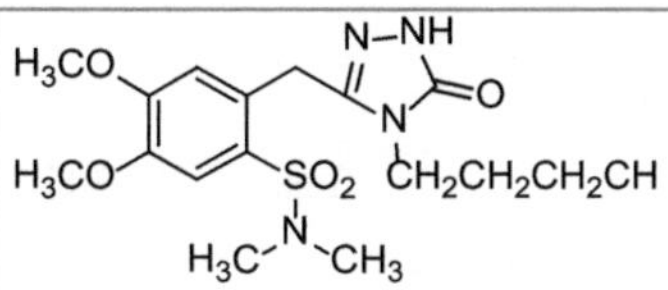	Training	3.202	3.196
SC50	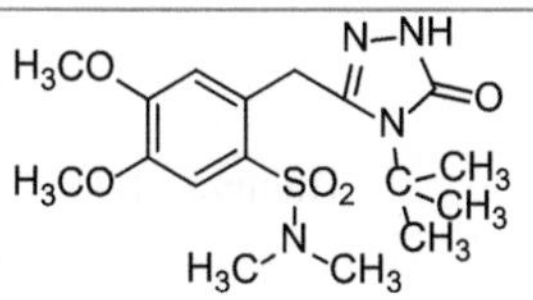	Training	3.009	3.027
SC51	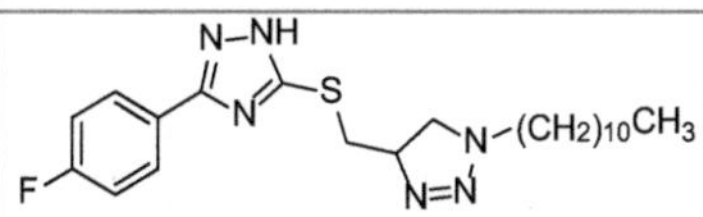	Training	1.432	1.449
SC52	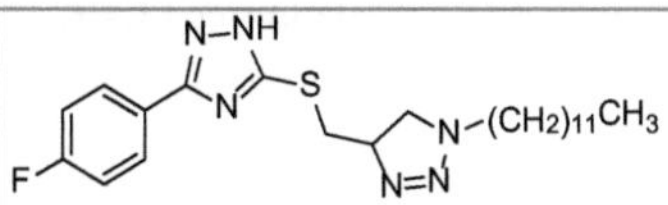	Training	1.446	1.456
SC53	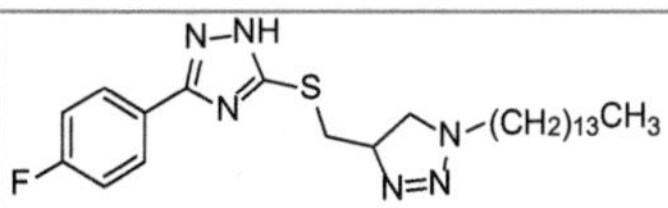	Training	1.773	1.825

(continued)

Table 2. (continued)

Compound no.	Chemical Structure	QSAR set	Observed activity	Predicted activity
SC54	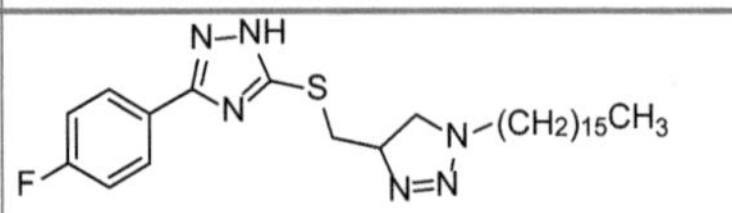	Training	1.798	1.842
SC55	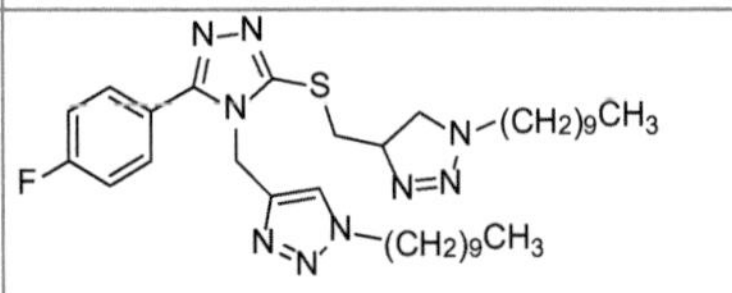	Training	1.602	1.601
SC56	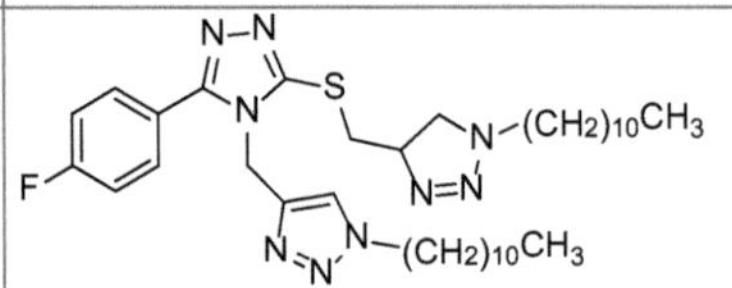	Training	1.922	1.929
SC57	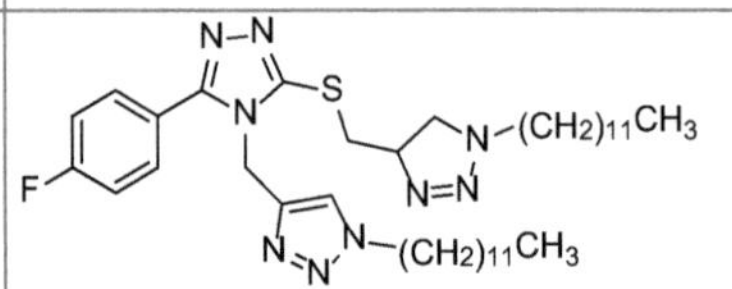	Training	2.241	2.303
SC58	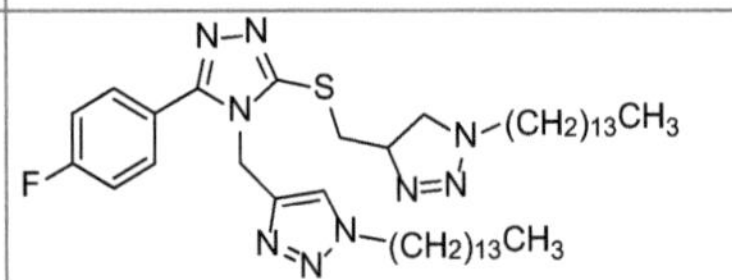	Training	2.274	2.316
SC59	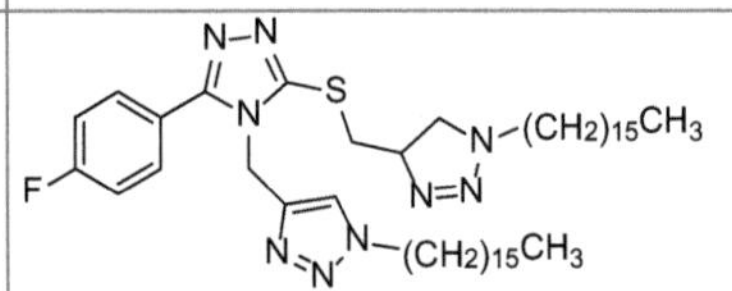	Test	2.305	2.013
SC60	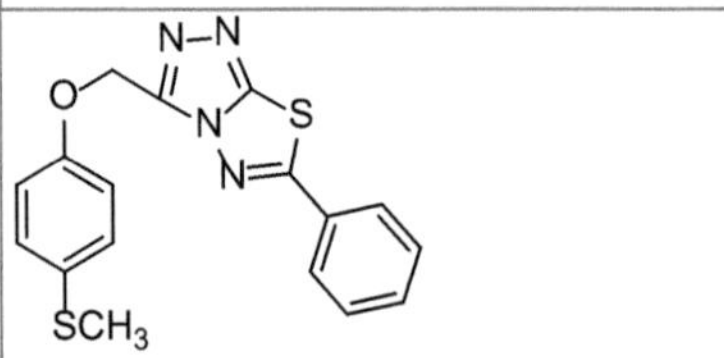	Training	1.753	1.713
SC61	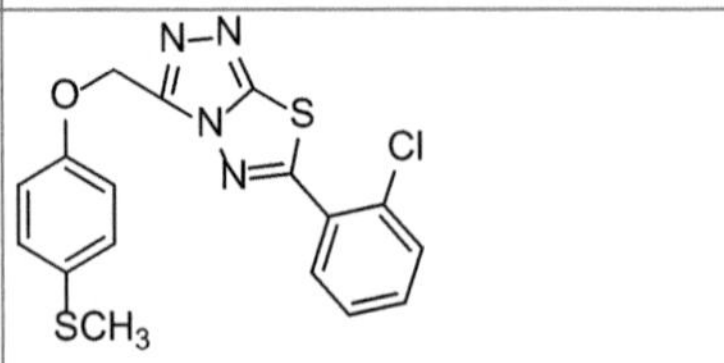	Training	1.192	1.413

(continued)

Table 2. (continued)

Compound no.	Chemical Structure	QSAR set	Observed activity	Predicted activity
SC62	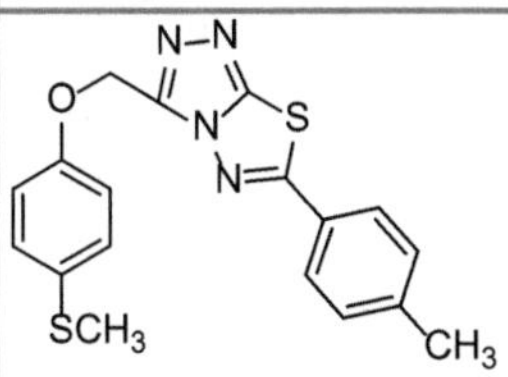	Training	1.770	1.750
SC63	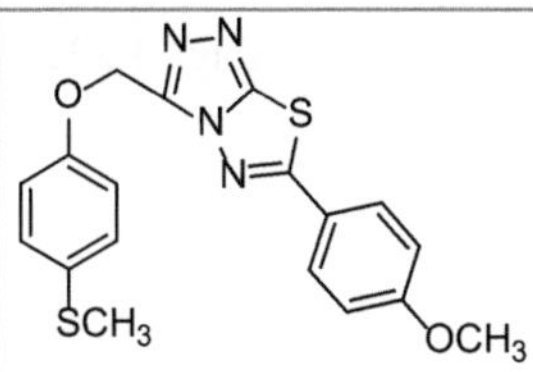	Test	1.788	1.306
SC64	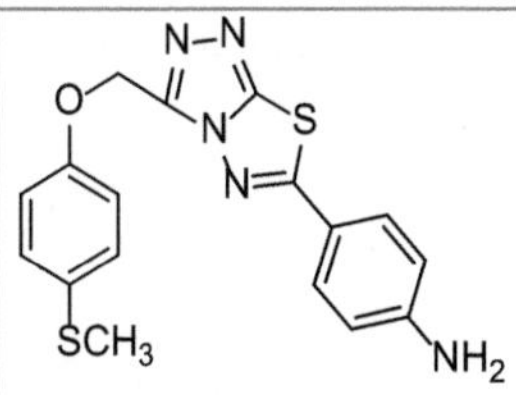	Training	1.771	1.760
SC65	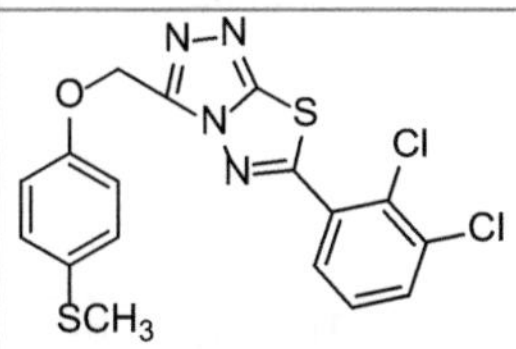	Training	1.774	1.561
SC66	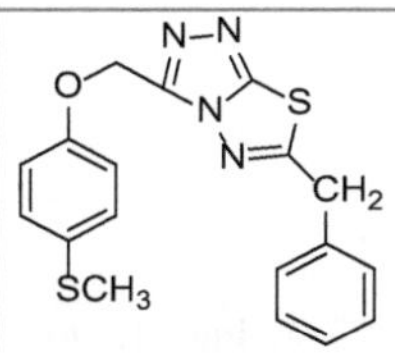	Training	1.770	1.732
SC67	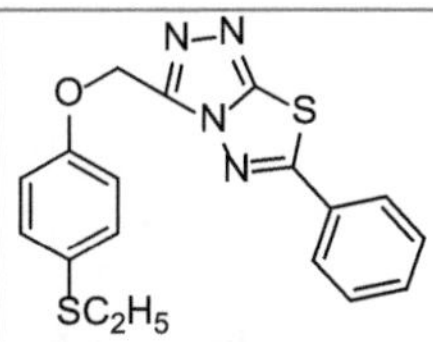	Test	1.770	1.431
SC68	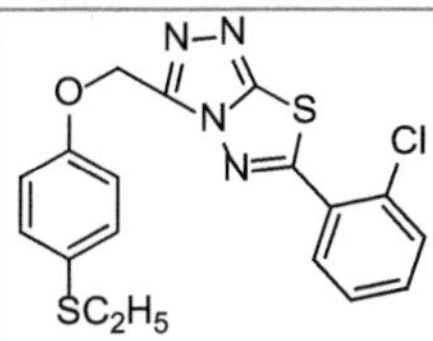	Training	1.206	1.265

(continued)

Table 2. (continued)

Compound no.	Chemical Structure	QSAR set	Observed activity	Predicted activity
SC69	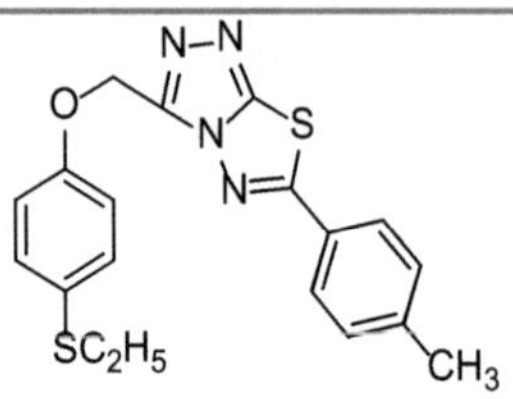	Training	1.786	1.821
SC70	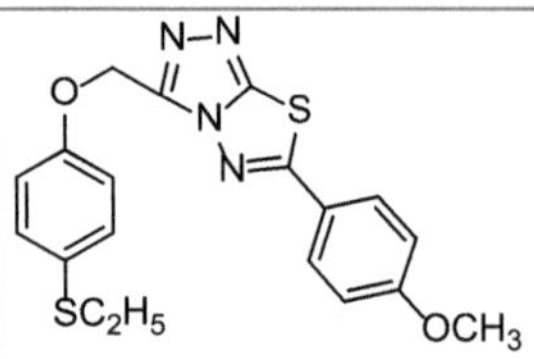	Training	1.202	1.181
SC71	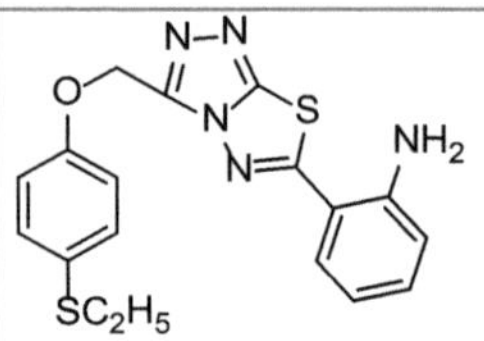	Training	1.185	1.204
SC72	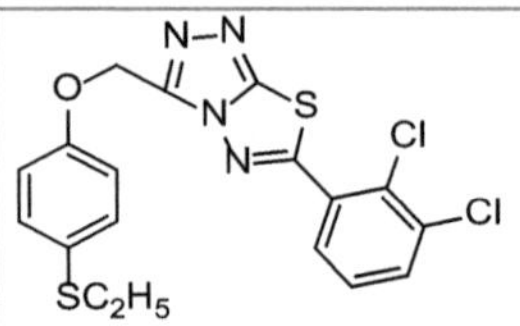	Training	1.243	1.238
SC73	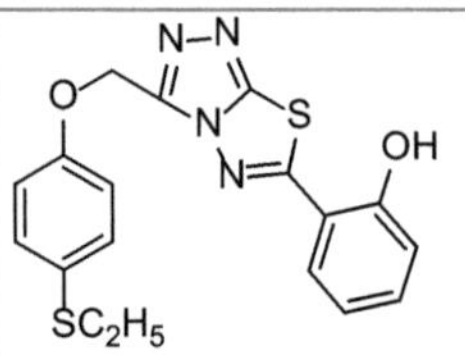	Training	1.186	1.171
SC74	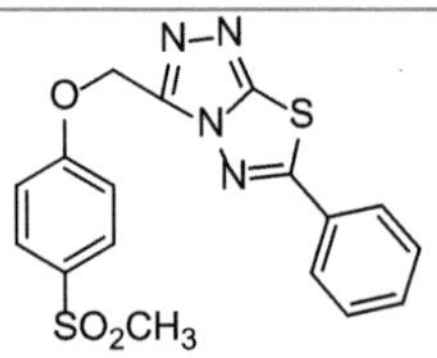	Training	1.490	1.484
SC75	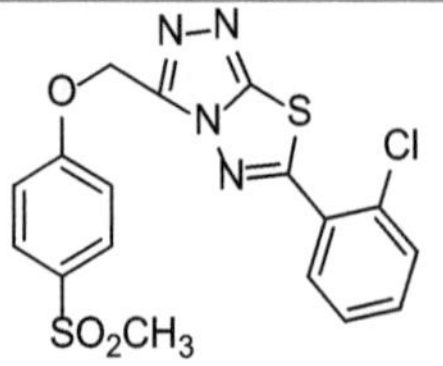	Training	1.526	1.611

(continued)

Table 2. (continued)

Compound no.	Chemical Structure	QSAR set	Observed activity	Predicted activity
SC76	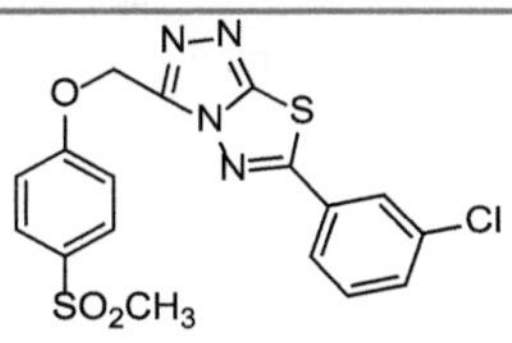	Training	1.526	1.488
SC77	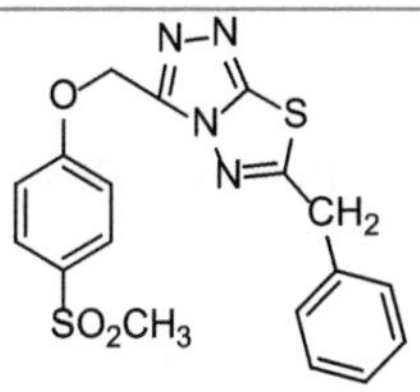	Training	1.505	1.508
SC78	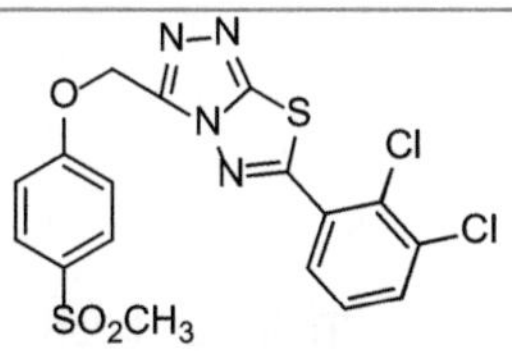	Training	1.806	1.675
SC79	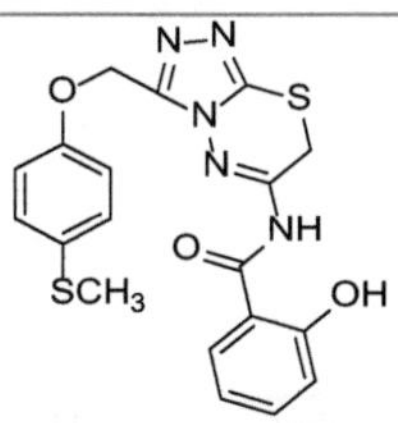	Training	1.232	1.235
SC80	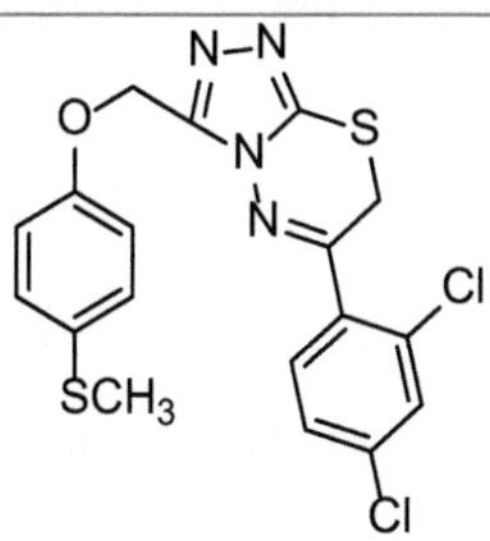	Training	1.485	1.704
SC81	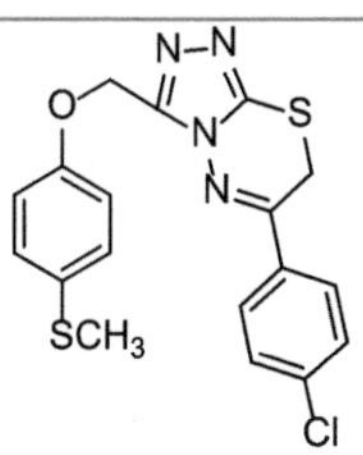	Test	1.808	1.394

(continued)

Table 2. (continued)

Compound no.	Chemical Structure	QSAR set	Observed activity	Predicted activity
SC82	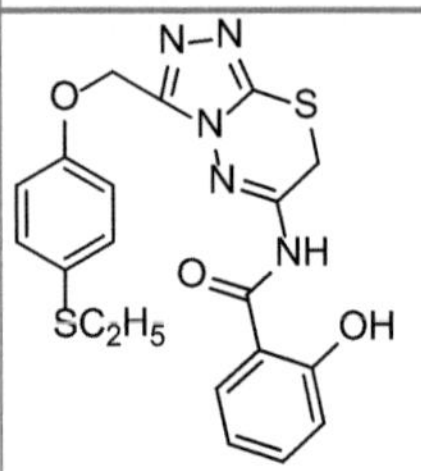	Training	1.849	1.854
SC83	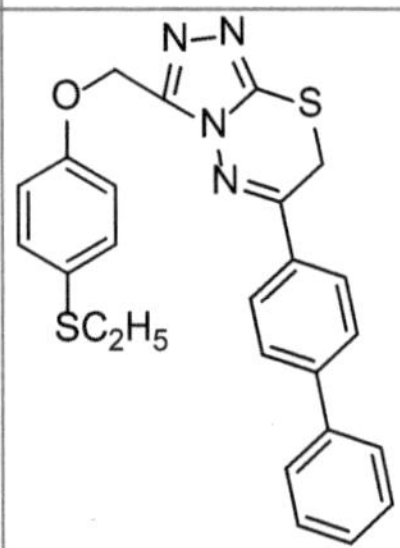	Training	1.263	1.263
SC84	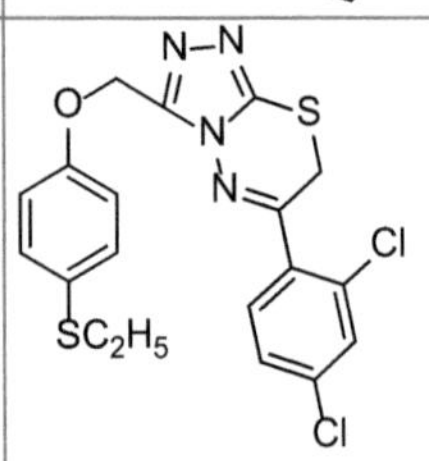	Training	1.256	1.337
SC85	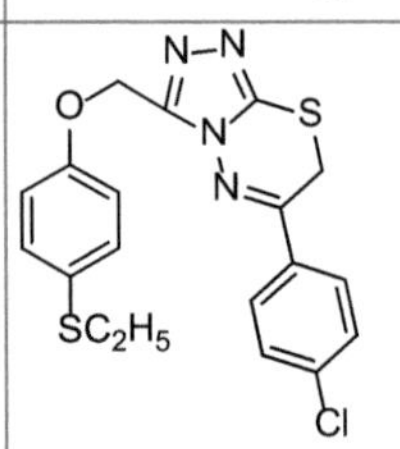	Test	1.221	1.651
SC86	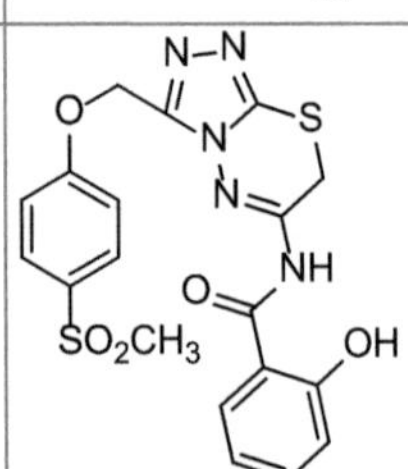	Training	1.565	1.587

(continued)

Table 2. (continued)

Compound no.	Chemical Structure	QSAR set	Observed activity	Predicted activity
SC87	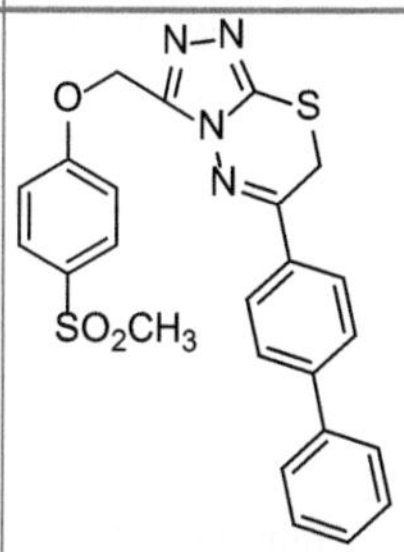	Training	1.581	1.553
SC88	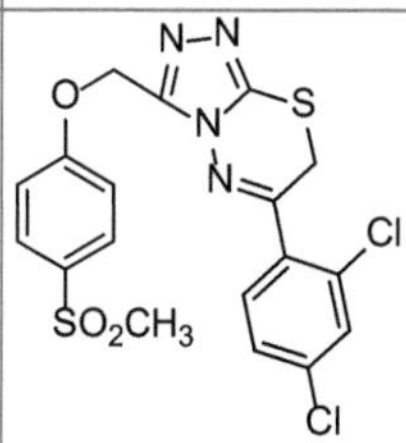	Test	1.574	1.715
SC89	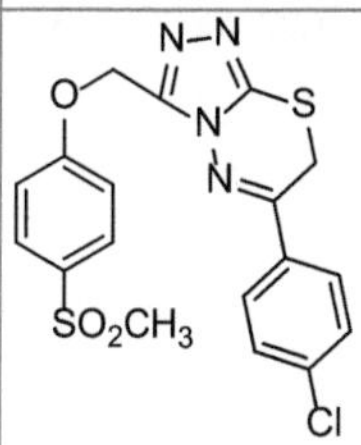	Test	1.240	1.573
SC90	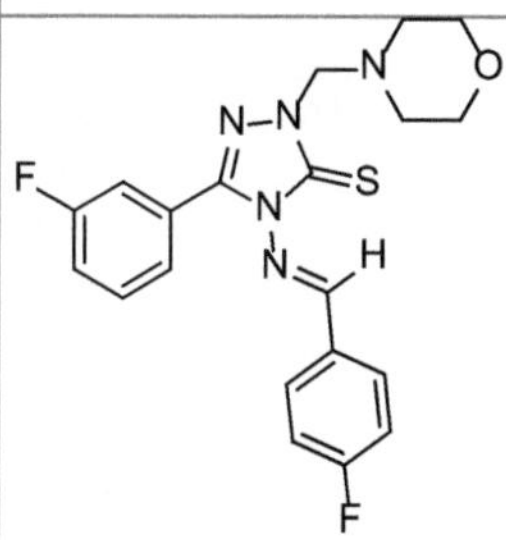	Test	1.715	1.669
SC91	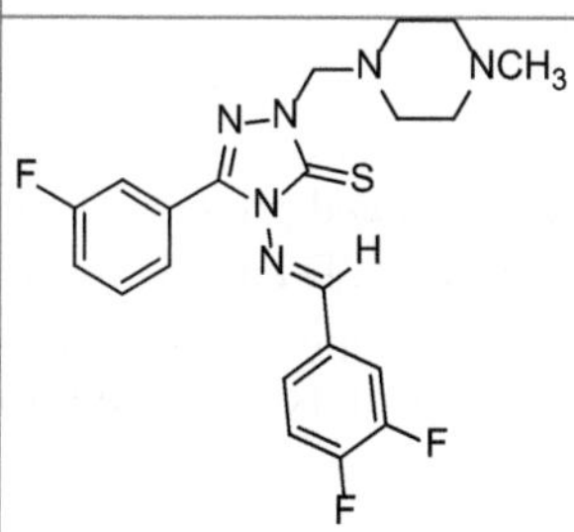	Training	1.747	1.722

(continued)

Table 2. (continued)

Compound no.	Chemical Structure	QSAR set	Observed activity	Predicted activity
SC92	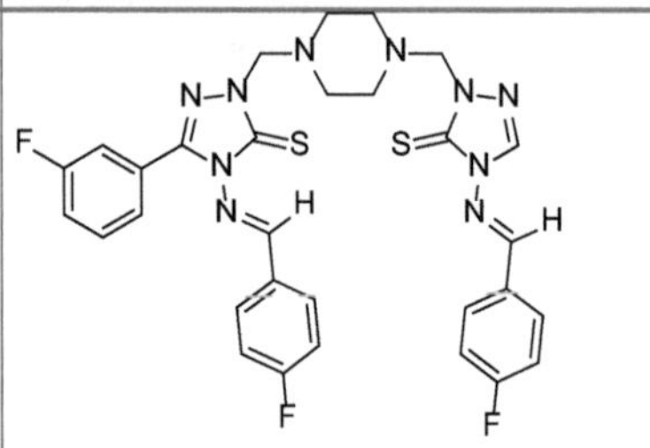	Training	2.210	2.219
SC93	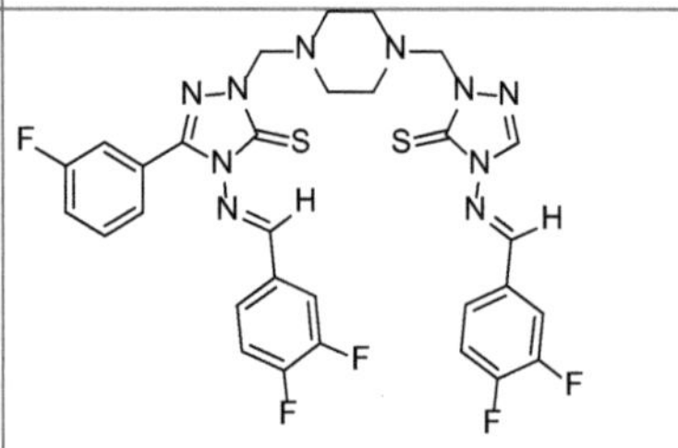	Training	2.083	2.111
SC94	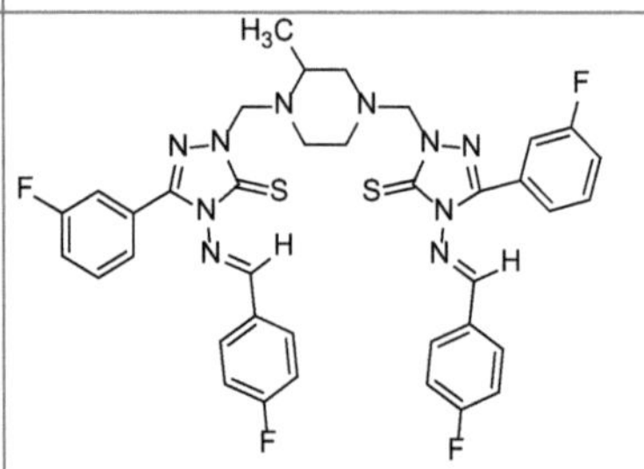	Training	2.278	2.274
SC95	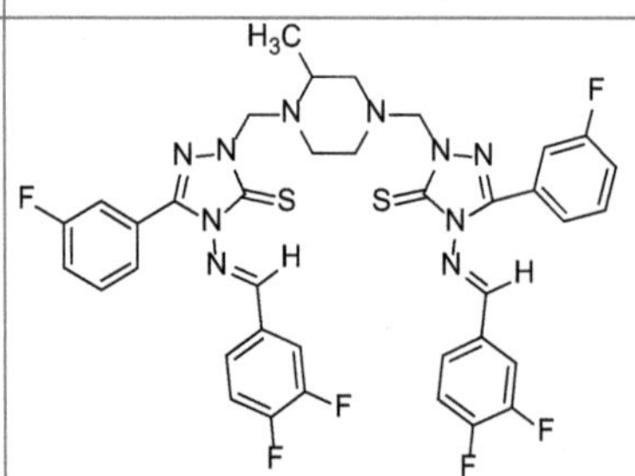	Training	2.297	2.282

module was used for the best orientation of ligand and protein. The best docked compound was chosen based on the docking scores and 2D and 3D ligand-protein interaction of the best docked compound was checked by employing the ligand interaction tool in the program [27]. Latterly, the best 5 hit compounds picked out for further ADME and toxicity studies (Fig. 6) (Table 5).

2.9 In Silico ADME Screening and Toxicity Predictions

To improve the safety, quality and efficacy of a drugs, Pharmacokinetics properties such as absorption, distribution, metabolism, excretion and toxicity profiles of the compounds were determined by using the ADMET Lab2.0 web tool. ADME profile depends upon the various factors includes intestinal absorption, blood brain barrier permeability, CYP inhibition and total clearance rate [28]. Toxicity profile can assess both qualitatively and quantitatively for determining the hepatotoxicity, mutagenicity, carcinogenicity and immunotoxicity [29]. Further, ADME and toxicity profile was performed on the top five virtually screened compounds which give us a good insight of how the drugs were absorbed, distributed, metabolized and excreted in the body with their toxicity effects.

3 Results and Discussion

3.1 Pharmacophore Hypothesis Analysis

Among the 7 generated hypotheses, the four point hypothesis HHRR_1 was selected as the best one based on their survival score (5.102), site score (0.3458), vector score (0.7539) and volume score (0.3363) (Table 1). The best selected hypothesis was next validated by using the decoy dataset (DUD-e database). The validated HHRR_1 hypothesis showed an Enrichment factor (1%) of 0.98, AUAC (area under accumulation curve) 1.00, robust initial enhancement (RIE) 19.50, BEDROC160.9 is 0.863, a ROC (receiver operating characteristics) 1.00. A ROC plot was generated for sensitivity (true positive rate) on X-axis against specificity (false positive rate) on Y-axis. If the curve shown sharp rise towards the upper left corner and flattened with the value closer to 1 depicts the ranking of active compounds higher than the inactive ones. The derived pharmacophore model HHRR_1 with ROC value 1.00 graphically represented the curve starting from upper left corner then flattened at the end shows the accuracy of the model (Fig. 3). The best hypothesis depicted that two hydrophobic bond and two aromatic ring features are required for an activity (Figs. 2 and 3).

3.2 Statistical Validation of Atom Based 3D-QSAR Model

As the 3D-QSAR model was generated, it was validated for their reliability, robustness and stability. For the regression analysis the leave-one-out cross internal validation method was used by keeping the partial least square factor (PLS-4). The internal validation report showed that the statistical parameters such as correlation coefficient for the training set ($Q^2 = 0.7995$), regression coefficient ($R^2 = 0.9808$), Scramble Fischer test ($F = 830.1$), a significance level of variance ratio ($P = 5.29e\text{-}5$), Standard deviation ($SD = 0.0728$), root mean square error ($RMSE = 0.26$) were found to be of good values for the generation of 3D-QSAR model (Tables 3 and 4). Graphically, the scattered plots were shown for the test and training set depicting the values of the generated QSAR

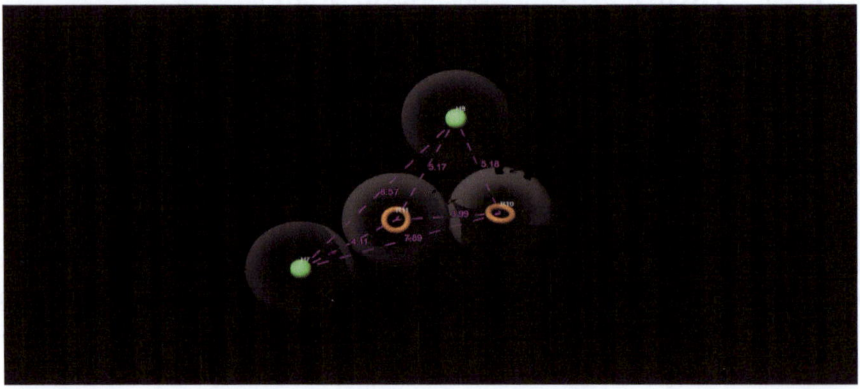

Fig. 2. Four features generated pharmacophore model (HHRR_1) by PHASE module.

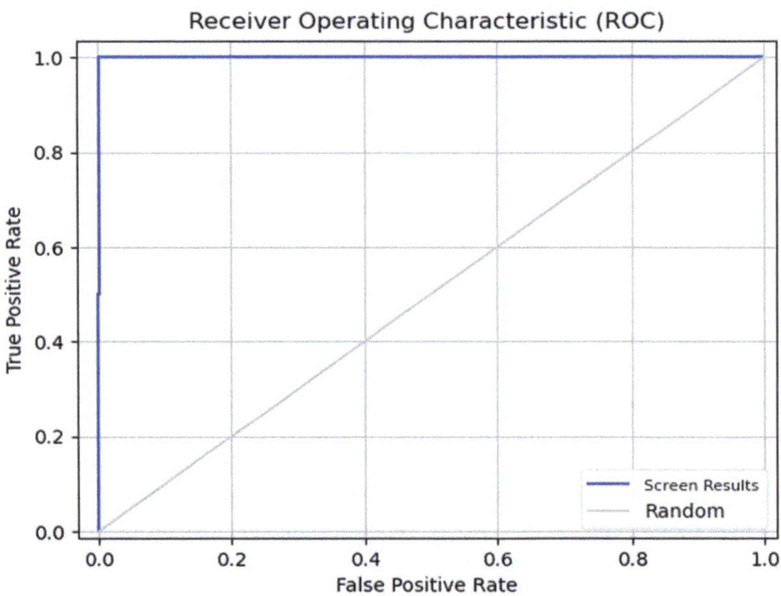

Fig. 3. A ROC plot of the best generated hypothesis HHRR- 1.

model. The training set depicted the development of 3D-QSAR model and test set was used for the QSAR superlative assessment. The best fit line equation for test set is $y = 0.75x + 0.46(R^2 = 0.81)$. where R^2 outlines the regression. The clustering of molecules towards the fit line represented the closeness between the actual and predicted activity by regression (Fig. 4).

Table 3. PLS parameters for the Atom based 3D-QSAR model (HHRR_1).

PLS Factor	SD	R^2	F	P	Stability	Q^2	RMSE	Pearson-R
1.	0.2646	0.7350	188.6	2.75e-21	0.893	0.7022	0.32	0.8506
2.	0.1678	0.8950	285.5	1.63e-33	0.811	0.7654	0.28	0.8772
3.	0.1029	0.9611	544.0	1.84e-46	0.711	0.7818	0.27	0.8870
4.	0.0728	0.9808	830.1	5.29e-55	0.663	0.7995	0.26	0.8973

Table 4. Statistical data of Atom based 3D-QSAR (HHRR_1).

Factors	H-bond donor	Hydrophobic/non-polar	Electron-withdrawing	Other
1	0.026875	0.588746	0.235253	0.148
2	0.035134	0.586819	0.228333	0.148
3	0.037117	0.591037	0.217642	0.154
4	0.039178	0.586953	0.218030	0.155

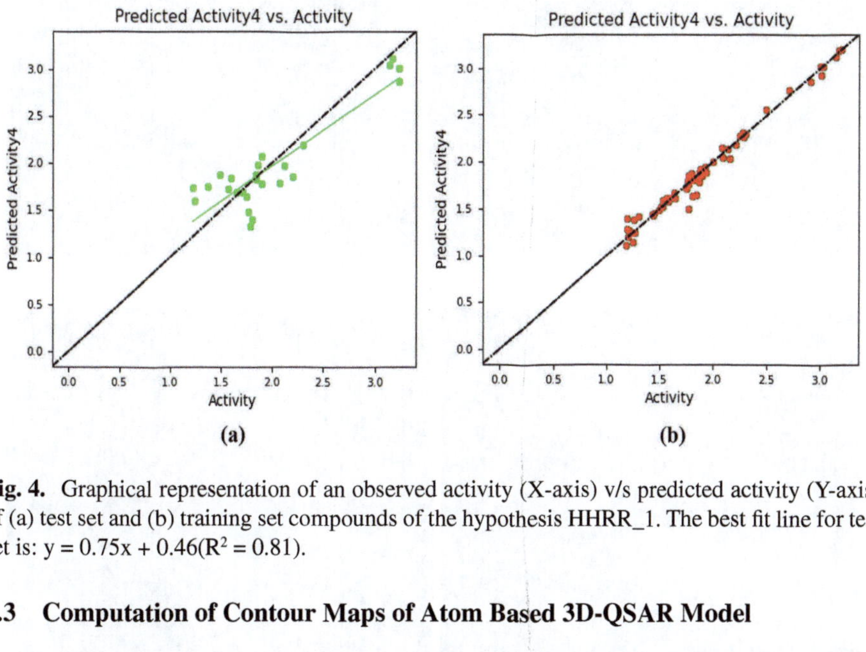

Fig. 4. Graphical representation of an observed activity (X-axis) v/s predicted activity (Y-axis) of (a) test set and (b) training set compounds of the hypothesis HHRR_1. The best fit line for test set is: $y = 0.75x + 0.46(R^2 = 0.81)$.

3.3 Computation of Contour Maps of Atom Based 3D-QSAR Model

3D-QSAR is an extensive approach to correlate the 3D properties of the ligands to predict their biological and physiological activities by using various statistical techniques. Biological activity prediction is necessitated by the presence of the different functional groups and moieties. The visualization of contour maps depicts the contribution of the functional groups or moieties to the biological activity of the compound. Blue regions of the maps show the favourable features that facilitated enzyme ligands interactions, while red one depicts the unfavourable interactions. In the atom-based 3D-QSAR contour map of Electron-withdrawing in Fig. 5 1(a) helps to compare and visualize the most and least potent group or moiety lay on the map.A blue region on almost all the

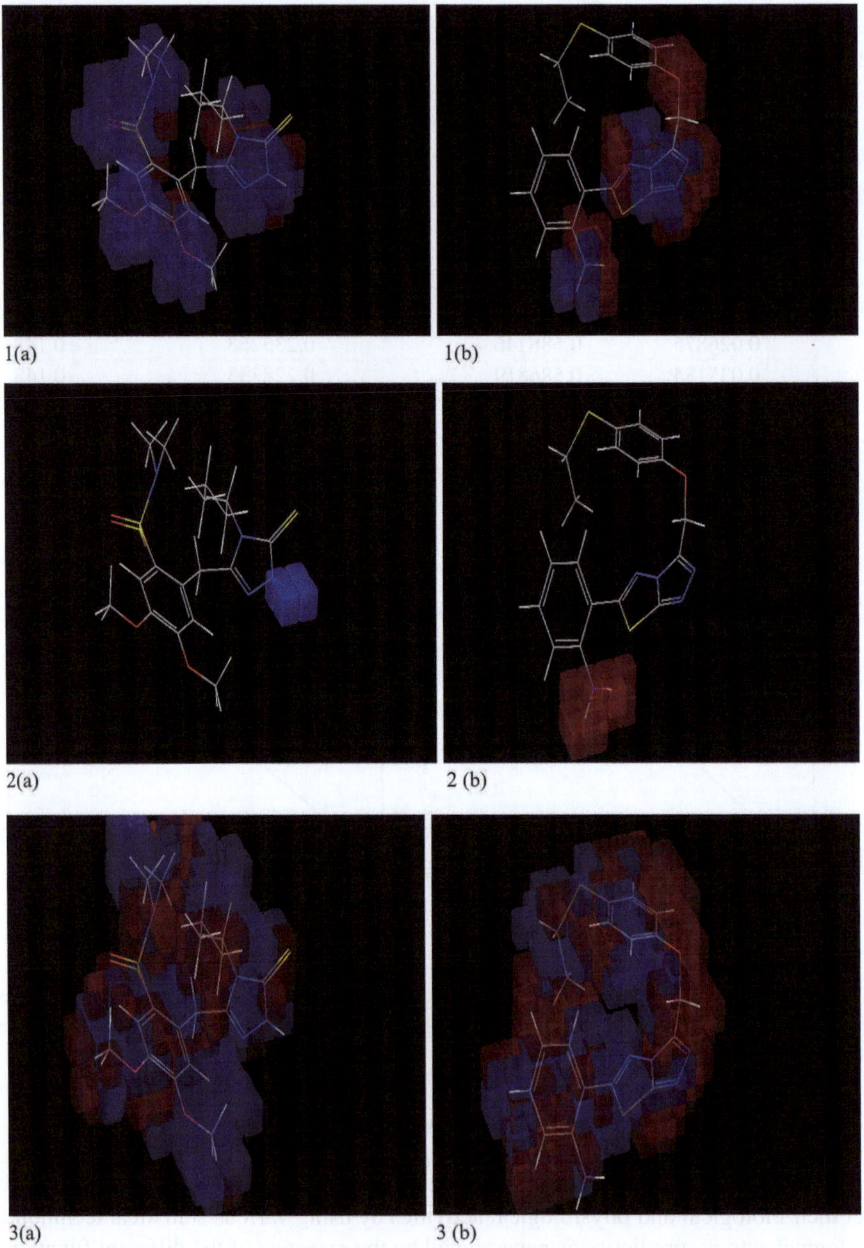

Fig. 5. Visualization of Atom based 3D-QSAR model of most active compound SC44-1(a) Electron withdrawing 2(a) Hydrogen bond donor 3(a) Hydrophobic and least active compound SC71-1(b) Electron withdrawing 2(b) Hydrogen bond donor 3(b) Hydrophobic withdrawing Blue coloured cubes shows increase in activity and red coloured cubes shows decrease in activity.

substituent that are attached to 1,2,4-triazole ring most active compound SC44 (Fig. 5(1a)) indicate that Electron-withdrawing is favourable for activity and, the red region around one carbon of alkyl group attached to the 1,2,4-triazole ring dictates that there is need for the substitution to increase the activity whereas, the blue cubes on the 1,2,4-triazole and attached ring of SC71 showed increment of an activity and red region around methoxy group attached to 1,2,4-triazole indicates that some modification should be done to improve the biological activity (Fig. 5 (1b)). Moreover, hydrogen bond donor map as shown in Fig. 5 (2a) states that blue region on Nitrogen at 1 position in 1,2,4-triazole ring is necessary for the biological activity however, red region on the amino group on benzene alongside the 1,2,4-triazole ring that it decreasing the biological activity of compound SC71 Fig. 5 (2b). Furthermore, in the non-polar contour map Blue region on all substituents of 1,2,4-triazole favouring in increase of the biological activity and red region over the oxygen group on 1,2,4-triazole dictates that there is need for the substitution to increase the biological activity of the compound SC44 Fig. 5 (3a) whereas, Blue region over benzene and its substituent attached to 1,2,4-triazole ring via carbon chain are favouring in increasing the biological activity, and red region over the other substituents dictates that there is need for the substitution to increase the biological activity of compound SC71 (Fig. 5 (3b)).

3.4 Molecular Docking

The docking was performed for 1,2,4-triazole derivatives and compared with the standard drugs used trimethoprim. Further, the best docked active compound was screened for drug similarity from the ChemBl database and the obtained 620 compounds were then put through for XP docking with the protein DHFR. The obtained result showed that, compound CID CHEMBL198820 had the best docking score (−8.761 kcal/mol) against the standard drugs trimethoprim (−7.675 kcal/mol). The compound CID CHEMBL198820 had one hydrogen bond with amino acid residue considered being the most active and has higher binding affinity among the all-hits compound thus outpacing the standard drugs. The top 5 hit lead ligand-amino acid residues were shown in Table 5, Fig. 6.

3.5 ADME and Toxicity Profile Prediction

We had performed the ADMET profile for the top 5 drug hits by using the ADMET Lab2.0 web tool. All the parameters were found to be acceptable ranges and didn't violate the Lipinski rule of five. For the better intestinal absorption and permeability, molecules with less than five hydrogen bond donors, less than 10 hydrogen bond acceptors, a molecular weight below 500 Da and LogP not greater than 5 are considered. Pharmacokinetic features properties such as partition coefficient (LogP o/w 3.454 to 5.614) used for an estimation of absorption of drugs within the body, the blood brain barrier permeability (Log BB 0.226 to −1.08), PPB, VD were considered for the selection of a best compound (Table 6).

Table 5. Docking scores of top 5 drug hits compared with the standard drugs.

S. No.	CHEMBL ID	Compounds Structure	Docking Score	Glide energy
1.	198820	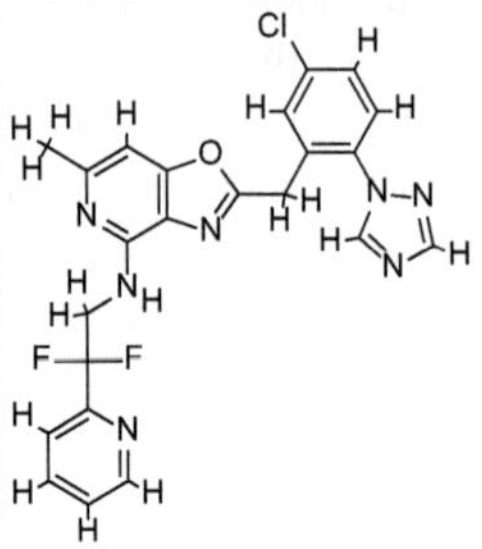	−8.761	−57.813
2.	275068	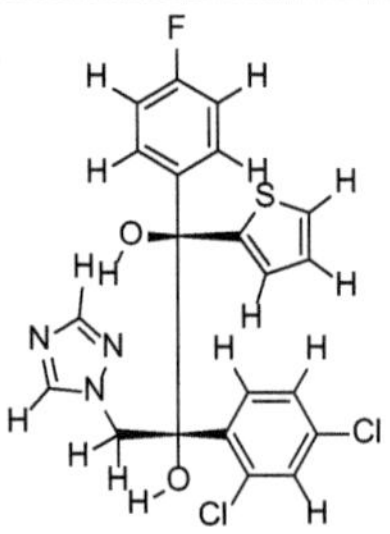	−8.674	−34.985
3.	198735	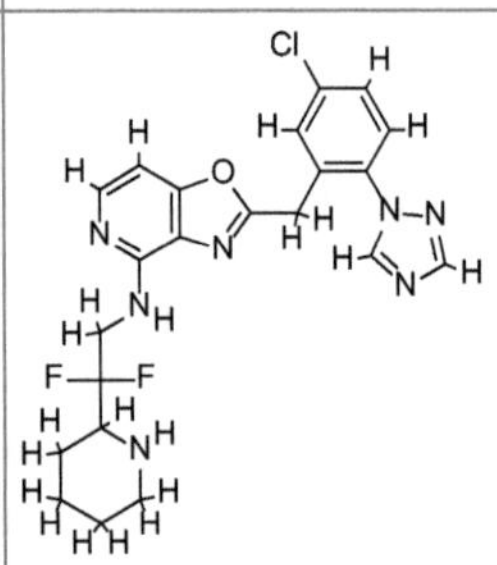	−8.542	−56.364
4.	196315	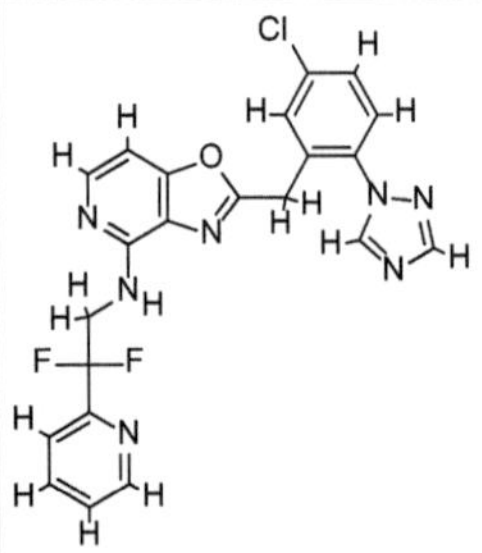	−8.419	−57.364
5.	208844	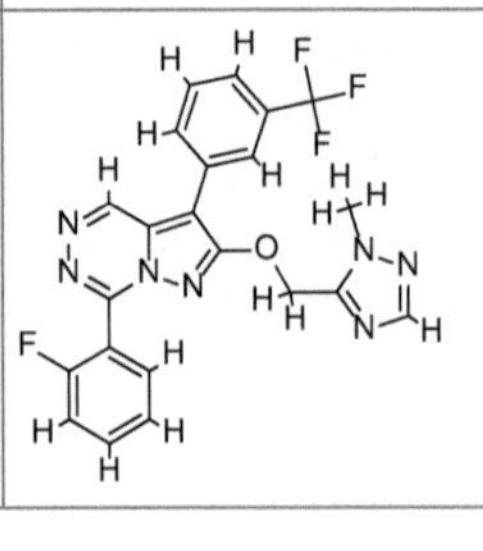	−8.341	−49.906

(continued)

Table 5. (continued)

S. No.	CHEMBL ID	Compounds Structure	Docking Score	Glide energy
6.	Trimethoprim	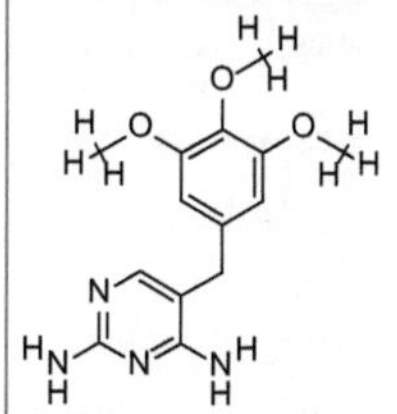	−7.675	−44.487
7.	Ciprofloxacin	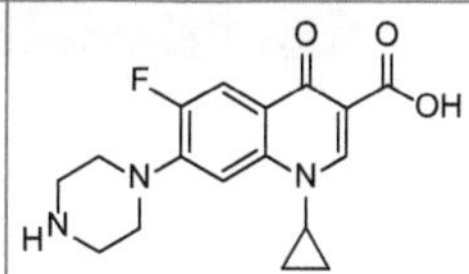	−7.886	−50.049
8.	Ampicillin	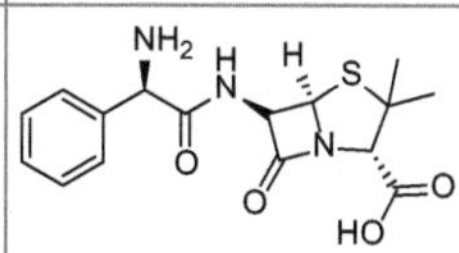	−7.458	−52.104

In silico toxicity studies were employed to evaluate the toxicity and adverse effects of the selected top 5 drug hits by utilizing the ADMET Lab2.0 online tool server which evaluated the results for hERG blocker, Carcinogenicity, Skin sensitivity and AMES Toxicity (Table 7)

4 Conclusion

A computational study on 1,2,4-triazoles derivatives consisting of dataset of 95 compounds was conducted for their antimicrobial activity. The protein ligand (PDB ID: 2W9G) were used for the model development against the dihydrofolate reductase. A ligand-based pharmacophore model was generated by the use of a PHASE module and the best generated hypothesis HHRR_1 was selected and validated with decoy dataset. The HHRR_1 hypothesis showed that the two hydrophobic bond and two aromatic ring features are good for the activity. The Atom based 3D-QSAR model was built by utilizing the best generated hypotheses HHRR_1 and the internal validation showed that statistical parameters such as (Q^2 = 0.7995), (R^2 = 0.9808), (SD = 0.0728), and (RMSE = 0.26) were found to be good values for the built model. The best docked compound was performed for drug similarity in the CheMBL database and molecular docking based virtual screening study was carried out and then preceded ADMET predictions. From the resulting data, it was clarified that CHEMBL 208844 and CHEMBL 275068 have more potential as antimicrobial agents and could be served for further studies in the battle against toxicity of drugs. It is concluded that various computational methods like molecular docking, pharmacophore hypothesis, 3D-QSAR model and ADME/T predictions are used to develop the more potent and effective antimicrobial

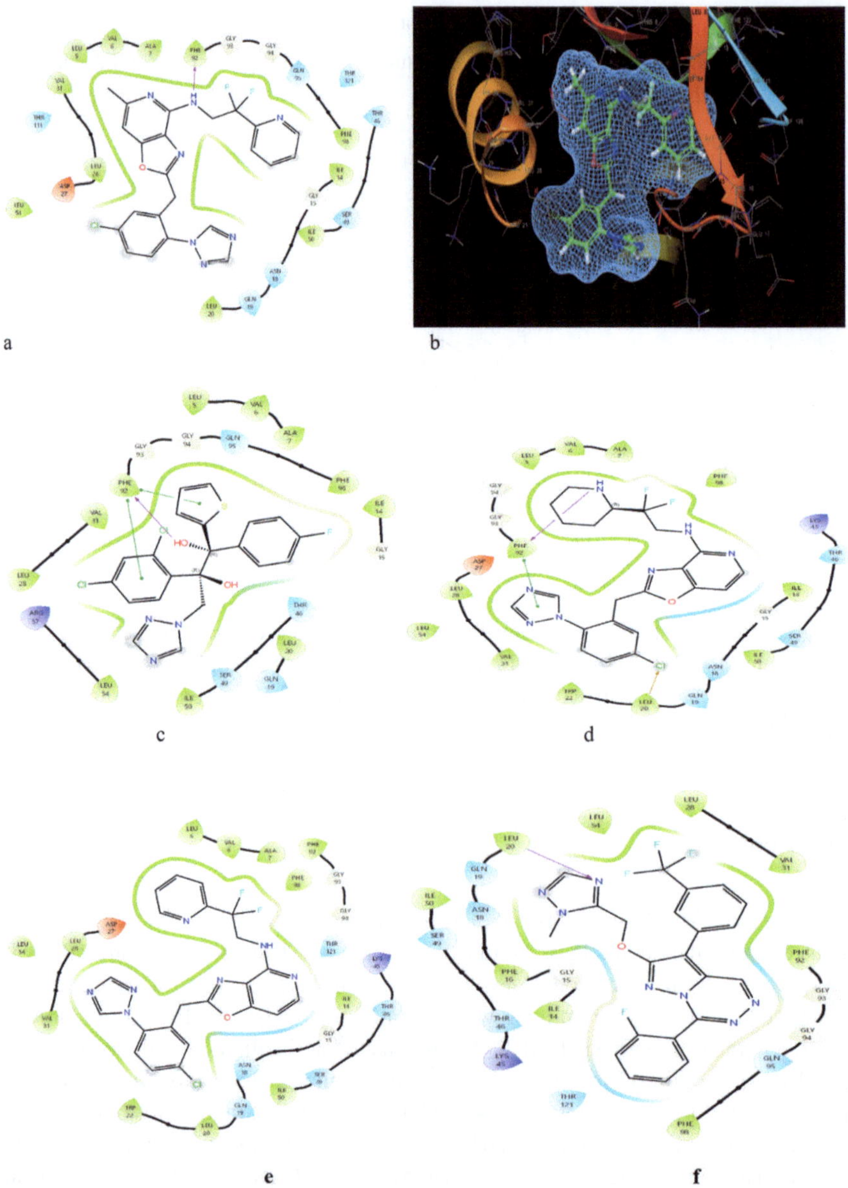

Fig. 6. 2D ligand interaction and surface binding of compound (a) CHEMBL198820 (c) CHEMBL275068 (d) CHEMBL198735 (e) CHEMBL196315 (f) CHEMBL208844 and 3D representation of binding modes of the best docked compound CHEMBL198820 (b).

Table 6. Top 5 Drug hits ADME predictions by ADMET Lab 2.0.

CID	198820	275068	198735	196315	208844	Acceptable values
Molecular weight	481.12	463.03	473.15	467.11	469.13	100–600
LogP	4.313	3.909	3.654	3.995	3.205	0–3
Lipinski Rule	Accepted	Accepted	Accepted	Accepted	Accepted	
Caco-2 permeability	−4.846	−4.578	−4.976	−4.837	−4.637	> − 5.15
HIA	0.001	0.007	0.004	0.005	0.006	0–0.3
BBB	0.14	0.151	0.231	0.556	0.125	
VD	4.712	1.865	3.451	4.824	2.616	0.04–20
PPB	98.32%	98.40%	92.68%	97.43%	93.49%	≤90%
CYP1A2 inhibitor	0.912	0.914	0.898	0.956	0.924	0–1
CYP2C19 inhibitor	0.977	0.948	0.963	0.98	0.927	
CYP2C9 inhibitor	0.969	0.932	0.896	0.974	0.808	
CYP2D6 inhibitor	0.907	0.971	0.922	0.91	0.019	
CYP3A4 inhibitor	0.975	0.957	0.973	0.973	0.453	
CL	5.884	5.125	7.038	5.597	9.823	≥5

Table 7. Toxicity parameters of the top 5 drug hits by ADMET Lab 2.0.

CID	hERG Blockers	Carcinogenicity	Skin sensitivity	AMES toxicity
198820	0.819	0.766	0.348	0.353
275068	0.461	0.343	0.43	0.11
198735	0.944	0.218	0.523	0.13
196315	0.776	0.11	0.477	0.17
208844	0.058	0.366	0.022	0.405

drug of 1,2,4-triazole derivatives. The presented results verified the significance of 1,2,4-triazole derivatives as a potent antimicrobial agent.

Acknowledgements. The authors are thankful to Head, Department of Pharmaceutical Sciences, Maharshi Dayanand University, Rohtak for providing software assistance.

Declaration of Interest Statement The authors declare that there are no competing financial interests or personal relationships that could have appeared to influence the work reported in this paper.

Disclosure Statement Not applicable

Data Availability Statement Not applicable

References

1. Prestinaci, F., Pezzotti, P., Pantosti, A.: Antimicrobial resistance: a global multifaceted phenomenon. Pathog. Glob. Health. **109**, 309–318 (2015)
2. Morrison, L., Zembower, T.R.: Antimicrobial Resistance. Gastrointest. Endosc. Clin. N. Am. **30**, 619–635 (2020)

3. Taneja, N., Sharma, M.: Antimicrobial resistance in the environment: the Indian scenario. Indian J. Med. Res. **149**, 119–128 (2019)
4. Almulla, A.F., Pharma, D., Al-Mulla, A.: ISSN 0975-413X CODEN (USA): PCHHAX a review: biological importance of heterocyclic compounds (2017)
5. Abdelli, A., et al.: Recent advances in the chemistry of 1,2,4-triazoles: synthesis, reactivity and biological activities. Tetrahedron Lett. **86**, 153518 (2021). https://doi.org/10.1016/J.TETLET.2021.153518
6. Gao, F., Wang, T., Xiao, J., Huang, G.: Antibacterial activity study of 1,2,4-triazole derivatives. Eur. J. Med. Chem. **173**, 274–281 (2019). https://doi.org/10.1016/J.EJMECH.2019.04.043
7. Langley, M.S., Clissold, S.P.: Brotizolam. A review of its pharmacodynamic and pharmacokinetic properties, and therapeutic efficacy as an hypnotic. Drugs. **35**, 104–122 (1988). https://doi.org/10.2165/00003495-198835020-00002
8. Bayrak, R., Bekircan, O., Durmuş, M., Değirmencioğlu, I.: The synthesis and structural characterization of new metallo phthalocyanines with 1,2,4-triazole fragments. J. Organomet. Chem. **767**, 101–107 (2014). https://doi.org/10.1016/J.JORGANCHEM.2014.06.003
9. Mathew, V., Keshavayya, J., Vaidya, V.P., Giles, D.: Studies on synthesis and pharmacological activities of 3,6-disubstituted-1,2,4-triazolo[3,4-b]-1,3,4-thiadiazoles and their dihydro analogues. Eur. J. Med. Chem. **42**(6), 823–840 (2007)
10. Al, A.A., Hassa, A.A., Makhlou, M.M., Bräse, S.: Chemistry and biological activities of 1,2,4- Triazolethiones-antiviral and anti-infective drugs. Molecules. **25** (2020)
11. Kazeminejad, Z., et al.: Novel 1, 2, 4-Triazoles as antifungal agents. Biomed. Res. Int. **2022** (2022)
12. Pei, T., et al.: Synthesis and anticancer activity of [1,2,4] triazole [4,3-b] [1,2,4,5] tetrazine derivatives. Phosphorus Sulfur Silicon Relat. Elem. **197**, 819–824 (2022). https://doi.org/10.1080/10426507.2022.2033743
13. Aggarwal, R., Sumran, G.: An insight on medicinal attributes of 1,2,4-triazoles. Eur. J. Med. Chem. **205**, 112652 (2020)
14. Kumari, M., et al.: Synthesis and biological evaluation of heterocyclic 1,2,4-triazole scaffolds as promising pharmacological agents. BMC Chem. **15** (2021). https://doi.org/10.1186/s13065-020-00717-y
15. Gomha, S., Abdel-Aziz, H., Badrey, M., Abdalla, M.: Efficient synthesis of some new 1,3,4-Thiadiazoles and 1,2,4-Triazoles linked to Pyrazolylcoumarin ring system as potent 5α-reductase inhibitors. J. Heterocycl. Chem. **56** (2019). https://doi.org/10.1002/jhet.3487
16. Ahmed, S.K., et al.: Antimicrobial resistance: impacts, challenges, and future prospects. J. Med. Surg. Public Health. **2**, 100081 (2024). https://doi.org/10.1016/J.GLMEDI.2024.100081
17. Hawser, S., Lociuro, S., Islam, K.: Dihydrofolate reductase inhibitors as antibacterial agents. BiochemPharmacol. **71**, 941–948 (2006). https://doi.org/10.1016/J.BCP.2005.10.052
18. Güner, O.F.: Pharmacophore Perception, Development, and Use in Drug Design. International University Line (2000)
19. Roy, K., Kar, S., Das, R.: Understanding the basics of QSAR for applications in pharmaceutical sciences and risk assessment (2015)
20. Kumar, M., et al.: Synthesis, antimicrobial, anticancer, antiviral evaluation and QSAR studies of 4-(1-aryl-2-oxo-1,2-dihydro-indol-3-ylideneamino)-N-substituted benzene sulfonamides. Arab. J. Chem. **7**, 396–408 (2014). https://doi.org/10.1016/j.arabjc.2012.12.005
21. Wan, Q., et al.: Toward resolving the catalytic mechanism of dihydrofolate reductase using neutron and ultrahigh-resolution X-ray crystallography. Proc. Natl. Acad. Sci. USA. **111**, 18225–18230 (2014). https://doi.org/10.1073/pnas.1415856111
22. Friesner, R.A., et al.: Glide: a new approach for rapid, accurate docking and scoring. 1. Method and assessment of docking accuracy. J. Med. Chem. **47**, 1739–1749 (2004). https://doi.org/10.1021/jm0306430

23. Gupta, S., Baweja, G.S., Gupta, G.D., Asati, V.: Identification of potential N-substituted 5-benzylidenethiazolidine-2,4-dione derivatives as α-amylase inhibitors: computational cum synthetic studies. J. Mol. Struct. **1287**, 135596 (2023). https://doi.org/10.1016/J.MOLSTRUC.2023.135596
24. Manal, M., et al.: Novel HDAC8 inhibitors: a multi-computational approach. SAR QSAR Environ. Res. **28**, 707–733 (2017). https://doi.org/10.1080/1062936X.2017.1375978
25. Dixon, S.L., Smondyrev, A.M., Rao, S.N.: PHASE: a novel approach to pharmacophore modeling and 3D database searching. Chem. Biol. Drug Des. **67**, 370–372 (2006)
26. Golbraikh, A., Tropsha, A.: Predictive QSAR modeling based on diversity sampling of experimental datasets for the training and test set selection. J. Comput.-Aided Mol. Des. **16**, 357 (2002)
27. Mali, S.N., Chaudhari, H.K.: Molecular modelling studies on adamantane-based Ebola virus GP-1 inhibitors using docking, pharmacophore and 3D-QSAR. SAR QSAR Environ. Res. **30**, 161–180 (2019). https://doi.org/10.1080/1062936X.2019.1573377
28. Sanapalli, B.K.R., Yele, V., Baldaniya, L., Karri, V.V.S.R.: Identification of novel protein kinase C-βII inhibitors: virtual screening, molecular docking and molecular dynamics simulation studies. J. Mol. Model. **28**, 117 (2022). https://doi.org/10.1007/s00894-022-05104-z
29. Khamouli, S., et al.: Multi-combined 3D-QSAR, docking molecular and ADMET prediction of 5-azaindazole derivatives as LRRK2 tyrosine kinase inhibitors. J. Biomol. Struct. Dyn. **40**, 1285–1298 (2022). https://doi.org/10.1080/07391102.2020.1824815
30. Maheshwari, K., et al.: In Silico examination of antimicrobial activity of Benzoxaole derivatives against Dihydrofolate reductase. Int. J. Res. Publ. Rev. **6**, 4944–4962 (2025)

23. Gupta, S., Bawejia, G.S., Gupta, G.D., Asati, V.: [illegible] Gupta, [illegible] 5-benzylidene[illegible] derivatives [illegible]: synthesis, computational and in [illegible] studies. J. Mol. Struct. 1287, 135590 (2023). https://doi.org/10.1016/j.molstruc.2023.135590
[illegible]

Rationale of Antifungal and Anti Bacterial Interventions of Imidazole Derivatives

Divya Negi[1,2] (✉), Akhilesh Kumar Mishra[1], and Narendra Kumar Singh[3]

[1] School of Pharmacy, Babu Banarsi Das University, Lucknow, Uttar Pradesh, India
[2] Faculty of Pharmaceutical Sciences, Amrapali University, Lamachaur, Haldwani, Nainital, Uttarakhand, India
divya917@amrapali.ac.in
[3] Veer Bahadur Singh Purvanchal, University Jaunpur, Uttar Pradesh, India

Abstract. Imidazole is a five-member heterocyclic molecule with two nitrogen atoms at positions 1 and 3 in the aromatic ring. It is categorized as a diazole. Clinical trials have utilized a variety of imidazole-based drugs for antifungal, antibacterial, anticancer, antiparasitic, antihypertensive, and anti-inflammatory properties. A remarkable class of chemicals with antimicrobial properties, including potent antifungal and antibacterial agents, are imidazole derivatives, which contain the imidazole ring. The biological, and therapeutic use of compounds containing imidazoles are the main topics of this review. Numerous studies have investigated the antifungal activity of imidazole derivatives, which interfere with the growth of cell membranes and the integrity of the walls of fungal cells. Likewise, their antibacterial properties have been noted in a variety of Gram-positive and Gram-negative bacteria, and they are linked to mechanisms such the suppression of nucleic acid replication, protein synthesis, and cell wall formation. We also investigate the structure-activity relationships (SAR) of imidazole derivatives, focusing on their pharmacokinetic, toxicological, and biological activities. In addition to increasing the number of effective medications available to treat bacterial and fungal diseases, the discovery of new imidazole-based treatments could contribute to combating the growing threat of drug resistance.

Keywords: Imidazole · Antifungal · Antibacterial

1 Introduction

Heterocyclic compounds, characterized by a cyclic structure with at least one non-carbon atom in the ring, represent crucial sources of bioactive molecules. These compounds can originate from natural sources or be synthesized organically, offering versatile interactions with therapeutic receptors and diverse biological effects [1, 2]. The most frequent hetero-atoms are nitrogen, oxygen, and sulphur, but heterocyclic rings including additional hetero-atoms i.e. phosphorus, iron, magnesium, selenium etc. are also common [3]. In recent years, especially in medical chemistry, the synthesis

A. K. Saxena, A. Saxena (Eds.): GTHTM 2025, pp. 117–142, 2026.
https://doi.org/10.1007/978-3-032-12320-6_6

of molecules based on heterocyclic scaffolds containing nitrogen, oxygen, and sulphur has become more and more significant. Due to their wide range of biological action, imidazole derivatives among nitrogen-heterocyclic compounds need special consideration [4–7]. These ring systems are also found as major skeleton in many biological molecules such as hemoglobin, DNA, RNA, vitamins, hormones, etc. It is also seen these structures in of many drugs approved by the FDA and used in the treatment of different diseases [8]. Furthermore, N-heterocyclic compounds have biological activities such as antifungal, anti-inflammatory, antibacterial, antioxidant, anticonvulsant, antiallergic, enzyme inhibitors, herbicidal activity, anti-HIV, antidiabetic, anticancer activity, insecticidal agents [9, 10].

1.1 Imidazole

Imidazoles are a significant group of heterocyclic compounds because of their wide range of medicinal and physiological effects, including their anti-inflammatory, antibacterial, anticonvulsant, antihypertensive, antiparasite, antioxidant, antifungal and anticancer properties [11, 12]. Numerous drugs have imidazole nucleus in their structure like cimetidine as antihistaminics,dacarbazine as anticancer, metronidazole as antifungal, and flumazenil as benzodiazepine antagonist (Table 1) [13–15].

The imidazole is a planar, five-membered heteroaromatic molecule having 3C and 2 N atoms in the 1 and 3 positions (Fig. 1). Its systemic name is 1,3-diaza-2,4-cyclopentadiene.The molecular formula of imidazole is $C_3H_4N_2$, which consists two hetero atoms (nitrogen) and two double bonds [16, 17].

Gloxaline was its original name because it was created using glyoxal and ammonia by Heinrich Debus in 1858. Attacks of an electrophilic and nucleophilic nature might affect amphoteric nature. Highly resistant to thermal, acid, base, oxidation, and reduction conditions [18]. Significant intramolecular hydrogen bonding is present. Because the hydrogen atom can be found on either of the two nitrogen atoms, it can exist in two identical tautomeric forms. The compound is classified as aromatic due to the existence of a sextet of π-electrons, made up of pair of electrons from the protonated nitrogen atom and one from each of the remaining four atoms in the ring [19–21]. The amino acid histidine, vitamin B12, a component of DNA base structure, and purines, histamine, and biotin are among the well-known components of human creatures that have an imidazole nucleus as their primary structural component. Cimetidine, azomycin, and metronidazole are a few examples of natural or synthesized medicine compounds that contain it in their structures [22–24].

2 History and Development

The discovery of imidazole is attributed to the German scientist ***Hermann Kolbe*** in 1858. His discovery of imidazole by reacting glyoxal with ammonia was a groundbreaking development in the field of heterocyclic compound research [25, 26]. ***Ernst J. Fisher*** identified imidazole as a component of purine-like compounds. Purines, such

Table 1. Pharmacological significance of imidazole nucleus.

S. No.	Name	Structure	Activity
1.	Clemizole	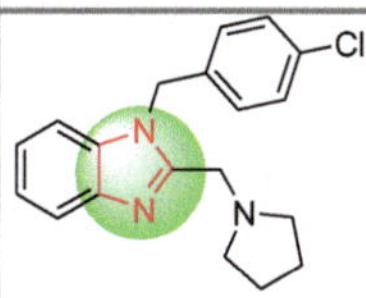 (1) **(1)**	Anti-histaminic Agent
2.	Etonitazene	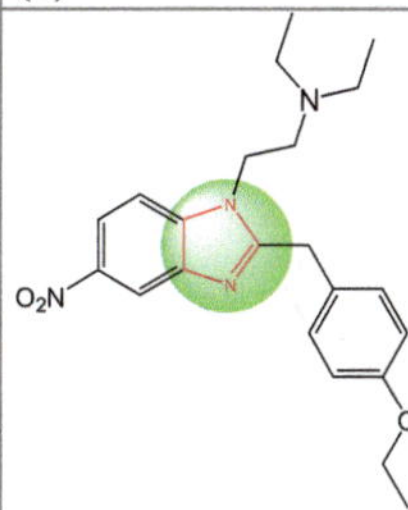 (2) **(2)**	Analgesic
3.	Enviroxime	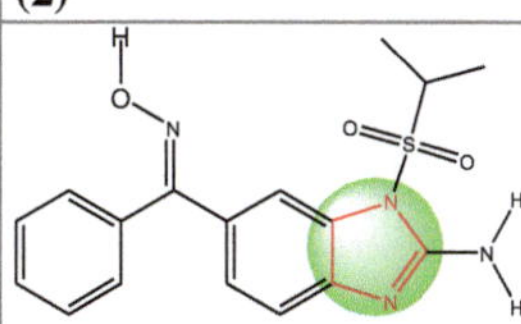 (3) **(3)**	Antiviral
4.	Irtemaazole	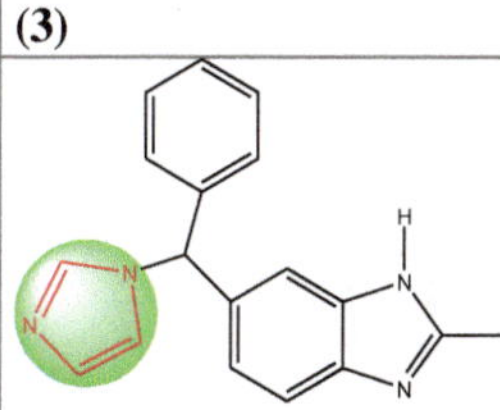 (4) **(4)**	For the promotion of excretion of uric acid
5.	Astemizole	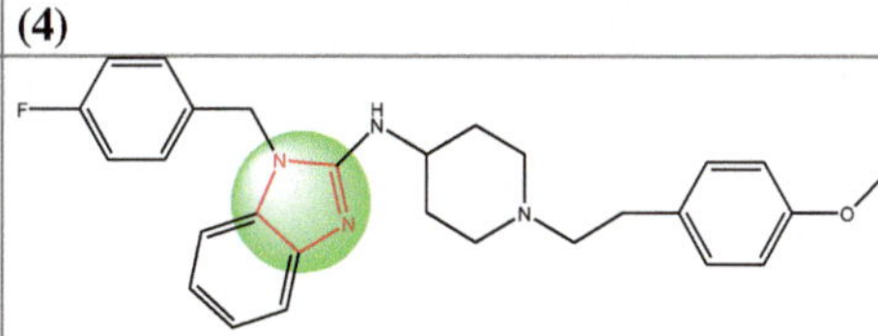 (5) **(5)**	Anti-histaminic Agent
6.	Omeprazole	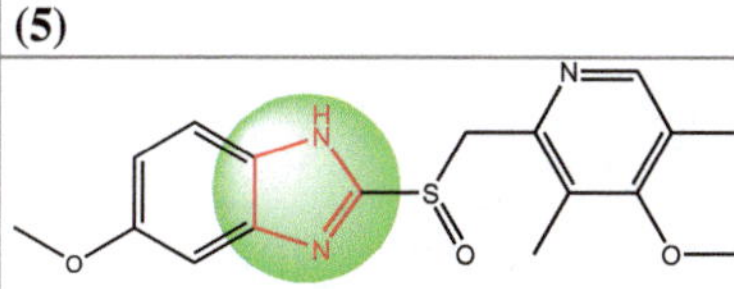 (6) **(6)**	Antiulcer

(continued)

Table 1. (continued)

S. No.	Name	Structure	Activity
7.	Pantoprazole	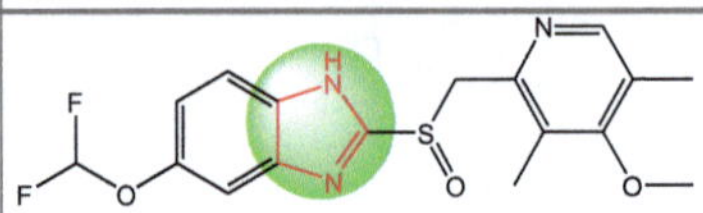 (7) **(7)**	Antiulcer
8.	Thiabendazole	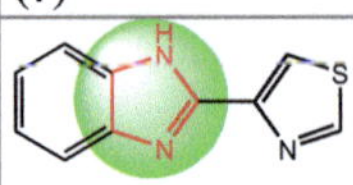 (8) **(8)**	Anti-helmintic
9.	Nocodazole	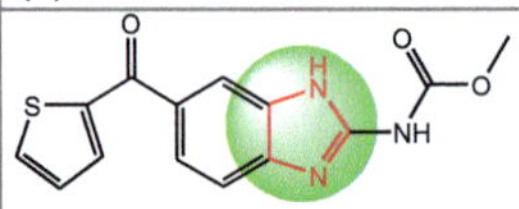 (9) **(9)**	Antinematodal
10.	Metronidazole	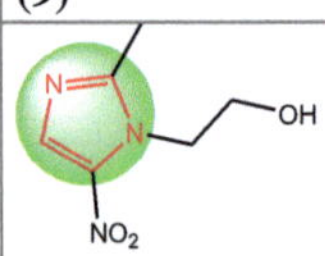 (10) **(10)**	Antibacterial
11.	Nitrosoimidazole	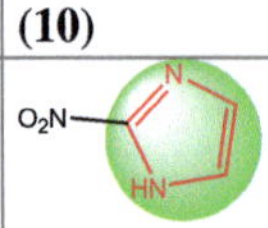 (11) **(11)**	Antibacterial
12.	Megazole	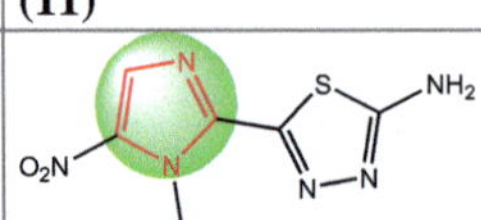 (12) **(12)**	Trypanocidal
13.	Azathioprine	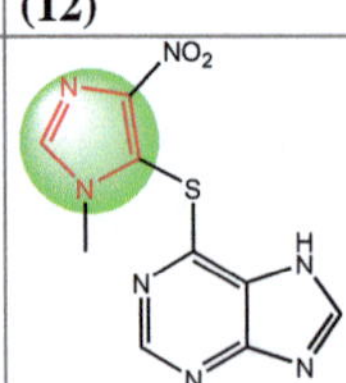 (13) **(13)**	Anti-rheumatoid Arthritis
14.	Tinidazole	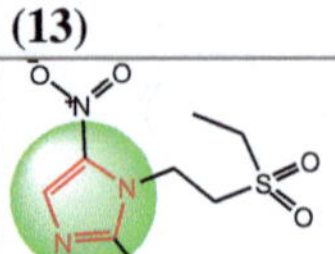 (14) **(14)**	Anti-protozoal and antibacterial

(continued)

Table 1. (continued)

S. No.	Name	Structure	Activity
15.	Ornidazole	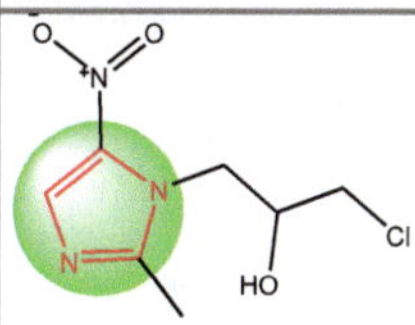 (15) **(15)**	Antiprotozoal & Antibacterial
16.	Satranidazole	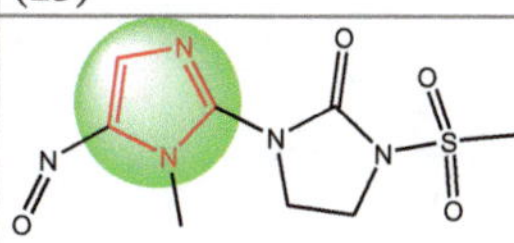 (16) **(16)**	Anti-amoebic
17.	Cimetidine	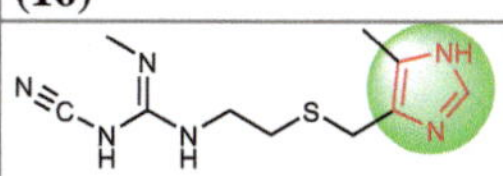 (17) **(17)**	Antiulcer
18.	Carbimazole	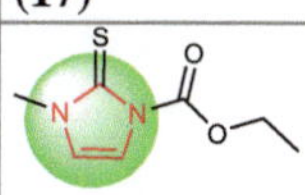 (18) **(18)**	Antithyroid
19.	Tolazoline	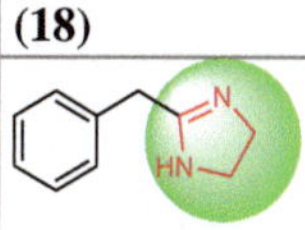 (19) **(19)**	Vasodilator
20.	Naphazoline	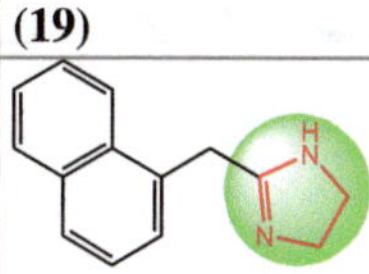 (20) **(20)**	Vaso-constrictor
21.	Tetra-hydrozoline	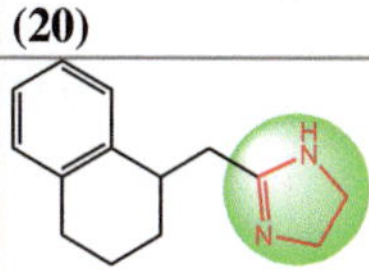 (21) **(21)**	Vaso-constrictor
22.	Etomidate	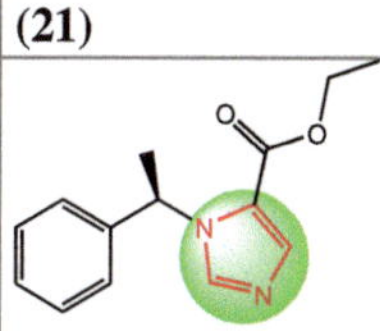 (22) **(22)**	Anesthetic agent

(continued)

Table 1. (continued)

S. No.	Name	Structure	Activity
23.	Lansoprazole	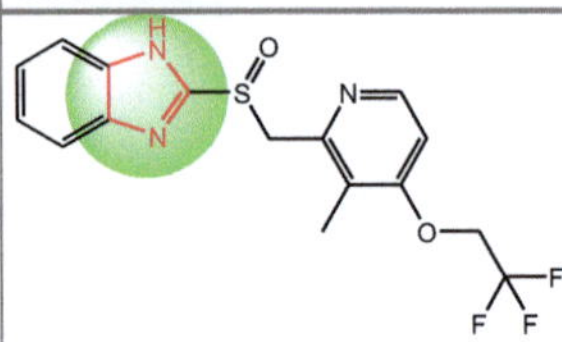 (23) **(23)**	Antiulcer
24.	Flumazenil	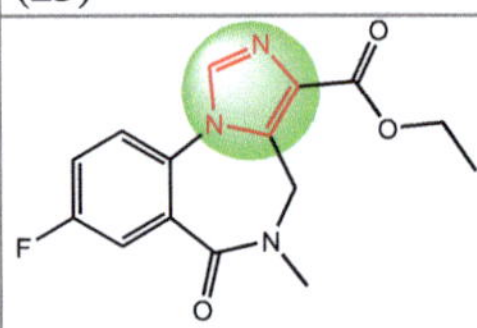 (24) **(24)**	Benzo-diazepine Antagonist
25.	Methimazole	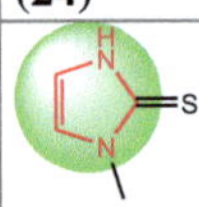 (25) **(25)**	Antithyroid
26.	Pilocarpine	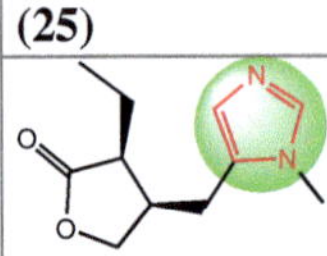 (26) **(26)**	Cholinergic agent
27.	Ketoconazole	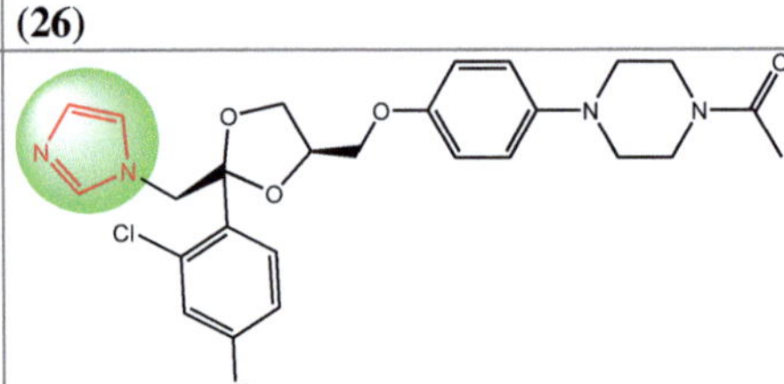 (27) **(27)**	Antifungal
28.	Dacarbazine	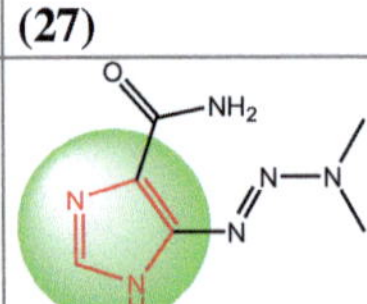 (28) **(28)**	Anticancer

(continued)

Table 1. (continued)

S. No.	Name	Structure	Activity
29.	Pimobendan	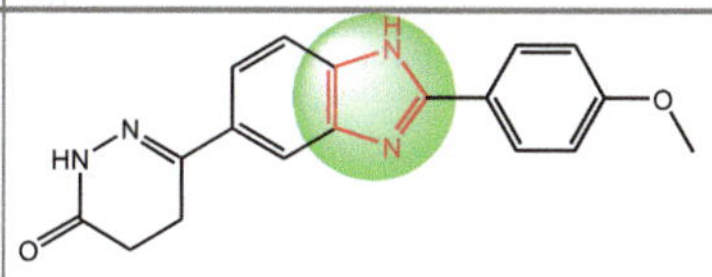 (29) **(29)**	Calcium sensitizer and phospho diesterase inhibitor
30.	Fenbendazole	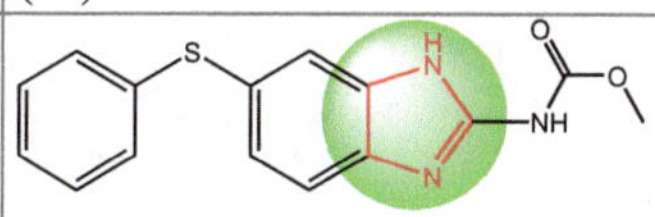 (30) **(30)**	Antihelmintic

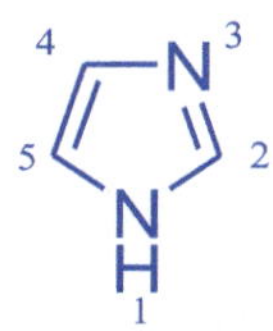

Fig. 1. Imidazole.

as guanine and adenine, are essential for biological functions like the creation of DNA and RNA [27]. Development on nitrogen-containing heterocycles emerged in the late 19th and early 20th centuries, leading to the clarification of imidazole's structure. The systematic synthesis and characterization of imidazole and its derivatives was carried out by ***Aldo Costa*** and other chemists in 1870s. The compound's ability to transfer electron pairs from its nitrogen atoms has been found to make it a highly reactive base [28]. The biological significance of imidazole was further demonstrated by the production of histidine, an amino acid with an imidazole group. ***Linus Pauling***, whose work on protein structures in the 1930s and 1940s highlighted the critical significance of imidazole in biological systems [29]. The amino acid histidine is the precursor of histamine, which has been found to function as a neurotransmitter and a channel for communication in allergic reactions (Fig. 2) [30].

In medicinal chemistry, imidazole derivatives became well-known, particularly after their potential as enzyme inhibitors and antifungal drugs was discovered. Several pharmacological agents were developed by using the imidazole structure to create medications that could interact with particular enzymes [31]. One of the earliest imidazole-based medications to be used in a clinical setting for fungal infections was ketoconazole, and its effectiveness prompted additional research into imidazole derivatives for use in pharmaceuticals [32]. Most commonly used medications that contain imidazoles are fluconazole and clotrimazole. Imidazole continues to play a key role in the development of selective enzyme inhibitors, which specifically target enzymes associated in autoimmune disorders, cardiovascular disorders, cancer, and viral infections [33]. Imidazole has also seen a lot of use in materials science lately, especially in the creation of metal-

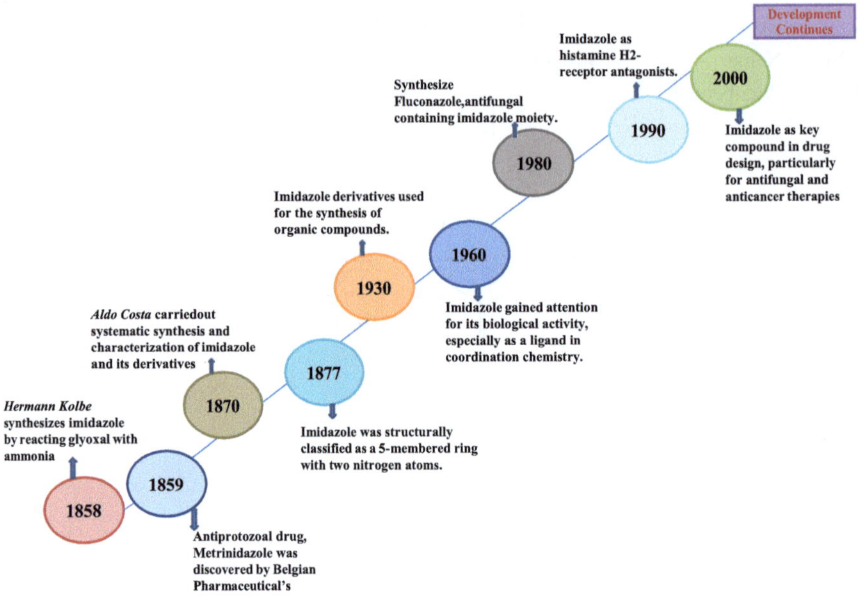

Fig. 2. Milestones in the development of imidazole and its derivatives.

organic frameworks (MOFs). Imidazole derivatives are employed as ligands in the production of MOFs, which are porous materials with uses in drug delivery, gas storage, and catalysis [34].

3 Structure Activity Relationship (SAR) of Imidazole

The relationship that exists between the atomic structure of imidazole-based compounds & their biological effects is referred to as the Structure Activity Relationship (SAR) of imidazole analogues. Imidazole is a five-membered aromatic heterocyclic ring comprised of two nitrogen atoms at positions one and three, respectively. Imidazole derivatives SAR serves as essential for interpreting how pharmacological attributes are influenced by modifications within the imidazole core. An overview of how structural modifications to imidazole compounds can impact their antimicrobial effectiveness is shown as follows (Fig. 3) [35].

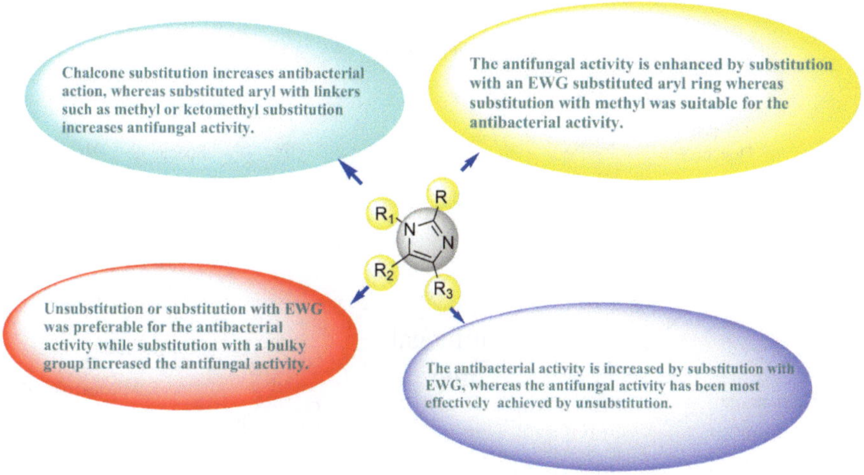

Fig. 3. Substituted imidazole as promising antifungal and antibacterial agents.

4 Anti Fungal Activity

4.1 Marketed Imidazole Containing Antifungal Drugs

In order to combat fungal infections, numerous kinds of imidazole derivatives have already been developed. Some of these include renowned antifungal imidazole medications such as Clotrimazole (31), Econazole (32), Miconazole (33), Eberconazole (34), Fenticonazole (35), Enilconazole (36), Ketoconazole (27), Flutrimazole (37), & Tioconazole (38) (Fig. 4). As compound 3 has applications for both topical administration & intravenous formulation, while compounds 1 & 2 are exclusively accessible for topical application only [36]. Compound 4 demonstrated exceptional antifungal properties against dermatophytes & yeasts of the genus Candida in various early in vitro investigations [37]. Drug 5 is used topically for managing mycotic skin infections which are superinfected with bacteria as well as to treat genital discharge [38]. Enilconazole 6 was developed for veterinary use in brain [39]. Compound 7 serves as a water-soluble imidazole that can be administered orally to combat fungal infections that may be superficial or deeply entrenched [40]. According to prior studies, compound 9 could potentially cure vaginal trichomoniasis, napkin rash triggered by Candida albicans, onychomycosis (in a particular nail composition), & impetigo [41].

4.2 Mechanism of Action of Imidazole as Antifungal Agent

Imidazole derivatives are very efficacious for treatment of a variety of superficial skin infections such as athlete's foot and other systemic damning fungal infections. The key interactions in the active site of fungal cell are these components: (i) the amidine nitrogen atom (N-3 in the imidazoles) to bind to the heme iron of enzyme; (ii) aromatic

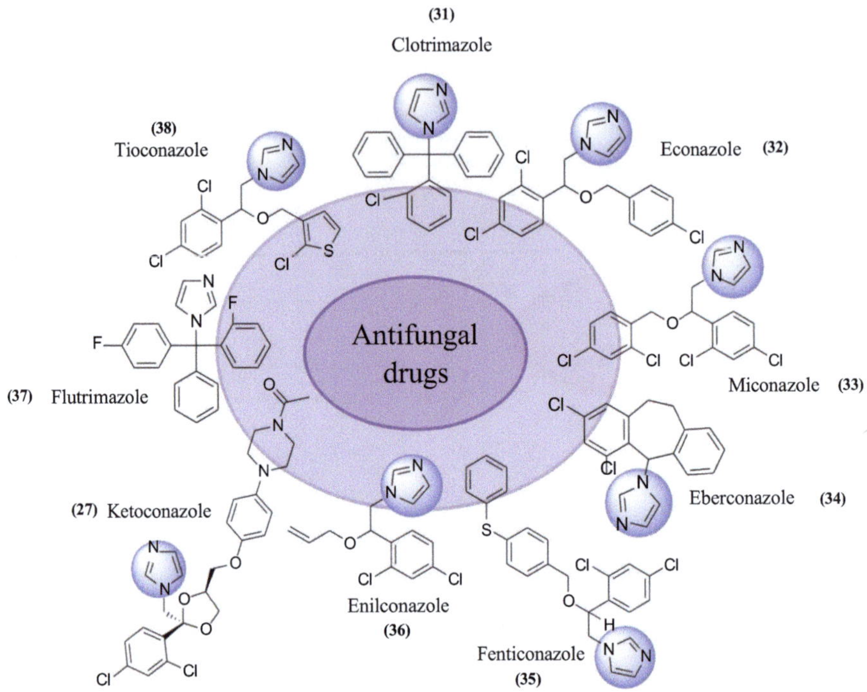

Fig. 4. Antifungal drugs with imidazole ring.

rings; (iii) the large nonpolar portion of molecule [42, 43]. Imidazole-derived compounds are a subgroup of antifungal agents that exert their effect primarily by hindering the stilling of ergosterol, a leading factor of the fungal membrane. Cytochrome P450 is an essential enzyme in the sterol biosynthetic pathway in eukaryotes. Imidazole derivatives successfully inhibit a fungal enzyme known as 14α-demethylase - a type of cytochrome P450. The enzyme has a major responsibility of converting lanosterol into ergosterol sterol, which is highly important for the growth and functional activity of the fungal cell membrane [44, 45]. Imidazole derivatives hamper the activity of 14α-demethylase and thereby dehydrate the fungus by way of allowing the polymorphonuclear osmiophilic granules to be synthesized. The lack of ergosterol leads to disruption and leakages of the fungal cell membrane – and subsequently disturbs the membrane functionalities needed to maintain the structure and integrity of the whole fuzzy mutant cell [46]. With the aggravation of the unstable and permeable menace of the fungus tainting the shelled cell enabling incapability of normal functionality, one may assume this leads to the scope of diseased cells aka cell demise. This is referred within the context as fungicide (destroying the fungus) or static (inhibiting fungal growth) depending on the degree as well as specific medicine (Fig. 5) [47].

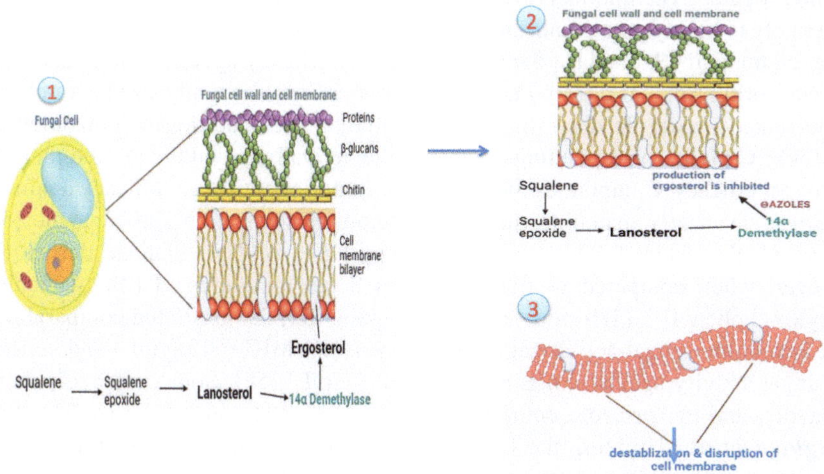

Fig. 5. Mechanism of action of azoles as antifungal agents.

4.3 Reported Imidazoles Having Antifungal Activity

***Z.A. Ketan* et al.** synthesized a series of novel heterocyclic compounds derived from imidazole and assessed their antifungal effects. Compounds **39 & 40** displayed outstanding antifungal activity towards *S.aureus & C.albicans* [48]. ***Zhengxiao H.* et al.** developed an assortment of novel azole derivatives containing a 1,2,3-triazole moiety & assessed their in-vitro antifungal efficacy. The findings demonstrated that compounds **41 & 42** had superior antifungal activity against *C.albicans* compared to the reference drug fluconazole by affecting the integrity of the cell structure, increasing permeability of the cell wall, in addition causing cell contents to leak out [49]. Several novel imidazole & triazole compounds have been developed by ***Emami L.* et al.** The synthetic compounds' antifungal effects were assessed in relation to fuconazole. Compounds **43 & 44** demonstrated the best activity against the *C.albicans* pathogen, with minimum inhibitory concentrations (MIC) ranging between 0.5 to 1 μM [50]. ***R. Ramachandran* et al.** produced imidazole derivatives which are substituted derivatives of piperidin-4-one. The synthetic compounds were tested against fungal strains to determine their in vitro antifungal activity. With IC50 values of 12.5 and 6.25 μg/mL, respectively, imidazole compounds **45 and 46** suppressed the development of *A.niger* and *C.neoformans* more effectively than the conventional medication Amphotericin-B [51].

Hussain* et al.** developed a number of di- and tri-substituted imidazole derivatives and screened them for antifungal activity against strains of the fungi *C. albicans*, *R. oryza*, *P. citrum* & *A. niger*. using ketoconazole as a reference medication (MIC: 12.5 μg/mL). Two of these, 2-(4-nitrophenyl)-4-(4-methoxyphenyl)-1- phenyl-1H-imidazole **(47)** and 2,4-di-(4-methoxyphenyl)-1-phenyl-1H-imidazole **(48),** stepped out as lead compounds with strong antifungal activity and reduced gastrointestinal irritation [52]. ***H. Mohammed Saleh Al-Jubori* et al.** created derivatives of α-aminophosphonate. The novel compounds' ability to combat the fungus *C.albicans* was investigated. The findings revealed that the majority of the novel compounds are moderately to highly potent antifungal agents. The medicine with the most significant value, compound **49**, displays the same inhibitory efficacy against *C.albicans* as the reference drug, Nezo-arm [53]. ***Padmavathi and colleagues synthesized a number of amido-linked imidazole derivatives by 1,3-dipolar cycloaddition of diazomethane and TosMIC to the corresponding cinnamamide derivatives, and then tested them for antifungal efficacy. Compounds **50 and 51** had superior antifungal efficacy against *P. chrysogenum,* nearly matching that of the common medication ketoconazole [54]. ***Emami*** **et al.** designed 1-[(2-benzyloxy)phenyl]-2-(azol-1-yl)ethanone derivatives. However, when compared to fluconazole as a standard drug, 1-[2-(2,4-dichloro benzyloxy) phenyl]-2-(1H-imidazol-1-yl)ethanone **(52)** demonstrated comparable or more potent antifungal activity against *S. cerevisiae* (MIC = 4 μg/mL) and superior antifungal activity against *A.niger* (MIC = 32 μg/mL) [55]. Some imidazole, benzimidazole, and benztriazole compounds were developed and synthesized by ***Sara Sadeghian*** **et al.**. Utilizing the Clinical & Laboratory Standards Institute (CLSI) technique, novel compounds were assessed for their antifungal effectiveness against a single species of fungus, *C.albicans*. As contrasted with Amphotericin B, the positive control, compound **53** was found to exhibit excellent anticandidal action (MIC = 1 μg/mL) [56].

Few novel imidazole compounds were created and synthesized by ***Verma*** **et al.** All synthetic compound were screened for antifungal effects against three distinct groups of fungi: *C. albicans, A. niger,* and *A. flavus.* The antifungal fluconazole was utilized as a standard medication for assessing the MIC (minimum inhibitory concentration) values. Among them, compounds **54 & 55** exhibited substantial antifungal activity [57]. ***Parameshwar*** **et al.** designed a novel family of imidazole derivatives associated with coumarins. The antifungal activity of newly developed compounds has been assessed against *C. krusei, C. albicans, A. niger, A. flavus, and P. notatum.* Compounds **56, 57 & 58** with o, p-chloro, & p-bromo substitutions as a result revealed powerful effects primarily against *C.krusei* and *C.albicans*, modest activity against *C.niger* and *C.flavus*, and inadequate activity against *C.notatum* respectively [58]. In addition, the ***F. Moraca*** **et al.** produced a new family of 2-(1H-imidazol-1-yl)-1-phenylethanol analogues. In-vitro testing was performed to assess the antifungal activity against *Candida albicans*, non-albicans Candida species, dermathophytes & *C.neoformans*. The physiological findings indicate that compound **59 & 60** exhibited antifungal activity that is on par with or greater than fluconazole [59].

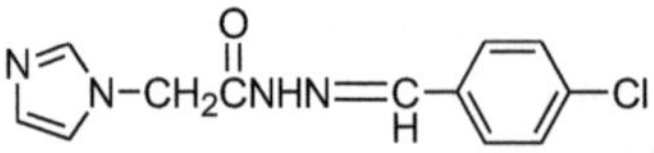

(39)

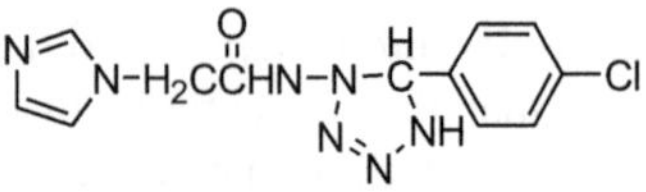

(40)

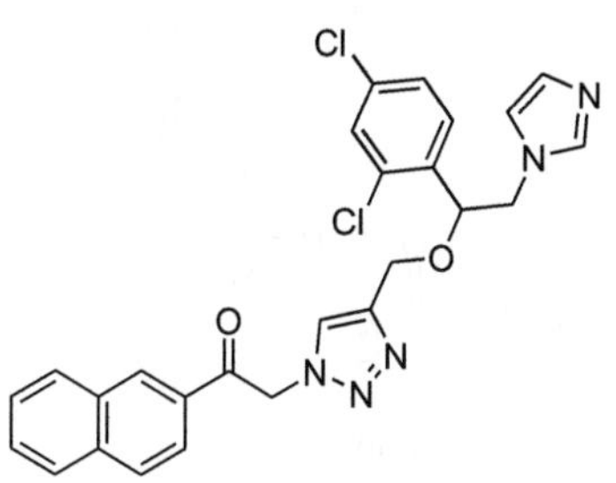

(41)

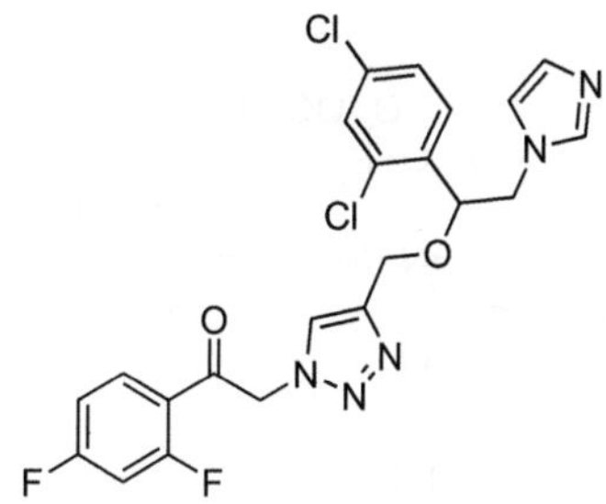

(42)

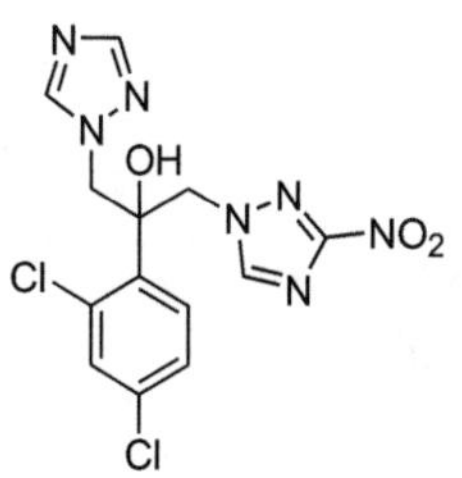

(43)

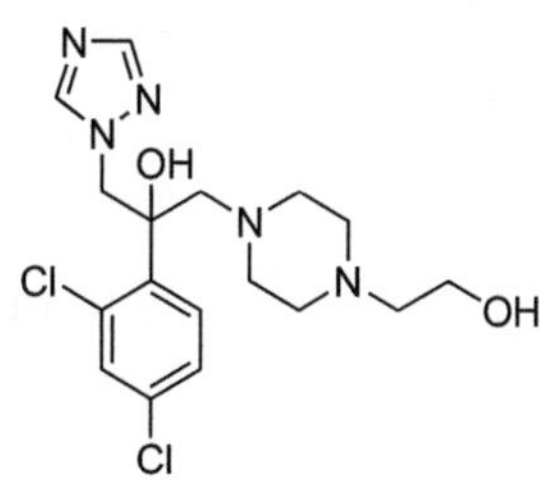

(44)

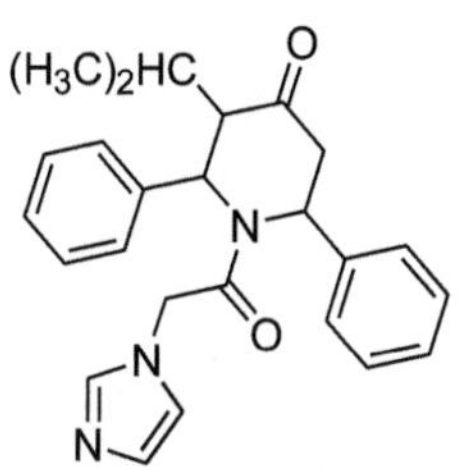

(45)

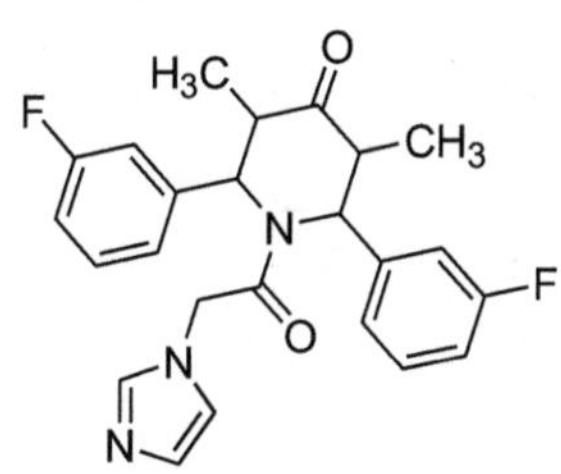

(46)

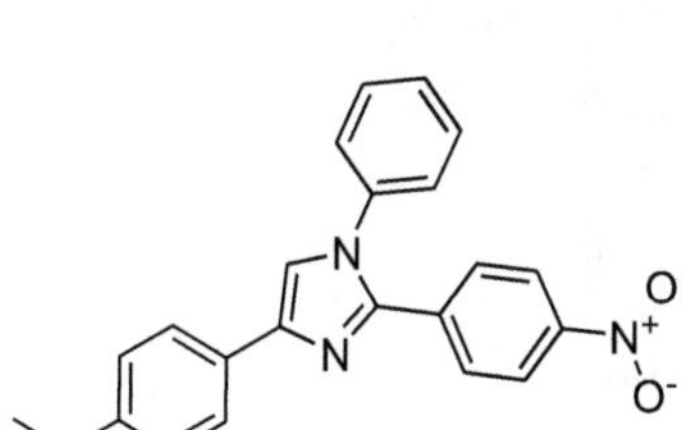

(47)

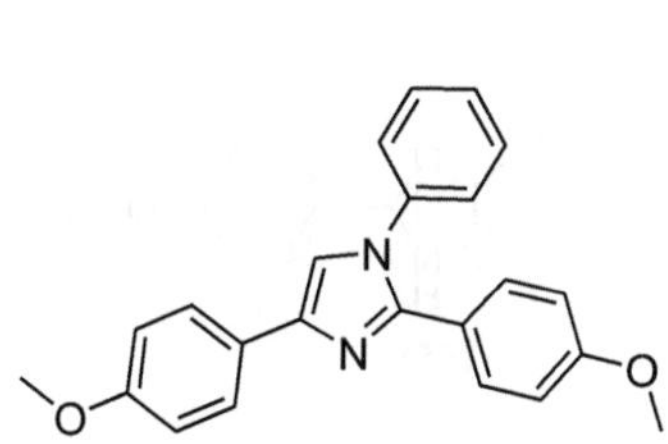

(48)

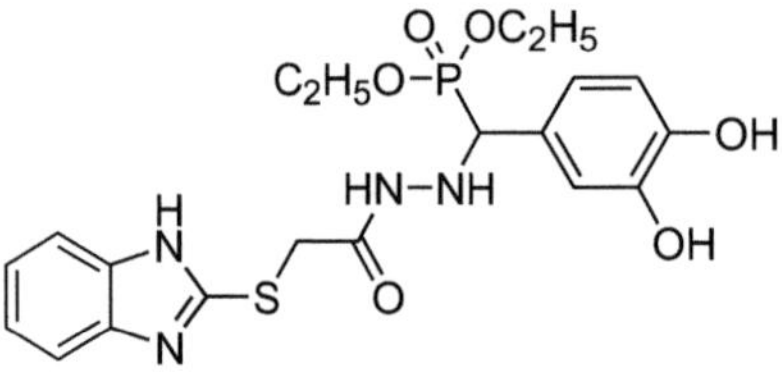

(49)

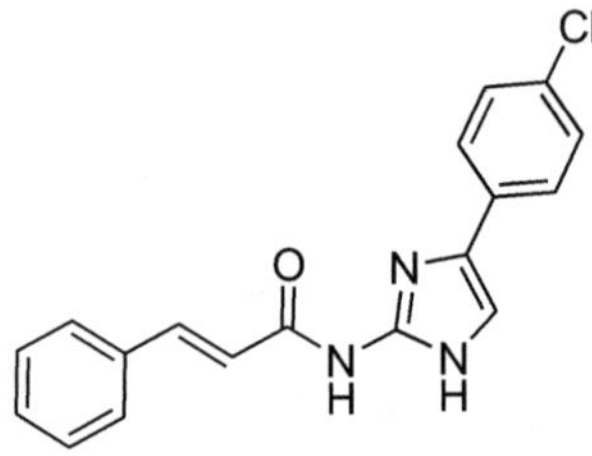

(50)

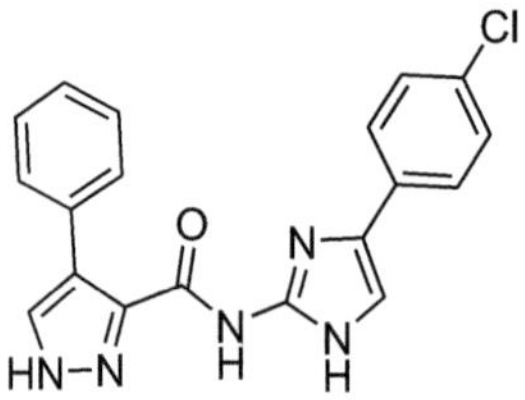

(51)

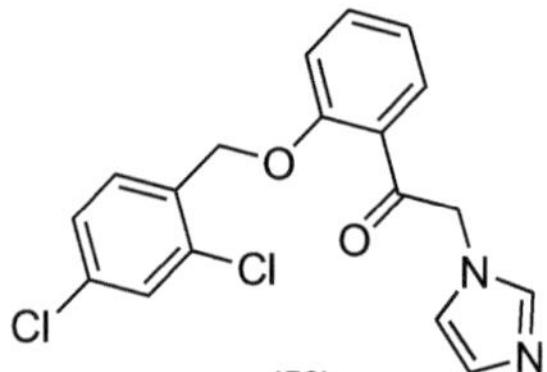

(52)

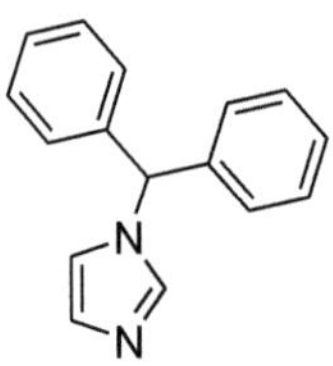

(53)

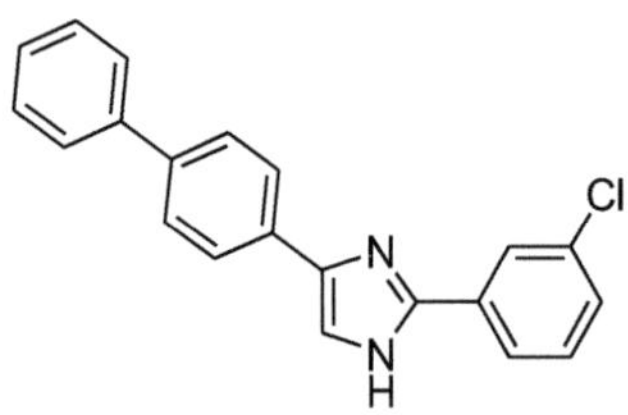

(54)

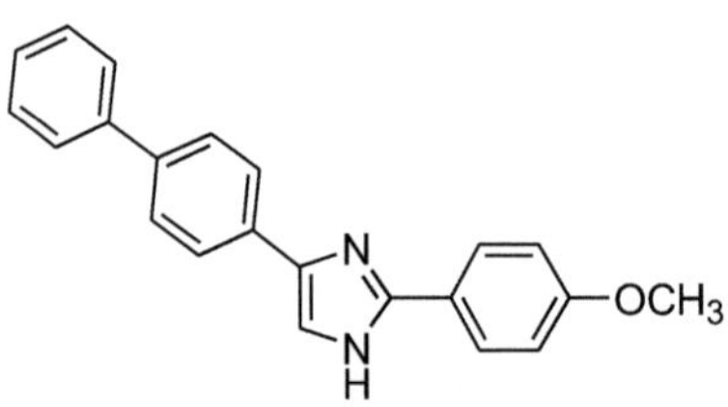

(55)

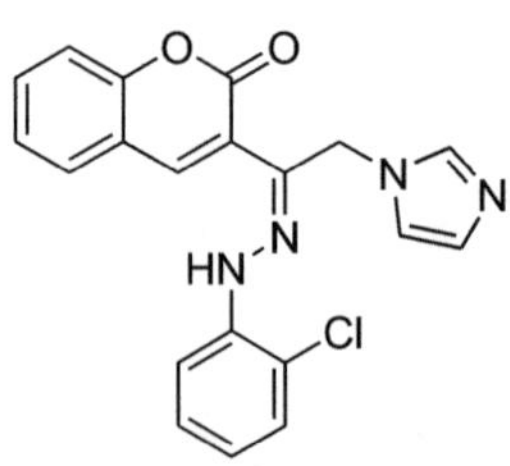

(56)

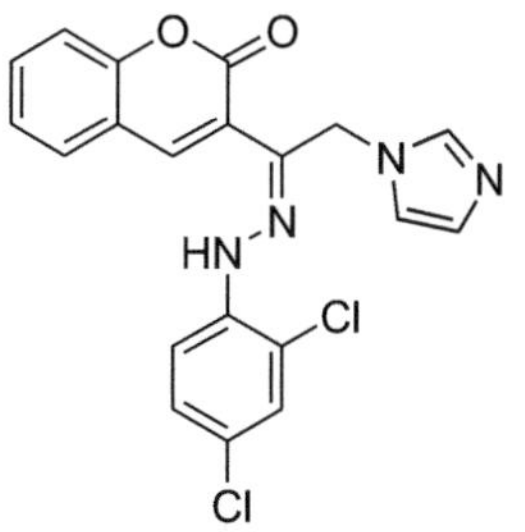

(57)

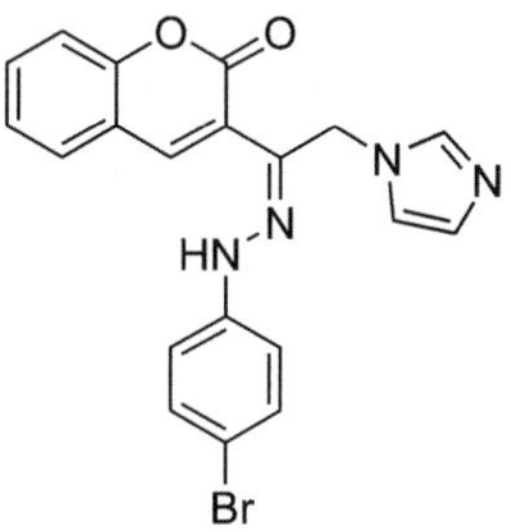

(58)

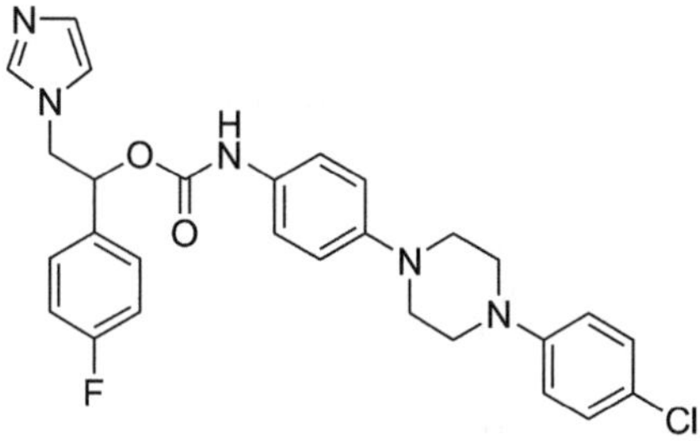

(59)

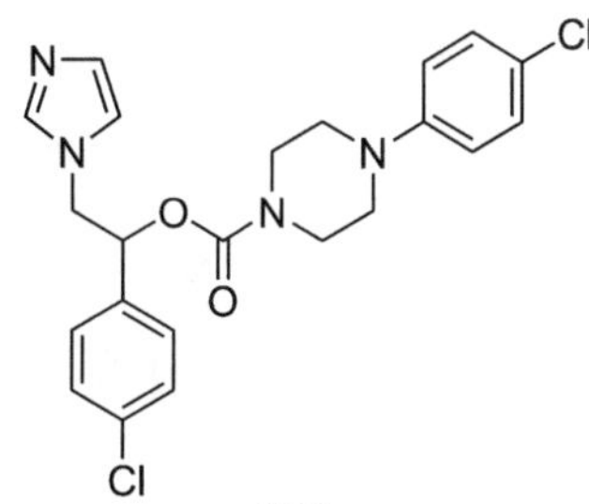

(60)

5 Antibacterial Activity

5.1 Marketed Imidazole Containing Antibacterial Drugs

Clinically, plenty of medications centered around imidazole rings are utilized for treating diverse bacterial & protozoal infections. Marketed medications based on nitroimidazole, such as Secnidazole (62), Nimorazole (61), Ornidazole (15), Tinidazole (14), Metronidazole (10) and Nitrosoimidazole (11), are employed for the treatment of bacterial infections (such as vaginosis caused by bacteria and endocarditis) along with protozoal infections (such as giardiasis, amoebiasis & trichomoniasis) (Fig. 6) [60, 61].

5.2 Mechanism of Action

Imidazole derivatives usually gain entrance into bacterial cells by means of passive diffusion or active transport. Within the cell, imidazole derivatives associate with DNA or any other enzymes that are important for DNA replication and repair. One of the principal targets is bacterial topoisomerase II (DNA gyrase) and topoisomerase IV, the enzymes responsible for supercoiling and the proper structure of DNA during its replication. Blocking action of these enzymes leads to improper unwinding of DNA, excessive supercoiling, leading to nicking of DNA strands at various sites, and degeneration of DNA [62]. This disturbance increases the permeability of the membrane and

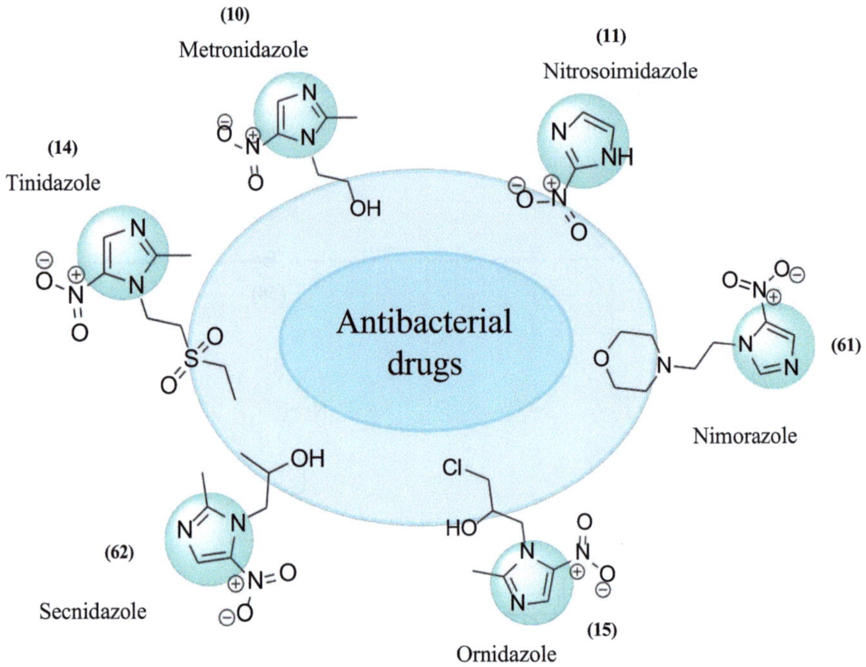

Fig. 6. Mechanism of action of azoles as antibacterial agents.

facilitates the ease with which ions, intracellular metabolites, and proteins can diffuse out. Deficiencies of the important cellular ions and nitrogen constituents which are vital for bacteria to synthesize proteins leads to the death of bacteria. This causes cessation of bacterial cell division and ultimately, death of the cells. Imidazole derivatives can also interfere with bacterial ribosomal function or the enzymes involved in protein biosynthesis. This inhibition prevents the bacterial ribosome from translating mRNA into proteins, halting bacterial growth and proliferation. Without proper protein synthesis, the bacterial cell cannot maintain its essential functions. Ultimately, the cumulative effects of DNA damage, membrane disruption, protein synthesis inhibition, and metabolic interference result in the death of the bacterial cell. This prevents bacterial growth, replication, and the spread of infection (Fig. 7) [63, 64].

5.3 Reported Imidazoles Having Antibacterial Activity

***Jain* et al.** synthesized 2-substituted-4,5-diphenyl-N- alkyl imidazole derivatives and assessed its antibacterial efficacy towards *E. coli, S. aureus & B. subtilis.* Out of all the derivatives, compounds **63 & 64** shown strong antibacterial capabilities using Norfloxacin as standard drug [65]. ***Vijesh* et al.** demonstrated the synthesis of novel imidazole derivatives containing substituted pyrazole moiety and the in-vitro antibacterial activity towards *S.aureus*, *S.typhimvrium*, *E.coli*, *B.subtilis*, *C.perfringens*, and *P.aeruginosa* using Streptomycin as standard drug. Compound **65** showed excellent

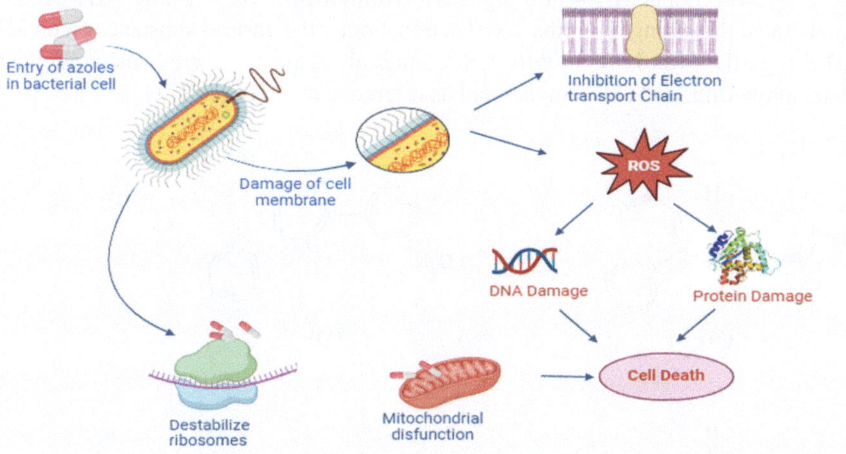

Fig. 7. Antibacterial drugs with imidazole ring.

antibacterial action towards *P.aeruginosa* & *C.perfringens* [66]. ***Narasimhan*** **et al.** created a number of [2-(substituted phenyl)-imidazol-1-yl]-pyridin-3-yl-methanones & examined for their antibacterial activity towards *B.subtilis*, *S.aureus* & *E.Coli.* A newly synthesized imidazole derivatives compounds **66 & 67** bearing a pyridine substituent had equivalent antibacterial activity efficacy (MIC = 0.002 μM/mL) in comparison with ciprofloxacin (MIC = 0.002 μM/mL) [67]. In order to develop strong antibacterial agents, ***Khabnadideh and associates*** used N-alkylated imidazoles, 2-methylimidazoles, and 2-methyl-4-nitroimidazoles. The antibacterial properties of the products were then examined against *P. aeruginosa, S. aureus,* and *E. coil.* The antibacterial potency was increased by substituting 2-methyl & 2-methyl-4-nitro groups on the imidazole ring. Using gentamycin as the benchmark medication, compound **68** was the most effective, with a MIC 10–39 g/ml towards *P. aeruginosa, S. aureus,* and *E. coli* [68].

Sapijanskaite **et al.** used p-aminobenzoic acid to synthesize substituted 1-phenyl-5-oxopyrrolidine derivatives containing a benzimidazole moiety in the structure. The in-vitro antibacterial activity of each of the compounds that were developed was assessed against the gram negative bacteria *E. coli, S. enteritidis,* & *P. aeruginosa*, as well as the gram positive bacteria *B. cereus, S. aureus,* & *L. monocytogenes.* The most promising compounds **69, 70 & 71** have antibacterial activity that was four times stronger than ampicillin [69]. According to ***Yang Hu*** **et al.,** coumarin compounds were designed, synthesized, and biologically evaluated to address antibacterial resistance by inhibiting the fatty acid production route. The antibacterial properties of these substances are evaluated against *E. Coli, S. aureus, S. agalactiae*, and *F. cloumnare*, & their mode of action against the primary enzyme in the bacterial fatty acid production pathway is examined. The findings demonstrate that, in comparison to the common medication norfloxacin, compounds **72, 73 & 74** exhibit strong and wide-ranging antibacterial action [70]. 3-(2,4-disubstituted phenyl)-1-(4-substituted phenyl)-4-(4,5-diphenyl-

1H-imidazol-2-yl)-1H-pyrazole was synthesized by ***Brahmbhatt* et al..** Compound **(75)** 3-(2-bromophenyl)-4-(4,5-diphenyl-1Himidazol-2-yl)-1-(p-tolyl)-1H-pyrazole demonstrated the strongest antibacterial action among the various substances with MIC of 31.0 mg/mL towards *B. subtilis* & 63.0 mg/mL against *S. aureus* using amikacin sulfate, ampicillin, and chloramphenicol as reference medications [71].

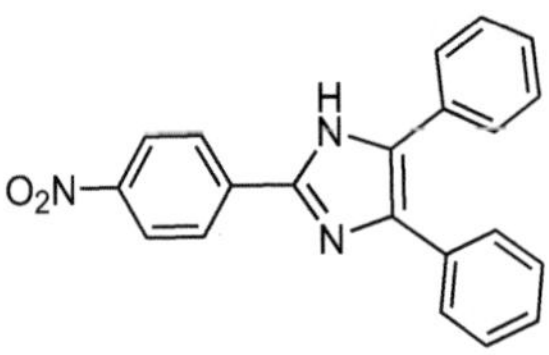

(63)

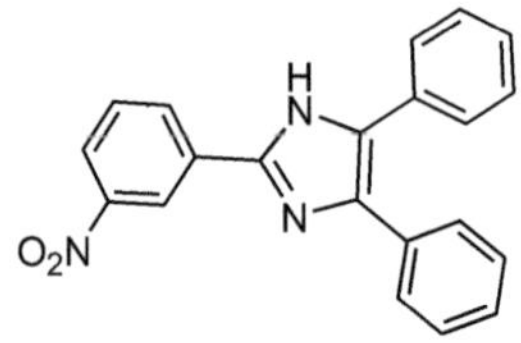

(64)

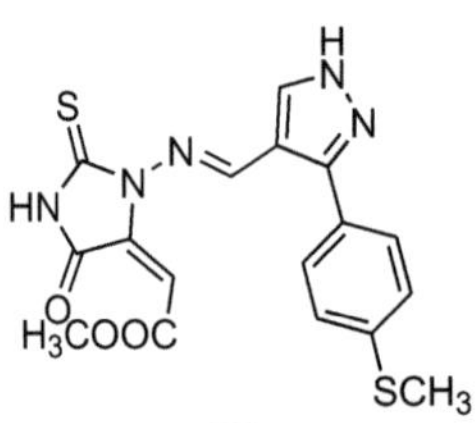

(65)

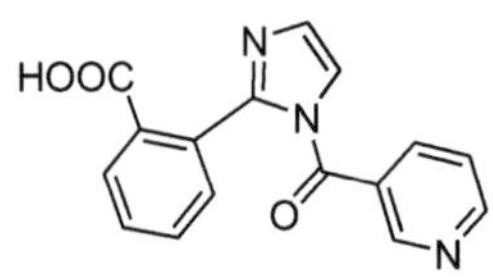

(66)

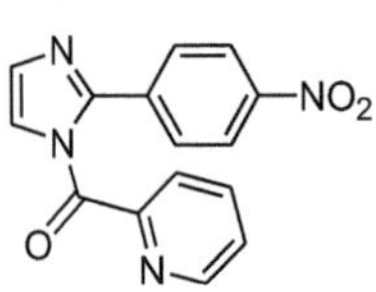

(67)

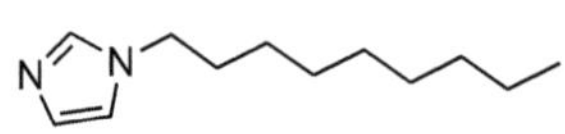

(68)

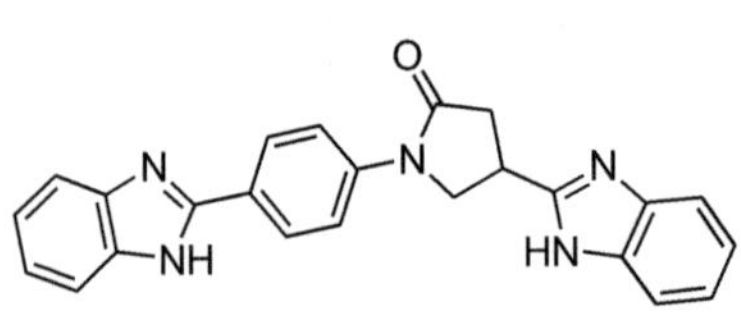

(69)

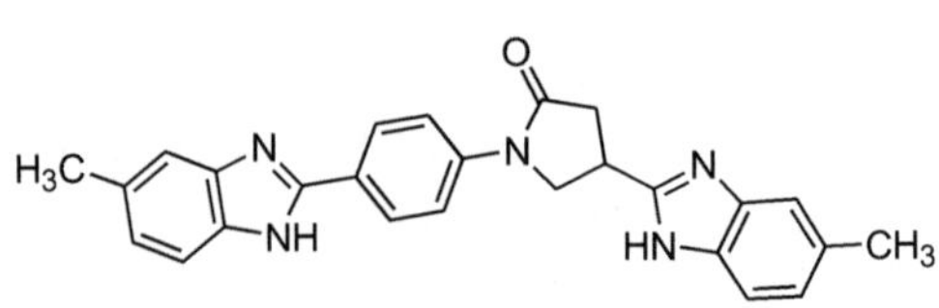

(70)

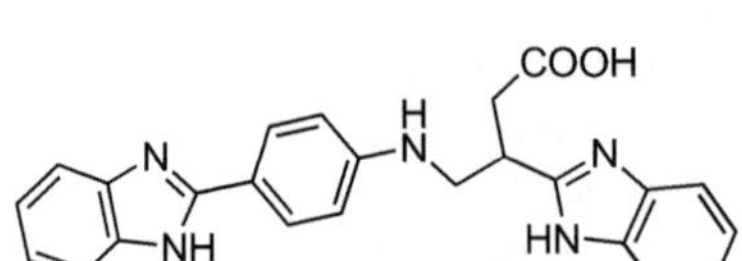

(71)

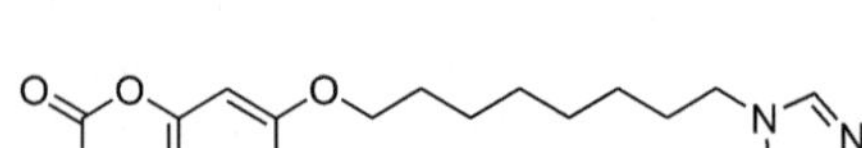

(72)

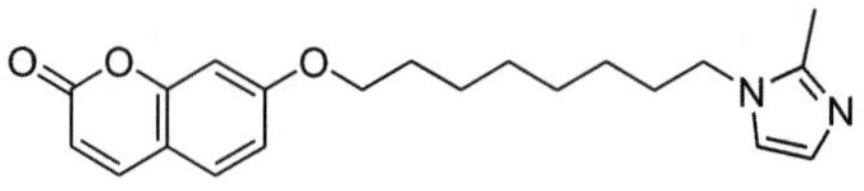

(73)

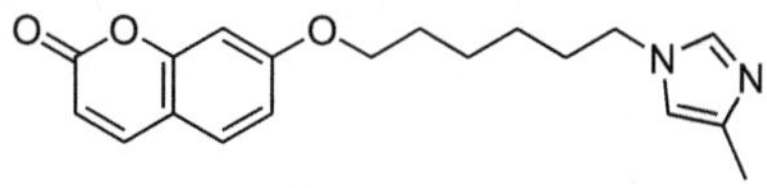

(74)

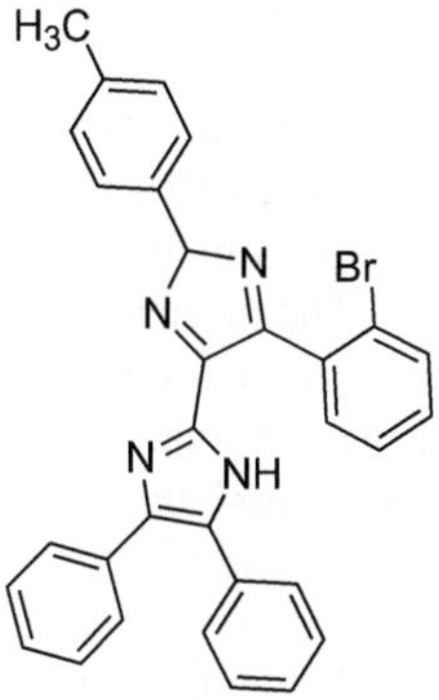

(75)

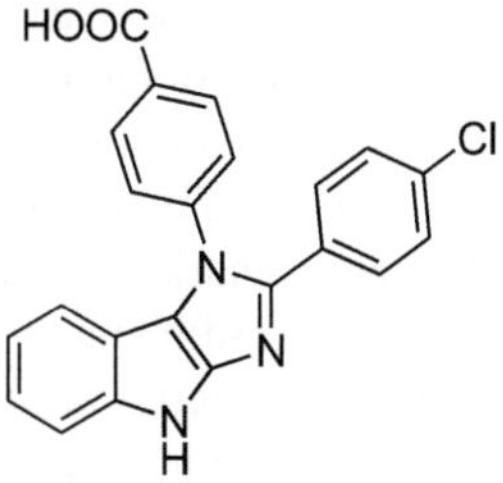

(76)

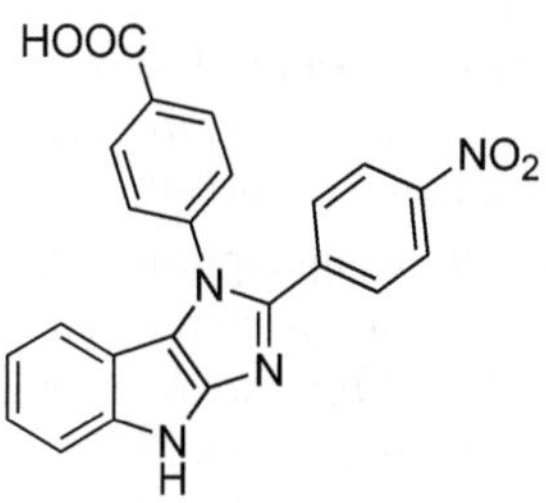

(77)

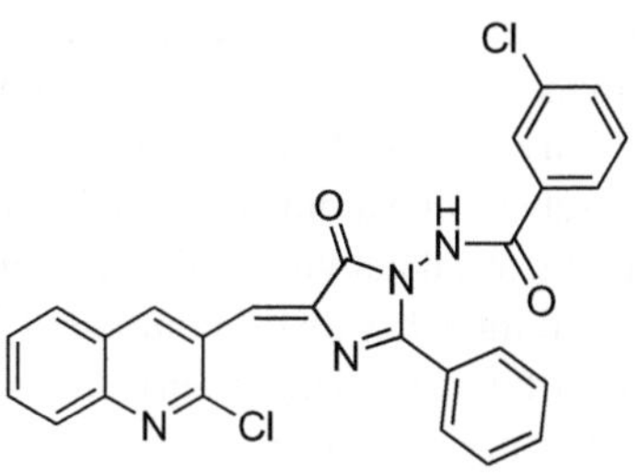

(78)

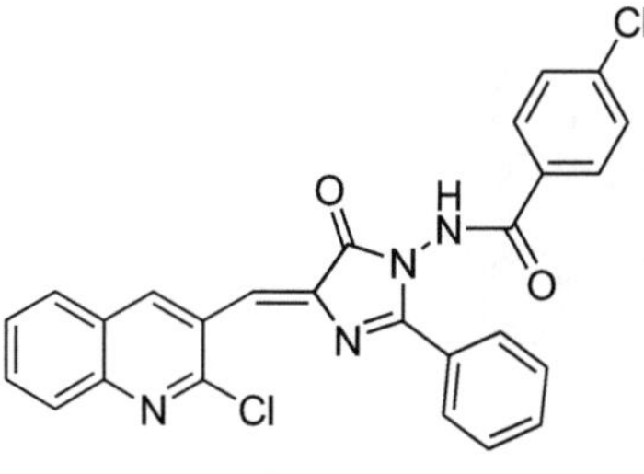

(79)

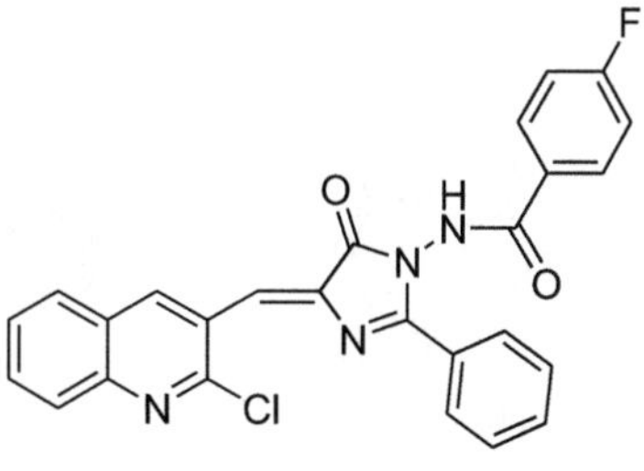

(80)

Sharma & colleagues created a number of new substituted imidazole derivatives using a solvent-free microwave method. The compounds were tested using a modified Kirby-Bauer disc diffusion method against two gram positve bacterial strains, *B. subti-*

lis & *S. aureus,* & two negative bacterial strains, *K. pneumonia* and *E. coli.* Compounds **76 & 77** demonstrated good antibacterial activity against gram negative bacterial strains, *K. pneumoniae* and *E. coli*, using the common medication ciprofloxacin [72]. Using conventional and microwave methods, ***Desai* et al.** synthesized N-(4-((2-chloroquinolin-3-yl)methylene)-5-oxo-2-phenyl-4,5-dihydro-1H-imidazol-1-yl)(aryl) amides and evaluated their antibacterial effects. Compared to the reference medication ampicillin, compounds **78, 79, 80, 81 & 82** which contain electron withdrawing groups, have a higher antibacterial activity (MIC 25 lg/mL) against the gram negative bacteria *E. coli* [73]. Novel 2-substituted aryl-4-fluoro phenyl-5-(2-Chloropyridinyl)-1H-imidazole derivatives have been produced via a four-step reaction strategy that starts with an ethanone-selenium dioxide reaction in dioxane and ends with aromatic aldehyde cyclization in the presence of ammonium acetate or acetic acid. The in-vitro antibacterial activity of these compounds against two gram positive *S. pyogenes and S. aureus* and two gram negative *P.aeruginosa and E. coli* bacteria was evaluated and compounds **83 & 84** shows remarkable anti bacterial activity [74]. ***Shobhashana* et al.** developed 6-substituted-3-(4,5-diphenyl-1H-imidazol-2-yl)-2-(4-substituted phenoxy) quinoline and tested its antibacterial efficacy against *B. subtilis, E. coli, C. tetani, S. pneumoniae*, and *S. typhi* employing the broth dilution method. Ciprofloxacin, ampicillin, and chloramphenicol were all utilized as positive controls. Compounds **85, 86 & 87** displayed substantial antibacterial action towards tested bacterial strains [75]. ***Selvan* et al.** discovered the N-(2-(1H-benzo[d]imidazol-2-yl)phenyl) substituted formimidoyl. The disc diffusion technique was used to assess the antibacterial effect against *S. aureus*, with ciprofloxacin serving as a positive control. Compounds **88 & 89** showed high antimycotic activity against *A. niger* using Nystatin as a reference drug [76]. ***Zala* et al.** synthesized 2-(substituted amino)-1-(2,4,5-triphenyl-1H-imidazol-1-yl) ethanone and examined its antibacterial activity against *S. aureus* and *E. coli* using ciprofloxacin as a standard. Out of all the derivatives, compound **90** shows strong antibacterial capability [77]. Tetra-substituted imidazole derivatives are created by ***Shahzad K.* et al.** employing a one-pot condensation process. Using Ciprofloxacin as a standard drug, antibacterial activity was assessed against *S. aureus, B. subtilis,* and *E. coli* and compound **91 & 92** presented promising results against all bacterial strains [78].

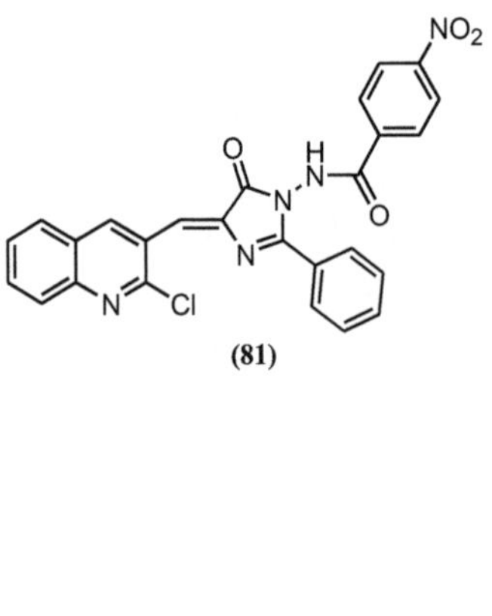

(81)

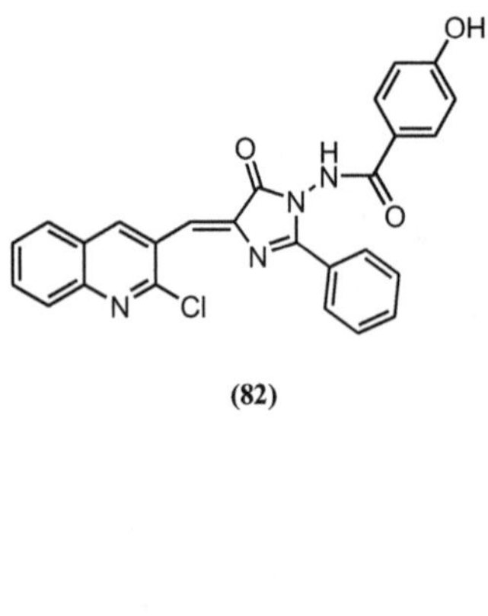

(82)

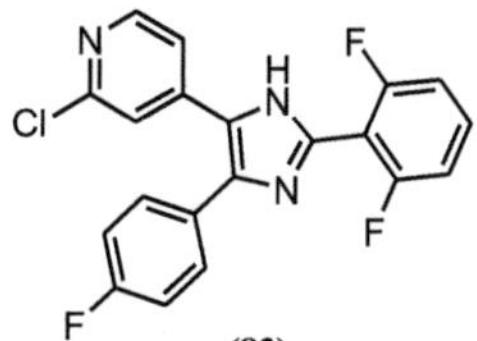
(83)

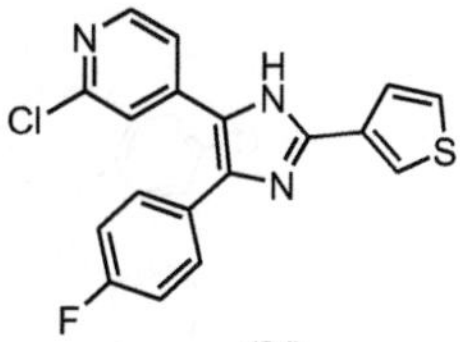
(84)

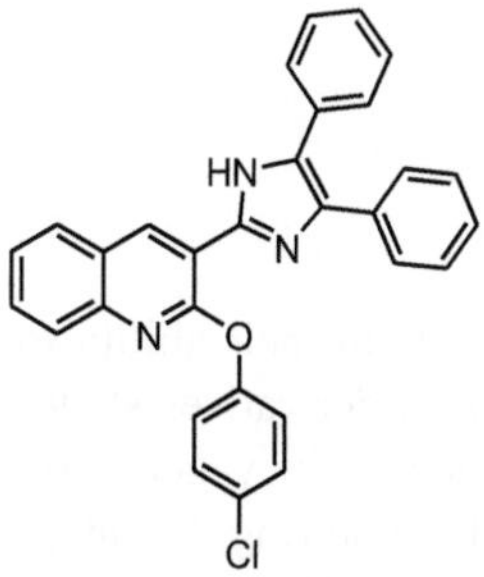
(85)

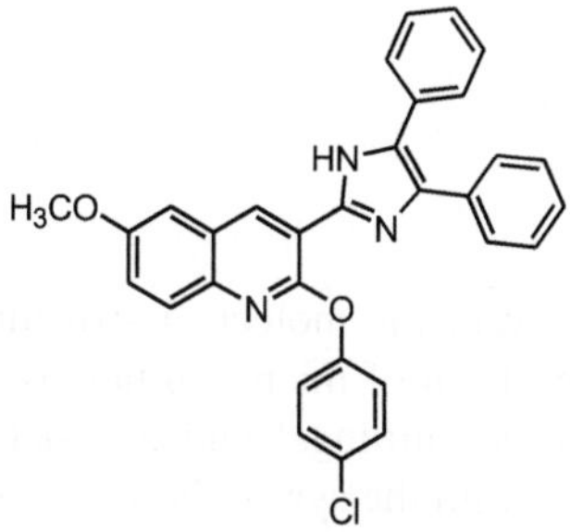
(86)

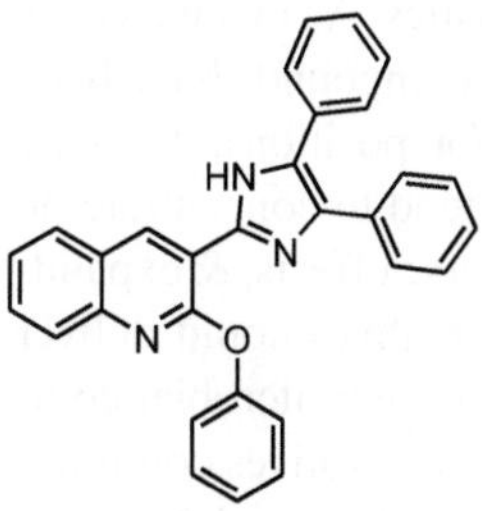
(87)

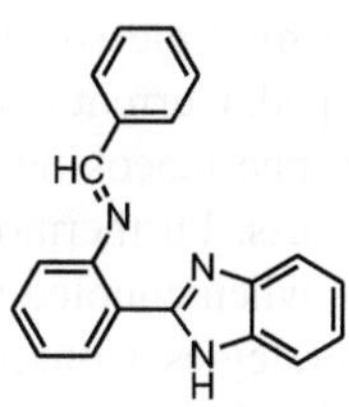
(88)

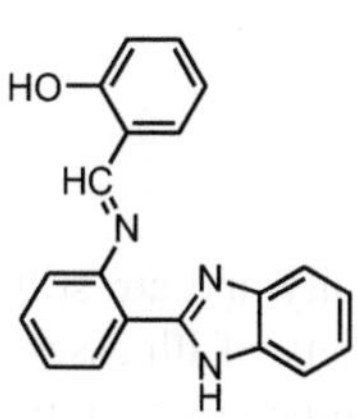
(89)

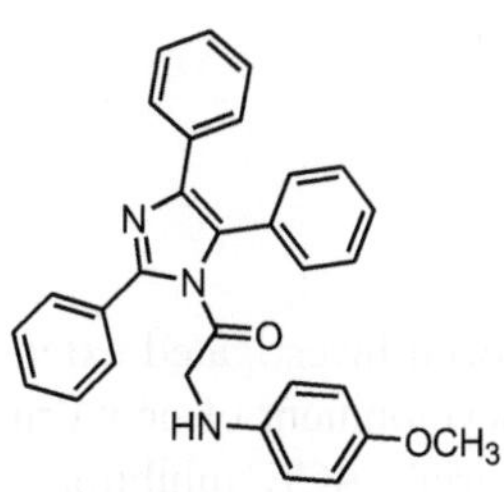
(90)

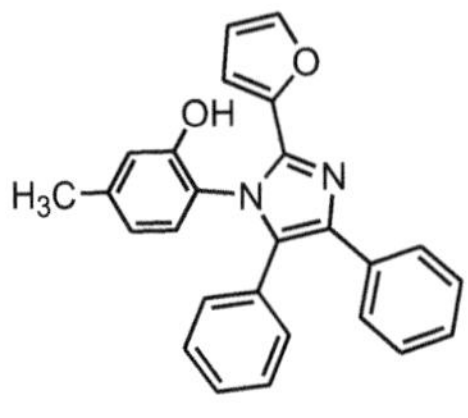

(91)

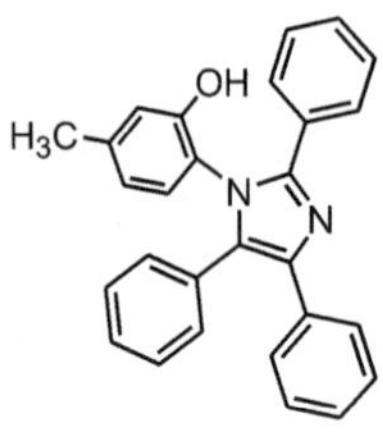

(92)

6 Conclusion

In conclusion, owing to their versatile molecular structure along with their ability to interact with an extensive spectrum of microbial targets, imidazole derivatives showcased considerable potential as antifungal and antibacterial agents. They generally accomplish this by interfering with the synthesis of nucleic acids, rupturing the integrity of cellular membranes, & preventing the functions of enzymes. The therapeutic management of fungal infections has led to extensive use of imidazole derivatives such clotrimazole, ketoconazole, & miconazole, whereas additional derivatives like econazole, metronidazole, & ornidazole offer antibacterial characteristics. As evidenced by the structure-activity relationship, a large number of remarkable compounds have been developed & assessed utilizing multiple substituent types and/or positions, & many more are expected to be developed. Current research endeavors tend to concentrate on enhancing imidazole derivatives pharmacokinetics, lowering adverse effects, & expanding their spectrum of effectiveness. Furthermore, imidazole derivatives could deliver enhanced therapeutic outcomes when coupled with supplementary antimicrobial compounds or novel drug delivery strategies. Consequently, imidazole derivatives continues to emerge as an intriguing family of pharmaceuticals for addressing bacterial & fungal conditions, nonetheless additional investigation must be conducted to maximize their potential in clinical practice.

7 Future Perspective

Imidazole compounds have been investigated extensively since antiquity and are still essential in facilitating the development of new drugs targeting a variety of illnesses, notably antifungal, antibacterial, ACE inhibitors, anticancer, anti-tubercular, anti-inflammatory and antiepileptic drugs. With the goal to accomplish more selectivity and minimal adverse effects, ongoing medicinal chemistry emphasizes on designing compounds with imidazole moieties. Economically accessible imidazole-based antibacterials that target topoisomerase II (DNA gyrase) and topoisomerase IV, the enzymes that supercoil and maintain the correct structure of DNA during replication, and imidazole-based antifungals that specifically target the 14-demethylase enzyme are gaining interest. Utilizing natural sources for the production of imidazole derivatives enhances accessibility, economy, and efficiency. By utilizing their varied physiological and pharmacological traits, medical researchers seek to develop new imidazole derivatives with remarkable pharmacodynamics and pharmacokinetics.

References

1. Sharma, V., Kumar, P., Pathak, D.: Biological importance of the indole nucleus in recent years: a comprehensive review. Heterocycl. Chem. **47**, 491–502 (2010)
2. Wan, Y., Li, Y., Yan, C., Yan, M., Tang, Z.: Indole: A privileged scaffold for the design of anticancer agents. Eur. J. Med. Chem. **183**, 111691 (2019)
3. Kabir, E., Uzzaman, M.: A review on biological and medicinal impact of heterocyclic compounds. Results Chem. **4**, 100606 (2022)
4. Jasiewicz, B., et al.: Indole derivatives bearing imidazole, benzothiazole-2-Thione or Benzoxazole-2-thione moieties—synthesis, structure and evaluation of their cytoprotective, antioxidant, antibacterial and fungicidal activities. Molecules. **28**, 708 (2023)
5. Dorababu, A.: Indole–a promising pharmacophore in recent antiviral drug discovery. RSC Med. Chem. **11**, 1335–1353 (2020)
6. Dadashpour, S., Emami, S.: Indole in the target-based design of anticancer agents: a versatile scaffold with diverse mechanisms. Eur. J. Med. Chem. **150**, 9–29 (2018)
7. Chen, Y., et al.: Synthesis and biological evaluation of indole-based peptidomimetics as antibacterial agents against Gram-positive bacteria. Eur. J. Med. Chem. **226**, 113813 (2021)
8. Wang, S., et al.: FDA-approved pyrimidine-fused bicyclic heterocycles for cancer therapy: Synthesis and clinical application. Eur. J. Med. Chem. **214**, 113218 (2021)
9. Mermer, A., Faiz, O., Demirbas, A., Demirbas, N., Alagumuthu, M., Arumugam, S.: Piperazine-azole-fluoroquinolone hybrids: conventional and microwave irradiated synthesis, biological activity screening and molecular docking studies. Bioorg. Chem. **85**, 308–318 (2019)
10. Mermer, A., et al.: Synthesis, biological activity and structure activity relationship studies of novel conazole analogues via conventional, microwave and ultrasound mediated techniques. Bioorg. Chem. **81**, 55–70 (2018)
11. Husain, A., Drabu, S., Kumar, N., Alam, M., Bawa, S.: Synthesis and biological evaluation of di-and tri-substituted imidazoles as safer anti-inflammatory-antifungal agents. J. Pharm. Bioallied Sci. **5**, 154 (2013)
12. Steel, H.C., Tintinger, G.R., Anderson, R.: Comparison of the anti-inflammatory activities of imidazole antimycotics in relation to molecular structure. Chem. Biol. Drug Des. **72**, 227–228 (2008)
13. Saini, M.S., Kumar, A., Dwivedi, J., Singh, R.: A review: biological significances of heterocyclic compounds. Int. J. Pharm. Sci. Res. **4**, 66–77 (2013)
14. Husain, A., Ajmal, M.: Synthesis of novel 1, 3, 4-oxadiazole derivatives and their biological properties. Acta Pharm. **59**, 223–233 (2009)
15. Chen, Q., Zhu, X.L., Jiang, L.L., Liu, Z.M., Yang, G.F.: Synthesis, antifungal activity and CoMFA analysis of novel 1, 2, 4-triazolo [1, 5-a] pyrimidine derivatives. Eur. J. Med. Chem. **43**, 595–603 (2008)
16. Kumar, J.R.: Review o f imidazole heterocyclic ring containing compounds with their biological activity. J. Hypertens. **10**, 11 (2010)
17. Reddy, D.R., Malleswari, K., Sreenivasulu, G., Sai, G.N., Rakesh, J.: Imidazole derivatives and its pharmacological activities. IJRAR. **9**, 98–107 (2022)

18. Debus, H.: Ueber die einwirkung des ammoniaks auf glyoxal. Justus Liebigs Ann. Chem. **107**, 199–208 (1858)
19. Romero, D.H., et al.: Synthesis of imidazole derivatives and their biological activities. J. Chem. Biochem. **2**, 45–83 (2014)
20. Wang, S., Zhao, L., Xu, Z., Wu, C., Cheng, S.: Novel nonlinearity–transparency–thermal stability trade-off of imidazole chromophores for nonlinear optical application. Mater. Lett. **56**, 1035–1038 (2002)
21. Forte, B., Malgesini, B., Piutti, C., Quartieri, F., Scolaro, A., Papeo, G.: A submarine journey: the pyrrole-imidazole alkaloids. Mar. Drugs. **7**, 705–753 (2009)
22. Verma, A., Joshi, S., Singh, D.: Imidazole: having versatile biological activities. J. Chem. **2013**, 329412 (2013)
23. Kleemann, A., Engel, J., Kutscher, B., Reichert, D.: Pharmaceutical substances. Thieme Verlag (2011)
24. Rautela, J., Bisht, A.S., Jakhmola, V., Ansori, A.N., Negi, D.: Microwave-assisted synthesis of new 4-amino acid substituted 1, 2, 4-triazole derivative derived from 1, 2, 3-oxadiazole nucleus and their anti-bacterial and anti-oxidant potential. RJPT. **16**, 2322–2326 (2023)
25. Kolbe, H.: Die Synthese der Imidazolverbindungen. Justus Liebigs Ann. Chem. **108**, 122–142 (1858)
26. Friedel, C.: Ueber sg gemischte Acetone. Justus Liebigs Ann. Chem. **108**, 122–125 (1858)
27. Fischer, E.: Zur Kenntniss der Purinstoffe. Ber. Dtsch. Chem. Ges. **2**, 106–110 (1869)
28. Griess, P.: Beiträge zur Chemie des Imidazols. J. Chem. Soc. **32**, 132–149 (1877)
29. Dale, H.H.: Histamine and its physiological effects. Pharmacol. Rev. **4**, 273–309 (1937)
30. Cataldi, M., Borriello, F., Granata, F., Annunziato, L., Marone, G.: Histamine receptors and antihistamines: from discovery to clinical applications. Chem. Immunol Allergy. **100**, 214–226 (2014)
31. Shepard, J.M., Bauer, G.E.: Imidazole Derivatives as Inhibitors of Enzyme Activity. J. Med. Chem. **8**, 670–675 (1965)
32. Borgers, M., Van den Bossche, H., De Brabander, M.: The mechanism of action of the new antimycotic ketoconazole. Am. J. Med. **74**, 2–8 (1983)
33. Anyanwu, E.C., Campbell, A.W., Ehiri, J.E.: Mycotoxins and antifungal drug interactions: implications in the treatment of illnesses due to indoor chronic toxigenic mold exposures. Sci. World J. **4**, 167–177 (2004)
34. Raptopoulou, C.P.: Metal-organic frameworks: Synthetic methods and potential applications. Materials. **14**, 310 (2021)
35. Rulhania, S., Kumar, S., Nehra, B., Gupta, G.D., Monga, V.: An insight into the medicinal perspective of synthetic analogs of imidazole. J. Mol. Struct. **1232**, 129982 (2021)
36. Khan, Z.K., Jain, P.: Antifungal agents and immunomodulators in systemic mycoses. Indian J. Chest Dis. Allied Sci. **42**, 345–356 (2000)
37. Torres-Rodríguez, J.M., et al.: In vitro susceptibilities of clinical yeast isolates to the new antifungal eberconazole compared with their susceptibilities to clotrimazole and ketoconazole. Antimicrob Agents Chemother. **43**, 1258–1259 (1999)
38. Jones, B.M., Geary, I., Lee, M.E., Duerden, B.: Comparison of the in vitro activities of fenticonazole, other imidazoles, metronidazole, and tetracycline against organisms associated with bacterial vaginosis and skin infections. Antimicrob Agents Chemother. **33**, 970–972 (1989)
39. Mahmoudabadi, A.Z., Drucker, D.B.: Effect of amphotericin B, nystatin and miconazole on the polar lipids of Candida albicans and Candida dubliniensis. Indian J. Pharmacol. **38**, 423–426 (2006)
40. Rani, N., Sharma, A., Kumar Gupta, G., Singh, R.: Imidazoles as potential antifungal agents: a review. Mini-Rev. Med. Chem. **13**, 1626–1655 (2013)

41. Clissold, S.P., Heel, R.C.: Tioconazole: a review of its antimicrobial activity and therapeutic use in superficial mycoses. Drugs. **31**, 29–51 (1986)
42. Lamb, D.C., Kelly, D.E., Baldwin, B.C., Kelly, S.L.: Differential inhibition of human CYP3A4 and Candida albicans CYP51 with azole antifungal agents. Chem Biol Interact. **125**, 165–175 (2000)
43. Borgers, M.: Mechanism of action of antifungal drugs, with special reference to the imidazole derivatives. Rev. Infect. Dis. **2**, 520–534 (1980)
44. Zhang, W., Ramamoorthy, Y., Kilicarslan, T., Nolte, H., Tyndale, R.F., Sellers, E.M.: Inhibition of cytochromes P450 by antifungal imidazole derivatives. Drug Metab Dispos. **30**, 314–318 (2002)
45. Wei, Q.L., Zhang, S.S., Gao, J., Li, W.H., Xu, L.Z., Yu, Z.G.: Synthesis and QSAR studies of novel triazole compounds containing thioamide as potential antifungal agents. Bioorg. Med. Chem. **14**, 7146–7153 (2006)
46. Rezaei, Z., Khabnadideh, S., Pakshir, K., Hossaini, Z., Amiri, F., Assadpour, E.: Design, synthesis, and antifungal activity of triazole and benzotriazole derivatives. Eur. J. Med. Chem. **44**, 3064–3067 (2009)
47. Khabnadideh, S., Rezaei, Z., Khalafinezhad, A., Pakshir, K., Heiran, M.J., Shobeiri, H.: Design and synthesis of 2-methyl and 2-methyl-4-nitro imidazole derivatives as antifungal agents, 31–36 (2009)
48. Ketan, Z.A., Hamed, Y.A., Saad, H.A.: Synthesis of some new heterocyclic compounds derived from imidazole and study of anti-microbial activity. Synthesis. **15**, 17 (2025)
49. Huang, Z., Chen, H., Zhang, X., Wang, R., Hu, C., Mao, Z.: Synthesis and antifungal evaluation of new azole derivatives containing 1, 2, 3-triazole. RSC Med. Chem. (2025)
50. Emami, L., et al.: In vivo antifungal activity and computational studies of some azole derivatives against Candida albicans. J. Chem. **2023**, 7834474 (2023)
51. Ramachandran, R., Rani, M., Senthan, S., Jeong, Y.T., Kabilan, S.: Synthesis, spectral, crystal structure and in vitro antimicrobial evaluation of imidazole/benzotriazole substituted piperidin-4-one derivatives. Eur. J. Med. Chem. **46**, 1926–1934 (2011)
52. Husain, A., Drabu, S., Kumar, N., Alam, M.M., Bawa, S.: Synthesis and biological evaluation of di-and tri-substituted imidazoles as safer anti-inflammatory-antifungal agents. J. Pharm. Bioallied Sci. **5**, 154–161 (2013)
53. Al-Ghamdi, H.A., et al.: Synthesis and biological evaluation of novel imidazole derivatives as antimicrobial agents. Biomolecules. **14**, 1198 (2024)
54. Padmavathi, V., Prema Kumari, C., Venkatesh, B.C., Padmaja, A.: Synthesis and antimicrobial activity of amido linked pyrrolyl and pyrazolyl-oxazoles, thiazoles and imidazoles. Eur. J. Med. Chem. **46**, 5317–5326 (2011)
55. Emami, S., et al.: Synthesis and antifungal activity of 1-[(2-benzyloxy)phenyl]-2-(azol-1-yl) ethanone derivatives: exploring the scaffold flexibility. Chem. Biol. Drug Des. **78**, 979–987 (2011)
56. Sadeghian, S., et al.: Evaluation of antibacterial and anticandidal activities of some imidazole, benzimidazole and benztriazole derivatives. TiPS. **8**, 75–84 (2022)
57. Verma, B.K., Kapoor, S., Kumar, U., Pandey, S., Arya, P.: Synthesis of new imidazole derivatives as effective antimicrobial agents. Indian J. Pharm. Biol. Res. **5**, 01–09 (2017)
58. Parameshwar, R., Ranganath, Y.S., Babu, V.H., Sandeep, G.: Synthesis and antifungal screening of some novel coumarin linked imidazole derivatives. Res. J. Pharm. Biol. Chem. Sci. **2**, 514–520 (2011)
59. Moraca, F., et al.: Synthesis, biological evaluation and structure–activity correlation study of a series of imidazol-based compounds as Candida albicans inhibitors. Eur. J. Med. Chem. **83**, 665–673 (2014)

60. Arana-Luna, L.L., et al.: Tinidazole efficacy for amebic colitis therapeutic prophylaxis in patients with de novo acute leukemia receiving intensive chemotherapy. Gac. Med. Mex. **155**, S22–SS7 (2019)
61. Changotra, R., Guin, J.P., Dhir, A., Varshney, L.: Decomposition of antibiotic ornidazole by gamma irradiation in aqueous solution: kinetics and its removal mechanism. Environ. Sci. Pollut. Res. **32591**, 602 (2018)
62. Andrei, G.S., Andrei, B.F., Roxana, P.R.: Imidazole derivatives and their antibacterial activity-a mini-review. Mini-Rev. Med. Chem. **21**, 1380–1392 (2021)
63. Verma, R.K., et al.: Green Synthesized Nanoparticles Targeting Antimicrobial Activities. Biointerface Res. Appl. Chem. **13**, 469 (2023)
64. Alghamdi, S.S., Suliman, R.S., Almutairi, K., Kahtani, K., Aljatli, D.: Imidazole as a promising medicinal scaffold: current status and future direction. Drug Des. Dev. Ther. **3289**, 312 (2021)
65. Jain, A.K., Ravichandran, V., Sisodiya, M., Agrawal, R.K.: Synthesis and antibacterial evaluation of 2–substituted–4, 5–diphenyl–N–alkyl imidazole derivatives. Asian Pac. J. Trop. Med. **3**, 471–474 (2010)
66. Vijesh, A.M., Isloor, A.M., Telkar, S., Peethambar, S.K., Rai, S., Isloor, N.: Synthesis, characterization and antimicrobial studies of some new pyrazole incorporated imidazole derivatives. Eur. J. Med. Chem. **46**, 3531–3536 (2011)
67. Narasimhan, B., Sharma, D., Kumar, P., Yogeeswari, P., Sriram, D.: Synthesis, antimicrobial and antimycobacterial evaluation of [2-(substituted phenyl)-imidazol-1-yl]-pyridin-3-yl-methanones. J. Enzyme Inhib. Med. Chem. **26**, 720–727 (2011)
68. Khabnadideh, S., Rezaei, Z., Khalafi-Nezhad, A., Bahrinajafi, R., Mohamadia, R., Farrokhroza, A.A.: Synthesis of *N*-alkylated derivatives of imidazole as antibacterial agents. Bioorg. Med. Chem. Lett. **13**, 2863–2865 (2003)
69. Sapijanskaitė-Banevič, B., Palskys, V., Vaickelionienė, R., Šiugždaitė, J., Kavaliauskas, P., Grybaitė, B., Mickevičius, V.: Synthesis and antibacterial activity of new azole, diazole and triazole derivatives based on p-aminobenzoic acid. Molecules. **26**, 2597 (2021)
70. Hu, Y., Shen, Y., Wu, X., Tu, X., Wang, G.X.: Synthesis and biological evaluation of coumarin derivatives containing imidazole skeleton as potential antibacterial agents. Eur. J. Med. Chem. **143**, 958–969 (2018)
71. Brahmbhatt, H., Molnar, M., Pavic, V.: Pyrazole nucleus fused tri-substituted imidazole derivatives as antioxidant and antibacterial agents. Karbala Int. J. Mod. Sci. **4**, 200–206 (2018)
72. Sharma, G.K., Pathak, D.: Microwave-assisted, solvent-free and parallel synthesis of some novel substituted imidazoles of biological interest. Chem. Pharm. Bull. **58**, 375–380 (2010)
73. Desai, N.C., Maheta, A.S., Rajpara, K.M., Joshi, V.V., Vaghani, H.V., Satodiya, H.M.: Green synthesis of novel quinoline based imidazole derivatives and evaluation of their antimicrobial activity. J. Saudi Chem. Soc. **18**, 963–971 (2014)
74. Singh, R.K., Bhatt, A., Kant, R.: Design and synthesis of some novel imidazole derivatives as potent antimicrobial & antimalarial agents. Pharm. Lett. **8**, 188–194 (2016)
75. Shobhashana, P.G., Prasad, P., Kalola, A.G., Patel, M.P.: Synthesis of imidazole derivatives bearing quinoline nucleus catalyzed by can and their antimicrobial, antitubercular and molecular docking studies. Res. J. Life Sci. bBoinform. Pharm. Chem. Sci. **4**, 175 (2018)
76. Selvan, S., Agalya, R., Maruthamuthu, S.R., Ruby, C., Stell, P.: Characterization and antimicrobial activity of newly synthesized benzimidazolyl compounds with docking studies. World J. Pharm. Res. **5**, 600–607 (2015)
77. Zala, S.P., Badmanaban, R., Sen, D.J., Patel, C.N.: Synthesis and biological evaluation of 2,4,5-triphenyl-1H-imidazole-1-yl derivatives. J. Appl. Pharm. Sci. **2**, 22 (2012)
78. Shahzad, K., Abbas, F., Pandey, D., Ajmal, S., Khadim, M., Tahir, M.U.: Synthesis, characterization and biological evaluation of novel tetrasubsituted Imidazole compounds (2020)

Quantitative Structure Pharmacophore–Activity Relationship of Tricyclic Antidepressant (TCAs) Utilizing Distance–Based Features

Sisir Nandi[1(✉)], Shuaib Hussain[1], Sandhya Dharkiya[1], Rukayya[1], and Ankita Sharma[1,2]

[1] Department of Pharmaceutical Chemistry, Global Institute of Pharmaceutical Education and Research (GIPER), Kashipur, India
sisir.iicb@gmail.com

[2] Sai Institute of Medical Sciences, Jaspur, Udham Singh Nagar, Uttarakhand, India

Abstract. Depression is a mental disorder which affects mood and pleasure and it is also affected by social, biological and psychological. Diagnosis and treatment of depression can improve the lives of sad people and their families and make them happier and more productive. At the time of COVID-19 pandemic which significantly increased the incidence and prevalence of anxiety and depression round the globe, there are significantly increased uses of these medications preoperatively. The current study aims is to develop quantitative structure-activity relationship (QSAR) and pharmacophore models based on compounds pharmacophoric distances, calculated using pharmacophoric features. In the present study, 36 Tricyclic antidepresent (TCA) compounds showing antidepressant activities were gathered from the published literature. Quantitative structure pharmacophore-activity relationship (QSPAR) was developed by taking calculated pharmacophoric distances and biological activity. Developed Triangular tropograph ΔABD (distance based) has shown significant distance based pharmacophore model responsible for antidepressant activities. The significant regression coefficient ($R = 0.53$) between biological activity and compounds pharmacophoric distances has been obtained. The proposed pharmacophoric designed model of effective congenric antidepressant drugs could benefit from using distance-based topography.

Keywords: Tricyclic antidepressants (TCAs) · 5-HT_2 receptor · Distance-based pharmacophore · Quantitative structure pharmacophore-activity relationship (QSPAR)

Type of Paper: Original Research

A. K. Saxena, A. Saxena (Eds.): GTHTM 2025, pp. 143–159, 2026.
https://doi.org/10.1007/978-3-032-12320-6_7

1 Introduction

A patient with depression experiences helplessness, hopelessness, tastelessness, and a sense of worthlessness *in* life. Depression is now become one of the most common conditions that renders a person immobile and is a condition that could be fatal. It is one of the disorders that is frequently ignored and causes many major issues, most individuals can benefit from the right therapy. Symptoms of depression include feeling of loneliness, sadness, difficulty in concentrating, sleep disturbaces, loss of interest or pleasure in hobbies or activities, lack of appetite, aching or pains, and making a person persistent sad. Depression can be classified as major depression, persistent depression, perinatal depression, seasonal depression, and depression with symptoms of psychosis. Typically medication, psychotherapy, or a combination of both can be used to treat depression [1]. According to the WHO report, an estimated 3.8% of people in the overall population suffer from depression, including 5.7% of people over 60 and 5% of adults (4% of males and 6% of women) population. Around 280 million people globally suffer from depression. Approximately 50% of women are more likely to experience depression than men. Every year, around 700,000 people commit suicide due to depression [2–4]. The reason for depression in children and adolescents may be due to various education-related difficulties, relationship issues with parents or at home, family-related issues, economic difficulties, and other factors [5]. Tricyclic antidepressants (TCAs), Selective Serotonin Reuptake Inhibitors (SSRIs), Serotonin/Norepinephrine Reuptake inhibitors (SNRIs), Atypical antidepressants, Serotonin modulators, Monoamine oxidase inhibitors (MAOIs), NMDA antagonists are given as antidepressant agents [6].

The present study is focused on the tricyclic antidepressant agents. Simply tricyclic antidepressants are made up of 2 phenyl rings joined to either a 7-membered or 6-membered central ring with a side chain. The central ring may contain one, several, or no heteroatoms, and it may or may not be saturated. The side chain having 2–3 carbon atoms with a terminal secondary, tertiary amino group may be attached to any one of the atoms of the central ring [7].

The targets of TCAs are monoamine oxidases such as norepinephrine, serotonin (5-HT), and dopamine. TCAs may antagonize target receptors or inhibit them in a non-selective manner. TCAs work by preventing serotonin and norepinephrine from being reabsorbed in the presynaptic terminals, which raises their concentrations in the synaptic cleft. The antidepressant action of the drug may be influenced by elevated amounts of norepinephrine and serotonin in the synapses. The mechanism of action of TCA in the treatment of headache and neuropathic pain is thought to be the suppression of norepinephrine and serotonin reuptake [8]. Serotonin and norepinephrine reuptake inhibitors may be utilized for a variety of clinical purposes, including as the first-line therapy for depression and anxiety as well as the management of chronic pain. Nor-epinephrine and serotonin reuptake inhibition frequently display dose and agent-dependent effects. Prescriptions for these drugs are being written much more frequently for perioperative use as the COVID-19. Epidemics significantly raised the incidence and prevalence of depression and anxiety across the nation [9].

TCAs have undergone SAR, however pharmacophore and QSAR based on structural features have largely remained unexplored. Therefore, the development of QSAR and pharmacophore models based on distances calculated using pharmacophoric features is the goal of the current study.

2 Material and Methods

2.1 Biological Activity Data

In the current investigation, several numbers of 36 TCA compounds showing antidepressant activities experimented on [3H] ketanserin labeled cloned at 5-HT_{2A} site were gathered from the published literature Human Clone Receptor experiments demonstrated the reduction of 5-HT reuptake and antagonistic effects on receptors. Table 1 contains the chemical structures of all the compounds as well as their biological activity ki (μM) values, which indicate how efficiently they impede re-uptake. Activity ranges for TCAs' amines were 0.7 to 4125 μM [10].

2.2 Computational Methods

2.2.1 Structure Optimization. Structural optimization was utilized to make the chemical structure as energetically stable as possible. Chem Draw software was used to draw the 2D structure of 36 TCAs. 2D structures were optimized into three-dimensional (3D) utilizing the MM2 force field and a 0.001 dielectric constant utilizing Chem3D Ultra [11]. These stable conformations with the lowest energy were taken into account when computing the theoretical structural descriptor for additional QSAR modelling.

2.2.2 Pharmacophoric Modelling. Pharmacophore means an assembly of 3D features responsible for the best supramolecular interaction with a specific biological target to obtain the biological action. A collection of 3-dimensional properties called pharmacophores are required for a group of compounds to be bioactive. It is useful to determine how many characteristics there are in, when conforming to variable spatial zones of binding shared by all active molecules but not the inactive compound [12–14]. As a result, the 3D pharmacophore is predicated on the idea that all of the structurally different compounds can attach to the biological target similarly. The six built-in pharmacophoric features such as hydrogen bond acceptor and donor, the moiety's hydrophobicity, aromatic rings, positive ionization (cation), and negatively ionization (anion) were used to define the chemical characteristics among the highly active ligands that could help non-covalent connecting the ligand to its target receptor. In the design of ligand-based drugs physiologically active compounds, pharmacophore shape can be a crucial strategy [15, 16].

The model was based on pharmacophoric features and created to concentrate on the structural characteristics of TCAs (cyproheptadine) congeners that are required for the active suppression of serotonin. In the present study, BIOVIA software has been utilized to develop pharmacophoric features [17, 18]. To explain how TCAs antagonists

Table 1. Biological Activity Data of 36 Tricyclic antidepressants.

Compound No.	Structure	Ki (μM)
1	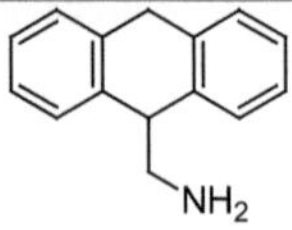	20
2	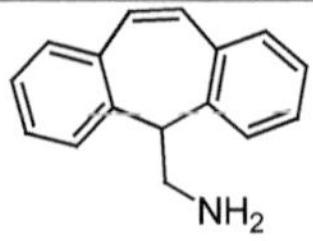	4125
3	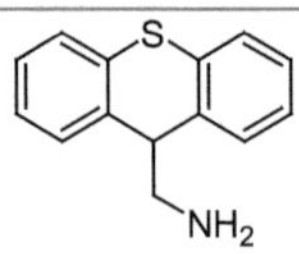	65
4	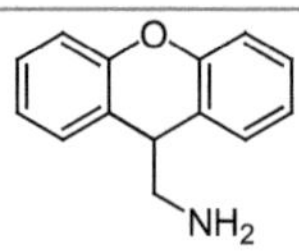	170
5	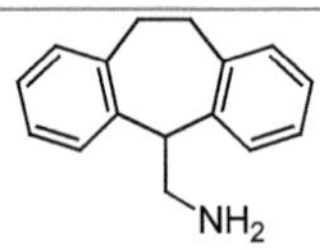	112
6	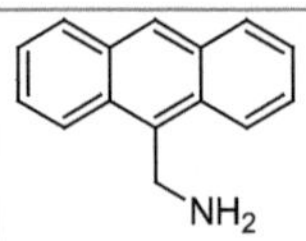	5700
7 (cyproheptadine) Reference compound	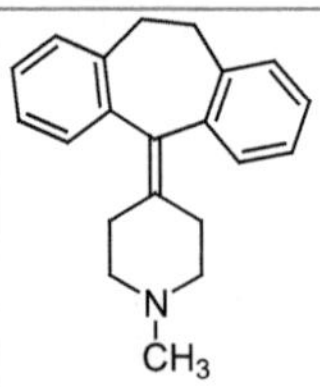	0.7
8	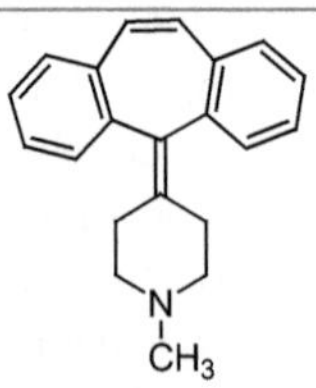	1.6
9	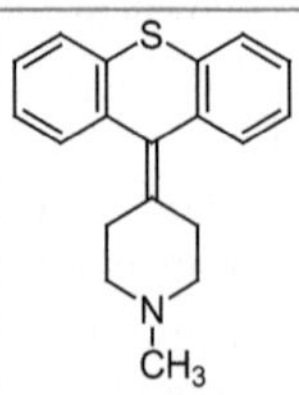	2.5

(continued)

Table 1. (continued)

Compound No.	Structure	Ki (μM)
10	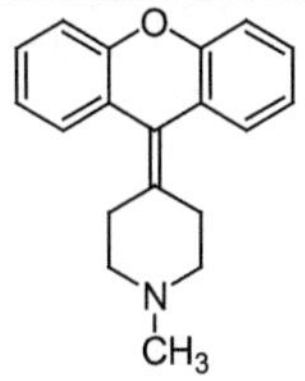	4.0
11	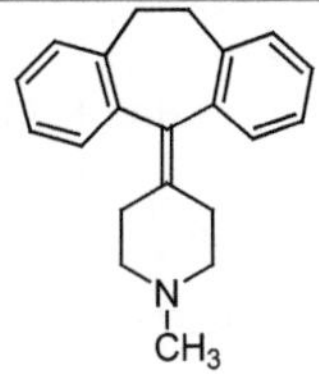	9.0
12	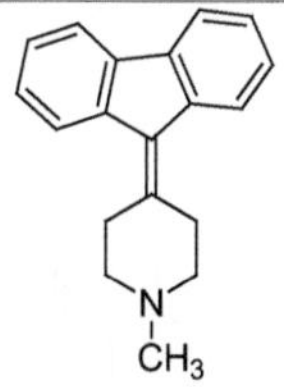	13
13	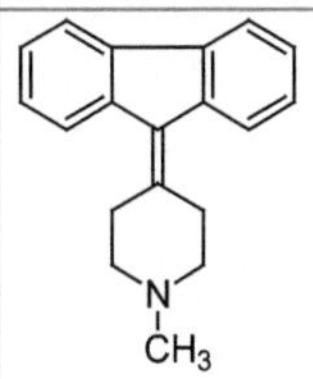	199
14	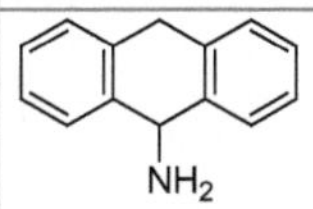	1200
15	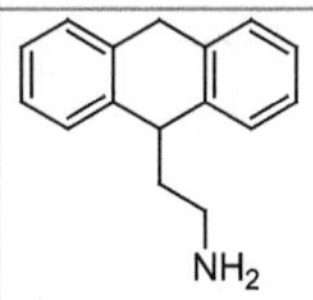	20
16	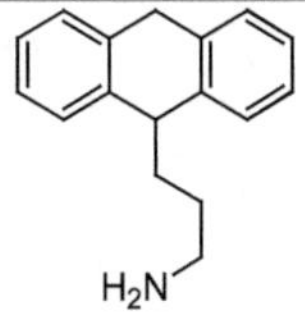	480
17	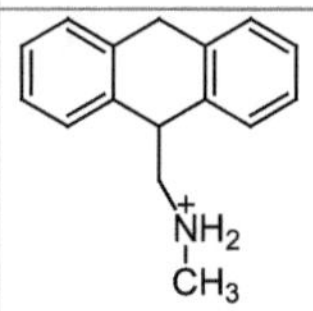	52

(continued)

Table 1. (continued)

Compound No.	Structure	Ki (μM)
18	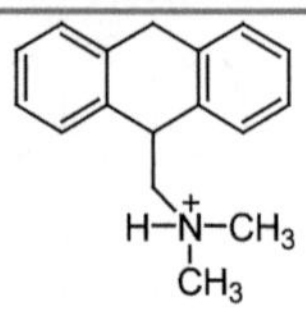	540
19	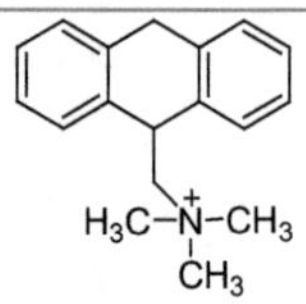	4000
20	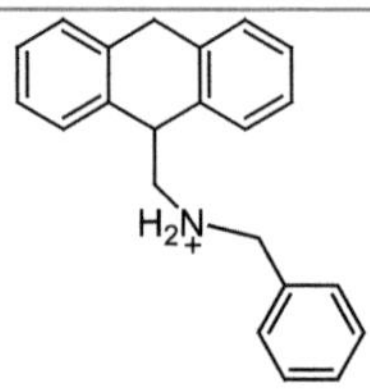	721
21	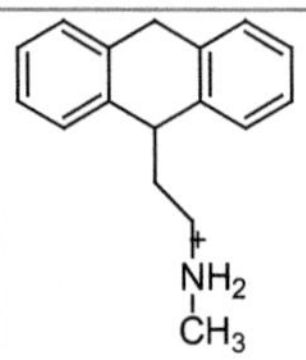	13
22	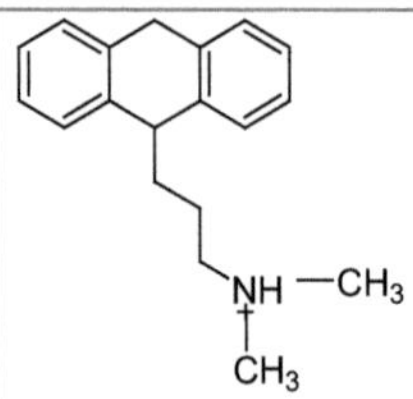	22
23	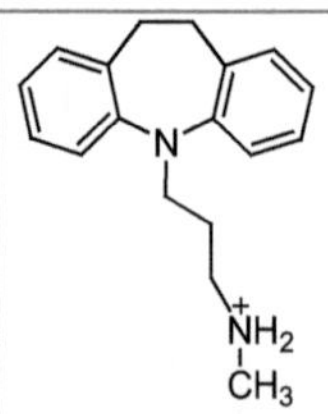	392
24	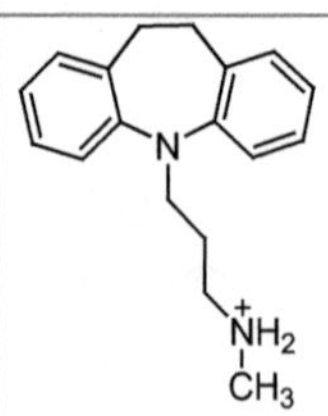	160

(continued)

Table 1. (continued)

Compound No.	Structure	Ki (μM)
25 (Imipramine)	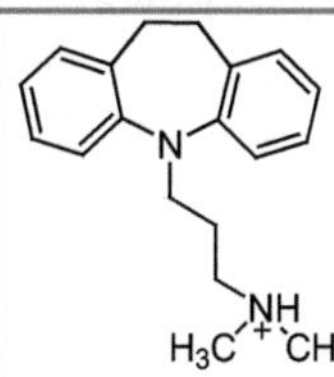	140
26	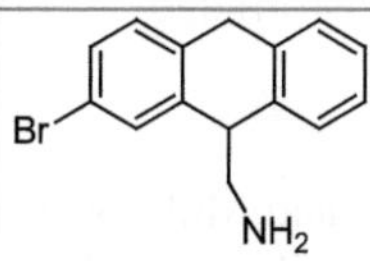	1.3
27	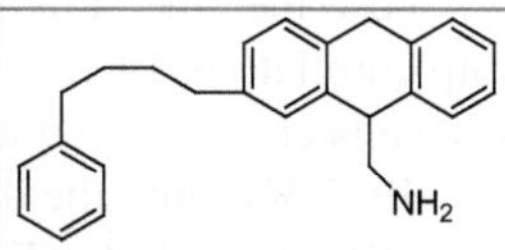	3.2
28	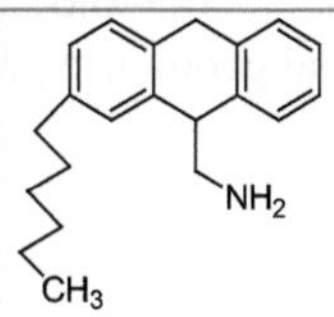	7.0
29	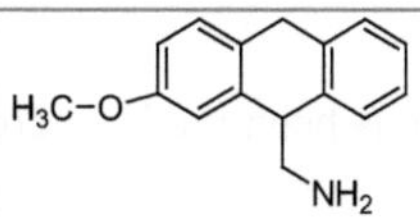	7.5
30	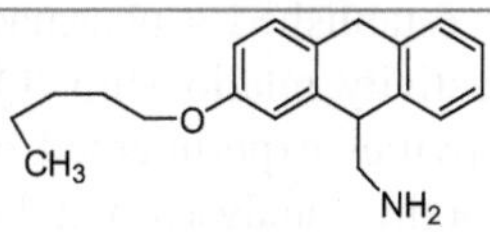	23
31	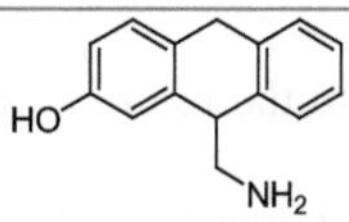	107
32	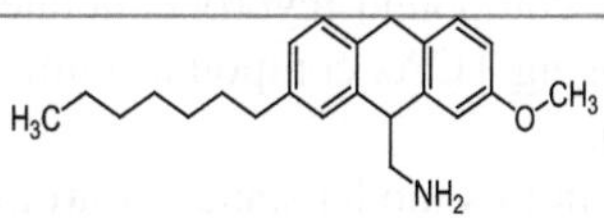	43
33	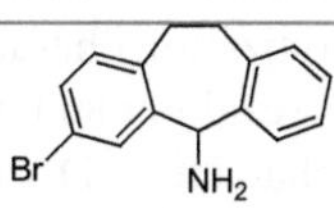	37
34	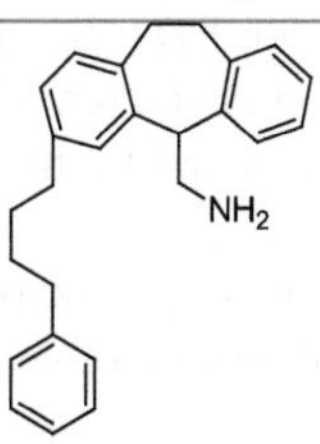	81

(continued)

Table 1. (continued)

Compound No.	Structure	Ki (μM)
35	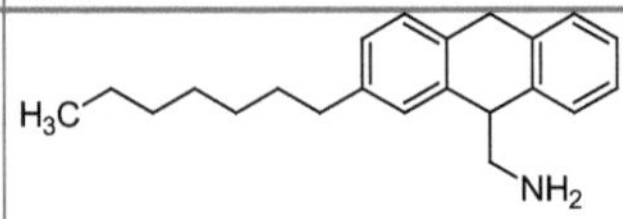	630
36	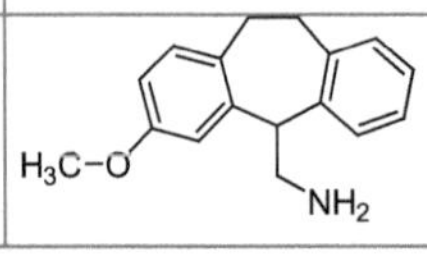	800

bind, Andersen et al. [1994] [19] and Mokrosz et al. [1995] [20] represented a general pharmacophoric model to describe the binding at 5-HT_2 receptors. For the structure-activity relationship correlations of well-known antagonists, several pharmacophore models for TCAs have been developed and the distance (angstrom) between two aromatic rings (AB = 4.149–5.360) and the distances between each aromatic ring and the basic amine nitrogen (BD = 3.429–6.672, DA = 3.686–7.685) are frequently used to characterize the fundamental geometric properties [21].

3 Results and Discussion

In the current work, an effort has been made to create general pharmacophores of various 36 TCAs congeners.

To this end, the author established a pharmacophoric distance-based quantitative structure pharmacophore-activity relationship (QSPAR) model using 36 compounds showing antidepressant activities experimented on [3H] ketanserin labeled cloned at 5-HT_{2A} site followed by outlier analyses and further development of the validated QSPAR model.

By matching the compounds to the references, a general pharmacophore based on alignment has been done.

The highest active compound (cyproheptadine) in the selected data was used as a reference when choosing TCAs compound numbers 1–36 for this study's pharmacophore model creation.

Vertices A, and B in this model stand for aromatic hydrophobic, whereas D stands for H-bond acceptor and positive ionization characters respectively.

According to the pharmacophore alignment model (Fig. 1), AB is the distance between 2 aromatic hydrophobic characters. DA is the distance between the first aromatic and positive ionization character. BD is the distance between the second aromatic character and the positive ionization character. The middle ring should be connected with 3 carbons followed by a tertiary amine in nature that gives positive ionization character.

For the selected 36 TCA compounds, specific pharmacophoric distances of AB, BD, and DA have been computed and Shown in Table 2 with their biological activity.

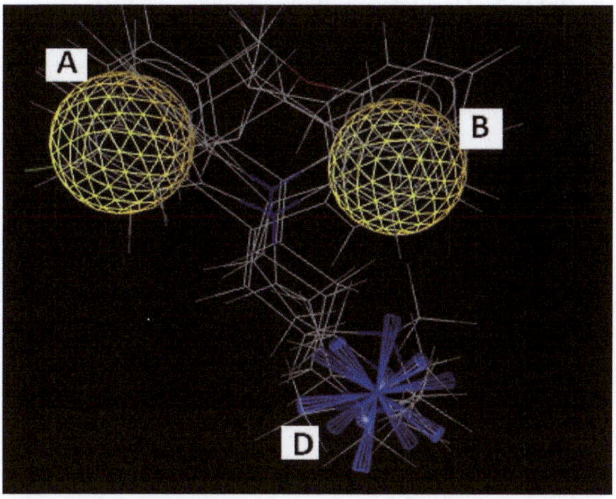

Fig. 1. TCAs (dibenzoazepine) pharmacophore alignment model.

3.1 Establishment of Pharmacophoric Distance Based Quantitative Structure Pharmacophore-Activity Relationship (QSPAR) Model

The pharmacophore features for all 36 compounds were generated with the help of BIOVIA software. Then distances between these 3D features are computed. It was also shown in the following Fig. 2.

In the present study, the calculated pharmacophoric distances for above 36 compounds are within the reference range as proposed by Andersen et al. [1994] and Mokrosz et al. [1995] i.e. the distance between two aromatic rings (AB = 4.6–7.3 Å) and each aromatic ring with the nitrogen in the basic amine (BD = 5.2–8.4 Å, DA = 5.7–8.5 Å)] [19, 20]. For 36 TCA compounds, the General Pharmacophore model can be represented by the separation between two aromatic rings as AB = 4.149–5.360 Å, the separation between the first aromatic ring and ionization character as BD = 3.429–6.672 Å, and the separation between the second aromatic ring and ionizing character as DA = 3.686–7.685 Å.

In our investigation, AB distance is almost within the reference range of Anderson but BD and DA are almost as per the standard reference studied by Nandi et al. [AB = 5.222–6.446 Å, BD = 3.695–7.257 Å, DA = 3.105–7.771 Å] which are congruent with these congeners' experimental observation] [22].

For the development of the Quantitative Structure Pharmacophore-Activity Relationship (QSPAR) model, these distances are taken as independent variables and biological activity (pKi) of the 36 compounds is selected as a dependent variable using MINITAB software utilizing multiple linear regressions [23].

The regression equation is as follows:

$$\mathrm{pKi} = -3.18 + 0.113\,\mathrm{AB} + 0.529\,\mathrm{BD} - 0.287\,\mathrm{DA} \tag{1}$$

$$N = 36, R = 0.38$$

Table 2. Computation of pharmacophoric distances.

Sr. No.	AB	BD	DA	pKi
1	4.906	3.750	5.156	−1.30103
2	4.870	3.893	5.220	−3.61542
3	5.145	3.694	5.165	−1.81291
4	4.766	3.804	5.161	−2.23045
5	4.510	4.174	5.267	−2.04922
6	4.944	4.222	4.713	−3.75587
7	5.219	6.579	6.438	0.154902
8	5.222	6.480	6.480	−0.20412
9	5.217	6.530	6.531	−0.39794
10	4.781	6.672	6.672	−0.60206
11	5.217	6.438	6.572	−0.95424
12	4.276	4.187	4.726	−1.11394
13	4.280	5.515	5.501	−2.29885
14	4.899	3.665	3.686	−3.07918
15	4.878	5.401	6.190	−1.30103
16	4.877	5.701	6.543	−2.68124
17	4.896	3.953	5.161	−1.71600
18	4.834	4.276	5.182	−2.73239
19	4.869	3.867	5.171	−3.60206
20	4.893	3.983	5.169	−2.85794
21	4.880	5.443	6.207	−1.11394
22	4.882	6.308	7.551	−1.34242
23	5.190	5.815	7.670	−2.59329
24	5.360	6.292	7.682	−2.20412
25	5.225	5.868	7.625	−2.14613
26	4.908	3.752	5.156	−0.11394
27	4.904	3.753	5.185	−0.50515
28	4.912	3.745	5.154	−0.8451
29	4.908	3.744	5.155	−0.87506
30	4.915	3.796	5.159	−1.36173
31	4.905	3.752	5.156	−2.02938
32	4.915	3.796	5.159	−1.63347
33	4.522	3.774	3.812	−1.56820
34	5.149	3.523	5.101	−1.90849
35	5.264	3.429	5.114	−2.79934
36	5.256	3.430	5.110	−2.90309

It model showed an R-value of 0.38, which is not good.

To detect nonlinearity, unequal error variances, and outliers present in the selected data set, standardized residual vs. fitted values were plotted. The plot (Fig. 3) showed that compounds having high residual values were predicted as 2, 6, 14, and 27 which were treated as outliers.

Compounds with high residual values were deleted and again regression model was developed as follows:

$$pKi = -4.32 + 0.733\,AB + 0.852\,BD - 0.868\,DA \tag{2}$$

$$N = 32, R = 0.53$$

It produces an R-value of 0.53, which is good.

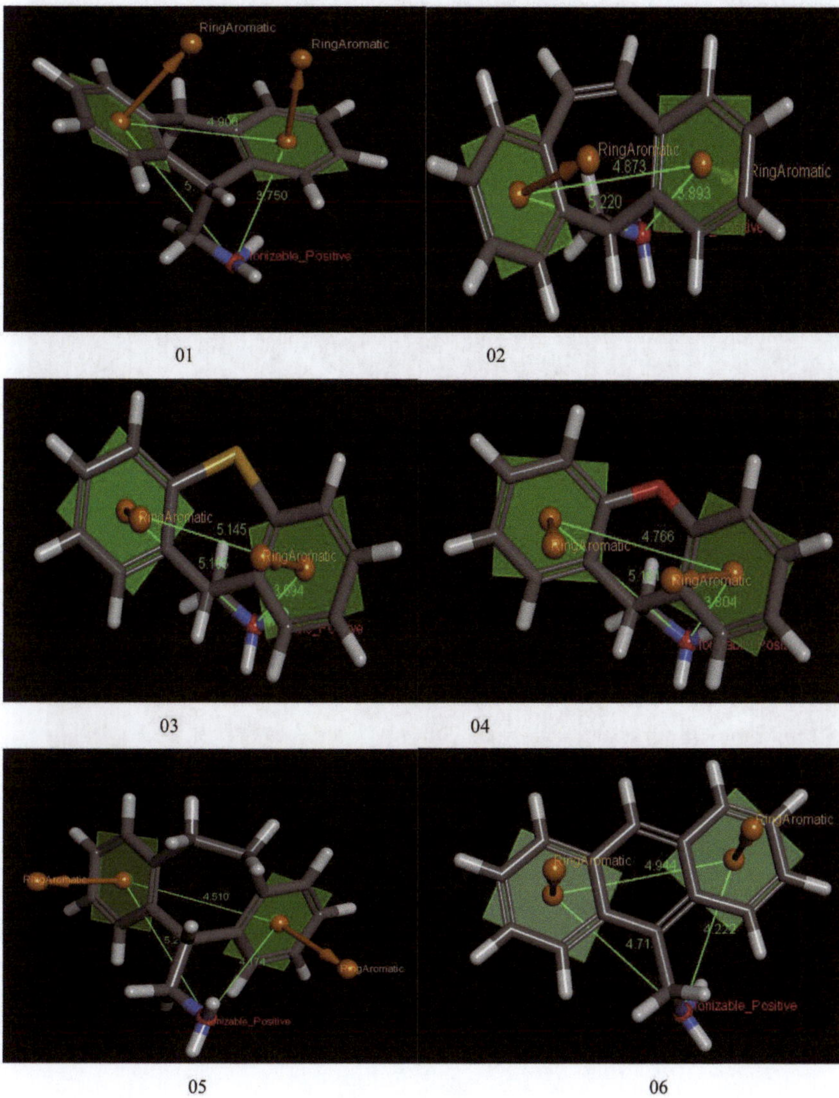

Fig. 2. Pharmacophore feature and distanced-based models for 36 compounds.

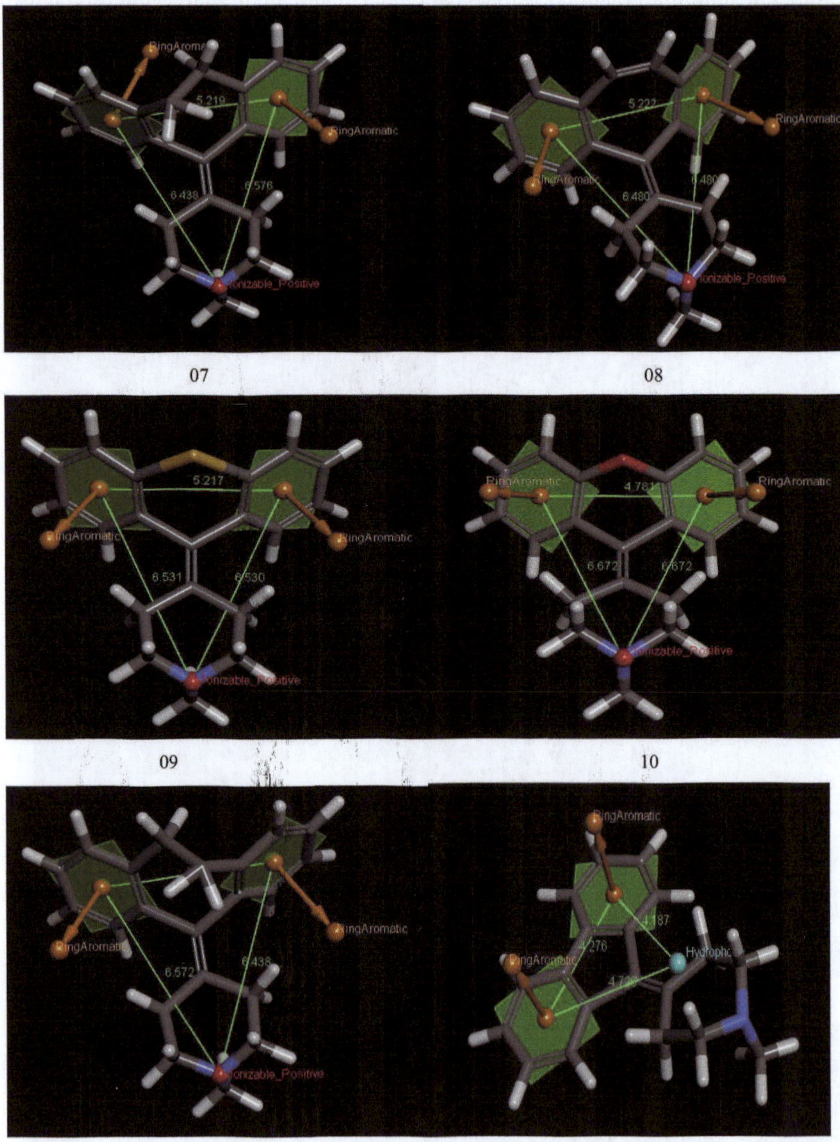

Fig. 2 (continued)

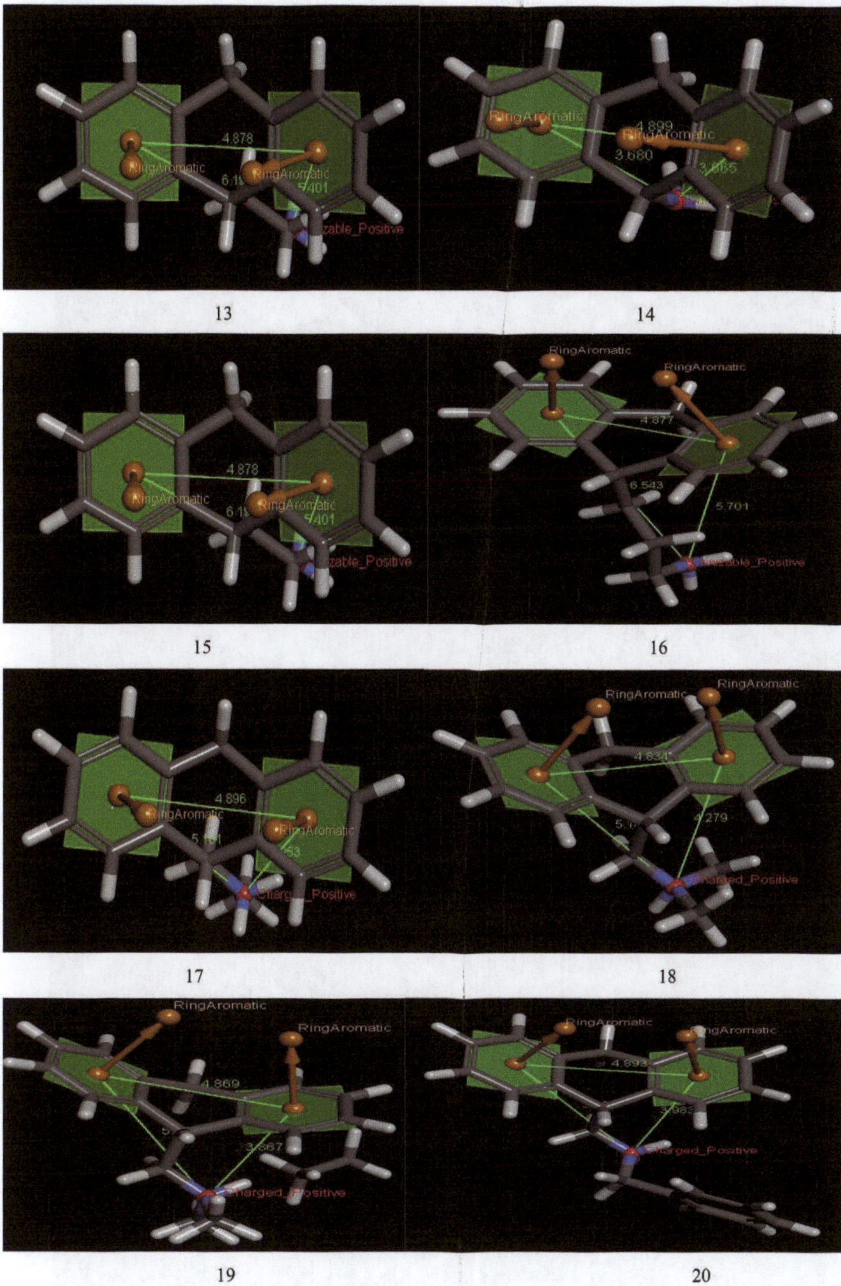

Fig. 2 (continued)

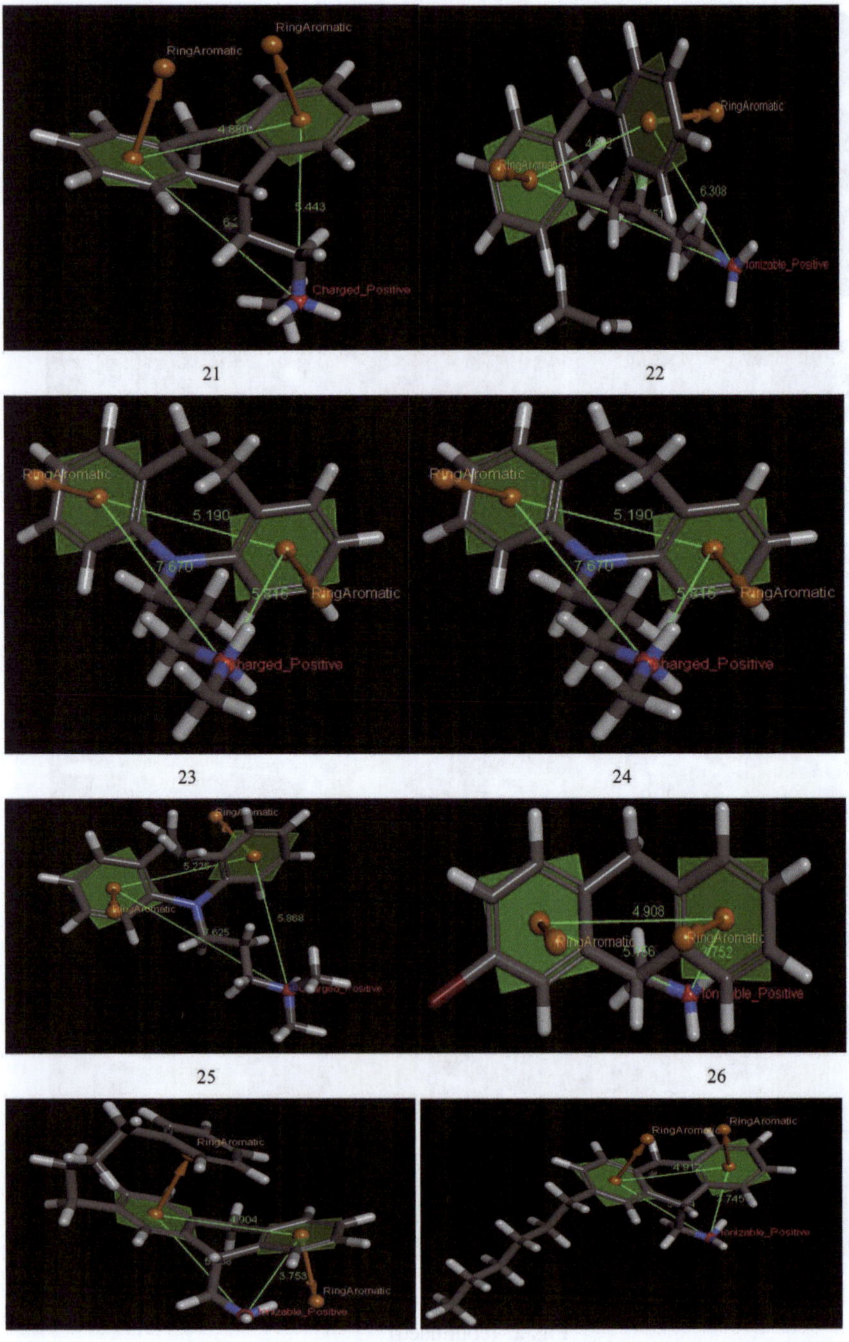

Fig. 2 (continued)

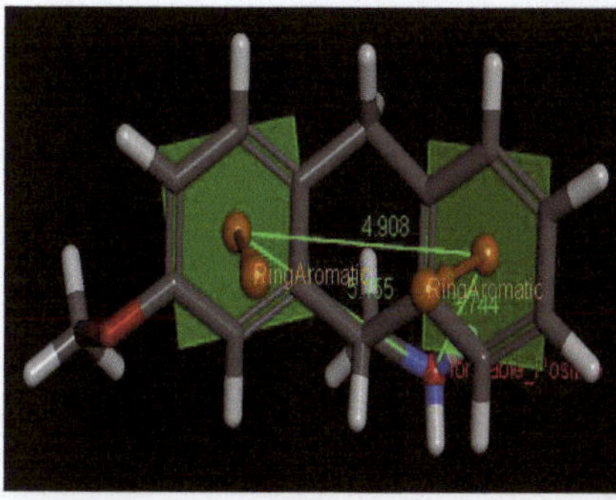

29

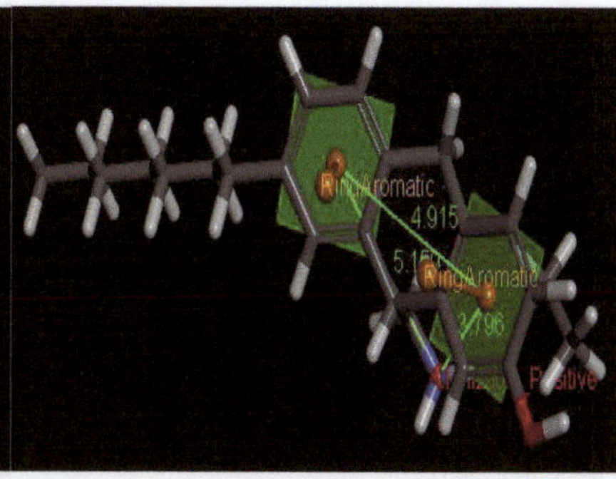

30

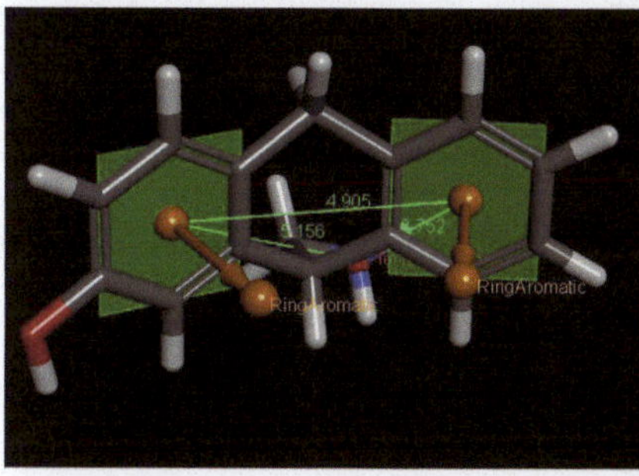

31

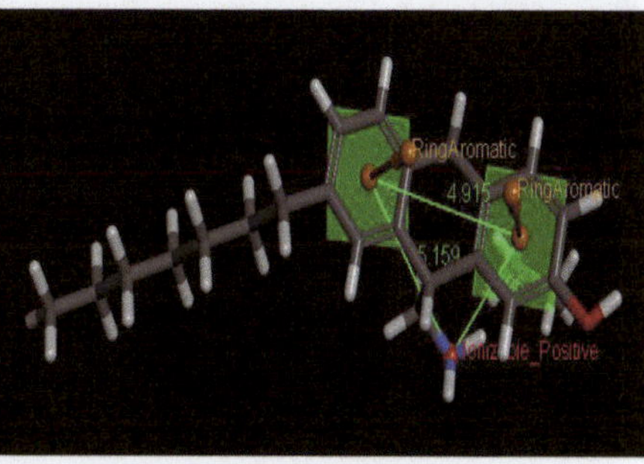

32

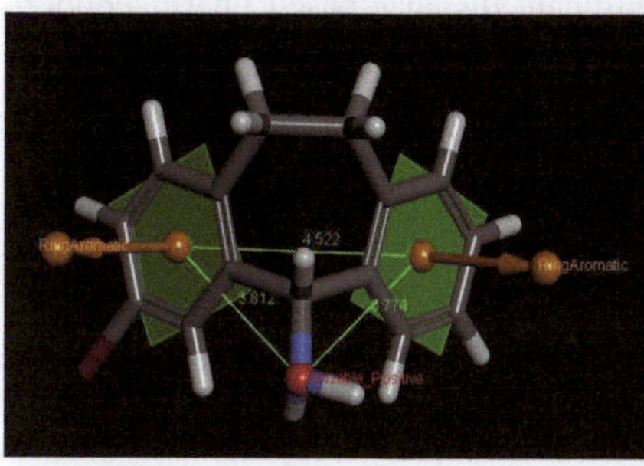

33

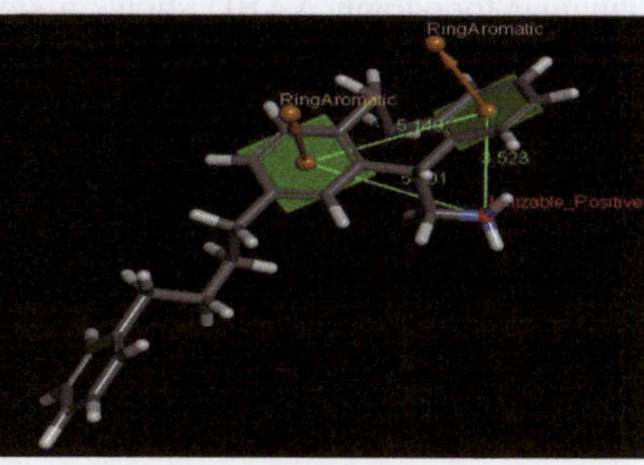

34

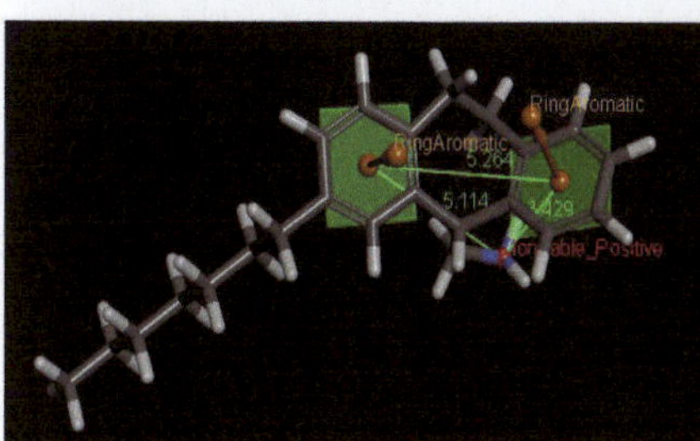

35

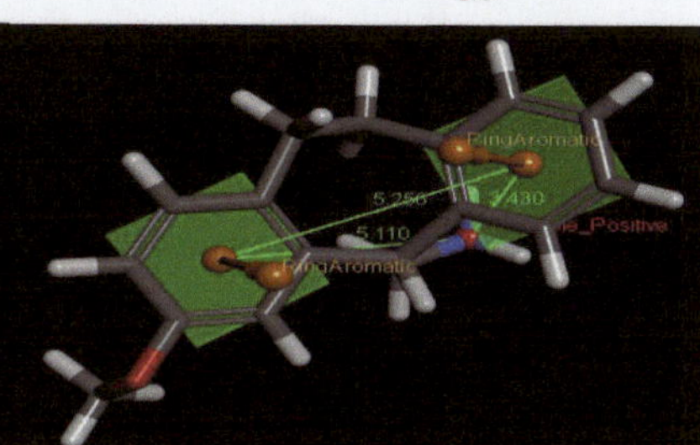

36

Fig. 2 (continued)

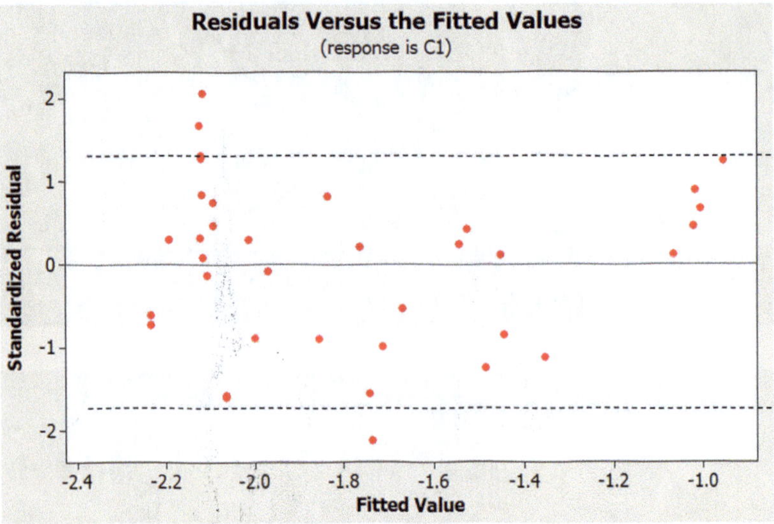

Fig. 3. Prediction of outlier by plotting fitted value versus standardized residual.

4 Conclusion

The triangular topograph ΔABD shows a significant distance-based pharmacophore model responsible for antidepressant activities. It has been statistically validated by developing a quantitative structure-pharmacophore activity relationship model showing a regression coefficient (R) value of 0.53, which is significant. Some compounds such as 2, 6, 14, and 27, were not fitted within the QSPAR model, so these were predicted as outliers.

It was seen that all studied compounds' pharmacophoric distances were within the reference range. The proposed pharmacophoric design for effective congeneric antidepressant drugs could benefit from using distance-based topograph.

References

1. NIMH: Depression: The National Institute of mental health. https://www.nimh.nih.gov/health/publications/depression. Accessed on: 18 Feb 2025 (2021)
2. Grover, S., Dutt, A., Avasthi, A.: An overview of Indian research in depression. Indian J. Psychiatry. **52**, S178–S188 (2010). https://doi.org/10.4103/0019-5545.69231
3. Beirão, D., Monte, H., Amaral, M., Longras, A., Matos, C., Villas-Boas, F.: Depression in adolescence: a review. Middle East Curr. Psychiatry. **27**, 50 (2020). https://doi.org/10.1186/s43045-020-00050-z
4. WHO: Depressive disorder. World Health Organisation (2023). https://www.who.int/health-topics/depression#tab=tab_1
5. Grover, S., Raju, V.V., Sharma, A., Shah, R.: Depression in children and adolescents: a review of Indian studies. Indian J. Psychol. Med. **41**, 216–227 (2019). https://doi.org/10.4103/IJPSYM

6. Sheffler, Z.M., Patel, P., Abdijadid, S.: Antidepressants. StatPearls [Internet]. https://www.ncbi.nlm.nih.gov/books/NBK538182/. Accessed on 22 Feb 2025 (2023)
7. Labrid, C., Moleyre, J., Poignant, J.C., Malen, C., Mocaër, E., Kamoun, A.: Structure-activity relationships of tricyclic antidepressants, with special reference to tianeptine. Clin. Neuropharmacol. **11**, S21–S31 (1988)
8. Moraczewski, J., Awosika, A.O., Aedma, K.K.: Tricyclic Antidepressants. StatPearls [Internet]. https://www.ncbi.nlm.nih.gov/books/NBK557791/. Accessed on: 02 Feb 2025 (2024)
9. Fanelli, D., Weller, G., Liu, H.: New serotonin-norepinephrine reuptake inhibitors and their anesthetic and analgesic considerations. Neurol. Int. **13**, 497–509 (2021). https://doi.org/10.3390/neurolint13040049
10. Wastkaemper, R.B., Glennon, R.A.: Application of ligand SAR, receptor modelling and receptor mutagenesis to the discovary and development of a new class of 5-HT2A ligand. Curr. Top. Med. Chem. **2**, 575–598 (2002). https://doi.org/10.2174/1568026023393741
11. Mills, N.: ChemDraw ultra 10.0. J. Am. Chem. Soc. **128**, 13649–13650 (2006)
12. Leach, A.R., Gillet, V.J., Lewis, R.A., Taylor, R.: Three-dimensional pharmacophore methods in drug discovery. J. Med. Chem. **53**, 539–558 (2009). https://doi.org/10.1021/jm900817u
13. Shoichet, B.K.: Virtual screening of chemical libraries. Nature. **432**, 862–865 (2004). https://doi.org/10.1038/nature03197
14. Mason, J.S., Good, A.C., Martin, E.J.: 3-D pharmacophores in drug discovery. Curr. Pharm. Des. **7**, 567–597 (2001). https://doi.org/10.2174/1381612013397843
15. Langer, T., Krovat, E.M.: Chemical feature-based pharmacophores and virtual library screening for discovery of new leads. Curr. Opin. Drug Discov. Dev. **6**(3), 370–376 (2003)
16. Koes, D.R., Camacho, C.J.: Pharmer efficient and exact pharmacophore search. J. Chem. Inf. Model. **51**, 1307–1314 (2011). https://doi.org/10.1021/ci200097m
17. Wolber, G., Langer, T.: LigandScout: 3-D pharmacophores derived from protein-bound ligands and their use as virtual screening filters. J. Chem. Inf. Model. **45**, 160–169 (2005). https://doi.org/10.1021/ci049885e
18. Schuster, D., Langer, T.: The identification of ligand features essential for PXR activation by pharmacophore modeling. J. Chem. Inf. Model. **45**, 431–439 (2005). https://doi.org/10.1021/ci049722q
19. Andersen, K., Liljefors, T., Gundertofte, K., Perregaard, J., Bogeso, K.P.: Development of a receptor interaction model for serotonin 5-HT2 receptor antagonists predicting selectivity with respect to dopamine D2 receptors. J. Med. Chem. **37**, 950–962 (1994). https://doi.org/10.1021/jm00033a013
20. Mokrosz, M.J., et al.: Structure-activity relationship studies of CNS agents, part 25. 4,6-di(heteroaryl)-2-(N-methylpiperazino)pyrimidines as new, potent 5-HT2A receptor ligands: a verification of the topographic model. Arch. Pharm. **328**, 659–666 (1995). https://doi.org/10.1002/ardp.19953280906
21. Westkaemper, R.B., Glennon, R.A.: Application of ligand SAR, receptor modeling and receptor mutagenesis to the discovery and development of a new class of 5-HT2A ligands. Curr. Top. Med. Chem. **2**, 575–598 (2002). https://doi.org/10.2174/1568026023393741
22. Kumar, A., Nandi, S., Saxena, A.K.: Antidepressant drug design on TCAs and Phenoxyphenylpropylamines utilizing QSAR and pharmacophore modeling. Comb. Chem. High Throughput Screen. **25**, 451–461 (2022). https://doi.org/10.2174/1386207323666200901104222
23. www.minitab.com

Computational Design of Pyrazole Derivatives for Management of Autoimmune Disease

Kriti Shukla[1], Jeevan Patra[1], and Neha Mathur[1](✉)

[1] Amity Institute of Pharmacy, Amity University Uttar Pradesh Lucknow Campus, Lucknow, Uttar Pradesh, India
neha07.mathur@gmail.com

Abstract. The discovery of an ideal, and effective autoimmune inhibitor is urgently required. The available therapies have substantial side effects, susceptibility to infections, metabolic disturbances, and organ toxicity. The solute carrier family 15 member 4 is a proton- coupled oligopeptide transporter that is mainly expressed in immune cells, according to several evidence sources. It has emerged as a viable target for investigating new regimens. The diverse pharmacological activities of amino-pyrazole-containing nuclei have made them noteworthy in the realm of medicinal chemistry. This study's primary goal is to investigate the effects of nuclei containing aminopyrazoles on the SLC15A4 protein. Receptor-based pharmacophore modeling was created and tested using the carefully selected amino-pyrazole libraries to test our hypothesis. The best aligned pharmacophoric featured hits were assessed for binding affinities and molecular interactions. Taking standard control as co-crystal of SLC15A4, three promising hits PubChem CID 17558339, 16191329, and 16188575 were confirmed based on their binding affinities, molecular interactions, molecular dynamics, and optimal ADMET profiles. Our findings represent a promising starting point for developing multi-modal regimens that act alleviate autoimmune disease syndromes.

Keywords: Autoimmune · SLC15A4 · Amino-pyrazole · Pharmacophore screening · Binding affinities

1 Introduction

A substantial global health burden, autoimmune illnesses impact between 7.6% and 9.4% of the world's population [1]. Current frontline therapeutics primarily consist of broadly immunosuppressive agents, biological response modifiers, and targeted immunotherapies [2]. Though these are effective with many patients, they often carry substantial side effects, such as increased susceptibility to infections, metabolic disturbances, and organ toxicity. Despite advances in biological therapies such as TNF-α inhibitors and B-cell depleting agents, a significant proportion of patients remain refractory to treatment.

A. K. Saxena, A. Saxena (Eds.): GTHTM 2025, pp. 161–170, 2026.
https://doi.org/10.1007/978-3-032-12320-6_8

The identification of novel molecular targets could modulate autoimmune pathogenesis effectively. Solute carrier family 15 member 4 (SLC15A4) is a proton-coupled oligopeptide transporter primarily expressed in immune cells and has emerged as a promising target for autoimmune disease [3]. SLC15A4 is a crucial modulator of endo lysosomal pH and is involved in the production of type I interferon in B lymphocytes and plasmacytoid dendritic cells as well as Toll-like receptor (TLR-7/9) signaling [4]. Mechanistically, SLC15A4 facilitates endosomal peptide transport and maintains the acidic pH required for optimal endosomal TLR function, thereby regulating downstream inflammatory signaling cascades implicated in autoimmunity [5]. Genetic study has identified SLC15A4 polymorphism associated with increased susceptibility to autoimmune conditions. In another study, SLC15A4 inhibition substantially attenuates lupus-prone mice [6]. Targeted inhibition of SLC15A4 could potentially disrupt pathogenic TLR7/9-driven interferon responses and autoreactive B cell functions without broadly compromising immune defense, offering a more selective approach to autoimmune disease management. Recently, a few studies elucidated the molecular mechanism of SLC15A4 inhibitors such as AJ2-30 [7] and Feeblin [8]. Many structural crystallography was resolved providing avenues for developing new regimens targeting against SLC15A4 [8–10].

Amino-pyrazoles are a promising class of compounds for medicinal chemistry due to their diverse biological activity, ease of synthesis, and potential for structural modification [11, 12]. The nucleus has a broad spectrum of pharmacological properties such as anti-inflammatory, anticancer, antiviral, antidepressant, analgesic, antimicrobial and selective enzyme inhibition [13, 14]. Several pyrazole containing clinical candidates are currently used such as anti-inflammatory and analgesic (Celecoxib, Tepoxalin, and Betazole), anticancer (Crizotimib), antiobesity (Surinabant and Difenamizole), antidepressant, and tranquilliser (Fezolamine and Mepiprazole), hence confirming their pharmacological importance and enhances their appeal [13]. They also possess several advantages as SLC15A4 inhibitors. The pyrazole ring provides a versatile scaffold for incorporating pharmacophoric features, optimizing binding affinity, and maintaining molecular stability. Furthermore, the favorable ADMET profiles of amino-pyrazole derivatives make them a promising candidates for further drug development. Thus, exploring a novel chemotypes containing amino-pyrazole nucleus drug candidates targeting against SLC15A4 protein would be open a new avenue for autoimmune drugs.

Encouraged by the facts we pioneered a first-line discovery of pyrazole-containing molecules targeting SLC15A4 computationally. By integrating pharmacophore modeling, and structure-based virtual screening, we identified novel pyrazole-based chemotypes capable of modulating SLC15A4 effectively. Our research represents a significant foremost step in the rational drug design, thus potentially opening new avenues for autoimmune therapeutic intervention.

2 Materials and Methods

2.1 Protein and Compound Database Preparation

From the RSCB PDB bank, a cryo-EM SLC15A4 structure with a good resolution of 2.50 Å was retrieved (PDB ID: 8JZX) [8]. To create the SLC15A4 protein, water molecules and HET groups were eliminated, hydrogen and Kollman charges were added, any missing loops were refined, and Chimera was used to minimize. In the current study, a virtual database bearing pyrazole-containing scaffold was explored from the chemical space of PubChem, ChemDiv, Enamine, and ChEMBL databases (Fig. 1).

The common molecules, ambiguous structures, and broken fragments were removed using the RDKit program. The remaining molecules underwent PAINS (Pan Assay Interferences), and REOS (Rapid Elimination of Swill) using False Positive Remover [15] followed by drug-likeness screening using DruLiTo v1.0 and prepared using MMFF94 of the Avogadro program.

2.2 Pharmacophore Development, and Virtual Screening

The pharmacophore model of SLC15A4 was developed using ZINCPharmer [16]. Pharmacophore features were identified as the chemical characteristics such as hydrophobic group, aromatic ring, hydrogen bond donor, and hydrogen bond acceptor (Fig. 2). The curated pyrazole databases were explored which can align accurately with the pharmacophore features.

2.3 Molecular Docking and Binding Free Energies Simulations

MzDOCK was used for molecular docking simulations [17]. The protein's active binding site was positioned 4 Å away from the co-crystallized ligand. MzDOCK is an integrated GUI-based pipeline which computes using various force fields using MMFF94, MMFF94s, UFF, GAFF, and Ghemical, thus, providing accurate and reliable docking results at significant lower RMSD while comparing poses with any standard. Following the execution of docking, an analysis image of the interactions was produced for additional PyMOL visualization. The g_mmpbsa program was used to determine the coulomb (ΔG_{coul}), hydrogen bonding (ΔG_{Hbond}), polar solvation (ΔG_{Solv}), van der Waals (ΔG_{vdW}), lipophilicity (ΔG_{lipo}), and Gibbs free energy (ΔG) [18].

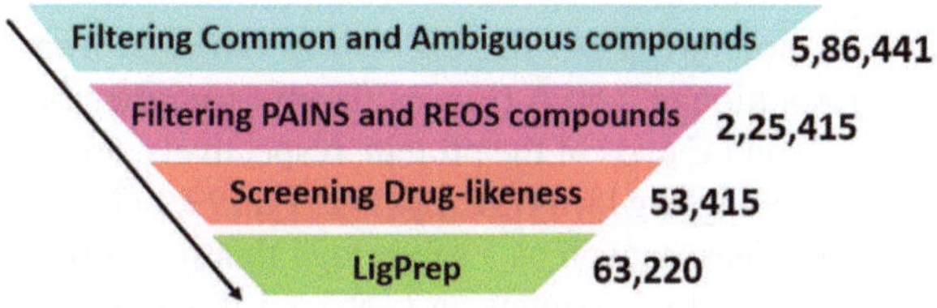

Fig. 1. Constructive workflow for amino-pyrazole preparation.

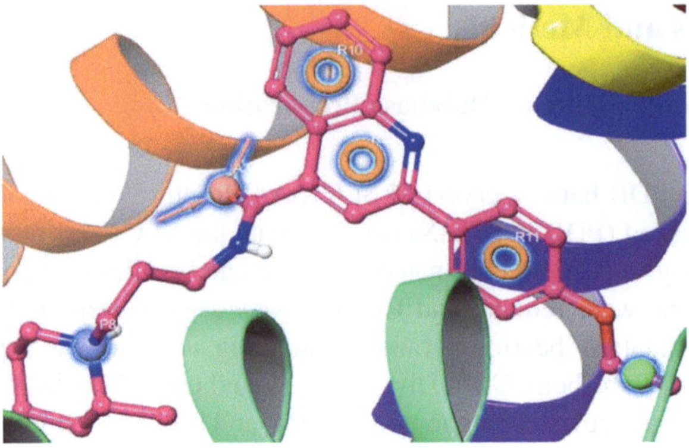

Fig. 2. Pharmacophore model of SLC15A4 complex featuring HBA, HBD, hydrophobic, and aromatic rings which are represented in red, blue, green, and orange respectively.

2.4 Molecular Dynamics Simulations

Academics at Desmond 2024 conducted molecular dynamics (MD) simulations of protein-ligand complexes [19]. The TIP5P model minimized in an orthorhombic box of 10 °C was used to solvate the best-docked conformers, and 0.15 M physiological salts were used to neutralize them. The systems relaxed at 300 K and 1.013 bar of pressure after reaching a gradient threshold of 25 kcal/mol/Å during the steepest fall. For every complex in which the trajectory frames were recorded, the MD simulations were run for 100 ns under the OPLS force field.

2.5 Physicochemical and ADMET Profiling

The physicochemical and virtual ADMET assessments are performed using ADMETlab 3.0 [20] and metabolic hotspot prediction using SuperCYPsPred [21]. The entry system was based on the canonical simplified molecular- input line-entry system (SMILES) of top-ranked molecules.

3 Results and Discussions

3.1 SLC15A4 Structural Crystallography

A cryo-EM SLC15A4 structure (PDB ID: 8JZX) was retrieved from the RSCB PDB bank having a good resolution of 2.50 Å (Fig. 3). Twelve transmembrane helices, numbered 19 to 569 at the N and C ends of the protein, respectively, are present. To interact with many residues, including Arg48, Tyr52, Lys172, Tyr199, Leu328, Tyr331, Leu473, Phe492, and Phe493 residues, as well as a salt bridge that forms between Arg200 and Glu482, the co-crystal "Feeblin" assumes an L-shaped binding stance projected outward opening conformation. A few polar

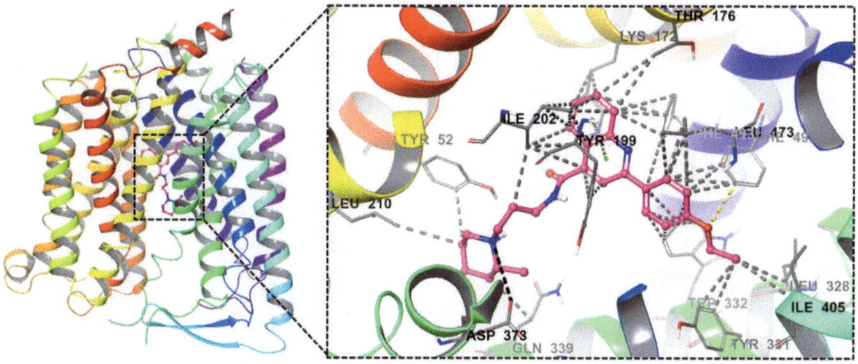

Fig. 3. Structural crystallography of SLC15A4 complexed with co-crystal interacting with the active residues. Here, aromatic- hydrogen bonding, hydrophobic, π-π stacking, and electrostatic interactions are represented in yellow, grey, green, and black respectively.

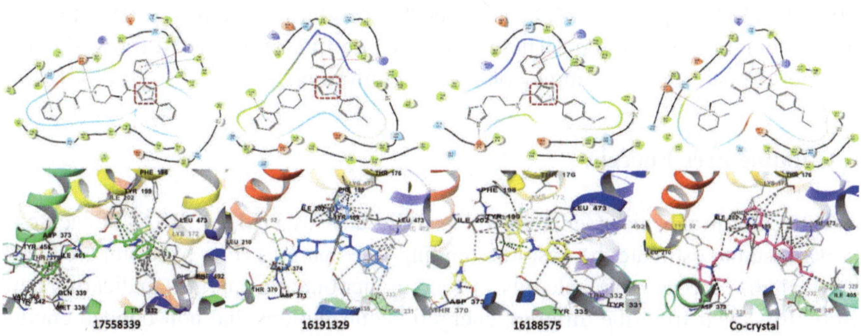

Fig. 4. Molecular interactions of best hits exhibiting higher affinity than co-crystal.

contacts, including Lys98-Gln493, Asp189-Tyr485, and Arg200 to the main chain carbonyl of Pro407, stabilize the conformation.

3.2 Molecular Docking Simulations

The initial phase before starting molecular docking investigations is to validate the co-crystal to replicate the precise docking poses of the built database. To compare pyrazole-based chemical libraries, the co-crystals were first re-docked and their binding affinities were calculated. The molecular interactions and binding pattern were identical in the re-docked co-crystal. To assess the aligned chemicals' molecular binding pattern, molecular docking simulations were performed. Every hit interacted with the essential active residues and produced precise docking conformers with the lowest RMSD. Amongst the screened database, best three hits—17558339, 16191329, and 16188575—were found to have higher docking scores than the co-crystal (Fig. 4). The docking scores of 17558339, 16191329, and 16188575 hits were − 9.388 kcal/mol, −9.516 kcal/mol, and − 8.567 kcal/mol, respectively whereas the co-crystal was

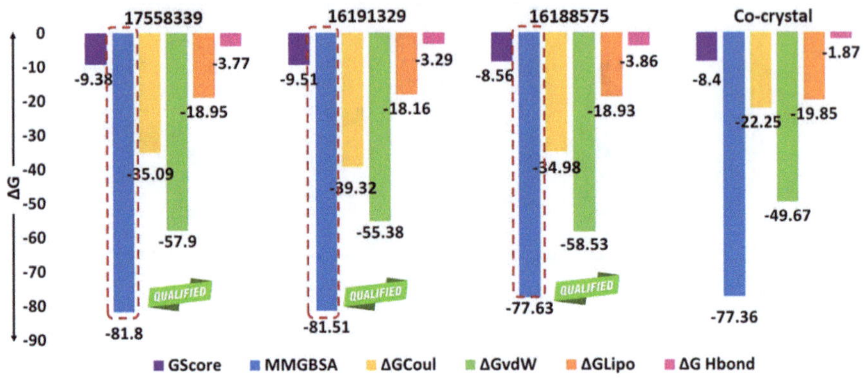

Fig. 5. MMGBSA of best hits demonstrating higher affinities than the co-crystal.

−8.409 kcal/mol. The crucial residues participating with the best hits are Lys172, Tyr199, Lys372, Asp373, Tyr458, and Phe493 residues exhibiting electrostatic and hydrogen bonding interactions. These three hits are further estimated for MM-GBSA parameters to validate the accuracy of docked conformers.

3.3 Binding Free Energies

The Molecular Mechanics Generalized Born Surface Area (MM-GBSA) of each complex was used to estimate the absolute binding free energies (Fig. 5). Compared to the co-crystal, the screened pyrazole hits showed higher binding energies, indicating better affinity. It is clear from the different energetic contributions that non-bonded energies (NBE) were mostly preferred for the binding process. Best hits 17558339, 16191329, and 16188575 had binding affinities of −81.8 kcal/mol, −81.51 kcal/mol, and – 77.63 kcal/mol, respectively. These values were higher than those of the co- crystal, which was −77.36 kcal/mol. The NBE, which likewise showed more free energy than the reference, is the sum of the coulombic and van der Waals energies.

3.4 Molecular Dynamics Simulations

Molecular dynamics (MD) simulations were conducted to validate the stability, conformational flexibility, and dynamic behavior of the docked SLC15A4-inhibitor complexes over time. A 100 ns simulation were performed for four complexes such as SLC15A4-17558339, SLC15A4-16191329, SLC15A4-16188575, and SLC15A4-Co-crystal. The root mean square deviation (RMSD), and root mean square fluctuation (RMSF) parameters were obtained and analyzed to assess the dynamic behavior and structural integrity of each complex. The RMSD represents overall conformational stability of protein-ligand complexes, relative to their initial docked poses. The RMSD trajectories of all complexes plateaued until equilibration of bounded ligands were achieved, and further ceased undergoing significant conformational changes (Fig. 6). Initially, there were a few distortions until the system plateaued which further stabilized

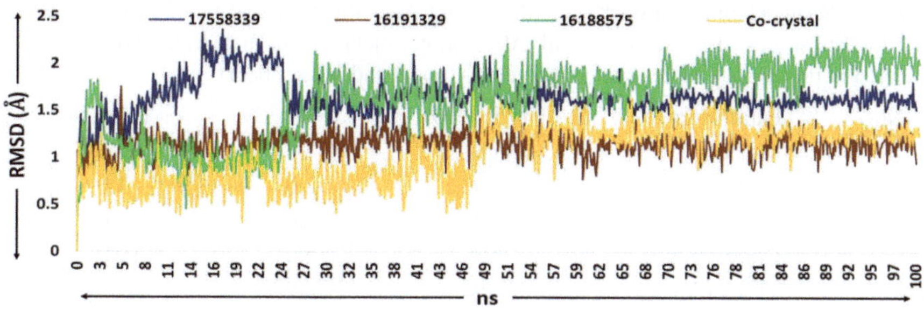

Fig. 6. All-atoms MDS of best hits showing non-significant distortions.

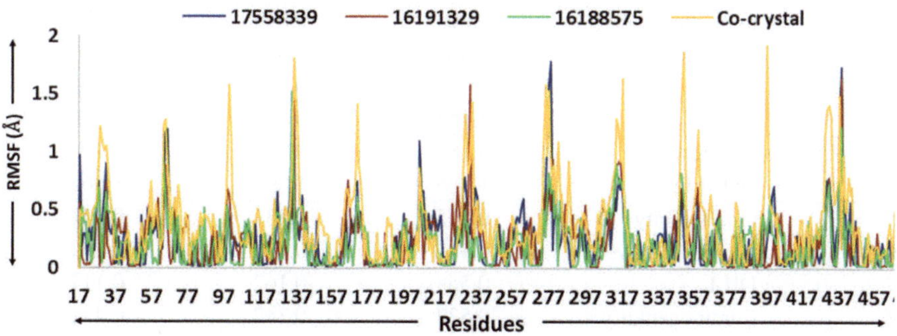

Fig. 7. RMSF of best hits against SLC15A4 protein.

and their deviations were non-significant. This could be due to the proper adjustment of the system inside the binding pocket. The average RMSD were 1.84 ± 0.23 Å, 1.86 ± 0.19 Å, 1.67 ± 0.23 Å, and 2.01 ± 0.61 Å for 17558339, 16191329, 16188575, and co-crystal respectively. The RMSF analysis was used to evaluate the flexibility of the protein backbone, particularly the α-carbon atoms, over the course of the simulation. The RMSF of all three hits produced a similar pattern of trajectories frames. The active residues involved around the binding pocket have minimal residues fluctuations due to the bounded optimized ligands whereas the reference molecules show a higher degree of residual fluctuations. The average RMSF of 17558339, 16191329, 16188575, and co-crystal were 0.13 ± 0.35 Å, 0.28 ± 0.32 Å, 0.13 ± 0.34 Å, and 1.4 ± 0.34 Å, respectively (Fig. 7). From the trajectories, the selected hits showed limited non-significant deviation and stabilized within the active pocket in contrast to the co-crystal (standard).

3.5 Physicochemical and Drug-Likeness Properties

The best hits were assessed for physicochemical and pharmacokinetic profiling (Table 1). All compounds exhibited predicted solubility within the acceptable limits (0.5 to −4 log mol/L). The partition coefficient shows higher lipophilicity, exceeding the optimal range of 0–3 for the selected molecules. All compounds demonstrated

Table 1. Physicochemical properties and ADMET assessment of best natural hits. Here, P-gp inhibitors/substrates, Human Intestinal Absorption, metabolism, half-life, and toxicity are predicted in probability values scaled between 0 (poor) and 1 (excellent).

Sr. No.	Parameters	17558339	16191329	16188575
1.	Mol Wt.	485.6	427.5	405.5
2.	HBD	2.0	2.0	1.0
3.	HBA	4.0	4.0	5.0
4.	Rot. Bonds	4.0	5.0	9.0
5.	TPSA (Å)	58.92	58.92	56.92
6.	Solubility	−4.029	−3.774	−3.529
7.	$LogP_{o/w}$	3.686	3.032	2.901
8.	Caco-2	−5.136	−4.873	−4.927
9.	p-gp substrate	0.003	0.049	0.004
10.	p-gp-inhibitor	0.06	0.592	0.004
11.	BBB	0.01	0.075	0.005
12.	PPB	84.143	90.482	86.926
13.	Clearance	16.186	9.812	8.102
14.	Half-life	1.286	1.39	1.563
15.	Oral toxicity	0.275	0.13	0.341
16.	Cardiotoxicity	0.381	0.502	0.209
17.	DILI	0.558	0.162	0.117
18.	Carcinogenicity	0.537	0.342	0.491
19.	Eye irritation	0.969	0.961	0.995
20.	Mutagenicity	0.739	0.561	0.646
21.	Hepatotoxicity	0.644	0.589	0.619
22.	Nephrotoxicity	0.298	0.393	0.131
23.	Hematotoxicity	0.115	0.233	0.062
24.	Genotoxicity	0.442	0.135	0.86
25	Neurotoxicity	0.298	0.4	0.333

favorable Caco-2 permeability. The estimated probability of being a P-gp inhibitor and substrate falls within ideal bounds. Within certain bounds, therapeutic window absorption (also known as human intestine absorption, or HIA) is permissible. This association is explained by blood plasma fractions and plasma protein binding (PPB), which have a direct impact on intestinal absorption in humans. A higher PPB (>90%) results in a lower volume of drug distribution in blood plasma leading to a lower absorption rate. All compounds showed optimal PPB which led to good drug distribution, and clearance time. In terms of toxicity, the selected hits showed very low probability values and had lower adverse risks.

3.6 Conclusion

In the present study, we unveiled the pyrazole-containing molecules that can effectively inhibit the SLC15A4 protein using receptor-based pharmacophore modeling and virtual screening. After the thorough virtual screening, three hits PubChem CID 17558339,

16191329, and 16188575 demonstrated better docking scores, MMGBSA, and optimal ADMET profiling. These were further subjected to 100 ns long all-atoms MD simulations to assess their molecular stability. However, after careful analysis, it was found that all hits retained a similar binding fashion forming stable molecular interactions. In comparison with the reference co-crystals, the screened hits can effectively modulate SLC15A4 in an autoimmune-mediated pathway and can alleviate myriads of pathogeneses. The rigorous testament is imperative in terms of biological evaluations which are still underway and will be reported once completed.

References

1. Zhao, M., et al.: Age-standardized incidence, prevalence, and mortality rates of autoimmune diseases in adolescents and young adults (15–39 years): an analysis based on the global burden of disease study 2021. BMC Public Health. **24** (2024). https://doi.org/10.1186/s12889-024-19290-3
2. Song, Y., Li, J., Wu, Y.: Evolving understanding of autoimmune mechanisms and new therapeutic strategies of autoimmune disorders. Signal Transduct. Target. Ther. **9** (2024). https://doi.org/10.1038/s41392-024-01952-8
3. Blasius, A.L., et al.: Slc15a4, AP-3, and Hermansky-Pudlak syndrome proteins are required for toll-like receptor signaling in plasmacytoid dendritic cells. Proc. Natl. Acad. Sci. USA. **107**, 19973–19978 (2010). https://doi.org/10.1073/PNAS.1014051107
4. Theofilopoulos, A.N., Kono, D.H., Baccala, R.: The multiple pathways to autoimmunity. Nat. Immunol. **18**, 716–724 (2017). https://doi.org/10.1038/NI.3731
5. Kobayashi, T., et al.: Human SLC15A4 is crucial for TLR-mediated type I interferon production and mitochondrial integrity. Int. Immunol. **33**, 399–406 (2021). https://doi.org/10.1093/INTIMM/DXAB006
6. Ban, T., et al.: Genetic and chemical inhibition of IRF5 suppresses pre-existing mouse lupus-like disease. Nat. Commun. **12**, 4379 (2021)
7. Chiu, T.-Y., et al.: Chemoproteomic development of SLC15A4 inhibitors with anti-inflammatory activity. Nat. Chem. Biol. **20**, 1000–1011 (2024)
8. Boeszoermenyi, A., et al.: A conformation-locking inhibitor of SLC15A4 with TASL proteostatic anti-inflammatory activity. Nat. Commun. **14**, 6626 (2023)
9. Chen, X., et al.: Structural basis for recruitment of TASL by SLC15A4 in human endolysosomal TLR signaling. Nat. Commun. **14** (2023)
10. Zhang, H., et al.: SLC15A4 controls endolysosomal TLR7–9 responses by recruiting the innate immune adaptor TASL. Cell Rep. **42**, 112916 (2023)
11. Secrieru, A., O'Neill, P.M., Cristiano, M.L.: Revisiting the structure and chemistry of 3(5)-substituted pyrazoles. Molecules. **25**, 42 (2019)
12. Alam, O., Naim, M., Nawaz, F., Alam, M.J., Alam, P.: Current status of pyrazole and its biological activities. J. Pharmacy Bioallied Sci. **8**, 2 (2016)
13. Lusardi, M., Spallarossa, A., Brullo, C.: Amino-pyrazoles in medicinal chemistry: a review. Int. J. Mol. Sci. **24**, 7834 (2023)
14. Li, G., Cheng, Y., Han, C., Song, C., Huang, N., Du, Y.: Pyrazole-containing pharmaceuticals: target, pharmacological activity, and their SAR studies. RSC Med. Chem. **13**, 1300–1321 (2022)

15. Baell, J.B., Holloway, G.A.: New substructure filters for removal of PAN assay interference compounds (pains) from screening libraries and for their exclusion in bioassays. J. Med. Chem. **53**, 2719–2740 (2010)
16. Koes, D.R., Camacho, C.J.: Zincpharmer: pharmacophore search of the zinc database. Nucleic Acids Res. **40**, W409 (2012)
17. Kabier, M., Gambacorta, N., Trisciuzzi, D., Kumar, S., Nicolotti, O., Mathew, B.: MzDOCK: a free ready-to-use GUI-based pipeline for molecular docking simulations. J. Comput. Chem. **45**, 1980–1986 (2024)
18. Kumari, R., Kumar, R., Lynn, A.: g_mmpbsa—a gromacs tool for high-throughput MM-PBSA calculations. J. Chem. Inf. Model. **54**, 1951–1962 (2014)
19. Bowers, K.J., et al.: Molecular dynamics—scalable algorithms for molecular dynamics simulations on commodity clusters. In: Proceedings of the 2006 ACM/IEEE Conference on Supercomputing – SC'06, p. 84 (2006)
20. Fu, L., et al.: ADMETlab 3.0: an updated comprehensive online admet prediction platform enhanced with broader coverage, improved performance, API functionality and decision support. Nucleic Acids Res. **52**, W422 (2024)
21. Banerjee, P., Dunkel, M., Kemmler, E., Preissner, R.: SuperCYPsPred—a web server for the prediction of cytochrome activity. Nucleic Acids Res. **48**, W580 (2020)

Designing, Synthesis and Characterization of Novel Heterocyclic Derivatives of-oxothiazolidine-3-carbonyl)-6- substituted-spiro[indoline-2,2′-thiazolidine]-3,4′-dione and -oxothiazolidine-3-carbonothioyl)-6- substituted-spiro [indoline-2,2′-thiazolidine]-3,4′-dione and their Bioevaluation

Ayushi Sahu[1], Shreya Dwivedi[2], Anoop K. Awasthi[3], Nitin Srivastava[2(✉)], and Krishna Srivastava[1(✉)]

[1] Faculty of Chemical Sciences, S R M University Lucknow-Dewa Road, Barabanki, Uttar Pradesh, India
Krishnajs2063@Gmail.com

[2] Department of Chemistry, Amity University Uttar Pradesh, Lucknow Campus, Noida, Uttar Pradesh, India
nitinsriv5@gmail.com; nsrivastava1@lko.amity.edu

[3] Department of Applied Science, Rajkiya Engineering College, Mainpuri, Uttar Pradesh, India

Abstract. Anovelseriesof3′-(2-(substituted-phenyl)-2-methyl-4-oxothiazolidine-3-carbonyl)-6- substituted-spiro[indoline-2,2′-thiazolidine]-3,4′-dione and 3′-(2-(substituted-phenyl)-2-methyl-4-oxothiazolidine-3-carbonothioyl)-6- substituted-spiro [indoline-2,2′-thiazolidine]-3,4′-dione derivatives were synthesized with good to excellent yield. The precursor imine has been formed by reaction of acetophenone with urea and thiourea then reacting the resulting products with thioglycolic acid to yields thiazolidine-Carboxamide and thiazolidine-carbothioamide which again formed imine by reaction with isatin to afford final spiro derivatives, further cyclization take place by the reaction of thioglycolic acid. The structures of the synthesized spiro derivatives were elucidated by their compatible spectral data given by U-Visible, Infra-Red-KBr, Proton Nuclear Magnetic Resonance and Mass. The prepared compounds depicted good antimicrobial activities with MIC ranged from 100 and 6.25 μg/mL. Among the synthesized compounds, **8b** exhibited promising antimicrobial activity against *A. clavatus MTCC1323* and *Staphylococcus pyogenes MTCC 442* with MIC value of 6.25 μg/mL for both these stains.

Keywords: Urea · Thiourea · Acetophenone · Isatin · Pyridine · Thioglycolic acid · Zinc chloride

A. K. Saxena, A. Saxena (Eds.): GTHTM 2025, pp. 171–188, 2026.
https://doi.org/10.1007/978-3-032-12320-6_9

1 Introduction

Heterocyclic-spiro compounds have been found in nature in many natural compounds like alkaloids, flavonoids, lactones, terpenoids, etc. proving their biocompatibility and justify their significance for development in novel biocompatible pharmaceutics [1–6]. The confrontational significance on spiro-heterocyclic provide them structural rigidity and specific bioactivity. The stearic strain in spiro compounds provides them with unexpected properties [7]. The single carbon adjoint to two hetero rings [8], is responsible for exceptional antimicrobial, analgesic, anti-inflammatory, antifungal, antitumor, and antiviral activities [9–12]. Among the spiral heterocycles, isatin (1H-indole-2,3-dione) containing pyrrole ring fused with the benzene ring can be employed for the synthesis of many derivatives [13]. The keto group at 2 and 3 positions, may be provoked for condensation and addition reactions, also the primary amine group present in the ring may be acetylated or alkylated for Michael, and Mannich reactions [14] and may be employed as a pre-cursor in pharmaceutics, for generating spiro derivatives with substitution at nitrogen and C2, C3 positions in the ring. These substituted isatin derivatives have previously shown potential bioactivities against various microbes [15–20].

Thiazolidinediones (TZDs) are another class of the heterocycles which are prominent component of many bioactive compounds specially pharmaceuticals. Thiazolidinediones moiety in the compounds shows antimicrobial, anticancer, antiarthritic, anti-inflammatory, aldose reductase inhibitors, and antidiabetic activities [21–24] e.g. Pioglitazone and rosiglitazone are clinically used drugs in the treatment of diabetes type 2 [25–27]. This necessitates the further exploration of the compounds containing spiro-thiazolidinedione moieties and their evaluation for the bioactivities [28–33]. Sunitinib and toceranib (as shown in Fig. 1) are isatin derivatives are approved as anticancer drug for different cancers while semaxanib, orantinib, nintedanib and esperidin are in the clinical testing as anticancer [34]. 1*H*-1,2,3-triazole-tethered ciprofloxacin-isatin semicarbazide/oxime have been found to possess antimycobacterial activities against *M. tuberculosis* (*Mtb*) $H_{37}Rv$ and MDR-TB [35]. Isatin derivative have also been reported as antituberculosis [35] and antibacterial against gram-positive and gram-negative bacteria [36]. Indole-thiazolidinones derivatives have also been found to exhibit antibacterial and antifungal activities against *S. aureus*, *P. aeruginosa*, *E. coli*, *M. tuberculosis*, *A. fumigates*, and *C. albicans* [37]. In our pursuit for the development of biologically potent scaffolds with cleaner greener and easier method [38–57], we get inspired by the structurally diverse and biologically potent scaffolds containing thiazolodins and isatin for the development of the hybrid of these two moieties and tried to explore them for the inherited bioactivities. In the reported research we have explored the synthesis of novel heterocyclic derivatives of-oxothiazolidine-3-carbonyl)-6-substituted-spiro[indoline-2,2′-thiazolidine]-3,4′-dione and -oxothiazolidine-3-carbonothioyl)-6- substituted-spiro [indoline-2,2′-thiazolidine]-3,4′-dione and their antimicrobial screening.

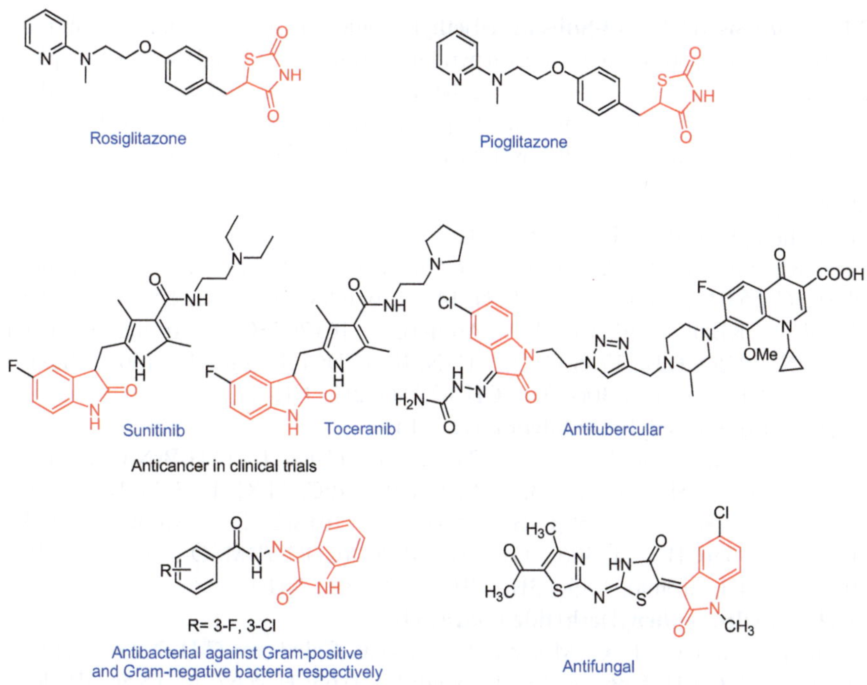

Fig. 1. Structurally diverse and biologically potent thiozolidones and isatin derivatives.

2 Materials and Method

2.1 Apparatus

The fusion temperatures were examined in the open capillary method. The purity and structure of the prepared derivatives were determined by using precoated TLC plates (Merk, 60F-254), iodine vapor as visualizing agent, eluent 5:2hexane/ethyl acetate). The ^{1}HNMR and ^{13}C NMR spectrum of the synthesized compounds were done on Bruker NMR spectrophotometer (300 MHz. ppm $CDCl_3$ and DMSO) using tetramethyl silane (TMS) as standard. Jasco FTIR-470 spectrophotometer for IR; MS-JEOL SX102 Mass spectroscopy for mass spectrum. Elemental analysis of the synthesized compounds done on CHN/O Analyzer.

2.2 Synthesis

Two parallel series of novel heterocyclic derivatives of-oxothiazolidine-3-carbonyl)-6-substituted-spiro[indoline-2,2′-thiazolidine]-3,4′-dione (**7a-f**) and -oxothiazolidine-3-carbonothioyl)-6- substituted-spiro [indoline-2,2′-thiazolidine]-3,4′-dione (**8a-f**) with the starting materials as given below

2.2.1 Synthesis of 1-(1-(4-Substituted)ethylidene)urea. A (0.02 mol) of substituted 1-Phenyl-ethanone and urea (0.02 mol) in pyridine (30 ml) was refluxed (6 h). On completion the cooled mixture drained into over ice cubes containing conc. hydrochloric acid (10 ml) giving precipitate in 1 h on decantation. The precipitated compound was cleaned with water and dried in vacuum desiccator. It was further recrystallized from ethanol.

Data Analysis

1-(1-(4-chlorphenyl)ethylidene)urea (1a)

Yield81%; mp87–88 °C; Mol. Wt: 196.63; Anal. Calcd. for $C_9H_9ClN_2O$, Elemental analysis: C, 55.01; H, 4.59; Cl, 17.99; N, 14.25; found: C, 54.93; H, 4.57; Cl, 17.98; N, 14.22.IR (KBr): 3051(C-H, Aroma), 1617(C=C, Aroma),935(C-Cl), 1231(C=N),2945(CH3), 1735(C=O). ^{1}H NMR ($CDCl_3$/DMSO-d6) (300 MHz): 6. 19–7.55(m,4H, Aroma), 1.40(s, 3H, -CH_3),2.28(s, 2H, -NH_2).

1-(1-(4-bromophenyl)ethylidene)urea (1b)

Yield 73%; mp93–94 °C; Mol. Wt: 241.09Anal. Calcd. for $C_9H_9BrN_2O$, Elemental analysis: C, 44.84; H, 3.76; Br, 33.14; N, 11.62found:C, 44.81; H, 3.73; Br, 33.10; N, 11.87, IR (KBr): 3059(C-H, Aroma), 1618(C=C, Aroma),735(C-Br), 1241(C=N),2955(CH_3), 1730(C=O). ^{1}H NMR ($CDCl_3$/DMSO-d6) (300 MHz): 6.09–7.59(m,4H, Aroma), 1.49(s,3H, -CH_3),2.22(s,2H, -NH_2).

1-(1-(4-hydroxyphenyl)ethylidene)urea (1c)

Yield 75%; mp80–81 °C; Mol.Wt: 178.19Anal. Calcd. for $C_9H_{10}N_2O_2$, Elemental analysis: C, 60.66; H, 5.66; N, 15.72; found: C, 60.63; H, 5.62; N, 15.56, IR (KBr): 3050(C-H,Aroma), 1619(C=C, Aroma), 3335 (C-OH), 1261(C=N), 2905(CH3), 1736(C=O). ^{1}H NMR ($CDCl_3$/DMSO-d6) (300 MHz): 6.39–7.66(m,4H, Aroma), 1.49(s,3H, -CH_3),2.28(s,2H, -NH_2), 5.49(s, 1H, -CH_3)

2.2.2 Synthesis of 1-(1-(4-Substituted)ethylidene)thiourea (2a). A (0.02 mol) of substituted 1-Phenyl-ethanone and thiourea (0.02 mol) in pyridine (30 ml) was refluxed for 6 h and the mixture was drained over ice cubes containing conc. hydrochloric acid (10 ml). The compound synthesized precipitated on standing for 1 h which was filtered and cleaned with water. The compound was recrystallized in ethyl alcohol.

Data Analysis

1-(1-(4-chlorophenyl)ethylidene)thiourea (2a)

Yield 68%; mp07–108 °C; Mol. Wt: 212.70Anal. Calcd. for $C_{15}H_{12}ClNO_2$, Elemental analysis: C, 50.82; H, 4.30; Cl, 16.70; N, 13.20, found: C, 50.77; H, 4.29; Cl, 16.68; N, 13.18. IR (KBr): 3055(C-H, Aroma), 1618(C=C, Aroma),945(C-Cl), 1235(C=N),2950(CH3), 1435(C=S). ^{1}H NMR ($CDCl_3$/DMSO-d6) (300 MHz): 6. 18–7.45(m,4H, Aroma), 1.38(s,3H, -CH3),2.18(s, 2H, -NH2).

1-(1-(4-bromophenyl)ethylidene)thiourea (2b)

Yield 59%; mp89–90 °C; Mol. Wt 257.15. Anal. Calcd. for $C_9H_9BrN_2S$, Elemental analysis: C, 41.98; H, 3.49; Br, 30.98; N, 10.92, found: C, 41.82; H, 3.50; Br, 31.03; N, 10.86.IR (KBr): IR (KBr): 3065(C-H, Aroma), 1622(C=C, Aroma), 1238(C=N), 2950(CH_3), 1439(C=S), 744(C-Br), ^{1}H NMR ($CDCl_3$/DMSO-d6) (300 MHz): 6.28–7.45(m, 4H, Aroma), 1.48(s, 3H, -CH_3), 2.23(s, 2H, -NH_2)

1-(1-(4-hydroxyphenyl)ethylidene)thiourea (2c)

Yield 63%; mp103–104 °C; Mol. Wt: 194.25Anal. Calcd. for $C_9H_{10}N_2OS$,Elemental analysis: C, 54.97; H, 5.20; N, 14.38; O, 8.26, found: C, 54.92; H, 5.15; N, 13.97,IR (KBr): IR (KBr): 3060(C-H, Aroma), 1629(C=C, Aroma), 1239(C=N),2958(CH_3), 1439(C=S), 3344(C-OH); ^{1}H NMR ($CDCl_3$/DMSO-d6) (300 MHz): 6.20–7.55 (m, 4H, Aroma), 1.40 (s, 3H,-CH_3), 2.28 (s, 2H,-NH_2), 5.40(s,1H,-OH).

2.2.3 Synthesis of 2-(substituted)-2-methyl-4-oxothiazolidine-3-carboxamide (3a). Cyclohexylidenehydrazine (I) (0.02 mol) of cyclohexylidenehydrazine and thioglycollic acid (0.01 mol) containing in trace $ZnCl_2$ (0.1 g) in DMF was heated under reflux for 4 h. It was poured into crushed ice and stirred vigorously. Solidification occurred after 15 min, which was filtered off and washed with cold water. Recrystallization from ethanol gave pure sample.

Data Analysis

2-(4-chlorophenyl)-2-methyl-4-oxothiazolidine-3-carboxamide (3a)

Yield 77%; mp109–110 °C; Mol.Wt: 270.73Anal. Calcd. for $C_{11}H_{11}ClN_2O_2S$,Elemental analysis: C, 48.80; H, 4.10; Cl, 13.09; N, 10.35,found:C, 48.76; H, 4.04; Cl, 13.06; N, 10.30, IR (KBr): 3048(C-H, Aroma), 1615(C=C, Aroma), 1738(C=O), 1037(C-N), 923(C-S), 968(C-Cl); ^{1}H NMR ($CDCl_3$/DMSO-d6) (300 MHz): 6.03–7.28 (m, 4H, Aroma), 1.08 (s, 3H, -CH_3), 2.08(s, 2H,-NH_2), 3.19 (s, 2H, CH_2-S, Spiro).

2-(4-bromophenyl)-2-methyl-4-oxothiazolidine-3-carboxamide (3b)

Yield 55%; mp 96 °C; Mol. Wt: 315.19; Anal. Calcd. for $C_{11}H_{11}BrN_2O_2S$; Elemental analysis: C, 41.92; H, 3.52; Br, 25.35; N, 8.89; found: C, 41.87; H, 3.48; Br, 25.31; N, 8.86. IR (KBr): 3044 (C-H, Aroma), 1625 (C=C, Aroma), 1748 (C=O), 1030 (C-N), 920 (C-S), 768(C-Br); ^{1}H NMR ($CDCl_3$ /DMSO-d6)(300 MHz): 6.09–7.38 (m, 4H, Aroma), 1.23 (s, 3H, -CH_3), 2.26 (s, 2H, -NH_2), 3.29 (s, 2H, CH_2-S, Spiro).CH_3).

2-(4-hydroxyphenyl)-2-methyl-4-oxothiazolidine-3-carboxamide (3c)

Yield 62%; mp109°C; Mol. Wt: 252.29; Anal. Calcd. for $C_{11}H_{12}N_2O_3S$, Elemental analysis: H, 4.80; C, 52.40; N, 10.98; found: H, 4.72; C, 51.99; N, 11.07; IR (KBr): 3048 (C-H, Aroma), 1622 (C=C, Aroma), 1743 (C=O), 1030 (C-N), 928 (C-S),762 (C-Br); ^{1}H NMR ($CDCl_3$/DMSO-d6) (300 MHz): 6.03–7.48 (m, 4H, Aroma), 1.27 (s, 3H, -CH_3), 2.29 (s, 2H, -NH_2), 3.39(s, 2H, CH_2-S, Spiro).CH_3), 5.27(s, 1H,-OH),

2.2.4 Synthesis of 2-(4-Substituted)-2-methyl-4-oxothiazolidine-3-carbothioa mide. Mixture of cyclohexylidene hydrazine (I) (0.02 mol) of cyclohexylidenehydrazine and thioglycolic acid (0.01 mol) containing in trace $ZnCl_2$ (0.02 g) in DMF was refluxed for 4 h. On completion of the reaction, the mixture was drained over ice cubes. Solidification took place after some time. After solidification the precipitated compound was filtered and cleaned with water. Recrystallization from ethyl alcohol gave pure sample.

Data Analysis

2-(4-chlorophenyl)-2-methyl-4-oxothiazolidine-3-carbothioamide (4a)

Yield 71%; mp112–113 °C; Mol.Wt: 286.79, Anal. Calcd. for $C_{11}H_{11}ClN_2OS_2$, Elemental analysis: H, 3.90; C, 46.07; Cl, 12.29; N, 9.80, found: H, 3.83; C, 46.04; Cl,

12.30; N, 9.72; IR (KBr): 3055 (C-H, Aroma), 1612 (C=C, Aroma), 1432 (C=S), 1035 (C-N), 912 (C-S), 945 (C-Cl); ^{1}H NMR ($CDCl_3$/DMSO-d6) (300 MHz): 6.18–7.15(m, 4H, Aroma), 1.12 (s, 3H, $-CH_3$), 2.12(s, 2H, $-NH_2$), 3.08 (s, 2H, CH_2-S, Spiro).

2-(4-bromophenyl)-2-methyl-4-oxothiazolidine-3-carbothioamide (4b)

Yield 82%; mp129–130 °C; Mol. Wt 331.25; Anal. Calcd. for $C_{11}H_{11}BrN_2OS_2$, Elemental analysis: H, 3.37; C, 39.40; Br, 23.98; N, 8.50, found: H, 3.36; C, 39.83; Br, 24.07; N, 8.42; IR (KBr): 3065 (C-H, Aroma), 1618 (C=C, Aroma), 1430 (C=S), 1039 (C-N), 910 (C-S), 745 (C-Br); ^{1}H NMR ($CDCl_3$/DMSO-d6) (300 MHz): 6.08–7.35 (m, 4H, Aroma), 1.02 (s, 3H, $-CH_3$), 2.22 (s, 2H, $-NH_2$), 3.28(s, 2H, CH_2-S, Spiro).

2-(4-hydroxyphenyl)-2-methyl-4-oxothiazolidine-3-carbothioamide (4c)

Yield 63%; mp 92–93 °C; Mol. Wt 268.35; Anal. Calcd. for $C_{11}H_{12}N_2O_2S_2$, Elemental analysis: C, 49.23; H, 4.51; N, 10.44, found: C, 49.17; H, 4.46; N, 10.40; IR (KBr): 3064 (C-H, Aroma), 1628 (C=C, Aroma), 1423 (C=S), 1045 (C-N), 918 (C-S), 749 (C-Br), 3365 (C-OH); ^{1}H NMR ($CDCl_3$/DMSO-d6) (300 MHz): 6.08–7.35 (m, 4H, Aroma), 1.02 (s, 3H, $-CH_3$), 2.22 (s, 2H, $-NH_2$), 3.28 (s, 2H, CH_2-S, Spiro), 5.02 (s, 1H, -OH).

2.2.5 Synthesis of 2-(4-Substituted)-2-methyl-N-(6-nitro-2-oxoindolin-3-ylidene)-4-oxothiazolidine-3-carboxamide. A mixture of 0.02 mol of **3a-c** and substituted-isatin (0.02 mol) in pyridine (30 ml) was refluxed for 6 h. On cooling the mixture was drained over ice cubes containing concentrated hydrochloric acid (10 ml). The precipitate formed on standing was filtered and cleaned with water. On drying the compound was recrystallized in ethyl alcohol.

Data Analysis

2-(4-chlorophenyl)-2-methyl-N-(6-nitro-2-oxoindolin-3-ylidene)-4-oxothiazolidine-3-carboxamide (5a)

Yield 56%; mp 120–121 °C; Mol. Wt: 444.85, Anal. Calcd. for $C_{19}H_{13}ClN_4O_5S$, Elemental analysis: C, 51.30; H, 2.95; Cl, 7.97; N, 12.59, found: C, 51.27; H, 2.91; Cl, 7.93; N, 12.54; IR (KBr): 3057 (C-H, Aroma), 1615 (C=C, Aroma), 1631 (N-C=O), 1351 (C=N), 1451, 1251 (NO_2), 976 (C-Cl); ^{1}H NMR ($CDCl_3$/DMSO-d6) (300 MHz): 6.03–7.55 (m, 7H, Aroma), 1.09 (s, 3H, $-CH_3$), 4.30 (s, 1H,-NH), 4.01 (s, 2H, CH_2-S, Spiro).

2-(4-chlorophenyl)-2-methyl-N-(6-methyl-2-oxoindolin-3-ylidene)-4-oxothiazolidine-3-carboxamide (5b)

Yield 57%; m.p. 118–119 °C; Mol. Wt: 413.88, Anal. Calcd. for $C_{20}H_{16}ClN_3O_3S$, Elemental analysis: C, 57.99; H, 3.90; N, 10.15, Cl, 8.57; found: C, 57.84; H, 3.85; N, 10.11; Cl, 8.51; IR (KBr): 3059(C-H, Aroma), 1635(C=C, Aroma), 1621(N-C=O), 1259 (C=N), 976 (C-Cl); ^{1}H NMR ($CDCl_3$/DMSO-d6) (300 MHz): 6.23–7.59 (m, 7H, Aroma), 1.39 (s, 3H, $-CH_3$), 4.38 (s, 1H, -NH), 4.05 (s, 2H, CH_2-S, Spiro).

2-(4-bromophenyl)-2-methyl-N-(6-nitro-2-oxoindolin-3-ylidene)-4-oxothiazolidine-3-carboxamide (5c)

Yield 60%; mp 93 °C; Mol. Wt 489.30; Anal. Calcd. for $C_{19}H_{13}BrN_4O_5S$; Elemental analysis: H, 2.68; C, 46.64; N, 11.45, Br, 16.33; found: H, 2.63; C, 46.60; N, 11.40; Br, 16.30; IR (KBr): 3066 (C-H, Aroma), 1618 (C=C, Aroma), 1633 (N-C=O), 1258 (C=N), 1350, 1459(NO_2), 770 (C-Br); ^{1}H NMR ($CDCl_3$/DMSO-d6) (300 MHz):

6.33–7.75 (m, 7H, Aroma), 1.29 (s, 3H, $-CH_3$), 4.39 (s, 1H, -NH), 4.08 (s, 2H, CH_2-S, Spiro).

2-(4-bromophenyl)-2-methyl-N-(6-methyl-2-oxoindolin-3-ylidene)-4-oxothiazolidine-3-carboxamide (5d)

Yield 74%; mp 123–124 °C; Mol. Wt, 458.33, Anal. Calcd. for $C_{20}H_{16}BrN_3O_3S$, Elemental analysis: H, 3.52; C, 52.41; N, 9.17, Br; 17.43; found: H, 3.47; C, 52.36; N, 9.14, Br, 17.40; IR (KBr): 3064 (C-H, Aroma), 1643(C=C, Aroma), 1629 (N-C=O), 1250 (C=N), 776 (C-Br); ^{1}H NMR ($CDCl_3$/DMSO-d6) (300 MHz): 6.03–7.79 (m, 7H, Aroma), 1.29 (s, 3H, $-CH_3$), 4.48 (s, 1H, -NH), 4.22 (s, 2H, CH_2-S, Spiro).

2-(4-hydroxyphenyl)-2-methyl-N-(6-nitro-2-oxoindolin-3-ylidene)-4-oxothiazolidine-3-carboxamide (5e)

Yield 70%; mp 131 °C; Mol. Wt: 426.40Anal. Calcd. for $C_{19}H_{14}N_4O_6S$, Elemental analysis: H, 3.31; C, 53.52; N, 13.14, found: H, 3.27; C, 53.46; N, 13.11; IR (KBr): 3069 (C-H, Aroma), 1638 (C=C, Aroma), 1653 (N-C=O), 1278 (C=N), 1360, 1450 (NO_2), 3370 (C-OH); ^{1}H NMR ($CDCl_3$/DMSO-d6) (300 MHz): 6.23–7.70 (m, 7H, Aroma), 1.29 (s, 3H, $-CH_3$), 4.30 (s, 1H, -NH), 4.03 (s, 2H, CH_2-S, Spiro), 5.30 (s, 1H, -NH).

2-(4-hydroxyphenyl)-2-methyl-N-(6-methyl-2-oxoindolin-3-ylidene)-4-oxothiazolidine-3-carboxamide (5f)

Yield 68%; mp116–117 °C; Mol.Wt:395.43Anal. Calcd. for $C_{20}H_{17}N_3O_4S$, Elemental analysis: C, 60.75; H, 4.33; N, 10.63, found: C, 60.72; H, 4.31; N, 10.60; IR (KBr): 3060 (C-H, Aroma), 1648 (C=C, Aroma), 1663 (N-C=O), 1258 (C=N), 3372(C-OH); ^{1}H NMR ($CDCl_3$/DMSO-d6) (300 MHz): 6.28–7.76 (m, 7H, Aroma), 1.27 (s, 3H, $-CH_3$), 4.38 (s, 1H, -NH), 4.09 (s, 2H, CH_2-S, Spiro), 5.38 (s, 1H, -NH), 5.08 (s, 1H, -OH).

2.2.6 Synthesis of 2-(4-substituted)-2-methyl-N-(6-nitro-2-oxoindolin-3-ylidene)-4-oxothiazolidine-3-carbothioamide:. A mixture of **4a-c** (0.02 mol) and substituted-isatin (0.02 mol) in pyridine (30 ml) was refluxed for 6 h. On completion of the reaction, the mixture was drained over ice cubes containing concentrated hydrochloric acid (10 ml). The desired compound precipitated which was filtered and cleaned with water and recrystallized in ethyl alcohol.

Data Analysis

2-(4-Chlorophenyl)-2-methyl-N-(6-nitro-2-oxoindolin-3-ylidene)-4-oxothiazolidine-3-carbothioamide (6a)

Yield 72%; mp 93–94 °C; Mol.Wt: 460.91 Anal. Calcd. for $C_{19}H_{13}ClN_4O_4S_2$C, 49.51; H, 2.84; Cl, 7.69; N, 12.16, found: C, 49.46; H, 2.81; Cl, 7.65; N, 12.12; IR (KBr): 3059 (C-H, Aroma), 1617 (C=C, Aroma), 1635 (N-C=O), 1255 (C=N), 1345, 1455(NO_2), 940(-C-Cl); ^{1}H NMR ($CDCl_3$/DMSO-d6) (300 MHz): 6.08–7.45 (m, 7H, Aroma), 1.08 (s, 3H, $-CH_3$), 4.2 (s, 1H, -NH), 3.70 (s, 2H, CH_2-S, Spiro).

2-(4-chlorophenyl)-2-methyl-N-(6-methyl-2-oxoindolin-3-ylidene)-4-oxothiazolidine-3-carbothioamide (6b)

Yield 69%; mp107–108 °C; Mol.Wt:429.94, Anal. Calcd. for $C_{20}H_{16}ClN_3O_2S_2$C, H, 3.76; 55.87; N, 9.77; Cl, 8.26; found: H, 3.72; C, 55.84; N, 9.73, Cl, 8.21; IR (KBr): 3069 (C-H, Aroma), 1637 (C=C, Aroma), 1655 (N-C=O), 1275 (C=N), 948(-C-Cl); ^{1}H

NMR ($CDCl_3$/DMSO-d6) (300 MHz): 6.09–7.48 (m, 7H, Aroma), 1.07 (s, 3H, -CH_3), 4.28 (s, 1H, -NH), 3.78(s, 2H, CH_2-S,Spiro).

2-(4-bromophenyl)-2-methyl-N-(6-nitro-2-oxoindolin-3-ylidene)-4-oxothiazolidine-3-carbothioamide (6c)

Yield 66%; mp 84–85 °C; Mol. Wt, 505.36, Anal. Calcd. for $C_{19}H_{13}BrN_4O_4S_2$, Elemental analysis: H, 2.59; C, 45.20; N, 11.09; Br, 15.81; found: H, 2.53; C, 45.12; N, 11.06; Br, 15.77; IR (KBr): 3065 (C-H, Aroma), 1637 (C=C, Aroma), 1645 (N-C=O), 1255 (C=N), 1355, 1467(NO_2); ^{1}H NMR ($CDCl_3$ /DMSO-d6) (300 MHz): 6.07–7.65 (m, 7H, Aroma), 1.03 (s, 3H, -CH_3), 4.29 (s, 1H,-NH), 3.60 (s, 2H, CH_2-S, Spiro).

2-(4-bromophenyl)-2-methyl-N-(6-methyl-2-oxoindolin-3-ylidene)-4-oxothiazolidine-3-carbothioamide (6d)

Yield 74%; mp 97 °C; Mol.Wt: 474.39, Anal. Calcd. for $C_{20}H_{16}BrN_3O_2S_2$, Elemental analysis: C, 50.64; H, 3.40; Br, 16.84; N, 8.86, found: C, 50.61; H, 3.36; Br, 16.81; N, 8.80; IR (KBr): 3078 (C-H, Aroma), 1622 (C=C, Aroma), 1645 (N-C=O), 1267 (C=N), 740 (-C-Br); ^{1}HNMR ($CDCl_3$/DMSO-d6) (300 MHz): 6.23–7.67 (m, 7H, Aroma), 1.32 (s, 3H, -CH_3), 4.34 (s, 1H, -NH), 3.69 (s, 2H, CH_2-S, Spiro).

2-(4-hydroxyphenyl)-2-methyl-N-(6-nitro-2-oxoindolin-3-ylidene)-4-oxothiazolidine-3-carbothioamide (6e)

Yield 59%; mp 98 °C; Mol.Wt: 442.46, Anal. Calcd. for $C_{19}H_{14}N_4O_5S_2$, Elemental analysis: C, 51.58; H, 3.19; N, 12.66, found: C, 51.53; H, 3.15; N, 12.62; IR (KBr): 3079 (C-H, Aroma), 1627 (C=C, Aroma), 1630 (N-C=O), 1275 (C=N), 1365, 1475 (NO_2), 3340 (-OH); ^{1}H NMR ($CDCl_3$/DMSO-d6) (300 MHz): 6.28–7.65 (m, 7H, Aroma), 1.28 (s, 3H, -CH_3), 4.28 (s, 1H, -NH), 3.79 (s, 2H, CH_2-S, Spiro), 5.28 (s, 1H, -OH).

2-(4-hydroxyphenyl)-2-methyl-N-(6-methyl-2-oxoindolin-3-ylidene)-4-oxothiazolidine-3-carbothioamide (6f)

Yield 76%; mp 106–107 °C; Mol.Wt: 411.49, Anal. Calcd. for $C_{20}H_{17}N_3O_3S_2$, Elemental analysis: H, 4.12; C, 58.40; N, 10.21, found: H, 4.10; C, 58.36; N, 10.16, IR (KBr): 3082 (C-H, Aroma), 1632 (C=C, Aroma), 1638 (N-C=O), 1275 (C=N), 3342 (-OH); ^{1}H NMR ($CDCl_3$/DMSO-d6) (300 MHz): 6.38–7.75(m, 7H, Aroma), 1.29 (s, 3H, -CH_3), 4.23(s, 1H, -NH), 3.70 (s, 2H, CH_2-S, Spiro), 5.29 (s, 1H, -OH).

2.2.7 Synthesis of 3′-(2-(4-Substituted)-2-methyl-4-oxothiazolidine-3-carbonyl)-6-nitrospiro[indoline-2,2′-thiazolidine]-3,4′-dione. Cyclohexylidene hydrazine (I) (0.02 mol) and thioglycolic acid (0.01 mol) containing in trace $ZnCl_2$ (0.02 g) in DMF was refluxed for 4 h. On completion of the reaction the mixture was drained over ice cubes. The desired compound precipitated which was filtered, cleaned with water and further recrystallized in ethyl alcohol.

Data Analysis

3′-(2-(4-chlorophenyl)-2-methyl-4-oxothiazolidine-3-carbonyl)-6-nitrospiro[indoline-2,2′-thiazolidine]-3,4′-dione (7a)

Yield 71%; mp 91 °C; Mol. Wt, 518.94, Anal. Calcd. for C_{21}H15Cl$N_4O_6S_2$, Elemental analysis: C, 48.60; H, 2.91; Cl, 6.83; N, 10.80, found: C, 48.55; H, 2.88; Cl, 6.80; N, 10.76, IR (KBr): 3061 (C-H, Aroma), 1612 (C=C, Aroma), 1656 (N-C=O), 934 (C-Cl), 967 (C-N), 3310 (N-H), 1365, 1475(NO_2); ^{1}H NMR ($CDCl_3$ /DMSO-d6) (300 MHz):

11–7.38 (m, 7H, Aroma), 1.38(s, 3H, -CH_3), 3.18 (s, 1H, -NH), 3.18 (s, 2H, CH_2-S, Spiro), 2.66 (s, 2H, -CH_2-S).

3′-(2-(4-chlorophenyl)-2-methyl-4-oxothiazolidine-3-carbonyl)-6-methylspiro[indoline-2,2′-thiazolidine]-3,4′-dione (7b)

Yield 65%; mp 91 °C; Mol. Wt, 487.97, Anal. Calcd. for $C_{22}H_{18}ClN_3O_4S_2$, Elemental analysis: H, 3.71; C, 54.15; N, 8.61, Cl, 7.28; found: H, 3.66; C, 54.11; N, 8.57, Cl, 7.21; IR (KBr): 3069 (C-H, Aroma), 1610 (C=C, Aroma), 1653(N-C=O), 963 (C-N), 3319 (N-H), 1365, 1475(NO_2), 932(C-Cl); ^{1}H NMR ($CDCl_3$/DMSO-d6) (300 MHz): 6.21–7.38(m, 7H, Aroma), 1.48 (s, 3H, -CH_3), 3.28 (s, 1H, -NH), 3.38 (s, 2H, CH_2-S, Spiro), 2.76 (s, 2H, -CH_2-S).

3′-(2-(4-bromophenyl)-2-methyl-4-oxothiazolidine-3-carbonyl)-6-nitrospiro[indoline-2,2′-thiazolidine]-3,4′-dione (7c)

Yield 59%; mp132°C; Mol. Wt: 563.40, Anal. Calcd. for $C_{21}H_{15}BrN_4O_6S_2$, Elemental analysis: C, 49.63; H, 3.41; Br, 15.01; N, 7.89, found: C, 49.60; H, 3.35; Br, 14.98; N, 7.83; IR (KBr): 3068 (C-H, Aroma), 1610 (C=C, Aroma), 166 6 (N-C=O), 987 (C-N), 3330 (N-H), 1360, 1470 (NO_2), 767(C-Br); ^{1}H NMR ($CDCl_3$/DMSO-d6) (300 MHz): 6.18–7.48 (m, 7H, Aroma), 1.58 (s, 3H, -CH_3), 3.38 (s, 1H, -NH), 3.28 (s, 2H, CH_2-S, Spiro), 2.69 (s,2H,-CH_2-S).

3′-(2-(4-bromophenyl)-2-methyl-4-oxothiazolidine-3-carbonyl)-6-methylspiro[indoline-2,2′-thiazolidine]-3,4′-dione (7d)

Yield 76%; mp116°C; Mol. Wt: 532.43, Anal. Calcd. for $C_{22}H_{18}BrN_3O_4S_2$, Elemental analysis: C, 49.63; H, 3.41; Br, 15.01; N, 7.89, found: C, 49.60; H, 3.37; Br, 14.99; N, 7.82; IR (KBr): 3068 (C-H, Aroma), 1632 (C=C, Aroma), 1676 (N-C=O), 968 (C-N), 3316 (N-H), 775(C-Br); ^{1}H NMR ($CDCl_3$ /DMSO-d6) (300 MHz): 6.21–7.48 (m, 7H, Aroma), 1.33 (s, 3H, -CH_3), 3.19 (s, 1H, -NH), 3.20 (s, 2H, CH_2-S, Spiro), 2.78 (s, 2H, -CH_2-S).

3′-(2-(4-hydroxyphenyl)-2-methyl-4-oxothiazolidine-3-carbonyl)-6-nitrospiro[indoline-2,2′-thiazolidine]-3,4′-dione (7e)

Yield 58%; mp89–90 °C; Mol. Wt: 500.50, Anal. Calcd. for $C_{21}H_{16}N_4O_7S_2$, Elemental analysis: H, 3.22; C, 50.40; N, 11.19, found: H, 3.18; C, 50.37; N, 11.16; IR (KBr): 3091 (H-C), 1652 (C=C), 1676 (O=C-N), 987 (N-C), 3380 (H-N), 1365, 1475 (NO_2), 3340 (-OH); ^{1}H NMR ($CDCl_3$/DMSO-d6) (300 MHz): 6.31–7.36 (m, 7H, Aroma), 1.48(s, 3H, -CH_3), 3.28 (s, 1H, -NH), 3.38 (s, 2H, CH_2-S, Spiro), 2.76 (s, 2H, -CH_2-S), 5.38 (s, 1H, -OH)

3′-(2-(4-hydroxyphenyl)-2-methyl-4-oxothiazolidine-3-carbonyl)-6-methylspiro[indoline-2,2′-thiazolidine]-3,4′-dione (7 f)

Yield 73%; mp 125–126 °C; Mol. Wt: 469.53, Anal. Calcd. for $C_{22}H_{19}N_3O_5S_2$, Elemental analysis: H, 4.10; C, 56.30; N, 8.95, found: H, 4.03; C, 55.98; N, 8.89; IR (KBr): 3081 (C-H, Aroma), 1662 (C=C, Aroma), 1670 (N-C=O), 983 (C-N), 3382 (N-H), 3344 (-OH); ^{1}H NMR ($CDCl_3$ /DMSO-d6) (300 MHz): 6.41–7.56 (m, 7H, Aroma), 1.58 (s, 3H, -CH_3), 3.38 (s, 1H, -NH), 3.48 (s, 2H, CH_2-S, Spiro), 2.70 (s, 2H, -CH_2-S), 5.33 (s, 1H, -OH).

2.2.8 Synthesis of 3′-(2-(4-chlorophenyl)-2-methyl-4-oxothiazolidine-3-carbono thioyl)-6-nitrospiro[indoline-2,2′-thiazolidine]-3,4′-dione. Cyclohexylidenehydrazine (I) (0.02 mol) and thioglycolic acid (0.01 mol) containing in trace $ZnCl_2$ (0.02 g) in DMF was refluxed for 4 h. On completion the reaction mixture was drained over ice cubes. The desired compound precipitated on standing which was filtered and cleaned with water and further recrystallized in ethyl alcohol.

Data Analysis

3′-(2-(4-chlorophenyl)-2-methyl-4-oxothiazolidine-3-carbonothioyl)-6-nitrospiro[indoline-2,2′-thiazolidine]-3,4′-dione (8a)

Yield 75%; mp107–108 °C; Mol. Wt: 535.00, Anal. Calcd. for $C_{21}H_{15}ClN_4O_5S_3$, Elemental analysis: C, 47.15; H, 2.83; Cl, 6.63; N, 10.47 found: C, 47.10; H, 2.80; Cl, 6.57; N, 10.43; IR (KBr): 3068 (C-H, Aroma), 1619 (C=C, Aroma), 1664 (N-C=O), 979 (C-N), 3319 (N-H),1367, 1478 (NO_2), 936 (C-Cl); ^1H NMR ($CDCl_3$/DMSO-d6) (300 MHz): 6.23–7.19(m, 7H, Aroma), 1.39 (s, 3H, -CH_3), 3.92 (s, 1H, -NH), 3.34 (s, 2H, CH_2-S, Spiro), 2.68 (s, 2H, -CH_2-S)

3′-(2-(4-chlorophenyl)-2-methyl-4-oxothiazolidine-3-carbonothioyl)-6-methylspiro[indoline-2,2′-thiazolidine]-3,4′-dione (8b)

Yield 71%; mp 89 °C; Mol. Wt: 504.03, Anal. Calcd. for $C_{22}H_{18}ClN_3O_3S_3$, Elemental analysis: H, 3.60; C, 52.43; N, 8.34, Cl, 7.03; found: H, 3.56; C, 52.40; N, 8.30; Cl, 7.01; IR (KBr): 3065 (C-H, Aroma), 1615 (C=C, Aroma), 1654 (N-C=O), 970 (C-N), 3318 (N-H), 938 (C-Cl); ^1H NMR ($CDCl_3$/DMSO-d6) (300 MHz): 6.29–7.18 (m, 7H, Aroma), 1.49 (s, 3H, -CH_3), 3.99 (s, 1H,-NH), 3.38 (s, 2H, CH_2-S, Spiro), 2.63(s, 2H, -CH_2-S).

3′-(2-(4-bromophenyl)-2-methyl-4-oxothiazolidine-3-carbonothioyl)-6-nitrospiro[indoline-2,2′-thiazolidine]-3,4′-dione (8c)

Yield 58%; mp104°C; Mol. Wt: 579.46, Anal. Calcd. for $C_{21}H_{15}BrN_4O_5S_3$, Elemental analysis: H, 2.61; C, 43.60; N, 9.67, Br, 13.79; found: H, 2.57; C, 43.50; N, 9.62; Br, 13.76; IR (KBr): 3052 (C-H, Aroma), 1614 (C=C, Aroma), 1668 (N-C=O), 972(C-N), 3315 (N-H), 1360, 1472 (NO_2), 736 (C-Br); ^1H NMR ($CDCl_3$/DMSO-d6) (300 MHz): 6.03–7.29 (m, 7H, Aroma), 1.49 (s, 3H, -CH_3), 3.82 (s, 1H, -NH), 3.39 (s, 2H, CH_2-S, Spiro), 2.58 (s, 2H, -CH_2-S)

3′-(2-(4-bromophenyl)-2-methyl-4-oxothiazolidine-3-carbonothioyl)-6-methylspiro[indoline-2,2′-thiazolidine]-3,4′-dione (8d)

Yield 66%; mp 112 °C; Mol. Wt: 548.49, Anal. Calcd. for $C_{22}H_{18}BrN_3O_3S_3$, Elemental analysis: H, 3.29; C, 48.20; N, 7.66; Br, 14.57; found: H, 3.27; C, 48.13; N, 7.63; Br, 14.52; IR (KBr): 3067 (C-H, Aroma), 1629 (C=C, Aroma), 1674 (N-C=O), 989 (C-N), 3319 (N-H), 736 (C-Br); ^1H NMR ($CDCl_3$/DMSO-d6) (300 MHz): 6.29–7.19 (m, 7H, Aroma), 1.38 (s, 3H, -CH_3), 3.90 (s, 1H, -NH), 3.32 (s, 2H, CH_2-S, Spiro), 2.69 (s, 2H, -CH_2-S)

3′-(2-(4-hydroxyphenyl)-2-methyl-4-oxothiazolidine-3-carbonothioyl)-6-nitrospiro[indoline-2,2′-thiazolidine]-3,4′-dione (8e)

Yield 74%; mp 97 °C; Mol. Wt, 516.56, Anal. Calcd. for $C_{21}H_{16}N_4O_6S_3$, Elemental analysis: C, 48.83; H, 3.12; N, 10.85, found: C, 48.78; H, 3.08; N, 10.81., IR (KBr): 3060 (C-H, Aroma), 1610 (C=C, Aroma), 1662 (N-C=O), 976 (C-N), 3310 (N-H), 1363, 1474 (NO_2), 3336(C-OH); ^1H NMR ($CDCl_3$/DMSO-d6) (300 MHz): 6.63–7.89 (m, 7H, Aroma), 1.49 (s, 3H, -CH_3), 3.99 (s, 1H, -NH), 3.38 (s, 2H, CH_2-S, Spiro), 2.63 (s, 2H, -CH_2-S), 4.99 (s, 1H, -OH).

3′-(2-(4-hydroxyphenyl)-2-methyl-4-oxothiazolidine-3-carbonothioyl)-6-methylspiro[indoline-2,2′-thiazolidine]-3,4′-dione (8 f)

Yield 77%; mp114–115 °C; Mol. Wt: 485.59, Anal. Calcd. for $C_{22}H_{19}N_3O_4S_3$, Elemental analysis: C, 54.42; H, 3.94; N, 8.65, found: C, 54.36; H, 3.90; N, 8.60; IR (KBr): 3067 (C-H, Aroma), 1620 (C=C, Aroma), 1660 (N-C=O), 978 (C-N), 3318 (N-H), 3336 (C-OH); ^1H NMR ($CDCl_3$/DMSO-d6) (300 MHz): 6.43–7.80 (m, 7H, Aroma), 1.40 (s, 3H, $-CH_3$), 3.96 (s, 1H, -NH), 3.37 (s, 2H, CH_2-S, Spiro), 2.60 (s, 2H, $-CH_2$-S), 5.09 (s, 1H, -OH).

2.3 Antimicrobial Activity

The serial dilution approach was used to assess the synthetic compounds' antifungal and antibacterial properties. The concentration of the test chemical in dimethyl sulfoxide (DMSO-d6) was determined to be 1 mg/mL. The culture was made by combining 20 mL of plane luria-bertani medium with 1 mL of broth that contained antibacterial growth in order to measure the antibacterial potency. Additionally, a control tube with no antibiotic sample was made. Finally, two control tubes containing fluconazole and ciprofloxacin were made. For antimicrobial examination, all prepared tube samples were incubated for 24 h at 37 °C; for antifungal testing, the tubes were incubated for 96 h at 28 °C. The absorbance value at 600 nm was calculated to verify growth for each conical tube. The MIC (corresponding to the drop in optical density) of the specific derivative was determined by plotting the compound concentration against the absorbance value.

3 Results and Discussion

3.1 Synthesis

In this research we undertook the parallel synthesis of two series namely heterocyclic derivatives of-oxothiazolidine-3-carbonyl)-6- substituted-spiro[indoline-2,2′-thiazolidine]-3,4′-dione **(7a-f)** and -oxothiazolidine-3-carbonothioyl)-6- substituted-spiro [indoline-2,2′-thiazolidine]-3,4′-dione (**8a-f**) as shown in Scheme 1.

For this synthesis we proceeded with refluxing the reaction mixture of substituted 1-Phenyl-ethanone with urea or thiourea in pyridine for 6 h for the formation of **1a-c** and **2a-c** respectively. The compound formed was ice cooled and recrystallized from ethanol. In the next step 2-(substituted)-2-methyl-4-oxothiazolidine-3-carboxamide **(3a-c)** and 2-(4-Substituted)-2-methyl-4-oxothiazolidine-3-carbothioamide **(4a-c)** were synthesized by refluxing **1a-c** and **2a-c** with thioglycolic acid (0.01 mol) containing in trace $ZnCl_2$ (0.1 g) in DMF respectively. All these compounds were characterized by IR NMR and elemental analysis.

Further **3a-c** and **4a-c** were separately refluxed with substituted-isatin in pyridine to get **5a-f** and **6a-f**. Again, the compound synthesized were characterized by IR, NMR and elemental analysis. In the final step of the synthesis 1-(1-(substituyted/

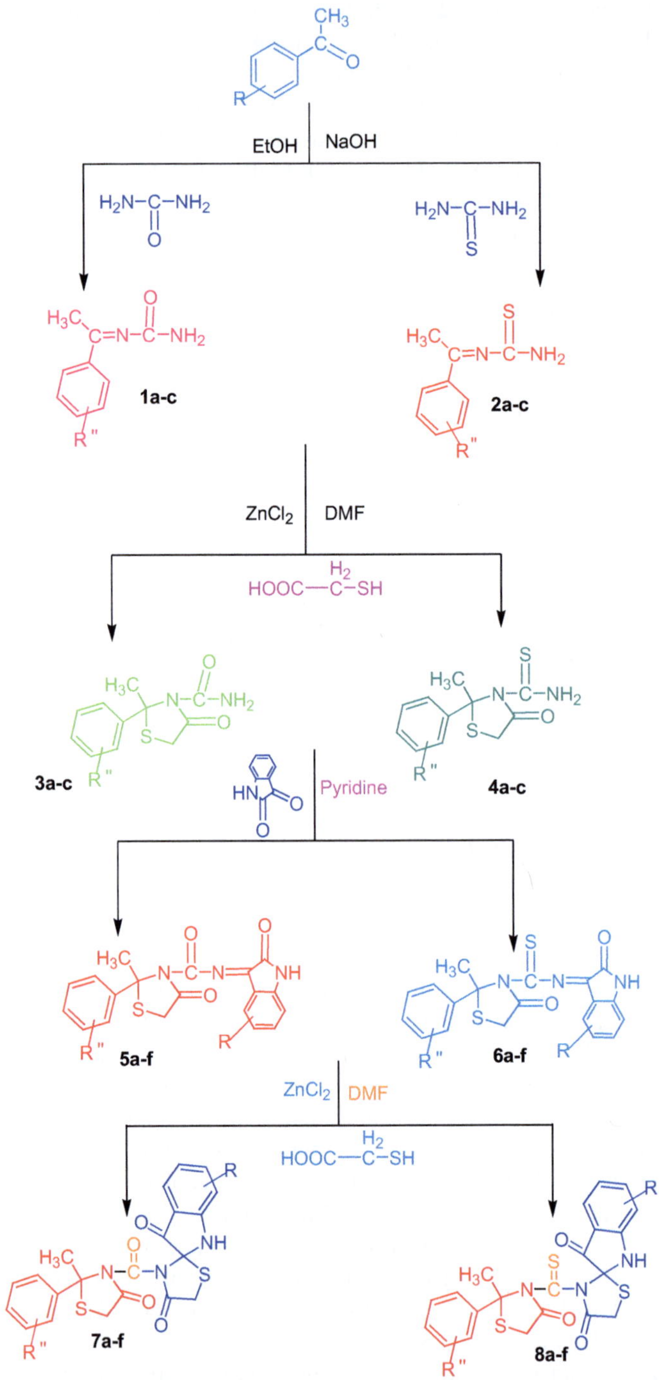

Scheme 1. Synthesis of heterocyclic derivatives of-oxothiazolidine-3-carbonyl)-6- substituted-spiro[indoline-2,2′-thiazolidine]-3,4′-dione and -oxothiazolidine-3-carbonothioyl)-6- substituted -spiro [indoline-2,2′-thiazolidine]-3,4′-dione.

unsubstitutedaryl)-1H-naphtho[1,2-e][1,3]oxazin-2(3H)-yl)-3-(2,4′-dioxospiro[indoline-3,2′-thiazolidin]-3′-yl)thiourea **(7a-f)** and1-(1-(substituyted/unsubstitutedaryl)-1H-naphtho[1,2-e][1,3]oxazin-2(3H)-yl)-3-(2,4′-dioxospiro[indoline-3,2′-thiazolidin]-3′-yl)urea (**8a-f**) were obtained by refluxing **5a-f** and **6a-f** with thioglycolic acid containing in trace $ZnCl_2$ in DMF. All synthesized compounds were characterized by IR, NMR and elemental analysis.

3.2 Antimicrobial Activity

The synthesized series of the compounds **7a-f** and **8a-f** were screened for in vitro antimicrobial action against the Gram-positive bacteria *Corynebacterium striatum MTCC 8963, Staphylococcus pyogenes MTCC 442* and fungus *Aspergillus clavatus MTCC 1323, A. clavatus MTCC1323.* Among the synthesized compound 7e exhibited prominent antimicrobial activity against *Aspergillus clavatus MTCC 1323* and Corynebacterium striatum MTCC 8963. Compound **8b** exhibited promising antimicrobial activity against *A. clavatus MTCC1323* and *Staphylococcus pyogenes MTCC 442.* In the antimicrobial evaluation the standard drug ciprofloxacin and fluconazole were taken as standard. The MIC values of the synthesized compounds in comparison to the standard drugs are given in the Table 1

The table provides values (presumably in micromolar concentrations or percentages) that represent the efficacy of the different compounds against the listed organisms. These values range from 6.25 to 100, indicating varying degrees of inhibition or effectiveness.

For *Aspergillus clavatus* (MTCC 1323), the best-performing compound is p-Cl-phenyl at 12.5 μg/mL (from 8b). Other compounds include p-OH-phenyl and p-CH3-phenyl (both 12.5 μg/mL from 7e and 5d respectively) also exhibit reasonable potency against *Aspergillus clavatus MTCC 1323*. The chlorine group (Cl) group is electron-withdrawing group that tends to make the aromatic ring more electrophilic.

This can enhance the compound reactivity, enabling it to more easily interact with bacterial cell membranes, proteins, or enzymes. The low MIC of indicates that this increased reactivity is beneficial in inhibiting fungal growth at a lower concentration. Chlorine is a moderately lipophilic group. This increases the compound's ability to cross fungal cell membranes, which is a critical step for antifungal activity. Since *Aspergillus clavatus* is a filamentous fungus, compounds that are lipophilic can penetrate its cell wall and membrane more easily, leading to higher efficacy at lower concentrations. The bromine group (Br), like chlorine, is an electron-withdrawing group but is slightly larger in size. While p-Br-phenyl is still potent (with a MIC of 50–100 μg/mL), the slightly larger size of the bromine atom compared to chlorine may decrease the overall interaction with the fungal cell, as its steric effect might not allow as much interaction with the target site.

The hydroxyl group (OH) is an electron-donating group, which makes the phenyl ring less electrophilic the presence of other electrophilic sites in the compound may be enhancing the interaction between the compound 7e and the fungi *Aspergillus clavatus MTCC 1323* and also with the gram-positive bacteria *Corynebacterium striatum MTCC*

Table 1. Antimicrobial activity: (MIC µg/mL) of compounds 7a-f and 8a-f.

			Fungus		Gram-positive Bacteria	
S.N	R	*R"*	*Aspergillus clavatus MTCC 1323*	*A. clavatus MTCC1323*	*Corynebacterium striatum MTCC 8963*	*Staphylococcus pyogenes MTCC 442*
7a.	p-Cl-aryl	6-nitro-isatin	50	50	100	100
7b.		6-methyl-isatin	25	100	100	25
7c.	p-Br-aryl	6-nitro-isatin	100	100	100	100
7d.		6-methyl-isatin	100	25	25	100
7e.	p-OH- aryl	6-nitro-isatin	12.5	50	12.5	50
7 f.		6-methyl-isatin	100	25	50	25
8a.	p-Cl-aryl	6-nitro-isatin	25	100	25	100
8b.		6-methyl-isatin	100	6.25	100	6.25
8c.	p-Br-aryl	6-nitro-isatin	100	50	100	50
8d.		6-methyl-isatin	100	100	50	100
8e.	p-OH- aryl	6-nitro-isatin	100	25	25	50
8 f.		6-methyl-isatin	50	100	50	100
	Ciprofloxacin		50	50		
	Fluconazole				25	25

8963. In other ways hydroxyl group (OH) enhances the compound reactivity, making it more potent against bacterial growth.

The hydroxyl group (OH) can also form hydrogen bonds, which might improve the compound's ability to interact with bacterial proteins, enzymes, or structural components. This could help the compound better penetrate the bacterial cell or disrupt cellular functions, increasing its effectiveness. Electron-Donating Groups (like OH in p-OH-phenyl) tend to increase the compound's ability to interact with bacterial enzymes, disrupt the cell membrane, and affect cell wall synthesis. The hydroxyl group increases the reactivity of the molecule, allowing it to bind more effectively to bacterial proteins or disrupt cellular functions. This is likely why p-OH-phenyl is the most effective compound, with the lowest MIC of 12.5 µg/mL.

4 Conclusion

Novel designing, synthesis and antimicrobial screening of heterocyclic derivatives of-oxothiazolidine-3-carbonyl)-6- substituted-spiro[indoline-2,2′-thiazolidine]-3,4′-dione and -oxothiazolidine-3-carbonothioyl)-6- substituted-spiro [indoline-2,2′-thiazolidine]-3,4′-dione resulted in the procurement of some prominent antimicrobials in the form of compound 7b, 7e, 8b, 8e with notable antimicrobial activity against *Corynebacterium striatum MTCC 8963, Staphylococcus pyogenes MTCC 442* and fungus *Aspergillus clavatus MTCC 1323, A. clavatus MTCC1323.* The compound needs to be further explored for *Insilco* studies and in vitro examinations.

References

1. Soleiman, H.A.: Some fused/isolated heterocyclic of pyrimidine, -lactam, Thiazolidine and Triazine derivatives. Open Catal J. **4**, 18–26 (2011)
2. Shukla, G., et al.: Effect of a novel series of benzothiazolo-quinazolones on epidermal growth factor receptor (EGFR) and biological evaluations. Chem. Biol. Drug Des. **72**, 533–539 (2008)
3. Srivastava, V., et al.: Disubstituted 4(3H) Quinazolones: a novel class of antitumor agents. Chem. Biol. Drug Des. **74**, 297–301 (2009)
4. Tiwari, A.K., et al.: Synthesis and pharmacological study of novel pyrido-quinazolone analogues as anti-fungal, antibacterial, and anticancer agents. Bioorg. Med. Chem. Lett. **16**, 4581–4585 (2006)
5. Deniz, E.A., et al.: Synthesis, characterization, and antimicrobial evaluation of some novel 3-(6- (substituted phenyl)-4-phenyl-6h-1,3-thiazine-2-yl)spiro[indoline3,3- isothiazolidine]-2,4-dione derivativesSynth. Ther. Communities. **48**, 2510–2521 (2018)
6. Aggarwal, S., et al.: Studies for development of novel quinazolinones: new biomarker for EGFR. Spectrochim. Acta A Mol. Biomol. Spectrosc. **143**, 309–318 (2015). https://doi.org/10.1016/j.saa.2015.01.069
7. Acosta, A.L., Rodríguez, A.D.: 11-Oxoaerothionin: a cytotoxic antitumor bromotyrosine-derived alkaloid from the Caribbean marine sponge Aplysina lacunose. J. Nat. Prod. **55**, 1007–1012 (1992)
8. Sannigrahi, M.: Stereocontrolled synthesis of spirocyclics. Tetrahedron. **55**, 9007–9071 (1999)
9. Youssef, M.M., Amin, M.A.: Microwave assisted synthesis of some new heterocyclic Spiro-derivatives with potential antimicrobial and antioxidant activity. Molecules. **15**, 8827–8840 (2010)
10. Shaker, R.M., et al.: Synthesis of novel 1-thia-5,9-diaza-spiro[5.5]undec-2-ene and its recyclization to 1,5,9-triaza-spiro[5.5]undec-2-ene and/or spiro(piperidine-4,2′-thieno[2,3-d]pyrimidine). ARKIVOC. **2**, 57–68 (2011)
11. Chin, Y.W., et al.: Potential anticancer activity of naturally occurring and semisynthetic derivatives of Aculeatins a and B from Amomum aculeatum. J. Nat. Prod. **71**, 390–395 (2008)
12. Wang, W.L., et al.: Three novel, structurally unique Spirocyclic alkaloids from the halotolerant B-17 fungal strain of aspergillus variecolor. Chem. Biodivers. **4**, 2913–2919 (2007)
13. Varun, S., Kakkar, R.: Isatin and its derivatives: a survey of recent syntheses, reactions, and applications. Med. Chem. Commun. **10**, 351–368 (2019)

14. Chinnasamy, R.P., Sundararajan, R.: Synthesis, characterization, and antimicrobial evaluation of some novel 3′-(6-(substituted phenyl)-4-phenyl-6H-1,3-Thiazine-2-yl)Spiro[Indoline-3,3′-Isothiazolidine]-2,4′-Dione derivatives. J. Saudi Chem. Soc. **17**, 337–343 (2016)
15. Kim, D.W., et al.: Quantitative analysis of phenolic metabolites from different parts of Angelica keiskei by HPLC–ESI MS/MS and their xanthine oxidase inhibition. Food Chem. **153**, 20–27 (2014)
16. Chen, X.B., Zhu, H.Y., Bao, K.: Bis-isatin derivatives: design, synthesis, and biological activity evaluation as potent dimeric DJ-1 inhibitors. Acta Pharmacol. Sin. **42**, 1160–1170 (2021)
17. Nath, S., et al.: Isatin containing heterocycles for different biological activities: analysis of structure activity relationship. J. Mol. Struct. **1222**, 128900 (2020)
18. Ferraz de Paiva, R.E., et al.: Anticancer compounds based on Isatin-derivatives: strategies to ameliorate selectivity and efficiency. Front. Mol. Biosci. **7**, 627272 (2021)
19. Zhang, Y.Z., et al.: Isatin dimers and their biological activities. Arch. Pharm. **353**, e1900299 (2020)
20. Ding, Y., et al.: Synthesis and Antiproliferatory activities evaluation of multi-substituted Isatin derivatives. Molecules. **26**, 176 (2020)
21. El-Adl, K., et al.: Design, synthesis, docking, ADMET profile, and anticancer evaluations of novel thiazolidine-2,4- dione derivatives as VEGFR-2 inhibitors. Arch. Pharm. **354**, e2000491 (2021)
22. Sever, B., et al.: A new series of 2,4-thiazolidinediones endowed with potent aldose reductase inhibitory activity. Open Chem. **19**, 347–357 (2021)
23. Oguchi, M., et al.: Molecular design, synthesis, and hypoglycemic activity of a series of Thiazolidine-2,4-diones. J. Med. Chem. **43**, 3052–3066 (2000)
24. Sucheta, Tahlan, S., Verma, P.K.: Biological potential of thiazolidinedione derivatives of synthetic origin. Chem. Cent. J. **11**(1), 130 (2017)
25. Long, N., Le Gresley, A., Wren, S.P.: Thiazolidinediones: an in–depth study of their synthesis and application to medicinal chemistry in the treatment of diabetes mellitus. Chem. Med. Chem. **16**, 1716–1735 (2021)
26. Lebovitz, H.E.: Thiazolidinediones: the forgotten diabetes medications. Curr. Diab. Rep. **19**, 151 (2019)
27. Huiying, Z., Guangying, C., Shiyang, Z.: Design, synthesis and biological activity evaluation of a new class of 2,4-thiazolidinedione compounds as insulin enhancers. J. Enzyme Inhib. Med. Chem. **34**, 981–989 (2019)
28. Kotha, S., et al.: Diversity-oriented synthesis of spirothiazolidinediones and their biological evaluation. Beilstein J. Org. Chem. **15**, 2774–2781 (2019)
29. Dhara, K., et al.: Diversity-oriented synthesis of spirothiazolidinediones and their biological evaluation. Org. Biomol. Chem. **9**, 3801–3807 (2011)
30. Apaydýn, C.B., et al.: Design, synthesis and anti-influenza virus activity of furan-substituted spirothiazolidinones. Bioorg. Chem. **112**, 104958 (2021)
31. Sun, J., et al.: Synthesis of Dihydrothiophenes or Spirocyclic compounds by domino reactions of 1,3- thiazolidinedione. J. Org. Chem. **74**, 3398–3401 (2009)
32. Rao, V.S., et al.: Synthesis of pyranothiazoles and spiro-thiazolidinediones. Ind. J. Heterocycl. Chem. **9**, 247–250 (2000)
33. Yang, W.L., et al.: Asymmetric construction of Spirocyclic Pyrrolidine-thia(oxa)zolidinediones via N,O-ligand/cu(I) catalyzed 1,3-dipolar cycloaddition of Azomethine Ylides with 5-Alkylidene Thia(oxa)zolidine-2,4-diones. Org. Lett. **17**, 4822–4825 (2015)
34. Puerta, A., et al.: Early pharmacological profiling of isatin derivatives as potent and selective cytotoxic agents. Biochem. Pharmacol. **222**, 116059 (2024). https://doi.org/10.1016/j.bcp.2024.116059

35. Xu, Z., Song, X., Hu, Y., Qiang, M., Lv, Z.: Design, synthesis and in vitro anti-mycobacterial activities of 8-OMe ciprofloxacin-1H-1,2,3-triazole-isatin-(thio) semicarbazide/oxime hybrids. J. Heterocyclic Chem. **55**, 192–198 (2018)
36. Lian, Z.M., Sun, J., Zhu, H.L.: Design, synthesis and antibacterial activity of isatin derivatives as FtsZ inhibitors. J. Mol. Struct. **1117**, 8–16 (2016)
37. Abo-Ashour, M.F., et al.: Novel indole-thiazolidinone conjugates: design, synthesis and whole-cell phenotypic evaluation as a novel class of antimicrobial agents. Eur. J. Med. Chem. **160**, 49–60 (2018)
38. Tyagi, R., Yadav, K., Srivastava, N., Sagar, R.: Applications of pyrrole and pyridine-based heterocycles in cancer diagnosis and treatment. Curr. Pharm. Des. **30**, 255–277 (2024). https://doi.org/10.2174/0113816128280082231205071504
39. Srivastava, N., et al.: Clean and green synthesis of organic isothiocyanates using zinc peroxide as desulfurizing agent and their antimicrobial activity. Russ. J. Org. Chem. **59**, 2205–2213 (2023). https://doi.org/10.1134/S1070428023120199
40. Tyagi, R., Singh, K., Srivastava, N., Sagar, R.: Recent advances in carbohydrate-based gelators. Mater. Adv. **4**, 3929–3950 (2023). https://doi.org/10.1039/D3MA00321C
41. Anisree, S., Shanmugavelan, P., Palanisamy, S., Kishore, R., Srivastava, N.: Synthesis and anticancer screening of substituted (phenyl)-1-((naphthalen-5-yl)methylene)thiosemicarbazide against PC3 prostate cancer cells. J. Indian Chem. Soc. **101**(9), 101229 (2024). https://doi.org/10.1016/j.jics.2024.101229
42. Yadav, S., Kishore, R., Srivastava, N.: Microwave-assisted phase transfer catalytic synthesis of aryl benzothiazole derivatives and their anticancer screening. Org. Prep. Proced. Int. **56**(6), 563–571 (2024). https://doi.org/10.1080/00304948.2024.2342692
43. Anisree, S., Shanmugavelan, P., Vijaylakshmi, P., Kishore, R., Srivastava, N.: Synthesis, characterization and anticancer screening of novel phenylbenzylidene thiosemicarbazone derivatives. Phosphorus Sulfur Silicon Relat. Elem. **199**(4), 267–276 (2024). https://doi.org/10.1080/10426507.2024.2330935
44. Srivastava, N., Saxena, A.K.: Caspase-3 activators as anticancer agents. Curr. Protein Pept. Sci. **24**(10), 783–804 (2023). https://doi.org/10.2174/1389203724666230227115305
45. Srivastava, N., Rathore, B.K., Singh, S., Singh, S.: Thermoacoustical study of intermolecular interactions in binary liquid mixtures of benzaldehyde with methanol and ethanol. Chem. Phys. Impact. **6**, 100144 (2022). https://doi.org/10.1016/j.chphi.2022.100144
46. Srivastava, N.: Efficient TBAI-CS_2 promoted synthesis of substituted hydrazinecarbodithiolates. Org. Prep. Proced. Int. **54**(6), 569–577 (2022). https://doi.org/10.1080/00304948.2022.2111171
47. Srivastava, N., Chaturvedi, D., Kishore, R., Kaur, M., Kaur, R.: Triton-B/CS_2 mediated novel synthesis of substituted thioureas: a novel class of anti-cancer agents. Indian J. Chem. B. **61**, 870–877 (2022). https://doi.org/10.56042/ijc.v61i8.65053
48. Srivastava, N.: Tetrabutylammonium iodide/I_2 mediated convenient and green synthesis of substituted organic isothiocyanates. Org. Prep. Proced. Int. **53**(6), 562–570 (2021). https://doi.org/10.1080/00304948.2021.1975487
49. Srivastava, N., Saxena, M.: Novel synthesis of symmetrical dialkyl/diaryl trithiocarbonates in non-aqueous medium at room temperature using CS_2, Cs_2CO_3, and alkyl/aryl halide. Rasayan J. Chem. **12**(1), 333–337 (2019). https://doi.org/10.31788/RJC.2019.1215038
50. Srivastava, N., Kishore, R.: Cleaner, greener synthesis, characterization, and anticancer evaluation of 2-thioxo-1,3-thiazolanes. Indian J. Heterocycl. Chem. **31**(2), 265–271 (2021)
51. Srivastava, N., Kishore, R.: Synthesis, characterization, anticancer, antibacterial, antifungal, and antimalarial activity of (aryl or heteroaryl) substituted-4-methyl-2,3-dihydro-2-thioxo-1H-imidazol-5-yl ethanone derivatives. Indian J. Heterocycl. Chem. **31**(3), 347–355 (2021)

52. Srivastava, N., Rathore, B.K., Vishnoi, R.K., Bajpai, S.: A simple and convenient method for synthesis of artemisinin dimer through aliphatic nitro compounds. Asian J. Chem. **33**(1), 49–52 (2020). https://doi.org/10.14233/ajchem.2021.22918
53. Kishore, R., Kamboj, M., Shukla, M., Chaturvedi, D., Srivastava, N.: Novel synthetic strategy of cyclic dithiocarbamates catalyzed by triton-B. Asian J. Chem. **31**(5), 1091–1094 (2019). https://doi.org/10.14233/ajchem.2019.21830
54. Srivastava, N., Saxena, M., Shukla, M.: Novel synthesis of 5-oxo-2-thioxo-2,5-dihydro-3-thiophenecarboxylate derivatives in non-aqueous medium. Asian J. Chem. **31**(1), 176–180 (2019). https://doi.org/10.14233/ajchem.2019.21622
55. Srivastava, N.: Key role of ionic liquids in the cleaner and greener synthesis of lactams. Res. J. Chem. Environ. **26**(1), 125–130 (2022). https://doi.org/10.25303/2601rjce125130
56. Vishnoi, R.K., Kishore, R., Chaturvedi, D., Shukla, M., Bajpai, S., Srivastava, N.: Synthesis and antimicrobial activity of cyclic dithiocarbamates employing triton-B/CS_2 system. Asian J. Chem. **33**(5), 1133–1136 (2021). https://doi.org/10.14233/ajchem.2021.23173
57. Srivastava, N., Kishore, R., Chaturvedi, D.: Novel and efficient method for the synthesis of cyclic trithiocarbonates employing Cs_2CO_3 and CS_2 system. Res. J. Chem. Environ. **25**(12), 142–148 (2021). https://doi.org/10.25303/2512rjce142148

Synthesis and Antimicrobial Evaluation of Novel Dioxo Spiro[indoline-3,2′ thiazolidin]-3′-yl)thiourea and -dioxaspiro[indoline-3,2′-thiazolidin]-3′-yl)urea

Tandarani Ghosh[1], Parthivi Awasthi[2], Ayushi Sahu[1], R. P. Tiwari[1], R. B. Singh[1], Krishna Srivastava[1](✉), and Nitin Srivastava[2](✉)

[1] Faculty of Chemical Sciences, S R M University, Lucknow-Dewa Road, Barabanki, Uttar Pradesh, India
Krishnajs2063@Gmail.com

[2] Department of Chemistry, Amity University, Lucknow Campus, Noida, Uttar Pradesh, India
nitinsriv5@gmail.com

Abstract. Novel synthesis and antimicrobial screening of 1-(1-(4-chlorophenyl)-1H-naphtho[1,2-e][1,3]oxazin-2(3H)-yl)-3-(2,4′-dioxospiro[indoline-3,2′-thiazolidin]-3′-yl)urea and 1-(1-(4-chlorophenyl)-1H-naphtho[1,2-e][1,3] oxazin-2(3H)-yl)-3-(2,4′-dioxospiro[indoline-3,2′-thiazolidin]-3′-yl)thiourea derivatives is reported in this manuscript. In this research a series of Schiff bases were formed initially by the reaction of isatin with semicarbazide and thiosemicarbazide which were then subjected to react with thioglycolic acid to form cyclized spiro derivatives. The cyclized derivatives were then reacted with substituted aryl aldehydes and β-naphthol under reflux conditions. Finally, the resulting cyclized derivatives of oxazine were formed by reaction of acetonitrile and formaldehyde. All newly synthesized derivatives conformation were characterized in detail by FT-IR and NMR spectroscopy including U-Visible and mass spectrometry which collated well with the proposed synthesis and the experimental observed data. All the prepared derivatives exhibited good to excellent antimicrobial activities with MIC ranging from 100 and 6.25 μg/mL.

Keywords: Isatin · Urea · β-naphthol · Thiourea · Acetonitrile · Formaldehyde · Thioglycolic acid · Substituted aromatic-aldehyde

1 Introduction

Schiff bases which are prominent moiety in the pharmaceutics and industry are reactions between amines and compounds containing carbonyl group [1–7]. The Schiff bases are azomethine and imines in organic chemistry [6, 8]. The presence of lone pairs of electrons makes the compound for exploration in the organic synthesis [9, 10]. The lability of the lone pair of electrons can be used for the formation of diverse class of organic compounds which may be explored for the variety of the synthetic protocols

A. K. Saxena, A. Saxena (Eds.): GTHTM 2025, pp. 189–206, 2026.
https://doi.org/10.1007/978-3-032-12320-6_10

and evaluation of the synthesized scaffolds for the biological activities and improvement in the efficacy and reducing the toxicity of the drug [11–15]. Schiff bases and their metal complexes have been extensively explored in material science, as catalyst, supramolecular chemistry, encapsulation chemistry, bioinorganic chemistry, biomedical chemistry [16–18]. The Schiff bases and their metal complexes have also been prominent antimicrobial and have been extensively explored in this field [19–22]. Schiff bases have also been reported recently to exhibit prominent anticancer activities [23, 24] which further need to be explored for the drug development phases. Moreover, the resistance developed in the microbes for the existing antimicrobial also necessitates the development of novel antimicrobial drugs with the help of involvement of other moieties in the Schiff bases. Antimicrobial resistance among bacteria, fungi, parasites and viruses, fungi and parasites change over time and the traditional drugs are no longer effective against the infections causing sudden rise in the risk of disease spread, severe illness and death. This needs further exploration of the novel drug as per World Health Organization [25]. Recent research reported the antibacterial action of the synthesized Schiff base scaffolds as antibacterial [26] and antifungal [27] (Fig. 1).

Advanced synthetic methodologies and computational methods have simplified the synthesis and facilitated the rational design and optimization of heterocyclic compounds with promising bioactivities using SAR. It has paved the way for novel drug candidates with better therapeutic effects [28]. Heterocyclic compounds bearing a five-membered ring are rich in several bioactivities. Among heterocyclics, thiazolidine is a five-membered with one N and S atom in the ring and plays a prominent role in therapeutics. Thiazolidine may be substituted at 2, 4 and 5 positions for the amelioration in the efficiency of the drug molecule. This makes thiazolidine as the main constituent of the natural products and many drugs, e.g. antimicrobial [29], anticancer [30], antidiabetic [31].

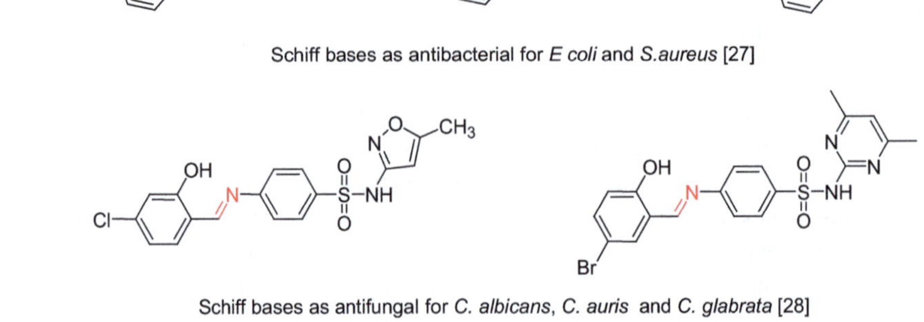

Fig. 1. Structurally diverse biologically potent Schiff bases for antimicrobial activities.

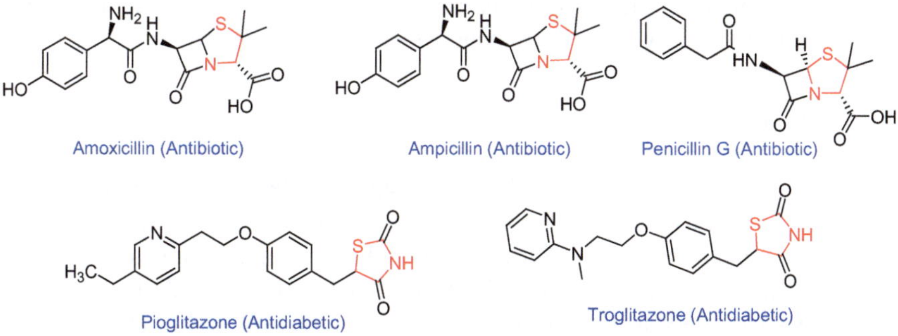

Fig. 2. Diverse clinical drugs containing thiazolidine moiety.

Many thiazolidine derived drugs are clinically used for different therapeutic effects as given in the Fig. 2.

1,3-oxazine has been considered as a privileged scaffold in many medicinal chemistry applications. Compounds having 1,3-oxazine moiety exhibit broad range of biological applications such as anticancer, antimicrobial, anti-inflammatory, antiplatelet, antitubercular and alpha-glucosidase inhibition activities [32]. So we have explored the novel synthesis of 1-(1-(4-chlorophenyl)-1H-naphtho[1,2-e][1,3]oxazin-2(3H)-yl)-3-(2,4′-dioxospiro[indoline-3,2′-thiazolidin]-3′-yl) urea and 1-(1-(4-chlorophenyl)-1H-naphtho[1,2-e][1,3]oxazin-2(3H)-yl)-3-(2,4′-dioxospiro[indoline-3,2′-thiazolidin]-3′-yl)thiourea derivatives which has both thiazolidine and oxazine moiety in the structure and evaluated them for the antimicrobial activities. In our pursuit for the development of biologically potent scaffolds through novel cleaner, greener and easier methods [22, 23, 33–46], here we are reporting the Synthesis and Antimicrobial Evaluation of Novel dioxospiro[indoline-3,2′ thiazolidin]-3′-yl)thiourea and -dioxospiro[indoline-3,2′-thiazolidin]-3′-yl)urea.

2 Materials and Methods

2.1 Instrumentation

The chemicals utilized are of LR grade. BRUKER 300-MHz spectrometer was employed for proton NMR spectral study in $CDCl_3$ or DMSO-d6 solutions. Chemical shifts were represented as part per million, δ values in relation to tetramethyl silane as the internal reference (TMS). A Perkin-Elmer 1650 FTIR spectrometer was used to collect the IR spectra v, cm^{-1} in KBr pellets. NBA served as the matrix in mass spectrometry, whereas Xenon/Argon (10 mA, 6Kv) served as the FAB gas. At CDRI Lucknow, India, elemental analyses were completed using a Thermo Fisher Scientific-11206100 Flash Smart Elemental Analyzer. The synthesized derivatives were determined by using precoated TLC plates as stationary (Merk, 60F-254) eluent 8:2 methanol: ethyl acetate as mobile phase, iodine vapor as visualizing agent. The melting point of all synthesized dioxospiro-thiourea and -dioxaspiro-urea derivatives were examined using the Thermo Scientific manual melting point instrument, which was not adjusted.

2.2 Synthesis

2.2.1 The Synthesis of (E)-N-(2-oxoindolin-3-ylidene)hydrazinecarbothioamide

0.02 mol isatin and 0.02 mol of thiourea in were slowly mixed in 10 mL of pyridine. The mixture was refluxed for 6 h. On completion the reaction mixture was poured into 100 mL of cold acidic water (0–5 °C) mixed with 10 mL of HCl. The desired compound separated out and settled in 1 h. The precipitate was filtered and washed repeatedly with water. The synthesized hydrazinecarbothioamide **1** was dried in vacuum desiccator and recrystallised from ethanol. Yield 86%; mp 93 °C; Mol. Wt: 293.30, Chemical Formula: $C_9H_8N_4OS$ (Scheme 1).

2.2.2 The Synthesis of (E)-N-(2-oxoindolin-3-ylidene)hydrazinecarboxamide

0.02 mol of isatin and semicarbazide was mixed in 10 mL of pyridine and refluxed for 6 h. The mixture was refluxed for 6 h. On completion the reaction mixture was poured into 100 mL of cold acidic water (0–5 °C) mixed with 10 mL of HCl. The desired compound separated out and settled in 1 h. The precipitate was filtered and washed repeatedly with water. The synthesized hydrazinecarbothioamide **2** was dried in vacuum desiccator and recrystallized from ethanol. Yield 88%; mp 89 °C; Mol. Wt: 204.18, Chemical Formula: $C_9H_8N_4O_2$ (Scheme 2)

2.2.3 TheSynthesisofN-(2,4′-dioxospiro[indoline-3,2′-thiazolidin]-3′-yl)hydrazinecarbothioamide

0.02 mol of (E)-N-(2-oxoindolin-3-ylidene)hydrazinecarbothioamide **1** and 0.01 mol of thioglycollic acid was taken in 10 mL of DMF with catalytic amount of $ZnCl_2$ and refluxed (4 h). After completion, the mixture was emptied over ice forming a solid product which was filtered off and repeatedly cleaned with cold water. The synthesized hydrazinecarbothioamide **3** was recrystallized from ethanol. Yield 84%; mp 79 °C; Mol. Wt: 309.36, Chemical Formula: $C_{11}H_{11}N_5O_2S_2$.

IR: 3060 (H-C, str., Aryl), 2856 (H-C, str.), 1685 (O=C, str.), 1652 (N=C, str.), 1545 (C=C,str.), 1360 (N-C, str.), 1022 (S=C, str.), 717 (Cl-C, str.); HNMR: δ 7.45–8.12 (m, 13H, Ar-H), 1.78 (s, 3H, CH_3) (Scheme 3).

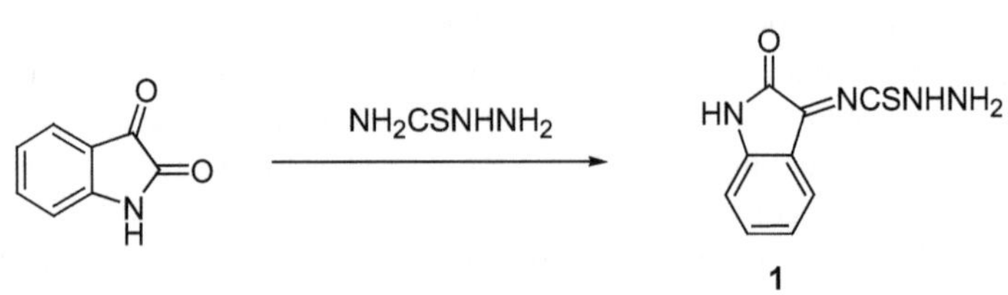

Scheme 1. Synthesis of (E)-N-(2-oxoindolin-3-ylidene)hydrazinecarbothioamide.

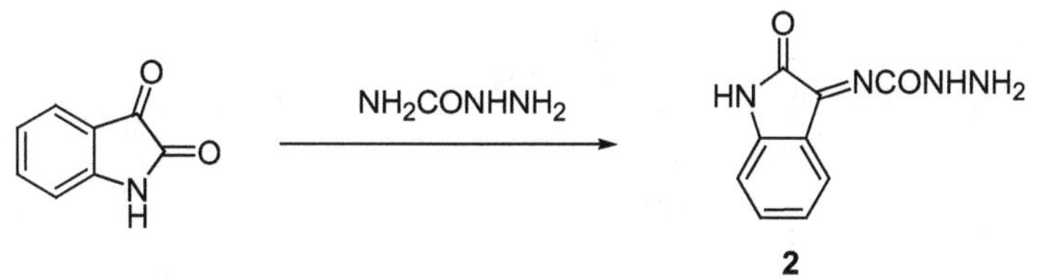

Scheme 2. Synthesis of (E)-N-(2-oxoindolin-3-ylidene)hydrazinecarboxamide.

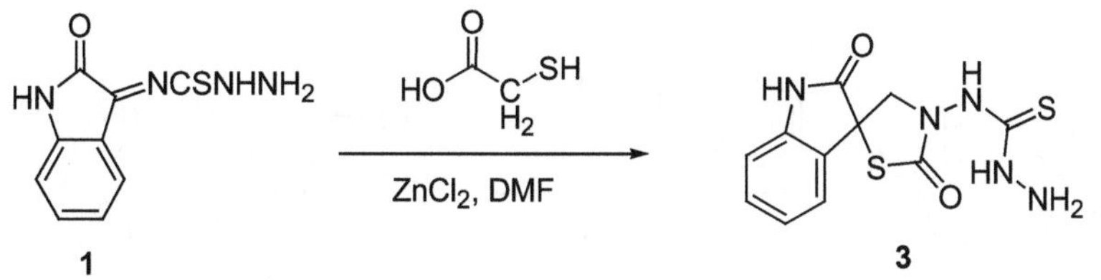

Scheme 3. Synthesis of N-(2,4′-dioxospiro[indoline-3,2′-thiazolidin]-3′-yl)hydrazinecarbothioamide.

2.2.4 The Synthesis of N-(2,4′-dioxospiro[indoline-3,2′-thiazolidin]-3′-yl)hydrazinecarboxamide

0.02 mol of (E)-N-(2-oxoindolin-3-ylidene)hydrazinecarboxamide **2** and 0.01 mol of thioglycollic acid in 10 mL of DMF containing in catalytic amount of $ZnCl_2$ and refluxed (4 h). After completion the reaction mixture was emptied on ice cubes. The product formed was solid, was filtered and cleaned repeatedly using cold water. The synthesized hydrazinecarboxamide **4** was recrystallized from ethanol. Yield 88%; mp 103 °C; Mol. Wt: 293.30 Anal. Calcd. for $C_{11}H_{11}N_5O_3S$, C, 45.05; H, 3.78; N, 23.88, S, 10.93 found: C, 45.01; N, 23.82. S, 10.91 (Scheme 4)

IR: 3050 (H-C, str.), 2852 (H-C, str.), 1680 (O=C, str.), 1650 (N=C, str.), 1550 (C=C,str.), 1360 (N-C, str.), 1020 (S=C str.), 719 (Cl-C, str.); HNMR: δ 7.49–8.10 (m, 13H, Ar-H), 1.78 (s, 3H, CH_3).

2.2.5 The Synthesis of 2-((substituyted/unsubstitutedaryl)(2-hydroxynaphthalen-1-yl)methyl)-N-(2,4′-dioxospiro[indoline-3,2′-thiazolidin]-3′-yl)hydrazine-1-carbothioamide

0.02 mol of N-(2,4′-dioxospiro[indoline-3,2′-thiazolidin]-3′-yl)hydrazinecarbothioamide **3**, 0.02 mol of aldehyde and 0.02 mol of β-naphthol were heated in oil bath at 110–120 °C for 7 h. The rection was observed on TLC and cooled on completion. The solidified compound was cleaned repeatedly with ethanol to get the desired products **5a-f**. The spectral analysis of **5a-f** are given below (Scheme 5).

5a. *2-((4-chlorophenyl)(2-hydroxynaphthalen-1-yl)methyl)-N-(2,4′-dioxospiro [indoline-3,2′-thiazolidin]-3′-yl)hydrazine-1-carbothioamide*

Yield 68%; mp 108–109 °C; Mol. Wt: 576.09 Anal. Calcd. for $C_{28}H_{22}ClN_5O_3S_2$ C, 58.38; H, 3.85; Cl, 6.15; N, 12.16 found: C, 58.32; H, 3.80; Cl, 6.11; N, 12.12.

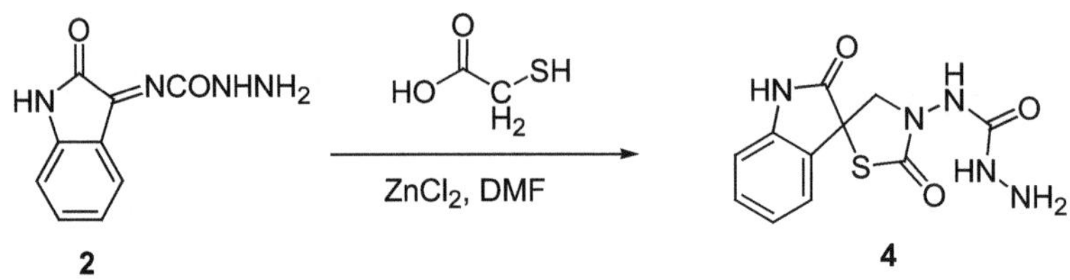

Scheme 4. Synthcsis of N-(2,4′-dioxospiro[indolinc-3,2′-thiazolidin]-3′-yl)hydrazinecarbox-amide.

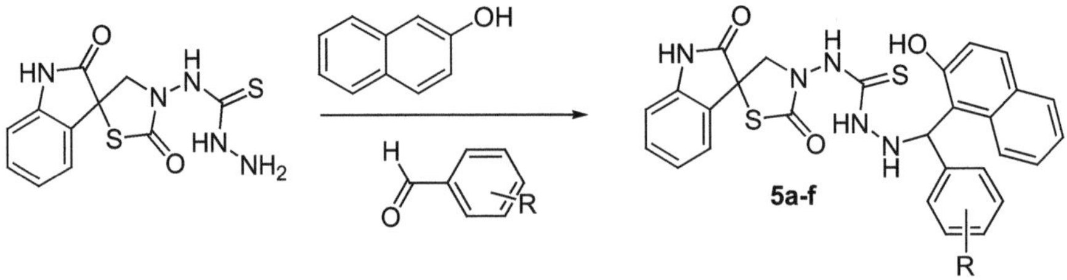

Scheme5. Synthesisof2-((substituyted/unsubstitutedaryl)(2-hydroxynaphthalen-1-yl)methyl)-N-(2,4′-dioxospiro[indoline-3,2′-thiazolidin]-3′-yl)hydrazine-1-carbothioamide.

IR (KBr): 3059 (H-C, str., Aryl), 2859 (H-C, str.), 1668 (O=C, str.), 1639 (N=C, str.), 1558 (C=C,str.), 1360 (N-C, str.), 1020 (S=C, str.), 719 per cm (Cl-C, str.); H NMR: δ7.60–8.10 (m, 13H, Ar-H), 1.78 (s, 3H, CH_3).

5b. *2-((4-bromophenyl)(2-hydroxynaphthalen-1-yl)methyl)-N-(2,4′-dioxospiro[indoline-3,2′-thiazolidin]-3′-yl)hydrazine-1-carbothioamide*

Yield 68%; mp 133 °C; Mol. Wt: 620.54 Anal. Calcd. for $_{28}H_{22}BrN_5O_3S_2$ C, 54.20; H, 3.57; Br, 12.88; N, 11.29 found: C, 54.16; H, 3.52; Br, 12.83; N, 11.25.

IR (KBr): 3070 (H-C, str.), 2871 (H-C str.), 1680 (O=C, str.), 1650 (N=C str.), 1570 (C=C,str.), 1362 (N-C str.), 1020 (S=C str.), 730 (Cl-C str.); H NMR: δ 7.49–7.98 (m, 13H, Ar-H), 1.90 (s, 3H, CH 3).

5c. *N-(2,4′-dioxospiro[indoline-3,2′-thiazolidin]-3′-yl)-2-((2-hydroxynaphthalen-1-yl)(4-hydroxyphenyl)methyl)hydrazine-1-carbothioamide*

Yield 61%; mp 83 °C; Mol. Wt: 557.64 Anal. Calcd. for $C_{28}H_{23}N_5O_4S_2$ C, 60.31; H, 4.20; N, 12.60 found: C, 60.29; H, 4.16; N, 12.52.

IR (KBr): 3072 (H-C str.), 2858 (H-Cstr.), 1682 (O=C str.), 1652 (N=C str.), 1555 (C=C,str.), 1360 (N-C str.), 1020 (S=C str.), 730 per cm (Cl-C str.); H NMR: δ7.60–8.15 (m, 13H, Ar-H), 1.78 (s, 3H, CH 3).

5d. *N-(2,4′-dioxospiro[indoline-3,2′-thiazolidin]-3′-yl)-2-((2-hydroxynaphthalen-1-yl)(phenyl)methyl)hydrazine-1-carbothioamide*

Yield 53%; mp 120–121 °C; Mol. Wt: 541.64 Anal. Calcd. for $C_{28}H_{23}N_5O_3S_2$ C, 62.09; H, 4.28; N, 12.93 found: C, 62.03; H, 4.23; N, 12.91.

IR (KBr): 3075 (H-C str.), 2876 (H-C str.), 1660 (O=C str.), 1650 (N=C str.), 1570 (C=C,str.), 1364 (N-C str.), 1020 (S=C str.), 750 per cm (Cl-C str.); HNMR: δ7.53–8.05 (m, 13H, Ar-H), 1.83 (s, 3H, CH_3).

5e. 2-((2-bromophenyl)(2-hydroxynaphthalen-1-yl)methyl)-N-(2,4′-dioxospiro[indoline-3,2′-thiazolidin]-3′-yl)hydrazine-1-carbothioamide

Yield 62%; mp 114–115 °C; Mol. Wt: 620.54 Anal. Calcd. for $C_{28}H_{22}BrN_5O_3S_2$ C, 54.20; H, 3.57; Br, 12.88; N, 11.29 found: C, 54.17; H, 3.53; Br, 12.82; N, 11.26.

IR (KBr): 3070 (H-C str), 2859 (H-C str.), 1668 (O=C str.), 1650 (N=C str.), 1572 (C=C,str.), 1364 (N-C str.), 1020 (S=C str.), 730 per cm (Cl-C, str.); HNMR: δ7.60–8.10 (m, 13H, Ar-H), 1.91 (s, 3H, CH_3).

5 f. N-(2,4′-dioxospiro[indoline-3,2′-thiazolidin]-3′-yl)-2-((2-hydroxynaphthalen-1-yl)(4-nitrophenyl)methyl)hydrazine-1-carbothioamide

Yield 77%; mp 92–93 °C; Mol. Wt: 586.64 Anal. Calcd. for $C_{28}H_{22}N_6O_5S_2$ C, 57.33 N, 11.29, S,10.93 found: C, 57.29 N, 11.27, S,10.90

IR (KBr): 3071 (H-C str.), 2856 (H-C str.), 1665 (O=C str.), 1650 (N=C str.), 1572 (C=C,str.), 1364 (N-C str.), 1020 (S=C str.), 730 (Cl-C, str.); HNMR: δ7.60–8.10 (m, 13H, Ar-H), 1.91 (s, 3H, CH_3).

2.2.6 The Synthesis of 2-((substituyted/unsubstitutedaryl)(2-hydroxynaphthalen-1-yl)methyl)-N-(2,4′-dioxospiro[indoline-3,2′-thiazolidin]-3′-yl)hydrazine-1-carboxamide

A (0.02 mol) of each N-(2,4′-dioxospiro[indoline-3,2′-thiazolidin]-3′-yl)hydrazinecarboxamide **4** and β-naphthol (0.02 mol) was heated for 9 h at 110-120 °C and monitored on TLC. After completion the cooled reaction led to the formation of solid which was cleaned with ethanol (3x10 mL) thrice resulted in the formation of **6a-f**. Characterization data of the derivatives (**6a-f**), thus synthesized, are given as (Scheme 6):

6a. 2-((4-chlorophenyl)(2-hydroxynaphthalen-1-yl)methyl)-N-(2,4′-dioxospiro[indoline-3,2′-thiazolidin]-3′-yl)hydrazine-1-carboxamide

Yield74%; mp 118–119 °C; Mol. Wt:560.03 Anal. Calcd. for $C_{28}H_{22}ClN_5O_4S$ C, 59.99; H, 4.00; Cl, 6.33; N, 12.49 found: C, 60.08; H, 3.98; Cl, 6.30; N, 12.46

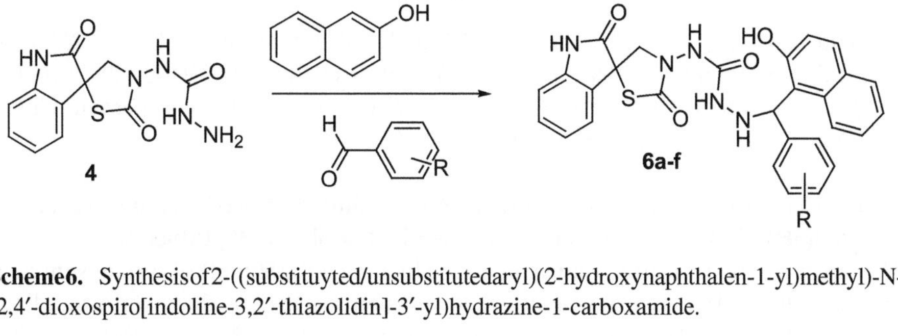

Scheme6. Synthesisof2-((substituyted/unsubstitutedaryl)(2-hydroxynaphthalen-1-yl)methyl)-N-(2,4′-dioxospiro[indoline-3,2′-thiazolidin]-3′-yl)hydrazine-1-carboxamide.

IR (KBr): 3072 (H-C str.), 2871 (H-C str.), 1680 (O=C str.), 1650 (N=C str.), 1570 (C=C,str.), 1362 (N-C str.), 1020 (S=C str.), 718 per cm (Cl-C, str.); HNMR: δ7.48–8.12 (m, 13H, Ar-H), 1.78 (s, 3H, CH_3).

6b. *2-((4-bromophenyl)(2-hydroxynaphthalen-1-yl)methyl)-N-(2,4′-dioxospiro[indoline-3,2′-thiazolidin]-3′-yl)hydrazine-1-carboxamide*

Yield 75%; mp 83–84 °C; Mol. Wt: 604.48 Anal. Calcd. for $C_{28}H_{22}BrN_5O_4S$ C, 55.64; H, 3.67; Br, 13.22; N, 11.59 found: C, 55.61; H, 3.62; Br, 13.18; N, 11.55.

IR (KBr): 3070 (H-C str.), 2874 (H-C str.), 1682 (O=C str.), 1638 (N=C str.), 1555 (C=C,str.), 1360 (N-C str.), 1020 (S=C str.), 730 (Cl-C, str.); HNMR: δ7.53–8.05 (m, 13H, Ar-H), 1.83 (s, 3H, CH_3).

6c. *N-(2,4′-dioxospiro[indoline-3,2′-thiazolidin]-3′-yl)-2-((2-hydroxynaphthalen-1-yl)(4-hydroxyphenyl)methyl)hydrazine-1-carboxamide*

Yield 54%; mp 117–118 °C; Mol. Wt: 541.58 Anal. Calcd. for $C_{28}H_{23}N_5O_5S$ C, 62.10; H, 4.28; N, 12.93 found: C, 61.97; H, 4.22; N, 12.90. IR (KBr): 3052 (H-C str.), 2875 (H-C str.), 1668 (O=C str.), 1639 (N=C str.), 1557 (C=C,str.), 1363 (N-C str.), 1022 (S=C str.), 719 per cm (Cl-C, str.); HNMR: δ7.49–7.98 (m, 13H), 1.83 (s, 3H).

6d. *N-(2,4′-dioxospiro[indoline-3,2′-thiazolidin]-3′-yl)-2-((2-hydroxynaphthalen-1-yl)(phenyl)methyl)hydrazine-1-carboxamide*

Yield 63%; mp 119 °C; Mol. Wt: 525.58 Anal. Calcd. for $C_{28}H_{23}N_5O_4S$ C, 64.01; H, 4.39; N, 13.29, found: C, 64.05; H, 4.36; N, 12.99.

IR (KBr): 3056 (H-C str.), 2872 (H-C str.), 1681 (O=C str.), 1650 (N=C str.), 1558 (C=C,str.), 1364 (N-C str.), 1012 (S=C str.), 718 (Cl-C str.); HNMR: δ7.48–7.99 (m, 13H), 1.78 (3H, s).

6e. *2-((2-bromophenyl)(2-hydroxynaphthalen-1-yl)methyl)-N-(2,4′-dioxospiro[indoline-3,2′-thiazolidin]-3′-yl)hydrazine-1-carboxamide*

Yield 55%; mp 109 °C; Mol. Wt: 604.48 Anal. Calcd. for $C_{28}H_{22}BrN_5O_4S$, C, 55.64; H, 3.67; Br, 13.22; N, 11.59 found: C, 55.60; H, 3.64; Br, 13.17; N, 11.56.

IR (KBr): 3055 (H-C), 2870 (H-C), 1680 (O=C), 1654 (N=C.), 1574 (C=C), 1365 (N-C), 1022 (S=C), 730 (Cl-C); HNMR: δ7.47–7.97 (m, 13H), 1.78 (3H, s).

6 f. *N-(2,4′-dioxospiro[indoline-3,2′-thiazolidin]-3′-yl)-2-((2-hydroxynaphthalen-1-yl)(4-nitrophenyl)methyl)hydrazine-1-carboxamide*

Yield 70%; mp 109 °C; Mol. Wt: 570.58 Anal. Calcd. for $C_{28}H_{22}N_6O_6S$ C, 58.94; N, 14.73; S, 5.62 found: C, 58.90; N, 14.71; S, 5.58IR (KBr): 3052 (H-C), 2876 (H-C), 1682 (O=C), 1651 (N=C), 1556 (C=C), 1364 (N-C), 1012 (S=C), 718 per cm (Cl-C); 1H NMR: δ7.54–8.06 (m, 13H, Ar-H), 1.92 (s, 3H, CH_3).

2.2.7 The Synthesis of 1-(1-(substituyted/unsubstitutedaryl)-1H-naphtho[1,2-e][1,3]oxazin-2(3H)-yl)-3-(2,4′-dioxospiro[indoline-3,2′-thiazolidin]-3′-yl)thiourea

The target compounds **7a-f** were synthesized by taking (0.02 mol) of 2-((substituyted/unsubstitutedaryl)(2-hydroxynaphthalen-1-yl)methyl)-N-(2,4′-dioxospiro[indoline-3,2′-thiazolidin]-3′-yl)hydrazine-1-carbothioamide (**5a-f**) and formaldehyde (37%,

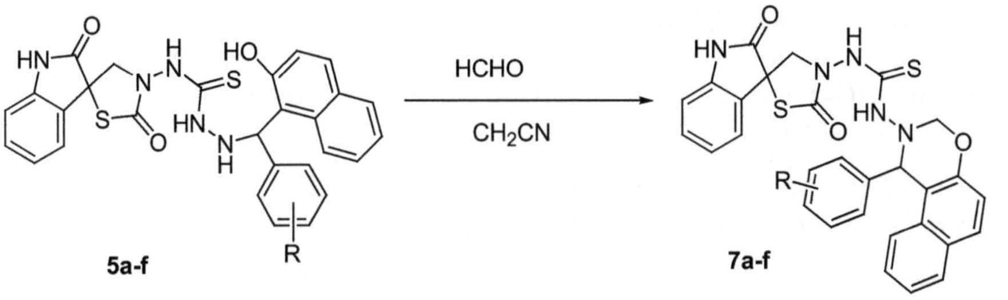

Scheme 7. Synthesis of 1-(1-(substituyted/unsubstitutedaryl)-1H-naphtho[1,2-e][1,3]oxazin-2(3H)-yl)-3-(2,4′-dioxospiro[indoline-3,2′-thiazolidin]-3′-yl)thiourea.

0.02 mol) were taken in 20 ml acetonitrile. The reaction Refluxing for 11–14 h led to the formation of the product as confirmed by TLC. The solvent in the reaction mixture was evaporated using rotatory evaporator and it was further added with 100 ml of distilled water. The resultant solution was extracted with ethyl acetate, and the solvent was distilled off under reduced pressure. The obtained residue on purification through recrystallization using ethanol yielded the product (Scheme 7).

7a. *1-(1-(4-chlorophenyl)-1H-naphtho[1,2-e][1,3]oxazin-2(3H)-yl)-3-(2,4′-dioxospiro[indoline-3,2′-thiazolidin]-3′-yl)thiourea*

Yield 61%; mp 124 °C; Mol. Wt: 588.10, Anal. Calcd. for $C_{29}H_{22}ClN_5O_3S_2$ C, 59.23; H, 3.77; Cl, 6.03; N, 12.00 found: C, 59.21; H, 3.73; Cl, 6.00; N, 12.01.

IR (KBr): 3056 (H-C), 2858 (H-C str.), 1680 (O=C), 1638 (N=C), 1573 (C=C), 1362 (N-C), 1020 (S=C), 730 (Cl-C); ^{1}H NMR: δ7.53–8.05 (m, 13H, Ar-H), 1.83 (s, 3H, CH_3).

7b. *1-(1-(4-bromophenyl)-1H-naphtho[1,2-e][1,3]oxazin-2(3H)-yl)-3-(2,4′-dioxospiro[indoline-3,2′-thiazolidin]-3′-yl)thiourea*

Yield 70%; mp 133–134 °C; Mol. Wt: 632.55 Anal. Calcd. for $C_{29}H_{22}BrN_5O_3S_2$ C, 55.07; H, 3.51; Br, 12.63; N, 11.07 found: C, 55.02; H, 3.46; Br, 12.60; N, 11.02.

IR (KBr): 3070 (H-C), 2876 (H-C), 1683 (O=C), 1654 (N=C), 1575 (C=C), 1372 (N-C), 1022 (S=C), 732 per cm (Cl-C); ^{1}HNMR: δ7.53–8.05 (m, 13H, Ar-H), 1.83 (s, 3H, CH_3).

7c. *1-(2,4′-dioxospiro[indoline-3,2′-thiazolidin]-3′-yl)-3-(1-(4-hydroxyphenyl)-1H-naphtho[1,2-e][1,3]oxazin-2(3H)-yl)thiourea*

Yield 77%; mp 92–93 °C; Mol. Wt: 569.65, Anal. Calcd. for $C_{29}H_{23}N_5O_4S_2$ C, 61.15; H, 4.07; N, 12.29 found: C, 61.12; H, 4.05; N, 12.26.

IR (KBr): 3070 (H-C), 2876 (H-C), 1686 (O=C), 1658 (N=C), 1587 (C=C), 1365 (N-C), 1024 (S=C), 732 (Cl-C); ^{1}HNMR: δ7.53–8.05 (m, 13H, Ar-H), 1.83 (s, 3H, CH_3).

7d. *1-(2,4′-dioxospiro[indoline-3,2′-thiazolidin]-3′-yl)-3-(1-phenyl-1H-naphtho[1,2-e][1,3]oxazin-2(3H)-yl)thiourea*

Yield 61%; mp 89 °C; Mol. Wt: 553.66, Anal. Calcd. For $C_{29}H_{23}N_5O_3S_2$: C, 63.01; H, 4.20; N, 12.65 found: C, 62.86; H, 4.16; N, 12.63; IR (KBr): 3054 (H-C), 2859 (H-C), 1668 (O=C), 1652 (N=C), 1572 (C=C), 1360 (N-C), 1020 (S=C), 734 (Cl-C); ^{1}H NMR: δ7.53–8.05 (m, 13H, Ar-H), 1.83 (s, 3H, CH_3).

7e. 1-(1-(2-bromophenyl)-1H-naphtho[1,2-e][1,3]oxazin-2(3H)-yl)-3-(2,4′-dioxospiro [indoline-3,2′-thiazolidin]-3′-yl)thiourea

Yield 52%; mp 128–129 °C; Mol. Wt: 632.55 Anal. Calcd. for $C_{29}H_{22}BrN_5O_3S_2$ C, 55.07; H, 3.51; Br, 12.63; N, 11.07 found: C, 55.00; H, 3.47; Br, 12.60; N, 11.02; IR (KBr): 3072 (H-C), 2858 (H-C), 1666 (O=C), 1635 (N=C), 1570 (C=C), 1368 (N-C), 1024 (S=C), 718 (Cl-C); ^{1}H NMR: δ7.48–7.98 (m, 13H, Ar-H), 1.78 (s, 3H, CH_3).

7 f. 1-(2,4′-dioxospiro[indoline-3,2′-thiazolidin]-3′-yl)-3-(1-(4-nitrophenyl)-1H-naphtho[1,2-e][1,3]oxazin-2(3H)-yl)thiourea

Yield 72%; mp 107–108 °C; Mol. Wt: 598.10 Anal. Calcd. for $C_{29}H_{22}N_6O_5S_2$ C, 58.18; N, 14.04 S, 10.71 found: C, 58.14; N, 14.00 S, 10.66; IR (KBr): 3070 (H-C), 2856 (H-C), 1664 (O=C), 1638 (N=C), 1572 (C=C), 1365 (N-C), 1022 (S=C), 718 (Cl-C); ^{1}H NMR: δ7.47–7.95 (m, 13H, Ar-H), 1.76 (s, 3H, CH_3).

2.2.8 The Synthesis of 1-(1-(substituyted/unsubstitutedaryl)-1H-naphtho[1,2-e][1,3] oxazin-2(3H)-yl)-3-(2,4′-dioxospiro[indoline-3,2′-thiazolidin]-3′-yl)urea:

The target compounds **8a-f** were synthesized by taking 0.02 mol of 2-((substituyted/ unsubstituted aryl)(2-hydroxynaphthalen-1-yl)methyl)-N-(2,4′-dioxospiro[indoline-3,2′-thiazolidin]-3′-yl)hydrazine-1-carboxamide(**6a-f**) and formaldehyde (37%,0.02 mol) were taken in 20 ml acetonitrile. The multicomponent reaction proceeded with refluxing for 10–14 h. Once the reaction finished, 1000 mL of water was added after the removal of the solvent. The content was extracted with ethyl acetate, and solvent distilled out under reduced pressure. The residue obtained was purified by crystallization from ethanol (Scheme 8).

8a. 1-(1-(4-chlorophenyl)-1H-naphtho[1,2-e][1,3]oxazin-2(3H)-yl)-3-(2,4′-dioxospiro [indoline-3,2′-thiazolidin]-3′-yl)urea

Yield 76%; mp 123–124 °C; Mol. Wt: 572.04 Anal. Calcd. for $C_{29}H_{22}ClN_5O_4S$ C, 60.89; H, 3.90; Cl, 6.18; N, 12.19 found: C, 60.82; H, 3.85; Cl, 6.16; N, 12.22; Exact Mass: 571.10810, Molecular Weight: 572.03600; m/z: 571.10810 (100.0%), 573.10515 (32.0%), 572.11146 (31.4%), 574.10851 (10.0%), 573.11481 (4.7%), 573.10390 (4.5%), 572.10514 (1.8%), 575.11186 (1.5%), 575.10095 (1.4%), 574.10725 (1.4%); IR (KBr): 3075 (H-C), 2856 (H-C), 1668 (O=C), 1653 (N=C), 1570 (C=C), 1363 (N-C), 1021 (S=C), 730 (Cl-C); 1H NMR: δ7.48–7.99 (m, 13H, Ar-H), 1.77 (s, 3H, CH_3).

8b. 1-(1-(4-bromophenyl)-1H-naphtho[1,2-e][1,3]oxazin-2(3H)-yl)-3-(2,4′-dioxospiro [indoline-3,2′-thiazolidin]-3′-yl)urea

Yield 64%; mp 108 °C; v Anal. Calcd. for $C_{29}H_{22}BrN_5O_4S$: C, 56.50; H, 3.60; Br, 12.96; N, 11.36 found: C, 56.42; H, 3.55; Br, 12.90; N, 11.33; Mol. Wt: 616.49000; m/z: 615.05759 (100.0%), 617.05554 (97.3%), 616.06094 (31.4%), 618.05890

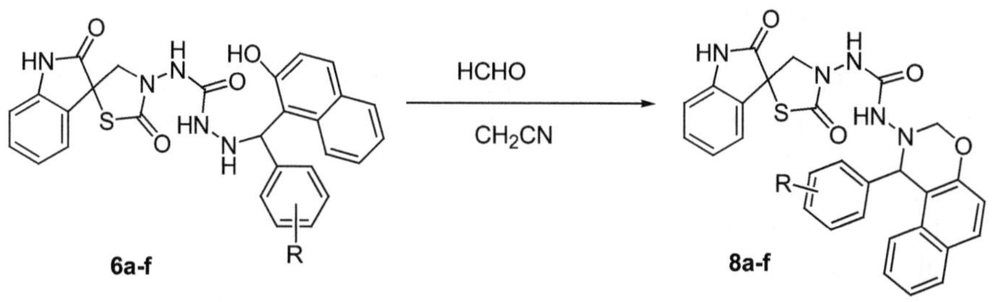

Scheme 8. Synthesis of 1-(1-(substituyted/unsubstitutedaryl)-1H-naphtho[1,2-e][1,3]oxazin-2(3H)-yl)-3-(2,4′-dioxospiro[indoline-3,2′-thiazolidin]-3′-yl)urea.

(30.5%), 617.06430 (4.7%), 619.06225 (4.6%), 617.05338 (4.5%), 619.05134 (4.4%), 616.05462 (1.8%), 618.05258 (1.8%), 618.05674 (1.4%), 620.05469 (1.4%); IR (KBr): 3075 (H-C), 2870 (H-C), 1680 (O=C), 1650 (N=C), 1556 (C=C), 1364 (N-C), 1021 (S=C), 718 per cm (Cl-C); ^{1}H NMR: δ7.48–7.98 (m, 13H, Ar-H), 1.79 (s, 3H, CH_3).

8c. *1-(2,4′-dioxospiro[indoline-3,2′-thiazolidin]-3′-yl)-3-(1-(4-hydroxyphenyl)-1H-naphtho[1,2-e][1,3]oxazin-2(3H)-yl)urea*

Yield 59%; mp 115 °C; Mol. Wt: 553.59 Anal. Calcd. for $C_{29}H_{23}N_5O_5S$; H, 4.20; C, 63.01; N, 12.59 found: C, 62.90; H, 4.14; N, 12.63; IR (KBr): 3072 (C-H, str., Aryl), 2870 (H-C), 1680 (O=C), 1652 (N=C), 1572 (C=C), 1365 (N-C), 1022 (S=C), 730 (Cl-C); ^{1}H NMR: δ7.62–8.15 (m, 13H, Ar-H), 1.90 (s, 3H, CH_3).

8d. *1-(2,4′-dioxospiro[indoline-3,2′-thiazolidin]-3′-yl)-3-(1-phenyl-1H-naphtho[1,2-e][1,3]oxazin-2(3H)-yl)urea*

Yield 56%; mp 79 °C; Mol. Wt:537.59 Anal. Calcd. for $C_{29}H_{23}N_5O_4S$: H, 4.31; C, 65.01; N, 12.99 found: H, 4.26; C, 63.98; N, 13.00; IR (KBr): 3058 (H-C), 2876 (H-C), 1680 (O=C), 1654 (N=C), 1572 (C=C), 1364 (N-C), 1020 (S=C), 732 (Cl-C); ^{1}H NMR: δ 7.53–8.05 (m, 13H, Ar-H), 1.83 (s, 3H, CH_3).

8e. *1-(1-(2-bromophenyl)-1H-naphtho[1,2-e][1,3]oxazin-2(3H)-yl)-3-(2,4′-dioxospiro[indoline-3,2′-thiazolidin]-3′-yl)urea*

Yield 60%; mp 103 °C; Anal. Calcd. for $C_{29}H_{22}BrN_5O_4S$: H, 3.60; C, 56.50; Br, 12.96; N, 11.36 found: H, 3.58; C, 56.44; Br, 12.95; N, 11.33; Mol. Wt: 616.49, m/z: 615.05759 (100.0%), 617.05554 (97.3%), 616.06094 (31.4%), 618.05890 (30.5%), 617.06430 (4.7%), 619.06225 (4.6%), 617.05338 (4.5%), 619.05134 (4.4%), 616.05462 (1.8%), 618.05258 (1.8%), 618.05674 (1.4%), 620.05469 (1.4%); IR (KBr): 3070 (H-C), 2872 (H-C), 1684 (O=C), 1653 (N=C), 1575 (C=C), 1360 (N-C), 1020 (S=C), 732 (Cl-C); ^{1}H NMR: δ7.49–7.95 (m, 13H, Ar-H), 1.76 (s, 3H, CH_3).

8 f. *1-(2,4′-dioxospiro[indoline-3,2′-thiazolidin]-3′-yl)-3-(1-(4-nitrophenyl)-1H-naphtho[1,2-e][1,3]oxazin-2(3H)-yl)urea*

Yield 62%; mp 103–104 °C; Mol. Wt: 582.59 Anal. Calcd. for $C_{29}H_{22}N_6O_6S$ C, 59.79; N, 14.43 S, 5.50 found: C, 59.75; N, 14.40; S, 5.45; IR (KBr): 3072 (H-C), 2870 (H-C), 1682 (O=C), 1654 (N=C), 1572 (C=C), 1362 (N-C), 1025 (S=C), 734 (Cl-C); ^{1}H NMR: δ7.47–7.92 (m, 13H, Ar-H), 1.74 (s, 3H, CH_3)

2.3 In Vitro Antimicrobial Activity

The antimicrobial evaluation was done using serial dilution approach for the antifungal and antibacterial properties. The concentration of the test chemical in dimethyl sulfoxide (DMSO-d6) was determined to be 1 mg/mL. The culture was made by combining 20 mL of plane luria bertani medium with 1 mL of broth that contained antibacterial growth in order to measure the antibacterial potency. Additionally, a control tube with no antibiotic sample was made. Finally, two control tubes containing Ampicillin and Gentamycin were made. For antimicrobial examination, all prepared tube samples were incubated for 24 h at 37 °C; for antifungal testing, the tubes were incubated for 96 h at 28 °C. The absorbance value at 600 nm was calculated to verify growth for each conical tube. The MIC (corresponding to the drop in optical density) of the specific derivative was determined by plotting the compound concentration against the absorbance value.

3 Results and Discussion

In this research novel synthesis of 1-(1-(4-chlorophenyl)-1H-naphtho[1,2-e][1,3] oxazin-2(3H)-yl)-3-(2,4′-dioxospiro[i doline-3,2′-thiazolidin]-3′-yl)urea and 1-(1-(4-chlorophenyl)-1H-naphtho[1,2-e][1,3]oxazin-2(3H)-yl)-3-(2,4′-dioxospiro[indoline-3,2′-thiazolidin]-3′-yl)thiourea derivatives which has both thiazolidine and oxazine moiety in the structure and evaluated them for the antimicrobial activities.

First of all we synthesized Synthesis of (E)-N-(2-oxoindolin-3-ylidene)hydrazinecarbothioamide with the help of the reaction between isatin and thiourea in the presence of pyridine. The mixture was refluxed and then poured into cold acidic water. The desired compound separated out in the form of (E)-N-(2-oxoindolin-3-ylidene)hydrazinecarbothioamide (**1**) which is a starting material for the synthesis of dioxospiro[indoline-3,2′ thiazolidin]-3′-yl)thiourea**.**

For the synthesis of second series of -dioxospiro[indoline-3,2′-thiazolidin]-3′-yl) urea, the Synthesis of starting material (E)-N-(2-oxoindolin-3-ylidene)hydrazinecarboxamide (**2**) was done by reacting isatin and semicarbazide in presence of pyridine on similar pattern as above.

The second parallel step for the Synthesis of N-(2,4′-dioxospiro[indoline-3,2′-thiazolidin]-3′-yl)hydrazinecarbothioamide and N-(2,4′-dioxo spiro[indoline-3,2′-thiazolidin]-3′-yl)hydrazinecarboxamide was carried out by refluxing (E)-N-(2-oxoindolin-3-ylidene)hydrazinecarbothioamide (**2**) and thioglycolic acid in DMF with catalytic amount of $ZnCl_2$. After completion, the mixture was poured over ice. The solid product was filtered off and repeatedly washed with cold water and recrystallization yielded (**3**) and (**4**) separately in both series. These compounds were characterized using IR, NMR and elemental analysis with appreciable collation in the observed and the theoretical values.

In the third parallel step we synthesized 2-((substituyted/unsubstitutedaryl) (2-hydroxynaphthalen-1-yl)methyl)-N-(2,4′-dioxospiro[indoline-3,2′-thiazolidin]-3′-yl)hydrazine-1-carbothioamide (**5a-f**) and 2-((substituted/unsubstitutedaryl)

(2-hydroxynaphthalen-1-yl)methyl)-N-(2,4′-dioxospiro[indoline-3,2′-thiazolidin]-3′-yl)hydrazine-1-carboxamide **(6a-f)** by reaction of **3** and **4** with different aldehyde and β-naphthol by heating over oil bath at 110–120 °C for 7 h. The solidified compound was washed repeatedly with ethanol to get the desired products. These compounds were also characterized using IR, NMR and elemental analysis with appreciable matching in the observed and the theoretical values.

In the fourth step, the target compounds were synthesized by taking **5a-f** and **6a-f** parallelly and refluxing them with formaldehyde in the presence of acetonitrile for 11–14 h. This resulted in the formation of the target compounds **7a-f** and **8a-f** for the two series (Scheme 9).

The two synthesized series of urea and thiourea derivatives were screened for their in-vitro for their anti-bacterial and anti-fungal activity against two gram positive, two gram-negative bacteria and one fungus namely *C. krusei*, *S.pneumonia*, *E.coli*, *Salmonella typhimurium* and *Aspergillus fumigatus* respectively.

The standard antibacterial Ampicillin and Gentamicin were used as standard for the antimicrobial evaluation. It was found that almost all of the synthesized compounds possessed appreciable antimicrobial activity. Among the synthesized compounds **7e** and **7 f** possessed the highest activity against gram-positive bacteria *C. krusei* while the compound **7b** and **7d** possessed the highest activity for the gram-positive bacteria *S. pneumonia*. The compound 7e and **7 f** were most prominent against the gram-negative bacteria *E coli* and *Salmonella typhimurium* respectively. In the second series synthesized compound **8e** possesses antimicrobial action against both gram-negative bacteria *Salmonella typhimurium* and fungus *Aspergillus fumigatus* as shown in Table 1.

4 Conclusion

Novel Synthetic series of 1-(1-(4-chlorophenyl)-1H-naphtho[1,2-e][1,3]oxazin-2(3H)-yl)-3-(2,4′-dioxospiro[indoline-3,2′-thiazolidin]-3′-yl)urea and 1-(1-(4-chlorophenyl)-1H-naphtho[1,2-e][1,3]oxazin-2(3H)-yl)-3-(2,4′-dioxospiro[indoline-3,2′-thiazolidin]-3′-yl)thiourea derivatives are prominent Schiff bases have been synthesized and found to exhibit significant antimicrobial activities with MIC ranging from 100 and 6.25 μg/mL. The study of the antimicrobial activities against two gram positive, two gram-negative bacteria and one fungus namely *C. krusei*, *S. pneumonia*, *E. coli*, *Salmonella typhimurium* and *Aspergillus fumigatus* indicates that the ortho and para substituted derivatives based on their electron withdrawing effects possess notable antimicrobial action. This research provides novel compounds bearing thiourea and urea moieties which are well known for their bioactivities. The reported compounds that are easily synthesized in the laboratory will surely provide another antimicrobial in the rising cases of antimicrobial resistance among the microbes. Such synthesized compound may further be explored for other bioactivities like anticancer, antioxidant, anti-malarial antituberculosis etc. The synthesized compound may be further explored for their in vivo action and may be developed as new antimicrobials in the event of the rising adaption of the clinical drugs among the discussed classes of the microbes.

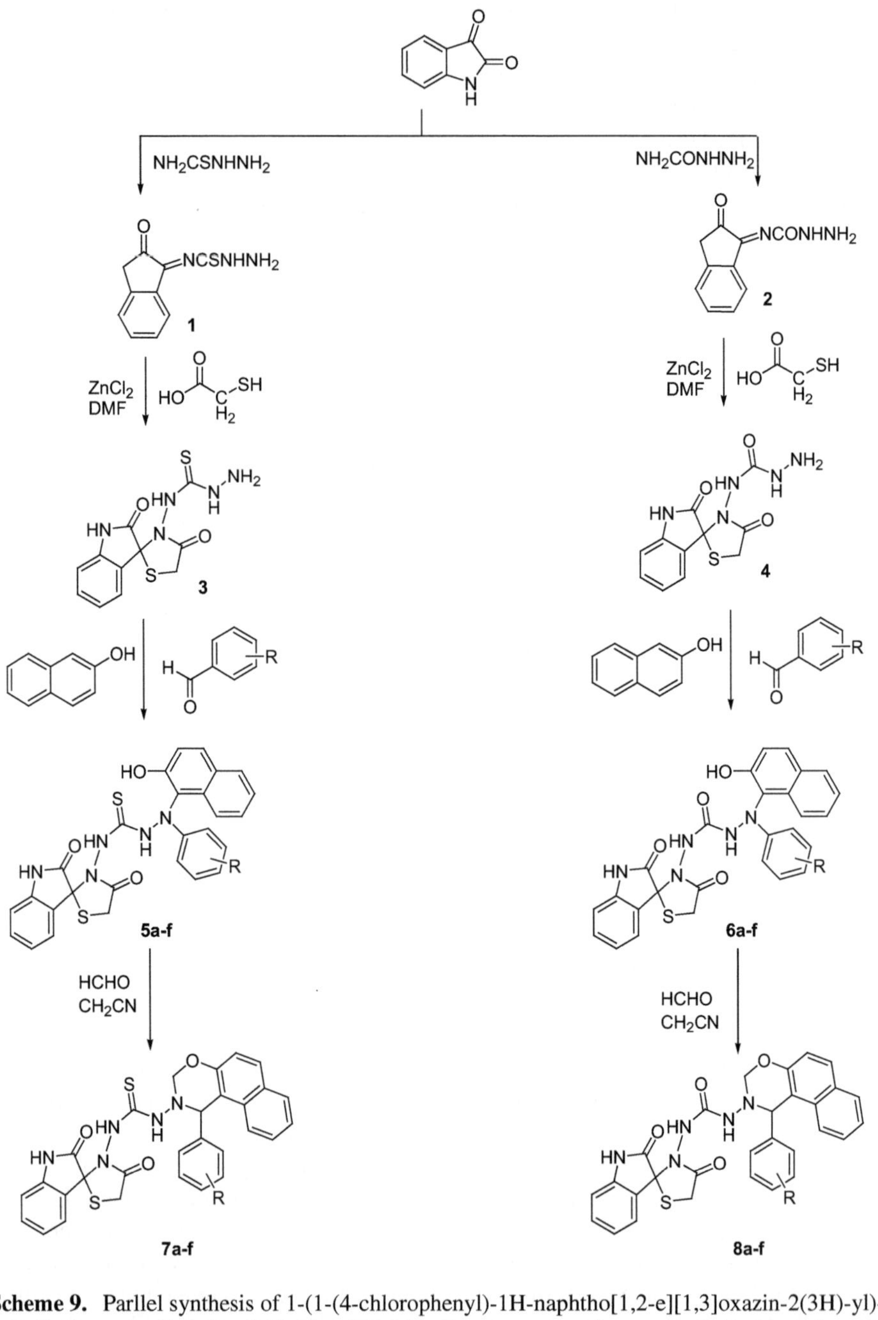

Scheme 9. Parllel synthesis of 1-(1-(4-chlorophenyl)-1H-naphtho[1,2-e][1,3]oxazin-2(3H)-yl)-3-(2,4′-dioxospiro[indoline-3,2′-thiazolidin]-3′-yl)urea and 1-(1-(4-chlorophenyl)-1H-naphtho[1,2-e][1,3]oxazin-2(3H)-yl)-3-(2,4′-dioxospiro[indoline-3,2′-thiazolidin]-3′-yl) thiourea derivatives.

Table 1. In vitro anti-microbial activity for the newly synthesized derivatives of dioxospiro-thiourea(**7a-e**) and -dioxospiro-urea(**8a-e**) against +ve, Gram -ve and fungus.

Entry		Gram positive MIC (μg/mL)		Gram negative MIC (μg/mL)		Fungus MIC μ(g/mL)
	Substituent	C. krusei	S. pneumonia	E. coli	Salmonella typhimurium	Aspergillus fumigatus
7a.	p-Cl-phenyl	50	100	50	25	100
7b.	p-Br- phenyl	100	25	50	100	50
7c.	p-OH-phenyl	50	100	25	25	50
7d.	Phenyl	50	25	100	100	100
7e.	o-Br-phenyl	12.5	100	12.5	25	25
7 f.	p-NO_2-phenyl	6.25	100	50	12.5	6.25
8a.	p-Cl-phenyl	100	50	50	100	100
8b.	p-Br- phenyl	50	25	25	25	100
8c.	p-OH-phenyl	100	50	50	100	50
8d.	Phenyl	100	100	25	25	50
8e.	o-Br-phenyl	50	100	25	12.5	12.5
8 f.	p-NO_2-phenyl	100	25	25	100	50
	Ampicillin	25	–	25	25	
	Gentamicin					25

References

1. Berhanu, A.L., et al.: A review of the applications of Schiff bases as optical chemical sensors. TrAC Trends Anal. Chem. **116**, 74–91 (2019)
2. Berhanu, A.L., Mohiuddin, I., Malik, A.K., Aulakh, J.S., Kumar, V., Kim, K.H.: A review of the applications of Schiff bases as optical chemical sensors. TrAC Trends Anal. Chem. **116**, 74–91 (2019)
3. Juyal, V.K., et al.: Schiff base metal complexes as a versatile catalyst: a review. J. Organomet. Chem. **999**, 122825 (2023). https://doi.org/10.1016/j.jorganchem.2023.122825
4. Vithal, A., Kishore, R., Dongare, S.J., Chundawat, N.S., Srivastava, N., Singh, G.P.: Novel thiosemicarbazone based sensors for transition metals. Sens. Diagn. **3**, 1822–1826 (2024). https://doi.org/10.1039/D4SD00266K
5. Ashraf, T., Ali, B., Qayyum, H., Haroone, M.S., Shabbir, G.: Pharmacological aspects of Schiff base metal complexes: a critical review. Inorg. Chem. Commun. **150**, 110449 (2023)
6. Li, M., et al.: Nitrogen-doped porous carbon nanosheets based on a Schiff base reaction for high-performance lithium-ion batteries anode. Energies. **16**(4), 1733 (2023)
7. Dubey, R., Pandey, J., Srivastava, N.: Terminal-substituted benzoxazole based Schiff-base mesogens and their Cu(II) complexes: synthesis and structural characterization. Chem. Phys. Impact. **8**, 100586 (2024). https://doi.org/10.1016/j.chphi.2024.100586
8. Uddin, M.N., Ahmed, S.S., Alam, S.R.: Biomedical applications of Schiff base metal complexes. J. Coord. Chem. **73**(23), 3109–3149 (2020)

9. Antony, R., Arun, T., Manickam, S.T.D.: A review on applications of chitosan-based Schiff bases. Int. J. Biol. Macromol. **129**, 615–633 (2019)
10. Yadav, M., Yadav, D., Singh, D.P., Kapoor, J.K.: Pharmaceutical properties of macrocyclic Schiff base transition metal complexes: urgent need in today's world. Inorg. Chim. Acta. **546**, 121300 (2023)
11. Zhang, J., Xu, L., Wong, W.Y.: Energy materials based on metal Schiff base complexes. Coord. Chem. Rev. **355**, 180–198 (2018)
12. Kakkassery, J.T., Raphael, V.P., Johnson, R.: In vitro antibacterial and in silico docking studies of two Schiff bases on Staphylococcus aureus and its target proteins. Fut. J Pharm Sci. **7**(1), 1–9 (2021)
13. El-Sonbati, A.Z., Mahmoud, W.H., Mohamed, G.G., Diab, M.A., Morgan, S.M., Abbas, S.Y.: Synthesis, characterization of Schiff base metal complexes and their biological investigation. Appl. Organomet. Chem. **33**(9), e5048 (2019)
14. Xu, J., Liu, Y., Hsu, S.H.: Hydrogels based on Schiff base linkages for biomedical applications. Molecules. **24**(16), 3005 (2019)
15. Ghanghas, P., Choudhary, A., Kumar, D., Poonia, K.: Coordination metal complexes with Schiff bases: useful pharmacophores with comprehensive biological applications. Inorg. Chem. Commun. **130**, 108710 (2021)
16. Verma, C., Quraishi, M.A.: Recent progresses in Schiff bases as aqueous phase corrosion inhibitors: design and applications. Coord. Chem. Rev. **446**, 214105 (2021)
17. Shakir, M., Hanif, S., Sherwani, M.A., Mohammad, O., Al-Resayes, S.I.: Pharmacologically significant complexes of Mn(II), Co(II), Ni(II), Cu(II) and Zn(II) of novel Schiff base ligand, (E)-N-(furan-2-yl methylene) quinolin-8-amine: synthesis, spectral, XRD, SEM, antimicrobial, antioxidant and in vitro cytotoxic studies. J. Mol. Struct. **1092**, 143–159 (2015)
18. Yousif, E., Majeed, A., Al-Sammarrae, K., Salih, N., Salimon, J., Abdullah, B.: Metal complexes of Schiff base: preparation, characterization and antibacterial activity. Arab. J. Chem. **10**, S1639–S1644 (2017)
19. Liu, X., Hamon, J.R.: Recent developments in penta-, hexa- and heptadentate Schiff base ligands and their metal complexes. Coord. Chem. Rev. **389**, 94–118 (2019)
20. El-Gammal, O.A., Mohamed, F.S., Rezk, G.N., El-Bindary, A.A.: Structural characterization and biological activity of a new metal complexes based on Schiff base. J. Mol. Liq. **330**, 115522 (2021)
21. Keshavarzian, E., Asadi, Z., Poupon, M., Dusek, M., Rastegari, B.: Heterodinuclear Cu–Gd (3d–4f) complex with di-compartmental Schiff base ligand in biological activity: synthesis, crystal structure, catecholase activity and DNA & BSA-binding studies. J. Mol. Liq. **345**, 117785 (2022)
22. Anisree, S., Shanmugavelan, P., Palanisamy, S., Kishore, R., Srivastava, N.: Synthesis and anticancer screening of substituted (phenyl)-1-((naphthalen-5-yl)methylene)thiosemicarbazide against PC3 prostate cancer cells. J. Indian Chem. Soc. **101**(9), 101229 (2024). https://doi.org/10.1016/j.jics.2024.101229
23. Anisree, S., Shanmugavelan, P., Vijaylakshmi, P., Kishore, R., Srivastava, N.: Synthesis, characterization and anticancer screening of novel phenylbenzylidene thiosemicarbazone derivatives. Phosphorus Sulfur Silicon Relat. Elem. **199**(4), 267–276 (2024). https://doi.org/10.1080/10426507.2024.2330935
24. Tyagi, M., Dubey, M.: A fight against cancer with advancement of Schiff base metal complexes: future prospects. Oral Oncol. Rep. **13**, 100692 (2025). https://doi.org/10.1016/j.oor.2024.100692
25. https://www.who.int/health-topics/antimicrobial-resistance#:~:text=AMR%20occurs%20when%20bacteria%2C%20viruses,spread%2C%20severe%20illness%20and%20death

26. Sumrra, S.H., et al.: Computational investigation of molecular structures, spectroscopic properties, cholinesterase inhibition and antibacterial activities of triazole Schiff bases endowed metal chelates. J. Mol. Struct. **1238**, 130382 (2021). https://doi.org/10.1016/j.molstruc.2021.130382
27. Hamad, A., et al.: Schiff bases of sulphonamides as a new class of antifungal agent against multidrug-resistant *Candida auris*. Microbiologyopen. **10**, e1218 (2021). https://doi.org/10.1002/mbo3.1218
28. Pibiri, I.: Recent advances: heterocycles in drugs and drug discovery. Int. J. Mol. Sci. **25**, 9503 (2024). https://doi.org/10.3390/ijms25179503
29. Sankar, P.S., Divya, K., Padmaja, A., Padmavathi, V.: Synthesis and antimicrobial activity of azetidinone and thiazolidinone derivatives from azolylindolyl Schiff's bases. Med. Chem. **7**, 340–347 (2017)
30. Osmaniye, D., Levent, S., Ardıç, C.M., Atlı, Ö., Özkay, Y., Kaplancıklı, Z.A.: Synthesis and anticancer activity of some novel benzothiazole-thiazolidine derivatives. Phosphorus Sulfur Silicon Relat. Elem. **193**, 249–256 (2017)
31. Datar, P.A., Aher, S.B.: Design and synthesis of novel thiazolidine-2,4-diones as hypoglycemic agents. J. Saudi Chem. Soc. **20**, S196–S201 (2016)
32. Gupta, N., Saini, V., Basavarajaiah, S.M., Dar, M.O., Das, R., Dahiya, R.S.: 1,3-Oxazine as a promising scaffold for the development of biologically active lead molecules. ChemSelect. **8**(39), e202301456 (2023). https://doi.org/10.1002/slct.202301456
33. Repale, A.V., Kishore, R., Chundawat, N.S., Srivastava, N., Singh, G.P.: A novel and easy synthesis of 1H-Tetrazoles employing p-Tolenesulfonyl azide. Res. J. Chem. Environ. **28**(10), 35–43 (2024). https://doi.org/10.25303/2810rjce035043
34. Yadav, S., Kishore, R., Srivastava, N.: Microwave-assisted phase transfer catalytic synthesis of aryl Benzothiazole derivatives and their anticancer screening. Org. Prep. Proced. Intern. **56**(6), 563–571 (2024). https://doi.org/10.1080/00304948.2024.2342692
35. Tyagi, R., Yadav, K., Srivastava, N., Sagar, R.: Applications of pyrrole and pyridine-based heterocycles in cancer diagnosis and treatment. Curr. Pharm. Des. **30**(4), 255–277 (2024). https://doi.org/10.2174/0113816128280082231205071504
36. Tyagi, R., Singh, K., Srivastava, N., Sagar, R.: Recent advances in carbohydrate based gelators. Mater. Adv. **4**, 3929–3950 (2023). https://doi.org/10.1039/D3MA00321C
37. Srivastava, N.: Efficient TBAI-CS2 Promoted Synthesis of Substituted Hydrazinecarbodithiolates. OPPI. **54**(6), 569–577 (2022). https://doi.org/10.1080/00304948.2022.2111171
38. Srivastava, N., Chaturvedi, D., Kishore, R., Kaur, M., Kaur, R.: Triton-B/CS_2 mediated novel synthesis of substituted thioureas: a novel class of anti-cancer agents. Indian J. Chem. 'B'. **61**, 870–877 (2022). https://doi.org/10.56042/ijc.v61i8.65053
39. Srivastava, N., Kishore, R.: Cleaner and greener synthesis of 3H-benzothiazole-2-thione and its derivatives. J. Sulfur Chem. **42**(1), 29–39 (2020). https://doi.org/10.1080/17415993.2020.1803321
40. Srivastava, N., Kishore, R.: Cleaner, greener synthesis, characterization, and anticancer evaluation of 2-Thioxo-1,3-Thiazolanes. Indian J. Heterocycl. Chem. **31**(2), 265–271 (2021)
41. Srivastava, N., Kishore, R.: Synthesis, characterization, anticancer, antibacterial, antifungal, and antimalarial activity of (aryl or hetero aryl) substituted-4-Methyl-2,3-dihydro- 2-thioxo-*1H*-imidazol-5-yl Ethanone derivatives. Indian J. Heterocycl. Chem. **31**(03), 347–355 (2021)
42. Srivastava, N., Rathore, B.K., Vishnoi, R.K., Bajpai, S.: A simple and convenient method for synthesis of Artemisinin dimer through aliphatic nitro compounds. Asian J. Chem. **33**(1), 49–52 (2020). https://doi.org/10.14233/ajchem.2021.22918

43. Kishore, R., Kamboj, M., Shukla, M., Chaturvedi, D., Srivastava, N.: Novel synthetic strategy of cyclic Dithiocarbamates catalyzed by triton-B. Asian J. Chem. **31**(5), 1091–1094 (2019). https://doi.org/10.14233/ajchem.2019.21830
44. Srivastava, N., Saxena, M., Shukla, M.: Novel synthesis of 5-Oxo-2-thioxo-2,5-dihydro-3-thiophenecarboxylate derivatives in non-aqueous medium. Asian J. Chem. **31**(1), 176–180 (2019). https://doi.org/10.14233/ajchem.2019.21622
45. Srivastava, N.: Key role of ionic liquids in the cleaner and greener synthesis of lactams. Res. J. Chem. Environ. **26**(1), 125–130 (2022). https://doi.org/10.25303/2601rjce125130
46. Vishnoi, R.K., Kishore, R., Chaturvedi, D., Shukla, M., Bajpai, S., Srivastava, N.: Synthesis and antimicrobial activity of cyclic dithiocarbamates employing triton-B/CS_2 system. Asian J. Chem. **33**(5), 1133–1136 (2021). https://doi.org/10.14233/ajchem.2021.23173

Synthesis, Characterization and Antimicrobial Screening of Novel Heterocyclic –2-oxo-2H-chromene-3-carbonyl) spiro[indoline-3,3′-pyrazolidin]-2-One Based Derivatives

Tandrani Ghosh[1], Geet Gupta[2], Vaishnavi Dwivedi[1], Nitin Srivastava[2(✉)], and Krishna Srivastava[1(✉)]

[1] Faculty of Chemical Sciences, SRM University, Lucknow-Dewa Road, Barabanki, India
krishnajs2063@Gmail.com

[2] Department of Chemistry, Amity University Uttar Pradesh, Lucknow Campus, Lucknow, Uttar Pradesh, India
nsrivastava1@lko.amity.edu

Abstract. A series of novel pyrazole-spiro based cyclized derivatives of acetyl, iminoethyl and ethanethioyl were synthesized and screened for antimicrobial activities. The final derivatives were procured by initiating the condensation of precursor molecules, salicylaldehyde and ethyl acetoacetate form acetyl coumarin. The obtained products were allowed to react with substituted-isatin which then yielded chalcone. Upon further reaction of chalcone with hydrazine hydrate, aminoguanidine HCl and thiosemicarbazide we got final products as 1′-acetyl-5- substituted –5′-(2-oxo-2H-chromene-3-carbonyl)spiro[indoline-3,3′-pyrazolidin]-2-one (**3a-c**), 5- substituted –1′-(1-iminoethyl)-5′-(2-oxo-2H-chromene-3-carbonyl)spiro[indoline-3,3′-pyrazolidin]-2-one (**4a-c**) and 5- substituted –1′-ethanethioyl-5′-(2-oxo-2H-chromene-3-carbonyl)spiro[indoline-3,3′-pyrazolidin]-2-one (**5a-c**). The structures of all synthesized derivatives were identified by their spectral data. The synthesized compounds exhibited promising in-vitro antimicrobial activity against Gram-positive bacteria (*S. Aureus*, *B. subtilis*), Gram-negative bacteria (*Klebsiella pneumoniae*), and Fungi (*Candida albicans*) as compared to the standard drug Chloramphenicol and Ketoconazole. Compound **3c** exhibited the highest antimicrobial activity against Gram-positive bacteria *S. Aureus*, *B. subtilis* and Fungi *Candida albicans* in comparison to the standard drug Chloramphenicol and Ketoconazole with MIC values of 12.5, 25.0 and 6.25 respectively. The other compounds **4a**, **5a**, **5b** have also exhibited promising antimicrobial activities with appreciable MIC values.

Keywords: Isatin · Thiourea · Salicylaldehyde · 1.4-dioxane · Semicarbazide · Hydrazine Hydrate · Antimicrobial activity

A. K. Saxena, A. Saxena (Eds.): GTHTM 2025, pp. 207–219, 2026.
https://doi.org/10.1007/978-3-032-12320-6_11

1 Introduction

Spirocyclic compounds are prominent bicyclic compounds in which two cycles are joined together by a common atom [1]. The organic Spiro-compounds are a significant class of compounds found in naturally occurring stuff [2, 3] illustrated via extremely prominent bioactivities [4]. These compounds have exhibited medicinal properties like, antimicrobial activity [5, 6], antifungal activity [7], anti-viral [8, 9] and anti-cancer [10], etc. Many natural and biologically active compounds and drug molecules contain the organic Spiro-compounds as one of the most important constituents. 4-Amino-1,2,4-triazole is one of the important heterocyclic compounds due to their pharmaceutical and biological activity like, antitubercular [11], antitumor [12], antileishmanial [13], anti-convulsant [14], anticancer [15], anti-inflammatory [16], antimicrobial activity [17], etc. As a result of these valuable activities, many researchers have considered the discovery of diverse and proficient methodologies to produce this class of compounds. Small ring heterocycles constitute valuable scaffolds in medicinal chemistry for generating biologically active derivatives. Remarkably, data indicates that at least 85% of all bioactive compounds incorporate a heterocyclic moiety into their structure [18–33]. Steroids have proved their significance in several therapeutic targets [34]. So, the heterocyclic spiro steroids were explored by the synthesis of spironolactone useful in curing edema [35]. Spironolactone has been developed as antagonist for aldosterone receptor which is helpful in combating the salt and water retention in heart failure, hypertension, nephrosis, liver cirrhosis etc. [36]. Spirorenone and drospirenone which exhibit anti-aldosterone activity have been developed and used as oral contraceptives [37]. Steroidal spiro heterocycles are promising drugs for diverse diseases varying from Hedgehog (HH) signaling inhibition for the treatment of cancers [38] and prevention of tumorigenesis [39]. Synthesis of spiro heterocyclic steroids is facilitated by stereochemistry of steroids for generating a single diastereomer but the creation of a new chiral center remains a constraint in their synthesis. Study of interaction between ligand and receptor have made easy to identify conformations with maximum binding affinity. Introduction of spiro moiety in the drug molecule has enhanced the bioactivity of the drug with enhanced pharmacokinetic and stability. Pyrazole moiety is another widespread scaffold in nature with diverse bioactivities [40] Pyrazole is a five-member heterocyclic compound containing two adjacent N-atoms. Pyrazole-scaffolds possess biological activity including pesticidal, herbicidal activities, anti-inflammatory activity by inhibiting COX-2 [41]. Taking advantage of the therapeutic significance of the pyrazole moiety and its regioselective synthesis for the functionalization at C4 or C5, due to electrophilicity of Two Carbon in the vicinity of N-atoms many syntheses have been reported [42]. In the reported research we have extended our research efforts for the novel synthesis Pyrazole-Spiro based Derivatives of Carboxamide, Carboximidamide and Carbothioamide derivatives and evaluated them for antimicrobial activities.

2 Materials and Methods

2.1 Experimental

The chemicals utilized are of laboratory grade. A BRUKER 300-MHz spectrometer was used to record ^{1}H NMR spectra in DMSO-d6 or $CDCl_3$ solutions. Chemical shifts were represented as part per million, δ values in relation to tetramethyl silane as the internal reference (TMS). A Perkin-Elmer 1650 FTIR spectrometer was used to collect the IR spectra v,cm-1 in KBr pellets. NBA served as the matrix in mass spectrometry, whereas Xenon/Argon (10 mA, 6Kv) served as the FAB gas. At CDRI Lucknow, India, elemental analyses were completed using a Thermo Fischer Scientific-11206100 Flash Smart Elemental Analyzer. The synthesized derivatives were determined by using pre-coated TLC plates as stationary (Merk, 60F-254) eluent 8:2 methanol: ethyl acetate as mobile phase, iodine vapor as visualizing agent. The melting point of all newly synthesized Pyrazole-Spiro based Derivatives of Carboxamide, Carboximidamide and Carbothioamide derivatives were examined using the Thermo Scientific manual melting point instrument, which was not adjusted.

2.2 Synthesis

2.2.1 The Synthesis of 3-acetylocoumarin

A 20 mmol. salicylaldehyde, 30 mmol ethyl acetoacetate, absolute alcohol (10 mL) and diethylamine (1 mL) were reflux at 120 °C with continuous stirring for 4–5 h. The product was filtered and washed with excess water and dried in air and recrystallized from ethanol. Yield 87%; mp 120–121 °C; Mol. Wt: Chemical Formula $C_{11}H_8O_3$ (Scheme 1).

2.2.2 Synthesis(E)-substituted-3-(2-oxo-2-(2-oxo-2H-chromen-3-yl)ethylidene) indolin-2-one(Chalcone 2a-c)

0.01 mol of each of **3-acetylocoumarin** and 5-substituted-isatin were mixed in 30 ml ethanol as solvent in a round bottom flask placed on ice bath with continuous stirring for 30 min. The reaction was kept in a refrigerator overnight. Then it was diluted with 50 ml ice-cold distilled water and washed well with cold water and recrystallized from rectified methanol. The synthesized compounds were characterized as (Scheme 2):

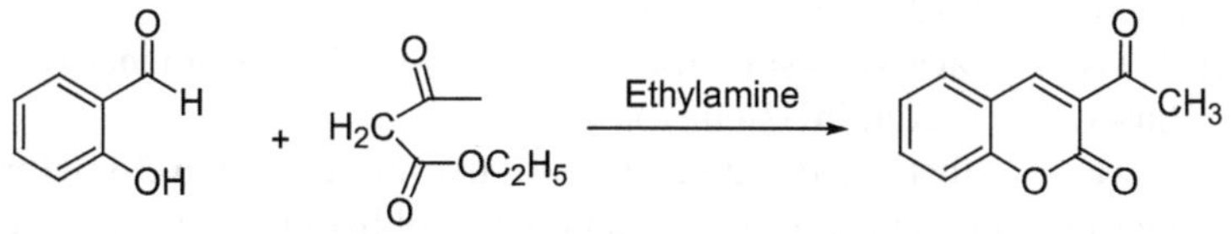

Scheme 1. Synthesis of 3-acetylocoumarin.

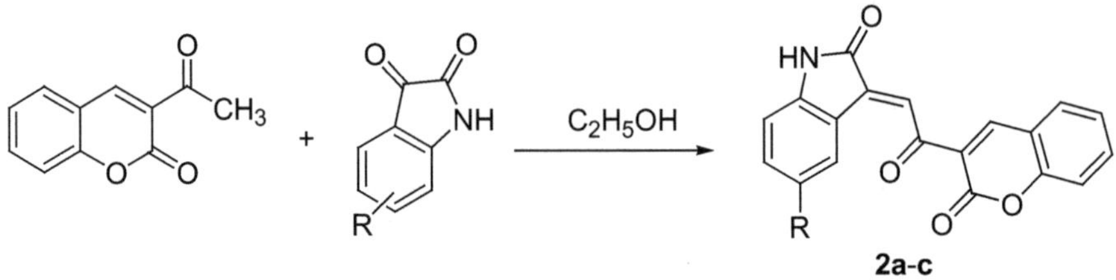

Scheme2. Synthesisof(E)-substituted-3-(2-oxo-2-(2-oxo-2H-chromen-3-yl)ethylidene)indolin-2-one(Chalcone.

i. **(E)-5-chloro-3-(2-oxo-2-(2-oxo-2H-chromen-3-yl)ethylidene)indolin-2-one (2a):**
Yield 74%; pale yellow, mp106 °C; Mol. Wt: 351.75 Anal. Calcd. for $C_{19}H_{10}ClNO_4$, Elemental analysis C-64.88 Cl- 10.08 N-3.98 found: C-64.83 Cl-10.05 N-3.94,IR (KBr): 3020 (C-H, Aroma), 1632(C=C, Aroma), 2933(CH_2. Cumarin), 1720(C=O, Conju), 1615(NH-CO),3312(N-H),965(C-Cl); ^{1}H NMR ($CDCl_3$/DMSO-d6) (300 MHz): δ 6.13–7.15(m,7H,Aroma),5.07(s,1H,CH,Cumarin),4.23(s, 1H,C=C, Conjug), 3.75(s, 1H,NH); ^{13}C NMR (100 MHz, $CDCl_3$): δ 63.8, 135.9, 187.0, 162.0, 150.8, 124.0, 127.8, 120.3.

ii. **(E)-5-bromo-3-(2-oxo-2-(2-oxo-2H-chromen-3-yl)ethylidene)indolin-2-one (2b):**
Yield 66%; pale yellow, mp 97–98 °C; Mol.Wt: 396.20 Anal. Calcd. for $C_{19}H_{10}BrNO_4$, Elemental analysis: C-57.60 Br- 20.17 N- 3.54 found: C- 57.57 Br-20.14 N-3.51; IR (KBr): 3024(C-H, Aroma), 1630(C=C, Aroma), 2939(CH_2. Cumarin), 1728(C=O, Conju), 1619(NH-CO),3312(N-H),763(C-Br), ^{1}H NMR ($CDCl_3$ /DMSO-d 6) (300 MHz):6.03–7.25(m,7H,Aroma),5.08(s,1H,CH,Cumarin),4.33(s, 1H,C=C, Conjug), 3.65(s, 1H,NH); ^{13}C NMR (100 MHz, $CDCl_3$): δ 63.9, 135.8, 187.1, 162.0, 150.8, 124.0, 127.8, 120.3.

iii. **(E)-5-hydroxy-3-(2-oxo-2-(2-oxo-2H-chromen-3-yl)ethylidene)indolin-2-one (2c):**
Yield 71%; pale yellow, mp105–106 °C; Mol. Wt: 333.30 Anal. Calcd. for $C_{19}H_{11}NO_5$, Elemental analysis: C-68.47 N-4.20 found: C-68.43 N-4.17. IR (KBr): 3030(C-H, Aroma), 1642(C=C, Aroma), 2943(CH_2.Cumarin), 1740(C=O, Conju), 1635(NH-CO), 3322(N-H),3443(-OH); ^{1}H NMR ($CDCl_3$/DMSO-d6) (300 MHz): 6.03–7.25(m, 7H, Aroma), 5.27(s,2H,CH,Cumarin), 4.33(s, 1H,C=C, Conjug), 3.65(s, 1H,NH), 4.95(s, 1H,OH); ^{13}C NMR (100 MHz, $CDCl_3$): δ 63.8, 135.9, 187.0, 162.0, 150.8, 124.0, 127.8, 120.3.

The synthesis of 1′-acetyl-5- substituted – 5′-(2-oxo-2H-chromene-3-carbonyl) spiro[indoline-3,3′-pyrazolidin]-2-one (3a-c):

0.01 mol of **(E)-5- substituted – 3-(2-oxo-2-(2-oxo-2H-chromen-3-yl)ethylidene) indolin-2-one (3a-c)** in 20 ml 1.4-dioxane and 0.01 mol hydrazinehydrate in presence of acetic acid as catalyst was refluxed at 120 °C on heating mental for 8–10 h. The cooled reaction mixture was emptied over ice having HCl for neutralizing the mixture. The precipitate thus formed was filtered, dried and recrystallized from methanol. The characterization data of the derivatives thus synthesized, are given as (Scheme 3)

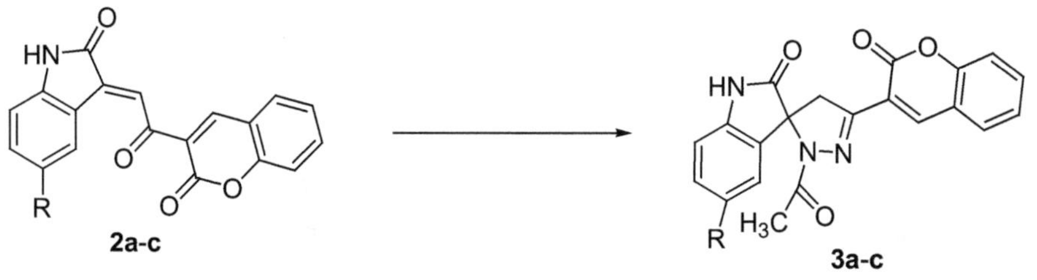

Scheme 3. Synthesis of 1′-acetyl-5- substituted –5′-(2-oxo-2H-chromene-3-carbonyl)spiro [indoline-3,3′-pyrazolidin]-2-one.

i. **2′-acetyl-5-chloro-5′-(2-oxo-2H-chromen-3-yl)-2′,4′-dihydrospiro[indoline-3,3′-pyrazol]-2-one (3a):**

Yield 59%; mp 121 °C; Anal. Calcd. for: $C_{21}H_{14}ClN_3O_4$,Mol. Wt: 407.81000; Elemental Analysis: C, 61.85; Cl, 8.69; N, 10.30;Found: C, 61.82; Cl, 8.64; N, 10.26;IR (KBr): 3018(C-H, Aroma), 1659(C=C, Aroma),948(N-N), 1651 (C=O), 1015(C-O), 1665(CON-H),992(C-Cl); ^{1}H NMR ($CDCl_3$/DMSO-d6) (300 MHz): 6.13–7.05(m,7H, Aroma),5.45(s,1H,CH,Cumarin),1.13(s,3H,-CH_3), 4.11(s, 1H,NH,Amide), 5.45(s,1H,N-H,N-N), 2.3(s,2H,diazine ring); ^{13}C NMR (100 MHz, $CDCl_3$): δ 17.8, 127.8, 142.1, 155.6, 162.0, 121.0, 125.8, 126.0.

ii. **2′-acetyl-5-bromo-5′-(2-oxo-2H-chromen-3-yl)-2′,4′-dihydrospiro[indoline-3,3′-pyrazol]-2-one (3b):**

Yield 73%; mp109–110 °C; Anal. Calcd. for: $C_{21}H_{14}BrN_3O_4$,Molecular Weight: 452.26400, Elemental Analysis: C, 55.77; Br, 17.67; N, 9.29;found: C, 55.70; Br, 17.61; N, 9.22; IR (KBr): 3023 (C-H, Aroma), 1649(C=C, Aroma), 968(N-N), 1657 (C=O), 1019(C-O), 1668(CON-H),752(C-Br),^{1}H NMR ($CDCl_3$ /DMSO-d6) (300 MHz): 6.03–7.25(m,7H, Aroma.), 5.45(s,1H,CH,Cumarin), 1.18(s,3H,-CH_3), 4.19 (s, 1H,NH,Amide), 5.48(s,1H,N-H,N-N), 2.40(s,2H,diazine ring); ^{13}C NMR (100 MHz, $CDCl_3$): δ 17.9, 126.8, 142.1, 155.6, 162.0, 121.0, 125.8, 126.0.

iii. **2′-acetyl-5-hydroxy-5′-(2-oxo-2H-chromen-3-yl)-2′,4′-dihydrospiro[indoline-3,3′-pyrazol]-2-one (3c):**

Anal. Calcd. for: $C_{21}H_{15}N_3O_5$, Yield 61%; mp102 °C; Molecular Weight: 389.36700, Elemental Analysis: C, 64.78; N, 10.79;Found: C, 64.72; N, 10.73,IR (KBr): 3032(C-H, Aroma), 1649(C=C, Aroma), 936(N-N), 1661 (C=O), 1018(C-O), 1660(CON-H),3392(C-OH); ^{1}H NMR ($CDCl_3$/DMSO-d6)(300 MHz): 6.12–7.35(m,7H,Aroma), 5.45(s,1H,CH,Cumarin), 1.19(s, 3H,-CH_3), 4.14 (s, 1H,NH,Amide), 5.53(s,1H,N-H,N-N), 2.33(s,2H,diazine ring), 5.18(s, 1H,OH); ^{13}C NMR (100 MHz, $CDCl_3$): δ 17.8, 127.8, 142.1, 155.6, 162.1, 121.2, 125.8, 126.1.

The synthesis of 5- substituted-1′-(1-iminoethyl)-5′-(2-oxo-2H-chromene-3--carbonyl)spiro[indoline-3,3′-pyrazolidin]-2-one (4a-c):

A 0.01 mol of **(E)-5- substituted – 3-(2-oxo-2-(2-oxo-2H-chromen-3-yl)ethylidene)indolin-2-one (2a-c),** 0.01 mol aminoguanidine HCl, 20 ml 1.4-dioxane was

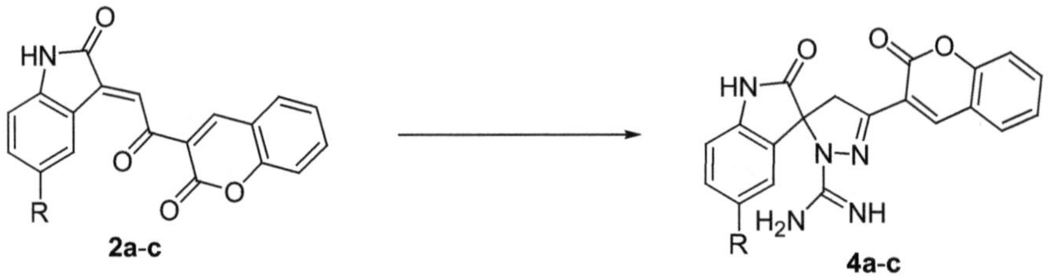

Scheme 4. Synthesis of 5- substituted –1′-(1-iminoethyl)-5′-(2-oxo-2H-chromene-3-carbonyl) spiro[indoline-3,3′-pyrazolidin]-2-one.

refluxed at 120 °C on heating mental for 12–13 h. The cooled reaction mixture was emptied over ice which was added with HCl for the neutralization. The precipitate was filtered, dried and recrystallized from methanol. The characterization data of the derivatives thus synthesized, are given as (Scheme 4):

(i) **5-chloro-1′-(1-iminoethyl)-5′-(2-oxo-2H-chromene-3-carbonyl)spiro[indoline-3,3′-pyrazolidin]-2-one (4a):**

Yield 66%; mp 98 °C; Mol. Wt 407.81400, Anal. Calcd. for $C_{20}H_{14}ClN_5O_3$, Elemental Analysis: C, 58.90; Cl, 8.69; N, 17.17; Found: C, 58.87; Cl, 8.64; N, 17.12; IR (KBr): 3021(C-H,Aroma), 1661 (C=C,Aroma), 1255(C=N), 951 (N-N), 1735(C=O), 1670(CON-H),969(C-Cl), ^{1}H NMR ($CDCl_3$ /DMSO-d6)(300 MHz): 6.18–7.15(m,7H,Aroma), 5.41(s,1H,CH,Cumarin), 2.18(s, 1H,-C=NH), 4.17 (s, 1H,NH,Amide), 5.45(s,1H,N-H,N-N); ^{13}C NMR (100 MHz, $CDCl_3$): δ 56.2, 142.1, 127.8, 142.1, 155.6, 162.0, 121.0, 125.8, 126.0, 172.7.

(ii) **5-bromo-2-oxo-5′-(2-oxo-2H-chromen-3-yl)spiro[indoline-3,3′-pyrazole]-2’(4’H)-carboximidamide (4b):**

Yield 74%; mp89–90 °C; Mol. Wt 452.26800, Anal. Calcd. for $C_{20}H_{14}BrN_5O_3$, Elemental Analysis: C, 53.11; Br, 17.67; N, 15.49; Found: C, 53.07;Br, 17.63; N, 15.42; IR (KBr): 3041(C-H,Aroma), 1668 (C=C,Aroma), 1259(C=N), 958 (N-N), 1745(C=O), 1679(CON-H),769(C-Br), ^{1}H NMR ($CDCl_3$/DMSO-d6) (300 MHz): 6.08–7.15(m,7H,Aroma), 5.51(s,1H,CH,Cumarin), 2.08(s, 1H,-C=NH), 4.18 (s, 1H,NH,Amide), 2.38 (s, 2H,NH_2), 2.17(s,2H,diazine); ^{13}C NMR (100 MHz, $CDCl_3$): δ 56.3, 142.2, 127.8, 142.1, 155.6, 162.0, 121.0, 125.8, 126.0, 172.9.

(iii) **5-hydroxy-2-oxo-5′-(2-oxo-2H-chromen-3-yl)spiro[indoline-3,3′-pyrazole]-2’(4’H)-carboximidamide (4c):**

Yield 66%; mp103°C; Mol. Wt: 389.37100, Anal. Calcd. For $C_{20}H_{15}N_5O_4$, Elemental Analysis: C, 61.69; N, 17.99;found: C, 61.64; N, 17.96,IR (KBr): 3027(C-H,Aroma), 1668 (C=C,Aroma), 1250(C=N), 961 (N-N), 1730(C=O), 1676(CON-H), 3369(C-OH); ^{1}H NMR ($CDCl_3$ /DMSO-d6) (300 MHz): 6.34–7.75(m,7H,Aroma), 5.31(s,1H,CH,Cumarin), 2.28(s, 1H,-C=NH), 4.38 (s, 1H,NH, Amide), 2.38 (s, 2H,NH_2), 5.38 (s, 1H,OH), 2.14(s,2H,diazine); ^{13}C NMR (100 MHz, $CDCl_3$): δ 56.2, 142.1, 127.8, 142.1, 155.6, 162.0, 121.3, 125.9, 126.1, 172.7.

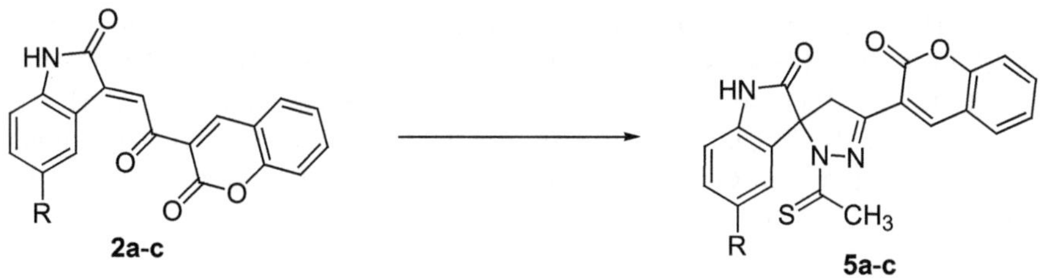

Scheme 5. Synthesis of 5- substituted –1′-ethanethioyl-5′-(2-oxo-2H-chromene-3-carbonyl) spiro[indoline-3,3′-pyrazolidin]-2-one.

The synthesis of 5- substituted – 1′-ethanethioyl-5′-(2-oxo-2H-chromene-3--carbonyl)spiro[indoline-3,3′-pyrazolidin]-2-one (5a-c):

A 0.01 mol of **(E)-5- substituted – 3-(2-oxo-2-(2-oxo-2H-chromen-3-yl)ethylidene)indolin-2-one (2a-c)** and 0.01 mol of thiosemicarbazide in 20 ml 1.4-dioxane was refluxed at 120 °C on heating mantle for 7–8 h. The cooled reaction mixture was emptied over ice which was added with HCl for the neutralization. The precipitate was filtered, dried and recrystallized from methanol. The characterization data of the derivatives thus synthesized, are given as (Scheme 5):

(i) **5-chloro-2′-ethanethioyl-5′-(2-oxo-2H-chromen-3-yl)-2′,4′-dihydrospiro[indoline-3,3′-pyrazol]-2-one (5a):**

Yield 61%; mp102–103 °C; Mol. Wt 423.87100, Anal. Calcd. for $C_{21}H_{14}ClN_3O_3S$, Elemental Analysis: C, 59.51; N, 9.91; S, 7.56, Found: C, 59.45; N, 9.87; S, 7.52; IR (KBr): 3023(C-H, Aroma), 1663 (C=C, Aroma), 1730(C=O, Ester), 953 (N-N), 1735(C=O), 1435(CSN-NH), 1032(C-O), 1690(NH-CO),953(C-Cl); ^{1}H NMR ($CDCl_3$ /DMSO-d6) (300 MHz): 6.30–7.32(m,7H,Aroma),2.37(s,2H,CH_2,diazine), 5.31(s,1H,CH,Cumarin), 2.10(s, 1H,-C=NH), 4.27 (s, 1H,NH,Amide), 2.14(s,2H,-NH_2); ^{13}C NMR (100 MHz, $CDCl_3$): δ 25.8, 36.1, 63.6, 120.4, 124.6, 130.0, 155.6, 162.0, 172.7, 198.8.

(ii) **5-bromo-2′-ethanethioyl-5′-(2-oxo-2H-chromen-3-yl)-2′,4′-dihydrospiro[indoline-3,3′-pyrazol]-2-one (5b)**

Yield 55%; mp 96 °C; Mol.Wt 468.32500, Anal. Calcd. for $C_{21}H_{14}BrN_3O_3S$, Elemental Analysis: C, 53.86; N, 8.97; S, 6.85; IR (KBr): 3028(C-H, Aroma), 1660 (C=C, Aroma), 1738(C=O, Ester), 958 (N-N), 1739(C=O), 1430(CSN-H), 1030(C-O), 1693(NH-CO),755(C-Br); ^{1}H NMR ($CDCl_3$/DMSO-d6) (300 MHz): 6.20–7.38(m,7H,Aroma), 3.39(s,2H,CH_2,Aroma), 5.55(s,1H,CH,Cumarin), 2.23(s, 1H,-C=NH), 4.30 (s, 1H,NH,Amide), 2.18(s,2H,-NH_2); ^{13}C NMR (100 MHz, $CDCl_3$): δ 25.9, 36.2, 63.8, 120.4, 124.6, 130.0, 155.6, 162.0, 172.7, 198.9.

(iii) **2′-ethanethioyl-5-hydroxy-5′-(2-oxo-2H-chromen-3-yl)-2′,4′-dihydrospiro[indoline-3,3′-pyrazol]-2-one (5c):**

Yield 62%; mp 127–128 °C; Mol. Wt 405.42800, Anal. Calcd. for $C_{21}H_{15}N_3O_4S$, Elemental Analysis: C, 62.21; N, 10.36; S, 7.91, found: C, 62.21; N, 10.36; S, 7.91; IR

(KBr): 3033(C-H, Aroma), 1667 (C=C, Aroma), 1730 (C=O, Ester), 950 (N-N), 1734(C=O, Ester), 1439(CSN-N-H), 1039(C-O), 1690 (NH-CO), 3345(C-OH); ^{1}H NMR ($CDCl_3$ /DMSO-d6) (300 MHz): 6.29–7.44(m, 7H, Aroma), 3.30 (s, 2H, CH_2, Aroma), 5.65(s,1H,CH,Cumarin), 2.33(s, 1H,-C=NH), 4.33 (s, 1H,NH, Amide), 2.19(s,2H,-NH_2); ^{13}C NMR (100 MHz, $CDCl_3$): δ 26.0, 36.2, 63.6, 120.4, 124.6, 130.0, 155.6, 162.1, 172.7, 198.9.

3 Results and Discussion

3.1 Chemistry

3-acetylocoumarin was first synthesized using salicylaldehyde, ethyl acetoacetate in absolute alcohol on refluxing in the presence of base diethylamine. The synthesized pale yellow coumarin (m.p. 119–120 °C) in solid state was dissolved in ethanol and reacted with 5-substituted-isatin in ice cooled conditions with final recrystallization in the presence of rectified methanol which led to the formation of (E)-substituted-3-(2-oxo-2-(2-oxo-2H-chromen-3-yl)ethylidene)indolin-2-one (Chalcone). These chalcones were employed for the synthesis of three series of acetyl, iminoethyl and ethanethioyl derivatives of novel pyrazole-spiro based compounds 1′-acetyl-5- substituted –5′-(2-oxo-2H-chromene-3-carbonyl)spiro[indoline-3,3′-pyrazolidin]-2-one (**3a-c**), 5- substituted –1′-(1-iminoethyl)-5′-(2-oxo-2H-chromene-3-carbonyl)spiro[indoline-3,3′-pyrazolidin]-2-one (**4a-c**) and 5- substituted –1′-ethanethioyl-5′-(2-oxo-2H-chromene-3-carbonyl)spiro[indoline-3,3′-pyrazolidin]-2-one (**5a-c**) In IR spectral analysis of the synthesized series showed vibrational bands at 3118–3500, 1560–1589, and 1318–1370 cm^{-1} which confirmed the presence of NH_2, C=N, and S=C groups respectively in the synthesized compounds. In the proton NMR spectral analysis, we observed CH_2 protons in the range of 2.98–3.30 ppm and 3.73–4.12 ppm as double doublet which confirmed pyrazoline ring in the synthesized compound. Proton of –CH_2 group which are not magnetically equivalent then the C-H protons showed double doublet in the range of 5.28–6.00 ppm. The aromaticity was confirmed by the signal in the range of 6.60–9.37 ppm.^{13}C NMR analysis showed peaks in the range of 40.33–45.54 ppm which confirmed the presence of CH_2 groups in the synthesized compound. The signal in the range of 154.85–178.58 confirmed the presence of C=O and C=S in carboxamide and carbothioamide formation. The scheme of the synthesis is given below (Scheme 6):

3.2 Antimicrobial Activity

The antimicrobial screening of the synthesized compounds was done Gram-positive bacteria *S. aureus* and *B. subtilis*, Gram-negative bacteria *Klebsiella pneumoniae.* The screening of the synthesized series of compounds was also done against fungi *Candida albicans.* It was found that all compounds of the synthesized series exhibited potency towards the considered bacteria and fungi in comparison to the standard drug

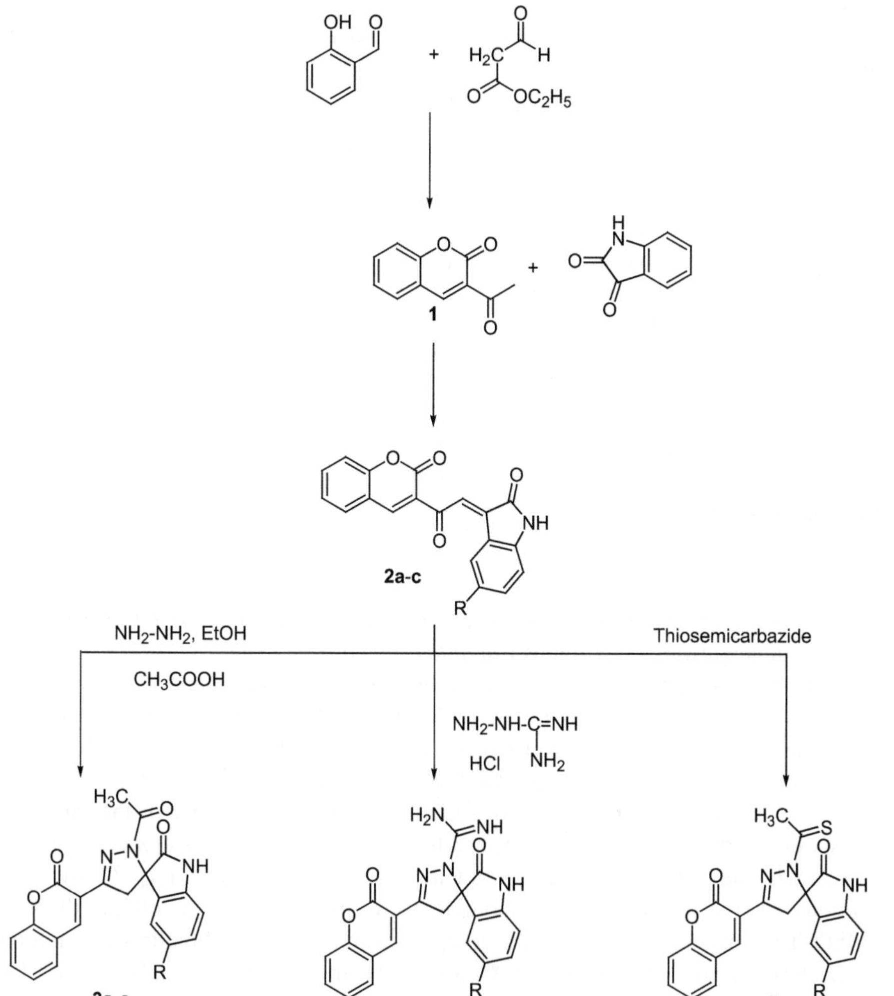

Scheme 6. Synthesis of 1′-acetyl-5- substituted –5′-(2-oxo-2H-chromene-3-carbonyl) spiro[indoline-3,3′-pyrazolidin]-2-one (**3a-c**), 5- substituted –1′-(1-iminoethyl)-5′-(2-oxo-2H-chromene-3-carbonyl)spiro[indoline-3,3′-pyrazolidin]-2-one (**4a-c**) and 5- substituted – 1′-ethanethioyl-5′-(2-oxo-2H-chromene-3-carbonyl)spiro[indoline-3,3′-pyrazolidin]-2-one (**5a-c**).

Chloramphenicol and ketoconazole. Among the derivatives of Carboxamide, compound **3c** having 5-OH-isatin moiety exhibited highest anti-bacterial and antifungal activity. Among carboximidamide derivatives, compound **4a** containing 5-Cl-isatin moiety exhibited highest potency towards both gram-positive and gram-negative bacteria, while compound **4c** of the series exhibited highest potency towards the gram-positive bacteria *B. subtilis*. Among the Carbothioamide derivatives compound **5a**, **5b**

Table 1. In vitro anti-microbial activity for the newly synthesized Pyrazole-Spiro based Derivatives of Carboxamide (**3a-c**), Carboximidamide (**4a-c**) and Carbothioamide (**5a-c**) against +ve, Gram -ve and fungus.

S. No.	R	Gram +ve MIC (μg/mL)		Gram -ve MIC (μg/mL)	Fungus MIC μ(g/mL)
		S. aureus	*B. subtilis*	*Klebsiella pneumoniae*	*Candida albicans*
3a	5-Cl-isatin	100	100	25	100
3b	5-Br-isatin	50	100	100	100
3c	5-OH-isatin	12.5	25	100	6.25
4a	5-Cl-isatin	25	25	25	100
4b	5-Br-isatin	50	100	25	25
4c	5-OH-isatin	50	6.25	100	50
5a	5-Cl-isatin	25	100	100	25
5b	5-Br-isatin	25	50	50	100
5c	5-OH-isatin	100	50	100	100
Chloramphenicol		50	–	50	–
Ketoconazole		–	–	–	50

exhibited the highest potency towards *S. aureus,* while the compound **5a** exhibited highest potency towards the fungus *Candida albicans* (Table 1).

Derivatives of Carboxamide (**3a-c**), Carboximidamide (**4a-c**) and Carbothioamide (**5a-c**) against +ve, Gram -ve and fungus with heterocyclic structures like spiro[indoline-3,3′-pyrazolidin], have exhibited promising antibacterial and antifungal activity. These compounds may have targeted specific bacterial enzymes, disrupted bacterial membranes, or may have inhibited biofilm formation. The susceptibility of the microbial stains were evaluated using broth microdilution assay in accordance with the procedures laid by the National Committee for Clinical Laboratory Standards [43]. The stock solutions of the synthesized compounds were prepared by dissolving them in DMSO. The final concentration of the compounds in 200 mm^3 of Mueller-Hinton broth (for bacterial strains) or in RPMI 1,600 (for fungi) ranged over 0.120–250 $\mu g/cm^3$. The *MIC* values of the compounds against the bacterial and fungal stains were determined as minimum inhibitory concentration preventing the growth of microbial strains with respect to the probe with no test compound.

4 Conclusion

Among the series of novel pyrazole-spiro based cyclized derivatives of acetyl, iminoethyl and ethanethioyl namely 1′-acetyl-5- substituted –5′-(2-oxo-2H-chromene-3--carbonyl)spiro[indoline-3,3′-pyrazolidin]-2-one (**3a-c**), 5- substituted –1′-(1-iminoethyl)-5′-(2-oxo-2H-chromene-3-carbonyl)spiro[indoline-3,3′-pyrazolidin]-2-one (**4a-c**) and 5- substituted –1′-ethanethioyl-5′-(2-oxo-2H-chromene-3-carbonyl)spiro[indoline-3,3′-pyrazolidin]-2-one (**5a-c**), we find that some of the compounds exhibit notable antimicrobial activities. Thus, the research adds some more compound to our fight against the resistant bacteria and fungus which may be further explored for the in vivo studies and clinical trials. Compound **3c** (2′-acetyl-5-hydroxy-5′-(2-oxo-2H-chromen-3-yl)-2′,4′-dihydrospiro[indoline-3,3′-pyrazol]-2-one) exhibited the highest antimicrobial activity against Gram-positive bacteria *S. Aureus*, *B. subtilis* and Fungi *Candida albicans* in comparison to the standard drug Chloramphenicol and Ketoconazole with MIC values of 12.5, 25.0 and 6.25 respectively. The other compounds **4a, 5a, 5b** have also exhibited promising antimicrobial activities with appreciable MIC values against the Gram-positive bacteria and fungi. Compounds **3a, 4a, 4b** have also exhibited promising antimicrobial activities against Gram-negative bacteria *Klebsiella pneumoniae*.

References

1. Gilles, L., Antoniotti, S.: Spirocyclic compounds in fragrance chemistry: synthesis and olfactory properties. ChemPlusChem. **87**(11), e202200227 (2022)
2. Chupakhin, E., et al.: Spirocyclic motifs in natural products. Molecules. **24**, 4165 (2019)
3. Quintavalla, A.: Spirolactones: recent advances in natural products, bioactive compounds and synthetic strategies. Curr. Med. Chem. **25**, 917–962 (2018)
4. Zheng, Y., Tice, C.M., Singh, S.B.: The use of spirocyclic scaffolds in drug discovery. Bioorg. Med. Chem. Lett. **24**, 3673–3682 (2014)
5. Gul, M., et al.: Synthesis, antimicrobial activity and molecular docking studies of spiroquinoline-indoline-dione and spiropyrazolo-indoline-dione derivatives. Sci. Rep. **13**, 1676 (2023). https://doi.org/10.1038/s41598-023-27777-z
6. Honmore, V.S., Natu, A.D., Khedkar, V.M., Arkile, M.A., Sarkar, D., Rojatkar, S.R.: Two antibacterial spiro compounds from the roots of Artemisia pallens wall: evidence from molecular docking. Nat. Prod. Res. **36**(10), 2465–2472 (2021). https://doi.org/10.1080/14786419.2021.1902325
7. Du, C., et al.: Design, synthesis and biological evaluation of novel spiro-quinazolinone derivatives as chitin synthase inhibitors and antifungal agents. Eur. J. Med. Chem. **255**, 115388 (2023). https://doi.org/10.1016/j.ejmech.2023.115388
8. Osolodkin, D.I., et al.: Phenotypic assessment of antiviral activity for spiro-annulated oxepanes and azepenes. Chem. Biol. Drug Des. **103**(5), e14553 (2024). https://doi.org/10.1111/cbdd.14553
9. Ma, S., et al.: Antiviral spirooliganones C and D with a unique spiro[bicyclo[2.2.2]octane-2,2′-bicyclo[3.1.0]hexane] carbon skeleton from the roots of Illicium oligandrum. Chin. Chem. Lett. **33**(9), 4248–4252 (2022). https://doi.org/10.1016/j.cclet.2022.02.050
10. Bora, D., Kaushal, A., Shankaraiah, N.: Anticancer potential of spirocompounds in medicinal chemistry: a pentennial expedition. Eur. J. Med. Chem. **215**, 113263 (2021). https://doi.org/10.1016/j.ejmech.2021.113263
11. Rangaswamy, V., Laddi, U.: A review on emerging impact of Antitubercular activity of 1,2,4-Triazole derivatives. Anti-Infect. Agents. **22**(1) (2024). https://doi.org/10.2174/0122113525268198230921071016
12. Jiang, G.W., Chang, Q., Liang, D.Y., Zhang, Y.T., Meng, Y.J., Yi, Q.Q.: Preparation and antitumor effects of 4-amino-1,2,4-triazole Schiff base derivative. J. Int. Med. Res. **48**(2), 300060520903874 (2020). https://doi.org/10.1177/0300060520903874

13. El-Saghier, A.M., et al.: Green synthesis, antileishmanial activity evaluation, and in silico studies of new amino acid-coupled 1,2,4-triazoles. Med. Chem. Res. **28**, 169–181 (2019). https://doi.org/10.1007/s00044-018-2274-x
14. Kaproń, B., et al.: 1,2,4-Triazole-based anticonvulsant agents with additional ROS scavenging activity are effective in a model of pharmacoresistant epilepsy. J. Enzyme Inhib. Med. Chem. **35**(1), 993–1002 (2020). https://doi.org/10.1080/14756366.2020.1748026
15. Emami, L., et al.: Design, synthesis and evaluation of novel 1,2,4-triazole derivatives as promising anticancer agents. BMC Chem. **16**, 91 (2022). https://doi.org/10.1186/s13065-022-00887x
16. Li, S.-M., et al.: Synthesis and anti-inflammatory activity evaluation of NO-releasing furoxan/1,2,4-triazole hybrid derivatives. Eur. J. Med. Chem. **257**, 115496 (2023). https://doi.org/10.1016/j.ejmech.2023.115496
17. Strzelecka, M., Świątek, P.: 1,2,4-Triazoles as important antibacterial agents. Pharmaceutics. **14**, 224 (2021). https://doi.org/10.3390/ph14030224
18. Repale, A.V., Ram, K., Chundawat, N.S., Srivastava, N., Singh, G.P.: A novel and easy synthesis of 1H-Tetrazoles employing p-Tolenesulfonyl azide. Res. J. Chem. Environ. **28**(10), 35–43 (2024). https://doi.org/10.25303/2810rjce035043
19. Anisree, S., Shanmugavelan, P., Palanisamy, S., Kishore, R., Srivastava, N.: Synthesis and anticancer screening of substituted (phenyl)-1-((naphthalen-5-yl)methylene)thiosemicarbazide against PC3 prostate cancer cells. J. Indian Chem. Soc. **101**(9), 101229 (2024). https://doi.org/10.1016/j.jics.2024.101229
20. Yadav, S., Kishore, R., Srivastava, N.: Microwave-assisted phase transfer catalytic synthesis of aryl Benzothiazole derivatives and their anticancer screening. Org. Prep. Proced. Int. **56**(6), 563–571 (2024). https://doi.org/10.1080/00304948.2024.2342692
21. Anisree, S., Shanmugavelan, P., Vijaylakshmi, P., Kishore, R., Srivastava, N.: Synthesis, characterization and anticancer screening of novel Phenylbenzylidene Thiosemicarbazone derivatives. Phosphorus Sulfur Silicon Relat. Elem. **199**(4), 267–276 (2024). https://doi.org/10.1080/10426507.2024.2330935
22. Tyagi, R., Yadav, K., Srivastava, N., Sagar, R.: Applications of pyrrole and pyridine-based heterocycles in cancer diagnosis and treatment. Curr. Pharm. Des. **30**(4), 255–277 (2024). https://doi.org/10.2174/0113816128280082231205071504
23. Tyagi, R., Singh, K., Srivastava, N., Sagar, R.: Recent advances in carbohydrate-based gelators. Mater. Adv. **4**, 3929–3950 (2023). https://doi.org/10.1039/D3MA00321C
24. Srivastava, N.: Efficient TBAI-CS2 promoted synthesis of substituted Hydrazinecarbodithiolates. Org. Prep. Proced. Int. **54**(6), 569–577 (2022). https://doi.org/10.1080/00304948.2022.2111171
25. Srivastava, N., Chaturvedi, D., Kishore, R., Kaur, M., Kaur, R.: Triton-B/CS2 mediated novel synthesis of substituted thioureas: a novel class of anti-cancer agents. Indian J. Chem. B. **61**, 870–877 (2022). https://doi.org/10.56042/ijc.v61i8.65053
26. Srivastava, N., Kishore, R.: Cleaner and greener synthesis of 3H-benzothiazole-2-thione and its derivatives. J. Sulfur Chem. **42**(1), 29–39 (2020). https://doi.org/10.1080/17415993.2020.1803321
27. Srivastava, N., Kishore, R.: Cleaner, greener synthesis, characterization, and anticancer evaluation of 2-Thioxo-1,3-Thiazolanes. Indian J. Heterocycl. Chem. **31**(2), 265–271 (2021)
28. Srivastava, N., Kishore, R.: Synthesis, characterization, anticancer, antibacterial, antifungal, and antimalarial activity of (aryl or hetero aryl) substituted-4-Methyl-2,3-dihydro-2-thioxo-1H-imidazol-5-yl Ethanone derivatives. Indian J. Heterocycl. Chem. **31**(03), 347–355 (2021)

29. Srivastava, N., Rathore, B.K., Vishnoi, R.K., Bajpai, S.: A simple and convenient method for synthesis of artemisinin dimer through aliphatic nitro compounds. Asian J. Chem. **33**(1), 49–52 (2020). https://doi.org/10.14233/ajchem.2021.22918
30. Kishore, R., Kamboj, M., Shukla, M., Chaturvedi, D., Srivastava, N.: Novel synthetic strategy of cyclic Dithiocarbamates catalyzed by triton-B. Asian J. Chem. **31**(5), 1091–1094 (2019). https://doi.org/10.14233/ajchem.2019.21830
31. Srivastava, N., Saxena, M., Shukla, M.: Novel synthesis of 5-Oxo-2-thioxo-2,5-dihydro-3-thiophenecarboxylate derivatives in non-aqueous medium. Asian J. Chem. **31**(1), 176–180 (2019). https://doi.org/10.14233/ajchem.2019.21622
32. Srivastava, N.: Key role of ionic liquids in the cleaner and greener synthesis of lactams. Res. J. Chem. Environ. **26**(1), 125–130 (2022). https://doi.org/10.25303/2601rjce125130
33. Vishnoi, R.K., Kishore, R., Chaturvedi, D., Shukla, M., Bajpai, S., Srivastava, N.: Synthesis and antimicrobial activity of cyclic dithiocarbamates employing triton-B/CS2 system. Asian J. Chem. **33**(5), 1133–1136 (2021). https://doi.org/10.14233/ajchem.2021.23173
34. Dembitsky, V.M.: Steroids bearing heteroatom as potential drugs for medicine. Biomedicine. **11**(10), 2698 (2023). https://doi.org/10.3390/biomedicines11102698
35. Romero-Hernández, L.L., Ahuja-Casarín, A.I., Merino-Montiel, P., Montiel-Smith, S., Vega-Báez, J.L., Sandoval-Ramírez, J.: Syntheses and medicinal chemistry of spiro heterocyclic steroids. Beilstein J. Org. Chem. **20**, 1713–1745 (2024). https://doi.org/10.3762/bjoc.20.152
36. Shim, J.Y., Garbo, G., Grimstad, F.W., Scatoni, A., Barrera, E.P., Boskey, E.R.: Use of the Drospirenone-only contraceptive pill in adolescents with endometriosis. J. Pediatr. Adolesc. Gynecol. **37**(4), 402–406 (2024). https://doi.org/10.1016/j.jpag.2024.02.003
37. Larik, F.A., et al.: Synthetic approaches towards the multi target drug spironolactone and its potent analogues/derivatives. Steroids. **118**, 76–92 (2017). https://doi.org/10.1016/j.steroids.2016.12.010
38. Heretsch, P., Tzagkaroulaki, L., Giannis, A.: Cyclopamine and hedgehog signaling: chemistry, biology, medical perspectives. Angew. Chem. Int. Ed. **49**, 3418–3427 (2010). https://doi.org/10.1002/anie.200906967
39. Gupta, S., Takebe, N., LoRusso, P.: Targeting the hedgehog pathway in cancer. Ther. Adv. Med. Oncol. **2**, 237–250 (2010). https://doi.org/10.1177/1758834010366430
40. Naim, M.J., Alam, O., Nawaz, F., Alam, M.J., Alam, P.: Current status of pyrazole and its biological activities. J. Pharm. Bioallied Sci. **8**(1), 2–17 (2016). https://doi.org/10.4103/0975-7406.171694
41. Muñoz, J.S., Zarate, X., Jaramillo, J.S., Bustos, C., Schott, E.: Synthesis and theoretical study of a new family of pyrazole derivatives. J. Mol. Struct. **1301**, 137267 (2024). https://doi.org/10.1016/j.molstruc.2023.137267
42. Faria, J.V., Vegi, P.F., Miguita, A.G.C., Dos Santos, M.S., Boechat, N., Bernardino, A.M.R.: Recently reported biological activities of Pyrazole compounds. Bioorg. Med. Chem. **25**(21), 5891–5903 (2017). https://doi.org/10.1016/j.bmc.2017.09.035
43. National Committee for Clinical Laboratory Standards.: Methods for dilution antimicrobial susceptibility test for bacteria that grow aerobically. 3rd edn. Approved standard NCCLS document M7-A3, No 24, vol 13, NCCLS, Villanova (1993)

Synthesis, Characterization and Antibacterial Screening of Novel Heterocyclic Derivatives of Oxazine as 3-(Substituted-phenyl)-4-(3-(Substituted-phenyl)-3,4-dihydro-2H-benzo[e][1,3]oxazin-6-yl)-3,4-dihydro-2H-naphtho[2,3-e][1,3]oxazine

Krishna Srivastava[1(✉)], Vanshika Gupta[2], Ajay Kumar Tripathi[1], R. K. Vishnoi[2], and Nitin Srivastava[2(✉)]

[1] Department of Chemistry, SRM University, Barabanki, Uttar Pradesh, India
krishnajs2063@gmail.com

[2] Department of Chemistry, Amity University Uttar Pradesh Lucknow Campus, Lucknow, Uttar Pradesh, India
nitinsriv5@gmail.com

Abstract. Innovative oxazine derivatives have been produced through one-pot, multicomponent condensation. The starting material oxazine-6-carbaldehyde was synthesized by the reaction of 1° amine and p-hydroxy benzaldehyde in toluene. **1a-c** further the condensation of carbaldehyde, substituted-1oamine and β-naphthol gives oxazin-6-yl)methyl)naphthalen-2-ol (**2a-l**) The obtained yields has been allowed to react with in acetonitrile to get target cyclized derivatives as 3-(4-bromophenyl)-4-(3-(4-chlorophenyl)-3,4-dihydro-2H-benzo[e][1,3]oxazin-6-yl)-3,4-dihydro-2H-naphtho[2,3-e][1,3]oxazine The structure of all the synthesized derivatives were elucidated by using FT-infrared(−ν max -cm −1 -KBr), H NMR-$CDCl_3$–300 MHz, mass spectral studies and antimicrobial activity in vitro studied by serial dilution method. The synthesized oxazin-derivatives exhibited significant antimicrobial properties against the tested microbes. Compound **3c**, exhibited notable activity against *Pseudomonas aeruginosa* and *Escherichia coli* with MIC values of 6.25 and 25 μg/ml respectively. Compound **3d** exhibited promising activity against *B. subtills* with MIC value of 12.5 μg/ml**.** Compound **3 f** was effective against *S. aureus* with MIC 12.5 μg/ml. Compound **3 g**, effectively reduced the growth of *Pseudomonas aeruginosa* with MIC value of 12.5 μg/ml. Compound **3i** showed notable antimicrobial activity against *B. subtills* MIC 12.5 μg/ml while compound **3j** showed its promising activity against *Klebsiella pneumoniae* with MIC value of 12.5 μg/ml.

Keywords: Substituted 1° amine · β-naphthol · Formaldehyde · Oxazine · P-hydroxy benzaldehyde · Toluene · Acetonitrile

A. K. Saxena, A. Saxena (Eds.): GTHTM 2025, pp. 221–236, 2026.
https://doi.org/10.1007/978-3-032-12320-6_12

1 Introduction

Oxazines are a class of a six-membered heterocyclic ring which contain one oxygen and one nitrogen atom [1]. There are four carbon atoms, one nitrogen atom, and one oxygen atom in the ring structure of oxazine. Because of their varied chemical characteristics and biological activity, oxazines are frequently employed in organic synthesis and medicinal chemistry [2, 3]. They play a crucial role as building blocks in the production of several materials, agrochemicals [4], and medications. The science of dye chemistry is one of the important areas in which oxazines are used. Vibrant and adaptable colorants, oxazine dyes are utilized in fabrics, inks, and other materials [5]. They are well-liked for a variety of industrial applications due to their vibrant colors and strong lightfastness. 1,3-oxazine derivatives have a variety of biological actions in medicinal chemistry, such as antiviral, anticancer, and antibacterial qualities [6]. Many molecules based on oxazine have been created by researchers as possible therapeutic candidates for a range of illnesses [7]. Due to their structural diversity, oxazines are interesting candidates for drug discovery since it is possible to modify their pharmacological characteristics.

1,3-oxazine has been considered as a privileged scaffold in many medicinal chemistry applications. Compounds having 1,3-oxazine moiety exhibit broad range of biological applications such as anticancer, antimicrobial, anti-inflammatory, antiplatelet, antitubercular and alpha-glucosidase inhibition activities. Trifluoromethyl-1,3-oxazin-2-one (Efavirenz) is a non-nucleoside reverse-transcriptase inhibitor and selective anti-HIV drug. Etifoxine is an anti-anxiety drug which is also used for the treatment of insomnia. In addition, 1,2,4-Oxadiazin-5(6 H)-One Derivatives have exhibited therapeutic potential for the treatment of Parkinson's disease [8]. Oxazines have also proved their significance as antibacterials [9]. Oxazine moiety has been found to possess anticonvulsant activities [10], analgesic [11], antimicrobial [12] etc. as shown in Fig. 1.

Oxazines are also used in the creation of fluorescent probes and sensors that are used to identify biological molecules and contaminants in the environment [13]. Their structural adaptability and fluorescence qualities make them useful instruments in chemical biology and environmental monitoring. In our pursuit for the development of biologically potent scaffold and exploring their specific bioactivities [14–33], we focussed on the oxazines for the novel and easier methods for the synthesis and bio-evaluation. In the present research we are reporting the synthesis and antibacterial screening of the new heterocyclic derivatives of Oxazine as 3-(Substituted-phenyl)-4-(3-(Substituted-phenyl)-3,4-dihydro-2H-benzo[e][1,3]oxazin-6-yl)-3,4-dihydro-2H-naphtho[2,3-e][1,3]oxazine.

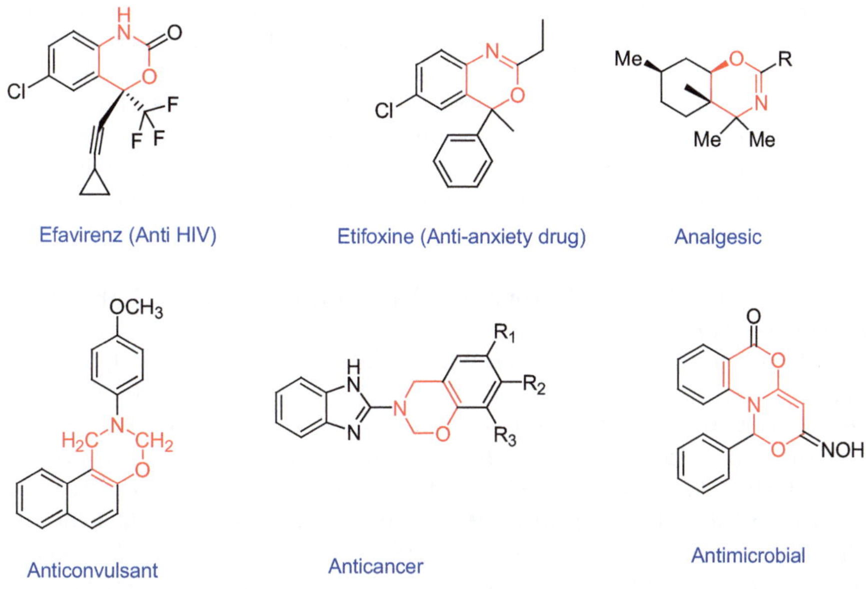

Fig. 1. Structurally diverse biologically potent 1,3-oxazin scaffolds.

2 Experimental

2.1 Instrumental

The purity and structure of the prepared derivatives were determined by using precoated TLC plates (Merk, 60F-254), iodine vapor as visualizing agent, eluent 5:2hexane/ethyl acetate). The ^{1}HNMR and ^{13}C NMR spectra were recorded in $CDCl_3$ and DMSO on Bruker NMR spectrophotometer at 300 MHz. Tetramethyl silane (TMS) was taken as the internal standard and chemical shift value (δ) were given in part per million(ppm). The melting points were examined in the open capillary method. Jasco FTIR-470 spectrophotometer (KBr) used for IR study of the synthesized compounds. MS-JEOL SX102 Mass spectrometer was used for mass spectrum of synthesized compounds. PerkinElmer 2400 CHNS Organic Elemental Analyzer was used for the elemental analysis of the synthesized compounds.

2.2 Synthesis of 3-(substituted-phenyl)-3,4-dihydro-2H-benzo[e][1,3]oxazine-6-carbaldehyde (1a-c)

A three-necked flask with stirrer was added with 0.4 mol aqueous formaldehyde solution and 50 ml toluene at room temperature. The mixture was cooled up to 5 °C by ice water with stirring and then added with 0.2 mol of substituted aromatic amine in 15 ml of toluene dropwise into flask under constant stirring at temperature below 5 °C. After 5 h stirring at same temperature, 0.2 mol of p-hydroxybenzaldehyde in 50 ml toluene was added into flask with continuous stirring. The reaction mixture was kept at room temperature for one hour with continuous stirring and then heated for 5 h at 95 °C. The residue removed under reduced pressure was added with extra chloroform for its dissolution. The resulting solution was washed with 1.0 ml of aqueous NaOH solution and repeatedly washed with water to remove to separate it from unreacted reagent and impurities. Further, the evaporation of solvent on rotary evaporator was followed by recrystallization of pure compound in toluene. Characterization data of the derivatives thus synthesized, are given as:

2.2.1 Data Analysis. 3-(4-chlorophenyl)-3,4-dihydro-2H-benzo[e][1,3]oxazine-6-carbaldehyde (1a)

Yield 84%; mp 123–124 °C; Mol. Wt: 273.72; Anal. Calcd. for $C_{15}H_{12}ClNO_2$, C, 65.82; H, 4.42; Cl, 12.95; N, 5.12 found: C, 65.82; H, 4.38; Cl, 12.90; N, 5.07; IR (KBr): 3063 (C-H, str., Aryl), 2867 (C-H, str.), 1675 (C=O, str.), 1643 (C=N, str.), 1560 (C=C, str., Aryl), 1358 (C-N, str.), 1016 (C=S, str.), 723 per cm (C-Cl, str.); 1 H NMR ($CDCl_3$ /DMSO-d 6) (300 MHz): δ7.53–8.05 (m, 13H, Ar-H), 1.83 (s, 3H, CH 3).

3-(4-bromophenyl)-3,4-dihydro-2H-benzo[e][1,3]oxazine-6-carbaldehyde (1b):

Yield 87%; mp 112–113 °C; Mol. Wt: 318.17 Anal. Calcd. for $C_{15}H_{12}BrNO_2$, H, 3.80; C, 56.63; N, 4.40; Br, 25.11; found: H, 3.874; C, 56.56; N, 4.33; Br, 25.09; IR(KBr): 3455 (H-N), 3076 (H-C), 1672(O=C), 1640 (N=C), 1579 (C=C,),1355 (N-C), 1022 (S=C), 680 cm^{-1} (Br-C); ^{1}H NMR: δ 7.49–8.02 (m, 12H, Ar-H), 8.10 (s, 1H, free NH).

3-(4-hydroxyphenyl)-3,4-dihydro-2H-benzo[e][1,3]Oxazine-6-carbaldehyde (1c)

Yield 78%; mp 111 °C; Mol. Wt: 255.27 Anal. Calcd. for $C_{15}H_{13}NO_3$, C, 70.58; H, 5.13; N, 5.49 found: C, 70.52; H, 5.10; N, 5.44. FT-infrared (ν_{max}, cm^{-1}, KBr): 1722 (C=O, str.), 1618 (C=N, str.), 1484 (C=C- aromatic), 1619 (C=S), 3470 (4-hygroxybenzene, br), 1181 (3-methoxybenzene), 2940 (C-H, str, aromatic), 3256 (1-naphthol, br) 3432 (H-N-C, br); P-NMR (d6, DMSO, 300 MHz, δ in ppm): 2.75 (b, 1H, C-NH), 5.3 (1H, bs, OH), 4.73 (1H, s,p-hydroxybenzene), 3.25(s, 3H, m-methoxybenzene), 6.71–7.70 (m, 18H, Ar-H).

2.3 Synthesis of 3-(((substituted-pheny)amino)(3-(substituted-pheny)-3,4-dihydro-2H-benzo[e][1,3]oxazin-6-yl)methyl)naphthalen-2-ol (2a-l)

A (0.02 mol) of 3-(4-hydroxyphenyl)-3,4-dihydro-2H-benzo[e][1,3]Oxazine-6-carbaldehyde(1a-c), substituted amine (0.02 mol) and β-naphthol (0.02 mol) was heated in preheated oil bath at 110-120 °C for 2 h. The reaction was allowed to cool after completion. The solid thus obtained on cooling was cleaned with ethanol (3x10 ml) to give pure products. The synthesized compound was characterized as given below:

2.3.1 Data Analysis. 3-(((3-bromophenyl)amino)(3-(4-chlorophenyl)-3,4-dihydro-2H-benzo[e][1,3]oxazin-6-yl)methyl)naphthalen-2-ol (2a):

Yield 72%; mp 107–108 °C; Mol. Wt: 571.90; Anal. Calcd. for $C_{31}H_{24}BrClN_2O_2$: C, 65.11; H, 4.23; Br, 13.97; Cl, 6.20; N, 4.90 found: C, 65.03S; H, 4.16; Br, 13.91; Cl,

6.17; N, 4.93; IR (KBr): 3061 (C-H, Aroma), 1612 (C=C, Aroma), 1656 (N-C-O), 700 (C-Cl), 650 (C-Br), 967 (C-N), 3310 (N-H); ^{1}H NMR: δ 7.20–7.60 (m, 4H), 6.92 (s, 1H), 7.41 (s, 1H), 5.60 (br, 1H OH), 4.80 (br, 1H, NH), 6.38–6.95 (m, 4H), 6.53–6.65 (dd, 2H), 6.75 (s, 1H), 4.61 (s, 2H), 6.02 (s, 2H), 6.53–7.10 (dd, 4H); ^{13}C NMR: 53.0, 123.3, 125.8, 127.1, 128.8, 132.5, 108.6, 155.5, 126.9, 146.8, 106.9, 132.6, 119.1, 114.4, 128.3, 135.1, 127.7, 114.4, 123.3, 159.1, 54.1, 90.5, 114.5, 142.6, 114.5, 129.8, 123.3, 129.8

3-((3-(4-chlorophenyl)-3,4-dihydro-2H-benzo[e][1,3]oxazin-6-yl)((4-chlorophenyl)amino)methyl)naphthalen-2-ol (2b):

Yield 71%; mp 119–120 °C; Mol. Wt: 527.45; Anal. Calcd. for $C_{31}H_{24}Cl_2N_2O_2$, C, 70.59; H, 4.59; Cl, 13.44; N, 5.31 found: C, 70.55; H, 4.54; Cl, 13.37; N, 5.26; IR (KBr): 3061 (C-H,Aroma), 1612 (C=C, Aroma), 1656 (N-C-O), 700 (C-Cl), 967 (C-N), 3310 (N-H)^{1}H NMR: δ 7.20–7.60 (m, 4H), 6.92 (s, 1H), 7.41 (s, 1H), 5.60 (br, 1H OH), 4.80 (br, 1H, NH), 6.40–6.99 (m, 4H), 6.53–6.65 (dd, 2H), 6.75 (s, 1H), 4.61 (s, 2H), 6.02 (s, 2H), 6.53–7.10 (dd, 4H); ^{13}C NMR: 53.0, 123.3, 125.8, 127.1, 128.8, 132.5, 108.6, 155.5, 126.9, 146.8, 106.9, 132.6, 119.1, 114.4, 128.3, 135.1, 127.7, 114.4, 123.3, 159.1, 54.1, 90.5, 114.5, 142.6, 114.5, 129.8, 123.3, 129.8

3-((3-(4-chlorophenyl)-3,4-dihydro-2H-benzo[e][1,3]oxazin-6-yl)((4-hydroxyphenyl)amino)methyl)naphthalen-2-ol (2c):

Yield 70%; mp 106–107 °C; Mol. Wt: 509.00; Anal. Calcd. For $C_{31}H_{25}ClN_2O_3$;, C, 73.15; H, 4.95; Cl, 6.96; N, 5.50 found: C, 73.108; H, 4.91; Cl, 6.89; N, 5.46; IR (KBr): 3061 (C-H, Aroma), 1612 (C=C, Aroma), 1656 (N-C-O), 700 (C-Cl), 967 (C-N), 3310 (N-H) ^{1}H NMR: δ 7.20–7.60 (m, 4H), 6.92 (s, 1H), 7.41 (s, 1H), 5.60 (br, 1H OH), 4.80 (br, 1H, NH), 6.40–6.99 (m, 4H), 6.53–6.65 (dd, 2H), 6.75 (s, 1H), 4.61 (s, 2H), 6.02 (s, 2H), 6.53–7.10 (dd, 4H); ^{13}C NMR: 53.0, 123.3, 125.8, 127.1, 128.8, 132.5, 108.6, 155.5, 126.9, 146.8, 106.9, 132.6, 119.1, 114.4, 128.3, 135.1, 127.7, 114.4, 123.3, 159.1, 54.1, 90.5, 114.5, 142.6, 114.5, 129.8, 123.3, 129.8

3-(((2-bromophenyl)amino)(3-(4-chlorophenyl)-3,4-dihydro-2H-benzo[e][1,3]oxazin-6-yl)methyl)naphthalen-2-ol (2d):

Yield 66%; mp 131–132 °C; Mol. Wt: 571.90; Anal. Calcd. for $C_{31}H_{24}BrClN_2O_2$, C, 65.11; H, 4.23; Br, 13.97; Cl, 6.20; N, 4.90 found: C, 65.06; H, 4.17; Br, 13.92; Cl, 6.14; N, 4.83; IR (KBr): 3061 (C-H, Aroma), 1612 (C=C, Aroma), 1656 (N-C-O), 700 (C-Cl), 650 (C-Br), 967 (C-N), 3310 (N-H); ^{1}H NMR: δ 7.20–7.60 (m, 4H), 6.90 (s, 1H), 7.41 (s, 1H), 5.70 (br, 1H OH), 4.78 (br, 1H, NH), 5.19 (d, 1H), 6.32–6.76 (m, 3H), 6.75–7.21 (m, 3H), 6.32–6.78 (m, 3H), 4.61 (s, 2H), 6.02 (s, 2H), 6.55–7.10 (dd, 4H); ^{13}C NMR: 53.0, 123.3, 125.8, 127.1, 128.8, 132.5, 108.6, 155.5, 126.9, 146.8, 106.9, 132.6, 119.1, 114.4, 128.3, 135.1, 127.7, 114.4, 123.3, 159.1, 54.1, 90.5, 114.5, 142.6, 114.5, 129.8, 123.3, 129.8

3-(((4-bromophenyl)amino)(3-(4-hydroxyphenyl)-3,4-dihydro-2H-benzo[e][1,3]oxazin-6-yl)methyl)naphthalen-2-ol (2e):

Yield 74%; mp 109–110 °C; Mol. Wt: 553.46; Anal. Calcd. for $C_{31}H_{25}BrN_2O_3$; C, 67.28; H, 4.55; Br, 14.44; N, 5.06 found: C, 67.23; H, 4.49; Br, 14.37; N, 5.01; IR (KBr): 3061 (C-H, Aroma), 1612 (C=C, Aroma), 1656 (N-C-O), 700 (C-Cl), 650 (C-Br), 967 (C-N), 3310 (N-H); ^{1}H NMR: 7.22–7.64 (m, 4H), 6.90 (s, 1H), 7.45(s, 1H), 5.76 (br, 1H, OH), 4.80(br,1H, NH), 6.32–7.21 (dd, 2H), 6.32 (s, 1H), 6.63–6.76 (dd, 2H), 4.61 (s, 2H), 6.00 (s, 2H), 6.42–6.55 (dd, 2H), 5.79 (s, 1H, OH); ^{13}C NMR: 125.2,

123.1, 125.9, 132.5, 128.8, 108.6, 155.2, 126.9, 53.7, 142.5, 129.0, 132.6, 115.5, 132.6, 114.5, 114.5, 135.1, 127.7, 114.4, 159.1, 90.5, 54.1, 114.5, 116.6, 146.8, 112.6

3-(((4-chlorophenyl)amino)(3-(4-hydroxyphenyl)-3,4-dihydro-2H-benzo[e][1,3]oxazin-6-yl)methyl)naphthalen-2-ol (2f):

Yield 62%; mp 117–118 °C; Mol. Wt: 509.00; Anal. Calcd. for $C_{31}H_{25}ClN_2O_3$; C, 73.15; H, 4.95; Cl, 6.96; N, 5.50 found: C, 73.110; H, 4.988; Cl, 6.92; N, 5.43; IR (KBr): 3061 (C-H, Aroma), 1612 (C=C, Aroma), 1656 (N-C-O), 700 (C-Cl), 967 (C-N), 3310 (N-H); ^{1}H NMR: 7.20–7.60 (m, 4H), 6.92 (s, 1H), 7.41(s, 1H), 5.70 (br, 1H, OH), 4.80(br,1H, NH), 6.32–7.21 (dd, 2H), 6.32 (s, 1H), 6.63–6.76 (dd, 2H), 4.61 (s, 2H), 6.00 (s, 2H), 6.42–6.55 (dd, 2H), 5.79 (s, 1H, OH); ^{13}C NMR: 125.3, 123.3, 125.8, 132.5, 128.8, 108.6, 155.2, 126.9, 53.7, 142.5, 129.0, 132.6, 115.5, 132.6, 114.5, 114.5, 135.1, 127.7, 114.4, 159.1, 90.5, 54.1, 114.4, 116.6, 146.8, 112.5

3-((3-(4-hydroxyphenyl)-3,4-dihydro-2H-benzo[e][1,3]oxazin-6-yl)((4-hydroxyphenyl)amino)methyl)naphthalen-2-ol (2 g)

Yield 74%; mp 118–119 °C; Mol. Wt: 490.56; Anal. Calcd. for $C_{31}H_{26}N_2O_4$: C, 75.90; H, 5.34; N, 5.71 found: C, 75.83; H, 5.27; N, 5.66; IR (KBr): 3061 (C-H, Aroma), 1612 (C=C, Aroma), 1656 (N-C-O), 967 (C-N), 3310 (N-H); ^{1}H NMR: 7.22–7.64 (m, 4H), 6.88 (s, 1H), 7.43(s, 1H), 5.70 (br, 1H, OH), 4.85 (br,1H, NH), 6.26–6.51 (dd, 2H), 6.75 (s, 1H), 6.63–6.76 (dd, 2H), 4.61 (s, 2H), 6.02 (s, 2H), 6.40–6.54 (dd, 2H), 5.76 (s, 1H, OH); ^{13}C NMR: 125.3, 123.3, 125.8, 132.5, 128.8, 108.6, 155.2, 126.9, 53.7, 142.5, 129.0, 132.6, 115.5, 132.6, 114.5, 114.5, 135.1, 127.7, 114.4, 159.1, 90.5, 54.1, 114.4, 116.6, 146.8, 112.5

3-(((2-bromophenyl)amino)(3-(4-hydroxyphenyl)-3,4-dihydro-2H-benzo[e][1,3]oxazin-6-yl)methyl)naphthalen-2-ol (2 h)

Yield 68%; mp 115–116 °C; Mol. Wt: 553.46; Anal. Calcd. for $C_{31}H_{25}BrN_2O_3$, H, 4.55; C, 67.28; N, 5.06 Br, 14.44; found: H, 4.49; C, 67.23; N, 5.01; Br, 14.37; IR (KBr): 3061 (C-H, Aroma), 1612 (C=C, Aroma), 1656 (N-C-O), 650 (C-Br), 967 (C-N), 3310 (N-H); ^{1}H NMR: 7.18–7.34 (m, 4H), 6.76 (s, 1H), 7.43(s, 1H), 5.70 (br, 1H, OH), 4.85 (br,1H, NH), 6.26–6.51 (dd, 2H), 6.75 (s, 1H), 6.63–6.76 (dd, 2H), 4.61 (s, 2H), 6.02 (s, 2H), 6.40–6.54 (dd, 2H), 5.76 (s, 1H, OH); ^{13}C NMR: 125.3, 123.3, 125.8, 132.5, 128.8, 108.6, 155.2, 126.9, 53.7, 142.5, 129.0, 132.6, 115.5, 132.6, 114.5, 114.5, 135.1, 127.7, 114.4, 159.1, 90.5, 54.1, 114.4, 116.6, 146.8, 112.6

3-((3-(2-bromophenyl)-3,4-dihydro-2H-benzo[e][1,3]oxazin-6-yl)((4-bromophenyl)amino)methyl)naphthalen-2-ol (2i)

Yield 55%; mp 93 °C; Mol. Wt: 616.35; Anal. Calcd. for $C_{31}H_{24}Br_2N_2O_2$, C, 60.41; H, 3.93; Br, 25.93; N, 4.55 found: C, 60.35; H, 3.87; Br, 25.86; N, 4.49; IR (KBr): 3061 (C-H, Aroma), 1612 (C=C, Aroma), 1656 (N-C-O), 650 (C-Br), 967 (C-N), 3310 (N-H); ^{1}H NMR: 7.20–7.54 (m, 4H), 6.92 (s, 1H), 7.41(s, 1H), 5.50 (br, 1H, OH), 4.76 (br,1H, NH), 6.32–7.21 (dd, 2H), 6.53–6.75 (dd, 2H), 4.61 (s, 2H), 6.00 (s, 2H), 6.48–7.25 (dd, 2H); ^{13}C NMR: 125.3, 123.3, 125.8, 132.5, 128.8, 108.6, 155.2, 126.9, 53.7, 142.5, 129.0, 132.6, 115.5, 132.6, 114.5, 114.5, 135.1, 127.7, 114.4, 159.1, 89.8, 53.4, 107.4, 115.6, 120.2, 128.4, 132.7, 147.8, 112.6

3-((3-(2-bromophenyl)-3,4-dihydro-2H-benzo[e][1,3]oxazin-6-yl)((4-chlorophenyl)amino)methyl)naphthalen-2-ol (2j)

Yield 74%; mp 93 °C; Mol. Wt: 571.90; Anal. Calcd. for $C_{31}H_{24}BrClN_2O_2$, C, 65.11; H, 4.23; Br, 13.97; Cl, 6.20; N, 4.90 found: C, 65.06; H, 4.16; Br, 13.88; Cl, 6.14; N,

4.83; IR (KBr): 3061 (C-H, Aroma), 1612 (C=C, Aroma), 1656 (N-C-O), 700 (C-Cl), 650 (C-Br), 967 (C-N), 3310 (N-H); ^{1}H NMR: 7.24–7.60 (m, 4H), 6.92 (s, 1H), 7.41(s, 1H), 5.50 (br, 1H, OH), 4.76 (br,1H, NH), 6.15–7.20 (dd, 2H), 6.53–6.75 (dd, 2H), 4.61 (s, 2H), 6.00 (s, 2H), 6.48–7.20 (dd, 2H); ^{13}C NMR: 125.3, 123.3, 125.8, 132.5, 128.8, 108.6, 155.2, 126.9, 53.7, 142.5, 129.0, 132.6, 115.5, 132.6, 114.5, 114.5, 135.1, 127.7, 114.4, 159.1, 89.8, 53.4, 107.4, 115.6, 120.2, 128.4, 132.7, 147.8, 112.6

3-((3-(2-bromophenyl)-3,4-dihydro-2H-benzo[e][1,3]oxazin-6-yl)((4-hydroxyphenyl)amino)methyl)naphthalen-2-ol (2 k)

Yield 69%; mp 127–128 °C; Mol. Wt: 553.46; Anal. Calcd. for $C_{31}H_{25}BrN_2O_3$, C, 67.28; H, 4.55; Br, 14.44; N, 5.06 found: C, 67.22; H, 4.50; Br, 14.38; N, 5.02; IR (KBr): 3061 (C-H, Aroma), 1612 (C=C, Aroma), 1656 (N-C-O), 650 (C-Br), 967 (C-N), 3310 (N-H); ^{1}H NMR: 7.24–7.60 (m, 4H), 6.92 (s, 1H), 7.41(s, 1H), 5.50 (br, 1H, OH), 4.76 (br,1H, NH), 6.15–7.20 (dd, 2H), 6.53–6.75 (dd, 2H), 4.61 (s, 2H), 6.00 (s, 2H), 6.48–7.20 (dd, 2H); ^{13}C NMR: 125.3, 123.3, 125.8, 132.5, 128.8, 108.6, 155.2, 126.9, 53.7, 142.5, 129.0, 132.6, 115.5, 132.6, 114.5, 114.5, 135.1, 127.7, 114.4, 159.1, 89.8, 53.4, 107.4, 115.6, 120.2, 128.4, 132.7, 147.8, 112.6

3-((3-(2-bromophenyl)-3,4-dihydro-2H-benzo[e][1,3]oxazin-6-yl)((2-bromophenyl)amino)methyl)naphthalen-2-ol (2 l)

Yield 70%; mp 121–122 °C; Mol. Wt: 616.35; Anal. Calcd. for $C_{31}H_{24}Br_2N_2O_2$:, C, 60.41; H, 3.93; Br, 25.93; N, 4.55 found: C, 60.36; H, 3.86; Br, 25.88; N, 4.47; IR (KBr): 3061 (C-H, Aroma), 1612 (C=C, Aroma), 1656 (N-C-O), 650 (C-Br), 967 (C-N), 3310 (N-H); ^{1}H NMR: 7.18–7.60 (m, 4H), 6.92 (s, 1H), 7.41(s, 1H), 5.30 (br, 1H, OH), 4.60 (br,1H, NH), 6.30–7.20 (m, 4H), 5.19 (d, 2H), 6.32 (s, 1H), 6.53–6.76 (dd, 2H), 4.64 (s, 2H), 6.48–7.25 (m, 4H); ^{13}C NMR: 125.3, 123.3, 125.8, 132.5, 128.8, 108.6, 155.2, 126.9, 53.7, 142.5, 129.0, 132.6, 115.5, 132.6, 114.5, 114.5, 135.1, 127.7, 114.4, 159.1, 89.8, 53.4, 147.4, 107.6, 132.2, 120.4, 128.4, 116.3

2.4 Synthesis of 3-(substituted-phenyl)-4-(3-(substituted-phenyl)-3,4-dihydro-2H-benzo[e][1,3]oxazin-6-yl)-3,4-dihydro-2H-naphtho[2,3-e][1,3]oxazine (3a-l)

The target compounds were synthesized by taking (0.02 mol) of 3-(((3-bromophenyl)amino)(3-(4-chlorophenyl)-3,4-dihydro-2H-benzo[e][1,3]oxazin-6-yl)methyl)naphthalen-2-ol (**2a-l**) and formaldehyde (37%,0.02 mol) were taken in acetonitrile (15 ml). The reaction was refluxed for 2 h. Once the reaction was completed, the solvent was removed with addition of 30 ml of water and extracted with ethyl acetate. After this the solvent was removed with the help of rotatory evaporator. The residue obtained was purified by recrystallization from ethanol to produce good yield, characterization data of the derivatives thus synthesized, are given as:

2.4.1 Data Analysis. 3-(4-bromophenyl)-4-(3-(4-chlorophenyl)-3,4-dihydro-2H-benzo[e][1,3]oxazin-6-yl)-3,4-dihydro-2H-naphtho[2,3-e][1,3]oxazine (3a)

Yield 62%; mp 94–95 °C; Mol. Wt: 583.91; Anal. Calcd. for $C_{32}H_{24}BrClN_2O_2$: C, 65.82; H, 4.14; Br, 13.68; Cl, 6.07; N, 4.80 found: C, 65.77; H, 4.11; Br, 13.61; Cl, 6.02; N, 4.72; IR (KBr): 3061 (C-H, Aroma), 1612 (C=C, Aroma), 1656 (N-C-O), 700

(C-Cl), 650 (C-Br), 967 (C-N), 3310 (N-H); ^{1}H NMR: δ 7.20–7.60 (m, 4H), 6.92 (s, 1H), 7.41 (s, 1H), 5.60 (br, 1H OH), 4.80 (br, 1H, NH), 6.38–6.95 (m, 4H), 6.53–6.65 (dd, 2H), 6.75 (s, 1H), 4.61 (s, 2H), 6.02 (s, 2H), 6.53–7.10 (dd, 4H); ^{13}C NMR: 53.0, 123.3, 125.8, 127.1, 128.8, 132.5, 108.6, 155.5, 126.9, 146.8, 106.9, 132.6, 119.1, 114.4, 128.3, 135.1, 127.7, 114.4, 123.3, 159.1, 54.1, 90.5, 114.5, 142.6, 114.5, 129.8, 123.3, 129.8

3-(4-chlorophenyl)-4-(3-(4-chlorophenyl)-3,4-dihydro-2H-benzo[e][1,3]oxazin-6-yl)-3,4-dihydro-2H-naphtho[2,3-e][1,3]oxazine (3b)

Yield 62%; mp 123–124 °C; Mol. Wt: 539.46; Anal. Calcd. for $C_{32}H_{24}Cl_2N_2O_2$: H, 4.48; C, 71.25; N, 5.19; Cl, 13.14; found: H, 4.43; C, 71.21; N, 5.13; Cl, 13.08; IR (KBr): 3061 (C-H,Aroma), 1612 (C=C, Aroma), 1656 (N-C-O), 700 (C-Cl), 967 (C-N), 3310 (N-H); ^{1}H NMR: δ 7.20–7.60 (m, 4H), 6.92 (s, 1H), 7.41 (s, 1H), 5.60 (br, 1H OH), 4.80 (br, 1H, NH), 6.40–6.99 (m, 4H), 6.53–6.65 (dd, 2H), 6.75 (s, 1H), 4.61 (s, 2H), 6.02 (s, 2H), 6.53–7.10 (dd, 4H); ^{13}C NMR: 53.0, 123.3, 125.8, 127.1, 128.8, 132.5, 108.6, 155.5, 126.9, 146.8, 106.9, 132.6, 119.1, 114.4, 128.3, 135.1, 127.7, 114.4, 123.3, 159.1, 54.1, 90.5, 114.5, 142.6, 114.5, 129.8, 123.3, 129.8

4-(4-(3-(4-chlorophenyl)-3,4-dihydro-2H-benzo[e][1,3]oxazin-6-yl)-2H-naphtho[2,3-e][1,3]oxazin-3(4H)-yl)phenol (3c)

Yield 67%; mp 116–1117 °C; Mol. Wt: 521.01; Anal. Calcd. for $C_{32}H_{25}ClN_2O_3$: H, 4.84; C, 73.77; N, 5.38; Cl, 6.80; found: H, 4.78; C, 73.73; N, 5.30; Cl, 6.74; IR (KBr): 3061 (C-H, Aroma), 1612 (C=C, Aroma), 1656 (N-C-O), 700 (C-Cl), 967 (C-N), 3310 (N-H); ^{1}H NMR: δ 7.20–7.60 (m, 4H), 6.92 (s, 1H), 7.41 (s, 1H), 5.60 (br, 1H OH), 4.80 (br, 1H, NH), 6.40–6.99 (m, 4H), 6.53–6.65 (dd, 2H), 6.75 (s, 1H), 4.61 (s, 2H), 6.02 (s, 2H), 6.53–7.10 (dd, 4H); ^{13}C NMR: 53.0, 123.3, 125.8, 127.1, 128.8, 132.5, 108.6, 155.5, 126.9, 146.8, 106.9, 132.6, 119.1, 114.4, 128.3, 135.1, 127.7, 114.4, 123.3, 159.1, 54.1, 90.5, 114.5, 142.6, 114.5, 129.8, 123.3, 129.8

3-(2-bromophenyl)-4-(3-(4-chlorophenyl)-3,4-dihydro-2H-benzo[e][1,3]oxazin-6-yl)-3,4-dihydro-2H-naphtho[2,3-e][1,3]oxazine (3d)

Yield 71%; mp 131–132 °C; Mol. Wt: 583.91; Anal. Calcd. for $C_{32}H_{24}BrClN_2O_2$: C, 65.82; H, 4.14; Br, 13.68; Cl, 6.07; N, 4.80 found: C, 65.76; H, 4.11; Br, 13.66; Cl, 6.03; N, 4.77; IR (KBr): 3061 (C-H, Aroma), 1612 (C=C, Aroma), 1656 (N-C-O), 700 (C-Cl), 650 (C-Br), 967 (C-N), 3310 (N-H); ^{1}H NMR: δ 7.20–7.60 (m, 4H), 6.90 (s, 1H), 7.41 (s, 1H), 5.70 (br, 1H OH), 4.78 (br, 1H, NH), 5.19 (d, 1H), 6.32–6.76 (m, 3H), 6.75–7.21 (m, 3H), 6.32–6.78 (m, 3H), 4.61 (s, 2H), 6.02 (s, 2H), 6.55–7.10 (dd, 4H); ^{13}C NMR: 53.0, 123.3, 125.8, 127.1, 128.8, 132.5, 108.6, 155.5, 126.9, 146.8, 106.9, 132.6, 119.1, 114.4, 128.3, 135.1, 127.7, 114.4, 123.3, 159.1, 54.1, 90.5, 114.5, 142.6, 114.5, 129.8, 123.3, 129.8

4-(6-(3-(4-bromophenyl)-3,4-dihydro-2H-naphtho[2,3-e][1,3] oxazin-4-yl)-2H-benzo[e][1,3]oxazin-3(4H)-yl)phenol (3e)

Yield 76%; mp 101–102 °C; Mol. Wt: 565.47; Anal. Calcd. for $C_{32}H_{25}BrN_2O_3$: C, 67.97; H, 4.46; Br, 14.13; N, 4.95 found: C, 67.92; H, 4.41; Br, 14.106; N, 4.88; IR (KBr): 3061 (C-H, Aroma), 1612 (C=C, Aroma), 1656 (N-C-O), 650 (C-Br), 967 (C-N), 3310 (N-H); ^{1}H NMR: 7.22–7.64 (m, 4H), 6.90 (s, 1H), 7.45(s, 1H), 5.76 (br, 1H, OH), 4.80(br,1H, NH), 6.32–7.21 (dd, 2H), 6.32 (s, 1H), 6.63–6.76 (dd, 2H), 4.61 (s, 2H), 6.00 (s, 2H), 6.42–6.55 (dd, 2H), 5.79 (s, 1H, OH); ^{13}C NMR: 125.2, 123.1,

125.9, 132.5, 128.8, 108.6, 155.2, 126.9, 53.7, 142.5, 129.0, 132.6, 115.5, 132.6, 114.5, 114.5, 135.1, 127.7, 114.4, 159.1, 90.5, 54.1, 114.5, 116.6, 146.8, 112.6

4-(6-(3-(4-chlorophenyl)-3,4-dihydro-2H-naphtho[2,3-e][1,3]oxazin-4-yl)-2H-benzo[e][1,3]oxazin-3(4H)-yl)phenol (3 f)

Yield 67%; mp106–107 °C; Mol. Wt: 521.01; Anal. Calcd. for $C_{32}H_{25}ClN_2O_3$: H, 4.84; C, 73.77; N, 5.38; Cl, 6.80; found: H, 4.78; C, 73.74; N, 5.27; Cl, 6.74; IR (KBr): 3061 (C-H, Aroma), 1612 (C=C, Aroma), 1656 (N-C-O), 700 (C-Cl), 967 (C-N), 3310 (N-H); ^{1}H NMR: 7.20–7.60 (m, 4H), 6.92 (s, 1H), 7.41(s, 1H), 5.70 (br, 1H, OH), 4.80(br,1H, NH), 6.32–7.21 (dd, 2H), 6.32 (s, 1H), 6.63–6.76 (dd, 2H), 4.61 (s, 2H), 6.00 (s, 2H), 6.42–6.55 (dd, 2H), 5.79 (s, 1H, OH); ^{13}C NMR: 125.3, 123.3, 125.8, 132.5, 128.8, 108.6, 155.2, 126.9, 53.7, 142.5, 129.0, 132.6, 115.5, 132.6, 114.5, 114.5, 135.1, 127.7, 114.4, 159.1, 90.5, 54.1, 114.4, 116.6, 146.8, 112.5

4-(4-(3-(4-hydroxyphenyl)-3,4-dihydro-2H-benzo[e][1,3]oxazin-6-yl)-2H-naphtho[2,3-e][1,3]oxazin-3(4H)-yl)phenol (3 g)

Yield 70%; mp 125–126 °C; Mol. Wt: 502.57; Anal. Calcd. for $C_{32}H_{26}N_2O_4$: H, 5.21; C, 76.48; N, 5.57 found: H, 5.16; C, 76.44; N, 5.49; IR (KBr): 3061 (C-H, Aroma), 1612 (C=C, Aroma), 1656 (N-C-O), 967 (C-N), 3310 (N-H); ^{1}H NMR: 7.22–7.64 (m, 4H), 6.88 (s, 1H), 7.43(s, 1H), 5.70 (br, 1H, OH), 4.85 (br,1H, NH), 6.26–6.51 (dd, 2H), 6.75 (s, 1H), 6.63–6.76 (dd, 2H), 4.61 (s, 2H), 6.02 (s, 2H), 6.40–6.54 (dd, 2H), 5.76 (s, 1H, OH); ^{13}C NMR: 125.3, 123.3, 125.8, 132.5, 128.8, 108.6, 155.2, 126.9, 53.7, 142.5, 129.0, 132.6, 115.5, 132.6, 114.5, 114.5, 135.1, 127.7, 114.4, 159.1, 90.5, 54.1, 114.4, 116.6, 146.8, 112.5

4-(6-(3-(2-bromophenyl)-3,4-dihydro-2H-naphtho[2,3-e][1,3] oxazin-4-yl)-2H-benzo[e][1,3]oxazin-3(4H)-yl)phenol (3 h)

Yield 68%; mp 103 °C; Mol. Wt: 565.47; Anal. Calcd. for$C_{32}H_{25}BrN_2O_3$:, C, 67.97; H, 4.46; Br, 14.13; N, 4.95 found: C, 67.94; H, 4.43; Br, 14.08; N, 4.88; IR (KBr): 3061 (C-H, Aroma), 1612 (C=C, Aroma), 1656 (N-C-O), 650 (C-Br), 967 (C-N), 3310 (N-H); ^{1}H NMR: 7.18–7.34 (m, 4H), 6.76 (s, 1H), 7.43(s, 1H), 5.70 (br, 1H, OH), 4.85 (br,1H, NH), 6.26–6.51 (dd, 2H), 6.75 (s, 1H), 6.63–6.76 (dd, 2H), 4.61 (s, 2H), 6.02 (s, 2H), 6.40–6.54 (dd, 2H), 5.76 (s, 1H, OH); ^{13}C NMR: 125.3, 123.3, 125.8, 132.5, 128.8, 108.6, 155.2, 126.9, 53.7, 142.5, 129.0, 132.6, 115.5, 132.6, 114.5, 114.5, 135.1, 127.7, 114.4, 159.1, 90.5, 54.1, 114.4, 116.6, 146.8, 112.6

3-(4-bromophenyl)-4-(3-(2-bromophenyl)-3,4-dihydro-2H-benzo[e] [1,3]oxazin-6-yl)-3,4-dihydro-2H-naphtho[2,3-e][1,3]oxazine (3i)

Yield 71%; mp 129–130 °C; Mol. Wt: 628.36; Anal. Calcd. for$C_{32}H_{24}Br_2N_2O_2$: H, 3.85; C, 61.17; N, 4.46; Br, 25.43; found: H, 3.80; C, 61.11; N, 4.42; Br, 25.37; IR (KBr): 3061 (C-H, Aroma), 1612 (C=C, Aroma), 1656 (N-C-O), 650 (C-Br), 967 (C-N), 3310 (N-H); ^{1}H NMR: 7.20–7.54 (m, 4H), 6.92 (s, 1H), 7.41(s, 1H), 5.50 (br, 1H, OH), 4.76 (br,1H, NH), 6.32–7.21 (dd, 2H), 6.53–6.75 (dd, 2H), 4.61 (s, 2H), 6.00 (s, 2H), 6.48–7.25 (dd, 2H); ^{13}C NMR: 125.3, 123.3, 125.8, 132.5, 128.8, 108.6, 155.2, 126.9, 53.7, 142.5, 129.0, 132.6, 115.5, 132.6, 114.5, 114.5, 135.1, 127.7, 114.4, 159.1, 89.8, 53.4, 107.4, 115.6, 120.2, 128.4, 132.7, 147.8, 112.6

4-(3-(2-bromophenyl)-3,4-dihydro-2H-benzo[e][1,3]oxazin-6-yl)-3-(4-chlorophenyl)-3,4-dihydro-2H-naphtho[2,3-e][1,3]oxazine (3j)

Yield 74%; mp 133–134 °C; Mol. Wt: 583.91 Anal. Calcd. for $C_{32}H_{24}BrClN_2O_2$: C, 65.82; H, 4.14; Br, 13.68; Cl, 6.07; N, 4.80 found: C, 65.76; H, 4.08; Br, 13.62; Cl,

6.01; N, 4.73; IR (KBr): 3061 (C-H, Aroma), 1612 (C=C, Aroma), 1656 (N-C-O), 700 (C-Cl), 650 (C-Br), 967 (C-N), 3310 (N-H); ^{1}H NMR: 7.24–7.60 (m, 4H), 6.92 (s, 1H), 7.41(s, 1H), 5.50 (br, 1H, OH), 4.76 (br,1H, NH), 6.15–7.20 (dd, 2H), 6.53–6.75 (dd, 2H), 4.61 (s, 2H), 6.00 (s, 2H), 6.48–7.20 (dd, 2H); ^{13}C NMR: 125.3, 123.3, 125.8, 132.5, 128.8, 108.6, 155.2, 126.9, 53.7, 142.5, 129.0, 132.6, 115.5, 132.6, 114.5, 114.5, 135.1, 127.7, 114.4, 159.1, 89.8, 53.4, 107.4, 115.6, 120.2, 128.4, 132.7, 147.8, 112.6

4-(4-(3-(2-bromophenyl)-3,4-dihydro-2H-benzo[e][1,3]oxazin-6-yl)-2H-naphtho[2,3-e][1,3]oxazin-3(4H)-yl)phenol (3 k)

Yield 60%; mp 140–141 °C; Mol. Wt: 565.47; Anal. Calcd. for $C_{32}H_{25}BrN_2O_3$: C, 67.97; H, 4.46; Br, 14.13; N, 4.95 found: C, 67.92; H, 4.42; Br, 14.06; N, 4.89; IR (KBr): 3061 (C-H, Aroma), 1612 (C=C, Aroma), 1656 (N-C-O), 650 (C-Br), 967 (C-N), 3310 (N-H); ^{1}H NMR: 7.24–7.60 (m, 4H), 6.92 (s, 1H), 7.41(s, 1H), 5.50 (br, 1H, OH), 4.76 (br,1H, NH), 6.15–7.20 (dd, 2H), 6.53–6.75 (dd, 2H), 4.61 (s, 2H), 6.00 (s, 2H), 6.48–7.20 (dd, 2H); ^{13}C NMR: 125.3, 123.3, 125.8, 132.5, 128.8, 108.6, 155.2, 126.9, 53.7, 142.5, 129.0, 132.6, 115.5, 132.6, 114.5, 114.5, 135.1, 127.7, 114.4, 159.1, 89.8, 53.4, 107.4, 115.6, 120.2, 128.4, 132.7, 147.8, 112.6

3 l. 3-(2-bromophenyl)-4-(3-(2-bromophenyl)-3,4-dihydro-2H-benzo[e][1,3]oxazin-6-yl)-3,4-dihydro-2H-naphtho[2,3-e][1,3]oxazine (3 l)

Yield 65%; mp 108–109 °C; Mol. Wt: 628.36; Anal. Calcd. for $C_{32}H_{24}Br_2N_2O_2$: C, 61.17; H, 3.85; Br, 25.43; N, 4.46 found: C, 61.12; H, 3.77; Br, 25.38; N, 4.40; IR (KBr): 3061 (C-H, Aroma), 1612 (C=C, Aroma), 1656 (N-C-O), 650 (C-Br), 967 (C-N), 3310 (N-H); ^{1}H NMR: 7.18–7.60 (m, 4H), 6.92 (s, 1H), 7.41(s, 1H), 5.30 (br, 1H, OH), 4.60 (br,1H, NH), 6.30–7.20 (m, 4H), 5.19 (d, 2H), 6.32 (s, 1H), 6.53–6.76 (dd, 2H), 4.64 (s, 2H), 6.48–7.25 (m, 4H); ^{13}C NMR: 125.3, 123.3, 125.8, 132.5, 128.8, 108.6, 155.2, 126.9, 53.7, 142.5, 129.0, 132.6, 115.5, 132.6, 114.5, 114.5, 135.1, 127.7, 114.4, 159.1, 89.8, 53.4, 147.4, 107.6, 132.2, 120.4, 128.4, 116.3

2.5 Antibacterial Activity Assessment

The newly designed heterocyclic derivatives of β-lactam and thiazole were screened for their bacterial activity against different bacterial species viz., *S. aureus*, *B. subtills*, *P. seudomonas, Klebsiella pneumoniae* and *Escherichia coli* in vitro. The pure isolates of the test bacterial species were obtained from the Department of Microbiology KGMU Lucknow. A reported methodology was used to confirm the identity of the working strains by gram staining and colony morphology. To evaluate the antibacterial activity, the culture was prepared by mixing 1 ml of anti-bacterial growth containing broth and 20 ml of plane luria-bertani medium. From the prepared mixture, 1.0 ml of the culture was taken in six different sterile tubes while one sterile tube contains 1.8 ml of culture. Subsequently, 0.2 ml of sample solution (**2 a-h** and **3a-h**) in EtOH (1 mg/ml) was inoculated in the tube containing 1.8 ml of culture. From this tube, 1 ml of culture was taken out and transferred into the second tube. Then again 1 ml of the culture was taken out from the second tube and transferred into the third tube; in this manner concentration of the sample in each successive tube was reduced to half. One control tube containing ciprofloxacin was prepared at last. All the prepared tube samples were incubated for 24 h at 37 o C. Bacterial growth for every conical tube was checked after 24 h by computing the absorbance value at 600 nm. The plot of compound concentration and absorbance value was used to obtain the MIC (corresponding to the drop in optical density) of the particular derivative. The MIC value of these derivatives was found between 12.5 and 100 µg/m.

3 Results and Discussion

3.1 Synthesis

Novel derivatives of Oxazine as 3-(Substituted-phenyl)-4-(3-(Substituted-phenyl)-3,4-dihydro-2H-benzo[e][1,3]oxazin-6-yl)-3,4-dihydro-2H-naphtho[2,3-e][1,3]oxazine (**3a-l**) were synthesized and the stepwise synthesis of the desired compounds is given in Scheme 1.

In the first step of synthesis compound **1a-c** were synthesized by reacting substituted aromatic amine with p-hydroxybenzaldehyde in toluene in ice cool conditions. Which led to the formation of 3-(4-hydroxyphenyl)-3,4-dihydro-2H-benzo[e][1,3]Oxazine-6-carbaldehyde (**1a-c**). In the next step these synthesized **1a-c** were reacted with substituted amine and β-naphthol to give the compounds **2a-l**. In the final step of the synthesis (**2a-l**) was reacted with formaldehyde in acetonitrile under reflux conditions to give compound **3a-l**. All synthesized compounds were characterized by IR, NMR and elemental analysis.

3.2 Antimicrobial Activity

The toxicity of the synthesized compounds was first determined against Human Embryonic Kidney 293 T (HEK293T) cells and none of the compounds exhibited toxicity to HEK293T cells, and the survival of the cells exceeded 80%. The synthesized series of the compounds 3**a-l** was then screened for in vitro antimicrobial action against the Gram-positive bacteria *S. aureus, B. subtills* and gram-negative bacteria *Pseudomonas aeuroginosa, Klebsiella pneumoniae, Escherichia coli*. In the antimicrobial evaluation the standard drug Ampicillin was taken as standard. All of the synthesized compound exhibited antimicrobial activities gains the considered gram-positive and gram-negative bacteria. Compound **3 f** exhibited promising antimicrobial activity against *S. aureus*. Compounds **3d** and **3i** exhibited promising antimicrobial activities against *B. subtills.* In case of gram-negative bacterial stain of *Pseudomonas aeruginosa,* the compound **3c** and **3 g** exhibited the notable antimicrobial activities. Compound **3j** exhibited the best antimicrobial activity against gram-negative bacteria *Klebsiella pneumoniae.* In case of the antimicrobial screening of the synthesized compounds against *Escherichia coli,* it was found that compound **3c**, **3 g**, **3 l** exhibited significant antimicrobial activities. The presence of the electron withdrawing groups on the phenyl group attached to oxazine rings are responsible for the enhanced activity of the compounds. The MIC values of the synthesized compounds in comparison to the standard drugs are given in the Table 1.

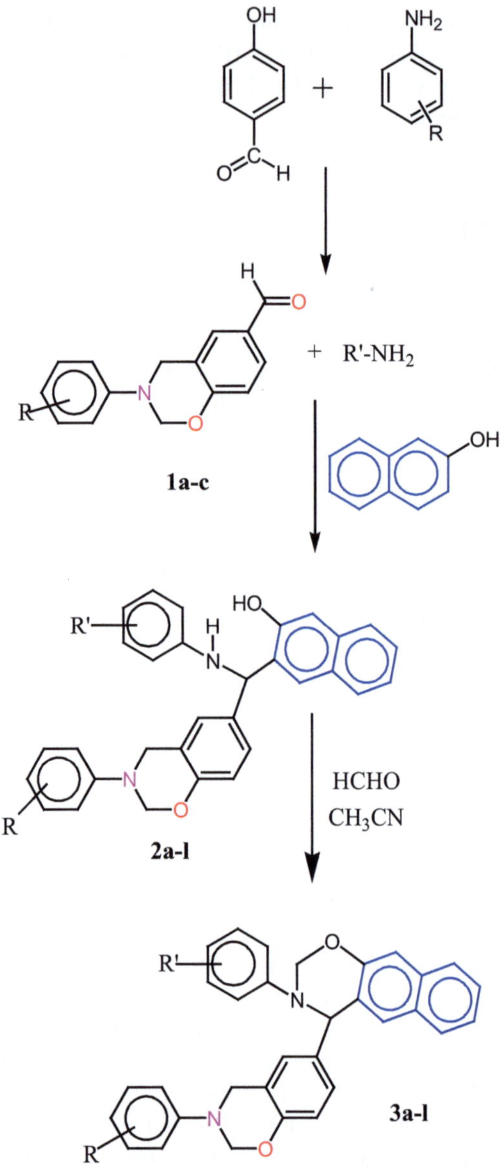

Scheme 1. Synthesis of derivatives of Oxazine as 3-(Substituted-phenyl)-4-(3-(Substituted-phenyl)-3,4-dihydro-2H-benzo[e][1,3]oxazin-6-yl)-3,4-dihydro-2H-naphtho[2,3-e][1,3]oxazine.

Compound **3c**, exhibited notable activity against *Pseudomonas aeruginosa* and *Escherichia coli* with MIC values of 6.25 and 25 μg/ml respectively. Compound **3d** exhibited promising activity against *B. subtills* with MIC value of 12.5 μg/ml**.** Compound **3 f** was effective against *S. aureus* with MIC 12.5 μg/ml. Compound **3 g**, effectively reduced the growth of *Pseudomonas aeruginosa* with MIC value of 12.5 μg/

Table 1. In vitro anti-microbial activity (MIC μg/ml) of compounds **3a-l.**

entry			Gram positive		Gram negative		
	R-with benzene Ring	R^1-with benzene ring	*S. aureus*	*B. subtills*	*Pseudomonas aeuroginosa*	*Klebsiella pneumoniae*	*Escherichia Coli*
3a.	*p*-Cl	*p*-Br-phenyl	100	25	100	50	100
3b.		*p*-Cl-phenyl	50	100	50	25	100
3c.		*p*-OH-phenyl	100	50	6.25	100	25
3d.		*o*-Br-phenyl	25	12.5	50	25	50
3e.	*p*-OH	*p*-Br-phenyl	25	25	25	100	100
3 f.		*p*-Cl-phenyl	12.5	50	25	100	50
3 g.		*p*-OH-phenyl	50	25	12.5	25	25
3 h.		*o*-Br-phenyl	25	100	50	50	100
3i.	*o*-Br	*p*-Br-phenyl	50	12.5	100	50	50
3j.		*p*-Cl-phenyl	100	25	50	12.5	50
3 k.		*p*-OH-phenyl	25	100	25	100	100
3 l.		*o*-Br-phenyl	100	100	25	50	25
	Ampicillin		–	25	25	–	–

ml. Compound **3i** showed notable antimicrobial activity against *B. subtills* MIC 12.5 μg/ml while compound **3j** showed its promising activity against *Klebsiella pneumoniae* with MIC value of 12.5 μg/ml. Innovative oxazine derivatives based on their electron withdrawing effects possess notable antimicrobial action The reported compounds that are easily synthesized in the laboratory will surely provide another antimicrobial in the rising cases of antimicrobial resistance among the microbes. Such synthesized compound may further be explored for other bioactivities like anticancer, antioxidant, anti-malarial antituberculosis etc. The synthesized compound may be further explored for their in vivo action and may be developed as new antimicrobials in the event of the rising adaption of the clinical drugs among the discussed classes of the microbes.

4 Conclusions

Novel designing, synthesis and antimicrobial screening of novel heterocyclic derivatives of oxazine as 3-(Substituted-phenyl)-4-(3-(Substituted-phenyl)-3,4-dihydro-2H-benzo[e][1,3]oxazin-6-yl)-3,4-dihydro-2H-naphtho[2,3-e][1,3]oxazineresulted in the procurement of some prominent antimicrobials in the form of compound **3c**, **3d**, **3 f**, **3 g**, **3i**, **3 l** with notable antimicrobial activity against *B. subtills, Pseudomonas aeruginosa, Klebsiella pneumoniae, Escherichia coli*. The compounds need to be further explored for their *in-silico* studies and in-vitro examinations.

References

1. Zhang, Y.-R., Xie, J.-W., Huang, X.-J., Zhu, W.-D.: Construction of functionalized 2,3-dihydro-1,4-benzoxazines via [5 + 1] annulations of 2-halo-1,3-dicarbonyl compounds with imines. Org. Biomol. Chem. **10**, 6554–6561 (2012). https://doi.org/10.1039/c2ob25927c
2. Zinad, D.S., Mahal, A., Mohapatra, R.K., Sarangi, A.K., Paramata, M.R.F.: Medicinal chemistry of Oxazines as promising agents in drug discovery. Chem. Biol. Drug Des. **95**, 16–47 (2020). https://doi.org/10.1111/cbdd.13633
3. Hannath, K.M., Chandra, M., Krishnakumar, K.: Biological potentials of Oxazines as promising agents for drug discovery - A short review. Int. J. Pharm. Sci. Rev. Res. **63**(2), 102–106 (2020)
4. Zhang, P., et al.: 6-Aryl-1,4-dihydro-benzo[*d*][1,3]oxazin-2-ones: a novel class of potent, selective, and orally active nonsteroidal progesterone receptor antagonists. J. Med. Chem. **45**, 4379–4382 (2002). https://doi.org/10.1021/jm025555e
5. Ziarani, G.M., Moradi, R., Lashgari, N., Kruger, H.G.: Chapter 12 - Oxazine Dyes. In: Ziarani, G.M., Moradi, R., Lashgari, N., Kruger, H.G. (eds.) Metal-Free Synthetic Organic Dyes, pp. 177–178. Elsevier (2018). https://doi.org/10.1016/B978-0-12-815647-6.00012-1
6. Gupta, N., Saini, V., Basavarajaiah, S.M., Dar, M.O., Das, R., Dahiya, R.S.: 1,3-Oxazine as a promising scaffold for the development of biologically active Lead molecules. Chem. Select. **8**(39), e202301456 (2023). https://doi.org/10.1002/slct.202301456
7. Khollat, A., Moradi, L.: Efficient synthesis of some [1,3]-oxazine derivatives in the presence of solid acid nano catalyst based on ferrierite and study on their activity against breast cancer through molecular docking calculation. Sci Rep **14**, 16211 (2024). https://doi.org/10.1038/s41598-024-67292-3; (b) Gali, S., Raghu, D., Mallikanti, V. et al. Design, synthesis of benzimidazole tethered 3,4-dihydro-2H-benzo[e] [1, 3] oxazines as anticancer agents. Mol Divers **28**, 1347–1361 (2024)
8. Presnukhina, S.I., et al.: Synthesis of 1,2,4-Oxadiazin-5(6*H*)-one derivatives and their biological investigation as monoamine oxidase inhibitors. Molecules. **29**(23), 5550 (2024). https://doi.org/10.3390/molecules29235550
9. Arshad, M., Khan, T.: 1-(substituted-phenylsulfonyl)-2Hthieno [2, 3-d][1, 3] oxazine-2, 4 (1H)-dione: drug likeness, physicochemical, synthesis, characterization, antibacterial and cytotoxicity assessment. Int. J. Pharm. Sci. Res. **9**, 12–19 (2018)
10. Khabnadideh, S., solhjoo, A., Heidari, R., et al.: Efficient synthesis of 1,3-naphtoxazine derivatives using reusable magnetic catalyst (GO-Fe_3O_4–$Ti^{(IV)}$): anticonvulsant evaluation and computational studies. BMC Chem. **16**, 44 (2022). https://doi.org/10.1186/s13065-022-00836-8

11. Li-Zhulanov, N.S., Pavlova, A.V., Korchagina, D.V., et al.: Synthesis of 1,3-Oxazine derivatives based on (−)-Isopulegol using the Ritter reaction and study of their analgesic activity. Chem. Heterocycl. Comp. **56**, 936–941 (2020). https://doi.org/10.1007/s10593-020-02753-x
12. Elsayed, D.A., Abdu, M.E., Marzouk, M.A., et al.: Bio-computational modeling, POM analysis and molecular dynamic simulation for novel synthetic quinolone and benzo[*d*][1,3]oxazine candidates as antimicrobial inhibitors. Sci. Rep. **14**, 28709 (2024). https://doi.org/10.1038/s41598-024-73972-x
13. Peng, L., Chen, H., Sun, M., Yu, H., Hou, J., Wang, S.: A benzooxazine-based ratiometric fluorescent probe for pH imaging in living cells and bacteria. Sensors Actuators B Chem. **335**, 129711 (2021). https://doi.org/10.1016/j.snb.2021.129711
14. Repale, A.V., Ram, K., Chundawat, N.S., Srivastava, N., Singh, G.P.: A novel and easy synthesis of 1H-Tetrazoles employing p-Tolenesulfonyl azide. Res. J. Chem. Environ. **28**(10), 35–43 (2024). https://doi.org/10.25303/2810rjce035043
15. Anisree, S., Shanmugavelan, P., Palanisamy, S., Kishore, R., Srivastava, N.: Synthesis and anticancer screening of substituted (phenyl)-1-((naphthalen-5-yl)methylene)thiosemicarbazide against PC3 prostate cancer cells. J. Indian Chem. Soc. **101**(9), 101229 (2024). https://doi.org/10.1016/j.jics.2024.101229
16. Yadav, S., Kishore, R., Srivastava, N.: Microwave-assisted phase transfer catalytic synthesis of aryl Benzothiazole derivatives and their anticancer screening. Org. Prep. Proced. Int. **56**(6), 563–571 (2024). https://doi.org/10.1080/00304948.2024.2342692
17. Anisree, S., Shanmugavelan, P., Vijaylakshmi, P., Kishore, R., Srivastava, N.: Synthesis, characterization and anticancer screening of novel Phenylbenzylidene Thiosemicarbazone derivatives. Phos. Sulf. Silicon Relat. Elem. **199**(4), 267–276 (2024). https://doi.org/10.1080/10426507.2024.2330935
18. Tyagi, R., Yadav, K., Srivastava, N., Sagar, R.: Applications of pyrrole and pyridine-based heterocycles in cancer diagnosis and treatment. Curr. Pharm. Des. **30**(4), 255–277 (2024). https://doi.org/10.2174/0113816128280082231205071504
19. Tyagi, R., Singh, K., Srivastava, N., Sagar, R.: Recent advances in carbohydrate-based gelators. Mater. Adv. **4**, 3929–3950 (2023). https://doi.org/10.1039/D3MA00321C
20. Srivastava, N.: Efficient TBAI-CS2 promoted synthesis of substituted Hydrazinecarbodithiolates. Org. Prep. Proced. Int. **54**(6), 569–577 (2022). https://doi.org/10.1080/00304948.2022.2111171
21. Srivastava, N., Chaturvedi, D., Kishore, R., Kaur, M., Kaur, R.: Triton-B/CS2 mediated novel synthesis of substituted thioureas: A novel class of anti-cancer agents. Indian J. Chem. B. **61**, 870–877 (2022). https://doi.org/10.56042/ijc.v61i8.65053
22. Srivastava, N., Kishore, R.: Cleaner and greener synthesis of 3H-benzothiazole-2-thione and its derivatives. J. Sulfur Chem. **42**(1), 29–39 (2020). https://doi.org/10.1080/17415993.2020.1803321
23. Srivastava, N., Kishore, R.: Cleaner, greener synthesis, characterization, and anticancer evaluation of 2-Thioxo-1,3-Thiazolanes. Indian J. Heterocycl. Chem. **31**(2), 265–271 (2021)
24. Srivastava, N., Kishore, R.: Synthesis, characterization, anticancer, antibacterial, antifungal, and antimalarial activity of (aryl or hetero aryl) substituted-4-Methyl-2,3-dihydro-2-thioxo-1H-imidazol-5-yl Ethanone derivatives. Indian J. Heterocycl. Chem. **31**(03), 347–355 (2021)
25. Srivastava, N., Rathore, B.K., Vishnoi, R.K., Bajpai, S.: A simple and convenient method for synthesis of Artemisinin dimer through aliphatic nitro compounds. Asian J. Chem. **33**(1), 49–52 (2020). https://doi.org/10.14233/ajchem.2021.22918
26. Kishore, R., Kamboj, M., Shukla, M., Chaturvedi, D., Srivastava, N.: Novel synthetic strategy of cyclic Dithiocarbamates catalyzed by triton-B. Asian J. Chem. **31**(5), 1091–1094 (2019). https://doi.org/10.14233/ajchem.2019.21830

27. Srivastava, N., Saxena, M., Shukla, M.: Novel synthesis of 5-Oxo-2-thioxo-2,5-dihydro-3-thiophenecarboxylate derivatives in non-aqueous medium. Asian J. Chem. **31**(1), 176–180 (2019). https://doi.org/10.14233/ajchem.2019.21622
28. Srivastava, N.: Key role of ionic liquids in the cleaner and greener synthesis of lactams. Res. J. Chem. Environ. **26**(1), 125–130 (2022). https://doi.org/10.25303/2601rjce125130
29. Vishnoi, R.K., Kishore, R., Chaturvedi, D., Shukla, M., Bajpai, S., Srivastava, N.: Synthesis and antimicrobial activity of cyclic dithiocarbamates employing triton-B/CS2 system. Asian J. Chem. **33**(5), 1133–1136 (2021). https://doi.org/10.14233/ajchem.2021.23173
30. Srivastava, N., Kishore, R., Chaturvedi, D.: Novel and efficient method for the synthesis of cyclic trithiocarbonates employing Cs_2CO_3 and CS_2 system. Res. J. Chem. Environ. **25**(12), 142–148 (2021). https://doi.org/10.25303/2512rjce142148
31. Srivastava, N., Chaturvedi, D., Kaur, M., Kalia, N.: Cleaner, greener synthesis and anticancer screening of asymmetrical Trithiocarbonates. Org. Prep. Proced. Intern. **56**, 606–616 (2024). https://doi.org/10.1080/00304948.2024.2326356
32. Srivastava, N., et al.: Clean and green synthesis of organic Isothiocyanates using zinc peroxide as desulfurizing agent and their antimicrobial activity. Russ. J. Org. Chem. **59**(12), 2205–2213 (2023). https://doi.org/10.1134/S1070428023120199
33. Srivastava, N., Saxena, M.: Novel synthesis of symmetrical Dialkyl/Diaryl trithiocarbonates in non-aqueous medium at room temperature using CS_2. CS_2CO_3 and alkyl/aryl halide. Rasayan J Chem. **12**(1), 333–337 (2019). https://doi.org/10.31788/RJC.2019.1215038

Small Molecules for Drug Development of Alzheimer's Disease

Nitin Srivastava[1(✉)] and Shivi Singh[1]

[1] Department of Chemistry, Amity University Uttar Pradesh, Lucknow Campus, Lucknow, Uttar Pradesh, India
nitinsriv5@gmail.com; nsrivastava1@lko.amity.edu

Abstract. Alzheimer's disease (AD) is the most common form of dementia and a challenging disease. It is characterized by cognitive decline, memory loss, and behavioral problems. The pathogenesis of AD is complex, and the molecular mechanisms and mechanisms involved in this devastating neurodegeneration are numerous. The development of small molecules targeting tauopathy and the search for antibodies are promising for the future of AD treatment. Here, we consider small-molecule therapeutics as a promising tool for the development of new drugs for the treatment of AD. Small molecules exhibit better shelf life, easier to handling and distribution, cheap, making them more suitable over the monoclonal antibodies. This review develops structural insight into the small molecules which have shown potential in drug development inhibiting acetylcholinesterase (AChE) in the treatment of AD. It may help the study of AD pharmacokinetics using small molecules, including new drug development, immunology, and clinical trial challenges.

Keywords: Alzheimer's disease · Small molecules · β-Amyloid · Acetylcholinesterase · Tauopathy

1 Introduction

Alzheimer's disease (AD) is the root cause of dementia. AD is the biological phenomena which leads to the formation of proteins in the form of amyloid plaques and neurofibrillary tangles in the brain leading to the death of brain cells with passing time. The World Health Organization report in 2022 says that an estimated 55.2 million people are affected by AD around the world. Europe reporting a prevalence of 6.5%, Southeast Asia at 2.9% [1]. The cases of dementia are increasing, apart from some high-income nations observing a decline [2]. By 2030, there will be a rise in people with Alzheimer disease which would be approximately 78 million. This will enhance the cost of medical treatment, care with global loss in the human resource which may exceed US$ 2.8 trillion. It will badly affect the individuals, families, and societies [1]. AD leads to cognitive deterioration, increasing memory loss with disease advancement, language impairment, behavior change, disorientation, advancing dementia and ultimately death. The dementia advances with growing age, So, intricacies of the disease need to be

A. K. Saxena, A. Saxena (Eds.): GTHTM 2025, pp. 237–253, 2026.
https://doi.org/10.1007/978-3-032-12320-6_13

understood molecular-level for the development precise diagnosis and treatment. Various aspects associated with AD include neurodegenerative disorder caused by aggregation and misfolding of proteins, oxidative stress, changes in the mitochondria and inflammation [3]. The onset of Alzheimer's disease is characterized by loss in memory of recent conversation and events. With the advancement of the disease memory loss increases and a person is unable to perform his routine work. Different theories have been proposed the development of Alzheimer disease. Among the proposed theories the cholinergic hypothesis is extensively studied. It studies the effect of neurotransmitter Acetylcholine (ACh) in AD, and which is responsible for cognitive functions. This theory focusses on degeneration of cholinergic neurons, which produce, regulate and are responsible for the production and regulation of the neurotransmitter acetylcholine, and plays a critical role in the pathogenesis of AD. According to the cholinergic hypothesis, due to the degeneration of cholinergic neurons, the activity of ACh and associated enzymes are reduced in some parts of the brain like the hippocampus and cortex resulting in learning and memory dysfunction. This reduction is believed to contribute to the hallmark symptoms of AD, including memory impairment and cognitive decline [4]. ACh, is a neurotransmitter which controls lung, heart, intestine, stomach, etc. is found at synapses and at the neuromuscular junction. Energy pumps make the vesicles acidic and facilitate the entry of Ach into the storage vesicle. In the process of neurotransmission, acetylcholinesterase is liberated into the synaptic cleft and interact with the acetylcholinesterase receptors in the post-synaptic membrane to pass on the nerve signal [5].

Cholinesterase are enzymes present in the postsynaptic membrane which hydrolyses neurotransmitter Ach into acetic acid and choline. Choline is necessary for the activation of cholinergic neurons. The released choline is lifted by choline carrier for production of ACh by using Acetyl CoA in ambience of choline acetyl transferase in the presynaptic neuron [6]. Acetylcholinesterase (AChE) and butyrylcholinesterase (BuChE) are two types of cholinesterase.

AChE and BuChE are serine hydrolases found at neuromuscular junctions and cholinergic brain synapses and have ellipsoidal shapes with Peripheral anionic site (PAS) at the entry of the active centre gorge and Catalytic active site (CAS) at the lower end of extended slender gorge. The CAS has esteratic subsite, anionic subsite, and catalytic machinery. The esteratic subsite is choline-binding pocket [7]. Binding of acetylcholine in the catalytic pathway and allosterically modulates catalysis is initiated by PAS. The anionic subsite gets bounded with the quaternary trimethyl ammonium choline [8]. The hydrolysis of ACh into choline acetates and occurs at esteratic site having amino acids Glu327, His440 and Ser 200 [9, 10]. Despite resemblance in structures AChE and BuChE are substrate-specific and possess different sensitivity to inhibitors. AChE and BuChE modulate the activity proteins, phosphorylation, amyloid cascade, blood flow and play specific role in neuronal development, differentiation [11]. AChE can be easily hydrolyzed rather than BuChE making it a prime focus of research for the development of therapeutics against AD [12].

Presence of AChE inhibitors prevent the hydrolysis of acetylthiocholine via AChE and help in managing the symptoms of AD. There are some FDA approved drugs which control the progression of AD and these drugs are basically AChE inhibitors like Donepezil, tacrine, rivastigmine and galantamine. Tacrine was the first AChE inhibitor

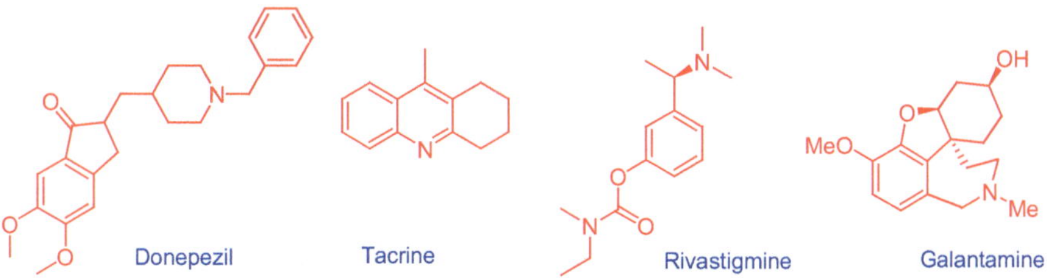

Fig. 1. Structurally diverse anti-Alzheimer drug molecules.

drug but later found to be hepatotoxic and so rarely used. Clump of proteins called amyloid plaques which are formed in the brains of people with Alzheimer's disease (AD). These are a symbol of AD progression and interfere with the normal brain functioning (Fig. 1).

Aducanumab and donanemab are FDA approved drugs for targeting Amyloid plaques but the manufacturer of aducanumab has withdrawn it. The FDA approved drugs which are AChE inhibitors can be modified, and their several derivatives can be made which may be explored for the anti-Alzheimer drugs using SAR studies, computational methods, drug repurposing, etc. Such research for drug development may improve potency, bioavailability metabolic stability, less toxicity and side effects. We have been exploring the synthesis and bioactivity of several drug molecules [13–32]. In this review we extend our efforts towards discovery and modification of existing small molecule-based AChE inhibitors drugs in the last ten years.

1. **Piperidine based derivatives**

Donepezil is a widely recommended piperidine-based drug for the treatment of Alzheimer. It is least hepatotoxic and presents a significant moiety for the development of novel anti-Alzheimer drugs.

Masitinib (**1**), a kinase inhibitor prevents fibroblast growth factor receptors and carries synaptic protection in a dual amyloid precursor protein (APP)/presenilin 1 (PSEN1) in Alzheimer's disease (AD) [33, 34]. In Phase 3 study and clinical trial (NCT01872598) it improved cognition [35]. It inhibited tau accumulation and proved potential for further development as anti-Alzheimer drug [35, 36].

Heterocyclic scaffolds indoline-2-one, quinoline and pyrrolidone-2-onewere explored for in-vitro and in-vivo anti-Alzheimer's drug in animal models [37]. Docking studies and molecular dynamics simulations were performed to evaluate the interaction with acetylcholinesterase enzyme (AChE) with further QSAR. The designed compounds inhibited AChE with micromolar IC_{50} values. N-(1-benzylpiperidin-4-yl)-2-(2-oxoindolin-3-yl)acetamide (**2**), 2-(1-benzylpiperidin-4-yl)-6,7-dimethoxy-4-(4-methoxyphenyl)quinoline (**3**) 6-chloro-N-(1-(1-(3,4-dimethoxybenzyl)-2-oxopyrrolidin-3-yl)piperidin-4-yl)pyridine-3-sulfonamide (**4**) from indolin-2-one, quinoline and pyrrolidone-2-one series respectively inhibited AChE with an IC50 value of 0.01 μM. Further lipid peroxidation, glutathione, superoxide dismutase, studies revealed that compound 9 inhibited AChE more effectively and may be further developed as lead compound for AD treatment (Fig. 2).

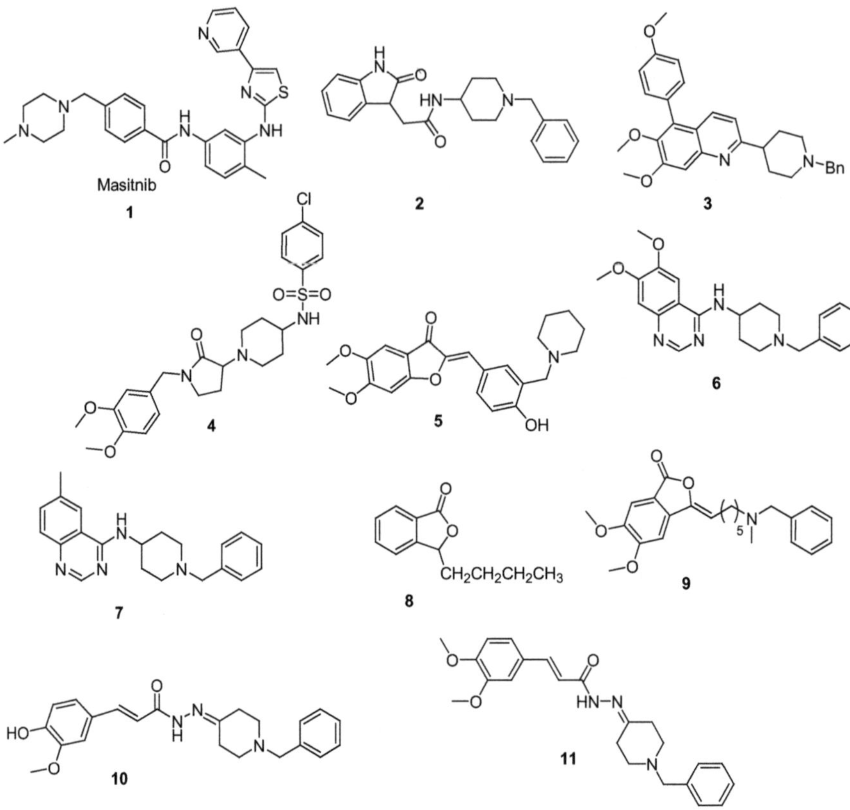

Fig. 2. Piperidine based derivatives.

Aurone-2-benzylidenebenzofuran-3(2H)-ones exhibited good interaction with Aβ aggregates, inhibition to MAO, and AChE inhibition to be used as MTDLs. It was found that benzylamine and Mannich base are pharmacophores which possess AChE inhibition, so novel aurone Mannich base derivatives have been synthesized and screened for antioxidant action, bio metal chelation activity anti-Aβ aggregation, and AChE inhibition [38]. Among the synthesized compound, **5** exhibited highest inhibition against *rat*AChE, *ee*AChE and *hu*AChE with IC_{50} values of 0.00878, 0.0212, and 0.0371 μM respectively which is better than donepezil. The synthesized compounds showed more selectivity to AChE than BuChE. Compound **5** exhibited mixed types of inhibition with Ki = 0.012 μM which was due to π-π interaction between the Tyr334, Trp279 and benzofuran-3(2H)-one moiety, Phe330 and benzene ring. The compound exhibited hydrophobic interaction between Trp84 and piperidine moiety, H-bond interactions with OH and Asp72. This compound showed the highest antioxidant effect with the least toxicity to the PC-12 cells. The compound **5** showed bio metal chelation and inhibition to Aβ1—42 aggregation and was able to cross blood brain barrier.

Oxoindolinone donepezil derivatives have exhibited potential for the anti-Alzheimer drug so quinazoline analogues of donepezil were studied through molecular docking, fragment hopping, and structural similarity approach [39]. The quinazoline moiety interacted with Tyr341 and Trp286. Compound **6** has exhibited H-bond interaction with Phe295 and Tyr124 while **7**, lacked H bonding due to higher distance between Phe295 and NH of quinazoline. Both **6** and **7** exhibited the highest activity against AChE and least toxicity. Further scavenging action was also shown by **6** for DPPH. The study favored the compounds for development as lead anti-AD compounds.

DL-3-n-butylphthalide and DL-NBP **8** which are used in treatment of stroke are useful in improving cognition in mice AD models. Hybrids of DL-NBP having indanone structure and secondary amines were prepared and screened for anti-Alzheimer action [40]. Among the synthesized compounds, **9** showed the highest inhibitory activity with IC_{50} value of 2.66 nM. Docking evaluation revealed that **9** covered AChE, PAS and CAS with hydrophobic interactions between the side chain and amino acids Phe331, Phe330 and Tyr334 along with H-bonding with Phe288. It also exhibited π-π interactions with Trp279 and phthalein, Trp84 and benzylamine. Compound **9** was able to cross the blood brain barrier with enhanced antioxidant activity and inhibited Aβ1-42 aggregation. In Vivo, examinations revealed that **9** possess higher potential than donepezil in reversing scopolamine-induced memory loss in mice and may be considered for multi-targeted drug design.

The availability of H acceptor and donor groups in N-acyl hydrazones (NAHs) makes them potential pharmacophore based on higher polarity. Hybridization of N-benzylpiperidine from donepezil and cinnamoyl from curcumin led to the formation of cinnamoyl-N-acylhydrazone-donepezil derivatives which are potential ligands [41]. Among the synthesized compounds, **10** and **11** are potential inhibitors for AChE with IC_{50} values of 13.04 and 9.10. Docking studies revealed that they exhibited mixed type inhibition. Interactions like π-stacking between Trp86 and the phenyl ring, cation-π interaction with piperidine Tyr337, H-bonding C=O of NAH and Tyr121 were found during docking studies. Also, some hydrophobic interactions between the PAS and 1,2-dimethoxybenzene, and methoxy with Tyr72 and Trp286 in PAS. The study also reported that donepezil, having a long NAH group, hindered dimethoxy benzene moiety to exhibit stacking interaction with Trp279. Although PAS area can undergo easy interaction with the compounds due to higher flexibility. Compound **10** also showed DPPH scavenging against oxidative damage caused by AD and least toxicity posing it to be a potential lead molecule in drug discovery against AD.

2. **Tetrahydroacridine derivatives**

These are the traditional compounds which have been employed for increasing aminotransferase. They inhibit acetylcholine esterase and induce the production and release of acetylcholine and modulate nicotinic and muscarinic receptors. The tacrine showed toxicity to liver but its structure has given a clue for the development of anti-Alzheimer drugs by modification of acridine moiety.

Among the synthesized 1,2,3-triazole derivatives, compound **12** and **13** exhibited significant activities anti-Alzheimer activities with an IC_{50} of 7.31 ± 0.03 μM and IC_{50} of 10.84 ± 0.07 μM. In comparison to the reference drug, rivastigmine [42]. Compound **12** showed H-bonding with the methoxy group present in the acridone moiety and OH group of Ser122.

Xanthenone derivatives with similar structure to tacrine were synthesized and evaluated for the anti-Alzheimer action. These synthesized compounds were found to interact in a similar way to tacrine with the acetylcholinesterase during molecular docking studies [43]. The in vitro evaluation revealed that compound **14**, (R)-9-mercapto-5-methoxy-3,3-dimethyl-3,4-dihydro-2H-xanthen-1(9H)-one exhibited the highest inhibition (94.35%), which is as good as tacrine. This compound was able to overcome the blood-brain barrier (BBB) which is a basic requirement for the anti-Alzheimer drug. The compounds also exhibited non-toxic nature when they were examined in vivo.

Binuclear condensed tacrine Naphthyridine-11-amine, are also found to possess anti = Alzheimer activities [44]. In this synthesized series one compound with para-fluoro substituted phenyl ring **15** exhibited prominent AChE inhibitory activity with IC_{50} = 0.091 μM and least toxicity on normal HepG2 cells in comparison to tacrine. This compound on further studies found to cross blood brain barrier easily.

Derivatives of Indole-3-acetic acid (IAA) have also been found to possess inhibition to acetylcholinesterase (AChE). In this series compound **16** showed highest inhibitory activity toward acetylcholinesterase (AChE) with least toxicity to normal HepG2 cells. The compound was further modified in silico methods by altering the -N(iPr)2 terminus with a tacrine moiety which resulted in the increase in the inhibition with IC_{50} = 6.62 μM 38-fold rise in the activity [45]. Compound **17** inhibited AChE and BChE both with IC_{50} values of 0.173 μM and 0.066 μM respectively and fits well inside both AChE and BChE. The compound exhibits effective π-π interactions between tacrine scaffold and Tyr84, Phe330 inside AChE. The π interactions were also found with the N and Phe330. The indole formed π-π interaction with Tyr279 and a hydrogen bond with Arg289. Similarly, in BChE inhibition, tacrine forms cation-π interaction with Tyr332 and H bond with Pro265 and nitrogen. This compound showed a promising score in docking (−9.3) which was much better than tacrine (docking score of −8.5). Molecular modelling showed that compound **17** inhibits linearly with inhibition constants of 0.065 μM and 0.198 μM for AChE and BChE respectively. The modified compound showed good blood brain barrier transfusion with higher lipophilicity and solubility and tolerable side effects (Fig. 3).

Schiff bases of tacrine also showed promising anti-Alzheimer activities by inhibiting AChE [46]. Among the synthesized compounds, **18** exhibited promising inhibition with IC_{50} = 22.1 nM (Ki = 2.14 mM).

Steric ligands of tacrine moiety and xanomeline have also been employed for the inhibition by activating M1 receptor at the orthosteric site by xanomeline and consequently inhibition of AChE allosterically [47]. Among the synthesized compound **19** with p IC_{50} value of 9.55 showed the highest inhibitory activities with enhanced lipophilicity.

Tacrine-benzofuran hybrids were developed as ligands for multitargets in inhibiting AChE, *B*enzofuran acted as biomimetic for indanone scaffold in donepezil. Molecular docking studies revealed π-π stacking interaction with Phe330 and Tyr84, resulting in insertion of tacrine in the Catalytic active site (CAS) [48], benzofuran may undergo π-π interaction with Peripheral anionic site (PAS) by Trp279 and Tyr70. Chlorine substituted tacrine derivatives exhibited effective fitting in the hydrophobic site and rendered highest inhibiting capacity to compound **20, 21** and **22** with IC_{50} values of 0.12, 0.13 and 0.13 μM respectively. Compound **23** which inhibited AChE with IC_{50} value of 0.67 μM and was found to be potential metal chelator for Fe and Cu. Compound **23**

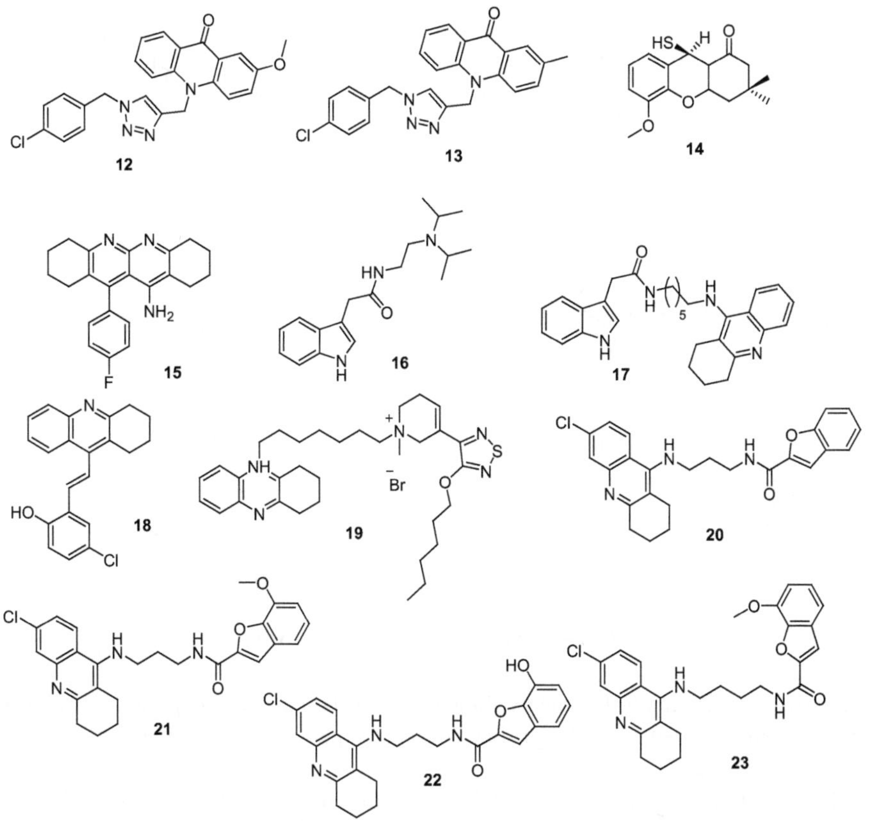

Fig. 3. Tetrahydroacridine derivatives.

showed inhibition with higher capacity to cross BBB and lipophilicity to become a potential drug for AD (Fig. 4).

AD is associated with the production of reactive oxygen species which cause apoptosis in the nerve cells. Antioxidants have been developed in the form of tacrinekojic acid scaffolds which have shown AChE inhibition which is helpful in reduction of ROS [49]. Among the synthesized compounds, cyclopentapyranopyridine-kojic acid derivative **24** possesses the highest inhibition activity with IC_{50} value of 4.18 μM. Compound **24** possessed the highest protective action against ROS H2O2 with higher drug-likeness and neuroprotection.

Pyridine is a basic pharmacophore in most of the anti-AD drugs. Simultaneously δ Sultone fused pyrazoles have shown strong inhibition to BuChE. The congregation of these two have been explored by synthesis of pyridinyl sultones and screened for their inhibitory activities to AChE [50]. This research concluded that among the synthesized compounds, compound **25, 26, 27** with 4-halophenyl substituents showed promising activities against AChE with IC_{50} values of 8.93, 11.51, and 13.06 μM respectively. The in vitro activity was further supported by the molecular docking study. Compound **25** showed the best fitting in the enzyme pocket, forming hydrogen bonding with δ sultone

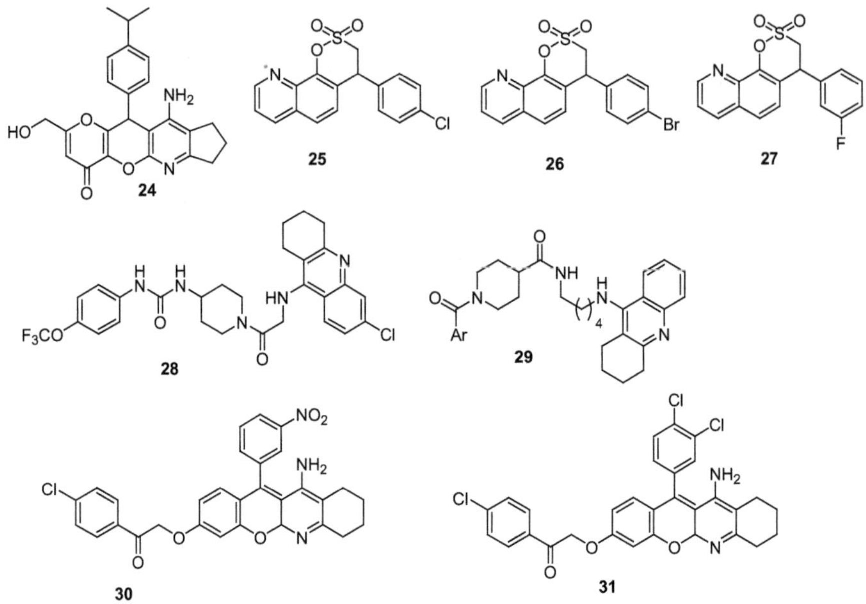

Fig. 4. Tetrahydroacridine derivatives.

and Asp72, Ser122 and NH of the pyridine ring with Trp84. The compound also showed a π-π stacking interaction with pyridine ring and Trp84, an anion π interaction among Phe288, Phe290 and Cl.

Neuronal inflammation is one of primary factor enhancing the Alzheimer. The oxidative stress causing inflammation may be reduced by Epoxyeicosatrienoic acids (EETs). But EETs are metabolized by epoxide hydrolase (sEH) in Alzheimer patients. So, AChE inhibitors which exhibit sEH inhibition may be more fruitful. sEH inhibition can be achieved by TPPU. So, TPPU-huprine and TPPU-chlorotacrine hybrids have been synthesized and evaluated for their AChE inhibition effects [51]. Among the synthesized compounds, compound **28** inhibited AChE and sEH both with IC_{50} value of 12.9 nM. The compound showed good pharmacokinetic stability to cross blood brain barrier with the least toxicity to nerve cells. In vivo studies revealed that compound **28** enhanced the memory function significantly affected synaptic, amyloid and tau dysfunction.

Novel tacrine having and *N*-aryl-piperidine-4-carboxamide moiety designed, synthesized and screened for inhibition of AChE and BuChE [52]. The synthesized compounds exhibited prominent ChE inhibition with IC_{50} in the range of 117.5–455 nM for AChE and 34–324 nM for BuChE. Among the synthesized compounds 2-chlorobenzoyl derivative **29** showed best AChE/BChE inhibition, high selectivity index, capable of binding to both CAS and PAS of the enzymes. The compound also showed neuroprotective against amyloid-β. Molecular dynamics (MD) simulation study reported that 2-chlorobenzoyl derivative, formed stable complex with AChE and BuChE. Evaluation of ADME properties showed good penetration across blood–brain barrier. The compound may be developed as a lead in the anti-Alzheimer drug development.

Among the synthesized tacrine-chromene derivatives, the compounds **30** and **31** exhibited the potential inhibition to *h*AChE with IC_{50} values of 0.44 and 0.25 μM, respectively [53]. They also exhibited an inhibition to BuChE with IC_{50} = 0.08 and 0.14 μM respectively which was 10 and 5 times more than rivastigmine. Compounds **30** and **31** bind to PAS, influenced Aβ aggregation. Compound **31** was eight-fold effective (76% inhibition) than curcumin with IC_{50} = 0.74 μM. Compounds **30** and **31** inhibited BACE-1 with IC_{50} = 0.38 and 0.44 μM, respectively better than curcumin. They also inhibited MAO-B with IC_{50} = 5.15 and 2.42 μM, respectively. These compounds possess antioxidant activities and metal-chelation for Fe^{+2}, Zn^{+2}, and Cu^{+2}, which may inhibit oxidative stress in AD brains. Compounds **30** and **31** were less toxic to HepG2 and SH-SY5Y. Compounds **30** and **31** are potential candidates for development of drugs against AD.

3. **Rivastigmine derivatives**

Another FDA approved drug for treatment of AD is Rivastigmine (Riva) which is a AChE inhibitor and is disintegrated by AChE and releases metabolite NAP226–90 which undergo chelation with metals in the brain and minimizes Aβ aggregation [54]. The Structure of Riva inspired the development of AChE inhibitors as anti-AD drugs. Synthesis of novel 1-phenyl-ethanone derivatives has been further screened for AChE inhibition and its hydrolysis products have been explored for metal chelation [54]. Among the synthesized **32** and **33** compounds effectively inhibited Aβ aggregation. Also, compound **34** inhibited AChE, an IC_{50} value of 31.9 μM.

Twenty-four Rivastigmine structure-based hybrids as potential multi-target anti-Alzheimer's drug candidates [55]. Among the synthesized compounds, **36** and **37** exhibited potential as multi-target agents as ChEs, MAO-B and FAAH inhibitors. Compound **35** (**ROS151**) was found to be a potent AChE inhibitor, as well as a chelator of Fe^{3+} and Cu^{2+}. Compounds **ROS151**, **36**, **37**, **38** exhibit similar pharmacokinetic properties as rivastigmine and donepezil.

Novel rivastigmine derivatives have been synthesized as selective BuChE inhibitors and evaluated as anti-Alzheimer [56]. Among the synthesized compounds **39** was identified to be a BuChE inhibitor on the basis of structural modification of rivastigmine. It proved to be a potential and selective inhibitor of BuChE inhibitor with IC_{50} value of 0.33 μM, and selectivity index of higher than 151.5. Compound **39** showed promising neuroprotective properties with drug-likeness and blood brain barrier penetration. It also ameliorated $AlCl_3$-induced zebrafish AD model and scopolamine-induced impairment in mice memory.

Rivastigmine–indole (RIV-IND) hybrids were designed, synthesized and screened for anti-Alzheimer potential [57]. Docking evaluation of the hybrids with acetylcholinesterase (AChE) and butyrylcholinesterase (BChE) these are inhibitor of both enzymes which is better than clinical drugs. Compounds **40** and **41** exhibited better inhibition of AChE than rivastigmine with IC_{50} values of 10.9 and 26.8 μM. drug. Hydroxyl substituted hybrids in indole possess significant antioxidant activity with EC_{50} values in the range of 7.8–20.7 μM. These compounds were effective inhibitors of $A\beta_{42}$ self-aggregation and compounds **42** and **43** prevented $A\beta_{1-42}$ induced toxicity to cells. The compounds also exhibited drug-likeness with potential oral availability. The hybrids, **40** and **41** appeared as lead compounds for anti-AD drug candidates (Fig. 5).

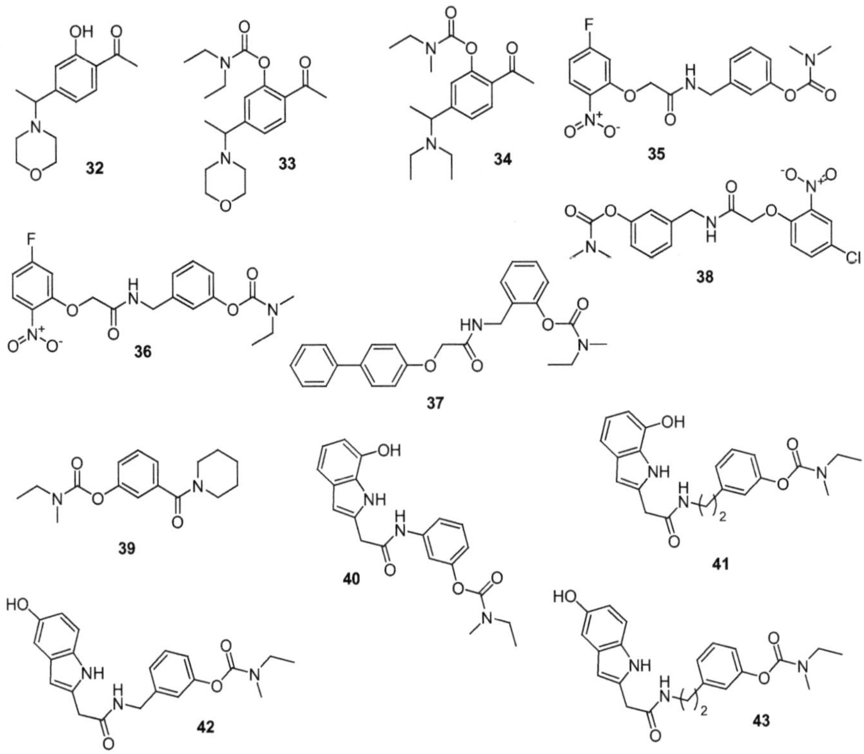

Fig. 5. Rivastigmine derivatives.

Novel rivastigmine derivatives developed and explored for AChE/BuChE inhibition and neuroprotective effect [58]. Compounds exhibited *ee*AChE inhibition with IC_{50} value of 1.78 μM and but weak BuChE inhibition The molecular docking study also confirmed the AChE inhibitory potency. Further studies revealed that one of the compounds was a neuroprotective on L-Glu-induced HT22 cells injury. It also possesses drug-like properties and represents a promising selective AChE inhibitor.

In order to explore for the cheaper treatment of Alzheimer's disease a cheaper synthesis of acetylcholinesterase (AChE) inhibitors has been suggested from the natural products. In this regards cashew nutshell liquid has been reported for synthesis of dual-binding site inhibitors [59]. Literature showed that cardanol moiety (found in cashew nutshell oil) binds to PAS, in the same way as rivastigmine, and heteroamines bind to CAS. In this regards various N-ethyl-N-(2-methoxybenzyl)amino derivatives were synthesized and among the synthesized compounds, the compound **44** was most effective in inhibiting eeAChE with an IC_{50} of 6.6 μM with least toxicity on human colon epithelial cell line HT-29. Compound **44** also inhibited *h*AChE with IC_{50} value of 5.65 μM. Compound **44** has potential bioavailability (Pe ± SEM = 16.8 ± 1 × 10^{-6} cm s^{-1}), making it a promising candidate for further development (Fig. 6).

Quinoline derivatives of rivastigmine were synthesized and screened for their anti-Alzheimer activities. Among the synthesized compounds, **45** inhibited AChE more

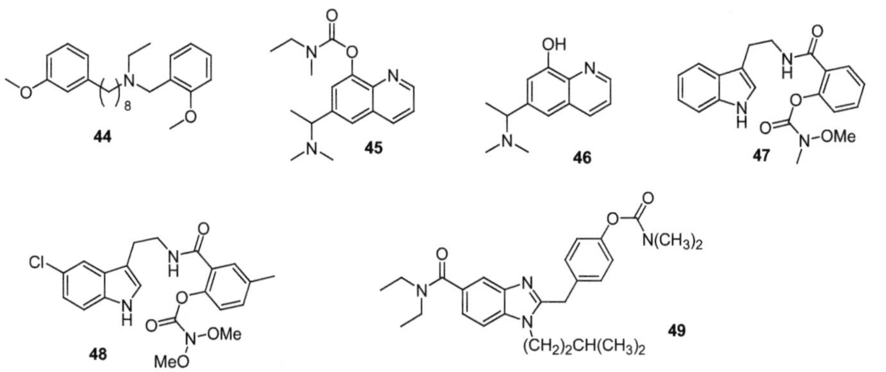

Fig. 6. Rivastigmine derivatives.

effectively than drug Riva [60]. The hydrolytic metabolite **46** of **45** chelated with metals more effectively. Compound **45** inhibited Aβ aggregation. During in vitro hydrolysis of compound by AChE, the enzyme eliminated the carbamoyl moiety. In vivo efficacy indicated that the compound effectively targeted AChE, performs metal chelation, and improves cognition. So, compound **45** may be developed as a drug for the treatment of Alzheimer's disease.

Multitarget drug molecules N-salicyloyl tryptamine derivatives were ssynthesized and screened for AD [61]. In this series, compound **47** exhibited dual inhibition of *ee*AChE and *eq*BChE with IC_{50} values of 3.24 μM and 1.84 μM respectively. This compound exhibited anti-inflammatory properties effectively suppressed Aβ1–42 aggregation. The compound is also nontoxic when studied over mice and able to cross blood-brain barrier with favorable pharmacokinetics requires further research for drug development.

Carbamate-based N-salicyloyl tryptamine derivatives have been synthesized and screened for activity against cholinesterase and neuroinflammation [62]. Compound **48** exhibited higher BChE inhibition but lesser AChE inhibition with IC_{50} value of 0.057 μM which was better than rivastigmine. Compound **48** also exhibited gradual decarbamylation as compared to rivastigmine. In vitro studies showed that the compound exhibited neuroprotection against oxidative stress and Aβ-induced cytotoxicity.

Design and synthesis of novel benzimidazole-carbamate compounds were achieved, and their structural modifications were also explored for interaction for BChE [63]. It was found that compound **49** exhibited pseudo irreversible binding for better selectivity over AChE therapeutic effect over long time. It inhibited hBChE with an IC_{50} value of 6.9 nM and retained selectivity over cannabinoid receptors and AChE. It enhanced cognitive function, and reverses Aβ25–35 induced learning impairments. Compound **49** has proved to be developed as an effective therapeutic agent for AD.

4. Galantamine

Alkaloid galantamine is a traditional AChE inhibitor which fits easily in CAS of AChE but owing to its small size does not interact with PAS. Galantamine also

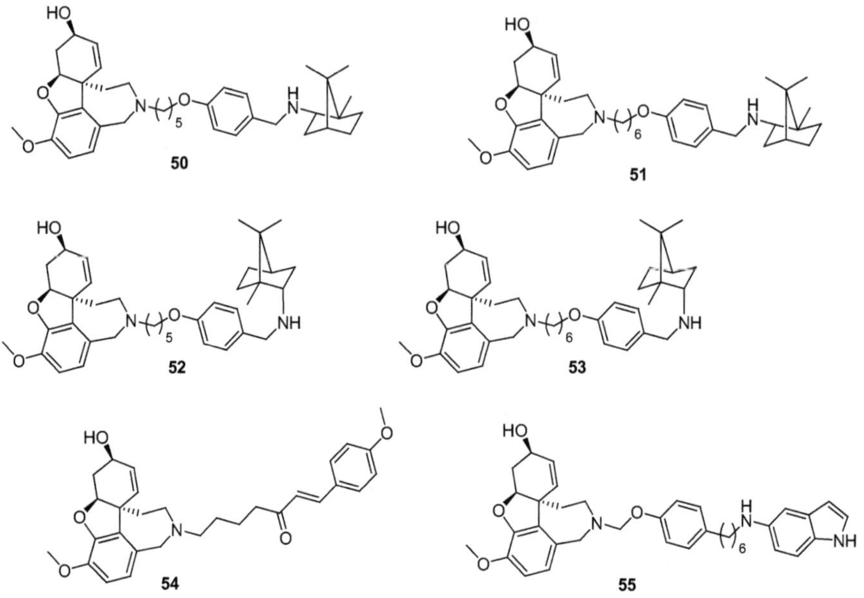

Fig. 7. Galantamine hybrids.

modulates nAChRs and increases Ca^{2+} level to induce release of noradrenaline which is responsible for cognitive ability [64] (Fig. 7).

Galantamine hybrids with long N side chains may be used as dual-binding AChE inhibitors. Keeping this in mind galantamine camphane (CAM) hybrids were synthesized and screened for inhibiting AChE [65]. Among the synthesized compounds, **50, 51, 52, 53** were highly potent with IC_{50} values of 0.0047, 0.0099, 0.0056 and 0.0029 μM respectively. Their inhibition action was 108–369 times more than that of galantamine. CAM inhibited Aβ from binding to AChE. MD simulation studies revealed that the AChE-**50**-Aβ complex possesses a higher distance between Aβ and AChE compared to AChE-Aβ complex. Hence it restricts the formation of amyloid plaques.

Similarly, the galantamine curcumin derivatives were also reported as AChE inhibitors with anti-Aβ aggregation. Here curcumin interacts with fibrils and Aβ oligomers inhibiting the formation of β sheets [64]. Virtually screening of GAL-CU derivatives for ADMET study reported 44 molecules to fit the criteria. Further on molecular docking studies, the compounds having best gold scores were synthesized and screened for neurotoxicity and their inhibitory action on AChE. It was found that compound **54** inhibited AChE far better than curcumin and galantamine with IC_{50} value of 0.02 μM as compared to curcumin and galantamine having IC_{50} values of 68 and 3.52 μM respectively. Further on the basis of MD stimulation the compound with best gold scores were synthesized and found that compound **55** was the most potent AChE inhibitor with IC_{50} value of 0.011 μM [66].

2 Conclusion

In conclusion it may be said that the small molecule-based acetylcholinesterase (AChE) inhibitors are very important in the treatment of Alzheimer's disease. In this review we focused on the recent advances in the development small molecules which are derivatives of the established drugs for AChE inhibition. These small molecules demonstrate better selectivity, potency, and more favorable pharmacokinetics. They enhance our understanding of the enzyme function and structure. Further research may concentrate on optimizing these compounds to reduce adverse effects, enhance blood brain barrier crossing, and overcome drug resistance. Exploring the known drug scaffolds and their modification adds to the novel drug design methodologies and plays a pivotal role in the development of next-generation AChE inhibitors which have better efficacy and safety.

Conflict of Interest. Authors declare no potential conflict of interest.

References

1. WHO: A Blueprint for Dementia Research. World Health Organization, Geneva (2022)
2. Knopman, D.S., et al.: Alzheimer disease. Nat. Rev. Dis. Primers. **7**, 33 (2021). https://doi.org/10.1038/s41572-021-00269-y
3. Hardy, J.: A hundred years of Alzheimer's disease research. Neuron. **52**, 3–13 (2006). https://doi.org/10.1016/j.neuron.2006.09.016
4. Terry Jr., A.V., Buccafusco, J.J.: The cholinergic hypothesis of age and Alzheimer's disease-related cognitive deficits: recent challenges and their implications for novel drug development. J. Pharmacol. Exp. Ther. **306**(3), 821–827 (2003). https://doi.org/10.1124/jpet.102.041616
5. Sobha, A., Ganapathy, A., Mohan, S., Madhusoodanan, N., Babysulochana, A.D., Alaganandan, K., Somappa, S.B.: Novel small molecule-based acetylcholinesterase (AChE) inhibitors: from biological perspective to recent developments. Eur. J. Med. Chem. Reports. **12**, 100237 (2024)
6. Colovic, M.B., Krstic, D.Z., Lazarevic-Pasti, T.D., Bondzic, A.M., Vasic, V.M.: Acetylcholinesterase inhibitors: pharmacology and toxicology. Curr. Neuropharmacol. **11**, 315–335 (2013). https://doi.org/10.2174/1570159x11311030006
7. Nicolet, Y., Lockridge, O., Masson, P., Fontecilla-camps, J.C., Nachon, F.: Crystal structure of human butyrylcholinesterase and of its complexes with substrate and products crystal structure of human butyrylcholinesterase and of its complexes with substrate and products. J. Biol. Chem. **278**(42), 41141–41147 (2023). https://doi.org/10.1074/jbc.M210241200
8. Johnson, G., Moore, S.: The peripheral anionic site of acetylcholinesterase: structure, functions and potential role in rational drug design. Curr. Pharmaceut. Des. **12**, 217–225 (2005). https://doi.org/10.2174/138161206775193127
9. Tripathi, A., Srivastava, U.C.: Acetylcholinesterase: a versatile enzyme of nervous system. Ann. Neurosci. **15**(4), 106–111 (2010)
10. Soreq, H.: Acetylcholinesterase—new roles for an old actor. Nat. Rev. Neurosci. **2**, 294–302 (2001). https://doi.org/10.1038/35067589

11. Reale, M., et al.: Butyrylcholinesterase and acetylcholinesterase polymorphisms in multiple sclerosis patients: implication in peripheral inflammation. Sci. Rep. **8**, 1–9 (2018). https://doi.org/10.1038/s41598-018-19701-7
12. Nunes, D., Loureiro, J.A., Pereira, M.C.: Drug delivery systems as a strategy to improve the efficacy of FDA-approved Alzheimer's drugs. Pharmaceutics. **14** (2022). https://doi.org/10.3390/pharmaceutics14112296
13. Repale, A.V., Ram, K., Chundawat, N.S., Srivastava, N., Singh, G.P.: A novel and easy synthesis of 1H-Tetrazoles employing p-Tolenesulfonyl azide. Res. J. Chem. Environ. **28**(10), 35–43 (2024). https://doi.org/10.25303/2810rjce035043
14. Anisree, S., Shanmugavelan, P., Palanisamy, S., Kishore, R., Srivastava, N.: Synthesis and anticancer screening of substituted (phenyl)-1-((naphthalen-5-yl)methylene)thiosemicarbazide against PC3 prostate cancer cells. J. Indian Chem. Soc. **101**(9), 101229 (2024). https://doi.org/10.1016/j.jics.2024.101229
15. Yadav, S., Kishore, R., Srivastava, N.: Microwave-assisted phase transfer catalytic synthesis of aryl Benzothiazole derivatives and their anticancer screening. Org. Prep. Proced. Int. **56**(6), 563–571 (2024). https://doi.org/10.1080/00304948.2024.2342692
16. Anisree, S., Shanmugavelan, P., Vijaylakshmi, P., Kishore, R., Srivastava, N.: Synthesis, characterization and anticancer screening of novel Phenylbenzylidene Thiosemicarbazone derivatives. Phosphorus. Sulfur. Silicon Relat. Elem. **199**(4), 267–276 (2024). https://doi.org/10.1080/10426507.2024.2330935
17. Tyagi, R., Yadav, K., Srivastava, N., Sagar, R.: Applications of pyrrole and pyridine-based heterocycles in cancer diagnosis and treatment. Curr. Pharm. Des. **30**(4), 255–277 (2024). https://doi.org/10.2174/0113816128280082231205071504
18. Tyagi, R., Singh, K., Srivastava, N., Sagar, R.: Recent advances in carbohydrate-based gelators. Mater. Adv. **4**, 3929–3950 (2023). https://doi.org/10.1039/D3MA00321C
19. Srivastava, N.: Efficient TBAI-CS2 promoted synthesis of substituted Hydrazinecarbodithiolates. Org. Prep. Proced. Int. **54**(6), 569–577 (2022). https://doi.org/10.1080/00304948.2022.2111171
20. Srivastava, N., Chaturvedi, D., Kishore, R., Kaur, M., Kaur, R.: Triton-B/CS2 mediated novel synthesis of substituted thioureas: a novel class of anti-cancer agents. Indian J. Chem. B. **61**, 870–877 (2022). https://doi.org/10.56042/ijc.v61i8.65053
21. Srivastava, N., Kishore, R.: Cleaner and greener synthesis of 3H-benzothiazole-2-thione and its derivatives. J. Sulfur Chem. **42**(1), 29–39 (2020). https://doi.org/10.1080/17415993.2020.1803321
22. Srivastava, N., Kishore, R.: Cleaner, greener synthesis, characterization, and anticancer evaluation of 2-Thioxo-1,3-Thiazolanes. Indian J. Heterocycl. Chem. **31**(2), 265–271 (2021)
23. Srivastava, N., Kishore, R.: Synthesis, characterization, anticancer, antibacterial, antifungal, and antimalarial activity of (aryl or hetero aryl) substituted-4-Methyl-2,3-dihydro-2-thioxo-1H-imidazol-5-yl Ethanone derivatives. Indian J. Heterocycl. Chem. **31**(03), 347–355 (2021)
24. Srivastava, N., Rathore, B.K., Vishnoi, R.K., Bajpai, S.: A simple and convenient method for synthesis of Artemisinin dimer through aliphatic nitro compounds. Asian J. Chem. **33**(1), 49–52 (2020). https://doi.org/10.14233/ajchem.2021.22918
25. Kishore, R., Kamboj, M., Shukla, M., Chaturvedi, D., Srivastava, N.: Novel synthetic strategy of cyclic Dithiocarbamates catalyzed by triton-B. Asian J. Chem. **31**(5), 1091–1094 (2019). https://doi.org/10.14233/ajchem.2019.21830
26. Srivastava, N., Saxena, M., Shukla, M.: Novel synthesis of 5-Oxo-2-thioxo-2,5-dihydro-3-thiophenecarboxylate derivatives in non-aqueous medium. Asian J. Chem. **31**(1), 176–180 (2019). https://doi.org/10.14233/ajchem.2019.21622

27. Srivastava, N.: Key role of ionic liquids in the cleaner and greener synthesis of lactams. Res. J. Chem. Environ. **26**(1), 125–130 (2022). https://doi.org/10.25303/2601rjce125130
28. Vishnoi, R.K., Kishore, R., Chaturvedi, D., Shukla, M., Bajpai, S., Srivastava, N.: Synthesis and antimicrobial activity of cyclic dithiocarbamates employing triton-B/CS2 system. Asian J. Chem. **33**(5), 1133–1136 (2021). https://doi.org/10.14233/ajchem.2021.23173
29. Srivastava, N., Kishore, R., Chaturvedi, D.: Novel and efficient method for the synthesis of cyclic trithiocarbonates employing Cs_2CO_3 and CS_2 system. Res. J. Chem. Environ. **25**(12), 142–148 (2021). https://doi.org/10.25303/2512rjce142148
30. Srivastava, N., Chaturvedi, D., Kaur, M., Kalia, N.: Cleaner, greener synthesis and anticancer screening of asymmetrical Trithiocarbonates. Org. Prep. Proced. Intern. **56**, 606–616 (2024). https://doi.org/10.1080/00304948.2024.2326356
31. Srivastava, N., et al.: Clean and green synthesis of organic Isothiocyanates using zinc peroxide as desulfurizing agent and their antimicrobial activity. Russ. J. Org. Chem. **59**(12), 2205–2213 (2023). https://doi.org/10.1134/S1070428023120199
32. Dubey, R., Pandey, J., Srivastava, N.: Terminal-substituted benzoxazole based Schiff-base mesogens and their Cu(II) complexes: synthesis and structural characterization. Chem. Phys. Impact. **8**, 100586 (2024). https://doi.org/10.1016/j.chphi.2024.100586
33. Li, T., et al.: Effects of chronic masitinib treatment in APPswe/PSEN1dE9 transgenic mice modeling Alzheimer's disease. J Alzheimer's Dis. **76**, 1339–1345 (2020)
34. Das, V., et al.: Antineoplastics for treating Alzheimer's disease and dementia: evidence from preclinical and observational studies. Med. Res. Rev. **44**, 2078–2111 (2024)
35. Hamad, A.A., Amer, B.E.: Safety of masitinib in patients with neurodegenerative diseases: a meta-analysis of randomized controlled trials. Neurol. Sci. **45**, 3503–3507 (2024)
36. Dubois, B., et al.: Masitinib for mild-to-moderate Alzheimer's disease: results from a randomized, placebo-controlled, phase 3, clinical trial. Alzheimer's Res Ther. **15**, 39 (2023)
37. Gupta, M., et al.: Investigation, scaffold hopping of novel donepezil-based compounds as anti-Alzhiemer's agents: synthesis, in-silico and pharmacological evaluations. Sci. Rep. **14**, 1687 (2024). https://doi.org/10.1038/s41598-024-51713-4
38. Li, Y., et al.: Graphical abstract SC. Eur. J. Med. Chem. (2016). https://doi.org/10.1016/j.ejmech.2016.12.009
39. Lan, T.T., et al.: Design, synthesis and bioevaluation of two series of 3-[(1-Benzyl-1 H -1,2,3-triazol-4-yl)methyl]quinazolin-4(3 H)-ones and N -(1-Benzylpiperidin-4-yl) quinazolin-4-amines. Chem. Biodivers. **17** (2020). https://doi.org/10.1002/cbdv.202000290
40. Luo, L., Song, Q., Li, Y., Cao, Z., Qiang, X.: Design, synthesis and evaluation of phthalide alkyl tertiary amine derivatives as promising acetylcholinesterase inhibitors with high potency and selectivity against Alzheimer's disease. Bioorg. Med. Chem., 115400 (2020). https://doi.org/10.1016/j.bmc.2020.115400
41. Juliet, C., et al.: Cinnamoyl - N - acylhydrazone - donepezil hybrids: synthesis and evaluation of novel multifunctional ligands against neurodegenerative diseases. Neurochem. Res. (2020). https://doi.org/10.1007/s11064-020-03148-2
42. Mohammadi-khanaposhtani, M., Saeedi, M., Zafarghandi, N.S., Mahdavi, M., et al.: Potent acetylcholinesterase inhibitors: design, synthesis, biological evaluation, and docking study of acridone linked to 1, 2, 3-triazole derivatives. Eur. J. Med. Chem. (2015). https://doi.org/10.1016/j.ejmech.2015.01.044
43. Figueroa-Villar, J.D.: Design, synthesis, structure, toxicology and in vitro testing of three novel agents for Alzheimer's disease. RSC Adv. **7**(38), 23457–23467 (2017)
44. Kurt, B.Z.: Synthesis and anticholinesterase activity of novel non-hepatotoxic naphthyridine-11-amine derivatives. Mol. Divers. (2018). https://doi.org/10.1007/s11030-018-9897-1

45. Cheng, Z., Zhu, K., Zhang, J., Song, J.: Bioorganic chemistry molecular-dockingguided design and synthesis of new IAA-tacrine hybrids as multifunctional AChE/BChE inhibitors. Bioorg. Chem. **83**, 277–288 (2019). https://doi.org/10.1016/j.bioorg.2018.10.057
46. Koyuncu, E., et al.: Synthesis of novel Schiff base derivatives of tacrine and investigation of their acetylcholinesterase inhibition potency. Maced. J. Chem. Chem. Eng. **38**(1), 75–84 (2019). https://doi.org/10.20450/mjcce.2019.156
47. Maspero, M., et al.: Bioorganic chemistry Tacrine-xanomeline and tacrine-iperoxo hybrid ligands: synthesis and biological evaluation at acetylcholinesterase and M 1 muscarinic acetylcholine receptors. Bioorg. Chem. **96**, 103633 (2020). https://doi.org/10.1016/j.bioorg.2020.103633
48. Fancellu, G., et al.: Novel tacrine–benzofuran hybrids as potential multi-target drug candidates for the treatment of Alzheimer's disease. J. Enzyme Inhib. Med. Chem. **35**, 211–226 (2020). https://doi.org/10.1080/14756366.2019.1689237
49. Babaee, S., Chehardoli, G., Akbarzadeh, T., Zolfigol, A.: Design, synthesis, and molecular docking of some novel tacrine based cyclopentapyranopyridine- and tetrahydropyranoquinoline- kojic acid derivatives as anti-acetylcholinesterase agents. Chem. Biodivers. (2021). https://doi.org/10.1002/cbdv.202000924
50. Zhang, H., et al.: Novel pyridine-containing sultones: structure-activity relationship and biological evaluation as selective AChE Inhibitors for the treatment of Alzheimer's disease. ChemMedChem. **16**(20), 3189–3200 (2021)
51. Codony, S., et al.: Discovery and in vivo proof of concept of a highly potent dual inhibitor of soluble epoxide hydrolase and acetylcholinesterase for the treatment of alzheimer's disease. J. Med. Chem. **65**(6), 4909–4925 (2022). https://doi.org/10.1021/acs.jmedchem.1c02150
52. Jevtić, I.I., et al.: Multi-target potential of newly designed tacrine-derived cholinesterase inhibitors: synthesis, computational and pharmacological study. Bioorg. Med. Chem. **101**, 117649 (2024). https://doi.org/10.1016/j.bmc.2024.117649
53. Fares, S., El Husseiny, W.M., Selim, K.B., Massoud, M.A.M.: Modified Tacrine derivatives as multitarget-directed ligands for the treatment of Alzheimer's disease: synthesis, biological evaluation, and molecular modeling study. ACS Omega. **8**(29), 2601–2603 (2023)
54. Liang, M., et al.: Design, synthesis and biological evaluation of 1-phenyl-Ethanone derivatives for multi-targeted treatment of Alzheimer's disease. Med. Chem. **7**, 285–293 (2017). https://doi.org/10.4172/2161-0444.1000469
55. Leuci, V., et al.: Rivastigmine structure-based hybrids as potential multi-target anti-Alzheimer's drug candidates. Bioorg. Chem. **153**, 107895 (2024). https://doi.org/10.1016/j.bioorg.2024.107895
56. Li, Y., et al.: Development of novel rivastigmine derivatives as selective BuChE inhibitors for the treatment of AD. Bioorg. Chem., 108245 (2025). https://doi.org/10.1016/j.bioorg.2025.108245
57. Bon, L., et al.: New multitarget Rivastigmine–indole hybrids as potential drug candidates for Alzheimer's disease. Pharmaceutics. **16**(2), 281 (2024). https://doi.org/10.3390/pharmaceutics16020281
58. Liu, Z., et al.: Development of novel fluoro-substituted rivastigmine derivatives as selective AChE inhibitors for the treatment of AD. Med. Chem. Res. **33**, 1195–1204 (2024). https://doi.org/10.1007/s00044-024-03250-y
59. Flavia, L., et al.: Eur. J. Med. Chem. (2015). https://doi.org/10.1016/j.ejmech.2015.12.024
60. Huang, W., et al.: Development of the "hidden" multifunctional agents for Alzheimer's disease. Eur. J. Med. Chem. (2019). https://doi.org/10.1016/j.ejmech.2019.05.051
61. Liu, D., et al.: Design, synthesis, and biological evaluation of carbamate derivatives of N-salicyloyl tryptamine as multifunctional agents for the treatment of Alzheimer's disease. Eur. J. Med. Chem. **229**, 114044 (2022). https://doi.org/10.1016/j.ejmech.2021.114044

62. Wang, Y., et al.: Discovery of carbamate-based N-salicyloyl tryptamine derivatives as novel pleiotropic agents for the treatment of Alzheimer's disease. Bioorg. Chem. **127**, 105993 (2022). https://doi.org/10.1016/j.bioorg.2022.105993
63. Spatz, P., Zimmermann, T., Steinmüller, S., Hofmann, J., Maurice, T., Decker, M.: Novel benzimidazole-based pseudo-irreversible butyrylcholinesterase inhibitors with neuroprotective activity in an Alzheimer's disease mouse model. RSC Med. Chem. **13**, 944–954 (2022). https://doi.org/10.1039/d2md00087c
64. Stavrakov, G., et al.: Galantamine-curcumin hybrids as dual-sitebinding acetylcholinesterase inhibitors. Molecules. (2020). https://doi.org/10.3390/molecules25153341
65. Stavrakov, G., Philipova, I., Zheleva-Dimitrova, D., Valkova, I., Salamanova, E., Konstantinov, S.: Docking-based design and synthesis of galantamine – camphane hybrids as inhibitors of acetylcholinesterase. Chem. Biol. Drug Des. **90**, 709–718 (2017). https://doi.org/10.1111/cbdd.12991
66. Atanasova, M., Stavrakov, G., Philipova, I., Zheleva, D., Yordanov, N., Doytchinova, I.: Bioorganic & Medicinal Chemistry Galantamine derivatives with indole moiety: docking, design, synthesis and acetylcholinesterase inhibitory activity. Bioorg. Med. Chem. **23**(17), 5382–5389 (2015). https://doi.org/10.1016/j.bmc.2015.07.058

Greener Synthesis of Quinolines Using Ionic Liquids

Nitin Srivastava[1(✉)]

[1] Department of Chemistry, Amity University Uttar Pradesh, Lucknow Campus, Lucknow, Uttar Pradesh, India
nitinsriv5@gmail.com; nsrivastava1@lko.amity.edu

Abstract. Sustainable development demands cleaner greener synthesis in the pharmaceutical sector. Cleaner, greener synthesis may be done with the help of ionic liquids which are versatile class of green solvents which have several advantages over traditional medium. Ionic liquids are "designer solvents" whose properties like solubility, density, refractive index, and viscosity can be ameliorated to fit better as per requirements. This is done by structural modification of its anionic or cationic part or both. Ionic liquids have been extensively employed for the several synthetic transformations desired in medicinal chemistry. Heterocycles, which possess a key role in medicinal chemistry, have been synthesized using a variety of structurally diverse ionic liquids. Among these heterocycles quinoline plays a prominent role in drug discovery research and its cleaner greener and efficient synthesis has to be explored via ionic liquids. In this review, I will discuss the recent advances in the synthesis of various kinds of quinolines using a variety of ionic liquids

Keywords: Heterocycles · Quinolines · Ionic liquids · Drug development · Medicinal chemistry

1 Introduction

Heterocycles are organic compounds which are significant biologically and industrially but also to acquire apex position in the functioning of a developed human society [1]. Most of the therapeutics that mimic natural products belong to heterocycles. Advances against disease and their diagnosis are achieved by designing and testing new structures and heterocycles are most important in this regard [2–6]. Many biologically potent glycosides, alkaloids, antibiotics, pesticides are heterocycles [7]. Therefore, heterocyclic scaffolds are extensively explored for better insecticides, pesticides, herbicides and therapeutics. Many such biologically active compounds are heterocycles. Heterocycles are employed as additives and modifiers in antioxidants, cosmetics, data storage, solvents etc. Recently, several review articles have been reported by various researchers on the different aspects in literature [8, 9]. Quinoline nucleus is part of many natural products and pharmacologically active molecules, which are used as drugs. These diverse classes of drugs are indispensable part of the human life. Quinoline motif is found antitubercular [10], antimalarial [11], anticancer [12], anti-HIV drug [13], antiviral

A. K. Saxena, A. Saxena (Eds.): GTHTM 2025, pp. 255–274, 2026.
https://doi.org/10.1007/978-3-032-12320-6_14

[14], anti-inflammatory [15], analgesic [16], antibacterial [17], antihypertensive [18], anti-Alzheimer drug [19], antibiotic [20], antileishmanial [21, 22], antioxidant [23], antiprotozoal [24], DNA binding [25], inhibition of tyrosine kinase properties [26, 27], and tyrokinase PDGF-RTK inhibition [28] drugs. Also, 2-trifluoromethyl quinoline scaffolds have also been used as organic ligands [29, 30] and fluorescence probes [31, 32]. Some of the drugs containing quinoline core are most reliable and best sold in the market. These drugs include antimalarial drugs like quinine, quinidine, chloroquine, and mefloquine, [33, 34], antimicrobial drugs like Sparfloxacin and ciprofloxacin [35, 36], anticancer drugs like Irinotecan and camptothecin [37–39], anti-HIV drugs like Vesnarinone and Saquinavir [40], antimycobacterial drugs like Bedaquiline [41]. Among other drugs Dibucaine as local anesthesia [42], Primaquine as antimalarial [43], Topotecan as Anticancer [44]. The structures of these drugs are given in the Fig. 1.

Quinoline contains two six-membered rings fused together with a nitrogen atom in one of the rings and named as benzo[b]pyridine or 1-azanaphthalene. The aromatic heterocyclic ring gives nucleophilic and electrophilic substitution [45, 46]. In an acidic

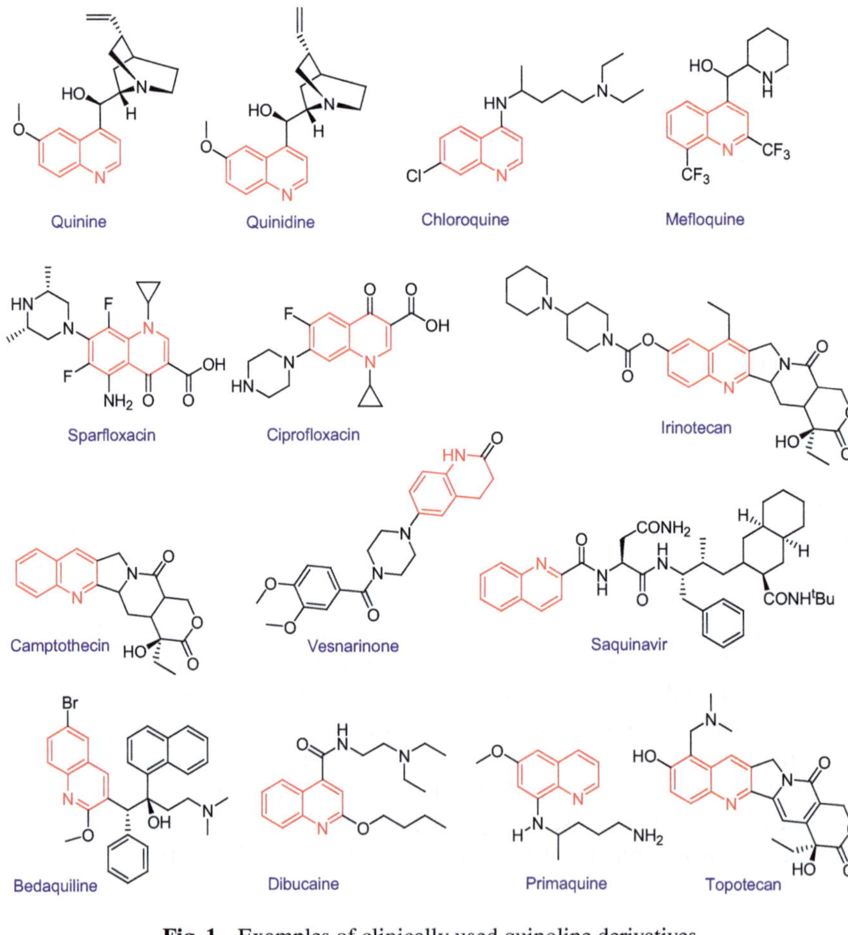

Fig. 1. Examples of clinically used quinoline derivatives.

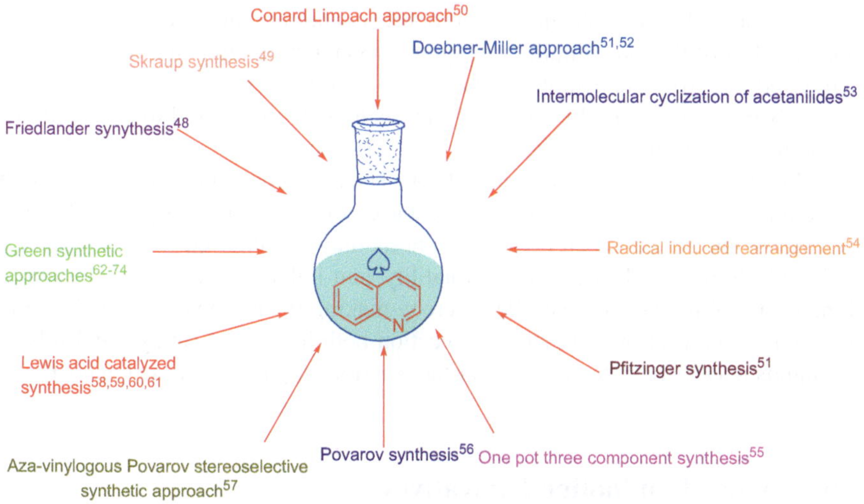

Fig. 2. Methods for the synthesis of Quinolines.

medium it forms salt and undergoes reactions like pyridine and benzene. It forms a weak base [45]. Quinoline hydrazones which contain $R_1R_2C{=}N{-}NH_2$ moiety exhibits polar and non-polar properties, which are helpful in transmission through the bacterial cells [47].

The traditional methods for the synthesis of quinoline have been given in Fig. 2 [48–61].

Traditional methods for synthesis of quinoline employed the use of costly starting materials and extreme temperature conditions with lower yields. Quinoline and its derivatives exhibit plethora of biological activities and quinoline-based drugs are employed for the treatment of variety of disorders. Owing to their therapeutic significance, scientists are exploring diverse one-pot synthetic methods employing various acidic/basic or metal catalysts and reagents. Such methods suffer several constraints like long reaction timing, extreme reaction conditions, the formation of unwanted byproducts, expensive reaction set up and hazardous solvents. Moreover, such processes have resulted in low yield and less chances of recovery of the catalyst after the reaction. The concept of sustainability in organic synthesis has paved way to effective methods which are environment-friendly for the synthesis of quinoline derivative.

In the greener synthesis of quinoline the research have reported the use of I2 in place of metal catalyst [62], water as solvent [63], solvent free condition using Indium chloride as catalyst [64], Irradiation methods in acetonitrile solvent [65], irradiation of the reaction mixture in water as solvent [66], irradiation in metal free condition [67], irradiation in solvent DMF [68], microwave radiations [69–74] and use of ionic liquids which act as both solvent and the catalyst.

This review discusses green methods involving ionic liquids for the synthesis of quinolines in the past 10 years. Ionic liquids are defined as pure compounds consisting of cations and anions (i.e. salts), which exist in liquid state at ambient temperature i.e. they are salts that do not normally melt by means of any external heat source [75]. The cation in the ionic liquids typically is heterocycle containing nitrogen while anions are inorganic. Most of the ionic liquids are liquid at room temperature that is why they are also referred to as room temperature ionic liquids (RTILs) [76].

The solvent depletion in the organic synthesis may be minimized by application of ionic liquids which are non-volatile in nature. Homogeneous catalysis can be best achieved in the presence of polar ionic solvent. The difference in the solubilities of the ionic liquids in the organic solvent and water have glorified them as organic solvent for the extraction purposes. Green organic synthesis may be achieved in the presence of ionic liquids. Ionic liquids possess non-volatility, non-inflammability, thermal stability, and their miscibility can be controlled thereby making them a potential candidate for gaining kinetic control over the organic reaction both in term of catalyst and solvent. Ionic liquids may be recycled easily after the reaction and may be reused many times.

2 Synthesis of Quinoline Derivatives

Synthesis of quinoline derivatives using novel ionic liquid-based nano-magnetic catalyst (MNPs@SiO_2-Pr-AP-tribromide) has been reported [77]. These ionic liquid-based magnetic nanoparticles of pyridinium-tribromide (MNPs@SiO_2-Pr-AP-tribromide) catalyzed the one pot synthesis of new quinoline derivatives employing ketones and 2-amino aryl ketones (Scheme 1). Bromine on the substrate of 4-aminopyridine and magnetic nano-Fe_3O_4 was used as a base. The Br_3- ions on the surface of catalyst enhances catalytic activity. The prepared nano catalyst helps with easy synthesis in less reaction time with higher efficiency, easier separation. Moreover, the process also permits recycling of the catalyst six times without compromise in the catalytic activity. The process also facilitates easy purification of products selectivity of the catalyst.

A practical one-pot synthesis of quinoline derivatives through protection-free chemo-selective oxidation of 2-aminobenzyl alcohols to give aldehydes. In the second step mild annulation with different 1,3-dicarbonyl compounds or nitriles in binary solvent water-acetonitrile in the presence of Cu as catalyst and glucose-based ionic liquids (GSILs) lead to the formation of quinoline derivatives (Scheme 2). This process was used for the synthesis of 40 functionalized quinolines with high yield up to 93% and followed green chemistry parameters. The glucose-based ionic liquids (GSILs) may be recycled without compromising the rate of the reaction and yields of the products. This research highlighted inexpensive catalyst, aqueous medium, Gram scale synthesis, Synergistic effect of Cu and GSILs and broad scope of the synthesis [78].

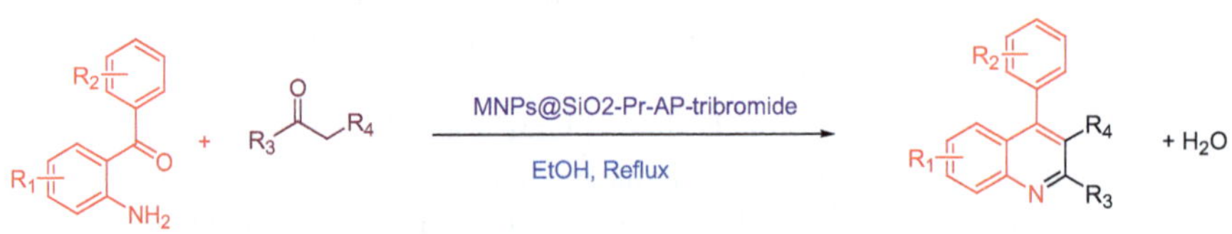

Scheme 1. Synthesis of quinolines with ionic liquid-based nano-magnetic catalyst.

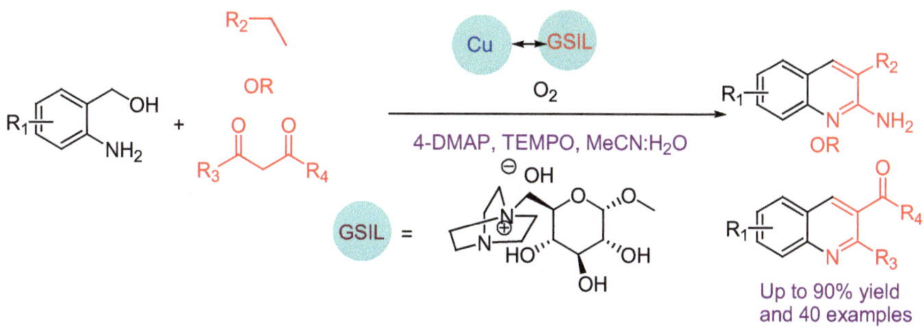

Scheme 2. Synthesis of quinoline derivatives using glucose-based ionic liquids (GSILs).

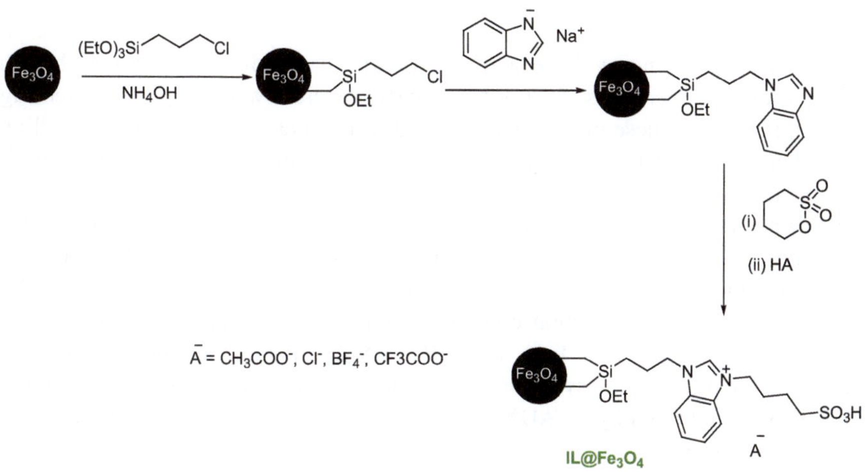

Scheme 3. Synthesis of the catalyst IL@Fe_3O_4.

Ionic liquid supporting magnetite nanoparticle (IL@Fe_3O_4) with different anions were synthesized through functionalization of Fe_3O_4 nanoparticles and employed as catalyst for the synthesis of quinolines [79] as shown in Scheme 3. The synthesized IL@Fe_3O_4 were characterized via FTIR, TEM, energy dispersive X-rays analysis, thermal gravimetry, X-ray diffraction. The efficiency of the catalysts was explored in the synthesis of fused polycyclic quinoline derivatives and bioactive substituted quinoline (Scheme 4). The mechanism of the synthesis has been explained on the basis of the different anion which affects the acidity of the catalyst. Among the various catalyst IL-1@ Fe_3O_4 having CH_3COO- ions was a potential catalyst for the Friedlander reaction which resulted in best yield of the desired products in the presence of mild reaction condition. The Ionic liquid supporting magnetite nanoparticle (IL@Fe_3O_4) with different anions which were employed here catalyst were easily separated through magnetic field and this catalyst may be reused for up to six times making the process eco-friendly.

The antimicrobial Resistance (AMR) among the microbes has jeopardized efficacy of the existing antibiotics against parasite fungi, bacteria, viruses threatening human health and their well-being. The drug discovery process with ionic Liquids (ILs) which

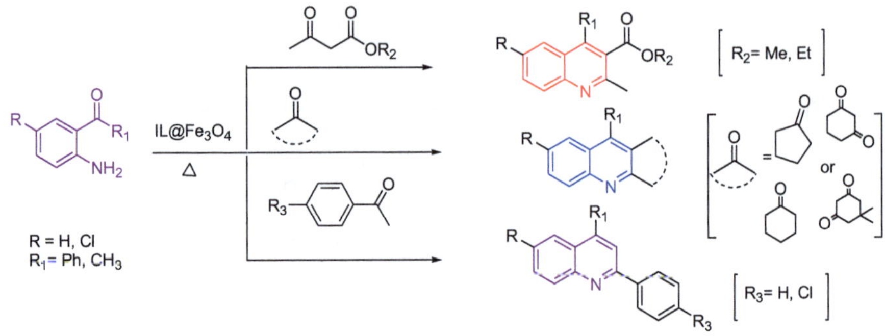

Scheme 4. Synthesis of quinoline and fused polycyclic quinoline derivatives via Friedlander reaction using IL@Fe_3O_4.

are thermally stable, non-volatile and eco-friendly has enthusiastically enriched the arsenal of the antimicrobials. Such ionic liquids possess low vapor pressure and high solvating potential. These entities have helped in making proteins stable as well and have enhanced enzymatic activity which has further enhanced their potential in the biomedical industry.

In this connection a green synthesis, characterization of quinoline-based IL, antimicrobial potency, has been reported [80]. These with quinoline-based IL exhibit low cytotoxicity and artificial protection. This one-pot synthetic approach in solvent-free and cleaner eco-friendly reaction conditions enhanced the efficiency of the reaction along with the yield of the reaction. ^{1}H NMR, ^{13}CNMR, and IR spectroscopy were used to ascertain the purity of the synthesized IL. The potential of the synthesized quinoline based-IL was confirmed by its ADMET analysis and further verified with the help of disc diffusion assay.

Nanoporous TiO_2 containing an ionic liquid bridge was employed for the synthesis of quinoline, quinazoline and spiro-quinazoline derivatives [81]. The synthesis of the Nanoporous TiO_2 containing an ionic liquid bridge is shown in Scheme 5. Synthesis of quinoline derivatives is represented in Scheme 6. This method resulted in the formation of the desired compounds under solvent free conditions in excellent yield and in shorter time. The catalyst used here can be reused for the repeated synthesis of the compounds.

DABCO based ionic liquid [H_2-DABCO][ClO_4]$_2$ has also synthesized and employed for the synthesis of pyrimido [4,5-*b*]-quinoline and pyrimido [4,5-*d*]pyrimidine derivatives as shown in Scheme 7 [82]. The synthesized catalyst was cheaper and can be recycled for further use in the synthesis of these compounds. This procedure was simple, high yielding, having short reaction times, facilitated easy work-up, and used a non-metal catalyst. The manuscript debated the effect of cations and anions on moisture-resistant property and its catalytic property in DABCO-based ionic liquids by comparing [H_2-DABCO][ClO_4]$_2$, [H_2-DABCO][HSO_4]$_2$, [H_2-DABCO][H_2PO_4]$_2$, [DABCO](SO_3H)$_2$(HSO_4)$_2$, and [DABCO](SO_3H)$_2$(Cl)$_2$.

A novel nanomagnetic material, Dicationic ionic liquid functionalized with Bisulfate and attached with silica-coated nano-Fe_3O_4 (nano-[Fe_3O_4@SiO_2@BDIL] or NFSBDI)

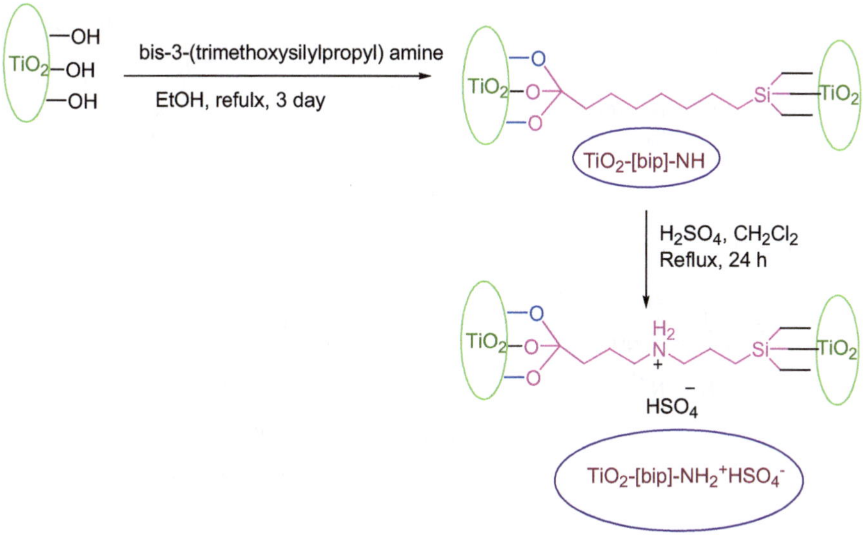

Scheme 5. Synthesis of Nanoporous TiO_2 containing an ionic liquid bridge (TiO_2-[bip]-NH_2^+ HSO_4^-).

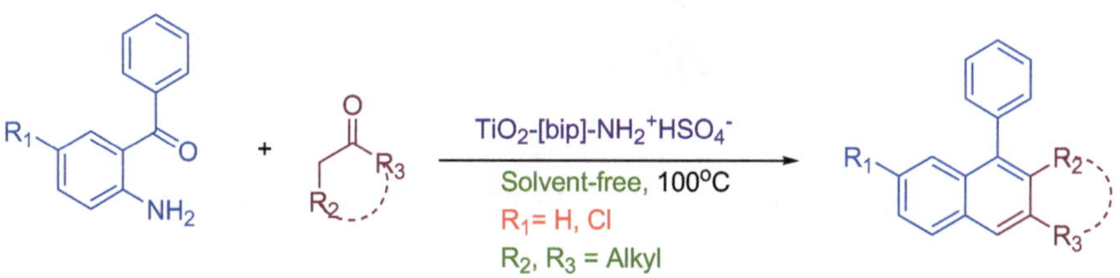

Scheme 6. Synthesis of quinolines catalyzed by TiO_2-[bip]-NH_2^+ HSO_4^-.

was synthesized and characterized with the help of TGA, XRD, VSM, FTIR, EDS, FE-SEM. Further it was employed as an efficient, magnetically recyclable and dual-functional catalyst for the synthesis of pyrimido[4,5-b]quinolines which are uracil-containing heterocycles as shown in Scheme 8 [83]. The synthesis was proposed to be one-pot multicomponent reaction among 6-amino-1,3-dimethyluracil, arylaldehydes and dimedone which was carried out in solvent free condition under the ambience of NFSBDI. This vital synthesis led to the synthesis of pyrimido[4,5-b]quinolines in less time with sufficient high yields.

Grohe cycloaracylation for the synthesis of biologically potent quinolones was carried out in the presence of ionic liquids rather than the toxic common solvent DMF as shown in Scheme 9 [84]. The advantages of the use of ILs in place of DMF include higher yields in shorter reaction times, 100% transformation of the reactants into the product with easy work-up than with DMF. Among the variety of ILs screened for the synthesis, tributylmethylammonium methanesulfonate ([TBMA][MsO]) served the purpose quite well. The process had a wide substrate scope, and the ionic liquid used

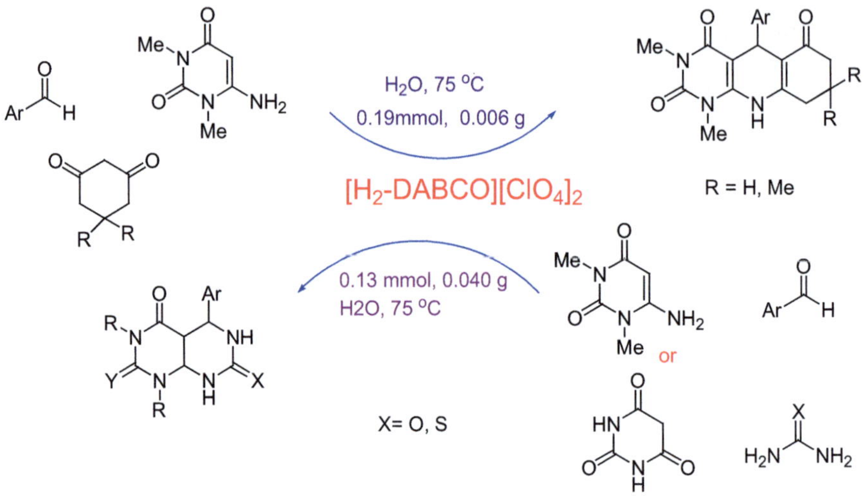

Scheme 7. Synthesis of quinolines using DABCO based ionic liquid $[H_2\text{-DABCO}][ClO_4]_2$.

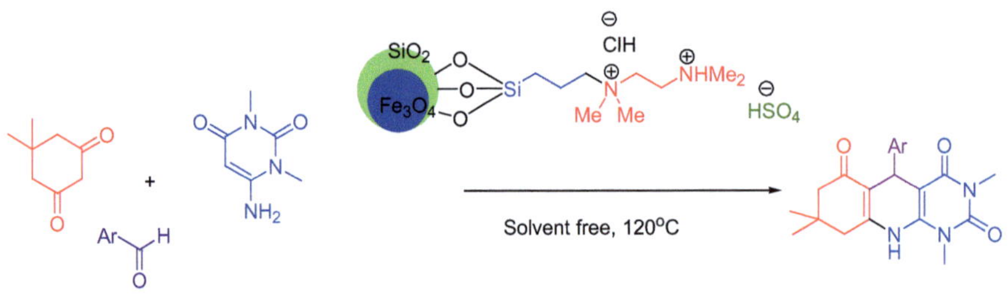

Scheme 8. Synthesis of quinolines using (nano-$[Fe_3O_4@SiO_2@BDIL]$ or NFSBDI).

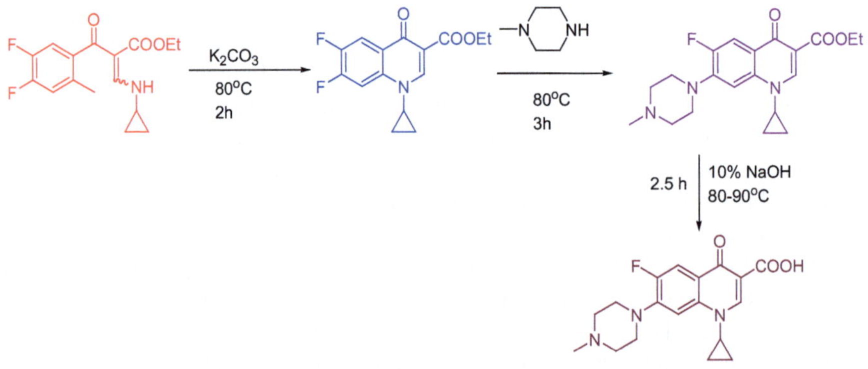

Scheme 9. Synthesis of quinolone employing [TBMA][MsO].

here may be recycled through a sustainable process. The ionic liquid [TBMA][MsO] was further employed for the synthesis of quinolone in a one-pot and three-step process. Among the various quinolones synthesized through the process the compound 3-carboxy-4-quinolone acid as given in scheme below, was obtained in high yield and in a shorter time. The clean, green and sustainable properties of ionic liquid and their non-volatility in comparison with DMF make this protocol an efficient and alternative greener and sustainable approach for synthesis of biologically significant quinolones.

Organic synthesis usually is done with organic solvents which are used up in large amounts during the synthetic process, separation and purification. The use of ionic liquids may resolve the problem of the use of organic solvent which is a big constraint of the organic synthesis in the way of the clean, green and sustainable synthesis. Ionic liquids may be employed for the synthesis of novel compound which act as solvent as well as catalyst for the reaction. In this regard one-pot synthesis of quinoline derivatives was done by reacting *β*-ketoester and substituted anilines in ethanol at 60 °C in in the presence of ionic liquid $[Et_3NH]^+[BF_4]^-$ which acted as catalyst for the reaction as shown in Scheme 10 [85]. The synthesis was reported to be of high purity and higher yields in the range of 78–93%. It has also been found that the synthesis occurred within 20–65 min which is much less than compared to the conventional methods. This protocol was screened for different solvents and recycling of the ionic liquid for further use. The structure of the synthesized compound was determined with the help of analytical and spectral data.

Homogeneous organocatalyst, triphenylmethyl chloride or trityl chloride (TrCl) neutral conditions were employed for the synthesis of pyrimido[4,5-*b*]quinolines derivatives [86]. The synthesis resulted in products in less time and high yields (Scheme 11). The trityl chloride is a cheap and easily available, catalyst which has been employed gives the product in less times and avoided the use of harsh acidic conditions.

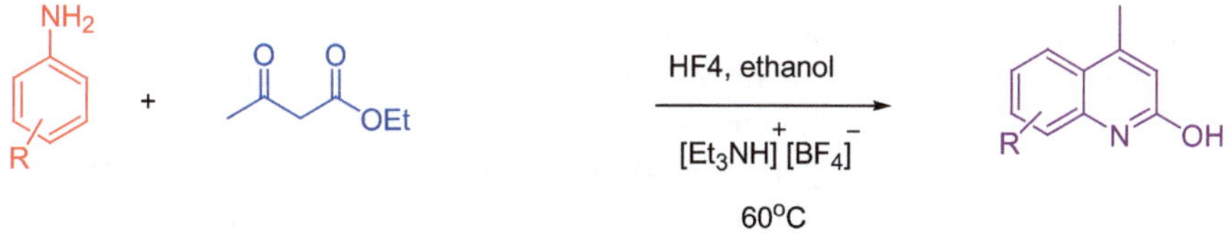

Scheme 10. Synthesis of substituted quinolines.

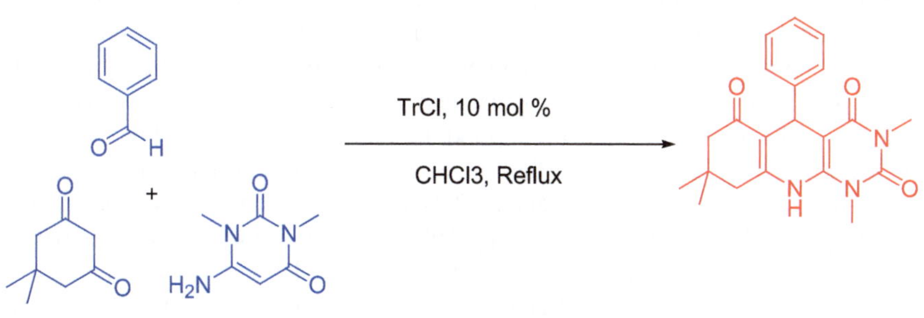

Scheme 11. Synthesis of quinolines using synthesis of pyrimido[4,5-*b*]quinolines derivatives.

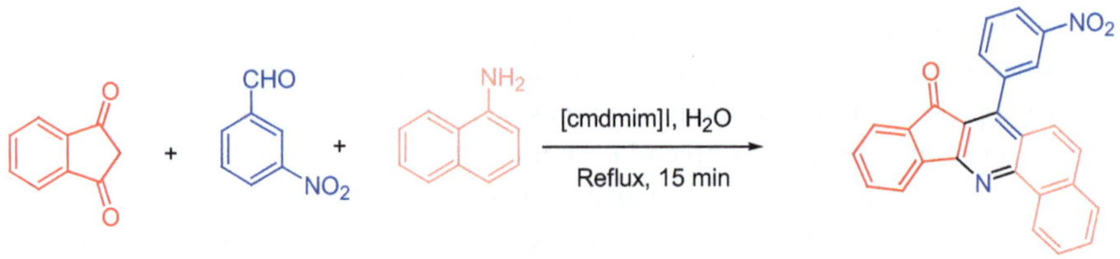

Scheme 12. Synthesis of fused quinoline derivatives in presence of 1-carboxymethyl-2,3-dimethylimidazolium iodide ([cmdmim]I) ionic liquid catalyst.

Hepatitis A virus (HAV) is the main cause of liver failure and mortality in old age or in the people with other health problems. So, the discovery of novel inhibitory compounds may help to treat the disease. A novel class of quinolines was brought forward with good inhibitory activities on HAV proteinase. The first step of the synthesis yields, fused quinoline derivatives were formed within 12.0 min and high yields up to 94% when ionic liquid 1-carboxymethyl-2,3-dimethylimidazolium iodide ([cmdmim]I) was used as catalyst (Scheme 12). Density functional theory (DFT) and molecular docking studies were also done to relate the chemical reactivity and inhibitory activity of synthesized quinolines [87]. Among the synthesized compounds, two possessed low $HOMO_{SWBNNT}$-$LUMO_{4a}$ gap (4.78 eV), more softness (0.67 eV), and higher adsorption energy (-87.9 kJ mol^{-1}). These compounds showed better delivery to the desired cells than other quinolines. Best cavity of the HAV has -134.2 kJ mol^{-1} as revealed in the docking studies with bonding and non-bonding interaction. The interactions were between fused quinolines and sidechain of the amino acid residues in the appropriate binding site of the HAV were reported to be electrostatic, steric, and H-bonding. The interactions confirmed that fused quinolines are potential inhibitor for HAV.

Graphene-oxide supported bronsted acidic ionic liquid [GrBenzImi]SO_3H have been employed for the synthesis of 2,4-disubstituted quinolines by coupling reactions of phenylacetylene anilines and aryl aldehydes, in a binary liquid medium of ethanol and water (1:1) and the synthesis is represented in Scheme 13 [88]. The protocol represented a green approach for the synthesis of quinolines at 80 °C.

Fe_3O_4 Nano particles supported by Brønsted acidic ionic liquid, 1-methyl-3-(3-trimethoxysilylpropyl)imidazolium hydrogen sulfate (Fe_3O_4-IL-HSO_4) has been designed and synthesized. The synthesized catalyst was employed for the single pot synthesis of polysubstituted quinolines by Friedlander reaction and shown in Scheme 14 [89]. This protocol utilized solvent free condensation of 2-aminoaryl ketones with 1,3-dicarbonyl compounds to give polysubstituted quinolines. The reaction was optimized for concentration of catalyst, temperature and solvent. The highest yield of the product was formed when 20 mg of catalyst was loaded at 90 °C in solvent-free conditions. The protocol resulted in good to high yield of quinoline derivatives (85–96%) in a very small time in the range of 15–60 min. Different carbonyls like β-ketoesters, cyclic ketones, and pentane-2,4-dione were employed for the reaction. The catalyst used in the reaction was recycled for further use up to six times without compromising its catalytic activity as shown in Scheme 14.

Magnetite (Fe_3O_4) nanoparticles supporting dodecylbenzenesulfonic acid (DDBSA@MNP) were designed and synthesized. This particle size of the synthesized

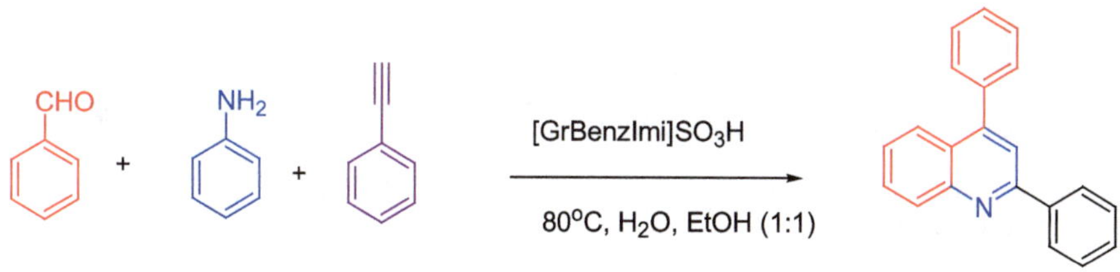

Scheme 13. Synthesis of 2,4-disubstituted quinolines employing [GrBenzImi]SO_3H.

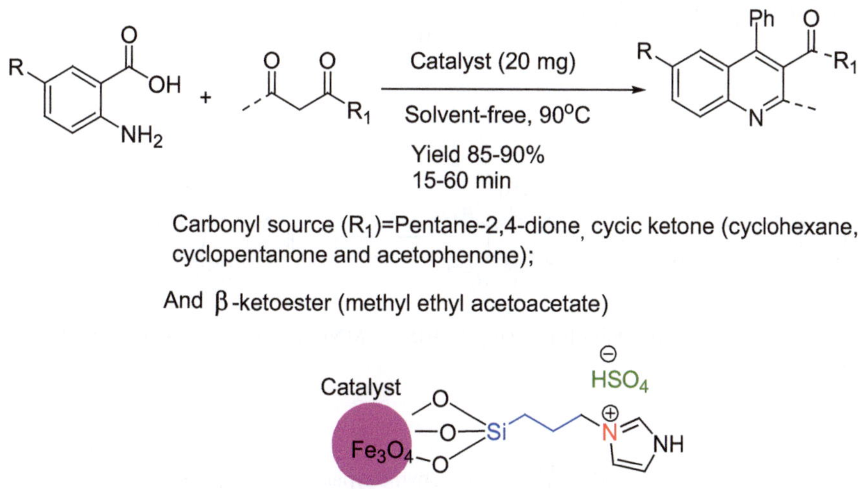

Scheme 14. Synthesis of Quinoline Using Fe_3O_4–IL-HSO_4 as a Catalyst.

material was approximately 16 nm and the catalyst had a core–shell structure and employed for the synthesis of substituted quinolines. The optimized synthesis used 0.15 g of catalyst and ethanol as a solvent and at 80 °C condensation of ketones and 2-aminobenzophenones for yielding substituted quinolines in one-pot as shown in Scheme 15 [90]. Higher yields, less reaction time, simple workup, and easy purification of the compound are advantages of this protocol. The catalyst employed here recovered easily for further use up to five consequent times. The mechanism of the synthesis was proposed likely that catalyst activates the carbonyl carbon of 2-aminoaryl ketone, facilitating active methylene of ketone or β-ketoester for nucleophilic attack and gives a cross-aldol product. β-hydroxy ketone thus formed dehydrates and yields α, β-unsaturated ketone which undergoes cyclization to give quinoline (Scheme 15).

Friedländer condensation by using α-chymotrypsin has been reported for the synthesis of quinoline derivatives in the presence of eco-friendly catalyst in an ionic liquid in aqueous medium (Scheme 16) [91]. In the reported protocol model reaction was carried between 2-aminoacetophenone and ethyl acetoacetate for the optimization of the reaction (Scheme 10). Considering the effect of the reaction medium on the activity of the enzyme the reactions were performed in six various dry ionic liquids which included three hexafluorophosphate and three tetrafluoroborate ionic liquids.

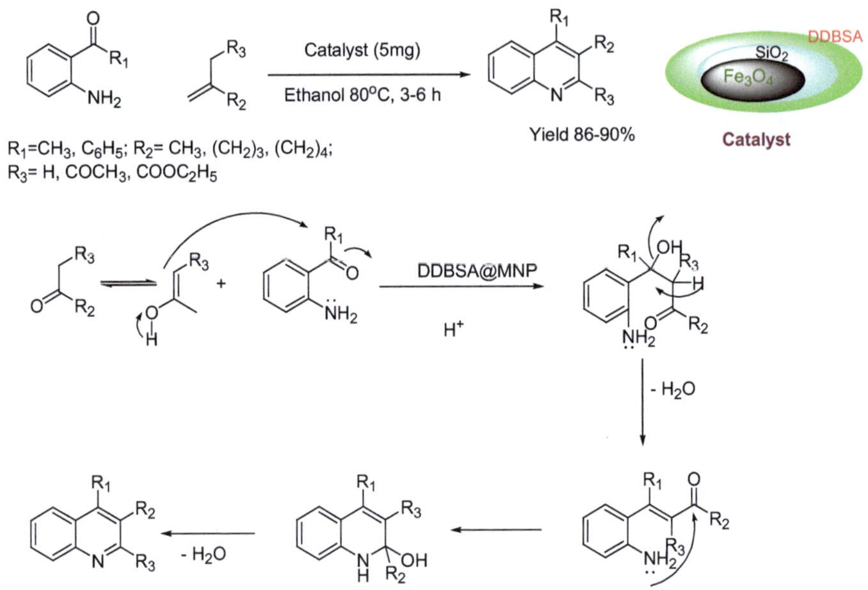

Scheme 15. Synthesis of Quinoline Using DDBSA@MNP as a Catalyst and Their Plausible Mechanism.

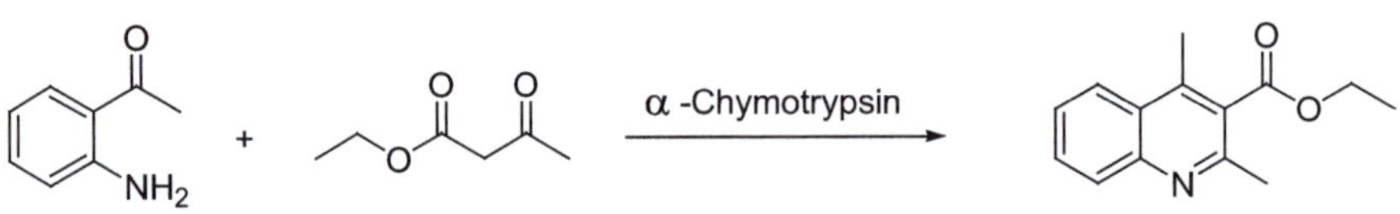

Scheme 16. Friedländer condensation for synthesis of quinolines in an ionic liquid in aqueous medium.

These ionic liquids were 1-butyl-3-methylimidazolium hexafluorophosphate, 1-hexyl-3-methylimidazolium hexafluorophosphate, and 1-ethyl-3-methylimidazolium hexafluorophosphate; 1-hexyl-3-methylimidazolium tetrafluoroborate, 1-butyl-3-methylimidazolium tetrafluoroborate and 1-ethyl-3-methylimidazolium tetrafluoroborate respectively. It was observed that the ionic liquids in which there was a shorter chain in the cation exhibited higher polarity and enhanced the enzyme and substrate solubility and helped in increasing the rate of the reaction. The reaction was found to be optimized in the case of ionic liquid 1-ethyl-3-methylimidazolium tetrafluoroborate) [EMIM][BF_4].

Friedländer synthesis of quinolines was achieved by using 3-methyl-1-sulfoimidazolium chloride ([Msim]Cl) and 3-methyl-1-sulfoimidazolium trichloroacetate ([Msim][$OOCCCl_3$]) at 100 °C (Scheme 17) [92]. These two acidic ionic liquids may be recycled for further application in the synthesis of the quinolines. This reaction yielded to a single product in less time with a high yield in the range of 95–100%. Moreover, the products were easily separated with easy work up process. These new ILs namely 1-methylimidazolium trichloroacetate ([Hmim][$OOCCCl_3$]) and [Msim][$OOCCCl_3$] were prepared from CCl_3COOH and characterized by spectral and analyti-

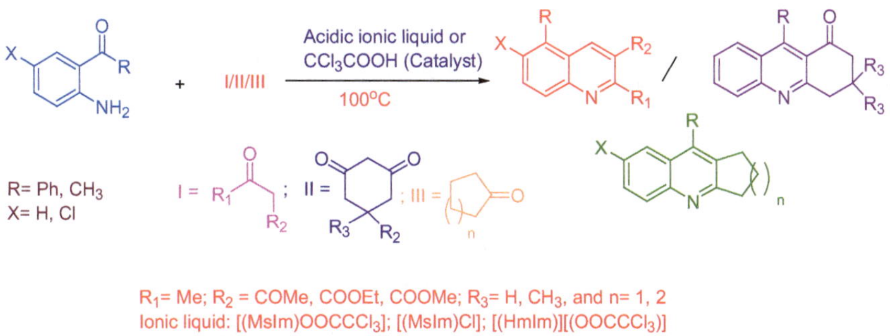

Scheme 17. Friedländer synthesis of quinolines was achieved by using 3-methyl-1-sulfoimidazolium chloride ([Msim]Cl) and 3-methyl-1-sulfoimidazolium trichloroacetate ([Msim][$OOCCCl_3$]).

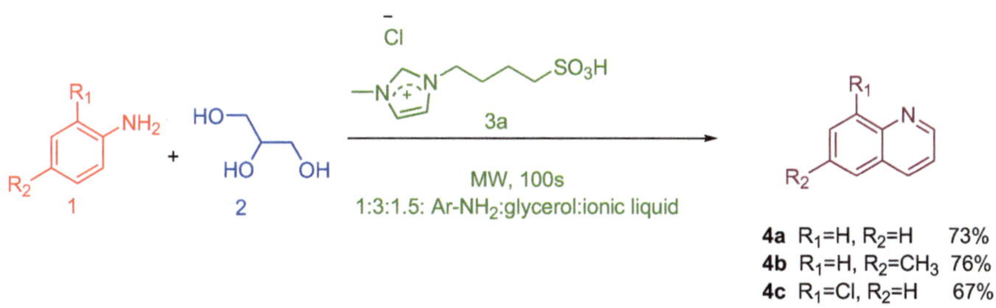

Scheme 18. Skraup reaction in ionic liquid medium under microwave irradiation.

cal data. During the synthesis of quinoline at 100 °C, the catalytic activity of acidic ILs and acid CCl_3COOH were also compared under solvent-free conditions.

Skraup synthesis quinolines were also explored under microwave conditions by reaction between glycerol and anilines (Scheme 18) [93]. Here during the reaction, it was found that if imidazolium cation-based sulfonic acid ionic liquid **3a** was used in place of conc. sulfuric acid for synthesis, there was significant rise in the yield of the reaction and there is no need for adding external oxidizing agent for the conduction of the reaction. This amelioration resulted in the formation of quinolines in appreciable yields.

3-substituted and 2,3-disubstituted quinolines have been synthesized with the help of aldehydes and anilines (Scheme 19) [94]. Inspired by the earlier reported work which employed trifluoromethanesulfonic acid in DMSO and copper(I) bromide for the synthesis of quinolines, this research concluded that ionic liquid ([Bmim]BF_4, **3b**) can replace the transition metal, acid and the solvent in the earlier work. Moreover, the application of the ionic liquid did not compromise with the yield of the reaction.

Brønsted-acidic ionic liquids have also been employed to catalyze the Friedländer reaction [95]. This research reported that the new developed ionic liquid catalyst [Msim][$OOCCCl_3$] was very effective in catalyzing the synthesis of quinolines resulting in almost up to 100% yield (Scheme 20) [95]. This novel IL was found to be comparable to other existing ILs in acidic medium for the synthesis of quinolines. This new ionic liquid possesses exceptional recycling and reuse capabilities.

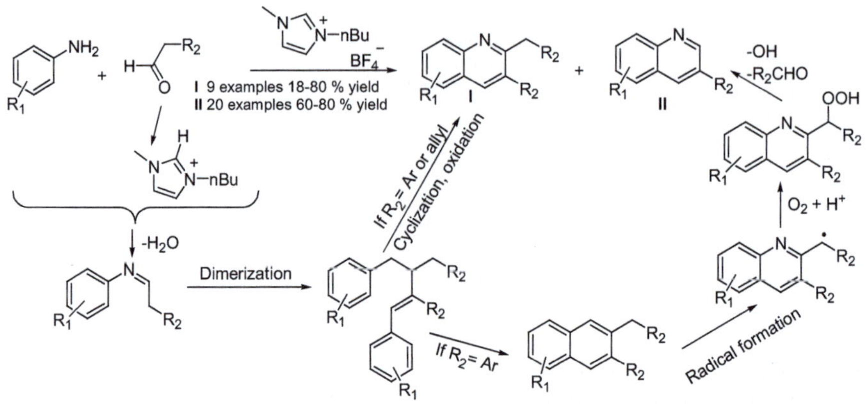

Scheme 19. Divergent reactivity in the Doebner reaction in ionic liquid medium.

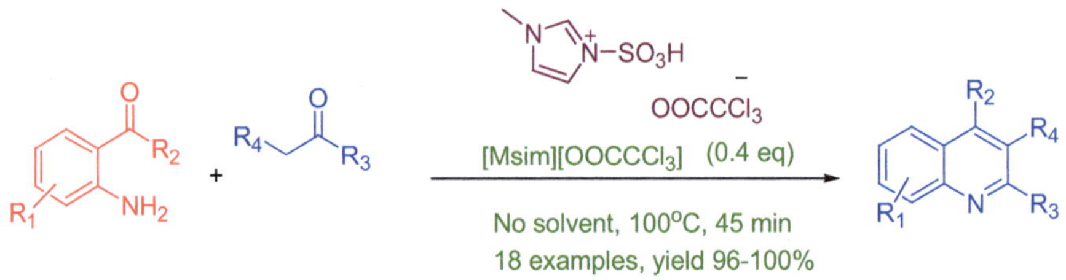

Scheme 20. Ionic liquid acid-catalyzed Friedländer quinoline synthesis.

3 Conclusion

The quinolines are prominent pharmacophore in drug discovery research as they possess a plethora of activities like antitumor, antiprotozoal, antimicrobial, antimalarial, antitumor, antitubercular, anti-inflammatory, antibacterial, antifungal, antihypertensive, anticancer, anti-HIV, DNA binding anti-inflammatory, antiproliferative, antioxidant etc. So, syntheses of this moiety have been extensively explored globally. The conventional routes for synthesis of quinolines suffer several constraints making it necessary for the development of novel protocols, particularly clean, green and sustainable discouraging the use of metal catalyst and organic solvents. The ionic liquids possess unique properties which make them an interesting green medium which need to be extensively employed for diverse reactions in organic synthesis. This review compiled and organized the diverse protocols for the synthesis of quinolines employing a variety of ionic liquids. We conclude from the discussion in this review that the synthesis of quinolines in the presence of ionic liquid has various advantages like high yield, mild reaction conditions, easy and affordable reaction workup, decrease in the reaction time, higher atom efficiency, safety, solvent free or the use of greener solvents, tolerance to functional groups, recycling of the ionic liquids and their reuse. Most of the approaches in the ionic liquids for the synthesis of quinolines ameliorated the traditional synthesis like Skraup synthesis, Friedlander synthesis, Doebner–Von Miller, Bronsted acid synthesis, one-pot multicomponent synthesis etc. The review will motivate synthetic chemists to explore more about the development of the cleaner greener and sustainable synthesis of the biologically potent scaffolds.

Conflict of Interest. The authors declare no potential conflict of interest.

References

1. Balaban, A.T., Oniciu, D.C., Katritzky, A.R.: Aromaticity as a cornerstone of heterocyclic chemistry. Chem. Rev. **104**, 2777–2812 (2004)
2. Tyagi, R., Yadav, K., Srivastava, N., Sagar, R.: Applications of pyrrole and pyridine-based heterocycles in cancer diagnosis and treatment. Curr. Pharm. Des. **30**(4), 255–277 (2024). https://doi.org/10.2174/0113816128280082231205071504
3. Anisree, S., Shanmugavelan, P., Palanisamy, S., Kishore, R., Srivastava, N.: Synthesis and anticancer screening of substituted (phenyl)-1-((naphthalen-5-yl)methylene)thiosemicarbazide against PC3 prostate cancer cells. J. Indian Chem. Soc. **101**(9), 101229 (2024). https://doi.org/10.1016/j.jics.2024.101229
4. Yadav, S., Kishore, R., Srivastava, N.: Microwave-assisted phase transfer catalytic synthesis of aryl Benzothiazole derivatives and their anticancer screening. Org. Prep. Proced. Intern. **56**(6), 563–571 (2024). https://doi.org/10.1080/00304948.2024.2342692
5. Srivastava, N., Kishore, R.: Cleaner, greener synthesis, characterization, and anticancer evaluation of 2-Thioxo-1,3-Thiazolanes. Indian J Heterocycl. Chem. **31**(2), 265–271 (2021)
6. Srivastava, N., Kishore, R.: Synthesis, characterization, anticancer, antibacterial, antifungal, and antimalarial activity of (aryl or hetero aryl) substituted-4-Methyl-2,3-dihydro- 2-thioxo-*1H*-imidazol-5-yl Ethanone derivatives. Indian J. Heterocycl. Chem. **31**(03), 347–355 (2021)
7. Matada, B.S., Pattanashettar, R., Yernale, N.G.: A comprehensive review on the biological interest of quinoline and its derivatives. Bioorg. Med. Chem. **32**, 115973 (2021). https://doi.org/10.1016/j.bmc.2020.115973
8. Qadir, T., Amin, A., Sharma, P.K., Jeelani, I., Abe, H.: A review on medicinally important heterocyclic compounds. Open Med. Chem. J. **16** (2022). https://doi.org/10.2174/18741045-v16-e2202280
9. Kabir, E., Uzzaman, M.: A review on biological and medicinal impact of heterocyclic compounds. Results Chem. **4**, 100606 (2022). https://doi.org/10.1016/j.rechem.2022.100606
10. Keri, R.S., Patil, S.A.: Quinoline: a promising antitubercular target. Biomed. Pharmacother. **68**, 1161–1175 (2014). https://doi.org/10.1016/j.biopha.2014.10.007
11. Michael, J.P.: Quinoline, quinazoline, and acridone alkaloids. Nat. Prod. Rep. **20**(5), 476–493 (2003). https://doi.org/10.1039/b208140g
12. Van de Walle, T., Cools, L., Mangelinckx, S., D'hooghe, M.: Recent contributions of quinolines to antimalarial and anticancer drug discovery research. Eur. J. Med. Chem. **226**(15), 113865 (2021). https://doi.org/10.1016/j.ejmech.2021.113865
13. Musiol, R.: An overview of quinoline as a privileged scaffold in cancer drug discovery. Expert Opin. Drug Discov. **12**, 583–597 (2017). https://doi.org/10.1080/17460441.2017.1319357
14. Chokkar, N., Kalra, S., Chauhan, M., Kumar, R.: A review on quinoline derived scaffolds as anti-HIV agents. Mini Rev. Med. Chem. **19**, 510–526 (2019). https://doi.org/10.2174/1389557518666181018163448

15. Kaur, R., Kumar, K.: Synthetic and medicinal perspective of quinolines as antiviral agents. Eur. J. Med. Chem. **215**, 113220 (2021). https://doi.org/10.1016/j.ejmech.2021.113220
16. Mukherjee, S., Pal, M.: Quinolines: a new hope against inflammation. Drug Discov. Today Off. **18**, 389–398 (2013). https://doi.org/10.1016/j.drudis.2012.11.003
17. Ghanim, A.M., et al.: Design and synthesis of ibuprofen-quinoline conjugates as potential anti-inflammatory and analgesic drug candidates. Bioorg. Chem. **119**, 105557 (2022). https://doi.org/10.1016/j.bioorg.2021.105557
18. Patel, K.B., Kumari, P.: A review: structure-activity relationship and antibacterial activities of quinoline based hybrids. J. Mol. Struct. **1268**, 133634 (2022). https://doi.org/10.1016/j.molstruc.2022.133634
19. Kumar, H., et al.: Antihypertensive activity of a quinoline appended chalcone derivative and its site specific binding interaction with a relevant target carrier protein. RSC Adv. **5**, 65496–65513 (2015). https://doi.org/10.1039/C5RA08778C
20. Mo, J., et al.: Design, synthesis, biological evaluation, and molecular modeling studies of quinoline-ferulic acid hybrids as cholinesterase inhibitors. Bioorg. Chem. **93**, 103310 (2019). https://doi.org/10.1016/j.bioorg.2019.103310
21. Mahamoud, A., Chevalier, J., Davin-Regli, A., Barbe, J., Pages, J.-M.: Quinoline derivatives as promising inhibitors of antibiotic efflux pump in multidrug resistant enterobacter aerogenes isolates. Curr. Drug Targets. **7**, 843–847 (2006). https://doi.org/10.2174/138945006777709557
22. Eswaran, S., Adhikari, A.V., Shetty, N.S.: Synthesis and antimicrobial activities of novel quinoline derivatives carrying 1,2,4-triazole moiety. Eur. J. Med. Chem. **44**, 4637–4647 (2009). https://doi.org/10.1016/j.ejmech.2009.06.031
23. Razzaghi-Asl, N., Sepehri, S., Ebadi, A., Karami, P., Nejatkhah, N., Johari-Ahar, M.: Insights into the current status of privileged N-heterocycles as antileishmanial agents. Mol. Divers. **24**, 525–569 (2020). https://doi.org/10.1007/s11030-019-09953-4
24. Prasad, M.V.V.V., Rao, R.H.R., Veeranna, V., Chennupalli, V.S., Sathish, B.: Novel quinolone derivatives: synthesis and antioxidant activity. Russ. J. Gen. Chem. **91**, 2522–2526 (2021). https://doi.org/10.1134/S1070363221120239
25. Dorababu, A.: Quinoline: a promising scaffold in recent antiprotozoal drug discovery. ChemistrySelect. **6**(9), 2164–2177 (2021). https://doi.org/10.1002/slct.202100115
26. Krstulović, L., et al.: New quinoline-arylamidine hybrids: synthesis, DNA/RNA binding and tumor activity. Eur. J. Med. Chem. **137**, 196–210 (2017). https://doi.org/10.1016/j.ejmech.2017.05.054
27. Ugale, V.G., Patel, H.M., Surana, S.J.: Molecular modeling studies of quinoline derivatives as VEGFR-2 tyrosine kinase inhibitors using pharmacophore based 3D QSAR and docking approach. Arab. J. Chem. **10**, S1980–S2003 (2017). https://doi.org/10.1016/j.arabjc.2013.07.026
28. Maguire, M.P., Sheets, K.R., McVety, K., Spada, A.P., Zilberstein, A.: A new series of PDGF receptor tyrosine kinase inhibitors: 3-substituted quinoline derivatives. J. Med. Chem. **37**, 2129–2137 (1994). https://doi.org/10.1021/jm00040a003
29. Pearson, D.M., Conley, N.R., Waymouth, R.M.: Oxidatively resistant ligands for palladium-catalyzed aerobic alcohol oxidation. Organ. **30**, 1445–1453 (2011). https://doi.org/10.1021/om101037k
30. Evans, P., et al.: 8-Methylquinoline palladacycles: stable and efficient catalysts for carbon–carbon bond formation. Tetrahedron. **61**, 9696–9704 (2005). https://doi.org/10.1016/j.tet.2005.06.118
31. Zhu, H., et al.: Rational design of a targetable fluorescent probe for visualizing H_2S production under golgi stress response elicited by Monensin. Anal. Chem. **92**, 1883–1889 (2020). https://doi.org/10.1021/acs.analchem.9b04009

32. Jia, P., et al.: A highly selective ratiometric fluorescence probe for bioimaging of hypobromous acid in living cells and zebrafish. Sensors Actuators B Chem. **320**, 128583 (2020). https://doi.org/10.1016/j.snb.2020.128583
33. Schlagenhauf, P.: Mefloquine for malaria chemoprophylaxis 1992–1998: a review. J. Travel Med. **6**, 122–133 (1999). https://doi.org/10.1111/j.1708-8305.1999.tb00843.x
34. Hein, J.E., Fokin, V.V.: Copper-catalyzed azide-alkyne cycloaddition (CuAAC) and beyond: new reactivity of copper(I) acetylides. Chem. Soc. Rev. **39**, 1302–1315 (2010). https://doi.org/10.1039/b904091a
35. Bisacchi, G.S.: Origins of the quinolone class of antibacterials: an expanded "discovery story". J. Med. Chem. **58**, 4874–4882 (2015). https://doi.org/10.1021/jm501881c
36. Zhanel, G.G., et al.: A critical review of the fluoroquinolones. Drugs. **62**, 13–59 (2002). https://doi.org/10.2165/00003495-200262010-00002
37. Fuchs, C., Mitchell, E.P., Hoff, P.M.: Irinotecan in the treatment of colorectal cancer. Cancer Treat. Rev. **32**, 491–503 (2006). https://doi.org/10.1016/j.ctrv.2006.07.001
38. Fallahi, P., et al.: Cabozantinib in thyroid cancer. J. Clin. Oncol. **10**, 259–269 (2015). https://doi.org/10.2174/1574892810666150708110816
39. Vaishampayan, U.: Cabozantinib as a novel therapy for renal cell carcinoma. Curr. Oncol. Rep. **15**, 76–82 (2013). https://doi.org/10.1007/s11912-012-0289-x
40. Souto-Padron, T., Lima, A.P., de Oliveira Ribeiro, R.: Effects of dibucaine on the endocytic/exocytic pathways in Trypanosoma cruzi. Parasit. Res. **99**, 317–320 (2006). https://doi.org/10.1007/s00436-006-0192-1
41. Yadav, S.: Bedaquiline: a novel antitubercular agent for the treatment of multidrug-resistant tuberculosis. J. Clin. Diag. Res. **10**(8), 01–02 (2016). https://doi.org/10.7860/JCDR/2016/19052.8286
42. Kang, C., Shin, S.C.: Preparation and evaluation of bioadhesive dibucaine gels for enhanced local anesthetic action. Arch. Pharm. Res. **33**, 1277–1283 (2010). https://doi.org/10.1007/s12272-010-0819-8
43. Camarda, G., et al.: Antimalarial activity of primaquine operates via a two-step biochemical relay. Nat. Commun. **10**, 3226 (2019). https://doi.org/10.1038/s41467-019-11239-0
44. Alshammari, M.K., et al.: Nanoplatform for the delivery of Topotecan in the cancer milieu: an appraisal of its therapeutic efficacy. Cancers (Basel). **5**(1), 65 (2022). https://doi.org/10.3390/cancers15010065
45. Ajani, O.O., et al.: Microwave-assisted synthesis and antibacterial propensity of N'-(s-benzylidene and s-1H-pyrrol-2-yl)methylene-2 propylquinol ine-4-carbohydrazide motifs. Arab. J. Chem. **13**(1), 1809–1820 (2020)
46. Poonam, S., Kamaldeep, K., Amit, C., Ranjodh, S., Dhawan, R.K.: A review on biological activities of quinoline derivatives. BEST: J. Manag. Inf. Technol. Eng. **2**(1), 1–14 (2016)
47. Mandewale, C.M., Patil, C.P., Shedge, S.V., Dappadwad, U.R., Yamgar, R.S.: A review on quinoline hydrazone derivatives as a new class of potent antitubercular and anticancer agents. Beni-Suef Univ. J. Basic Appl. Sci. **6**(4), 354–361 (2017). https://doi.org/10.1016/j.bjbas.2017.07.005
48. (a) Wang, G. W., Jia, C. S., Dong, Y. W.: Benign and highly efficient synthesis of quinoline from 2-aminoarylketone or 2-aminoarylaldehyde and carbonyl compounds mediated by hydrochloric acid in water, Tetrahedron Lett., **47**, 1059–1063 (2006); (b) Hasaninejad, A. R. et al.: Synthesis of poly-substituted quinolines via Friedl¨ander heteroannulation reaction using silica-supported P2O5 under solvent-free conditions. Iran. J. Chem. Chem. Eng. **30**(1), 73–81 (2011).
49. Yamashkin, S., Oreshkina, E.: Traditional and modern approaches to the synthesis of quinoline systems by the Skraup and Doebner- miller methods. Rev. Chem. Heterocycl. Compd. **42**, 701–718 (2006)

50. Brouet, J.C., Gu, S., Peet, N.P., Williams, J.D.: A survey of solvents for the Conrad Limpach synthesis of 4-hydroxyquinolones. Synth. Commun. **39**(9), 5193–5196 (2009)
51. Madapa, S., Tusi, Z., Batra, S.: Advances in the syntheses of quinoline and quinoline-annulated ring systems. Curr. Org. Chem. **12**, 1116–1183 (2008)
52. Wu, Y.C., Liu, L., Li, H.J., Wang, D., Chen, Y.J.: Skraup-Doebner-Von miller: Quinoline synthesis revisited: reversal of the regiochemistry for g-aryl-b,g-unsaturated a-keto esters. J. Org. Chem. **71**, 6592–6595 (2006)
53. Abdel-Wahab, B.F., Khidre, R.E., Farahat, A.A., El-Ahl, A.S.: 2-Chloroquinoline-3-carbaldehydes: synthesis,reactions and applications. ARKIVOC. **1**, 211–276 (2012)
54. Becker, A., Kohfeld, S., Pies, T., Wieking, K., Preu, L., Kunick, C.: Synthesis of 11H-indolo[3,2-c]quinoline-6-carboxylic acids by cascade autoxidation-ring contractions. Synthesis. **7**, 1185–1189 (2009)
55. Jumade, P.P., Wadher, S.J., Chourasia, A.J., Kharabe, U.V., Mude, D., Yeole, P.G.: Synthesis of newer Mannich bases of quinoline derivative for antimicrobial activity. Int. J. Chem. Sci. **7**(3), 1518–1530 (2009)
56. Almansour, A.I., Arumugam, N., Kumar, R.S., Menendez, J.C., Ghabbour, H.A., Fun, H.K., Kumar, R.R.: Straight forward synthesis of pyrrolo[3,4-b] quinolines through intramolecular Povarov reactions. Tetrahedron Lett. **56**(49), 6900–6903 (2015). https://doi.org/10.1016/j.tetlet.2015.10.107
57. Sridharan, V., Perumal, P., Avendano, C., Menendez, C.: The first aza Diels-Alder reaction involving an a,bunsaturated hydrazone as the dienophile: Stereoselective synthesis of C-4 functionalized 1,2,3,4-tetrahydroquinolines containing a quaternary stereocenter. J. Org. Biomol. Chem. **5**, 1351–1353 (2007)
58. Zhou, X.-Y., Chen, X., Wang, L.-G.: Highly efficient Brønsted acid and Lewis acid catalysis systems for the Friedländer Quinoline synthesis. Synth. Commun. **48**, 830–837 (2018)
59. Zou, L.-H., Zhu, H., Zhu, S., Shi, K., Yan, C., Li, P.-G.: Copper-catalyzed ring-opening/reconstruction of anthranils with oxo-compounds: synthesis of quinoline derivatives. J. Org. Chem. **84**, 12301–12313 (2019)
60. Huang, C., Li, A., Chao, Z.-S.: Heterogeneous catalytic synthesis of quinoline compounds from aniline and C1–C4alcohols over zeolite-based catalysts. RSC Adv. **7**, 48275–48285 (2017)
61. Gao, W.-Y., et al.: Investigation of prototypal MOFs consisting of polyhedral cages with accessible Lewis-acid sites for quinoline synthesis. Chem. Commun. **51**, 4827–4829 (2015)
62. Xiang, J.-C., et al.: Divergent synthesis of functionalized quinolines from aniline and two distinct amino acids. J. Org. Chem. **82**, 9210–9216 (2017)
63. Mou, R.-Q., Zhao, M., Lv, X.-X., Zhang, S.-Y., Guo, D.-S.: An efficient and green synthesis of ferrocenyl-quinoline conjugates via a TsOH-catalyzed three-component reaction in water. RSC Adv. **8**, 9555–9563 (2018)
64. Chanda, T., Verma, R.K., Singh, M.S.: InCl3-driven regioselective synthesis of functionalized/annulated quinolines: scope and limitations. Chem. Asian J. **7**, 778–787 (2012)
65. Hakki, A., Dillert, R., Bahnemann, D.W.: Arenesulfonic acidfunctionalized mesoporous silica decorated with titania: a heterogeneous catalyst for the one-pot photocatalytic synthesis of quinolines from nitroaromatic compounds and alcohols. ACS Catal. **3**, 565–572 (2013)
66. Kaur, M., Pramanik, S., Kumar, M., Bhalla, V.: Polythiopheneencapsulated bimetallic Au-Fe3O4nano-hybrid materials: a potential tandem photocatalytic system for nondirected C(sp2)–H activation for the synthesis of quinoline carboxylates. ACS Catal. **7**, 2007–2021 (2017)
67. Naruto, H., Togo, H.: Preparation of 2-arylquinolines from β-arylpropionitriles with aryllithiums and NIS through iminyl radical-mediated cyclization. Org. Biomol. Chem. **17**, 5760–5770 (2019)

68. Sun, X., Li, J., Ni, Y., Ren, D., Hu, Z., Yu, S.: Synthesis of fused Quinoline and Quinoxaline derivatives enabled by domino radical triple bond insertions. Asian J. Org. Chem. **3**, 1317–1325 (2014)
69. Oliver, K.C., Stadler, A., Dallinger, D.: Microwaves in Organic and Medicinal Chemistry, vol. 52. Wiley (2012). https://doi.org/10.1002/9783527647828
70. Zhang, X., et al.: One-pot and catalyst-free synthesis of pyrroloquinolinediones and quinolinedicarboxylates. Green Chem. **19**, 3851–3855 (2017)
71. Fedoseev, P., Van Der Eycken, E.: Temperature switchable Brønsted acid-promoted selective syntheses of spiro-indolenines and quinolines. Chem. Commun. **53**, 7732–7735 (2017)
72. Xiao, M., Sun, Q., Sun, J., Yan, C.G.: An efficient synthesis of spiropyrroloquinolines by the domino reaction of α-dicarbonyl compounds and anilinosuccinimides: an efficient synthesis of spiropyrroloquinolines by the domino reaction of α-dicarbonyl compounds and anilinosuccinimides. Eur. J. Org. Chem. **2017**, 6861–6866 (2017)
73. Yu, Y., Tu, M.-S., Jiang, B., Wang, S.-L., Tu, S.-J.: Multicomponent synthesis of polysubstituted dihydroquinoline derivatives. Tetrahedron Lett. **53**(38), 5071–5075 (2012)
74. Peng, J.H., et al.: A facile and expeditious microwave-assisted synthesis of furo[3,4-b]indeno[2,1-f]quinolin-1-one derivatives via multicomponent reaction. J. Heterocycl. Chem. **50**, 899–902 (2013)
75. Wasserchied, P., Keim, W.: Ionic liquids: new solutions for transition metal catalysis. Angew. Chem. Int. Ed. **39**, 3772–3789 (2000)
76. Larshen, A.S., Holbrey, J.D., Than, F.S., Reed, C.A.: Designing ionic liquids: imidazolium melts with inert carbonate anions. J. Am. Chem. Soc. **122**, 7264–7272 (2000)
77. Kharazmi, A., et al.: Synthesis of multiple quinoline derivatives using novel ionic liquid-based nano-magnetic catalyst (MNPs@SiO2-Pr-AP-tribromide). Res. Chem. Intermed. **48**, 1313–1329 (2022). https://doi.org/10.1007/s11164-022-04675-z
78. Jayakumara, J., Reddy, S.R.: Molecular oxygen-promoted sustainable synthesis of functionalized quinolines using catalytic glucose-derived ionic liquids and copper org. Biomol. Chem. **22**, 8472–8479 (2024). https://doi.org/10.1039/D4OB01354A
79. Dadhania, H., Raval, D., Dadhania, A.: A highly efficient and solvent-free approach for the synthesis of Quinolines and fused polycyclic Quinolines catalyzed by magnetite nanoparticle-supported acidic ionic liquid. Polycycl. Aromat. Compd. **41**(2), 440–453 (2021). https://doi.org/10.1080/10406638.2019.1595057
80. Sharma, K., Sharma, M.: Green synthesis of Quinoline-based ionic liquid. J. Vis. Exp. **211** (2024). https://doi.org/10.3791/67345
81. Mazloumi, M., Shirini, F.: Synthesis of Quinolines, Quinazolines and Spiro-Quinazolines using Nanoporous TiO_2 containing an ionic liquid bridge as an efficient and reusable catalyst. Polycycl. Aromat. Compd. **42**(5), 2087–2106 (2020). https://doi.org/10.1080/10406638.2020.1827271
82. Shirini, F., Langarudi, M.S.N., Daneshvar, N., Jamasbi, N., Irankhah-Khanghah, M.: Preparation and characterization of [H2-DABCO][ClO_4]$_2$ as a new member of DABCO-based ionic liquids for the synthesis of pyrimido[4,5-*b*]-quinoline and pyrimido[4,5-*d*]pyrimidine derivatives. J. Mol. Struct. **1161**, 366–382 (2018). https://doi.org/10.1016/j.molstruc.2018.02.069
83. Zare, A., Barzegar, M.: Dicationic ionic liquid grafted with silica-coated nano-Fe3O4 as a novel and efficient catalyst for the preparation of uracil-containing heterocycles. Res. Chem. Intermed. **46**, 3727–3740 (2020)
84. Cannalire, R., Tiecco, M., Cecchetti, V., Germani, R., Manfroni, G.: Advantageous use of ionic liquids for the synthesis of pharmaceutically relevant quinolones. Eur. J. Org. Chem. **2018**(23), 2977–2983 (2018). https://doi.org/10.1002/ejoc.201800415

85. Rajendran, A., Zarthikeyan, C., Rajathi, K., Ragupathy, D.: An environmentally benign one pot synthesis of substituted quinolines catalysed by fluoroboric acid based ionic liquid. J. Chem. Sci. **124**(4), 877–881 (2012)
86. Moosavi-Zare, A.R., Najafi, R.: Multicomponent synthesis of pyrimido[4,5-b] quinolines over a carbocationic catalytic system. Sci. Rep. **13**, 16501 (2023). https://doi.org/10.1038/s41598-023-43793-5
87. Jorshari, M.R., Mamaghani, M., Jahanshahi, P.: Synthesis, delivery, and molecular docking of fused quinolines as inhibitor of hepatitis a virus 3C proteinase. Sci. Rep. **11**, 18970 (2021). https://doi.org/10.1038/s41598-021-98529-0
88. Gajare, S., Patil, A., Hangirgekar, S., Dhanmane, S., Rashinkar, G.: Graphene-oxide supported ionic liquid catalyst-based approach for the synthesis of quinolines. Res. Chem. Intermed. **46**, 2417–2436 (2020)
89. Rezayati, S., Jafroudi, M.T., Nezhad, E.R., Hajinasiri, R., Abbaspour, S.: Imidazole-functionalized magnetic Fe3O4 nanoparticles: an efficient, green, recyclable catalyst for one-pot Friedländer quinoline synthesis. Res. Chem. Intermed. **42**, 5887–5898 (2016). https://doi.org/10.1007/s11164-015-2411-9
90. Katheriya, D., Patel, N., Dadhania, H., Dadhania, A.: Magnetite (Fe_3O_4) nanoparticles-supported dodecylbenzenesulfonic acid as a highly efficient and green heterogeneous catalyst for the synthesis of substituted quinolines and 1-amidoalkyl-2-naphthol derivatives. J. Ira. Chem. Soc. **18**, 805–816 (2021). https://doi.org/10.1007/s13738-020-02069-9
91. Le, Z.-G., Liang, M., Chen, Z.-S., Zhang, S.-H., Xie, Z.-B.: Ionic liquid as an efficient medium for the synthesis of Quinoline derivatives via α-chymotrypsin-catalyzed Friedländer condensation. Molecules. **22**, 762 (2017). https://doi.org/10.3390/molecules22050762
92. Sarma, P., et al.: 3-Methyl-1-sulfoimidazolium ionic liquids as recyclable medium for efficient synthesis of quinoline derivatives by Friedländer annulation. Monatsh Chem. **146**, 173–180 (2015). https://doi.org/10.1007/s00706-014-1305-7
93. Amarasekara, A.S., Hasan, M.A.: 1-(1-Alkylsulfonic)-3-methylimidazolium chloride Bronsted acidic ionic liquid catalyzed Skraup synthesis of quinolines under microwave heating. Tetrahedron Lett. **55**, 3319–3321 (2014)
94. Bharate, J.B., Bharate, S.B., Vishwakarma, R.A.: Metal-free, ionic liquid-mediated synthesis of functionalized quinolines. ACS Comb. Sci. **16**, 624–630 (2014)
95. Sarma, P., Dutta, A.K., Gogoi, P., Sarma, B., Borah, R.: 3-Methyl-1-sulfoimidazolium ionic liquids as recyclable medium for efficient synthesis of quinoline derivatives by Friedlander annulation. Monatsh. Chem. **146**, 173–180 (2015)

Working in a Good Laboratory Practice (GLP) Environment: The Positive Outcomes

Srikanta Kumar Rath[1,2] (✉)

[1] Genotoxicity Laboratory, Division of Toxicology and Experimental Medicine, CSIR-Central Drug Research Institute, Lucknow, Uttar Pradesh, India
[2] Academy of Scientific and Innovative Research (AcSIR), Ghaziabad, Uttar Pradesh, India
skrath@cdri.res.in

Abstract. Good Laboratory Practices (GLP) are guidelines that ensure the quality and integrity of data generated in the laboratory. There are several benefits to practicing it. The data generated in such an environment ensures credibility and reliability as this data is used to assess the safety of products and pass through stringent regulatory decisions to protect human health and the environment. GLP ensures animal healthcare and ethical treatment of animals. Wasteful research is also controlled and efficient data collection is in place which overall reduces the cost of research and also facilitates international acceptance. In this way, it helps to build trust in the scientific process and in the products that are developed using scientific research. The data generated in the GLP environment helps to ensure that scientific findings can be reproduced anywhere in the globe and hence duplication of studies is not required which saves time. GLP helps remove technical trade barriers. Therefore, working in a GLP environment is quite distinguished and dignified. The author will share his own experience at a test facility in the country.

Keywords: GLP · Non-clinical safety study · OECD · Data reliability · Mutual acceptance of data

1 Introduction

Good Laboratory Practices (GLP) is a set of established guidelines designed to ensure accuracy, reliability, and integrity in non-clinical safety studies. These standards make laboratory data trustworthy for regulatory bodies overseeing public health and environmental safety. By promoting ethical animal use, reducing unnecessary experiments, and enhancing data efficiency, GLP not only streamlines research but also lowers costs. Moreover, it strengthens the global acceptance of scientific findings by ensuring reproducibility and preventing redundant studies. By adhering to these standards, research conducted in one country can be confidently accepted by other nations, fostering mutual acceptance of data (MAD) [1, 2].

A. K. Saxena, A. Saxena (Eds.): GTHTM 2025, pp. 275–282, 2026.
https://doi.org/10.1007/978-3-032-12320-6_15

The Organization for Economic Cooperation and Development (OECD) defines GLP as a quality system that governs the planning, execution, monitoring, recording, reporting, and archiving of non-clinical health and environmental safety studies. The core principles of GLP focus on ensuring data quality and reliability, preventing research duplication, protecting human health and the environment, facilitating worldwide data acceptance, and eliminating technical trade barriers. Institutions conducting regulatory studies adhere to OECD GLP guidelines when assessing the safety of various chemicals, including pharmaceuticals, pesticides, cosmetics, veterinary drugs, food additives, and industrial chemicals. While these test items are often synthetic chemicals, they may also be derived from natural or biological sources and, in some cases, may even include living organisms [3–5].

2 Advantages of the GLP Study

2.1 Ensures Data Integrity and Reliability

GLP enforces strict documentation, meticulous record-keeping, and rigorous quality control, minimizing errors and preventing data manipulation. By maintaining transparency and traceability at every stage of a study, GLP enhances the credibility of research findings, ensuring they are reliable and trustworthy for regulatory approvals.

2.2 Enhances Study Reproducibility

GLP mandates the use of Standard Operating Procedures (SOPs), ensuring that experiments are conducted consistently under controlled conditions. This standardization enhances laboratory reproducibility, allowing studies to be reliably replicated and validated.

2.3 Facilitates Regulatory Compliance

GLP compliance is essential for regulatory submissions to authorities such as the U.S. FDA, OECD, EMA, and other global agencies. By ensuring that data meets international safety and efficacy standards, GLP streamlines the approval process for pharmaceuticals, chemicals, and other regulated products.

2.4 Improves Laboratory Efficiency and Organization

GLP establishes clear roles and responsibilities for laboratory personnel, ensuring a structured workflow and minimizing confusion during study execution. A well-organized laboratory infrastructure, coupled with properly maintained equipment and standardized procedures, reduces delays, prevents errors, and enhances overall research efficiency.

2.5 Enhances Quality Assurance and Auditability

GLP mandates independent Quality Assurance (QA) units to conduct regular audits and inspections, ensuring strict compliance with regulatory standards. This proactive oversight minimizes the risk of fraudulent, inconsistent, or unreliable data, reinforcing the credibility and integrity of scientific research.

2.6 Supports Global Acceptance of Research Data

GLP-compliant studies adhere to internationally recognized standards, enabling regulatory authorities worldwide to accept research data with confidence. This global recognition facilitates product approvals across multiple countries, reducing the need for duplicate testing and accelerating market entry.

2.7 Promotes Ethical Research Practices

GLP enforces the responsible handling of test substances, biological samples, and animal models, ensuring compliance with ethical guidelines in scientific research. By prioritizing animal welfare, environmental safety, and human health, GLP upholds the highest standards of integrity and responsibility in laboratory studies.

2.8 Reduces Research Costs and Time in the Long Run

While GLP implementation requires an initial investment, it prevents costly errors, rework, and regulatory rejections by ensuring studies are conducted correctly from the start. This efficiency minimizes delays, optimizes resource utilization, and ultimately saves both time and money in the research and development process.

2.9 Strengthen Consumer and Environmental Safety

GLP ensures the generation of reliable and high-quality safety data, enabling regulatory agencies to make informed decisions about pharmaceuticals, chemicals, and environmental hazards. By assessing potential risks accurately, GLP helps protect public health, consumer safety, and the ecosystem from harmful substances.

3 Positive Outcomes of the GLP Study

A key positive outcome of adhering to OECD GLP principles is the assurance of data integrity, reliability, and reproducibility in non-clinical safety studies. By enforcing standardized procedures for study execution, data recording, and quality assurance, GLP enhances the credibility and regulatory acceptability of research findings worldwide. This structured framework fosters transparency, minimizes experimental bias, and promotes international recognition of study results, ultimately contributing to safer pharmaceuticals, chemicals, and environmental protections for human health and the ecosystem.

4 Pillars of GLP: Non-clinical Safety Testing Under OECD GLP Guidelines

The pillars of GLP play a vital role in ensuring the reliability, accuracy, and credibility of scientific research. By establishing standardized procedures for study planning, execution, monitoring, reporting, and archiving, GLP enhances data integrity, regulatory compliance, and international collaboration. These principles not only improve laboratory efficiency and risk management but also contribute to the safety of personnel and the environment. Ultimately, adherence to GLP fosters scientific excellence, facilitates smoother regulatory approvals, and supports advancements in research and innovation. Non-clinical safety testing under the OECD GLP guidelines follows a structured approach to ensure data reliability, integrity, and regulatory compliance. The pillars of GLP uphold the reliability and consistency of research data by ensuring meticulous study planning, execution, and documentation. The key pillars of GLP include planning, performing, monitoring, reporting, recording, and archiving of study-related records.

4.1 Planning

A well-planned GLP study is the foundation of regulatory compliance and high-quality research. Each study must be incorporated into a master schedule that outlines timelines, responsibilities, and resource allocation. The process begins with defining clear study objectives, selecting test compounds and biological models, and establishing key evaluation criteria while adhering to OECD and FDA guidelines [5]. A robust study protocol is essential, detailing everything from the study title and funding source to the experimental design, dosing strategy, data collection methods, and ethical considerations.

Before the study begins, approval from the study director and QA personnel is mandatory. Selecting the appropriate test system, whether cell lines or animal models, requires detailed documentation of its source, species, strain, and genetic background. Proper handling, labelling, and tracking test substances are crucial for maintaining consistency and reliability. Laboratory facilities must meet GLP standards, ensuring a controlled environment, preventing contamination, and maintaining properly calibrated and validated equipment that follows SOPs. Clearly defined roles, responsibilities, and proper training of personnel ensure accountability and competency in research activities. Proper study design, protocol approval, and adherence to guidelines ensure systematic and unbiased research execution.

4.2 Performing

Conducting a GLP-compliant study demands unwavering precision, strict adherence to a predefined protocol, and meticulous documentation to meet regulatory standards. The process begins with an approved study plan, overseen by the study director, while QA personnel ensure compliance through regular inspections. Independent QA units monitor adherence to GLP principles, verifying compliance through audits and inspections. Every team member must be thoroughly trained in GLP principles, and all test substances must be handled according to standardized procedures. From dosing and environmental conditions to sample processing, every step follows established SOPs, with any deviations carefully documented and justified.

QA personnel routinely inspect procedures to verify compliance, address deviations, and implement corrective actions where necessary. Once the study concludes, data analysis is conducted according to a predefined statistical plan to uphold scientific integrity. Additionally, all equipment undergoes validation to confirm proper calibration, ensuring reliable outcomes. By maintaining these rigorous standards, GLP studies produce credible, reproducible results that support both regulatory approvals and scientific advancements.

4.3 Monitoring

Monitoring a GLP study is crucial for ensuring compliance with regulatory standards and maintaining the accuracy, integrity, and reliability of the study. It helps identify and correct deviations early, preventing errors and ensuring data integrity. Regular monitoring improves regulatory compliance, facilitating smoother approvals and audits. Additionally, it enhances laboratory efficiency, minimizes risks, and ensures the safety of personnel, test subjects, and the environment. This process involves routine inspections of test facilities and study audits to confirm adherence to GLP guidelines. Continuous monitoring is essential for preserving data integrity and credibility.

QA personnel oversee compliance by enforcing SOPs and maintaining detailed documentation during all study phases. QA teams play a key role in ensuring that every stage of the study strictly follows the approved protocol and GLP guidelines. Regular inspections and audits are scheduled to evaluate laboratory facilities, personnel practices, equipment calibration, test system management, and data recording. Monitoring also extends to laboratory equipment and environmental conditions. Instruments must undergo routine calibration and maintenance, with performance records maintained for accuracy. Also, environmental factors such as temperature, humidity, and light exposure are closely monitored and documented to maintain study integrity and reliability.

A well-designed Quality Assurance Program (QAP) is essential for maintaining compliance with GLP standards. It involves verifying study plans, ensuring adherence to standardized procedures, and conducting routine inspections. GLP guidelines categorize inspections into three types: study-based, facility-based, and process-based, each playing a crucial role in upholding research credibility and reliability. Effective monitoring strengthens data accuracy, regulatory compliance, and overall laboratory efficiency while minimizing risks. Additionally, it ensures the safety of personnel, test subjects, and the environment, reinforcing ethical research practices. By integrating a robust QAP, laboratories can enhance the reliability of their findings and streamline regulatory approval processes.

4.4 Recording

Accurate data recording is the backbone of GLP studies, ensuring data integrity, traceability, and compliance with regulatory standards. Comprehensive documentation enhances transparency, reproducibility, and the overall reliability of study findings. Real-time data entry must be meticulously recorded in laboratory notebooks or validated electronic systems, with each entry properly written, signed, and dated. Proper handling, storage, and documentation of test substances ensure study integrity and prevent contamination or misidentification.

Raw data includes observations, measurements, test system details, environmental conditions, and protocol deviations, each meticulously documented to maintain consistency and accuracy. To uphold data integrity, QA personnel conduct audits to verify accuracy and completeness, promptly addressing discrepancies or missing entries. Once the study concludes, data is systematically compiled for analysis. The final report integrates recorded data, statistical evaluations, and study conclusions, with all materials securely archived for regulatory review.

4.5 Reporting

The reporting of a GLP study is a critical step in transparently documenting study outcomes while ensuring regulatory compliance. A well-structured final report serves as an official record, allowing regulatory authorities and stakeholders to assess the validity and reliability of the results. Accurate and detailed reporting in GLP ensures that all study-related data, observations, and results are systematically documented. This enhances data integrity, making it easier to verify findings and ensure consistency across research. Proper reporting also facilitates regulatory compliance by providing clear and structured records that meet the requirements of authorities such as the FDA, OECD, and EPA. Additionally, well-prepared reports improve communication between researchers, sponsors, and regulatory bodies, ensuring transparency and trust in scientific studies.

The report begins with a title page that includes key details such as the study title, test substance identification, funding source, test facility, study director, and essential personnel. The introduction and study objective section provides context by outlining the study rationale, key research questions, and primary goals. The materials and methods section thoroughly describes the test system, study design, experimental procedures, dosing regimens, and analytical methods, ensuring reproducibility and independent verification. The results section systematically presents findings through tables, graphs, and statistical analyses, offering a clear and structured interpretation of the data. The discussion and conclusion section then evaluates these findings in relation to study objectives, emphasizing key observations and their broader implications. Additionally, a QA statement confirms adherence to GLP principles, summarizing QA inspections conducted throughout the study. Once finalized, the report undergoes thorough review and approval by the study director and relevant personnel before submission to regulatory authorities.

4.6 Archiving

Archiving a GLP study is essential for preserving study data and ensuring long-term accessibility for regulatory audits and inspections. Archiving plays a crucial role in preserving research data for future reference, validation, and compliance. Secure storage of study records allows researchers and regulatory bodies to verify findings, replicate studies, and conduct audits when necessary. Archiving helps prevent data loss due to accidental deletion, system failures, or unauthorized alterations, ensuring the longevity and reliability of research information. Furthermore, well-maintained archives support future studies by providing a valuable database of past experiments, reducing the need for redundant testing and saving both time and resources. Proper storage safeguards all study-related documents, raw data, and materials, protecting them from unauthorized access, deterioration, or loss. A GLP-compliant archive must be housed in a secure, controlled environment with access strictly limited to authorized personnel.

The archived materials include the final report, raw data, laboratory notebooks, electronic records, QA reports, SOPs, study protocols, test substance characterization data, and regulatory correspondence. Additionally, biological samples, specimens, or test systems must be stored according to specific regulatory guidelines to maintain their integrity. For electronic records, a validated GLP-compliant electronic document management system (EDMS) with robust security measures and backup protocols is required. Paper records should be carefully indexed and stored in fire, pest, and earthquake-resistant cabinets. The retention period for GLP study records varies by regulatory requirements, typically 10 to 15 years or longer, depending on the study type and the governing authority (e.g., FDA, OECD, EMA). In summary, proper archiving is critical for maintaining data integrity, ensuring regulatory compliance, and guaranteeing long-term access to essential study records.

5 Conclusion

GLP establishes a structured framework for conducting high-quality, reliable, and reproducible laboratory studies. By enforcing strict compliance, ensuring data integrity, and upholding ethical research practices, GLP enhances scientific credibility and facilitates regulatory approvals. This ultimately benefits researchers, industries, and public health by promoting safer, more efficient, and globally accepted scientific advancements. The positive outcome of GLP studies is evident in the enhanced transparency, reliability, and international recognition of scientific data. By preventing research duplication, promoting ethical practices, and safeguarding human and environmental health, GLP stands as a fundamental pillar of non-clinical safety testing. Moving forward, continued adherence to GLP principles will remain essential in ensuring scientific progress, regulatory compliance, and public safety.

Acknowledgments. The author gratefully acknowledges the contributions of Gourav Jha. Sakshi Mishra, Divyansh Sharma, and Bhawana Tomar in the form of critical reading, suggestions etc., during preparation of this manuscript.

References

1. Turnheim, D.: Benefits of good laboratory practice as a tool to improve testing. Hum. Exp. Toxicol. **12**, 528–532 (1993). https://doi.org/10.1177/096032719301200611
2. Canton, H.: Organisation for economic co-operation and development—OECD. In: The Europa Directory of International Organizations 2021, pp. 677–687. Routledge (2021)
3. Turnheim, D.: The OECD policy for the implementation of the principles of good laboratory practice. Ann. Ist. Super. Sanita. **30**, 395–400 (1994)
4. OECD: OECD series on principles of good laboratory practice and compliance monitoring. Ann Ist Super Sanita. **33**, 1–172 (1997)
5. Huntsinger, D.W.: OECD and USA GLP applications. Ann. Ist. Super. Sanita. **44**, 403–406 (2008)

BIMP: Unveiling the Phytochemical Richness of Indian Medicinal Plants as Potential Therapeutic Agents

Dheeraj Kumar Chaurasia[1], Raushan Anjum[3], Aman Sharma[2], Madhvi Mishra[2], Shashank Shekhar[2], Ashok Kumar Patel[4], Aditya Mittal[4], and B. Jayaram[2] [3](✉)

[1] School of Interdisciplinary Research, Indian Institute of Technology Delhi, Hauz Khas, New Delhi, Delhi, India

[2] Department of Chemistry & Biochemistry, The University of Arizona, Tucson, AZ, USA

[3] Supercomputing Facility for Bioinformatics and Computational Biology, Indian Institute of Technology Delhi, Hauz Khas, New Delhi, Delhi, India

bjayaram@chemistry.iitd.ac.in

[4] Kusuma School of Biological Sciences, Indian Institute of Technology Delhi, Hauz Khas, New Delhi, Delhi, India

Abstract. The historical importance of Indian medicinal plants is well documented in traditional systems of medicine such as Ayurveda and Unani. These plants are renowned for their medicinal effects that enhance general well-being. With modern scientific progresses, there has been an increased interest in exploring the bioactivity of these plants to uncover the active compounds responsible for their pharmacological effects. This has given rise to a growing field of research focused on Indian medicinal plants, which extends beyond traditional practices to contemporary pharmaceutical breakthroughs. In this context, a manually curated database called BIMP (Bioactivity information of the phytochemicals of Indian Medicinal Plants) is introduced here. This database provides information on 105,909 Phytochemicals found in 6,209 Indian medicinal plants. It includes information on both known and predicted targets for the compounds, along with specifics on the diseases and the target genes/proteins with information of organisms. Additionally, the database assesses physicochemical properties and drug similarity using cheminformatics tools, aiding drug discovery efforts. Of the 105,909 Phytochemicals, 72,318 comply with at least one of the major drug-likeness rules. The BIMP database represents a significant advancement in facilitating *in-silico* drug discovery processes by providing a comprehensive platform for discovering the therapeutic potential of Indian medicinal plant components. It also offers a protocol for identifying targets and molecules, streamlining the drug discovery process. It is openly accessible and offers researchers a valuable database for advancing drug development based on natural products. Database URL: https://scfbio-iitd.res.in/bimp

Keywords: Indian medicinal plants · BIMP · Phytochemicals · Bioactivity · Therapeutics · Virtual Screening · Similarity Search · Scaffolds Search

A. K. Saxena, A. Saxena (Eds.): GTHTM 2025, pp. 283–298, 2026.
https://doi.org/10.1007/978-3-032-12320-6_16

1 Introduction

Indian medicinal plants have a deep historical significance owing to their extensive use in Ayurveda, Unani, and other traditional healing methods [1–4]. India's rich legacy of herbal remedies highlights significant contributions that go beyond Ayurveda and Siddha, encompassing the realm of modern pharmaceutical breakthroughs and pharmacological investigations [5, 6, https://bsi.gov.in/page/en/medicinal-plant-database]. Over time, these plants have gained great respect for their therapeutic attributes, providing remedies for various ailments and contributing to overall well-being. With advancing scientific research, there is a growing interest in exploring the bioactivity of these plants to identify and understand the active compounds responsible for their remarkable medicinal effects.

Natural products remain a pivotal resource in the pharmaceutical industry, serving as valuable reservoirs for novel drugs [4]. India hosts approximately 45,000 plant species [https://www.cbd.int], with a substantial portion of these plants, numbering in thousands, being attributed with medicinal potential. Recent research spanning the past few decades has focused on plants referenced in ancient texts or employed in traditional practices related to diabetes management, revealing notable anti-diabetic attributes [7].

Another example is Ayush-64, a poly-herbal formulated by CCRAS, Ministry of AYUSH, Government of India and its constituents are *Alstonia scholaris, Picrorhiza kurrooa, Swertia chirayita,* and *Ticanto crista* [8]. These plants have a great significance in treating conditions like *Jvara* (disease conditions associated with pyrexia), *Kasa* (cough), *Swasa* (dyspnea), *Pravahika* (dysentry), *Visamajvara* (disease conditions associated with recurrent pyrexia), and *Sopha* (inflammation), as per the Ayurvedic indications. Extensive pharmacological, toxicological and clinical studies have proven its safety and effectiveness in infectious febrile diseases like malaria [9] and influenza [10]. AYUSH-64 possessing immunomodulatory, anti-inflammatory, antipyretic, antioxidant, and antiviral properties [11–16], has been repurposed during the COVID-19 outbreak [17] to handle asymptomatic and mild to moderate cases of the viral infection [16, 18, 19]. It has been used as a solo therapy or in addition to standard care in various clinical trials, showing positive results as it hindered the severe inflammatory responses in COVID-19 that led to the development of a life-threatening sickness.

Medicinal plants traditionally used have emerged as abundant repositories of biologically active Phytochemicals. The pharmaceutical sector has undergone a significant transformation, favouring the utilization of these Phytochemicals over synthetic compounds for drug development, intending to reduce potential side effects [20]. The presence of a carefully organized collection of data about plants, along with details about their corresponding natural compounds and an archive of their molecular arrangements, holds the potential to facilitate *in-silico* drug discovery processes. Progress in this domain has been noteworthy lately, particularly in database creation [4, 20–31] centred around natural compounds, primarily emphasizing the chemical composition of edible and herbaceous plants. Nonetheless, when viewed through the lens of traditional Indian medicine, there have been comparatively limited endeavours to construct digital repositories encompassing Indian medicinal plants, their phytochemical constituents, known and predicted drug targets, and the purposes they serve therapeutically such as CVDHD

[21], DISPEL [22], Dr. Duke's Phytochemical and Ethnobotanical Databases [23], IMPDB [20], IMPPAT [4], KNApSACK [24], MED-PDB database [25], NPACT [26], OSADHI [27], Phytochemica [28], and PlantMolecularTasteDB [29], TCM Database@ Taiwan [30], and TCMID [31]. The focus of establishing online databases for traditional Indian medicine has predominantly revolved around documenting medicinal plants and their corresponding therapeutic applications rather than encompassing the essential phytochemical components crucial for drug discovery and drug targets.

Consequently, we developed a well-curated BIMP database containing 105,909 Phytochemicals sourced from 6,209 Indian Medicinal Plants. The BIMP database offers a chemical library of Phytochemicals, each accompanied by downloadable chemical structures presented in one-dimensional notations such as InChI-Key and InChI [32], SMILES [33] (1D), two-dimensional (2D), and three-dimensional (3D) formats. Additionally, we performed computations on physicochemical properties and drug-likeness using the cheminformatics tool RDKit [http://www.rdkit.org]. Drug-likeness for each phytochemical was assessed using a range of scoring rules, including the Lipinski Rule of 5 (RO5) [34], Ghose Rule [35], Veber Rule [36], Egan Rule [37], and Muegge Rule [38]. Furthermore, we manually extracted data on Phytochemicals that are known to be beneficial in various diseases and their target genes and proteins across many organisms. BIMP offers an integrated platform that facilitates the application of cheminformatic [39] methodologies to speed up drug discovery based on natural products. The BIMP database is publicly available at https://scfbio-iitd.res.in/bimp. The database is also embedded into *Sanjeevini*, a comprehensive drug-discovery suite, as a phytochemical library for virtual screening [40, 41].

2 Methods

2.1 Indian Medicinal Plants List Curation

During the initial stages of constructing the database, we assembled an extensive list of over 15,000 Indian medicinal plants. This compilation was primarily derived from data sourced from the Indian Medicinal Plants Database [https://www.medicinalplants.in/] and the Environmental Information System (ENVIS), Centre on Medicinal Plants [https://envis.frlht.org/]. In addition, we collected comprehensive details about each plant from established databases [24], https://envis.frlht.org/, https://indiaflora-ces.iisc.ac.in/, http://www.flowersofindia.net/] and literature [42], including vernacular names in 10 Indian languages, including Gujarati, Hindi, Kannada, Malayalam, Marathi, Punjabi, Sanskrit, Tamil, Telugu, and Urdu. Also, we gathered extensive information on each plant's habits, habitats, and nationwide distribution. Please note that our all-inclusive list covers a wide range of Indian medicinal plants. This collection includes Ayurveda and other traditional systems of Indian medicine such as Siddha, Homeopathy, Sowa-Rigpa, Folk and Unani.

2.2 NCBI Taxonomy Retrieval

Employing the botanical names of the plants, we utilized NCBI Entrez with a focus on the Taxonomy [43, 44] Database as the primary text search. This method facilitated the retrieval of NCBI taxonomy information for each plant, streamlining the data collection process.

2.3 Phytochemicals List Curation

We obtained detailed phytochemical information for each plant from the pre-existing databases [4, 20, 24, 26, 28], increasing the comprehensiveness and accuracy of our research. We used the PubChem [45] and IMMPAT databases to search for natural products from each plant by providing the NCBI taxonomy. Plant metabolite extraction was done from PubChem and KNApSAcK databases. This enabled us to obtain PubChem ID, KNApSAcK C_ID, SMILES, InChI and InChIKey information for all Phytochemicals associated with each plant specified in our search.

2.4 Annotation of Phytochemicals

The main objective of this effort is to establish a platform for understanding the phytochemistry of Indian medicinal plants. Evaluating these phytochemicals for their therapeutic potential will aid in pinpointing promising compounds for target/biochemical pathway based future drug discovery efforts. To annotate each molecule, we extracted the InChI and InChIKey, and SMILES for each phytochemical from IMMPAT, KNApSAcK, and PubChem, the extracted data supplied to PubChem Identifier Exchange Service for the extraction of synonyms and standard PubChem chemical identifiers. Additionally, we made a manual interpretation of traditional nomenclature and additional details about Phytochemicals originating from Indian medicinal plants. These descriptions were obtained from literature references and enriched with documented synonyms and established chemical identifiers obtained from ChEBI [46], ChEMBL [47], ChemSpider [48], CompTox Chemicals Dashboard v2.2.1 [49], DrugBank [50], IMMPAT, KEGG [51], KNApSAcK, PubChem, SureChEMBL [52], and ZINC [53]. Furthermore, we employed ClassyFire [54] to determine the chemical classification of each phytochemical.

2.5 Structure Generation of Phytochemicals

The 3-dimensional (3D) structure of small molecules is one of the fundamental aspects of cheminformatics and drug discovery. It involves the generation of three-dimensional representations of the compounds that precisely depict the spatial arrangement of atoms and bonds. These generated structures are crucial in understanding molecular properties, predicting activities, and for designing new molecules with specific desired characteristics. We first extracted the existing 3D structures of Phytochemicals from the PubChem repository. For the remaining compounds, we generated their well-optimized

3D structures using external software, utilizing Balloon [55] and RDKit [https://www.rdkit.org/], employing the Merck Molecular Force Field (MMFF94) [56].

2.6 Physicochemical Properties Calculation

RDKit is an open-source toolkit for cheminformatics and molecular modelling. One of its commonly used features is the calculation of the physicochemical properties of molecules. These properties play a pivotal role in drug discovery, aiding in the identification, optimization, and development of small molecules with therapeutic potential. These properties also serve as indispensable parameters for characterizing the structural composition, intermolecular interactions, and potential functionalities of the small compounds. Here, we have used an in-house Python script using the RDKit module to calculate the physicochemical properties mentioned in Table 1.

2.7 Drug-Likeness of Phytochemicals

Drug-likeness pertains to the set of properties that increase the likelihood of a small molecule being transformed into an effective drug. These properties are important indicators for predicting the potential of a compound to become a safe and effective pharmaceutical product. Drug likeness is assessed through various molecular and physicochemical properties, which are often evaluated using computational methods. Here, we applied multiple drug-likeness rules to each phytochemical, such as Lipinski Rule of 5 (RO5), Ghose Rule, Veber Rule, Egan Rule and Muegge Rule. Each rule provides different criteria for assessing various aspects of a compound's drug-likeness.

Table 1. List of physicochemical properties calculated for each phytochemical.

S. No.	Physicochemical Properties
1	Number of Heavy Atoms
2	Number of Hydrogen Bond Donors
3	Number of Hydrogen Bond Acceptors
4	Molecular Weight
5	WlogP
6	Molar Refractivity
7	Molecular Formula
8	Number of Rings
9	Number of Rotatable Bonds
10	Topological Polar Surface Area
11	Fraction of sp3 Cs
12	Number of NH or OH groups
13	Number of N or O atoms
14	Number of Atoms
15	Number of Heteroatoms
16	Number of Carbons
17	Formal Charge

Table 2. List of drug-likeness rules with physicochemical properties and their range.

Rule Name	Physicochemical Properties	Range
Lipinski Rule of 5	Molecular Weight	≤ 500 Da
	logP	≤ 5
	Hydrogen Bond Donors	≤ 5
	Hydrogen Bond Acceptors	≤ 10
Ghose Rule	Molecular Weight	160 to 480 Da
	Number of Atoms	20 to 70
	logP	−0.4 to 5.6
	Molar Refractivity	40 to 130
Veber Rule	Number of Rotatable Bonds	≤ 10
	Topological Surface Area	≤ 140 $Å^2$
Egan Rule	logP	≤ 5.88
	Topological Surface Area	≤ 131.6 $Å^2$
Muegge Rule	Molecular Weight	200 to 600 Da
	logP	−2 to 5
	Topological Surface Area	≤ 150 $Å^2$
	Number of Rings	≤ 7
	Number of Carbons	> 4
	Number of Heteroatoms	> 1
	Number of Rotatable Bonds	≤ 15
	Hydrogen Bond Donors	≤ 5
	Hydrogen Bond Acceptors	≤ 10

Here is a concise summary of the rules outlined in Table 2. Apart from drug-likeness rules, Gastrointestinal absorption (GIA) and blood-brain barrier (BBB) access are two critical pharmacokinetic behaviours that play a significant role in determining the success of a drug candidate during the drug discovery and development process. The BOILED-Egg [57] model, which stands for Brain Or IntestinaL EstimateD permeation method, is put forward to predict the GIA and BBB for the BIMP Phytochemicals.

2.8 Phytochemical-Disease-Organism Correlation

Analyzing the therapeutic activity of Phytochemicals from a diverse range of plants provides a holistic understanding of their potential health benefits. The therapeutic activity of each phytochemical has been manually extracted from the comprehensive PubChem database. By examining the relationships between these compounds, diseases, and target genes in organisms, researchers can discover novel treatments and strengthen the scientific support for herbal medicine. This meticulous exploration not only facilitates the identification of promising natural remedies but also contributes to the advancement of personalized healthcare approaches tailored to individual needs.

3 Results and Discussion

We have prepared a curated list of ~15,000 Indian medicinal plants sourced from existing databases [45, https://envis.frlht.org, https://www.medicinalplants.in]. Of these, 2,559 are classified under Ayurveda, 2,267 under Siddha, 1,049 under Unani, 460 under Homeopathy, 671 under Sowa-Rigpa and 6,403 under Folk Indian Systems of Medicine. A list of 3,206 Indian Plants list was retrieved from the KNApSAcK database. Proceeding with the curation of the Phytochemicals list, we have effectively compiled a comprehensive list of 105,909 Phytochemicals from a selection of 6,209 Indian medicinal plants. Additionally, we acquired chemical identifiers from a diverse range of reputable sources, including ChEBI, ChEMBL, ChemSpider, CompTox Chemicals Dashboard v2.2.1, DrugBank, IMMPAT, KEGG, KNApSAcK, PubChem, SureChEMBL, and ZINC. Furthermore, we utilized ClassyFire to determine the chemical classifications of the Phytochemicals present in the BIMP database. It provides a hierarchical classification for each chemical compound, starting with the kingdom classification (organic or inorganic compounds). This is followed by superclass, class, and sub-class classification, creating a detailed and structured classification system for chemical compounds. The distribution of the Phytochemicals based on the kingdom and superclass has been represented in Fig. 1a and b respectively.

The 3D structures of the compounds, which do not have their 3D structures available in the PubChem database, were generated using Balloon and RDKit in conjunction with the application of the Merck Molecular Force Field (MMFF94) for the optimization process. Our initial step involved predicting the optimal conformer 3D structure for each phytochemical using Balloon, utilizing the MMFF94 force field and saving the results in SDF format. Subsequently, we used RDKit to convert the SDF files into various other formats, including PDB, MOL2, and XYZ, facilitating compatibility and versatility in the representation of the structures.

We have then calculated all the physicochemical properties listed in Table 1. The heavy atom count, molecular weight (Fig. 2a), and molecular formula offer valuable information on the compound's size and basic elemental composition, which are essential for understanding its pharmacokinetic and pharmacodynamic characteristics. Structural properties like ring count and rotatable bonds provide information about its spatial structure and flexibility, which is essential for optimizing its binding affinity and selectivity to target biomolecules. Properties such as the number of hydrogen bond donors (Fig. 2b) and acceptors (Fig. 2c), WlogP (Fig. 2d), and topological polar surface area influence the compound's solubility, membrane permeability, and distribution within the body, impacting bioavailability and efficacy. Molar refractivity (Fig. 2e) reveals details about the compound's steric volume and electron cloud density, impacting its lipophilicity and reactivity. A balance between hydrophilicity and lipophilicity, as indicated by the fraction C_{sp3}, is critical for achieving favourable drug-like properties. Furthermore, the presence of specific functional groups and aromatic systems influences the compound's interaction with biological targets, affecting its potency and specificity. Overall, understanding and leveraging these physicochemical properties are integral to the rational design and optimization of small molecules for therapeutic intervention in drug discovery.

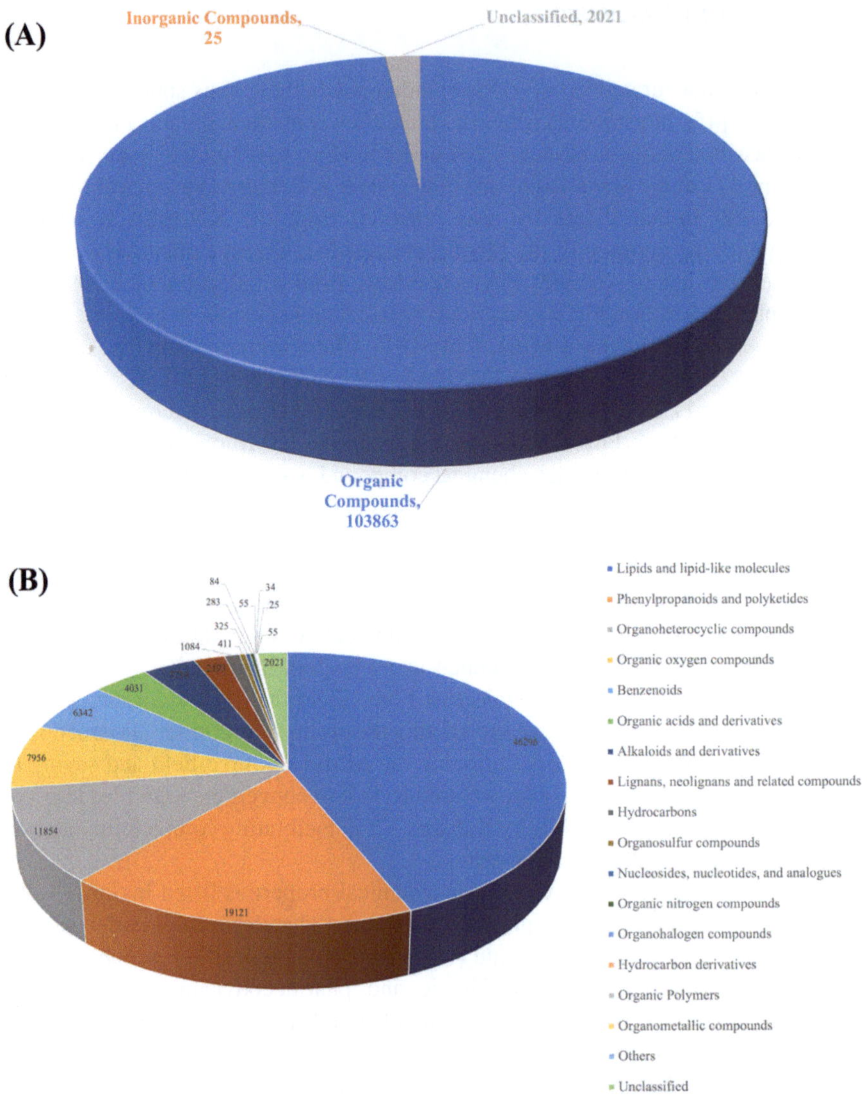

Fig. 1. Pie charts representing a distribution of compounds based on (A) Kingdom and (B) Super Classes.

We have then classified all the Phytochemicals within our dataset according to the drug-likeness criteria based on the calculated physicochemical properties. The pie chart (Fig. 3a) visually depicts two distinct categories of Phytochemicals within the dataset. The first category (Fig. 3a) comprises Phytochemicals that adhere to various drug-likeness rules, such as the Lipinski Rule of 5, Ghose Rule, Veber Rule, Egan Rule, and Muegge Rule. These rules serve as criteria to assess the potential suitability of these molecules for drug development. The second category represents Phytochemicals that do not align with these drug-likeness rules. A total of 33,591 molecules that did not

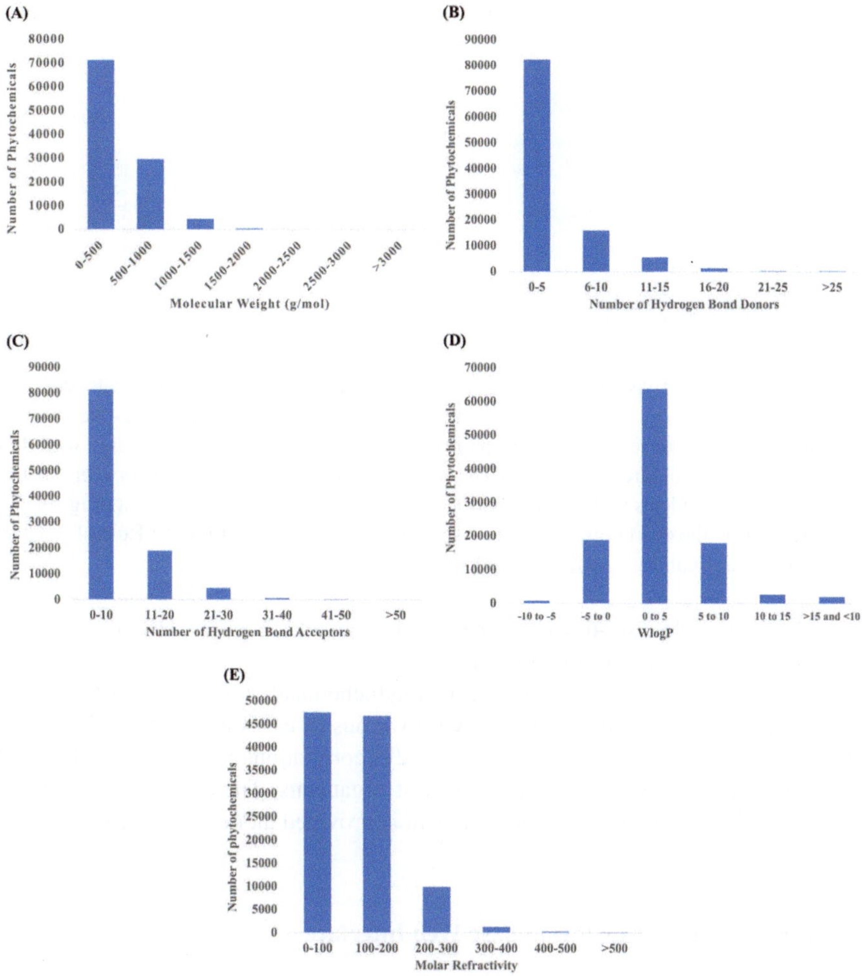

Fig. 2. Distribution of molecular descriptors of phytochemicals in BIMP (A) Molecular Weight, (B) Hydrogen Bond Donors (C) Hydrogen Bond Acceptors, (D) WLogP, and (E) Molar Refractivity.

conform to any of the established drug-likeness rules were identified. This subgroup of molecules may have properties that differ from the standard criteria used to assess their potential as drug candidates. Further investigation of the characteristics of these molecules may shed light on their unique features and potential applications.

The bar chart (Fig. 3b) presents a comparison between the total number of Phytochemicals and the distribution of Phytochemicals across different drug-likeness rules. Each bar on the chart corresponds to a specific rule. Specifically, we found that a total of 52,058 molecules adhered to Lipinski's rule of 5 (RO5), 45,817 to Ghose's rule, 66,020 to the Veber rule, 55,964 to Egan's rule, and 41,513 to Muegge's rule, with 33,031 molecules demonstrating alignment with all these rules, indicating broad com-

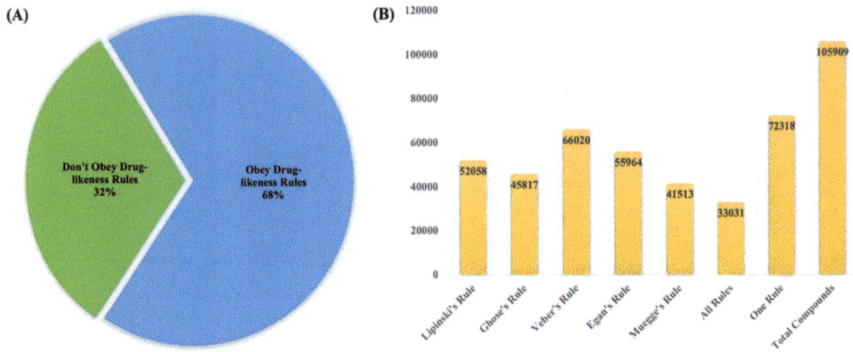

Fig. 3. (A) A pie chart visually depicting two distinct categories of phytochemicals within the dataset. The first category comprises phytochemicals that adhere to various drug-likeness rules (coloured sky blue). The second category represents phytochemicals that do not align with these drug-likeness rules (coloured green). (B) The bar chart presents a comparison between the total number of phytochemicals and the distribution of phytochemicals across different drug-likeness rules. Each bar on the chart corresponds to a specific rule, such as the Lipinski Rule of 5, Ghose Rule, Veber Rule, Egan Rule, and Muegge Rule.

pliance with multiple drug-likeness criteria, while an additional 72,318 molecules followed at least one of the drug-likeness rules.

We extracted the therapeutic data of each phytochemical present in BIMP. We found 30,429 relationships of Phytochemicals with various diseases involving 2,829 distinct disorders and diseases. Likewise, there are 750,299 correlations between Phytochemicals and target genes, including 611 different host organisms. The action of each phytochemical on the targets found in the organism is provided along with references to the research publications.

3.1 Introducing the BIMP Database Web Interface

An informative and user-friendly web interface (Fig. 4) utilizing PHP (5.5.2) [58], JavaScript [59], and HTML5 is presented. Apache (2.3) is employed as the foundational component to deliver webpage access services. The database encompassing all 105,909 Phytochemicals is named BIMP, which stands for Bioactivity information of the phytochemicals of Indian Medicinal Plants. The BIMP database offers two primary modes of exploration via the web server: the Dataset section (http://scfbio-iitd.res.in/bimp/Dataset.php) and the Search section (http://scfbio-iitd.res.in/bimp/Search.php). These subdivisions enable users to navigate the database based on Plant Family and Plant Name. Three additional subsections are available in the Search section: BIMP ID, Plant Name, and Physicochemical Properties. These subdivisions offer users distinct avenues to explore the database based on BIMP ID, Plant Name, and Physicochemical Properties, enhancing the precision and versatility of your search experience. Within the Plant Names subsection, users have the option to input queries such as botanical names or vernacular names in ten different Indian languages. These queries can be used to search within the database, providing a comprehensive and multilingual approach to

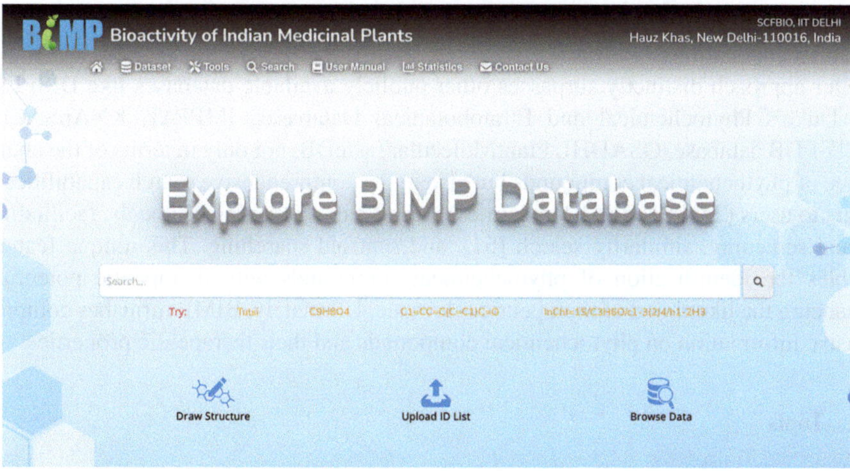

Fig. 4. Screenshot of the Homepage of the BIMP web server representing all the available features.

Table 3. The comparison of BIMP database over other publicly available databases on the basis of number of plant species, phytochemicals, compound-disease associations, and compound-target associations.

Database	Plants Species	Number of Phytochemicals	Compound-Disease Associations	Compound-Target Associations
BIMP	6,209	1,05,909	30,429	6,90,463
KNApSAck	3,685	~23,500	10,758	30,825
IMPPAT	1,742	9,596	-NA-	27,365
OSADHI	6,959	21,434	2,477	-NA-
IMPDB	113	1,737	-NA-	-NA-
NPACT	-NA-	1,574	-NA-	1,980
Phytochemica	5	963	-NA-	2,432
DISPEL	~5,500	-NA-	-NA-	-NA-
PlantMolecularTasteDB	-NA-	1,527	1,526	-NA-
MED-PDB database	147	369	-NA-	-NA-

searching for Phytochemicals. In the Descriptors subsection, users can search for Phytochemicals by specifying a range of physicochemical properties (Table 1).

The BIMP database comprehensively consolidates data pertaining to each phytochemical found in Indian medicinal plants across nine distinct association categories: Overview, Structure, Chemical Identifiers, Chemical and Physical Properties, Compound Classification, GIA and BBB, Druglikeness, Taxonomy and Similar Compounds. In addition to these functionalities, we have integrated 2D structure image representation of Phytochemicals. Additionally, we have successfully incorporated Mol* 3D Viewer [60] as the molecular visualizer for these Phytochemicals. These implementations enhance the user experience by providing visual insight into molecular structures, promoting a deeper understanding of compounds' characteristics and potential applications.

Our approach distinctly surpasses other publicly available databases like DISPEL, Dr. Duke's Phytochemical and Ethnobotanical Databases, IMPPAT, KNApSACK, MED-PDB database, OSADHI, PlantMolecularTasteDB, not only in terms of the abundance of phytochemical compounds but also in the comprehensive search capabilities it offers to users (Table 3). BIMP integrates three distinct screening protocols, facilitating virtual screening, similarity search [61], and scaffold searching. This unique feature enables the identification of phytochemical compounds with therapeutic potential, enhancing the likelihood of therapeutic relevance. Ultimately, BIMP furnishes comprehensive information on phytochemical compounds and their therapeutic properties.

3.2 Tools

We have dedicated a separate section to tools (Fig. 5a, b, and c) to highlight their significance in facilitating various computational analyses crucial for drug discovery and compound exploration within the BIMP database. We have implemented three tools: 1. Virtual Screening, 2. Similarity Search, and 3. Scaffolds Search. For Virtual Screening, we implemented our in-house RASPD+ [62, 63] software. Virtual screening is a computational technique used in drug discovery to identify potential drug candidates from large libraries of compounds. Rather than physically testing each compound, which can be time-consuming and costly, virtual screening uses computer programs and algorithms to predict which compounds are most likely to interact with the target proteins of interest [64]. In Virtual Screening, a protein with an active site or reference ligand is required for screening against BIMP Phytochemicals. In Similarity Search, similar molecules can be fetched from the entire BIMP database. In Scaffolds Search, BIMP Phytochemicals with the desired structural scaffold can be screened. Detailed user manual can be found on the web server (https://scfbio-iitd.res.in/bimp/Manual.php).

4 Conclusions

In conclusion, the BIMP database represents a carefully curated, organized, comprehensive and valuable resource in the field of medicinal plants for drug discovery. BIMP provides a rich repository of information covering a wide range of Phytochemicals, their properties, associations, and potential applications. Its diverse sections, including general information, physicochemical properties, drug analogues, known targets, and predicted targets, provide a comprehensive view of these compounds and their interactions. The BIMP database is expected to contribute significantly to advancing medicinal plant research, drug discovery and related areas. It provides a powerful tool for researchers, scientists and enthusiasts who are ready to uncover the potential of Indian medicinal plants and their profound impact on human health and well-being.

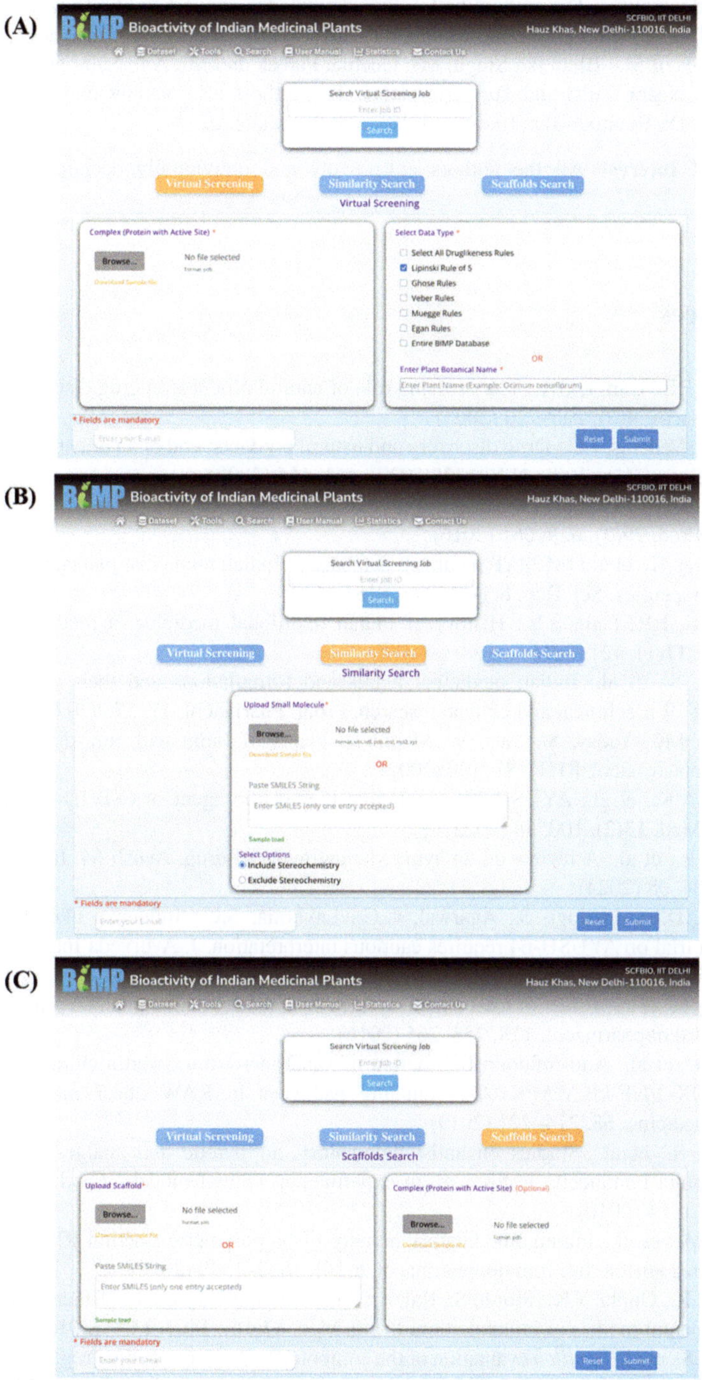

Fig. 5. Screenshots of tools integrated in BIMP (A) Virtual Screening, (B) Similarity Search, and (C) Scaffolds Search.

Acknowledgements. The authors gratefully acknowledge support to SCFBio, IIT Delhi from NSM, MEITY & DST, Govt of India and DBT, Govt. of India. The authors also acknowledge the initial studies of Ms. Bhumika Singh, Ms. Geetika Pareek and Mr. Anil Saini and Ms. Vandana Shekhar on Neem, Tulsi and Turmeric plants. The authors also acknowledge Ms. Khushboo Sharma and Dr. Ruchika Bhat for the validation of the database.

Conflict of Interest All the authors collectively and individually declare no conflict of interest.

References

1. Koehn, F.E., Carter, G.T.: The evolving role of natural products in drug discovery. Nat. Rev. Drug Discov. **4**(3), 206–220 (2005)
2. Li, J.W., Vederas, J.C.: Drug discovery and natural products: end of an era or an endless frontier? Science (New York, N.Y.). **325**(5937), 161–165 (2009)
3. Newman, D.J., Cragg, G.M.: Natural products as sources of new drugs from 1981 to 2014. J. Nat. Prod. **79**(3), 629–661 (2016)
4. Mohanraj, K., et al.: IMPPAT: a curated database of Indian medicinal plants, Phytochemistry and therapeutics. Sci. Rep. **8**(1), 4329 (2018)
5. Adhikari, P.P., Paul, S.B.: History of Indian traditional medicine: a medical inheritance. History. **11**(1), 421 (2018)
6. Ahmad, S., et al.: Indian medicinal plants and formulations and their potential against COVID-19-preclinical and clinical research. Front. Pharmacol. **11**, 578970 (2021)
7. Grover, J.K., Yadav, S., Vats, V.: Medicinal plants of India with anti-diabetic potential. J. Ethnopharmacol. **81**(1), 81–100 (2002)
8. Panda, A.K., et al.: AYUSH-64: a potential therapeutic agent in COVID-19. J. Ayurveda Integr. Med. **13**(2), 100538 (2022)
9. Allam, R., et al.: A review on an ayurvedic antimalarial drug: Ayush-64. Int. J. Mosq Res. **11**(1), 04–08 (2024)
10. Pandey, D.N., Rastogi, S., Agarwal, G.G., Lakhotia, S.C.: Influenza like illness related clinical trial on AYUSH-64 requires cautious interpretation. J. Ayurveda Integr. Med. **13**(1), 100346 (2022)
11. Zhao, Y.L., et al.: Effect of total alkaloids from Alstonia scholaris on airway inflammation in rats. J. Ethnopharmacol. **178**, 258–265 (2016)
12. Hu, T.Y., et al.: Anti-inflammation action of xanthones from Swertia chirayita by regulating COX-2/NF-κB/MAPKs/Akt signaling pathways in RAW 264.7 macrophage cells. Phytomedicine. **55**, 214–221 (2019)
13. Shukla, S., et al.: Studies on anti-inflammatory, antipyretic and analgesic properties of Caesalpinia bonducella F. Seed oil in experimental animal models. Food Chem. Toxicol. **48**(1), 61–64 (2010)
14. Gupta, A., et al.: Immunomodulatory activity of biopolymeric fraction RLJ-NE-205 from Picrorhiza kurroa. Int. Immunopharmacol. **6**(10), 1543–1549 (2006)
15. Kumar, R., Gupta, Y.K., Singh, S., Raj, A.: Anti-inflammatory effect of Picrorhiza kurroa in experimental models of inflammation. Planta Med. **82**(16), 1403–1409 (2016)
16. Ram, T.S., et al.: *In silico* evaluation of the compounds of the ayurvedic drug, AYUSH-64, for the action against the SARS-CoV-2 main protease. J. Ayurveda Integr. Med. **13**(1), 100413 (2022)-->

17. Srikanth, N., et al.: Disease characteristics, care-seeking behavior, and outcomes associated with the use of AYUSH-64 in COVID-19 patients in home isolation in India: a community-based cross-sectional analysis. Front. Public Health. **10**, 904279 (2022)
18. Singh, H., et al.: AYUSH-64 as an adjunct to standard care in mild to moderate COVID-19: an open-label randomized controlled trial in Chandigarh, India. Complement. Ther. Med. **66**, 102814 (2022)
19. Reddy, R.G., et al.: AYUSH-64 as an add-on to standard care in asymptomatic and mild cases of COVID-19: a randomized controlled trial. AYU. **41**(2), 107–116 (2020)
20. Sinha, S., et al.: IMPDB: Indian medicinal phytochemical database curated for drug designing. J. Comput. Biophys. Chem. **21**, 709 (2022)
21. Gu, J., et al.: CVDHD: a cardiovascular disease herbal database for drug discovery and network pharmacology. J. Cheminform. **5**(1), 51 (2013)
22. Singh, K., et al.: DISPEL: database for ascertaining the best medicinal plants to cure human diseases. Database (Oxford). **2023**, baad073 (2023)
23. Lans, C., van Asseldonk, T.: Dr. Duke's phytochemical and ethnobotanical databases, a cornerstone in the validation of Ethnoveterinary medicinal plants, as demonstrated by data on pets in British Columbia. In: Máthé, Á. (ed.) Medicinal and Aromatic Plants of North America. Medicinal and Aromatic Plants of the World, vol. 6, pp. 219–246. Springer, Cham (2020)
24. Nakamura, K., et al.: KNApSAcK-3D: a three-dimensional structure database of plant metabolites. Plant Cell Physiol. **54**(2), e4 (2013)-->
25. Gahlot, L.K., et al.: MED-PDB: an online database of medicinal plants. J. Adv. Pharm. Res. **7**(3), 204–207 (2017)
26. Mangal, M., et al.: NPACT: naturally occurring plant-based anti-cancer compound-activity-target database. Nucleic Acids Res. **41**, D1124–D1129 (2013)
27. Kiewhuo, K., et al.: OSADHI - an online structural and analytics based database for herbs of India. Comput. Biol. Chem. **102**, 107799 (2023)
28. Pathania, S., Ramakrishnan, S.M., Bagler, G.: Phytochemica: a platform to explore phytochemicals of medicinal plants. Database (Oxford). **2015**, bav075 (2015)
29. Gradinaru, T.C., Petran, M., Dragos, D., Gilca, M.: PlantMolecularTasteDB: a database of taste active phytochemicals. Front. Pharmacol. **12**, 751712 (2022)
30. Chen, C.Y.: TCM database@Taiwan: the world's largest traditional Chinese medicine database for drug screening in silico. PLoS One. **6**(1), e15939 (2011)
31. Xue, R., et al.: TCMID: traditional Chinese medicine integrative database for herb molecular mechanism analysis. Nucleic Acids Res. **41**, D1089–D1095 (2013)
32. Heller, S.R., et al.: InChI, the IUPAC international chemical identifier. J. Cheminform. **7**, 23 (2015)
33. Weininger, D.: SMILES, a chemical language and information system. 1. Introduction to methodology and encoding rules. J. Chem. Inf. Comput. Sci. **28**(1), 31–36 (1988)
34. Lipinski, C.A., Lombardo, F., Dominy, B.W., Feeney, P.J.: Experimental and computational approaches to estimate solubility and permeability in drug discovery and development settings. Adv. Drug Deliv. Rev. **46**(1–3), 3–26 (2001)
35. Ghose, A.K., Viswanadhan, V.N., Wendoloski, J.J.: A knowledge-based approach in designing combinatorial or medicinal chemistry libraries for drug discovery. 1. A qualitative and quantitative characterization of known drug databases. J. Comb. Chem. **1**(1), 55–68 (1999)
36. Veber, D.F., et al.: Molecular properties that influence the oral bioavailability of drug candidates. J. Med. Chem. **45**(12), 2615–2623 (2002)
37. Egan, W.J., Merz Jr., K.M., Baldwin, J.J.: Prediction of drug absorption using multivariate statistics. J. Med. Chem. **43**(21), 3867–3877 (2000)
38. Muegge, I.: Selection criteria for drug-like compounds. Med. Res. Rev. **23**(3), 302–321 (2003)

39. Olsson, T., Oprea, T.I.: Cheminformatics: a tool for decision-makers in drug discovery. Curr. Opin. Drug Discov. Devel. **4**(3), 308–313 (2001)
40. Jayaram, B., et al.: Sanjeevini: a freely accessible web-server for target directed lead molecule discovery. BMC Bioinformatics. **17**(Suppl 17), S7 (2012)
41. Jayaram, B., et al.: Sanjeevini: a comprehensive active-site directed lead design software. Indian J. Chem. **45**, 21–33 (2006)
42. Khare, C.P.: Indian Medicinal Plants: an Illustrated Dictionary. Springer (2007)
43. Schoch, C.L., et al.: NCBI taxonomy: a comprehensive update on curation, resources and tools. Database (Oxford). **2020**, baaa062 (2020)
44. Federhen, S.: The NCBI taxonomy database. Nucleic Acids Res. **40**, D136–D143 (2012)
45. Kim, S., et al.: PubChem 2023 update. Nucleic Acids Res. **51**(D1), D1373–D1380 (2023)
46. Degtyarenko, K., et al.: ChEBI: a database and ontology for chemical entities of biological interest. Nucleic Acids Res. **36**, D344–D350 (2008)
47. Gaulton, A., et al.: ChEMBL: a large-scale bioactivity database for drug discovery. Nucleic Acids Res. **40**, D1100–D1107 (2012)
48. Pence, H.E., Williams, A.: ChemSpider: an online chemical information resource. J. Chem. Educ. **87**(11), 1123–1124 (2010)
49. Williams, A.J., et al.: The CompTox chemistry dashboard: a community data resource for environmental chemistry. J. Cheminform. **9**(1), 61 (2017)
50. Wishart, D.S., et al.: DrugBank 5.0: a major update to the DrugBank database for 2018. Nucleic Acids Res. **46**(D1), D1074–D1082 (2018)
51. Kanehisa, M., Goto, S.: KEGG: Kyoto encyclopedia of genes and genomes. Nucleic Acids Res. **28**(1), 27–30 (2000)
52. Papadatos, G., et al.: SureChEMBL: a large-scale, chemically annotated patent document database. Nucleic Acids Res. **44**(D1), D1220–D1228 (2016)
53. Irwin, J.J., Shoichet, B.K.: ZINC - a free database of commercially available compounds for virtual screening. J. Chem. Inf. Model. **45**(1), 177–182 (2005)
54. Djoumbou Feunang, Y., et al.: ClassyFire: automated chemical classification with a comprehensive, computable taxonomy. J. Cheminform. **8**, 61 (2016)
55. Vainio, M.J., Johnson, M.S.: Generating conformer ensembles using a multiobjective genetic algorithm. J. Chem. Inf. Model. **47**(6), 2462–2474 (2007)
56. Halgren, T.A.: Merck molecular force field. I. Basis, form, scope, parameterization, and performance of MMFF94. J. Comp. Chem. **17**, 490–519 (1996)
57. Daina, A., Zoete, V.: A BOILED-egg to predict gastrointestinal absorption and brain penetration of small molecules. ChemMedChem. **11**(11), 1117–1121 (2016)
58. Bakken, S.S., Suraski, Z., Schmid, E.: PHP Manual: Volume 1. iUniverse, Incorporated (1997)
59. Flanagan, D.: JavaScript: the Definitive Guide. Reilly Media, Inc (2006)
60. Sehnal, D., et al.: Mol* viewer: modern web app for 3D visualization and analysis of large biomolecular structures. Nucleic Acids Res. **49**(W1), W431–W437 (2021)
61. Tanimoto, T.T.: IBM internal report. Nov, 17 (1957)
62. Mukherjee, G., Jayaram, B.: A rapid identification of hit molecules for target proteins via physico-chemical descriptors. Phys. Chem. Chem. Phys. **15**(23), 9107–9116 (2013)
63. Holderbach, S., et al.: RASPD+: fast protein-ligand binding free energy prediction using simplified physicochemical features. Front. Mol. Biosci. **7**, 601065 (2020)
64. Kattuparambil, A.A., et al.: Exploring chemical space for "druglike" small molecules in the age of AI. Front. Mol. Biosci. **12**, 1553667 (2025)

Current Trends in Healthcare Technology

Alok Kumar Saxena[1] (✉)

[1] AA Consultants,
1419-Milano, Mahagun Mascot, Crossings Republik, Ghaziabad, Uttar Pradesh, India[1]
[2] Consultant GTC Vorro Technology, 900 Threadneedle St Suite 500
Houston, TX, USA
alok.saxena@aaconsultants.in

Abstract. With the rise of Artificial Intelligence (AI), all sectors are poised for significant transformation and evolution of new business models. Pharmaceutical and healthcare industries are no exception to it. While the pharma sector has faced challenges in recent years, which have slowed the adoption of new technologies, it is now positioning itself for substantial change. These changes are necessary to address current issues such as the need for cost-effective medicines, their shortages, sudden demand fluctuations, and the sustainability of long-term business models.

Emerging trends in healthcare technology include:

- **Remote Consultation and Monitoring** through telehealth and telemedicine platforms.
- **Internet of Medical Things (IoMT)**, driven by wearable devices and connected sensors for continuous patient monitoring.
- **Artificial Intelligence (AI)** enabling predictive analytics and early detection of health conditions.
- **Advanced Technologies** such as simulations for medical procedures, pills with internal monitoring capabilities, and targeted treatments using nanomedicines.
- **Remote Patient Monitoring** systems designed to ease the burden on healthcare facilities.

This paper provides an overview of these key trends reshaping the pharmaceutical and healthcare sectors.

Keywords: Health Care · Artificial Intelligence · Health System; NLP

1 Introduction

A major transformation is underway in the pharmaceutical and healthcare industries, driven by advanced computing technologies, faster access to vast datasets, and the adoption of blockchain technology. While the sector has traditionally been cautious in embracing new technologies, it is now poised to shape a new future, leveraging these innovations to drive change and progress.

A. K. Saxena, A. Saxena (Eds.): GTHTM 2025, pp. 299–309, 2026.
https://doi.org/10.1007/978-3-032-12320-6_17

In the current scenario the industry is experiencing significant growth, making it an increasingly attractive space for investment, with a surge in start-ups and collaborations driving innovation. While the sector has faced numerous challenges in recent years, these obstacles have pushed companies to adopt more innovative approaches and respond quickly to market demands and pressures.

Recent challenges have included the global pandemic, which placed exceptional demands on pharmaceutical companies, particularly in terms of rapidly scaling vaccine production despite limited clinical trial opportunities. Furthermore, the worldwide shortage of personal protective equipment (PPE) exposed significant vulnerabilities within the supply chain, a situation that was further exacerbated by the crisis.

Amid such pressures and challenges, the pharmaceutical and healthcare industries are increasingly adapting to new technologies and processes, responsible for current trends. They are actively developing solutions to address medicine shortages and respond to sudden demand surges in a sustainable manner.

2 Factors Influencing Pharma and Healthcare Industry Current Trends

Having faced recent challenges, the pharmaceutical industry is now under external pressure to prove its resilience in time to come. Such pressures are rather obvious with mutating viruses and reporting of new viruses every alternate year. For example, China steps up emergency measures amid new Human metapneumo virus (HMPV) leading to respiratory issues specifically in relatively younger population. This comes five years after the world was first alerted to the emergence of a novel coronavirus in Wuhan, China, which later turned into a global pandemic with seven million deaths reported. Under such concerns and pressures the key issues to address and re-shape the pharma and health-care industry are as below:

2.1 Patients Expectations

According to the 2024 ZS Future of Health Report [1], which is based on a survey of 9,500 healthcare consumers and 1,400 healthcare providers across various countries, there are common challenges and opportunities within healthcare, despite the significant differences in healthcare systems. The survey reveals that, regardless of location or system, there are people's similar concerns and expectations. Consumers and providers alike are calling for solutions that create a more accessible, equitable, affordable, connected, and human-centered healthcare system.

Therefore, healthcare must be accessible, affordable, and equitable, which has become the cornerstone of emerging trends in healthcare systems. This emphasis stems from the common perceptions and attitudes observed among the surveyed population, where accessibility, cost, and fairness are seen as critical factors in shaping overall views of the healthcare system. These elements play a pivotal role in driving the need for transformative change within the sector. This level of exception is addressable by following key points as pin-pointed by surveyed people.:

- Preventive care that's accessible for everyone
- Greater coordination of care between providers
- Seamless access to medical records and histories
- Transparent billing practices
- Easier appointment scheduling

Many individuals across the countries believe that connected health tools have the potential to help prevent illness and promote better overall health. Moreover, there is a shared understanding between healthcare providers and consumers regarding the importance of preventive care, with both groups agreeing that it will be essential in enhancing the healthcare experience for individuals moving forward.

2.2 Reform of Health Systems

For enhanced affordability of healthcare system, it is rather imperative to ensure drugs availability at affordable prices. Generic Medicines in India is the buzz trend to meet this objective. Dr. Reddy's Laboratory has set targets of ~130 Billion US$ in 2025 which is expected to grow to 350 Billion US$ by 2047 for such medicines. This is happening in a context where drug manufacturers are facing increasingly stringent access and pricing regulations across key regions, as health systems implement reforms aimed at containing healthcare spending. In the U.S., the government has announced the first 10 products including 'healthcare costs' that will be subject to price negotiations under the Inflation Reduction Act (IRA). Proposed legislation in the EU seeks to reduce the exclusivity window for drug companies from 10 years to 8 years for new drugs that are not launched in all 27 member states within two years of their initial release. In Germany, the government is working to address perceived gaps in its national pharmaceutical regulations ("AMNOG 2.0") through additional reforms aimed at controlling drug costs. Similarly, in Italy, the long-awaited reorganization of the Italian Medicines Agency (AIFA) is currently in progress. In Japan and China as well, governments are pursuing comparable initiatives to reduce drug prices. These widespread, similar initiatives contribute to growing uncertainty within the pharmaceutical industry, prompting companies to reevaluate their business models.

2.3 Demographic Shifts

In the present era when mankind has found various means to interfere with nature, many countries are facing significant socioeconomic challenges due to a combination of rapidly aging populations and declining birth rates. India's population growth is slowing rapidly, with the annual growth rate dropping from 2.5% in 1971–81 to approximately 1.3% today. The population is projected to peak at 1.7 billion in 2060 before beginning to decline. These demographic shifts are profoundly impacting India's healthcare system. The population is aging rapidly, with the elderly population expected to increase from 139 million in 2021 to 319 million by 2050. This transition presents several critical challenges for healthcare such as:

2.3.1 Double Burden of Disease. India's healthcare system faces a complex challenge due to:

- A rising prevalence of non-communicable diseases (NCDs)
- The continued presence of communicable diseases
- Increasing chronic illness rates, with 23% of the elderly suffering [2] from chronic conditions

2.3.2 Healthcare System Challenges. This demographic shift places significant pressure on India's healthcare infrastructure, marked by:

- Uneven healthcare access between rural and urban [3] areas
- Limited health insurance coverage, with only 37% of the population insured [3]
- Insufficient public healthcare funding
- A growing demand for geriatric and palliative [4] care

2.3.3 Policy Implications. Focus on prioritizing the following actions to address such demographic shifts:

- **Reform Healthcare Systems**: Establish and implement robust national policies aimed at strengthening healthcare delivery, ensuring efficiency, and promoting equitable access for all.
- **Ensure Universal Healthcare Access**: Adopt comprehensive strategies to guarantee healthcare services reach every citizen, especially marginalized and underserved communities.
- **Develop Community-Based Health Insurance Schemes**: Create innovative and affordable health insurance programs tailored to local needs, empowering communities to access necessary care without financial strain.
- Invest **in Healthcare Infrastructure**: Significantly boost investments in healthcare infrastructure, including hospitals, clinics, and telemedicine facilities, ensuring quality care is available across urban and rural areas.
- **Leverage Emerging Technologies**: Integrate cutting-edge technologies, such as AI-driven monitoring and predictive analytics, to enhance healthcare delivery, improve diagnostics, and enable better decision-making.
- The demographic transition requires a coordinated effort, involving policymakers, the private sector, and local communities. Only through collaboration can we build a sustainable healthcare future that meets the needs of India's growing and aging population.

Though India's demographic shifts are very different from other countries, the challenges and policies are similar. Spain, for example, expects a complete reversal in the contribution of its labor force to GDP growth—from a positive 0.6% between 2000 and 2018 to a projected negative 0.6% by 2040—shifting from a historical labor-driven boost to a future economic burden.; Meanwhile, Switzerland also anticipates a 30% increase in public health expenditures by 2050, primarily driven by the rising costs associated with long-term care. Countries like South Korea and China are already facing the economic challenges of a shrinking working-age population, which is putting pressure on their ability to generate the tax revenues, required to support healthcare services for their rapidly growing elderly populations.

These demographic shifts highlight the pressing need to invest in preventive measures and proactive "well care" to mitigate future healthcare costs associated with treating illness.

3 Emerging Trends in Pharma and Healthcare Industry

In view of the Factors highlighted in Sect. 2 the current trends in Pharma and Healthcare [5] Industry are:

3.1 Increased R&D Focus and Investments

In India, pharmaceutical sector has experienced significant growth [6], driven by a rising population, substantial R&D investments, a modernized manufacturing base, and the increasing prevalence of chronic diseases. At the heart of this success lies a strong commitment to research and development, which has played a pivotal role in recent years. In the wake of COVID-19, pharmaceutical companies have further amplified their R&D investments, accelerating innovation and shaping the future of healthcare.

Several leading pharmaceutical companies have set up R&D centres in India, positioning the country as a global innovation hub. Pfizer's R&D centre in Chennai specializes in drug discovery and development, while Novartis' facility in Hyderabad is advancing novel drug research, backed by a planned $100 million investment. These initiatives underscore India's growing role in pharmaceutical innovation and development.

Meanwhile, Merck has inaugurated its Healthcare R&D Excellence Centre in Bengaluru, leveraging innovation to enhance healthcare solutions. Similarly, Bayer's Data Science and Analytics Centre in Hyderabad plays a crucial role in global pharmaceutical R&D, driving the development of new, innovative therapies to address unmet patient needs.

To boost the pharmaceutical industry, the Government of India introduced the Production Linked Incentive (PLI) scheme for pharmaceuticals in recent past, with a financial outlay of ₹15,000 crore over six years. In the latest budget, the government further reinforced its commitment to innovation by allocating ₹3,201 crore ($419.2 million) for research.

Notably, four pharmaceutical companies—Dr. Reddy's Laboratories, Biocon, Strides Pharma Science, and Premier Medical Corporation—have received the first tranche of incentives, totalling up to ₹165.74 crore, under the Department of Pharmaceuticals' PLI scheme. This initiative aims to strengthen domestic production of active pharmaceutical ingredients (APIs), which are essential raw materials for drug manufacturing.

As opined by pharma experts [6], in recent years, India has emerged as a prime destination for pharmaceutical investment in innovation. Despite some challenges, pharma companies continue to make significant R&D investments, particularly in drug discovery and development. With its strong capabilities in research, supportive government policies, and a vast pool of skilled professionals, India is well-positioned to become a global leader in pharmaceutical innovation in the years ahead.

On Global arena too, the pharmaceutical industry is undergoing a transformative phase, driven by key investment trends that are reshaping its future. These trends highlight a growing focus on innovative technologies, personalized medicine, and strategic collaborations, all aimed at accelerating drug development and improving delivery systems. Investors are increasingly favouring companies that specialize in targeted therapies and personalized treatments designed to meet the unique needs of individual patients. This shift is largely driven by breakthroughs in genetic research and a rising demand for more effective healthcare solutions. Notably, investment in biopharmaceuticals—especially in areas such as gene therapy and regenerative medicine—has rebounded strongly after a slowdown in 2022 and early 2023 [7].

3.2 Big Data Analytics in Pharma Innovation

Big Data analytics is basically systematic process of examining large, complex data sets to uncover valuable insights and patterns. Data analytics is revolutionizing the pharmaceutical industry by driving innovation at every stage of drug development, from discovery to manufacturing and marketing. By harnessing the power of big data, pharmaceutical companies can make informed decisions, optimize processes, and ultimately improve patient outcomes. This transformation is enabling greater efficiency and precision in developing new therapies.

Data analytics now plays a key role in accelerating drug discovery [8] by allowing researchers to analyse vast datasets, including scientific publications and clinical trial results. This capability helps identify promising drug candidates and predict their efficacy, significantly speeding up the selection process while minimizing the risks associated with investing in less viable compounds. Predictive analytics further refines this process by guiding researchers toward the most promising avenues, ultimately reducing both the time and cost of bringing new drugs to market.

3.3 Artificial Intelligence

Artificial Intelligence is a branch of computer science focused on creating systems that can perform tasks typically requiring human intelligence. This Enables machines to simulate human abilities like learning, problem solving, decision making and comprehension.

Artificial Intelligence (AI) is revolutionizing drug discovery and development by accelerating processes and enhancing efficiency. AI enables the automation and optimization of manufacturing, while also aiding in the strategic planning of product launches. It plays a crucial role at every stage of drug design, minimizing health risks associated with clinical trials and significantly reducing costs.

3.4 NLP Aided Advanced Communication

Natural Language Processing (NLP) is a crucial subfield of **Artificial Intelligence (AI)** that bridges the gap between human communication and machine understanding. It empowers computers to comprehend, interpret, and generate human language in a meaningful way, enabling more natural and seamless interactions between humans and technology.

Natural Language Processing (NLP) is revolutionizing the pharmaceutical industry by enabling the extraction and analysis of vast amounts of unstructured data. This advancement enhances critical processes, from drug discovery to clinical trials, driving efficiency and innovation.

3.5 Precision Medicine

Precision medicine, also known as personalized or individualized medicine, is a cutting-edge approach to healthcare that customizes medical treatment based on each patient's unique characteristics. This methodology integrates genetic information, environmental factors, and lifestyle choices to enhance disease prevention, diagnosis, and treatment, leading to more effective and targeted care.

Key features of precision medicine are:

- **Individualized treatment** which is not an approach like one size fits all. Precision medicine acknowledges that patients with the same condition may respond differently to the same treatment. By analyzing genetic and molecular data, healthcare providers can identify targeted therapies that are more likely to be effective for specific patient groups, leading to more personalized and efficient treatment strategies.
- **Use of Big Data** which means precision medicine harnesses vast datasets from diverse sources, including genomic information, electronic health records, and lifestyle data. This comprehensive analysis enables healthcare providers to identify patterns, refine treatment decisions, and improve risk assessments, leading to more precise and effective patient care.
- **Focus on Prevention**: A key goal of precision medicine is to enhance preventive care by identifying individuals at risk for certain diseases based on their genetic predispositions. This proactive approach enables earlier interventions, personalized prevention strategies, and improved health outcomes, ultimately reducing the burden of disease.

- This way **precision medicine** marks a paradigm shift in healthcare, focusing on more personalized approaches that account for the unique attributes of each patient. By integrating insights from genomics, environmental science, and lifestyle factors, this approach aims to enhance health outcomes through tailored care strategies. As technology evolves and more data becomes accessible, the potential for precision medicine to revolutionize treatment continues to expand, offering new possibilities for more effective and individualized healthcare.

3.6 Integration of Blockchain Technology

Blockchain technology is basically a decentralized, distributed digital ledger system that securely records transactions across a network of computers. It ensures data integrity by making information immutable—once a transaction is recorded, it cannot be altered or deleted, providing a high level of security and transparency. This it facilitates decentralization and at the same time ensures Immutability and Transparency.

These key features of the Blockchain technology are gaining recognition for its potential to transform the pharmaceutical industry, tackling key challenges such as supply chain integrity, data security, and regulatory compliance. By leveraging its decentralized and transparent nature, blockchain offers innovative solutions to improve trust and efficiency across the sector. Here are the key roles blockchain plays in pharma industry [9]:

- Enhancing supply chain transparency and security.
- Improving Clinical Trials and Data Integrity
- Facilitating Regulatory Compliance
- Addressing prescription misuse or abuse

3.7 Digital Transformation Across Businesses

Digital transformation is a strategic approach that enables organizations to integrate digital technologies across all facets of their operations. This shift aims to modernize workflows, enhance customer experiences, and drive business growth by leveraging innovation, automation, and data-driven decision-making. Key components and implications of digital transformation across businesses are:

- Integration **of Technologies** (AI, Machine Learning, IoT). This integration helps organizations adapt to changing market conditions and customer demands [10].
- **Cultural Shifts**: Digital transformation extends beyond technology; it demands a cultural shift within organizations. It requires a fundamental rethinking of how businesses operate, engage with customers, and deliver value. For successful implementation, employees at all levels must embrace this change, fostering a culture of innovation, collaboration, and agility.

Digital transformation is revolutionizing the pharmaceutical industry too, reshaping how companies develop, manufacture, and deliver medications. By integrating advanced technologies, the industry is enhancing efficiency, improving patient outcomes, and driving innovation. Here are some notable facts of this transformation:

- **Response to COVID-19:** The pandemic acted as a catalyst for digital transformation in the pharmaceutical industry, highlighting the need for rapid adaptation in drug development and distribution. Companies leveraged digital tools [11] to accelerate vaccine development, streamline supply chains, and enhance remote interactions with healthcare professionals and patients, ensuring continued access to critical treatments.

- Technological **Advancements:** The adoption of cutting-edge technologies such as **Artificial Intelligence (AI), Machine Learning (ML), Cloud Computing, and Big Data Analytics** is revolutionizing pharmaceutical operations [12]. These innovations enhance data management, improve decision-making, and streamline processes, ultimately driving efficiency, accuracy, and innovation across the industry.

This digital transformation has brought benefits of increased efficiency, enhanced drug development and improved patient engagement [13], though the challenges are cultural resistance and data security concerns [13].

3.8 Requirement of Reduced Carbon Footprint

The pharmaceutical industry is placing greater emphasis on reducing its carbon footprint in response to growing environmental concerns and regulatory pressures. To achieve sustainability, companies are adopting innovative strategies and initiatives aimed at mitigating carbon emissions. Here are the key approaches being implemented:

- **Continuous Manufacturing:** Shifting from traditional batch manufacturing to continuous manufacturing is a highly effective strategy for reducing carbon emissions in the pharmaceutical industry. By integrating multiple production stages into a seamless process, continuous manufacturing significantly lowers energy consumption and minimizes waste. Leading companies such as **Amgen and Sanofi** have adopted this approach, with reports indicating up to an **80%** [14] **reduction in carbon emissions** compared to conventional methods.
- **Adoption of Renewable Energy:** Pharmaceutical companies are increasingly committing to renewable energy sources such as **solar, wind, and geothermal power** to reduce their environmental impact. Initiatives like **Schneider Electric's Energize program** are helping pharma firms gain better access to sustainable energy solutions [14]. Industry leaders such as **Johnson & Johnson and Pfizer** have set ambitious goals for achieving **carbon neutrality**, with plans to transition to **100% renewable energy** within specified timeframes.
- **Sustainable Sourcing and Green Chemistry:** Prioritizing the sustainable sourcing of raw materials and integrating **green chemistry principles** can significantly reduce the environmental footprint [15] of pharmaceutical production. This approach focuses on **minimizing waste, reducing solvent use, and optimizing resource efficiency** throughout the supply chain, leading to more eco-friendly and sustainable manufacturing practices.
- **Circular Economy Practices:** Transitioning from a **linear** to a **circular economy model** promotes the **reuse and recycling** of materials, reducing waste and conserving valuable resources [16]. By adopting this sustainable approach, pharmaceutical companies can **lower carbon emissions, minimize environmental impact, and achieve long-term cost savings**, making their operations more efficient and eco-friendlier.

4 Conclusion

The pharmaceutical and healthcare industries are set for major changes soon, spurred by the COVID-19 pandemic. Rapid digitization, the integration of blockchain technology, big data analytics, and artificial intelligence (AI) are the primary forces driving this shift. The foundation for this digital evolution in the pharmaceutical sector was laid during the pandemic under immense pressure and is now steadily taking shape with the aid of AI.

This transformation is further fuelled by key factors such as changing patient expectations, demographic shifts, and the implementation of more stringent healthcare systems. These elements are leading to increased investments in research and development, which are expected to further accelerate the evolution of healthcare systems worldwide.

Innovative approaches, such as precision medicine and remote consultations, are becoming a reality and are aligned with broader objectives, such as reducing the carbon footprint of healthcare.

References

1. Future of healthcare insights: Health and patient experience. In: Www.zs.com. https://www.zs.com/zs-future-of-health-report-on-what-people-want-from-healthcare. Accessed 12 July 2025
2. Rangan, R.K.: India's demographic shift: Redefining retirement in the 21st century. In: The Sunday Guardian Live. https://sundayguardianlive.com/business/indias-demographic-shift-redefining-retirement-in-the-21st-century. Accessed 12 July 2025 (2025)
3. Ghia, C., Rambhad, G.: Implementation of equity and access in Indian healthcare: current scenario and way forward. J. Mark Access Health Policy. **11**, 2194507 (2023). https://doi.org/10.1080/20016689.2023.2194507
4. Sahoo, P.M., Rout, H.S., Jakovljevic, M.: Consequences of India's population aging to its healthcare financing and provision. J. Med. Econ. **26**, 308–315 (2023). https://doi.org/10.1080/13696998.2023.2178164
5. Innopharma Education.: Incredible pharma industry trends and innovations that are shaping the industry. In: Innopharma Education. https://www.innopharmaeducation.com/blog/incredible-pharma-industry-trends-and-innovations-that-are-shaping-the-industry. Accessed 12 July 2025 (2023)
6. Could 7.5% R&D Spends Make India Pharma Innovation Hub? In: Biospectrumindia.com. https://www.biospectrumindia.com/features/73/23024/could-7-5-rd-spends-make-india-pharma-innovation-hub.html. Accessed 12 July 2025b
7. What are the current investment trends in the pharmaceutical industry, and how are they influencing the sector's direction? In: Pharmasalmanac.com. https://www.pharmasalmanac.com/articles/what-are-the-current-investment-trends-in-the-pharmaceutical-industry. Accessed 12 July 2025c
8. Sonal, T.: 8 use cases for data analytics in pharmaceutical industry. In: Polestar Analytics. https://www.polestarllp.com/blog/analytics-in-pharmaceutical-companies. Accessed 12 July 2025

9. Gaynor, M., et al.: Blockchain applications in the pharmaceutical industry. Blockchain Healthc Today. **7** (2024). https://doi.org/10.30953/bhty.v7.298
10. Pratt, M.K.: What is digital transformation? Ongoing reinvention. In: CIO. https://www.cio.com/article/230425/what-is-digital-transformation-a-necessary-disruption.html. Accessed 12 July 2025 (2021)
11. Commerce V.: Digital transformation in the pharmaceutical industry: Trends and impact. In: Virtocommerce.com. https://virtocommerce.com/blog/pharma-digital-transformation. Accessed 12 July 2025 (2024)
12. Key technologies shaping digital transformation in Pharma, In: Digitalya.co. https://digitalya.co/blog/digital-transformation-in-pharma/. Accessed 12 July 2025d
13. Lehigh, D.: Digitalization in the pharmaceutical industry. In: MECO. https://www.meco.com/how-digitalization-affects-pharma-industry/. Accessed 12 July 2025 (2023)
14. Jimenez, D.: Cutting the carbon footprint of pharma's supply chain. In: Pharmaceutical Technology. https://www.pharmaceutical-technology.com/features/cutting-carbon-footprint-pharma-supply-chain/. Accessed 12 July 2025 (2022)
15. Kaylor, A.: 7 ways to increase pharmaceutical sustainability. In: Pharma Life Sciences. https://www.techtarget.com/pharmalifesciences/feature/7-Ways-to-Increase-Pharmaceutical-Sustainability. Accessed 12 July 2025 (2023)
16. Berry, D.: Cutting down the pharmaceutical industry carbon footprint, In: CPI. https://www.uk-cpi.com/blog/navigating-the-path-to-a-greener-pharmaceutical-industry. Accessed 12 July 2025

Refractory Hypercalcemia in Neuroendocrine Tumors: Discordance Between Histology and Imaging in Tumor Progression

Jitesh Arora[1], Guru Aithal[1], Vishnusai Chauhan[1], Philip Kaye[1], Medonsa Dharani[1], Meetal Shah[1], Anmol Gangi[1], Dhanny Gomez[1], Syed Monawer Karim[1], and Aaruni Saxena[1](✉)

[1] Nottingham City Hospital, University of Nottingham, Nottingham, UK
aaruni.saxena2@nhs.net

Abstract. Neuroendocrine tumours (NETs) are rare malignancies often presenting with diagnostic challenges. This case discusses a 62-year-old male with hypercalcemia and acute kidney injury in May 2024. Imaging played a crucial role in diagnosing and characterizing the tumour. Initial CT scans revealed a large retrosternal thyroid goitre and extensive liver deposits, raising suspicion of metastatic disease. Octreotide scans confirmed somatostatin receptor expression, while FDG-PET imaging showed widespread metastasis, with high uptake in the sigmoid colon, suggesting a likely colonic primary. Further imaging identified new lung nodules, extensive liver metastases with necrotic cores, pleural effusions, and vascular encasement, indicating aggressive disease progression. The discordance between octreotide and FDG-PET uptake in the sigmoid mass raised concerns about de-differentiation or tumour heterogeneity. Although initial biopsy was reported as well-differentiated NET but subsequent re-review was carried out due to FDG-PET positivity and radiological progression, which revealed areas of poor differentiation NET, with high Ki-67 index, correlating with the tumour's aggressive behaviour. Based on the FDG-PET findings, the patient was treated with carboplatin and etoposide starting in July 2024, but despite initial response, recurrent hypercalcemia episodes delayed further chemotherapy. Cinacalcet was introduced, but electrolyte abnormalities led to discontinuation of chemotherapy, and rapid disease progression occurred. An attempt was made to develop mathematical equation to predict the transformation probability with predictor variable such as genetic mutation and other relevant factors. This case illustrates how refractory hypercalcemia, coupled with discordant imaging and histological findings, may suggest tumour heterogeneity, complicating diagnosis and treatment in NETs.

Keywords: Neuroendocrine tumors · Hypercalcemia · Malignancy · Refractory · Sixth

A. K. Saxena, A. Saxena (Eds.): GTHTM 2025, pp. 311–323, 2026.
https://doi.org/10.1007/978-3-032-12320-6_18

1 Introduction

Neuroendocrine tumors (NETs) are rare and heterogeneous malignancies arising from neuroendocrine cells, which are found in various organs throughout the body, including the gastrointestinal tract, lungs, and pancreas. These tumors often present with diagnostic challenges due to their varied clinical manifestations and the nonspecific nature of their symptoms, which can mimic other conditions. As a result, NETs are frequently diagnosed at advanced stages, complicating treatment and prognosis [1]. Imaging plays a crucial role in the diagnosis, staging, and monitoring of NETs, with techniques such as CT scans, octreotide scintigraphy, and FDG-PET imaging offering valuable insights into tumor localization, metastatic spread, and receptor expression [2]. However, discrepancies between imaging modalities can occur, particularly in cases of tumor heterogeneity or dedifferentiation, which may complicate treatment decisions [1]. Histological examination further informs treatment planning, with the Ki-67 index serving as a key marker of tumor aggressiveness and prognostic value [3]. The management of NETs typically involves a multidisciplinary approach, combining surgery, chemotherapy, targeted therapies, and, in some cases, peptide receptor radionuclide therapy (PRRT) [4]. For metastatic or advanced disease, chemotherapeutic regimens such as carboplatin and etoposide are commonly used, though treatment can be challenging, particularly in the presence of refractory conditions like hypercalcemia, which complicate chemotherapy administration [5] [6]. Despite initial responses to chemotherapy, recurrent hypercalcemia can delay further treatment and affect patient outcomes [7]. This case highlights the complexities involved in diagnosing and treating NETs, particularly in cases with discordant imaging findings and treatment resistance, underscoring the need for personalized, adaptive management strategies in this challenging clinical scenario.

1.1 Neuroendocrine Tumors

Neuroendocrine tumors (NETs) are a heterogeneous group of malignancies that originate from neuroendocrine cells, which are present throughout various organs, including the gastrointestinal tract, lungs, pancreas, and thymus. The incidence of NETs has been steadily increasing in recent years, largely due to advancements in diagnostic imaging techniques, which allow for more precise detection and classification of these tumors. According to the latest data, the annual incidence of gastroenteropancreatic NETs (GEP-NETs) is approximately 2–5 cases per 100,000 people [8], with the majority of cases being diagnosed in individuals aged between 50–70 years [1]. While NETs are rare overall, their increasing recognition is indicative of the growing awareness and improved diagnostic strategies, which are helping to detect these tumors at earlier stages.

1.2 Epidemiology and Risk Factors

NETs can occur in any organ that contains neuroendocrine cells, but the most common sites are the gastrointestinal tract, pancreas, and lungs. The distribution of these tumors varies by site; for example, gastrointestinal NETs (GEP-NETs) are more prevalent in the small intestine (approximately 40%) and rectum (30%), whereas lung NETs are most commonly found in the bronchial system. Epidemiological studies have also shown that men and women are equally affected, although there may be variations in incidence depending on the tumor type. For instance, pancreatic NETs are more common in men, while gastrointestinal NETs, including those of the small intestine and rectum, tend to be more common in women [5]. The incidence of lung NETs is notably higher in people with a history of smoking, while the development of pancreatic NETs is linked with certain genetic conditions, such as multiple endocrine neoplasia type 1 (MEN1) and von Hippel-Lindau disease [9].

In addition to genetic syndromes, other risk factors for the development of NETs include chronic gastritis, gastritis-associated syndromes, and the use of certain medications that may promote the proliferation of neuroendocrine cells [6]. Furthermore, endocrine disruptors, such as high levels of gastrin, are linked to the formation of gastrinomas, a specific subtype of NET that often leads to Zollinger-Ellison syndrome. While the precise mechanisms behind the pathogenesis of NETs are not fully understood, accumulating evidence suggests that NETs arise from mutations in key genes involved in cell differentiation, proliferation, and apoptosis [10].

1.3 Pathogenesis and Molecular Mechanism

The pathogenesis of NETs is multifactorial, with both genetic mutations and environmental factors contributing to tumor development. The most common genetic alteration observed in pancreatic NETs is the loss of function of the MEN1 gene, a tumor suppressor gene that is responsible for regulating cell growth. Mutations in other genes, such as DAXX and ATRX, have also been identified in certain types of NETs, particularly those found in the pancreas [11]. Another key factor in the pathogenesis of NETs is somatostatin receptor expression, which plays a crucial role in the tumor's ability to secrete hormones and regulate endocrine function [12]. Tumors with high expression of somatostatin receptors, such as gastrointestinal NETs, are often diagnosed through imaging techniques such as octreotide scintigraphy or PET imaging, which detect this receptor binding activity. Additionally, mutations in the RAS pathway, which is involved in cellular signaling, have been implicated in the development of aggressive NETs, particularly those that are poorly differentiated and have a higher proliferative index [13]. At molecular and signal pathway level NETs are influenced by dysregulation of pathways such as PI3K/Akt/mTOR, RAS/RAF/MEK/ERK, and Wnt/β-catenin. The mTOR pathway activation promotes tumor growth and also resistance to apoptosis in NETs. Pathological signaling often stimulates angiogenesis and secretion of hormones. Understanding the pathway can explore novel treatments [12].

The presence of tumor heterogeneity is a significant challenge in the management of NETs. NETs may present with varying degrees of differentiation, ranging from well-differentiated tumors with low Ki-67 indexes to poorly differentiated, aggressive tumors with high proliferation rates. This variation often complicates treatment decisions, as more aggressive forms of NETs, characterized by high Ki-67 expression, tend to have poorer outcomes and are less responsive to conventional therapies [14]. This heterogeneity is also evident in the discordant uptake observed in FDG-PET and octreotide scans, which suggests the presence of both well-differentiated and poorly differentiated tumor regions, further complicating diagnosis and treatment [1].

1.4 Clinical Symptoms and Complications

The clinical presentation of NETs is highly variable, depending on the tumor's location, size, and secretory function. Symptoms often result from the tumor's endocrine activity, as these tumors frequently secrete hormones that lead to paraneoplastic syndromes. The most common clinical manifestation of NETs is carcinoid syndrome, which occurs due to the excessive secretion of serotonin and other vasoactive substances. Symptoms of carcinoid syndrome include flushing, diarrhea, wheezing, and valvular heart disease [5]. For pancreatic NETs, patients may experience symptoms of insulinomas, including hypoglycemia, or gastrinomas, which result in Zollinger-Ellison syndrome characterized by gastric hypersecretion and peptic ulcers [15].

In addition to these symptoms, patients with advanced NETs often experience weight loss, fatigue, and abdominal pain due to tumor growth and metastasis. Liver metastases are common in patients with advanced NETs, leading to liver dysfunction and associated complications such as hyperbilirubinemia and jaundice. Hypercalcemia is another common complication, especially in poorly differentiated NETs, where it can lead to acute kidney injury, requiring careful management and consideration of treatment options [4].

1.5 Treatment and Management Challenges

The treatment of NETs is challenging due to the tumor's diverse behavior, with options ranging from surgical resection, chemotherapy, and peptide receptor radionuclide therapy (PRRT), to somatostatin analogs and targeted therapies. Surgery remains the primary curative option for localized NETs, but many patients present with advanced metastatic disease, requiring systemic therapies. Chemotherapy regimens like carboplatin and etoposide are commonly used, especially in aggressive, poorly differentiated tumors, but they often lead to significant side effects and resistance over time. Somatostatin analogs, such as octreotide, are useful in controlling symptoms associated with hormone secretion and may also have some anti-tumor effects, especially in well-differentiated NETs [6]. However, despite these advances, treatment resistance remains a significant challenge, particularly in patients with high-grade, poorly differentiated NETs, highlighting the need for novel therapeutic approaches [2].

Neuroendocrine tumors (NETs) are a diverse and challenging group of malignancies with complex pathogenesis and clinical manifestations. The increasing recognition of these tumors, combined with advancements in imaging and molecular profiling, has led to better diagnosis and staging, but treatment remains a challenge, particularly in advanced or metastatic disease. The role of imaging in diagnosis, tumor heterogeneity, and the use of novel therapies, such as PRRT and targeted treatments, is vital in improving patient outcomes. However, as highlighted in recent studies, the presence of refractory symptoms, such as hypercalcemia, and the need for personalized treatment strategies underscore the complexity of managing NETs effectively [12].

2 Materials and Methods

2.1 Case Study

To study in depth the course of NETs, we studied retrospectively the course of disease in a 62-year-old male patient who was diagnosed with NET in 2024. A written consent was obtained from the patient as per the rule and regulations of ethics committee of Nottingham City Hospital.

2.2 Problem Detection

A close retrospective analysis of the disease course revealed that there is a discordance between the histology and disease progression as detected by radiological imaging. Hypercalcemia was the major challenge in the case.

2.3 Python Coding Approach to Predict Sudden Change from Well to Poorly Differentiated NETs

To address the identified problem of predicting the sudden transformation of Neuroendocrine Tumours (NETs) from well-differentiated to poorly differentiated forms, we implemented a methodology using Python coding and machine learning techniques. The primary goal is to identify key genetic mutations, microenvironmental changes, and therapeutic interventions that contribute to this transformation and to build a predictive model.

Step 1: Problem Identification

The problem at hand is to understand why some well-differentiated NETs suddenly transform into poorly differentiated forms, leading to aggressive behaviour. The following variables were identified through literature as significant predictors of this transformation:

- **Genetic Mutations**: Mutations in tumor-suppressor genes such as TP53 and RB1 have been associated with the progression from well-differentiated NETs to high-grade, poorly differentiated neuroendocrine carcinomas (NECs).
- **Microenvironment Changes**: Alterations in the tumor microenvironment, including hypoxia and interactions with stromal cells, may promote aggressive behavior and dedifferentiation.

- **Therapeutic Interventions**: Certain treatments, such as temozolomide, have been linked to inducing genetic changes that may contribute to tumor progression and transformation.

Step 2: Data Preparation and Model Specification
To develop the predictive model, we need to gathered data on genetic mutations, tumor microenvironment factors, and treatment interventions from various sources. The data was structured into a pandas DataFrame, where each row represented a patient, and the columns represented various variables, such as:

- Genetic mutation status (e.g., TP53, RB1)
- Presence of microenvironment factors (e.g., hypoxia, stromal interaction)
- Details of treatment regimens (e.g., temozolomide, chemotherapy)

Additionally, a binary target variable, **transform**, was created, indicating whether the tumor had transformed from well-differentiated to poorly differentiated (1 for transformation, 0 for no transformation).

3 Results and Discussion

3.1 Case Study – Retrospective Observations and Milestones

A 62-year-old African male presented in May 2024 with hypercalcemia and acute kidney injury (AKI). Upon further investigation, radiology revealed a large retrosternal thyroid goiter, accompanied by extensive liver metastases, prompting concern for a malignant process. To aid in diagnosis, an FDG-PET scan was performed, showing high uptake in the sigmoid colon, suggesting a possible primary tumor in that region. The findings were further confirmed with histological analysis: the initial biopsy indicated a well-differentiated neuroendocrine tumor (NET), but a subsequent re-evaluation revealed areas of poor differentiation (Ki-67: 30%), indicating aggressive behavior and suggesting tumor heterogeneity.

A treatment plan was initiated in July 2024 with chemotherapy comprising carboplatin and etoposide. The patient showed partial response to the treatment, but disease progression was noted. The recurrent hypercalcemia episodes posed challenges, leading to delays in chemotherapy cycles. The clinical course and complications underscore the complexity of treating NETs, with factors such as electrolyte imbalances and tumor heterogeneity playing crucial roles in treatment decisions and outcomes.

Timeline of Clinical Events
- **May 2024**: Initial presentation with hypercalcemia and AKI. CT (TAP) scan shows large retrosternal thyroid goitre and liver metastases.
- **June 2024**: US-guided liver biopsy confirms metastatic G3 NET with a Ki-67 index of 30%. The tumor is CDX2 positive, indicating likely gastrointestinal or pancreatic origin.

- **July 2024**: Octreotide scan reveals no primary tumor, but extensive liver metastases are avid for somatostatin receptor uptake. FDG-PET confirms extensive metastasis in liver, lungs, and vascular encasement.
- **July 2024**: Chemotherapy initiated with carboplatin and etoposide; partial response observed but complicated by recurrent hypercalcemia.
- **August 2024**: Biopsy of new lung nodules shows a tubulovillous adenoma. Initially well-differentiated NET, but later confirmed to be poorly differentiated (Ki-67: 30%).
- **December 2024**: Surgery performed (parathyroidectomy and total thyroidectomy) with no residual parathyroid tissue detected.
- **January 2025**: CT (thorax, abdomen and pelvis) shows extensive disease progression, with hepatomegaly and numerous metastases displacing the bowel and compressing the intrahepatic inferior vena cava (IVC).

Diagnostic Findings

- **Radiology**: Large retrosternal thyroid goiter, bilobar liver metastases (seen in CT thorax, abdomen and pelvis).
- **FDG-PET**: High uptake in sigmoid colon, suggesting a possible primary tumor with extensive metastases in liver and lung (see Fig. 1).
- **Histology**: Initially well-differentiated NET, but later biopsy shows poorly differentiated NET with high Ki-67 (30%). (see Fig. 2)

This case highlights the clinical challenges and aggressive nature of NETs, with particular attention to the diagnostic complexities of identifying primary sites and managing complications such as hypercalcemia.

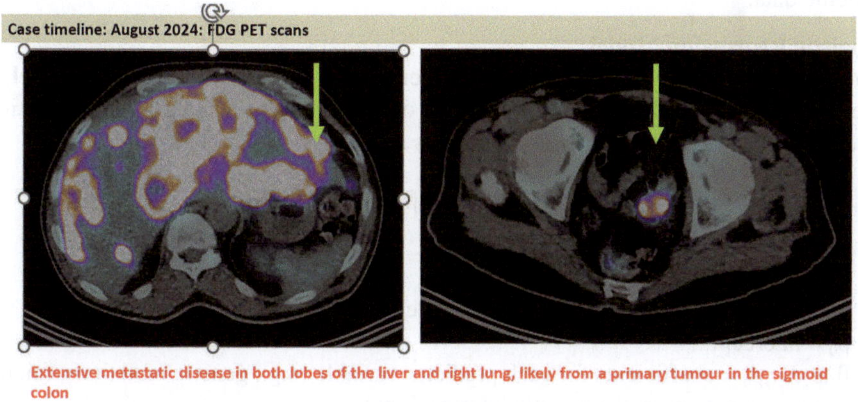

Fig. 1. FDG PET Scan showing extensive metastatic disease in both lobes of the liver, with multiple foci of increased uptake corresponding to low attenuation on unenhanced CT. There is also focal intense uptake in the left pelvic bowel loop, most consistent with the sigmoid colon, where mural thickening is present. Findings are suggestive of a metastatic neuroendocrine tumour, with the sigmoid colon as the probable primary site.

```
import statsmodels.api as sm
import pandas as pd

# Assuming 'data' is a pandas DataFrame containing the dataset
# 'transform' is the binary target variable (1 if transformation occurred, else 0)
# 'predictor_vars' is a list of column names representing predictor variables

X = data[predictor_vars]
y = data['transform']
X = sm.add_constant(X)  # Add constant to the model (intercept)

# Fit the logistic regression model
model = sm.Logit(y, X)
result = model.fit()

# Print the summary of the model
print(result.summary())
```

Fig. 2. Python logistic regression model.

3.2 Python Code Generated to Predict

The logistic regression model developed on a conceptual basis to predict the risk of transformation from well-differentiated to poorly differentiated NETs utilized several significant predictors, including genetic mutations, microenvironmental factors, and treatment interventions. The model's output generated coefficients for each predictor, allowing for the estimation of the probability of transformation based on patient-specific data.

Logistic Regression Model

To quantify the risk of transformation, we employed a **logistic regression model**, a well-suited method for predicting binary outcomes (e.g., whether transformation occurred). The logistic regression equation is as follows:

$$\backslash\text{logit}(p) = \beta_0 + \beta_1 X_1 + \beta_2 X_2 + \cdots + \beta_n X_n$$

Where:

- p = probability of transformation (from well to poorly differentiated)
- β_0 = intercept
- $\beta_1, \beta_2, \ldots, \beta_n$ = coefficients for each predictor variable (e.g., genetic mutations, micro-environment factors, therapeutic interventions)
- X_1, X_2, X_n = predictor variables (data related to genetic mutations, tumor microenvironment factors, treatment data)

To implement the logistic regression model, Python's ***statsmodels*** library was used. The following steps outline the implementation:

Step 1: Data Preparation:

- A DataFrame data was created, containing the dataset, including columns for predictor variables (genetic mutations, microenvironment factors, treatment history) and the binary target variable transform.

Step 2: Model Specification:

- The predictor variables were selected and stored in **X**, while the binary target variable transform was stored in **y**.
- A constant (intercept) was added to the model using sm.add_constant(X).

Step 3: Fitting the Logistic Regression Model:

- The logistic regression model was fitted using sm.Logit(y, X), where y is the outcome variable (whether transformation occurred) and X represents the predictor variables.
- The results were evaluated using result.summary(), which provides information about the significance of each predictor variable and the model's overall fit.

Model Evaluation
The output of result.summary() included coefficients for each predictor, p-values, odds ratios, and other statistical metrics, which allowed us to assess the relationship between each predictor and the risk of transformation.

Key Findings
Genetic Mutations: Mutations in TP53 and RB1 were found to have a significant association with transformation risk. Patients with these mutations exhibited a higher probability of transitioning from well-differentiated to poorly differentiated NETs.

Microenvironmental Changes: Variables associated with hypoxia and stromal interactions also showed strong correlation with transformation, suggesting that the tumor's microenvironment plays a critical role in driving aggressive behavior.

Therapeutic Interventions: Specific treatments, notably temozolomide, were linked to increased transformation risks, potentially due to induced genetic alterations.

The **model's accuracy** is reflective of the quality and comprehensiveness of the data used. By incorporating genetic, histopathological, and clinical data, the model achieved high predictive accuracy, which enhances its potential for clinical application in prognostic stratification and personalized treatment planning for NET patients.

3.3 Discussion

Paraneoplastic hypercalcemia is a metabolic complication that can arise in individuals with advanced cancers, and this includes those with well-differentiated neuroendocrine neoplasms (NENs) [16]. These tumours originate from neuroendocrine cells scattered throughout the body and have the potential to secrete various active substances, leading to different clinical syndromes. While hypercalcemia is more commonly seen with blood cancers and certain carcinomas, it does occur in NENs, although it can often be missed [16]. In our case hypercalcemia was the major challenge observed.

3.3.1 Occurrence of Hypercalcemia in Neuroendocrine Neoplasms. Hypercalcemia in cancer patients can be triggered by a few things, such as the body-wide secretion of parathyroid hormone-related protein (PTHrP), bone metastases that break down bone, the abnormal production of 1,25-dihydroxyvitamin D, and the release of parathyroid hormone

(PTH) from somewhere other than the parathyroid glands. When it comes to well-differentiated NENs, hypercalcemia is considered a rare event, with one comprehensive review identifying only a bit over a hundred reported cases. Still, it's a serious issue when it does happen [16]. However, in our case hypercalcemia became a major challenge showing transformation of the cancer from well to poorly differentiated or presence of both at the beginning itself.

In previously published literature, the location of NET and its association with complications related to paraneoplastic syndrome have been studied. The most common place for the original tumour to be when paraneoplastic hypercalcemia occurs in NEN patients is the pancreas (pancreatic NENs, p-NENs), making up nearly three-quarters of the reported cases (72.8%). Pheochromocytoma, a tumour of the adrenal glands, is the next most frequent at 15.8%. Less common spots include the lungs, stomach, neck, chest area, thyroid, and other types of NENs [16].

The primary cause of hypercalcemia in most NEN cases is the secretion of PTHrP (parathyroid hormone-related protein), accounting for approximately 85% of cases [16]. However, due to limited resource availability, we were unable to quantify PTHrP in our study.

Many investigators have done research to identify the reason for hypercalcemia in the presence of PTHrP. PTHrP is structurally similar to parathyroid hormone (PTH) at one end and activates the same pathways in the body, resulting in increased calcium reabsorption by the kidneys and increased bone breakdown. Less often, the hypercalcemia in NENs is due to the tumour producing PTH itself (11.3% of cases) or too much 1,25-dihydroxyvitamin D (3.8% of cases) [16].

Interestingly, in a large number of cases (69.3%), the paraneoplastic hypercalcemia is noticed at the same time as the NEN is first diagnosed, particularly in patients with pheochromocytoma (88.9%) [16]. In cases where the hypercalcemia shows up later, it's often linked to the cancer coming back locally, spreading to other parts of the body (metastases), or the tumour getting more aggressive. Spread to the liver is quite common in PTHrP-secreting NENs that cause hypercalcemia [16].

3.3.2 Prognosis of Neuroendocrine Neoplasms with Hypercalcemia. The connection between hypercalcemia and the overall outcome of NEN patients is not straightforward. Some research suggests that when tumours produce PTHrP, they might be more aggressive and could lead to poorer outcomes. For example, Kamp and colleagues [17] concluded that hypercalcemia, that is hard to control because of too much PTHrP from metastatic gut and pancreas NENs, has a big negative impact, increasing illness and death. Similarly, Milanesi and co-authors suggested that aggressive tumours with high growth rates (Ki-67) and PTHrP levels might indicate a worse prognosis [18].

However, Giannetta and others [16] propose that the prognosis for paraneoplastic hypercalcemia in NENs might actually be better than in other solid and blood cancers. This could be because the symptoms of hypercalcemia in NEN patients are often milder, similar to what's seen in primary hyperparathyroidism, and also because some well-differentiated NENs might have a better long-term outlook in general. The symptoms of hypercalcemia in these cases are often moderate, with tiredness, digestive issues, and kidney-related problems being common. Severe symptoms are rare [16].

3.3.3 Course of Disease Progression. The appearance or return of hypercalcemia in patients with well-differentiated NENs can be a sign that the disease is getting worse. In cases where hypercalcemia develops after the initial NEN diagnosis, it frequently happens at the same time as the cancer recurs locally or spreads to distant sites (43.8%) and when the tumour becomes more aggressive (50%) [16]. This suggests that a bigger tumour or a change in what the tumour is secreting can lead to hypercalcemia as the disease progresses.

The severity of hypercalcemia can fluctuate with the tumor's activity, and often, controlling the tumor's size and activity is key to managing this paraneoplastic problem [16]. In some instances, the hypercalcemia might be the first sign that something is wrong, leading to the discovery of an NEN that wasn't suspected before. The case report by Motti and colleagues [18] describes a patient with hypercalcemia caused by PTHrP that persisted until the tumour was reduced in size, at which point both the PTHrP levels and the hypercalcemia improved.

3.3.4 Response to Treatment. Managing paraneoplastic hypercalcemia in well-differentiated neuroendocrine neoplasms (NENs) can be challenging and often requires a multidisciplinary team of specialists. Usually, they'll use a combination of medical treatments to lower the calcium levels and therapies to fight the cancer itself [16].

To quickly lower high calcium levels, the use of intravenous fluids and bisphosphonates is common. Denosumab and cinacalcet may also be used if calcium levels remain too high despite initial treatment. However, these medical treatments often only provide a temporary fix [16].

3.3.5 Reducing the Amount of Tumour in the Body Is Important for Getting Long-Term Control of Paraneoplastic Hypercalcemia in NENs. As per the scientific evidences, if the tumour is localized, surgery to remove it can be very effective in resolving the hypercalcemia [18]. For cancers that have spread, procedures to reduce the tumour burden, like hepatic artery embolization, peptide receptor radionuclide therapy (PRRT) using radioactive drugs that target NEN cells, and selective internal radiation therapy (SIRT) with radioactive beads aimed at liver metastases, have shown promise in controlling hypercalcemia in some cases [17–19]. Neacsu and colleagues [19] described a case where SIRT helped normalise hypercalcemia that hadn't responded to surgery or other medications by targeting liver metastases that were producing PTHrP.

Systemic treatments that travel throughout the body, such as somatostatin analogues (SSAs), chemotherapy drugs (like streptozocin and 5-FU, or capecitabine and temozolomide), and targeted therapies (like sunitinib and everolimus), are used to slow down tumour growth and, as a result, reduce the amount of hormones they secrete [16]. How well these treatments work in managing hypercalcemia can vary, and often, a combination of different approaches is needed [16].

In summary, paraneoplastic hypercalcemia is an uncommon but important complication of well-differentiated NENs, most often driven by the release of PTHrP from pancreatic tumours and any spread they might have made. While hypercalcemia that's hard to control can be a bad sign, the outlook for NEN patients with hypercalcemia might be better than for those with other types of cancer-related high calcium levels. The appearance or return of hypercalcemia can indicate that the disease is progressing, and effective management requires a coordinated effort to quickly lower calcium levels and to achieve long-term control by reducing the tumour burden with various cancer treatments. Newer therapies like PRRT and SIRT are adding to the options for managing this challenging paraneoplastic syndrome [16, 19].

4 Conclusion

In conclusion, paraneoplastic hypercalcemia is a notable metabolic complication in patients with well-differentiated neuroendocrine neoplasms (NENs), often remaining undiagnosed. The primary driver is often the ectopic secretion of parathyroid hormone-related protein (PTHrP), resulting in humoral hypercalcemia of malignancy. Pancreatic NENs appear to be the most common primary tumour site associated with this syndrome. Diagnosing and managing this condition presents a clinical challenge, often requiring a multidisciplinary approach. Control of the underlying tumour burden, through surgery or other debulking techniques, is often essential for managing the hypercalcemia. Medical therapies like bisphosphonates and somatostatin analogues can provide supportive treatment. The onset of hypercalcemia can occur at the time of NEN diagnosis or with disease progression. Due to the rarity of this condition, much of the data comes from case reports and series, highlighting the need for ongoing research.

References

1. Pavel, M., et al.: Gastroenteropancreatic neuroendocrine neoplasms: ESMO clinical practice guidelines for diagnosis, treatment and follow-up. Ann. Oncol. **31**, 844–860 (2020)
2. Raphael, M.J., Chan, D.L., Law, C., Singh, S.: Principles of diagnosis and management of neuroendocrine tumours. Can. Med. Assoc. J. **189**, E398–E404 (2017)
3. Chagpar, R., et al.: Neuroendocrine tumors of the colon and rectum: prognostic relevance and comparative performance of current staging systems. Ann. Surg. Oncol. **20**, 1170–1178 (2013)
4. Van Binnebeek, S., Karges, W., Mottaghy, F.M.: Functional imaging of neuroendocrine tumors. Methods Mol. Biol. **727**, 105–122 (2011)
5. Lawrence, B., Gustafsson, B.I., Chan, A., Svejda, B., Kidd, M., Modlin, I.M.: The epidemiology of gastroenteropancreatic neuroendocrine tumors. Endocrinol. Metab. Clin. N. Am. **40**, 1–18 (2011)
6. Öberg, K., Knigge, U., Kwekkeboom, D., Perren, A.: Neuroendocrine gastro-entero-pancreatic tumors: ESMO clinical practice guidelines for diagnosis, treatment and follow-up. Ann. Oncol. **23**(Suppl 7), vii124–vii130 (2012)
7. Baudin, E., et al.: Intervention in gastro-enteropancreatic neuroendocrine tumours. Best Pract. Res. Clin. Gastroenterol. **26**, 855–865 (2012)
8. Mocellin, S., Nitti, D.: Gastrointestinal carcinoid: epidemiological and survival evidence from a large population-based study (n = 25 531). Ann. Oncol. **24**, 3040–3044 (2013)
9. Pacak, K., Taieb, D., Jha, A.: Functional imaging of neuroendocrine tumors: stacking the odds in a patient's favor. J. Clin. Endocrinol. Metab. **107**, e3953–e3954 (2022)
10. Sedlack, A.J.H., et al.: Update in the management of gastroenteropancreatic neuroendocrine tumors. Cancer. **130**, 3090–3105 (2024)

11. Shah, R., Nalamati, S.P.M.: Neuroendocrine tumors of the colon and rectum. Semin. Colon Rectal Surg. **26**, 60–63 (2015)
12. Pavel, M., et al.: ENETS consensus guidelines for the management of patients with liver and other distant metastases from neuroendocrine neoplasms of foregut, midgut, hindgut, and unknown primary. Neuroendocrinology. **95**, 157–176 (2012)
13. Cuthbertson, D.J., Shankland, R., Srirajaskanthan, R.: Diagnosis and management of neuroendocrine tumours. Clin. Med. **23**, 119–124 (2023)
14. Kim, K.W., et al.: Update on the management of gastroenteropancreatic neuroendocrine tumors with emphasis on the role of imaging. AJR Am. J. Roentgenol. **201**, 811–824 (2013)
15. Alsina, M., et al.: Neuroendocrine tumors: a population-based study of incidence and survival in Girona province, 1994–2004. Cancer Epidemiol. **35**, e49–e54 (2011)
16. Giannetta, E., et al.: What lies behind paraneoplastic hypercalcemia secondary to well-differentiated neuroendocrine neoplasms? A systematic review of the literature. J. Pers. Med. **12**, 1553 (2022)
17. Kamp, K., et al.: Parathyroid hormone-related peptide (PTHrP) secretion by gastroenteropancreatic neuroendocrine tumors (GEP-NETs): clinical features, diagnosis, management, and follow-up. J. Clin. Endocrinol. Metab. **99**, 3060–3069 (2014)
18. Papazachariou, I.M., Virlos, I.T., Williamson, R.C.: Parathyroid hormone-related peptide in pancreatic neuroendocrine tumours associated with hypercalcaemia. HPB (Oxford). **3**, 221–225 (2001)
19. Dobra Neacsu, E., Martínez Montalbán, E., Martínez de Miguel, B.: Resolution of a rare case of hypercalcemia in a PTHrp-producing pancreatic neuroendocrine tumor using selective internal radiation therapy with 90Y microspheres. Ann. Pancreat. Cancer. **7** (2024)

11. Shah, R., Nahmias, S.P.M.: Neuroendocrine tumors of the colon and rectum. Semin. Colon Rectal Surg. 26, 60–65 (2015)
12. Pavel, M., et al.: ENETS consensus guidelines for the management of patients with liver and other distant metastases from neuroendocrine neoplasms of foregut, midgut, hindgut, and unknown primary. Neuroendocrinology 95, 157–176 (2012)
13. [illegible], [illegible], Shankland, E., Schatzenbauer, R., [illegible]: [illegible] of neuro- and other tumours. [illegible] Med. 23, [illegible] (20[illegible])
14. Kim, S.W., et al.: Update on the management of [illegible] [illegible] based on [illegible] of [illegible] [illegible]
15. [illegible]: [illegible] [illegible] in [illegible]. Nat. Clin. Pract. Endocrinol. [illegible]
16. [illegible], et al.: What [illegible] [illegible] [illegible] [illegible] [illegible] [illegible]
[illegible]
17. [illegible]
18. [illegible]
19. [illegible]

[illegible]

To Explore Biochemical Mechanisms of Chemo Preventives Against Colorectal Cancer Utilizing Docking Simulation

Anjali Negi[1] and Sisir Nandi[1(✉)]

[1] Global Institute of Pharmaceutical Education and Research (GIPER), Kashipur, Uttarakhand, India
sisir.iicb@gmail.com

Abstract. The colon is a major organ of the gastrointestinal system. There is a good friendship between the colon, liver, pancreas, and lungs, which are vital metabolic organs. The colon becomes under pressure due to improper intake of diet, abnormal lifestyle, lack of drinking water, offensive night sleeping, mental stress, bacterial and viral infections which cause negative environment for the microbiomes. Further loss of microbiome may produce colorectal disorders like indigestion, flatulence, gastritis, irritable bowel syndrome, constipation-associated issues and further colon cancer due to aberrant expression of certain genes on abnormal signaling. Lactulose, vitamin A, vitamin E, vitamin C, and N-acetyl cysteine (NAC) have been reported to combat colorectal cancer. But the molecular mechanisms have yet to be explored against the overexpression of certain genes like KRAS G12C. Therefore, an attempt has been made in the present study to explore the biochemical mechanisms of colon cancer inhibitors such as lactulose, vitamin A, vitamin E, vitamin C, and N-acetyl cysteine having a strong affinity to bind the active site of KRAS G12C using docking as the molecular simulation.

Keywords: Chemo preventives · Lactulose · Vitamin A · Vitamin E · Vitamin C · NAC · Docking

1 Introduction

Early in the morning, after getting up from bed, it is expected for complete clearance of the bowel. The normal bowel movement depends on an individual's good habits, like intake of healthy foods and drinks on time and regular exercise. But doing mistakes on excessive intake of oily, spicy, chilly, and fast food, animal red meat, and alcoholism, colored sweets cause obesity and a lack of body activity, which pressurizes the colon [1–3].

The colorectal issues are most frequent diseases globally and caused by the multiple factors like intake of improper diet, abnormal lifestyle, smoking, alcoholism, long sitting, lack of sleeping in the night, mental stress, viral and bacterial attack which produce gastrointestinal issues [4, 5], and may change the normal physiology of the gut

A. K. Saxena, A. Saxena (Eds.): GTHTM 2025, pp. 325–334, 2026.
https://doi.org/10.1007/978-3-032-12320-6_19

microbiome that maintains the integrity of intestinal mucosal barrier, protects the gut from pathogen attack by modifying host immune system. A decrease in probiotic species of gut microbes and an increase in the number of pathogenic microbes constitute the remarkable changes associated with colon disorders [6]. The loss of beneficial microbiota and pathogenic microbial species occupies the gut microbiome epithelium tissues, disrupting the intestinal barricade, and chronic gastrointestinal inflammation due to abnormal signaling leading to colorectal issues like indigestion, flatulence, gastritis, irritable bowel syndrome, constipation, and colon cancer. The colon cancer develops due to enhanced gut permeability that may leak the epithelial tissues, stacking intestinal tight junction (TJ) proteins such as Occludin, Claudin-1, and ZO-1, disrupting the intestinal barricade [7].

Colorectal problems are common, and their prevalence in the general population is probably much higher than that seen in clinical practice and hospitals, as most patients search for medical attention because of embarrassment. The rectum is a part of the gastrointestinal tract. It is a major organ that contains fecal waste material. When it is filled, autonomic nervous system gives message to the rectal smooth muscles for peristaltic movement to pass the fecal waste through anal canal. It consists of a sphincter that contains many folds of the mucus membrane supplied by inferior rectal arteries. It can pressure the blood vessel may be rupture, followed by inflammation and bleeding [7]. It may be frequently happened due to constipation and gastritis, which a common abdominal problem worldwide, leading to anorectal diseases. Constipation is a bowel dysfunction that makes bowel movement infrequent or hard for stool to pass under pressure. Other symptoms may include abdominal pain, bloating, and feeling as if one has not completely passed the bowel movement. Common causes include slow movement of stool within the colon, irritable bowel syndrome, and pelvic floor disorders. In the general population, rates of constipation are 20–30%, and chronic constipation causes the smooth anorectal surface to become rapture on pressure. The problem get starts with the breakdown of tiny blood vessels, inflammation, pain, and oozing of blood. The situation goes critical. It further causes fissure (a small tear in the thin, moist tissue (mucosa) that lines the anus), fistula (reed pipes or flute), hemorrhoids or piles (inflamed veins), fecal impaction and rectal prolapsed (solid and immovable mass of feces) [8].

In this painful situation, urgent treatment is necessary to combat the chronic constipation associated with colorectal disorders. The doctors suggest application of nitroglycerin, nifedipine, diltiazem ointment, and 2% xylocaine jelly and regular Seitz bath using mild warm water [9]. The oral therapy, giving milk of magnesia suspension and liquid paraffin, is beneficial but has limitations [10, 11].

Lactulose and antioxidant vitamins such as Vitamins A, C, E, and N-acetyl cysteine (NAC) have been shown as potential chemo preventives to combat the recurrence of colorectal disorders [12, 13].

Lactulose, chemically known as 4-O-β-D-Galactosyl-D-fructose, is a non-absorbable sugar that reduces blood ammonia and is used in the treatment of constipation and hepatic encephalopathy [14]. Lactulose was shown to modulate the gut microbiota and thus be helpful for the treatment of inflammatory bowel disease and to reduce colorectal polyps and cancer [15].

Vitamins A maintain gut health by increasing the production of mucin and Bifidobacterium responsible for the development of a proper immune response and reducing colon cancer. Vitamin C, also known as L-ascorbic acid, is a water-soluble vitamin that produces an antioxidant effect to inhibit colorectal cancer. Vitamin E consists of tocopherols and tocotrienols function as fat-soluble antioxidants that may help protect cell membranes from reactive oxygen species, thus reducing the inflammatory bowel disorder and colorectal cancer [16].

N-acetylcysteine (NAC) is a derivative of the amino acid L-cysteine, with an acetyl group attached, known for its antioxidant reducing oxidative stress and inflammation, and may even have a role in preventing colon cancer [17].

But their mode of binding at a molecular level is not yet explained. Therefore, it is our prime focus to expedite the molecular mechanisms of these ligands against specific gene KRAS G12C overexpression linked with the PIK3CA mutations responsible for the development of gastric and colorectal cancer. While PIK3CA mutations often arise alongside inducing mutations in KRAS-BRAF cancer, they typically manifest as isolated events in gastric carcinoma, particularly in cases of mismatch repair deficiency. An attempt has been made in the present study to explore the mode of binding (MOB) and affinity prediction of these chemo preventives against KRASG$_{12}$C overexpression of which may cause colorectal cancer.

2 Materials and Methods

2.1 Preparation of Bioactive Ligands

The present small molecule database considers lactulose, vitamins A, C, E, and N-acetyl cysteine (NAC) as ligands. The 3D drawn structures of lactulose, vitamins A, C, E, and N-acetyl cysteine (NAC) were energetically minimized, taking MMFF as force field and 0.1 as RMS gradient [18] using Chem3Dultra [19]. The optimized ligands having minimal surface energy are taken as bioactive ligands for targeting the protein active site binding.

2.2 Preparation of the Target Protein

In the present study, KRAS G12C has been taken as a target protein. The KRAS G12C indicates Kirsten rat sarcoma viral oncogene homolog mutations glycine 12 to cysteine. The KRAS G12C mutations are the most common in pancreatic (2/3 of KRAS mutations) and colorectal (almost 50% of KRAS mutations) cancers [20, 21].

The target protein has been downloaded from the Protein Data Bank with the PDB ID of 6OIM. The water and heteroatoms were removed from the protein. Small-molecule ligands are known to explore the Switch I/II and Switch II pockets of the KRASG12C protein and inhibit its function. The active site of KRASG12C was shown to bind to the Switch II pocket, inhibiting GTP nucleotide exchange and rendering the protein in a GDP-bound inactive conformation. This pocket is occupied by amino acid

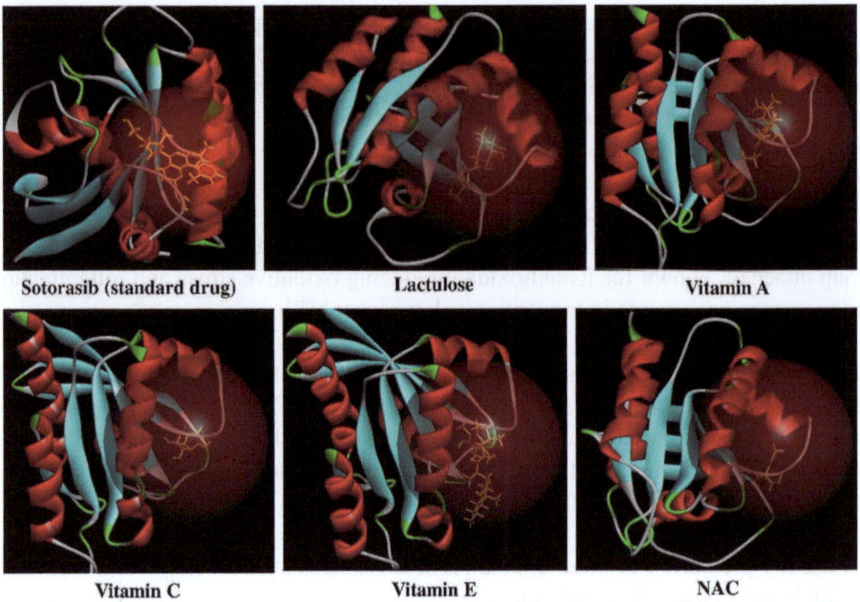

Fig. 1. Ligands perfectly cleft at the active site (red circle) of the target KRASG12C (PDB: 6OIM).

residues such as VAL/CYS12, GLN61, GLU62, GLU63, ARG68, ASP69, MET72, ASP92, HIS95, and TYR96 which have been taken to define the binding pocket of the co-crystal sotorasib [22].

2.3 Molecular Docking Simulation

The CHARMM energy minimization was applied using Discover Studio Client, v25.1.0.24284, using the standard smart minimizer protocol that considers 2000 steps associated with a gradient threshold of 0.01 kcal/mol. The active site of the energy-minimized protein molecule was defined by taking the sotorasib co-crystal docked amino acids surrounded by the active cavity grid box. The docking of all optimized ligands was performed by taking their MDL Mol files using the CDOCKER function of BIOVIA Discovery Studio [23]. The docked poses of all compounds are shown in Fig. 1.

3 Results and Discussion

Molecular docking is an in silico simulation to check the binding affinity of the ligand inside the cavity of the target. It helps to obtain the optimal geometry of the ligand-receptor complex having minimal interaction energy, which is expressed in terms of a score that indicates binding affinity [24, 25]. The present docking simulation is a flexible process [26] where the ligand can rotate inside the cavity to generate a number of

50 conformer complexes. The lowest score of the docked complex conformation has been selected for the analyses of the mode of binding at the molecular level, and it can be taken as the most bioactive conformation, having maximal affinity to bind the target site.

3.1 Mode of Binding Analyses

3.1.1 Sotorasib (Standard Drug). Sotorasib has been taken as the standard because it is co-crystallized at the active site of the target. The X-ray co-crystal cleft site has been analyzed. It was found that the residues such as GLY10, CYS12, LYS16, GLN61, GLU63, ARG68, MET72, TYR96, and GLN99 have been associated as the defined active site [27].

To validate the docking study, the cocrystal was removed. The sotorasib was drawn, and a 3D mol file with optimised structure was redocked into the same defined target site. It also shows interactions with different amino acid residues such as GLY10, CYS12, LYS16, ALA59, GLY60, ARG68, MET72, and TYR96, which match with the co-crystal docked residues having a docking score value of −61.97 kcal/mol (Fig. 2).

The pyrido-pyrimidine aromatic molecule is stacked with TYR96 by pi-pi. The propan-2-yl of the N1 pyridine moiety can produce an alkyl hydrophobic interaction with TYR96. The 4-prop-2-enoyl group produces a hydrogen bond with LYS16. The 4-piprazine can produce non-classical hydrogen bonding with GLY10 and GLY60. The same can produce alkyl hydrophobic interactions with CYS12 and ALA59. The 6-hydroxyl moiety of 7 phenyl can make a hydrogen bond with ARG 68, and it can also produce pi-alkyl hydrophobic interaction with Met72.

3.1.2 Lactulose. The 3,4,5-hydroxyl moiety of pyran can produce hydrogen bonding with GLN61 and GLU62. The 5-hydroxyl of furan can make the H bonding with ALA59. The non-classical hydrogen bonding interactions are produced by GLY10, CYS12, THR58, GLY60, GLU63, and TYR96 (Fig. 3).

3.1.3 Vitamin A. Vitamin A can make hydrogen bonding with PRO34 and alkyl hydrophobic interaction with CYS12, ALA59, and TYR96 (Fig. 4).

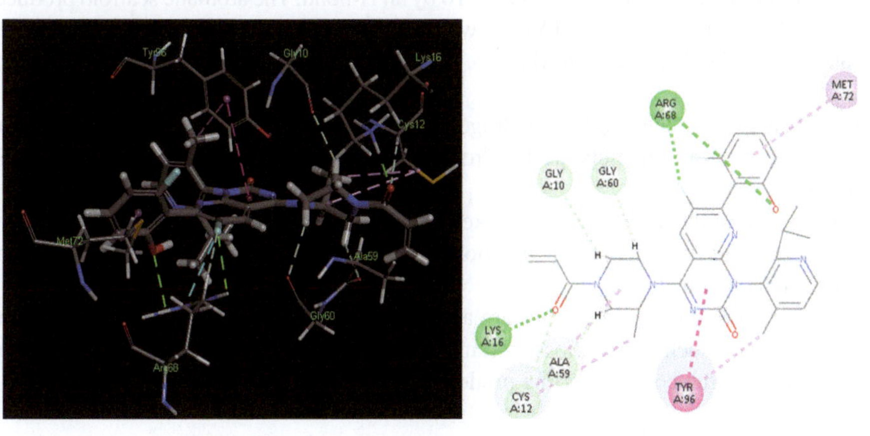

Fig. 2. Best pose conformer complex of sotorasib.

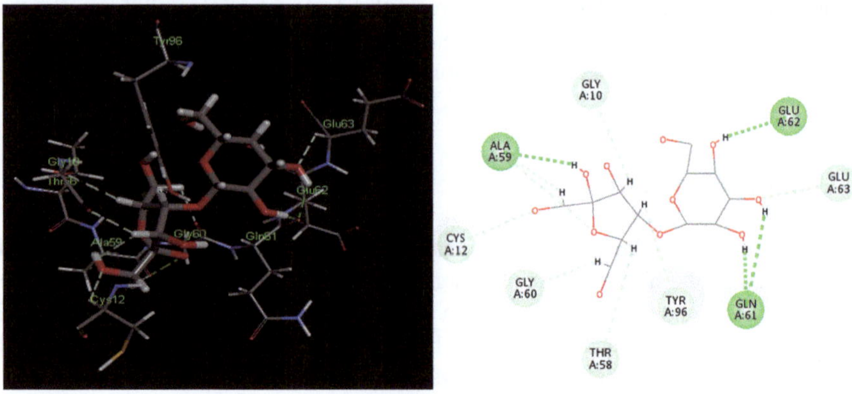

Fig. 3. Best pose conformer complex of lactulose.

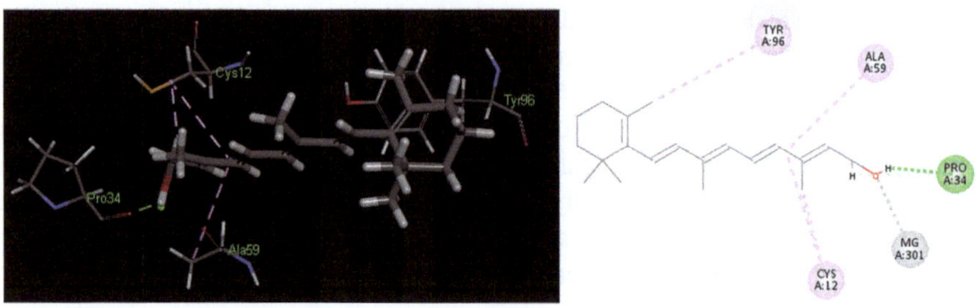

Fig. 4. Best pose conformer complex of Vitamin A.

3.1.4 Vitamin C. The 6-hydroxyl and 2-one moieties produce hydrogen bond interaction with GLN61 and ARG68, whereas the 1 and 2 dihydroxyl moieties make H-bonding with GLY10, TYR96, and THR58. The GLY 60 can produce a non-classical hydrogen bond (Fig. 5).

3.1.5 Vitamin E. It is bonded with GLY10 by an H-bond. The aromatic scaffold produces pi-hydrophobic interaction with TYR96, whereas CYS12, VAL9, and HIS95 are connected by alkyl hydrophobic interactions (Fig. 6).

3.1.6 NAC. NAC can produce hydrogen bond interaction with GLY60, TYR96, and LYS16 where whereas non-classical hydrogen bonding is produced by CYS12 and ALA59 (Fig. 7).

On analysing the above data of docked complexes of all ligands, the list of interacting residues and dock scores of the chosen ligand molecules in the present study has been summarised in Table 1.

All best docked poses of ligands are aligned on the reference sotorasib. It was found that the ligands such as lactulose, vitamins. A, C, and NAC (stick model) have been well aligned on the co-crystal (line model) at the target binding cavity (Fig. 8a).

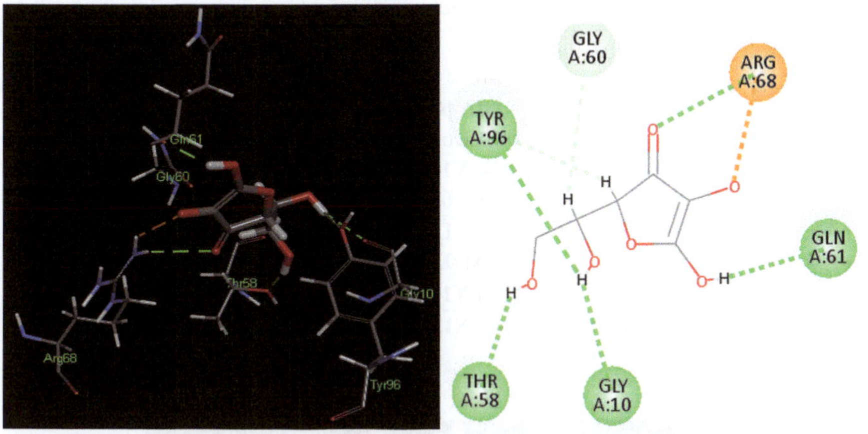

Fig. 5. Best pose conformer complex of Vitamin C.

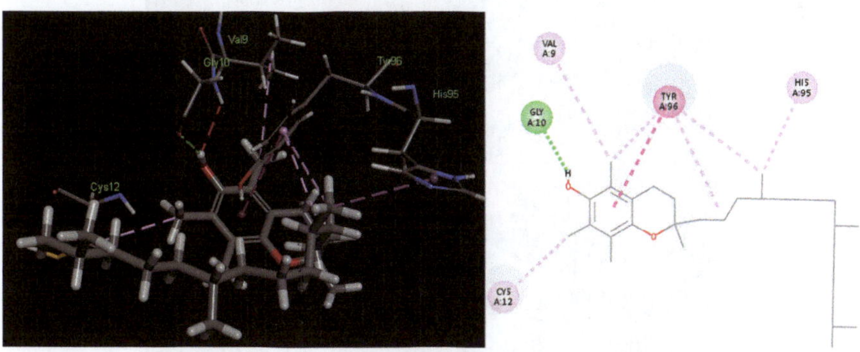

Fig. 6. Best pose conformer complex of Vitamin E.

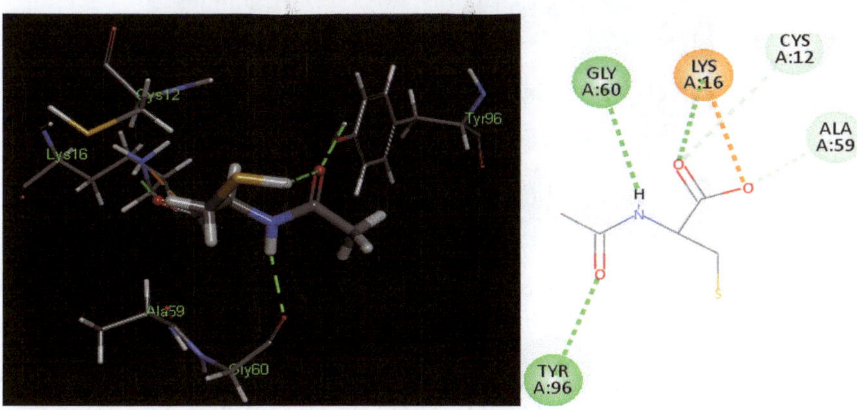

Fig. 7. Best pose conformer complex of NAC.

Table 1. Scores and modes of all ligand interactions.

Ligand	Binding Affinity (score)	Mode of binding towards KRASG12C target
Sotorasib (AMG510)	−45.05	GLY10, CYS12, LYS16, PRO34, ALA59, GLY60, GLU63, ARG68, MET72, TYR96
Lactulose	−44.44	**GLY10, CYS12**, THR58, **ALA59, GLY60,** GLN61, GLU62, **GLU63**
Vit A	−39.00	**CYS12, PRO34, ALA59, TYR96**
Vit E	−48.83	VAL9, **GLY10, CYS12**, HIS95, **TYR96**
Vit C	−28.28	**GLY10,** THR58, **GLY60,** GLN61, **ARG68, TYR96**
N-acetyl cysteine (NAC)	−29.49	**CYS12, LYS16, ALA59, GLY60, TYR96**

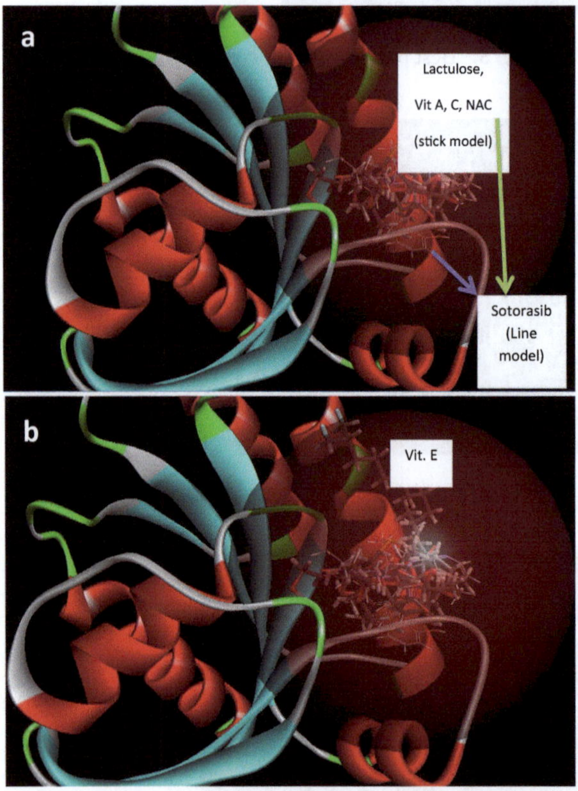

Fig. 8. Best docked pose conformer complexes of all ligands except Vit. E (a) and including Vit. E (b).

The *benzopyran ring of* Vit E was aligned with the reference molecule, and some portion of the *4,8,12-trimethyltridecyl long chain* has been deflected from the aligned zone, so it produces an uncommon binding a bit with a greater number of amino acid residues, such as VAL9 and HIS95 (Fig. 8b).

4 Conclusion

It was found that Vitamin E and lactulose show promising dock scores of −48.83 Kcal/mol and − 44.44 Kcal/mol. which are comparable with the standard drug sotorasib, having a dock score of −45.05 Kcal/mol. with a common binding residue of GLY10, CYS12, and TYR96, while vitamin A produces a dock score of −39 Kcal/mol. having common interactions of CYS12, PRO34, ALA59, and TYR96 towards KRAS G12C inhibition. The vitamin C and NAC have also shown satisfactory binding inhibition with dock scores of −28.28 Kcal/mol and − 29.49 Kcal/mol. Vitamin C produced interactions with GLY10, THR58, GLY60, GLN61, ARG68, and TYR96 whereas NAC produced interaction with CYS12, LYS16, ALA59, GLY60, and TYR96 towards KRAS G12C inhibition.

References

1. Amitay, E.L., et al.: Smoking, alcohol consumption and colorectal cancer risk by molecular pathological subtypes and pathways. Br. J. Cancer. **22**(11), 1604–1610 (2020)
2. Frezza, E.E., Wachtel, M.S., Chiriva-Internati, M.: Influence of obesity on the risk of developing colon cancer. Gut. **55**(2), 285–291 (2006)
3. Aykan, N.F.: Red meat and colorectal cancer. Oncol. Rev. **9**(1), 288 (2015)
4. Lin, C.L., Liu, T.C., Wang, Y.N., Chung, C.H., Chien, W.C.: The association between sleep disorders and the risk of colorectal cancer in patients: a population-based nested case-control study. In Vivo. **33**(2), 573–579 (2019)
5. Li, X., et al.: Unhealthy lifestyle factors and the risk of colorectal cancer: a Mendelian randomization study. Sci. Rep. **14**, 13825 (2024)
6. Chen, Y., Zhou, J., Wang, L.: Role and mechanism of gut microbiota in human disease. Front. Cell. Infect. Microbiol. **11**, 625913 (2021)
7. Loke, Y.L., Chew, M.T., Ngeow, Y.F., Lim, W.W.D., Peh, S.C.: Colon carcinogenesis: the interplay between diet and gut microbiota. Front. Cell Infect Microbial. **10**, 603086 (2020)
8. Kovatcheva-Datchary, P., Arora, T.: Nutrition, the gut microbiome and the metabolic syndrome. Best Pract. Res. Clin. Gastroenterol. **27**, 59–72 (2013)
9. Sanat, Z.M., Ganjaroudi, N.M., Mansouri, M.: The effect of topical Nifedipine versus diltiazem on the acute anal fissure: a randomized clinical trial. Middle East J. Dig. Dis. **15**(2), 121–125 (2023)
10. Mori, H., Tack, J., Suzuki, H.: Magnesium oxide in constipation. Nutrients. **13**(2), 421 (2021)
11. Tedsco, F.J., Dipiro, J.T.: Laxative use in constipation. Am. J. Gastroenterol. **80**(4), 303–309 (1985)
12. Roncucci, L., et al.: Antioxidant vitamins or lactulose for the prevention of the recurrence of colorectal adenomas. Dis. Colon Rectum. **36**(3), 227–234 (1993)
13. Ponz de Leon, M., Roncucci, L.: Chemoprevention of colorectal tumors: role of lactulose and of other agents. Scand. J. Gastroenterol. Suppl. **222**, 72–75 (1997)
14. Prasad, S., Dhiman, R.K., Duseja, A., Chawla, Y.K., Sharma, A., Agarwal, R.: Lactulose improves cognitive functions and health-related quality of life in patients with cirrhosis who have minimal hepatic. Encephalopathy. **45**(3), 549–559 (2007)
15. De Leon, M.P., Roncucci, L.: Chemoprevention of colorectal Tumors: role of lactulose and of other agents. Scand. J. Gastroenterol. **32**, 72–75 (1997)

16. Leenders, M., et al.: Plasma and dietary carotenoids and vitamins A, C and E and risk of colon and rectal cancer in the European prospective investigation into cancer and nutrition. Int. J. Cancer. **135**, 2930–2939 (2014)
17. Zuhra, K., et al.: *N*-Acetylcysteine serves as substrate of 3-Mercaptopyruvate Sulfurtransferase and stimulates Sulfide metabolism in colon. Cancer Cells. **8**(8), 828 (2019)
18. Halgren, T.A.: Merck molecular force field. III. Molecular geometries and vibrational frequencies for MMFF94. J. Comput. Chem. **17**, 553–586 (1996)
19. Mills, N.: Chemdrawultra10.0. **128**, 13649–13650 (2006)
20. Neumann, J., Zeindl-Eberhart, E., Kirchner, T., Jung, A.: Frequency and type of KRAS mutations in routine diagnostic analysis of metastatic colorectal cancer. Pathol. Res. Pract. **205**(12), 858–862 (2009)
21. Forbes, S.A., et al.: COSMIC: mining complete cancer genomes in the catalogue of somatic mutations in cancer. Nucleic Acids Res. **39**, 945–950 (2011)
22. Bröker, J., et al.: Fragment optimization of reversible binding to the switch II pocket on KRAS leads to a potent, in vivo active KRASG12C inhibitor. J. Med. Chem. **65**(21), 14614–14629 (2022)
23. Biovia, D.S., Dsme, R.: Dassault Systèmes, Discover Studio Client, v25.1.0.24284. Dassault Systèmes, San Diego (2025)
24. Nandi, S., Bagchi, M.C.: 3D-QSAR and molecular docking studies of 4-anilinoquinazoline derivatives: a rational approach to anticancer drug design. Mol. Divers. **14**, 27–38 (2010)
25. Stahl, M., Rarey, M.: Detailed analysis of scoring functions for virtual screening. J. Med. Chem. **44**, 1035–1042 (2001)
26. Teague, S.J.: Implications of protein flexibility for drug discovery. Nat. Rev. Drug Discov. **2**, 527–541 (2023)
27. Canon, J., et al.: The clinical KRAS(G12C) inhibitor AMG 510 drives anti-tumour immunity. Nature. **575**(7781), 217–223 (2019)

Pathogenesis, Mode of Treatment and to Explore Current Targets of Trypanosomiasis

Astha Saha[1], Rukayya[2], Debapriya Ghosh[1], Ashaporna Biswas[1], Sujan Murmu[1], Amar Faruk Saheb[1], Tamanna Khatun[1], Barsha Roy[1], Asmita Samadder[1,3](✉), and Sisir Nandi[2](✉)

[1] Cell and Developmental Biology Special, Department of Zoology, University of Kalyani, Kalyani, Nadia, West Bengal, India

[2] Department of Pharmaceutical Chemistry, Global Institute of Pharmaceutical Education and Research (GIPER), Kashipur, Uttarakhand, India
sisir.iicb@gmail.com

[3] Cytogenetics and Molecular Biology Lab., Department of Zoology, University of Kalyani, Kalyani, Nadia, West Bengal, India
asmita.samadder@gmail.com

Abstract. *Trypanosoma brucei* and *Trypanosoma cruzi* are the etiological agent of Human African Trypanosomiasis (sleeping sickness), and American Trypanosomiasis (Chagas disease) respectively. Trypanosomiasis has been the prime concern for global health issues and has been designated as world's 17th most neglected tropical diseases (NTD) by the World Health Organization (WHO). Trypanosomiasis is the most common among low-income countries and communities and claim immediate addressing to save lacks of lives. However, there are no specific synthetic drugs known to cure the Trypanosomiasis. Existing chemotherapeutics are unable to kill the protozoa and therefore, the causative microorganisms become modified and transformed to more pathogenic forms. Therefore, this issue calls for prime exploration of the disease pathogenesis, and current targets for *Trypanosoma brucei* and *Trypanosoma cruzi* for the design and development of potent specific drugs targeting the protozoa.

Keywords: Trypanosomiasis · Drug targets · Pathobiology · Mode of treatment

1 Introduction

Trypanosomiasis is caused by the bite of hematophagous insects and spread through the saliva of tsetse fly vectors. More than eleven subspecies of *Trypanosoma* are found that cause many acute diseases. Due to its lack of diagnostic paradigm and insufficient

The authors Astha Saha, Rukayya, Debapriya Ghosh, and Ashaporna Biswas have been equally contibuted to this work.

A. K. Saxena, A. Saxena (Eds.): GTHTM 2025, pp. 335–355, 2026.
https://doi.org/10.1007/978-3-032-12320-6_20

treatment strategies pertaining to medical advancement, and with low preventive action, WHO identified this disease as a neglected tropical disease. Present review elaborates the disease pathogenesis, various targets and possibilities to combat trypanosomiasis using synthetic chemotherapeutics [1].

Trypanosoma is a genus of order- Trypanosomatida that are particularly of both clinical and veterinary significance. The Trypanosomatidae are obligatory pathogenic protozoans that affect numerous vertebrates as well as insects and plants [2]. The organism belongs to the phylum Euglenozoa. Trypanosomatids are uniflagellate invertebrates, like all the other species in the class kinetoplastida, and are distinguished by something like a specialized mitochondrial genetic material also referred to as kinetoplast. This is constituted of DNA "minicircles" as well as "maxicircles," the quantity and catenation or linking of which varies in accordance with the species concerned [3]. Besides kinetoplast, they also have acidocalcisomes and glycosomes which are the main site used for storing minerals within the cell. Additionally, it has been demonstrated that the cytoskeleton of trypanosome cells is specialized and primarily made up of microtubules. However, centrosomes, which are crucial for cellular activities, are absent in them [4]. Because of this, parasite cells' ill-defined components are in charge of producing the microtubule spindles which aid in all these pathogens' closed divisions. Regarding the parasite's peculiar overall architecture, it is characterized by specific tubulin that works against the cellular membranes. Other distinguishing feature of trypanosomes includes their solitary flagellum, which can range in size between 2 and 20 μm and gets stabilized via basal and probasal structures [5]. The flagella of trypanosomes have a 9 + 2 arrangement made of parallel microtubules, similar to the motile ciliary structure and flagella found in numerous eukaryotes. All along the length of such flagella, there is a tubular structure known as paraflagellar rod (PFR) that has been proven to assist the flagellum's axoneme through increasing firmness and also to encourage movement [6]. There is a space called a flagellar pouch where the flagellum penetrates into the cell. It is thought that the location of vesicular transportation as well as nutrient absorption is within this space [7]. Members of salivaria category of trypanosomes are *Trypanosoma brucei*, *Trypanosoma cruzi, Trypanosoma vivax* and *Trypanosoma congolense* as they spread through the saliva of tsetse fly vectors. Amongst them, extensively researched salivarian trypanosome are *Trypanosoma brucei and Trypanosoma cruzi* [8].

1.1 *Trypanosoma brucei* Causing Sleeping Sickness

Nowadays, human African Trypanosomiasis (HAT) or sleeping sickness is becoming the world's most falling apart and ignored diseases. The parasites known to cause the disease are the African salivarian trypanosomes, transmitted by the tsetse flies (*Glossina* sp.) which spread the disease to human and other livestock by biting them. There are two forms of trypanosomes behind human disease- *T.b.rhodesiense* causing HAT in eastern and southern Africa and *Trypanosoma brucei gambiense* causing HAT in central and western Africa. *T. b. gambiense* is responsible for 92–97% of the reported cases of the disease. The disease not only causes acute morbidity and mortality if remain untreated but also create hindrance against production of livestock. 50 million people of

36 countries are facing the risk of getting the disease as per the record of 25000–50000 new cases reported annually [9].

Tsetse flies, the main vector of the disease although prevalent in Sub-Saharan African countries, but only few species carry this infection. Sleeping sickness is indeed uncommon in many places where tsetse fly is widespread for unspecified reasons [10]. People in rural areas who are dependent on agriculture, fishery, livestock farming, or poaching in areas where transmission prevails are particularly vulnerable to the tsetse flies and hence to the infection. The incidence of *Trypanosomiasis* as well as an increase in sleeping sickness, is inextricably connected to poverty, immigration, starvation, military conflicts, deforestation, and pushing people and animals to be exposed to tsetse flies (*Glossina* sp.) in tsetse-infested regions. On the contrary, socioeconomic insecurity as well as geopolitical tensions jeopardize disease surveillance networks. The disease can spread from just a single province to a whole nation. The burden of the disease may differ from one village to another in the core of an infected area [11].

1.1.1 Pathogenesis of *T. brucei* Infection. As trypanosomes depend on blood, they may be transmitted from one organism to another. During the blood meal, an infested tsetse fly (*Glossina* sp.) introduces metacyclic trypomastigotes inside the epidermal tissue of the human host [12]. The parasite penetrates the lymph nodes and afterwards reaches the bloodstream. After that if the healthy tsetse fly bite to the infected person, it ingests the blood conatinig trypomastigote convert into the procyclic trypomastigote followed by epimastigote and polycyclic trypomastigote. Within the recipient, they develop into circulatory trypomastigotes and are transported to different areas all throughout the body. They penetrate various body fluids (for example, lymphatic, cerebrospinal fluid), and complete multiplication by binary fission [13]. Extracellular stages comprise the total African trypanosome life cycle (Fig. 1).

Two clinical phases of the disease include an initial or preliminary phase in which the pathogen is present only in the bloodstream, and a second or final phase during which the pathogen penetrates the blood-brain boundary and enters the central nervous system (CNS) [14]. The very first phase includes generating an immunological reaction inside the host, which is avoided by the parasites through rounds of parasitic infection, which aid certain pathogens to persist and avoid the immunogenic reaction [15, 16]. Trypanosomes escape the immunogenic reaction through a procedure termed as antigenic variation by transforming their specific surface glycoprotein coat to a new variant surface glycoprotein (VSG) coat that cannot be recognized by the host's immune system. This persistent activity diminishes the patient's overall defense mechanisms against the parasitic infestation [17]. Endocytosis of VSG immunoglobulin aggregates occurs during the escape procedure, enabling them to avoid identification by the antibodies involved for complement-mediated death. Antigenic mutation allows trypanosomes to remain longer inside the recipient, resulting in greater multiplication as well as transfer to many other animals through the tsetse vector. Finally, this approach permits trypanosomes to attack organisms that are already familiar with VSGs from previous attacks [16, 18, 19]. The initial symptoms in the host consist of fever, headache, joint pain, dermatitis at the area of infection, etc., in approximately 19% of the total number of instances. The much more noticeable characteristic of this phase is fever, which lasts from a single day to approximately seven days, with gaps between occur-

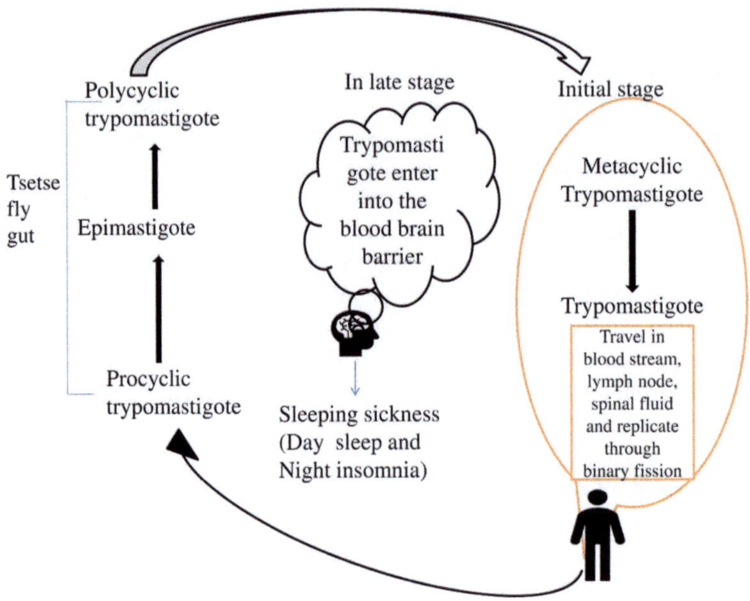

Fig. 1. Life cycle of *T. brucei*.

rences varying from a few days to almost a month. The disease progresses in two phases: a haemolymphatic phase, followed by the meningoencephalitis phase when the trypanosomes penetrate the CNS. Nonetheless, a large number of symptoms and indications are common in both phases, considering potential aspects alone difficult to identify in between two. Between 2 days to 14 days of just being stung by an infectious fly, the emergence of a trypanosomal pathological condition just at location of inoculation can elicit the beginning of first-stage signs [20]. The first stage is distinguished by ambiguous, generalized symptoms that begin 7 to 21 days after one tsetse fly attack with *Trypanosoma brucei rhodesiense*. The incubation time for *Trypanosoma brucei gambiense* is poorly defined, although illness proceeds relatively slower than that induced by *Trypanosoma brucei rhodesiense* infection.

1.1.2 Host Immune Invasion Strategies Used by the Parasite. Parasite evade the host immune system by complement system inhibition, antigenic variations and causing challenges in the development of vaccine. Parasite entered in to the host body and increase its number by division but escape from host immune systems by changing in variants surface glycoproteins (VSG) coat. Parasite have approx 100 VSG gene in their genome. Only one VSG is expressed by the parasite in a time from there telomeric expression site. Immune system couldn't recognize the parasite because of uniform surface coat. when host body try to generate antibodies against the pathogen, then parasite switch to another, this switching occurs by changing gene reengagement like gene conversion, reciprocal recombination. Again, host body try to generate antibody and ultimately cause immune exhaustion. GPI anchors help so that VSG are anchored to the membrane. To controlling chagas disease, targeting VSG switching mechanism s a potential strategy [21]

1.2 *Trypanosoma cruzi* Causing American Trypanosomiasis (Chagas Disease)

The American trypanosomiasis collectively known as Chagas disease is caused due to *Trypanosoma cruzi* infection which is transmitted through the bloodsucking bugs. Vector of Chagas disease is triatomine or 'kissing bug'. There are more than 100 species of Triatomine bugs which possess the capacity to harbor *T. cruzi*. The primary host of *T. cruzi* is humankind. Beside human, other animals such as palm tree-dwelling opossums, bat, guinea pig, rat, dog and various other small mammals serves as reservoir host. In the United States, opossums and raccoons are naturally found reservoir host of *T. cruzi*. Human are exposed to *T. cruzi* via transfusion, vectorial, congenital, organ transplant, oral and accidental ways which manifests the infection to occur with an initial phase of 6–8 weeks post infection.

Chagas disease is a concern with regard to health problems for multiple reasons like fever, headache, vomiting, abdominal pain, diarrhea and digestive hemorrhage, hepatosplenomegaly, jaundice, skin hemorrhages, neurological signs, myocarditis and meningoencephalitis for acute cases whereas chronic symptoms are dysphagia, odynophagia, circulatory issue, regurgitation, ptyalism, malnutrition, bradyarrhythmia, tachyarrhythmia, cardiac failure, epigastric pain, thromboembolism, apical aneurysm and sudden death [22].

1.2.1 Pathogenesis of *T. cruzi* Infection. When an infected triatomine bug with metacyclic trypomastigote in its saliva bite and sucks blood from the host, the parasites come into the contact with the wound surface and penetrate the host tissue. The metacyclic trypomastigote converted into amastigote followed by trypomastigote. After that if the healthy tsetse fly bites the infected person, it ingests the blood containing trypomastigote converts into the trypomastigote followed by epimastigote and metacytclic trypomastigote in the gut the tsetse fly (Fig. 2) [23]. Another way of infection is when the infected bug feeds on blood from vertebrates host, some of its blood meal enter in its stomach which helps in release of accumulated excreta including feces and urine which gets deposited on the skin and metacyclic trypomastigote penetrate inside the human body through the releasing of histolytic enzyme [24]. Then, by using various factors, *T cruzi* pathogen can enter the host body, which is a very complex process [25]. All phagocytic and nonphagocytic nucleated cells of *Trypanosoma cruzi* have tropism, and their surface is covered by the dense glycocalyx, with varying composition in each stage of the life cycle [25] [26]. Chagas disease has 2 clinical phases, one is the acute phase known as acute chagas disease (ACD), and another is the chronic phase known as Chronic chagas disease (CD) [27].

Acute chagas disease (ACD): It occurs 8–10 days after infection where 90% of ACD remains asymptomatic except for certain unspecific mild symptoms occurrence at instances [28] displayed feebly in adult and children after 7 to 14 days of incubation which includes fever, nausea, hepatosplenomegaly, anorexia, lymphadenopathy [29]. Sometimes the symptoms also seem similar to viral infections unless Romana's sign or chagoma appear. Chagoma is a very important clinical feature where the inflammation occurs at the bug bite site after conjunctival infection [30]. In this phase, very highly efficient control occurs as a result of the inflammatory response, which elicits the pro-

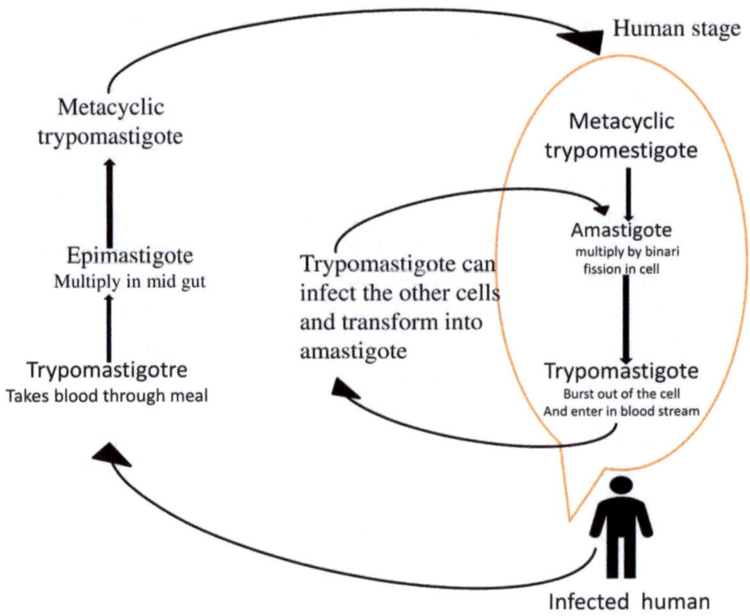

Fig. 2. Life cycle of *T. cruzi*.

duction of antibody activation and innate immune system activation by Th1 proinflammatory cytokines such as tumor necrosis factor alpha [31].

Chronic chagas disease: After one year of the acute phase, the chronic phase starts with low-intensity inflammatory responses as a marker, followed by pathological changes in the size and shape of the heart, the ventricular wall becoming thin, dense, and fibrous scar tissue occurring, atrophy, of heart failure. Especially in the heart, it has various effects such as sinus node dysfunction, bradycardia, ventricular arrhythmia, left ventricular aneurysm, enlargement of the heart, and many types of heart dysfunction. Failure of the heart and death is most common in patients with this phase. Besides these various types of cardiomyopathy, the function of the autonomic nervous system is also altered. The amount of cardiac acetylcholine and choline acetyltransferase is reduced in experimental animals and harms the parasympathetic nervous system [32]. The overflow of acetylcholine was greatly reduced in the case of infected rats. Due to the alternation of autonomic nervous system function in patients who is affected with *T. cruzi* sudden reduction of plasma norepinephrine, diastolic blood pressure has been repeated [29].

1.2.2 Complement System Inhibition by Parasite to Evade Host Immune System. Parasite have various complement system inhibitory proteins that allow it to evade the host immune response through various mechanisms. Some protein such as gp58/68, TcCRP, T-DAF, FH, CRIT, PMV these all proteins have main role in the Inhibition of C5 and C3 convertase and results are down regulation opsonization, reduce the formation of C3a & C5a, inhibit the complement mediated lysis.

Glycoprotein 58/68: A special type of protein present on trypomastigote. C3 convertase of alternative pathway is inhibited by this protein. *T cruzi* has fibronectin/collagen receptor and gp58/68 is the part of this receptor, have crucial role in host parasite interaction by facilitating adhesion to mammalian cells and aiding immune evasion through the inhibition of FB/C3b interaction, thereby preventing alternative pathway complement activation.

Trypanosoma cruzi Complement Regulatory Protein (TcCRP): A protein also known as gp160 found on the surface of *T cruzi* and inhibits alternative pathway & classical pathway. This protein is expressed by trypomastigote but not by epimastigote. TcCRP-transfected epimastigotes are protected from complement-mediated lysis.

Trypomastigote decay accelerating factor (T-DAF): A 70 kDa glycophospholipid present on the membrane of trypomastigote regulates CP and AP by C3 and C5 convertase. Epimastigote does not Express this protein. It alters C3 convertase in CP on trypomastigote. It gives protection to parasite from mammalian immune system [32, 33].

1.3 Treatment of Trypanosomiasis

It has been a big challenge to treat the Trypanosomiasis infection. Exploration of the efficacy of therapy in the second phase is based on therapeutics that can get through the blood-brain barrier and gain access to the pathogen [34]. Six major medications, such as suramin, pentamidine, melarsoprol, eflornithine, acorziborole and pafuramidine are given to treat sleeping sickness [35] whereas the nifurtimox and benznidazole are majorly given for the treatment of Chagas disease. Fexinidazole is a medicine used to treat the disease in both phases of the disease [36]. Suramin and pentamidine are two medications that are prescribed to treat the disease in the initial phase. Table 1 summarizes the current treatments for trypanosomiasis, their mechanisms of action, and valuable side effects.

1.3.1 Treatment of HAT. Suramin, the first WHO-approved drug given as injectable form, is a polysulphonatednaphthyl urea. It is a symmetrical molecule containing the urea functional group. Naphthaline moiety and 4-amide functional group. Suramin consists of 4 benzene rings responsible for aromaticity. The drug has more than five hydrogen bond donors and more than ten hydrogen bond acceptors. The mechanism of action is not clear, but it is thought that the parasite uptake suramin via receptor receptor-mediated endogenous pathway and the drug bound to the lipoprotein and other serum inside the parasite may inhibit the enzyme metabolism. Suramin has a number of negative adverse effects, such as nephrotoxicity and hypersensitivity responses [37].

Pentamidine is an antiprotozoal agent having antiparasitic activity against *T. b. rhodesiense* and used to early stage of Gambiense without involvement in CNS. It is also used to prevent and treat pneumocystis pneumonia (PCP) and visceral leishmaniasis that is also called kala azar caused by *Leishmania* parasite. It is available in injection form and given into vein or inhalation. This diamidine compound is used in the initial stage of the disease especially in haemolymphatic stage. Diamidine interferes with the DNA double helix, particularly by binding with AT rich sequence of DNA of these parasites, which inhibits the replication and transcription process of DNA inside the

Table 1. Drugs along with their structure, MOA and side effects.

Sl. No.	Drug	MOA	Structure	Side effects
1.	Suramin	Parasite uptake drug via receptor and drug bound to lipoprotein and other serum protein	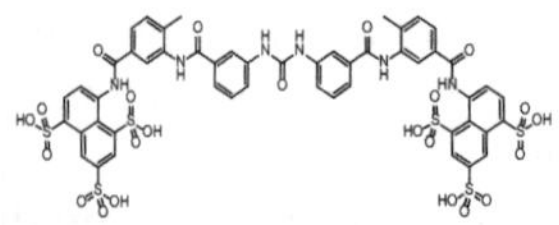	Loss of consciousness, fatigue, malasise
2.	Pentamidine	Drug's diamidine compound binds with AT rich sequence of DNA and prevent replication of parasite	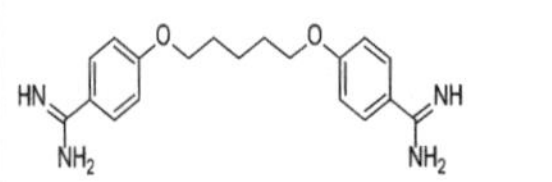	Hypo-glycemia, dysgeusia, syncope
3.	Melarsoprol	Drug enters into parasite body via aquaglyceroporin2 trans-protein and disrupt energy production	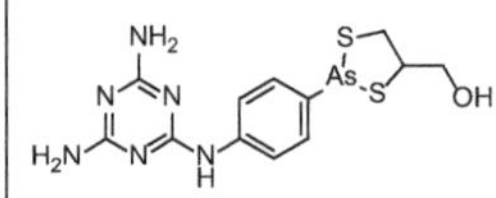	Presence of albumin kidney, increase blood pressure
4.	Eflornithine	Inhibit ornithine decarboxylase enzyme, so that process of formation of DNA, RNA and protein gets inhibited	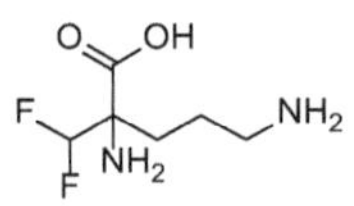	–
5.	Acoziborole	It inhibit the activity of the proteosome	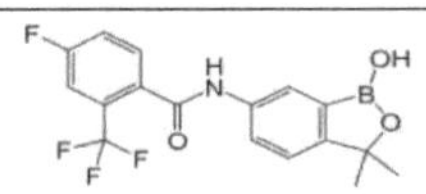	Mile side effect with fever and asthenia
6.	Nifurtimox	Generates oxidative stress by producing ROS such as superoxide anions via various oxidation reduction reaction of its nitro constituent	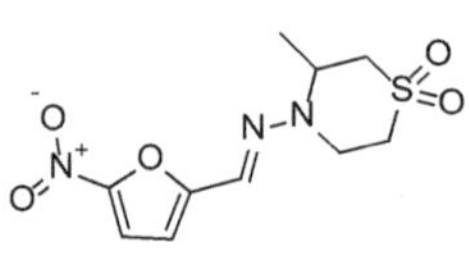	GI tract complication, nervous irritability, insomnia
7.	Benznidazole	Same as nifurtimox damage parasite's DNA by free radical generation	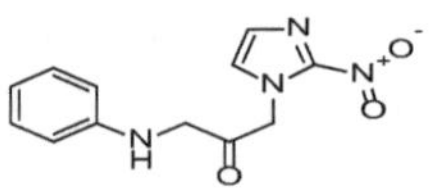	Polyneuropathy, bone marrow suppression

parasite, which may lead to the death of the parasite. It is given intravenously and due to its inability to pass the blood-brain barrier it has no activity against the late stage of sleeping sickness. The exact mechanism of pentamidine for pathogen clearance is not clear, but it is believed that it inhibits DNA, RNA, phospholipid and protein synthesis. The main adverse effect is mild nephrotoxicity.

The melarsoprol was first used in 1949 in the treatment of late stage of HAT having antiparasitic activity against *Trypanosoma brucei rhodesiense*. It is an arsenic containing medicament administered by intravenously. Administration of melarsoprol is painful and highly toxic. It is used to treat later stages of HAT when the parasite invades CNS. The drug has a large number of side effects as it contains arsenic, such as- brain dysfunction, numbness, rashes, kidney and liver problems, glucose-6-phosphate dehydrogenase (G6PD) deficiency, breakdown of red blood cell etc. Due to the high toxicity of these drugs, they are used in most severe parasitic infested cases, while other agents having low toxicity are used in the first stage of the disease. Melarsoprol drug enter the trypanosome cell with the help of some specific transporters via Aquaglyceroporin 2 (AQP 2) and the P2 adenosine transporters, thereby disrupting the energy production in the parasite [38].

Eflornithine (DFMO) is also called alpha difluoromethylornithine, used for chronic sleeping sickness treatment. Eflornithine shows better efficacy against *Trypanosoma brucei gambiese*, responsible for Gambian sleeping sickness, where it is more effective than melarsoprol to treat later stages, but not effective against Rhodesian (East African) sleeping sickness cause by *T. brucei rhodesiense*. The parasite contains the ornithine decarboxylase enzyme that is responsible for DNA, RNA, and protein synthesis. Eflornithine is an irreversible inhibitor, i.e., it inhibit the ornithine decarboxylase enzyme, whereby the parasites fail to synthesize DNA, RNA and protein.

Acoziborole, benzoxaborole derivative, novel boron containing oral drug and used to treat human African trypanosomiasis (HAT) caused by *Trypanosoma brucei gambiese* infestation. DNDi program run for the development of acoziborole. It assesses the safety efficacy of acoziborole for the patient with acute and chronic stages of the HAT. It is a safe and effective antiparasitic drug against *Trypanosoma brucei* along with *T. rhodesiense* and the S427 strain of *T. gambiense* [39].

Pafuramidine (DB289) is an active agent under clinical trials, which have shown their activity against trypanosomes. It is furamidine derivatives, which are in phase II clinical trials and also active against *Pneumocystis* pneumonia and malaria [40].

1.3.2 Treatment of Chagas Disease. Nifurtimox is a nitroimidazole derivative widely used in the treatment of Chagas disease, that is caused by *Trypanosoma cruzi* protozoan parasite. Nifurtimox is a prodrug and is administered orally. It undergoes nitro reduction for the activation of drugs. Two types of nitro reductase enzymes act for the activation of drugs, such as type I NADPH-dependent and type II which includes FAD and FMN cofactors. Though the proper mechanism of action is questionable, it is believed that it undergoes various reduction and oxidation reactions of its nitro constituent, which produces reactive oxygen species such as superoxide anions and generates oxidative stress within the cells. These free radicals accumulate and react with the parasite and cause DNA breakdown that may lead to the death of the parasite. The prescribed dose of nifurtimox is 8–10 mg/kg daily in three doses for 90 days, including some side effects, like gastrointestinal-tract complications, harmful effects on the central nervous system, like irritability, insomnia, and psychic alterations [41].

Benznidazole is nitroimidazole family of drugs having antitrypanosomal activity against *Trypanosoma cruzi* used in early and late stages. The mode of action is the same of nifurtimox. It is a prodrug and generates toxic free radicals such as reactive oxygen species after activation by the nitro reductase enzyme. These free radicals cause DNA damage inside the parasite. It is administered orally, and prolonged use causes toxicity and side effects such as dermatitis and hepatitis. The effective dose of benznidazole is

10 mg/kg daily for children or adults with acute infection. In case of chronic infection, the dose is 2.5 mg/kg twice daily for 60 days in case of adults. A maximum dose of about 300 mg/kg can be used for treatment; however associated with some side effects, like dermatitis can be seen in the first week of treatment among almost 30% of the patients. Polyneuropathy, bone-marrow suppression are other adverse effects found during treatment with this drug [42]. With the idea of more efficiency and safety, some new drugs have been discovered, though currently they are under observation. For example, several triazole derivatives TAK-187, are currently used. Posaconazole146 and E1224,147 a pro-drug of ravuconazole, are in phase II trials. Dronedarone, an amiodarone derivative, is used as for its anti-parasitic property. An anti-arrhythmic drug named as amiodarone may be used, though clinical efficiency is under observation [43].

1.4 Exploration of Trypanosomiasis Targets

Few drugs are available to combat *T. brucei* and *T. cruzi*. But these old drugs cannot withstand against chronic and acute forms of HAT and Chagas disease. To explore this issue, quest of potential targets is essential for the design of potent antitrypanosomal compounds. Figure 3 and Table 2 summarizes major treatments and various potential targets and their functions which focus the importance of this study for future perspectives.

1.4.1 *T. brucei* Targets

Cysteine Protease. Cysteine proteases (CP) activity was first recognized in culture of *T. brucei* during 1980s. The identification of cathepsin-L like peptidase was

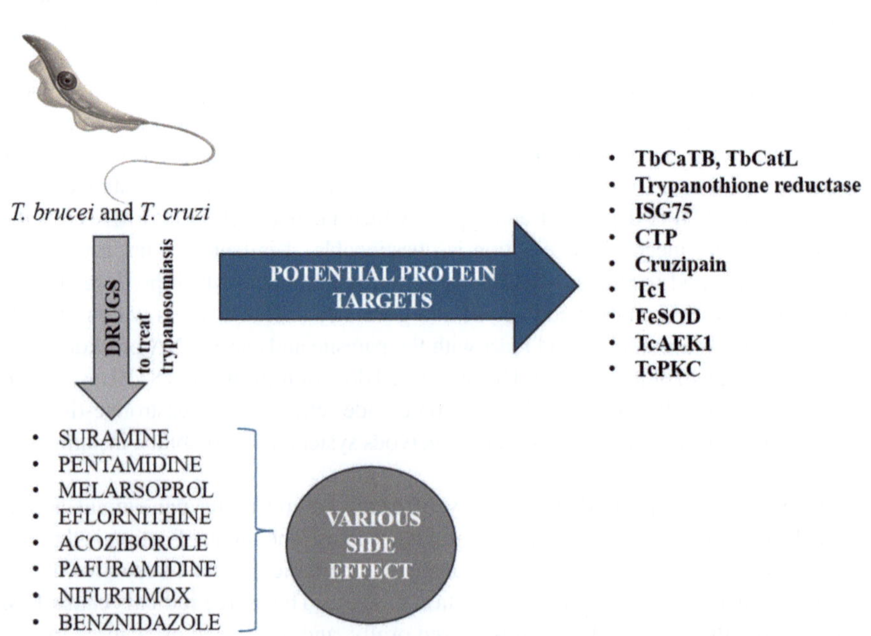

Fig. 3. Major treatments and various potential targets to combat trypanosomiasis.

Table 2. Targets of Trypanosomiasis and their functions.

Sl No.	Parasite	Target protein	Function
1.	*T. bruci*	TbCatB (cysteine protease)	*T.brucei* uptake iron from host body; It facilitates internalization of transferrin
2.		TbCatL (cysteine protease)	It cross the blood brain barrier and cause BBB failure; It breaks the host immunoglobulin
3.		Trypanothione reductase	It prevents oxidative damage derived from host immune system
4.		Intravariant surface glycoprotein (ISG75)	It binds with IgG with host and prevent the activation of classical complement system
5.		Cytidine triphosphate synthetase	It plays de-novo synthesis of CTP and maintain RNA and DNA level
6.	*T. cruzi*	Cruzipain (cystine protease)	Proper differentiation, host cell invasion; Host immune system inactivate
7.		Surface protein (Tc1)	Host cell invasion, activation of cascade essential for parasite infection survival
8.		Antioxidant defense enzyme (FeSOD)	Provide resistance against apoptosis, protect from superoxide radicals, drug resistance
9.		Protein kinases TcAEK1 and TcPKC)	Proper metaclyclogenesis, amastigote replication; Regulation of DNA synthesis, differentiation of epimastigote.

demonstrated through partial sequencing and cloning methodologies. By using immunoelectron microscopy [44], Rhodesain (TbCatL) was shown to be the major CP. Later, another lysosomal cathepsin-B-like enzyme was recognized and designated as TbCatB [45]. Nevertheless, the knockdown of rhodesain did not yield any observable phenotypic alterations; on the other hand, even a partial knockdown of TbCatB resulted in the mortality of the parasite, leading to the conclusion that the latter enzyme represents a more promising drug target for a new drug discovery strategy. According to the research conducted in vitro RNAi rhodesain knockdown (by 60%) [46] did not stop splenomegaly in mice or cause parasite mortality. Nonetheless, it improved mouse survival, indicating a crucial part in the onset of illness [47].

TbCatB

Clan CA cysteine proteases are integral to the developmental stages of numerous parasitic organisms. Bloodstream *T. Brucei* parasites exhibit the expression of two cysteine proteases belonging to the papain family, specifically rhodesain (also known as brucipain or trypanopain), which functions as a cathepsin L-like enzyme, alongside TbCatB, a cathepsin B-like enzyme [48]. TbCatB exhibit hallmark characteristics including triad (cys/His/Asn) and a substrate binding site like "occluding loop" composed of 20 amino acid residues on the surface of *Trypanosoma brucei* which provides with a quality of the cysteine protease with additional exopeptidase activity [49]. Conversely, it has been established that the bloodstream form (BSF) of *T. brucei* takes iron from the host organism through the process of endocytosis which facilitates the internalization of transferrin [50] subsequently, transferrin undergoes swift degradation within the endosomal/lysosomal compartments of the parasite, mediated by TbCatB. Remarkably human cathepsin B contain an invariant "GEGD" theme in this particular position [51]. The glycine residues gives extra empowerment in this region thereby enabling the negatively charged build-up to wave in and out of the dynamic location [47].

TbCatL

Rhodesain (TbCatL) is a cathepsin L-like cysteine protease belonging to Clan CA, family C1 (papain-family), and it is made up of a single 215 amino acid polypeptide chain. With its two domains, and the catalytic triad situated in a cleft between them, protease takes on the classic papain-like folding [52]. Rhodesain is important because it serves multiple purposes, inducing the disease's neurological stage because it helps the parasite to cross blood-brain barrier. This event is brought on by the activation of Gq-protein linked receptors called protease activated receptors (PARs), which raise intracellular calcium levels and cause endothelial cell cytoskeletal alterations as well as BBB failure [53]. By using antigenic variation, the parasite can easily dodge the host's immune system. Additionally, cysteine protease exhibits pertinent proteolytic activity in the lysosome, participating in the breakdown of host proteins that are carried intracellularly as well as parasite proteins. Its involvement in the breakdown of host immunoglobulin as an extracellular action is also important for drug development strategies [54].

Trypanothione Reductase. It is classified as the most potent drug target in *T. brucei* and absent in host. TR is a homodimer protein and consists of three distinct domains [55]. It shows more than 40% similarity with human glutathione reductase except the substrate binding site. It has molecular weight with 52 kDa [56]. TR functions as an NADPH-dependent flavoproteindisulphide oxidoreductase that is essential for the proliferation of trypanosomatids. TR prevents oxidative damage derived from the host immune system [57]. Dehydroascorbate and hydrogen peroxide are unpromptedly reduced by the trypanothione. Trypanothione reductase shields the parasite body by converting trypanothionedisulphide (TS2) to its active reduced form, trypanothione (T(SH)2), The process is done by using NADPH as an electron donor. The resulting form of trypanothione also helps in the synthesis of DNA precursor [58].

Invariant Surface Glycoprotein. The ISGs are categorized as type 1 transmembrane proteins, and their intracellular trafficking is connected with the ubiquitination of their cytoplasmic tails. Moreover, the ISGs are predominantly expressed in the parasitic forms, where they interact with the host's immune system. Although the expression levels of ISGs are significantly lower than those of the VSGs, the ISGs constitute the second most prevalent category of trypanosome surface proteins.

The invariant surface glycoprotein ISG75 is present in approximately 50,000 molecules per cell respectively. ISG75 is characterized as a 58-kDa surface protein, comprising a substantial ectodomain (residues 294–468), a transmembrane helix (residues 469–489), and a brief cytoplasmic tail (residues 490–523). The ectodomain of ISG75 possesses a singular predicted N-linked glycosylation site (Asn134) and features a coiled-coil motif (residues 280–450). The expression of ISG75 appears to be exclusive to the Trypanozoon subgenus [59]. ISG75 remains accessible for antibody recognition, rendering it a promising candidate for HAT antibody-based diagnostic applications. Given its invariant characteristics and elevated expression levels, ISG75 has also been considered a potential target for vaccine development [60]. Via interaction with IgG, ISG75 can prevent C1q and CD32/CD16 from binding to IgG, which interferes with the classical pathway of the complement system, Ab-mediated cellular phagocytosis (ADCP), and cytotoxicity (ADCC). This suggests that ISG75 protects the parasite from the host's immune system. With an estimated KD of 0.12 mM, ISG75's binding to IgG3 seems to be the strongest. With KD values between 5 and 6 mM, ISG75 interacts with the other IgG isoforms with comparable affinity [48, 61].

Cytidine Triphosphate Synthetase (CTP). It plays an important role in de novo synthesis of CTP, by combining the dephosphorylation of ATP with the deamination of glutamine to glutamate, which provides the necessary amino group. CTP synthetase catalyzes the conversion of UTP into CTP. The enzyme is very crucial in the synthesis of cytosine nucleotides, which are very important to maintain RNA and DNA levels. It is composed of separate synthetase and glutaminase domains. CTP synthetase is a possible target for drug discovery. Afterward, the parasite has a low quantity of CTP and cannot recover cytosine from the human host; the enzyme may be a good target for drug discovery [62].

1.4.2 *T. cruzi* Targets

T. cruzi CP. According to various studies of metazoan and protozoan papain family of cystine protease function, *T. cruzi* cystine protease Cruzipain, is an essential protease that is necessary for pathogen survival.

Cruzipain

In case of *T. cruzi*, cruzipain is one of the most well-characterized type I cysteine proteases [63] amongst its seventy different types of cysteine proteases. Cruzipain, a thiol-dependent cathepsin L-like cysteine protease [64], is a member of papain family which is involved not only in host cell invasion but also in activation of host immune response [63]. Cruzain is a glycoprotein, with a molecular mass of 51 kDa [65] expressed as a mix of isoforms [66]. The name cruzipain designates the native form of the enzyme and cruzain designates the recombinant form [63]. The enzyme produced as a pre-pro enzyme form has a pro-domain [67], a catalytic domain homologous to papain and some mammalian cathepsina C-terminal extension, comprised of 130 amino acids. The C-terminal part, along with the catalytic domain, remains in the main functional form of the enzyme [68]. From the three N-glycosylation sites of cruzipain two sites are present in the catalytic domain (Asn33, Asn169) and the remaining one in the C-terminal antigenic extension (Asn255) region [63]. Major amount of the enzyme is stored in lysosomal vesicles near flagellar pocket [69], but other localization sites are also there, like in epimastogote forms, it is located inside reservosomes [69] or attached to the membrane [70]. A single-chained protein, named chagasin binds tightly to cystine proteases and inhibits its function [71]. In epimastigotes cruzipain-chagasin complex is formed, and here chagasin controls the catalytic activity of cystinyl-proteases, resulting in proper differentiation and host cell invasion of the pathogen [72]. Overexpression of cruzipain leads to enhanced metacyclogenesis [73]. The trypomastigotes secrete cysteine proteases with high molecular weight and cruzain is one of them [74, 75]. Secreted cruzipain converts plasma kallikrein to active kallikrein and lyses kininogen and producing a proinflammatory peptide called Lys-bradykinin or kallidin [76]. Activation of kinin cascade activates G-protein coupled kinin receptors. There is a relationship between the amount of released cruzipain and the host cell invasion capacity of the parasite. In smooth muscle cell invasion depends on the cumulative effect of cruzipain and chagasin [63]. Cruzipain, the major cystine protease of *T. cruzi* also plays a key role to protect the parasite from the host immune system as it can partially hydrolyze Fab region and completely degrade Fc region to smaller peptides of host IgG [77]. The enzyme is essential for the persistence, growth, and infectivity of the pathogen as it induces several molecular pathways such as PKA, MAPK-mediated arginase induction [78] and Ca^{2+}signalling [79].

T. cruzi Surface Protein (Tc-1). Tc-1 is may be the first described, trypomastigote specific transmembrane surface protein of *T. cruzi* which acts as a substrate for casein kinase2 of both the parasite and the host and mediates the process of host cell interaction and invasion of the parasite. Tc1 is a single-pass transmembrane protein with a molecular mass of 62 kDa and pI-6.1. Among the total 550 amino acids that forms the protein first 343 amino acids form the extracellular, 344 to 364 forms the transmembrane and 365 to 550 form the cytosolic domain respectively. Tc1 is present as a single copy gene in the parasite, which is only expressed in the invasive trypomastigotes. The protein bears total 14 phosphorylation sites, among which 8 are extracellular and 6 are intracellular. These sites get phosphorylated by exogenous and endogenous CK2, cAMP, PKC, tyrosine kinases. In intact trypomastigotes the exogenous Ck2 phosphorylates the serine residues. CK2 phosphorylation sites bears S-x-x-D/E consensus sequence whereas phosphosites of other proteins are present near this site. Studies on the structure of the Tc-1 protein suggest that may be during infection, Tc-1 gets phosphorylated on its ectodomain by the host's casein kinase2, causing host cell invasion and intracellular domain by its own CK2, leading to the activation of several other pathways essential for parasite infection and survival. As mentioned earlier, Tc-1 is also phosphorylated by cAMP, tyrosine kinases and PKC which plays a role in mobilization of Ca2+ inside the cell. It is speculated that invasion of trypomastigotes increase host free calcium concentration of the host cell [80].

Antioxidant Defense Enzyme

Iron Superoxide Dismutase (FeSOD). Trypanosomatids have their own antioxidant defense system to protect themselves from the host's ROS as these highly reactive molecules react with the pathogen's cellular components like proteins, lipids, and nucleic acid and adversely affect them [81]. Antioxidant enzymes of *T. cruzi* are listed below.

SOD is a metalloenzyme (EC 1.15.1.1) [82] having iron (Fe) as its cofactor which provide resistance against apoptosis and protect the pathogen from external superoxide radical (O_2^-) by converting them to H_2O_2 and O_2 [83]. FeSOD-A and C are localized in mitochondria, whereas FeSOD-B1 and B2 are localized in the glycosome. In *T. cruzi*, overexpression of FeSOD is related to the drug resistance property of the parasite inside the host cell. Mammals do not have FeSOD; instead, they have CuZn-SOD [84].

Protein Kinases

TcAEK1

AGC essential kinase 1 (*TcAEK1*) is a protein kinase that plays a vital role in cell cycle regulation, metabolism, differentiation, and survival of trypanosomaids. TcAEK1 is generally found in three main life stages i.e., amastigote, epimastigote, and trypomastigote of *T. cruzi*. In epimastigotes it is present in a scattered fashion throughout the cell. Inhibition of TcAEK1 causes morphological changes in *T. cruzi,* like the cells aggregate and form a mass of undivided cells where cell components, associated peripheral flagella, large, double-membrane-bound vacuoles can be identified, indicating impaired cytokinesis. Some of the mitochondria of fused cells show dilated cristae, though mitochondrial membrane potential remains unchanged. In trypomastigote, TcAEK1 level is lower than epimastigote and amastigote, suggesting that this downregulation might help the pathogen to enter from epimastigote, a replicative form, to trypomastigote, a non-replicative form. TcAEK1 helps in survival as well as host cell invasion of trypomastigotes. Both in trypomastigote and amastigote, the protein is localized all over the cytoplasm but is more abundant near the nucleus and kinetoplast. It helps in amastigote replication. Inhibition of this protein in zooid (cells that have one or two kinetoplasts) results in cell cycle defects along with nuclear disjunction and degradation, polyploidy, and higher DNA content in cells [85].

TcPKC

One of the major roles of Protein kinase C(PKC) of *Trypanosoma cruzi* is to regulate DNA synthesis by phosphorylating Thr137 and Ser286 of the H form of Tc DNA polymerase beta, which in turn is also involved with the kDNA replication and repair mechanism of the parasite. In *T. cruzi* genome, three isoforms of the PKC gene are identified- TcPKC1, TcPKC2, and TcPKC3. TcPKC transports signals via second messengers. Phorbol esters, along with DAG, phosphatidylserine and stimulate the PKC activity, necessary for the regulation of metacyclogenesis and differentiation of the organism especially differentiation of the epimastigote form. Though unlike other eukaryotic PKC, the TcPKC1 and TcPKC2 lack calcium-binding domain inferring that the protein does not require calcium for its functioning. Also, protamine sulfate, a substrate of eukaryotic PKC, is not phosphorylated by either TcPKC1 or 2 suggests that these proteins use different substrates. TcPKC phosphorylates the histone H1 protein along with other microsomal and cytosolic functions. Isoforms of PKC contain regulatory domains that bind to phosphorinositols and other ligands at the N-terminus and at the C-terminus, and have a catalytic domain. The TcPKC1 with a MW of 120 kDa, is an 858-residue polypeptide, bears a domain homologous to Phox (PX) from 37 to 147 residues that binds to phosphoinositol and a fatty acid binding C-terminal lipocalin domain, which undergoes in vivo proteolysis. The MW of TcPKC2 is 90 kDa and it is a polypeptide with 705 amino acid residues. At the N-terminus, it has a zinc finger binding domain located from 86 to 143 amino acids with FYVE-type amino acid sequences. This part interacts with phospholipids. TcPKC3 is composed of 458 amino acid residues. At the N-terminus, it has phospholipid binding pleckstrin homology (PH) domain spanning from 3 to 98 amino acid residues, which also interacts with different proteins of signal transduction pathways, and at the C-terminus, it has STKX domain spanning residues 397 to 438 [86].

2 Conclusion

Over time immemorial it has been a very difficult task to treat trypanosomiasis infection because the pathogens can acquire various mechanisms to outlive within the host body. Antiparasitic drugs are used to treat infections that are somehow incapable of curing chronic infections completely, as the parasites quickly grow resistance against these long-established drugs. Further, these drugs have some adverse effects on human health. Drugs used against African trypanosomiasis, such as pentamidine can cause cardiovascular disease; suramine can cause proteinuria and renal toxicity; melarsoprol, used during the neurological treatment phase, is highly toxic. On the other hand, Nifurtimox and Benznidazole that are used for the treatment of Chagas, have some side effects like gastrointestinal-tract complications; they also affect the central nervous system, causing insomnia, irritability, etc. Therefore, it has become the call of the hour for a detailed study of essential endogenous proteins of the parasites, which would definitely be beneficial to develop drugs capable of inhibiting these targets while ensuring lower toxicity, minimal or no side effects, and high efficacy. Additionally, certain phytocompounds could be utilized either independently or in combination at lower doses to mitigate the adverse effects associated with conventional medicines. Further research is crucial in this area, emphasizing the need to identify potent targets that can effectively combat the pathogenesis and severity of trypanosomiasis, ultimately improving the quality of life for those affected by the disease.

References

1. Weng, H.-B., Chen, H.-X., Wang, M.-W.: Innovation in neglected tropical disease drug discovery and development. Infect. Dis. Poverty. **7**(1), 67 (2018)
2. Zeledón, R.A.: Hemoflagellates. In: Baron, S. (ed.) Medical Microbiology. University of Texas Medical Branch at Galveston, Galveston (1996)
3. Kaufer, A., Ellis, J., Stark, D., Barratt, J.: The evolution of Trypanosomatid taxonomy. Parasites Vectors. **10**(1), 287 (2017). https://doi.org/10.1186/s13071-017-2204-7
4. de Souza, W.: Structural organization of trypanosoma cruzi. Mem. Inst. Oswaldo Cruz. **104**(suppl 1), 89–100 (2009). https://doi.org/10.1590/S0074-02762009000900014
5. Vedrenne, C., et al.: Two related Subpellicular cytoskeleton-associated proteins in Trypanosoma Brucei stabilize microtubules. Mol. Biol. Cell. **13**(3), 1058–1070 (2002). https://doi.org/10.1091/mbc.01-06-0298
6. Ralston, K.S., Hill, K.L.: The flagellum of Trypanosoma Brucei: new tricks from an old dog. Int. J. Parasitol. **38**(8–9), 869–884 (2008). https://doi.org/10.1016/j.ijpara.2008.03.003
7. Perdomo, D., Bonhivers, M., Robinson, D.R.: The trypanosome flagellar pocket collar and its ring forming protein-TbBILBO1. Cells. **5**(1), E9 (2016). https://doi.org/10.3390/cells5010009
8. Van Den Abbeele, J., Caljon, G., De Ridder, K., De Baetselier, P., Coosemans, M.: Trypanosoma Brucei modifies the tsetse salivary composition, altering the Fly feeding behavior that favors parasite transmission. PLoSPathog. **6**(6), e1000926 (2010). https://doi.org/10.1371/journal.ppat.1000926

9. Caffrey, C.R., Scory, S., Steverding, D.: Cysteine proteinases of trypanosome parasites: novel targets for chemotherapy. Curr. Drug Targets. **1**(2), 155–162 (2000). https://doi.org/10.2174/1389450003349290
10. Solomon Ngutor, K., Idris, L.A., OluseyiOluyinka, O.: Silent human *Trypanosoma Brucei Gambiense* infections around the old Gboko sleeping sickness focus in Nigeria. J. Parasitol. Res. **2016**, 1–5 (2016). https://doi.org/10.1155/2016/2656121
11. Kasozi, K.I., et al.: Epidemiology of trypanosomiasis in wildlife—implications for humans at the wildlife Interface in Africa. Front. Vet. Sci. **8**, 621699 (2021). https://doi.org/10.3389/fvets.2021.621699
12. Silvester, E., McWilliam, K., Matthews, K.: The cytological events and molecular control of life cycle development of Trypanosoma Brucei in the mammalian bloodstream. Pathogens. **6**(3), 29 (2017). https://doi.org/10.3390/pathogens6030029
13. Matthews, K.R.: The developmental cell biology of *Trypanosoma Brucei*. J. Cell Sci. **118**(2), 283–290 (2005). https://doi.org/10.1242/jcs.01649
14. Greenwood, B.M., Whittle, H.C.: The pathogenesis of sleeping sickness. Trans. R. Soc. Trop. Med. Hyg. **74**(6), 716–725 (1980). https://doi.org/10.1016/0035-9203(80)90184-4
15. Stich, A.: Human African trypanosomiasis. BMJ. **325**(7357), 203–206 (2002). https://doi.org/10.1136/bmj.325.7357.203
16. Stockdale, C., Swiderski, M.R., Barry, J.D., McCulloch, R.: Antigenic variation in Trypanosoma Brucei: joining the DOTs. PLoSBiol. **6**(7), e185 (2008). https://doi.org/10.1371/journal.pbio.0060185
17. Mugnier, M.R., Stebbins, C.E., Papavasiliou, F.N.: Masters of disguise: antigenic variation and the VSG coat in Trypanosoma Brucei. PLoSPathog. **12**(9), e1005784 (2016). https://doi.org/10.1371/journal.ppat.1005784
18. Kato, C.D., Matovu, E., Mugasa, C.M., Nanteza, A., Alibu, V.P.: The role of cytokines in the pathogenesis and staging of Trypanosoma brucei rhodesiense sleeping sickness. Allergy, Asthma Clin. Immunol. **12**(1), 4 (2016)
19. Magez, S., Pinto Torres, J.E., Obishakin, E., Radwanska, M.: Infections with extracellular trypanosomes require control by efficient innate immune mechanisms and can result in the destruction of the mammalian humoral immune system. Front. Immunol. **11**, 382 (2020). https://doi.org/10.3389/fimmu.2020.00382
20. Lundkvist, G.B., Kristensson, K., Bentivoglio, M.: Why trypanosomes cause sleeping sickness. Physiology. **19**(4), 198–206 (2004)
21. Onyilagha, C., Uzonna, J.E.: Host immune responses and immune evasion strategies in African trypanosomiasis. Front. Immunol. **10** (2019). https://doi.org/10.3389/fimmu.2019.02738
22. Tarleton, R.L.: Trypanosoma cruzi and chagas disease: cause and effect. World Class Parasites, 107–115 (2003) Available at: https://doi.org/10.1007/978-1-4419-9206-210
23. De Souza Gonzaga, B.M., Ferreira, R.R., Coelho, L.L., Carvalho, A.C.C., Garzoni, L.R., Araujo-Jorge, T.C.: Clinical trials for Chagas disease: etiological and pathophysiological treatment. Front. Microbiol. **14** (2023). https://doi.org/10.3389/fmicb.2023.1295017
24. Onyekwelu, K.C.: Life cycle oftrypanosomacruziin the invertebrate and the vertebrate hosts. In IntechOpen eBooks. https://doi.org/10.5772/intechopen.84639 (2019)
25. Tanowitz, H.B., Kirchhoff, L.V., Simon, D., Morris, S.A., Weiss, L.M., Wittner, M.: Chagas' disease. Clin. Microbiol. Rev. **5**(4), 400–419 (1992). https://doi.org/10.1128/cmr.5.4.400
26. Rodríguez-Bejarano, O.H., Avendaño, C., Patarroyo, M.A.: Mechanisms associated with Trypanosoma cruzi host target cell adhesion, Recognition and Internalization. Life. **11**(6), 534 (2021). https://doi.org/10.3390/life11060534
27. Pérez-Molina, J.A., Molina, I.: Chagas disease. Lancet. **391**(10115), 82–94 (2017). https://doi.org/10.1016/s0140-6736(17)31612-4

28. Bonney, K.M., Luthringer, D.J., Kim, S.A., Garg, N.J., Engman, D.M.: Pathology and pathogenesis of Chagas heart disease. Ann. Rev. Pathol. Mech. Dis. **14**(1), 421–447 (2018). https://doi.org/10.1146/annurev-pathol-020117-043711
29. Echavarría, N., Echeverría, L.E., Stewart, M., Gallego, C., Saldarriaga, C.: Chagas disease: chronic Chagas cardiomyopathy. Curr. Probl. Cardiol. **46**(3), 100507 (2021). https://doi.org/10.1016/J.CPCARDIOL.2019.100507
30. Campi-Azevedo, A., et al.: Etiological treatment of Chagas disease patients with benznidazole lead to a sustained pro-inflammatory profile counterbalanced by modulatory events. Immunobiology. **220**(5), 564–574 (2015). https://doi.org/10.1016/j.imbio.2014.12.006
31. Lage, T.A.R., et al.: Stroke in Chagas disease: from pathophysiology to clinical practice. Revista Da SociedadeBrasileira De Medicina Tropical. **55** (2022). https://doi.org/10.1590/0037-8682-0575-2021
32. Ramírez-Toloza, G., Ferreira, A.: Trypanosoma cruzi evades the complement system as an efficient strategy to survive in the mammalian host: the specific roles of host/parasite molecules and Trypanosoma cruzi Calreticulin. Front. Microbiol. **8** (2017). https://doi.org/10.3389/fmicb.2017.01667
33. Lidani, K.C.F., Bavia, L., Ambrosio, A.R., De Messias-Reason, I.J.: The complement system: a prey of Trypanosoma cruzi. Front. Microbiol. **8** (2017). https://doi.org/10.3389/fmicb.2017.00607
34. Kennedy, P.G.: Clinical features, diagnosis, and treatment of human African trypanosomiasis (sleeping sickness). Lancet Neurol. **12**(2), 186–194 (2013)
35. Steverding, D.: The development of drugs for treatment of sleeping sickness: a historical review. Parasites Vectors. **3**(1), 15 (2010). https://doi.org/10.1186/1756-3305-3-15
36. Wéry, M.: Drug used in the treatment of sleeping sickness (human African trypanosomiasis: HAT). Int. J. Antimicrob. Agents. **4**(3), 227–238 (1994). https://doi.org/10.1016/0924-8579(94)90012-4
37. Maxmen, A.: Sleeping sickness can now be cured with pills. Nature. **550**(7677), 441–441 (2017). https://doi.org/10.1038/nature.2017.22856
38. Rodgers, J., et al.: Melarsoprolcyclodextrin inclusion complexes as promising oral candidates for the treatment of human African trypanosomiasis. PLoSNegl Trop Dis. **5**(9), e1308 (2011). https://doi.org/10.1371/journal.pntd.0001308
39. Kumeso, V.K.B., et al.: Efficacy and safety of acoziborole in patients with human African trypanosomiasis caused by Trypanosoma brucei gambiense: a multicentre, open-label, single-arm, phase 2/3 trial. Lancet Infect. Dis. **23**(4), 463–470 (2023)
40. Pohlig, G., et al.: Efficacy and safety of Pafuramidine versus Pentamidine maleate for treatment of first stage sleeping sickness in a randomized, comparator-controlled, international phase 3 clinical trial. PLoSNegl Trop Dis. **10**(2), e0004363 (2016). https://doi.org/10.1371/journal.pntd.0004363. PMID: 26882015; PMCID: PMC4755561
41. Castro, J.A., de Mecca, M.M., Bartel, L.C.: Toxic side effects of drugs used to treat Chagas' disease (American trypanosomiasis). Hum. Exp. Toxicol. **25**(8), 471–479 (2006). https://doi.org/10.1191/0960327106het653oa
42. Viotti, R., et al.: Side effects of benznidazole as treatment in chronic Chagas disease: fears and realities. Expert Rev. Anti-Infect. Ther. **7**(2), 157–163 (2009). https://doi.org/10.1586/14787210.7.2.157
43. Ribeiro, A.L., Nunes, M.P., Teixeira, M.M., Rocha, M.O.: Diagnosis and management of Chagas disease and cardiomyopathy. Nat. Rev. Cardiol. **9**(10), 576–589 (2012). https://doi.org/10.1038/nrcardio.2012.109
44. Caffrey, C., et al.: Active site mapping, biochemical properties and subcellular localization of rhodesain, the major cysteine protease of Trypanosoma brucei rhodesiense. Mol. Biochem. Parasitol. **118**(1), 61–73 (2001)

45. O'Brien, T.C., et al.: A parasite cysteine protease is key to host protein degradation and iron acquisition. J. Biol. Chem. **283**(43), 28934–28943 (2008)
46. Mackey, Z., O'Brien, T., Greenbaum, D., Blank, R., McKerrow, J.: A Cathepsin B-like protease is required for host protein degradation in Trypanosoma brucei. J. Biol. Chem. **279**(46), 48426–48433 (2004)
47. Abdulla, M., O'Brien, T., Mackey, Z., Sajid, M., Grab, D., McKerrow, J.: RNA interference of Trypanosoma brucei Cathepsin B and L affects disease progression in a mouse model. PLoS Negl. Trop. Dis. **2**(9), e298 (2008)
48. Chung, W.L., Carrington, M., Field, M.C.: Cytoplasmic targeting signals in transmembrane invariant surface glycoproteins of trypanosomes. J. Biol. Chem. **279**(52), 54887–54895 (2004). https://doi.org/10.1074/jbc.M409311200
49. Uzureau, P., et al.: Mechanism of Trypanosoma brucei gambiense resistance to human serum. Nature. **501**(7467), 430–434 (2013). https://doi.org/10.1038/nature12516
50. Redecke, L., et al.: Natively inhibited Trypanosoma brucei Cathepsin B structure determined by using an X-ray laser. Science. **339**(6116), 227–230 (2012). https://doi.org/10.1126/science.1229663
51. Mott, B.T., et al.: Identification and optimization of inhibitors of trypanosomal cysteine proteases: cruzain, Rhodesain, and TBCATB. J. Med. Chem. **53**(1), 52–60 (2009). https://doi.org/10.1021/jm901069a
52. Steverding, D., Sexton, D.W., Wang, X., Gehrke, S.S., Wagner, G.K., Caffrey, C.R.: Trypanosoma brucei: chemical evidence that cathepsin L is essential for survival and a relevant drug target. Int. J. Parasitol. **42**(5), 481–488 (2012). https://doi.org/10.1016/j.ijpara.2012.03.009
53. Steverding, D., Da Nóbrega, F.R., Rushworth, S.A., De Sousa, D.P.: Trypanocidal and cysteine protease inhibitory activity of isopentyl caffeate is not linked in Trypanosoma brucei. Parasitol. Res. **115**(11), 4397–4403 (2016). https://doi.org/10.1007/s00436-016-5227-7
54. Rico, E., Rojas, F., Mony, B.M., Szoor, B., MacGregor, P., Matthews, K.R.: Bloodstream form pre-adaptation to the tsetse fly in Trypanosoma brucei. Front. Cell. Infect. Microbiol. **3** (2013). https://doi.org/10.3389/fcimb.2013.00078
55. Krieger, S., Schwarz, W., Ariyanayagam, M.R., Fairlamb, A.H., Krauth-Siegel, R.L., Clayton, C.: Trypanosomes lacking trypanothione reductase are avirulent and show increased sensitivity to oxidative stress. Mol. Microbiol. **35**(3), 542–552 (2000). https://doi.org/10.1046/j.1365-2958.2000.01721.x
56. Krauth-Siegel, R.L., Enders, B., Henderson, G.B., Fairlamb, A.H., Schirmer, R.H.: Trypanothione reductase from Trypanosoma cruzi. Purification and characterization of the crystalline enzyme. Eur. J. Biochem. **164**(1), 123–128 (1987). https://doi.org/10.1111/j.1432-1033.1987.tb11002.x
57. Dormeyer, M., Reckenfelderbäumer, N., Lüdemann, H., Krauth-Siegel, R.L.: Trypanothione-dependent synthesis of Deoxyribonucleotides by Trypanosoma brucei ribonucleotide reductase. J. Biol. Chem. **276**(14), 10602–10606 (2001). https://doi.org/10.1074/jbc.m010352200
58. Reckenfelderbaumer, N., Lüdemann, H., Schmidt, H., Steverding, D., Krauth-Siegel, R.L.: Identification and functional characterization of thioredoxin from Trypanosoma brucei brucei. J. Biol. Chem. **275**(11), 7547–7552 (2000)
59. Mikkelsen, J.H., et al.: Trypanosoma brucei invariant surface glycoprotein 75 is an immunoglobulin fc receptor inhibiting complement activation and antibody-mediated cellular phagocytosis. J. Immunol. **212**(8), 1334–1344 (2024). https://doi.org/10.4049/jimmunol.2300862
60. Sullivan, L., Wall, S.J., Carrington, M., Ferguson, M.A.J.: Proteomic selection of immunodiagnostic antigens for human African trypanosomiasis and generation of a prototype lateral flow immunodiagnostic device. PLoS Negl. Trop. Dis. **7**(2), e2087 (2013). https://doi.org/10.1371/journal.pntd.0002087A

61. Ziegelbauer, K., Multhaup, G., Overath, P.: Molecular characterization of two invariant surface glycoproteins specific for the bloodstream stage of Trypanosoma brucei. J. Biol. Chem. **267**(15), 10797–10803 (1992). https://doi.org/10.1016/s0021-9258(19)50089-8
62. Hofer, A.: Targeting the nucleotide metabolism of Trypanosomabrucei and other trypanosomatids. FEMS Microbiol. Rev. **47**, 1–20 (2023). https://doi.org/10.1093/femsre/fuad020
63. Duschak, V., Couto, A.: Cruzipain, the major cysteine protease of Trypanosoma cruzi: a sulfated glycoprotein antigen as relevant candidate for vaccine development and drug TargetA review. Curr. Med. Chem. **16**(24), 3174–3202 (2009c). https://doi.org/10.2174/092986709788802971
64. Engel, J.C., Doyle, P.S., Palmer, J., Hsieh, I., Bainton, D.F., McKerrow, J.H.: Cysteine protease inhibitors alter Golgi complex ultrastructure and function in Trypanosoma cruzi. J. Cell Sci. **111**(5), 597–606 (1998). https://doi.org/10.1242/jcs.111.5.597
65. Eakin, A.E., McKerrow, J.H., Craik, C.S.: A cysteine protease is a target for the enzyme structure-based design of Antiparasitic drugs. Drug Inf. J. **29**(1_suppl), 1501S–1517S (1995b). https://doi.org/10.1177/00928615950290s102
66. De La Cruz Pech-Canul, Á., Monteón, V., Solís-Oviedo, R.: A brief view of the surface membrane proteins fromTrypanosomacruzi. J. Parasitol. Res. **2017**, 1–13 (2017b). https://doi.org/10.1155/2017/3751403
67. Eakin, A., Mills, A., Harth, G., McKerrow, J., Craik, C.: The sequence, organization, and expression of the major cysteine protease (cruzain) from Trypanosomacruzi. J. Biol. Chem. **267**(11), 7411–7420 (1992). https://doi.org/10.1016/s0021-9258(18)42533-1
68. Cazzulo, J.J., Stoka, V., Turk, V.: Review. Biol. Chem. **378**(1), 1–10 (1997). https://doi.org/10.1515/bchm.1997.378.1.1
69. Souto-Padrón, T., Campetella, O.E., Cazzulo, J.J., De Souza, W.: Cysteine proteinase in Trypanosoma cruzi: Immunocytochemical localization and involvement in parasite-host cell interaction. J. Cell Sci. **96**(3), 485–490 (1990c). https://doi.org/10.1242/jcs.96.3.485
70. Soares, M.J., Souto-Padrón, T., De Souza, W.: Identification of a large pre-lysosomal compartment in the pathogenic protozoon Trypanosoma cruzi. J. Cell Sci. **102**(1), 157–167 (1992). https://doi.org/10.1242/jcs.102.1.157
71. Monteiro, A.C.S., Abrahamson, M., Lima, A.P.C.A., Vannier-Santos, M.A., Scharfstein, J.: Identification, characterization and localization of chagasin, a tight-binding cysteine protease inhibitor in Trypanosoma cruzi. J. Cell Sci. **114**(21), 3933–3942 (2001). https://doi.org/10.1242/jcs.114.21.3933
72. Santos, C.C., et al.: Chagasin, the endogenous cysteine-protease inhibitor of Trypanosoma cruzi, modulates parasite differentiation and invasion of mammalian cells. J. Cell Sci. **118**(5), 901–915 (2005). https://doi.org/10.1242/jcs.01677
73. Tomas, A.M., Miles, M.A., Kelly, J.M.: Overexpression of Cruzipain, the major cysteine proteinase of Trypanosoma cruzi, is associated with enhanced Metacyclogenesis. Eur. J. Biochem. **244**(2), 596–603 (1997). https://doi.org/10.1111/j.1432-1033.1997.t01-1-00596.x
74. Yokoyama-Yasunaka, J.K., Pral, E.M., Oliveira, O.C., Alfieri, S.C., Stolf, A.M.S.: Trypanosoma cruzi: identification of proteinases in shed components of trypomastigote forms. Acta Trop. **57**(4), 307–315 (1994). https://doi.org/10.1016/0001-706x(94)90076-0
75. Duschak, V.G., Barboza, M., García, G.A., Lammel, E.M., Couto, A.S., Isola, E.L.D.: Novel cysteine proteinase inTrypanosomacruzimetacyclogenesis. Parasitology. **132**(3), 345–355 (2005). https://doi.org/10.1017/s0031182005009030
76. Del Nery, E., Juliano, M.A., Lima, A.P.C., Scharfstein, J., Juliano, L.: Kininogenase activity by the major Cysteinyl proteinase (Cruzipain) from Trypanosoma cruzi. J. Biol. Chem. **272**(41), 25713–25718 (1997). https://doi.org/10.1074/jbc.272.41.25713

77. Bontempi, E., Cazzulo, J.J.: Digestion of human immunoglobulin G by the major cysteine proteinase (cruzipain) from Trypanosoma cruzi. FEMS Microbiol. Lett. **70**(3), 337–342 (1990). https://doi.org/10.1111/j.1574-6968.1990.tb14000.x
78. Stempin, C.C., Tanos, T.B., Coso, O.A., Cerbán, F.M.: Arginase induction promotes Trypanosoma cruzi intracellular replication in Cruzipain-treated J774 cells through the activation of multiple signaling pathways. Eur. J. Immunol. **34**(1), 200–209 (2003). https://doi.org/10.1002/eji.200324313
79. Scharfstein, J., et al.: Host cell invasion by TRYPANOSOMA CRUZI is potentiated by activation of bradykinin B2 receptors. J. Exp. Med. **192**(9), 1289–1300 (2000). https://doi.org/10.1084/jem.192.9.1289
80. Augustine, S.A.J., et al.: Molecular cloning of a Trypanosoma cruzi cell SurfaceCasein kinase II substrate, Tc-1, involved in cellular infection. Infect. Immun. **74**(7), 3922–3929 (2006)
81. Pineyro, M., et al.: Crystal structure of the tryparedoxin peroxidase from the human parasite. J. Struct. Biol. **150**(1), 11–22 (2005). https://doi.org/10.1016/j.jsb.2004.12.005
82. Abreu, I.A., Cabelli, D.E.: Superoxide dismutases—a review of the metal-associated mechanistic variations. Biochim. Et Biophys. (BBA) Proteins Proteomics. **1804**(2), 263–274 (2009). https://doi.org/10.1016/j.bbapap.2009.11.005
83. Turrens, J.F.: Oxidative stress and antioxidant defenses: a target for the treatment of diseases caused by parasitic protozoa. Mol. Asp. Med. **25**(1–2), 211–220 (2004b). https://doi.org/10.1016/j.mam.2004.02.021
84. Dufernez, F., et al.: The presence of four iron-containing superoxide dismutase isozymes in Trypanosomatidae: characterization, subcellular localization, and phylogenetic origin in Trypanosoma brucei. Free Radic. Biol. Med. **40**(2), 210–225 (2005). https://doi.org/10.1016/j.freeradbiomed.2005.06.021
85. Chiurillo, M.A., Jensen, B.C., Docampo, R.: Drug target validation of the protein kinase AEK1, essential for proliferation, host cell invasion, and intracellular replication of the human pathogen Trypanosoma cruzi. Microbiol. Spectr. **9**(2) (2021). https://doi.org/10.1128/spectrum.00738-21
86. Maldonado, E., Rojas, D.A., Urbina, F., Valenzuela-Pérez, L., Castillo, C., Solari, A.: Trypanosoma cruzi DNA polymerase β is phosphorylated in vivo and in vitro by protein kinase C (PKC) and casein kinase 2 (CK2). Cells. **11**(22), 3693 (2022). https://doi.org/10.3390/cells11223693

Biofilm Formation in Gastric Mucosa by *Helicobacter pylori* Infection and Its Role in Drug Resistance

Pratyusa Biswas[1] and Laishram Pradeepkumar Singh[1(✉)]

[1] Laboratory of Translational Biology and Molecular Medicine, Department of Zoology, University of Kalyani, Kalyani, Nadia, West Bengal, India
laishrampk@gmail.com

Abstract. *Helicobacter pylori* (*H. pylori*) is one of the most common pathogenic agents of human which infects approximately 50% of the world's population. The colonization of this gram-negative bacterium in gastric mucosa often has an implication on several gastrointestinal (GI) diseases including gastro-duodenal ulcer and gastric cancer. The eradication of infection is the most effective strategy to decrease the infection rate and prevalence of gastro-duodenal complications. However, development of antibiotic-resistant strain throughout the world hindered the therapeutic efficacies against the bacterium. Extensive studies on the molecular mechanisms underlying antibiotic resistance indicated that upon infection, *H. pylori* enter and occupy the gastric epithelium and encase themselves in an extracellular polymeric substance known as biofilm. The biofilm structure protects the bacterial colony from any external threat and contributes to several chronic and recurrent infection of GI tract. Furthermore, bacterial aggregates in biofilm are 1000 times more resistant to any type of antibiotics as compared to solitary bacteria. Thus, currently anti-biofilm agents are being explored as a novel therapeutic alternative of antibiotics to combat the drug-resistance strategies of *H. pylori* infection. This review focuses on the molecular mechanisms of biofilm formation of *H. pylori* in gastric mucosa and the associated drug resistance strategy of biofilm producing *H. pylori*. Additionally, efforts are made to compile the existing knowledge of the antibiofilm agents which are designed for the treatment of *H. pylori* induced gastric infection.

Keywords: Antibiofilm agents · Biofilm · Drug-resistance · Extracellular polymeric substance · *Helicobacter pylori*

1 Introduction

Helicobacter pylori, a leading pathogenic agent that colonizes gastric epithelium in human is a microaerophilic, spiral-shaped gram-negative bacterium [1]. It was isolated initially from the patient suffering from antral gastritis in the year 1983 [2]. Later, it was characterized as a potential pathogen associated with chronic gastritis, gastro-intestinal ulcer and mucosa-associated lymphoid tissue (MALT) lymphoma [3, 4]. It was classified as group I carcinogen in 2017 by International Agency for Research on Cancer

A. K. Saxena, A. Saxena (Eds.): GTHTM 2025, pp. 357–374, 2026.
https://doi.org/10.1007/978-3-032-12320-6_21

(IARC) for its strong association with gastric cancer [5]. *Helicobacter pylori* possess several harmful components like flagella, urease, lipopolysaccharide (LPS), vacuolating cytotoxin A (VacA) and cytotoxin-associated gene A (CagA) which contributes to damage of the stomach's mucosal lining. The flagella assist in movement (motility) and chemotaxis and help the bacteria to drill into the mucosal layer. Urease enzyme secreted from the bacteria breaks down urea to produce ammonia and neutralizes the acidic environment of the stomach. In addition, lipopolysaccharides help the bacteria attach to the host cells. *H. pylori* also release enzymes like mucinase, protease and lipase which break down the protective mucous layer of the stomach. Moreover, the VacA toxin harms the cells of the gastric mucosa by forming vacuole. The bacterium also has a Type IV secretion system, which is similar to a pili-like injector. This system injects harmful effectors such as CagA and induces interleukin-8 (IL-8) production that triggers an inflammatory response in the stomach's epithelial cells [6]. Infection with *Helicobacter pylori* often remains dormant in the stomach, showing no noticeable symptoms in most individuals. However, in some cases, it can give rise to a variety of gastric disorders ranging from chronic gastritis to gastric cancer. The severity of these conditions is closely linked to the virulence of the bacterial strain and the genetic or physiological susceptibility of the host [7]. Currently, it is estimated that a two-third of global population is infected with *H. pylori* and the infection is most common among the developing countries [8]. The rate of *H. pylori* infection has exceeded 80% in Asia [9] while it crossed 79.1% and 63% of infection prevalence in Latin America and Africa, respectively [10]. Under such situation, *H. pylori* eradication is regarded as the most prudent step to decrease the rate of gastro-intestinal ulcer and the gastric cancer prevalence [11]. But emergence of antibiotic-resistant strain becomes a huge barrier in the path of *H. pylori* eradication [12]. In current scenario, a combination of proton pump inhibitor (PPI) with two antimicrobial classes (mainly, fluroquinolones or penicillin-based combined with metronidazole and macrolides), known as triple therapy, is considered effective standard treatment for *H. pylori* infection [13]. However, according to the report of World Health Organization (WHO) there is a rapid increase of antibiotic-resistant strain of *H. pylori* particularly against clarithromycin and metronidazole [13]. Additionally, the multi-drug resistance strain rose up to 40% in some region of the world, endangering the eradication procedure of the bacterium [14].

Various factors account for the rise in antibiotic-resistant strain of *H. pylori* which includes enhanced efflux pump expression, reduction of membrane permeability, targeted gene mutation and biofilm formation [4, 15]. *Helicobacter pylori* establish a quorum sensing mechanism to develop antibiotic resistance. In general, gram-negative bacteria release certain extracellular signalling molecules known as autoinducers which initiate intercellular communications. *H. pylori* produce autoinducer-2 (AI-2) which is a quorum sensing molecule generated through LuxS-enzyme-mediated cleavage of S-ribosyl-L homocysteine (SRH). The enzymatic cleavage produces 4,5-dihydroxy-2,3-pentanedione (DPD) and homocysteine, which serve as the precursor of AI-2. Chemically AI-2 is furanosyl borate diester and addition of naturally occurring borate to its precursor, can generate active AI-2. Several microorganisms utilize boron for inducing Quorum sensing (QS) mechanism. AI-2-mediated QS system increases virulence and development of antibiotic resistance in *H. pylori*. This happens because antibiotics are unable to break down the dense molecular barrier known as biofilm formed by QS activity [6]. The biofilm of *Helicobacter pylori* has also been observed in the gastric mucosa of the patients suffering from peptic ulcer [1]. Bacteria that produce biofilm could persist in the stomach and show slow growth rates [16]. After the colonization of the gastric mucosa, *H. pylori* secret polysaccharides, proteins, extracellular DNA (eDNA) and other molecules to form an extracellular polymeric substance (EPS) to connect and held the bacterial colony [17]. The EPS protects the bacterial colony from harsh external environment and antibiotic exposure [18]. Report suggests that, biofilm producing bacteria possess 10–1000 times more antibiotic resistance capacity than the planktonic bacteria. The biofilm formation of *H. pylori* has been shown as an important step for chronic gastric infection, antibiotic resistance and the primary reason for the failure of the conventional therapies [19]. As a consequence, biofilm could be a novel therapeutic target to combat against *H. pylori* drug resistance and the future studies should be aimed to generate novel anti-biofilm molecules without any side-effects. The current review explores the process of biofilm formation of *H. pylori* in gastric mucosa and the associated drug resistance strategy mediated by it. Additionally, it further highlights the existing anti-biofilm molecules or agents targeted against the drug-resistant *H. pylori* strain colonizing the gastric mucosa.

2 Biofilm and Bacterial Survivability

Biofilm is defined as a structured consortium made up of adhered microbial cells which are encased in their own extracellular polymeric matrix. It is considered as an adaptive process of microorganisms to fight against pugnacious environment [20]. The biofilm may consist of a single type of microbe or a combination of several microorganisms including bacteria, archaea, fungi, yeasts and protozoa. It contains channel proteins which regulate gaseous exchange, nutrients and antimicrobial uptake [21]. Anthony von Leeuwenhoek first observed the structure in bacteria but the word "biofilm" was later introduced by Bill Costerton and their team [22]. In 1993, American Society for Microbiology acknowledged the importance of biofilm in various infections and health-related issues [23].

The biofilm of bacteria is made up of 10% mass of microbes and 90% of EPS [24]. The polysaccharides of the matrix contribute to about 50–90% of entire organic mass of biofilm [25]. A thick, mesh-like structure is created by weaving chains of polysaccharides together. The hydroxyl groups of polysaccharides interact with one another to strengthen its mechanical properties [26]. The biofilm structure may have cationic charges such as Ca^{2+} and Mg^{2+} to form cross bridges of polymers and supports the matrix of up to 300 μm thickness. In other cases, the polysaccharides from EPS of gram-negative bacteria may be neutral or polycationic. Uronic acids including D-glucuronic, D-galactorunic and mannuronic acids, as well as ketal-linked pyruvates, which provide anionic charge, could be associated with the biofilm [27]. The anionic charge of uronic acids associates with the divalent positive charges of Ca^{2+} and Mg^{2+} to produce stronger inter-linked polymer strands in matured biofilm. The bacterial communities when covered with biofilm remain sessile in nature and show different adaptive behaviour during extreme osmolarity, decreased nutrient condition and enhanced cellular density. This final architecture of biofilm is incredibly viscoelastic and has rubbery characteristics [23]. This helps them to evade the immune defence system of hosts and causes persistent and chronic infections which are rarely ameliorated by any antimicrobials. The microbes in the biofilm remain in dormant state, change their metabolism, gene expression and protein synthesis in response to ambient anoxia and nutritional limitations. This can result in decreased metabolic rate and a slower rate of cell division. Additionally, these adaptations help the bacteria to become more antibiotic resistant through inactivation of antimicrobial targets, decreasing the cellular activities and reducing interference with the cellular components [28].

Microbial biofilm affects every aspect of human existence from public health to perturbed industrial efficiency leading to loss in energy and economy, equipment deterioration, contaminated products and infections [23]. Microorganisms forming biofilm colonize either on the living tissues or on the inert objects like medical devices have raised serious risk on human health and patient's life. The biofilms containing microbes acts as a biological container to modulate host inflammatory responses and increase implant failure [21]. To address the complexity of biofilm activity and behaviour, it demands multidisciplinary research to devise an effective solution to mitigate the damages associated with such structures.

3 Biofilm Formation Contributes to Bacterial Infections

It is estimated that 70% of all microorganisms induced-infections are linked to biofilm formation and healthcare-associated infections (HAIs) in human [23]. A number of illnesses are both medical device and non-medical device related, such as endocarditis, fibrosis, periodontitis, non-healing chronic wounds, meningitis and osteomyelitis kidney infections. The most common bacteria found in biofilm include *Staphylococcus epidermidis*, *Staphylococcus aureus*, *Enterococcus faecalis*, *Escherichia coli*, *Klebsiella pneumoniae* and *Pseudomonas aeruginosa*. Periodontitis is a biofilm-related infection of teeth caused by *P. aerobicus* and *Fusobacterium nucleatum* [23, 29–31]. The biofilm which forms on the tooth surface interferes with calcium transport of epithelial cells leading to mineralized biofilm [23]. Biofilm also associates with the pathogenesis of otitis media (OM), an inflammation of the middle ear cavity. The endocarditis is caused by biofilm formation of host and bacterial components on cardiac valve. Costerton et al. had identified endocarditis as a biofilm-linked infection and acknowledged viridans group streptococci as the causative agent [28]. The open wound is mostly colonized by *Staphylococcus aureus* species which develops biofilm structure and decreases the rate of wound healing process [32]. The aerobic bacteria form biofilm structure on the wound surface and help the anaerobic bacteria to invade the interior tissue [33]. The open wounds of diabetic patients are often infected with biofilm of bacteria which contribute to chronic inflammatory stages [34]. Bacterial vaginosis (BV) is the most prevalent infection of genital tract of women caused by the presence of dense, polymicrobial and organised biofilm of *Gardnerella vaginalis* that firmly attach to the vaginal epithelium [35].

4 *Helicobacter pylori* Biofilm Formation in Gastric Epithelium

Gastric ulcer, also known as peptic ulcer affects about 5–10% of global population. It is characterized by sore of mucous lining of digestive system especially the stomach. Various factors such as alcohol, tobacco, cigarettes, non-steroidal anti-inflammatory drugs (NSAIDs) associate with the gastro-intestinal damages. However, studies suggest *H. pylori* infection has been identified as the key contributor of the gastric ulcer pathogenesis where it accounts for about 90–95% of non-NSAIDs gastric ulcer and 50–80% of patient with peptic ulcer [36]. It colonizes gastric mucosa within a small range of 25 μm of gastric epithelial cell surface. The wild-type pathogen mainly infects two parts of the stomach: the antrum and the corpus [12]. The biofilm formation of *H. pylori* commences from the planktonic stage of bacteria when it adheres both biotic and abiotic surfaces [1, 37]. When *H. pylori* enter the antrum region, it encounters several stress factors that modulate various gene expressions to enhance biofilm formation. Elevated stress conditions lead to increased expression of several genes including hrcA (HP0111), crdR (HP1365), pgdA (HP0310) and hspR (HP1025) [37]. crdR is a stress regulatory gene that increases the expression of iron transporter (fecA), flagellar components and efflux pumps (glnP) to enable the bacterium survive in stressed condition. Additionally, the elevated expression of the RsfS (HP1414) enhances biofilm forming stage by lowering metabolic activity and reduction in protein synthesis [12, 38] (Fig. 1). Hence, the formation of a complex and mature biofilm structure is a complicated multi-stage phenomenon that necessitates the interaction of a single bacterium with the surface and other bacteria [39] (Fig. 2).

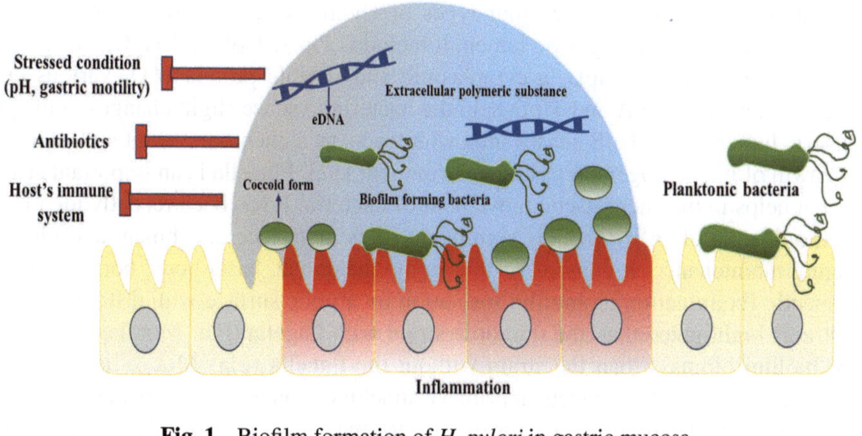

Fig. 1. Biofilm formation of *H. pylori* in gastric mucosa.

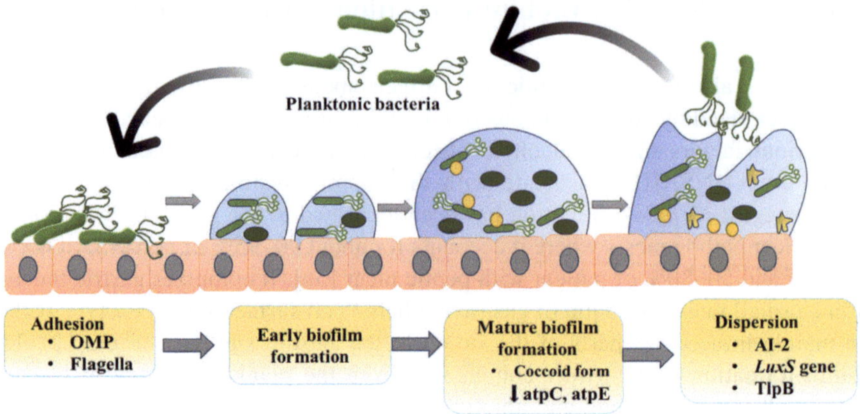

Fig. 2. Molecular mechanism of biofilm formation of *H. pylori.*

5 Molecular Mechanism of Biofilm Formation of *H. pylori*

5.1 *H. pylori* Adhesion on Gastric Mucosa

The biofilm formation of *H. pylori* starts from the adhesion of the planktonic bacteria on the surface of gastric epithelial cells. The chemotactic motility and outer membrane proteins (OMPs) play significant role in adhesion state of the pathogen [1, 12]. The pathogen is directed to the proper location to build a stable biofilm primarily by chemotaxis. *H. pylori* upregulate various chemotactic receptors, flagellar proteins and other accessory molecules to modulate their motilities during biofilm formation [14, 40]. This pathogen mainly possesses four types of chemotactic receptors to regulate for specific orientations inside gastric lumen. It includes TlpA, TlpB and TlpC, the integral membrane proteins and TlpD, a cytoplasmic and soluble protein [41]. Various evidences suggest that, TlpA and TlpD help the bacteria to sense slight changes in the pH of gastric lumen while TlpB senses urea which acts as a chemoattractant and supports the growth of the pathogen on gastric environment [12]. Flagella is an important structure that helps in the chemotactic movement of bacteria. It has been recently identified that, the flagella of *H. pylori* play significant roles in surface attachment and biofilm structure maintenance. It consists of four components viz. basal body, hook, filament and sheath. Examination on biofilm formation on abiotic surface with different bacterial strains highlighted that, the immotile strain with flagella (Fla^{+} Mot^{-}) accumulates more biofilm biomass than the strain without the flagella (Fla^{-} Mot^{-}). Interestingly, Fla.$^{-}$ Mot^{-} strain exhibits defects in primary attachment on gastric epithelial cells, suggesting that flagella are necessary for biofilm formation of *H. pylori*. Additionally, flagella filaments were seen in the biofilm and seemed to be creating a matrix [37, 42].

Outer membrane proteins (OMPs) are another important molecule which promote bacterial adherence. It is either a cell surface protein or outer membrane vesicles (OMVs) protein with extracellular DNA (eDNA) [1, 43]. These proteins play significant roles in adaptation of bacteria in gastric environment and modulate various stages of life cycle including maintenance of structure, host-microbial interactions and transportation of substances [44, 45]. Scanning electron microscopy (SEM) imaging of OMVs of *H. pylori* biofilm concluded that it promotes both cell to cell and cell to abiotic surface adherence. The common structural feature of OMP of *H. pylori* include anti-parallel β sheet with extensively stable pore-like conformation [1, 46]. The OMPs of *H. pylori* are mainly grouped under a family of four protein members encoded by the genes namely *homA*, *homB*, *homC* and *homD*. These four proteins play critical role in the pathogenesis of peptic ulcer and have been used as a marker for peptic ulcer and gastric cancer diseases [46–49]. Moreover, another family of OMP called Hop is involved in the regulation of biofilm formation. One of the members AlpB is associated with antibiotic resistance and biofilm formation. It is also known as adhesion protein which helps in attachment of bacteria on epithelial cell surface. AlpB has recently been studied as a therapeutic target to eliminate biofilm of *H. pylori* as it is highly conserved among the species [1].

5.2 *H. pylori* Microcolony Formation and Biofilm Maturation

The early growth of microcolony is the crucial step for biofilm formation where about 100 cells group together without specific boundaries, overlapping with the primary attachment state [50]. These microcolonies are composed of single or multiple strains of bacteria that provide protection, optimal condition and nutrients for survival. The biofilm produces a different microenvironment where diffusion of oxygen is limited and creates a suitable habitat for facultative anaerobes, aerobes and obligate anaerobic bacteria [51]. The biofilm maturation of *H. pylori* signifies a notable morphological change which culminates with the increased synthesis of extracellular polymeric substances (EPS) [52]. Studies have revealed that, flagella associated with pili maintain structural integrity of biofilm both on biotic and abiotic surfaces [37, 42]. The mature biofilm structure of *H. pylori* on abiotic surface mainly consists of coccoid form (0.4–0.6 μm) of bacteria with a small percentage of rod or bacillus shape (2–3 μm). The coccoid shapes of bacteria help to survive in environmental stress condition but detailed mechanism of this structure formation is still not known [1].

The main components of mature biofilm include extracellular polysaccharides, outer membrane vesicles and extracellular DNA. A mannose-related proteoglycan known as proteomannans is the primary component which supports EPS formation [53]. This is revealed by Nuclear Magnetic Resonance (NMR) study of EPS which identified the presence of 1–4 mannosyl linkages in mature and as well as in developing biofilm. Various monosaccharides like galactose, fucose, glucose, N-acetylmuramic acid, N-acetylglucosamine and glyceromannoheptose have been isolated from mature biofilm. Some lipids mainly tetradecanoic acid (C14) and hexadecanoic acid (C16) along with amino acids are also involved in biofilm formation [39].

OMVs are the essential components for pathogenesis, survival and intercellular communication of gram-negative bacteria like *H. pylori* [54]. Additionally, OMVs have been linked to a variety of processes such as conveying virulence factors, aiding adherence to host cells and inducing inflammatory responses in hosts. Various studies have displayed the existence of OMVs in the matrix of biofilm, further indicating that these structures play significant roles in biofilm maturations [39].

The other component of a mature biofilm is extracellular DNA which play significant role in their stabilization and higher recombination processes. Furthermore, eDNA play critical role in transformation, recombination and genomic variability. Studies on extracellular and intracellular DNA highlighted that eDNA may be purposefully released into extracellular areas rather than originating from cell lysis [39, 55].

5.3 Dispersion of *H. pylori* Biofilm

After reaching maximum growth conditions, the *H. pylori* biofilm disperse like other bacterial biofilm do. Although little is known about the signals that cause *H. pylori* biofilms to spread, there are evidences of bacteria using the quorum-sensing protein AI-2 as a signalling molecule to control the formation and spreading of biofilm [1].

6 Quorum Sensing and *H. pylori* Biofilm Formation

Quorum sensing is a biochemical process that facilitates cell-to-cell communication through the production of signalling molecule known as autoinducers. There are at least two factors which are essential for quorum sensing – autoinducer (AI)-2 and diffusible signal factors (DSFs). The threshold level of these factors controls the bacterial adhesion, aggregation, spatial organisation and even biofilm dispersal [56]. The quorum sensing process follows the following steps - 1) Production of autoinducer (AI), 2) Release of AI into the surrounding environment, 3) Detection and binding of AI to its receptor at higher cellular density 4) Separation of the receptor-signal complex from the cells and binding to the promoter of specific genes, 5) Gene expression activation [57]. In *H. pylori*, LuxS is the fundamental enzyme which catalyses reaction for the generation of autoinducer-2 (AI-2) [58]. AI-2 or furanosyl borate diesters are produced by recycling of S- adenosyl-homocysteine through methyl cycle activation [59]. After infecting the gastric mucosa, *H. pylori* respond to various chemical signals with the help of their chemoreceptors. The quorum sensing pathway of *H. pylori* involves a chemoreceptor known as TlpB, which binds to AI-2 and initiates various gene expression. The signalling of TlpB-AI-2 complex is regarded as chemorepellent where two periplasmic proteins AibA and AibB play significant roles in repulsive chemotactic response [57]. Various studies indicated that quorum sensing control every phase of biofilm formation, from the first attachment of cells to the detachment of biofilm. AI-2 signalling is essential for coordinating bacterial communication and controlling gene expression in relation to cell density during *H. pylori* microcolony development. AI-2 also regulates structure and functions of bacterial flagella. It is a well-established fact that flagella play a crucial role in biofilm formation by recruiting planktonic bacteria to the biofilm [39, 57].

Diffusible signal factor on the other hand is also suggested as a signalling molecule in biofilm production. DSFs are a type of fatty acid derivatives in quorum sensing system involved in inter and intra-species communication. DSF signalling is reported to stimulate *H. Pylori* transform into a sedentary state which helps the bacteria to transform into a more resistant coccoid form and initiates biofilm formation. Lauric acid and cis-7-tetradecenic acid were demonstrated to be diffusible signal factor of *H. pylori*. These DSFs has bacteriostatic effect and cause *H. pylori* to transform from spiral into coccoid form [39].

7 Biofilm Mediated Drug-Resistance Induction During *H. pylori* Infection

According to a recent meta-analysis, several antibiotic-resistant strains of *H. pylori* have emerged extensively in many parts of the world from 2013 to 2023. The metronidazole resistance was observed the most prevalent strain which is followed by clarithromycin resistant strain in the region of Europe, Africa and Southeast Asia [39]. According to the meta-analysis report of Boyanova et al., strains of *H. pylori* in Bulgaria showed 42% resistance to metronidazole and 30% to clarithromycin [2]. Globally, amoxicillin resistant strain of *H. pylori* is very uncommon, but in certain region the trend showed increased pattern especially in Mongolia and the Democratic Republic of the Congo [60–62]. The prevalence of antibiotic resistance varies significantly among industrialized country and developing nations too [2]. Investigations revealed that the biofilm producing strains display higher level of antibiotic resistance than those of free-floating planktonic bacteria [39, 61]. Incidentally, biofilm producing *H. pylori* strain tolerance level is given at 16, 8, 31.25 and 1000-fold higher for levofloxacin, metronidazole, clarithromycin and amoxicillin respectively in comparison to planktonic strain [39]. Though the detailed mechanism is still under investigation, a permanent alteration in cells such as mutation or horizontal gene transfer has been attributed for promoting antibiotic resistance in bacteria [39].

7.1 Extracellular Polysaccharide (EPS) Mediated Drug-Resistance

The extracellular polymeric substance (EPS) that surrounds the biofilm of *H. pylori* preserves the biofilm's structural integrity, encourages adhesion and promotes cell-to-cell communications [1]. EPS plays a vital role in protecting bacterial colony from antibiotic exposures [26, 63]. EPS blocks antibiotic penetration and creates a barrier that hinder the host's immune cell from recognition and phagocytosis [39]. Exposure to antibiotics had little effects on EPS itself corroborating with the inability to completely remove *H. pylori* biofilm through antibiotic therapy [1]. The proteinaceous components of EPS play crucial role in antibiotic resistance as treatment with proteinase K decreases antibiotic resistance and induces biofilm dispersion. Thus, this observation indicates that disruption of EPS could successfully hinder *H. pylori* colony and biofilm formation [37, 64].

7.2 Morphology of *H. pylori* in Drug-Resistance

Unlike the spiral shape of *H. pylori*, which is prevalent in planktonic stage of the bacteria, coccoid morphology is frequently identified in mature biofilm of *H. pylori* [37, 42]. The nonculturable but viable and metabolically less active coccoid form has the potential to revive and trigger the recurrence of *H. pylori* infection [65]. This is understandable that antibiotic treatments are most effective against the lag and log phases of the bacterium. Therefore, one of the primary reasons for the ineffectiveness of antibiotics against bacterial biofilm is the existence of dormant cells [1, 39]. Various studies have indicated that, remarkable changes in the cell wall of *H. pylori* are associated with transition of the bacteria from spiral to coccoid forms [65, 66]. Certain genes have been identified whose upregulation promotes coccoid formation and antibiotic resistance. UppS, also known as undecaprenyl pyrophosphate synthase regulate coccoid formation by modifying peptidoglycan of the *H. pylori* cell wall [1]. Peptidoglycan deacetylase (PdgA) is another component of the bacterial cell wall which preserves biofilm structure. The *pdga* gene is found to be upregulated during the oxidative stress of host and it induces biofilm formation in the bacteria [1, 37, 67]. Additionally, *hp0421* gene encodes cholesteryl-α-glucoside transferase which is capable of modifying coccoid morphology of biofilm. Hence, genes that control *H. pylori* morphologies and antibiotic resistance could be an important area of research to understand the biofilm related drug-resistance pathways [1].

7.3 Metabolic Gene Expression and Drug-Resistance

A positive correlation persists between powerful biofilm former and reduction of metabolic rate. Transcriptomic analysis of *H. pylori* in biofilm producing state shows downregulation of metabolic gene expression [37, 68]. *H. pylori* also resist antibiotics possibly by upregulating enzymes to metabolize them. For example, the people of South Africa consume Combretum molle extract to relieve stomach ailments, since it possesses bactericidal properties against *H. pylori*. The plant extract contains acetone which stimulates the expression of acetone carboxylase enzyme encoded by *acxA* gene of *H. pylori*. Furthermore, mutation of *acxA* gene is also associated with defective biofilm formation signifying that the biofilm state of bacteria modulates expression of metabolic gene to confer drug-resistance [42, 69].

7.4 Upregulation of Efflux Pump in Drug-Resistance

Efflux pump is mainly located in the cell membrane of *H. pylori* and plays significant role in the external transportation of various drugs [70]. Studies have identified that there is strong association between expression of efflux pumps and the antibiotic-resistant strains generation. Various genes encoding efflux pump such as HPG27_715, HP1165, gluP and HP1118 are upregulated during biofilm state of the bacteria [1]. TK1402, a clarithromycin-resistant strain exhibits upregulation of efflux pump expressing *hefA*, *hefG*, *hefD* and HP1489 genes [18]. Moreover, *H. pylori* upregulates *gluP* efflux gene expression to combat the effect of antibiotics, while its mutation is associated with defective biofilm formation and increased antibiotic susceptibility [71].

8 Therapeutic Strategies against Biofilm-Associated Drug-Resistant *H. pylori*

Chronic infection with *H. pylori* is associated with severe gastrointestinal disorders including peptic ulcers and gastric cancers. It has been found that, the biofilm formation of *H. pylori* reduces and impedes the effectiveness of widely used antibiotics. The fight against the infections has sparked creative investigation into new forms of treatment. Therefore, it has necessitated incorporating alternate drug molecules which has minimal adverse effects with enhanced antibiofilm activities for the eradication of infection of various bacterial strains including *H. pylori* infections.

8.1 Phytochemical Compounds as Antibiofilm Agents

Several natural products have been identified which show antibiofilm properties against the *H. pylori*. The phytocompounds from *Vitex trifolia*, *Acorus calamus* and *Azadirachta indica* have shown significant antibiofilm properties against *H. pylori* [72, 73]. Nimbolide is a phytocompound from *Azadirachta* which is effective against both biofilm and planktonic form of *H. pylori* [72]. The volatile oil of *Antractylodes lancea* has the properties to hinder the biofilm stage of *H. pylori* [1]. Moreover, the shoot extract of *Rubus idaeus* when combined with levofloxacin and doxycycline, shows synergistic effects against biofilm formation [74]. Eugenol is an essential oil extracted from cloves that could inhibit biofilm formation and shows anti-inflammatory activities [75]. Curcumin of curcuma longa possess antibacterial and antibiofilm properties against several strains of *H. pylori*. Notably, aloe-emodin extracted from *aloe vera* targets outer membrane protein-6 (OMP-6) of *H. pylori* and destroys the biofilm structures. On the hand Myricetin derived from Myrica tree decreases the rate of transformation of *H. pylori* into coccoid form and inhibits biofilm formation. Despite the fact that many plant extracts contain antibiofilm properties, extensive studies are required to identify the bioactive ingredients and understand their molecular mechanisms of actions. Furthermore, environmental variables may result in changes in plant compositions, altering the bioactive content and hamper the effectiveness of the medications [14].

8.2 Probiotics Therapy against Drug-Resistant *H. pylori*

Various probiotics exhibit significant antibacterial and antibiofilm properties thus play supplementary effect to antibiotic therapies. Initially, *Lactobacillus fermentum* (UCO-979C) has been investigated which inhibits biofilm formation of *H. pylori* [76]. Later, *Lactobacillus plantarum* LN66 cell-free supernatant (CFS) has been shown to reduce the production of *H. pylori* biofilm. Since LN66 CFS assist antibiotics to block EPS release, combinatorial treatment improves the effectiveness of levofloxacin [77]. Armeniaspirol A (ARM1), collected from *Streptomyces armeniacus* exhibits significant antibacterial properties and found to inhibit biofilm formation of *H. pylori* in dose dependent manner. Dual therapy with ARM1 and omeprazole demonstrated effective killing efficacy as compared to conventional triple therapy in a mouse model. ARM1 demonstrated significantly stronger anti-*H. pylori* action than metronidazole at a higher acidic pH. These benefits increase the likelihood of ARM1 application in several therapeutic settings [1].

8.3 Antimicrobial Peptides Against Drug-Resistant *H. pylori*

Antimicrobial peptides (AMPs) play auxiliary roles in the conventional antibiotic treatments of *H. pylori*. The benefit of employing AMPs is that, they could be used to target several bacterial components thereby lowering the generation of resistance in the bacteria. These small structures allowed them to penetrate the extracellular matrix shields of biofilm forming bacteria. Cathelicidin-like peptide (Cbf-K16) has recently been investigated for their significant antimicrobial properties against clarithromycin and amoxicillin resistant *H. pylori* in both in vitro and in vivo condition. Cbf-K16 potentially downregulates the adhesion associated gene, *alpA* and *alpB* expression and hinder biofilm formation [1]. MSI-78A or Pexiganan is also another antimicrobial peptide composed of 22 amino acids that possess antimicrobial properties in solution [1, 78]. Moreover, various synthetic peptides have been discovered which enhance biofilm dispersion either alone or in combination with host antimicrobial peptide. For example, IDR-1018 and DJK-5 are synthetic peptides that can increase the susceptibility of *H. pylori* biofilm to host-derived antimicrobial peptides [64].

8.4 Nanodrug Formulation as Novel Antibiofilm Candidates

Several small molecule drugs are found to play significant role in the treatment of bacterial infection. These compounds are generally inorganic and organic monomers or polymers which can target essential structure, enzymes and metabolic pathways of bacteria. For instance, Thymol and carvacrol inhibits carbonic anhydrase which is essential for biofilm growth [79]. Encapsulation of specific antibiotic like clarithromycin within lipid polymer nanoparticles could effectively diminish the structural integrity and viability of *H. pylori* biofilm by avoiding EPS and mucous layer of biofilm [1]. Moreover, synthesized silver ultra-nano clusters (SUNCs) have shown synergistic effect against *H. pylori* biofilm formation when combined with metronidazole [80]. Furthermore, mannose-modified nanoparticles were effective against multidrug resistant *H. pylori* and their associated biofilm formation [1].

8.5 Quorum Quencher (QQ) Molecules against Multidrug Resistant *H. pylori*

As mentioned earlier, the quorum sensing (QS) plays significant role in the generation of multidrug resistant strain of *H. pylori* through upregulation of the genes associated with EPS and efflux pumps [81]. Moreover, the biofilm development is hindered by inhibition of QS by making the bacteria more vulnerable to antibiotic exposure. Quorum sensing inhibitors also known as quorum quenchers (QQ) are either synthetic or naturally produced from plants, bacteria and animals. Since the core aspect of QS involves the production and detection of signalling molecules, QQ can disrupt the process through various intracellular and extracellular mechanisms (Table 1). Some of the mechanisms include inhibiting signal synthesis, blocking signal transmission, enzymatic degradation of AI and prevention of signal detection [57]. A study evaluating the effect of *Acorus calamus* plant extract on *H. pylori* revealed a significant antibiofilm and anti-adhesive properties. Molecular docking studies later revealed that β-sitosterol, a bioactive compound of *A. calamus* interacts with QS proteins ToxB, PhnB, Sip and DnaA [73]. The most direct and effective approach to inhibit the QS system is the enzymatic degradation of QS molecules to halt signal transduction. In gram-negative bacteria, two types of hydrolases have been identified viz. AHL-lactonase and AHL-acylase. Notably, *Bacillus licheniformis*-derived N-acylhomoserine lactonase degraded AHL produced by *H. pylori*, leading to reduced biofilm formation and attenuated virulence [57, 82]. Another effective approach to QS inhibition is disrupting the signalling cascade to inactivate downstream response regulators. AI-2 synthesis relies on the precursor SRH, which is derived from the action of 5′-methylthioadenosine/*S*-adenosylhomocysteine nucleosidase (MTAN) on S-adenosylhomocysteine (SAH). Inhibiting

Table 1. Different quorum quenchers and their mode of action against *H. pylori* biofilm formation.

Sl No.	Quorum quencher molecules (QQ)	Mechanism of actions	Effects on *H. pylori*	Test	References
1.	β-sitesterol (*Acorus calamus*)	It has high binding affinity with AI-2 receptor of *H. pylori* and acts as AI-2 antagonist	Antiadhesive, Antibiofilm, Antibacterial	In silico and in vitro	[57, 73]
2.	N-acylhomoserinelactonase (*Bacillus licheniformis*)	Enzymatic degradation of AHL, produced by *H. pylori*	Antibiofilm, Antibacterial	In vitro	[57, 82]
3.	Methylthio-DADMe-immucillin-A (DADMe-ImmA)	Inhibiting MTAN and thereby supressing AI-2 production	MTAN inhibitor	In silico	[57]
4.	-SH Furanosyl Borate Diester	Competitive inhibitor of AI-2	Antibiofilm	In silico	[57]

MTAN leads to the accumulation of 5-methylthioadenosine (MTA) and SAH, thereby suppressing AI-2 production. In silico testing of DADMe-ImmA derivatives demonstrated strong MTAN inhibition by tightly binding to the receptor. Furthermore, computational studies explored the potential of designing furanosyl borate diester derivatives through pharmacophore modelling, where replacing the –OH groups of AI-2 and DPD with –SH could derive a potent competitive inhibitor of AI-2 [6].

9 Conclusion

Significant progress has been made by the scientific communities in deciphering the complex nature of the *H. pylori* and its biofilm formation mechanism. Notably, research utilizing clinically isolated strains has been instrumental in enhancing our understanding and has laid the foundation for developing promising antibiofilm-based strategies to eradicate *H. pylori*. However, future research should adopt a "bench-to-bedside" approach, combining laboratory studies with clinical trials to bridge the gap between theoretical knowledge and practical applications. While several factors responsible for *H. pylori* biofilm formation have been identified, the exact mechanisms and key targets for eradication remain uncertain. The advances in omics studies, sequencing, bioinformatics, and machine learning could expedite the discovery of potential drug targets by analysing interactions between antibiotics or antibiofilm agents and bacterial proteins. These predicted targets can then be validated through experimental lab-based studies. Moreover, the complexity of the in vivo environment and the limitations of existing techniques pose challenges in characterizing bacterial biofilms within the host and to evaluate their potential treatments. Future research must extend beyond in vitro experiments by incorporating comprehensive animal models and rigorous clinical trials. Such studies will provide a more accurate assessment of the therapeutic potential of these treatments and their effectiveness on both the host and the pathogen. Additionally, examining the long-term effects of these innovative approaches is essential to ensure their safety and effectiveness of applications. Moreover, studying potential resistance mechanisms that *H. pylori* may develop in response to biofilm-targeting therapies will be crucial for designing more resilient treatment strategies.

Acknowledgement. The authors express their gratitude to the University of Kalyani for the Personal Research Grants (PRG) financial assistance to carry out this research work.

References

1. Elshenawi, Y., Hu, S., Hathroubi, S.: Biofilm of helicobacter pylori: life cycle, features, and treatment options. Antibiotics (Basel, Switzerland). **12**(8), 1260 (2023)
2. Elbehiry, A., et al.: Helicobacter pylori: routes of infection, antimicrobial resistance, and alternative therapies as a means to develop infection control. Diseases (Basel, Switzerland). **12**(12), 311 (2024)

3. Correa, P., Piazuelo, M.B.: Natural history of helicobacter pylori infection. Dig. Liver Dis. **40**(7), 490–496 (2008)
4. Yonezawa, H., Osaki, T., Kamiya, S.: Biofilm formation by helicobacter pylori and its involvement for antibiotic resistance. Biomed. Res. Int. **2015**, 914791 (2015)
5. Vogiatzi, P., Cassone, M., Luzzi, I., Lucchetti, C., Otvos Jr., L., Giordano, A.: Helicobacter pylori as a class I carcinogen: physiopathology and management strategies. J. Cell. Biochem. **102**(2), 264–273 (2007)
6. Kumar, M., Naaz, A., Sharma, A., Teotia, D., Nandi, S.: Pharmacophore modeling of Furanosyl borate Diester: an attempt to combat antibiotic resistance induced helicobacter pylori quorum sensing mediated peptic and duodenal ulcer. J. Bioanal. Biomed. **10** (2018)
7. Di Fermo, P., et al.: Helicobacter pylori dormant states are affected by vitamin C. Int. J. Mol. Sci. **24**(6), 5776 (2023)
8. Helicobacter pylori CDC Yellow Book.: https://wwwnc.cdc.gov/travel/yellowbook/2024/infections-diseases/helicobacter-pylori (2024)
9. Tao, H., et al.: Transcriptomic and functional approaches unveil the role of tmRNA in zinc acetate mediated levofloxacin sensitivity in helicobacter pylori. Microbiol Spectr. **10**(6), e0115222 (2022)
10. Mesquita, A., Rocha-Castro, C., Guimarães, D., Costa, J., Soutinho, J., Taveira-Gomes, T.: Multicentric study to assess helicobacter pylori incidence, patient reported adverse events, compliance and effectiveness, in real-world setting. Int. J. Environ. Res. Public Health. **19**(19), 12847 (2022)
11. Yamaoka, Y., Graham, D.Y.: Helicobacter pylori virulence and cancer pathogenesis. Future Oncol. **10**(8), 1487–1500 (2014)
12. Gupta, N., Kumar, A., Verma, V.K.: Strategies adopted by gastric pathogen helicobacter pylori for a mature biofilm formation: antimicrobial peptides as a visionary treatment. Microbiol. Res. **273**, 127417 (2023)
13. Savoldi, A., Carrara, E., Graham, D.Y., Conti, M., Tacconelli, E.: Prevalence of antibiotic resistance in helicobacter pylori: a systematic review and meta-analysis in World Health Organization regions. Gastroenterology. **155**(5), 1372–1382.e17 (2018)
14. Savitri, C.M.A., et al.: Opportunities for helicobacter pylori eradication beyond conventional antibiotics. Microorganisms. **12**(10), 1986 (2024)
15. Blair, J.M., Webber, M.A., Baylay, A.J., Ogbolu, D.O., Piddock, L.J.: Molecular mechanisms of antibiotic resistance. Nat. Rev. Microbiol. **13**(1), 42–51 (2015)
16. Cellini, L., et al.: Dynamic colonization of helicobacter pylori in human gastric mucosa. Scand. J. Gastroenterol. **43**(2), 178–185 (2008)
17. Rather, M.A., Gupta, K., Mandal, M.: Microbial biofilm: formation, architecture, antibiotic resistance, and control strategies. Braz. J. Microbiol. **52**(4), 1701–1718 (2021)
18. Yonezawa, H., Osaki, T., Hojo, F., Kamiya, S.: Effect of helicobacter pylori biofilm formation on susceptibility to amoxicillin, metronidazole and clarithromycin. Microb. Pathog. **132**, 100–108 (2019)
19. Hou, C., Yin, F., Wang, S., Zhao, A., Li, Y., Liu, Y.: Helicobacter pylori biofilm-related drug resistance and new developments in its anti-biofilm agents. Infect. Drug Resist. **15**, 1561–1571 (2022)
20. Wu, H., Moser, C., Wang, H.Z., Høiby, N., Song, Z.J.: Strategies for combating bacterial biofilm infections. Int. J. Oral Sci. **7**(1), 1–7 (2015)
21. Zhao, A., Sun, J., Liu, Y.: Understanding bacterial biofilms: from definition to treatment strategies. Front. Cell. Infect. Microbiol. **13**, 1137947 (2023)
22. Costerton, J.W., Geesey, G.G., Cheng, K.-J.: How Bacteria Stick. Sci. Am. **238**(1), 86–95 (1978)

23. Sharma, S., Mohler, J., Mahajan, S.D., Schwartz, S.A., Bruggemann, L., Aalinkeel, R.: Microbial biofilm: a review on formation, infection, antibiotic resistance, control measures, and innovative treatment. Microorganisms. **11**(6), 1614 (2023)
24. Donlan, R.M.: Biofilms: microbial life on surfaces. Emerg. Infect. Dis. **8**(9), 881–890 (2002)
25. Sutherland, I.: Biofilm exopolysaccharides: a strong and sticky framework. Microbiology (Reading). **147**(Pt 1), 3–9 (2001)
26. Singh, S., Datta, S., Narayanan, K.B., Rajnish, K.N.: Bacterial exo-polysaccharides in biofilms: role in antimicrobial resistance and treatments. J. Genet. Eng. Biotechnol. **19**(1), 140 (2021)
27. Vandana, Das, S.: Structural and mechanical characterization of biofilm-associated bacterial polymer in the emulsification of petroleum hydrocarbon. 3 Biotech. **11**(5), 239 (2021)
28. Vestby, L.K., Grønseth, T., Simm, R., Nesse, L.L.: Bacterial biofilm and its role in the pathogenesis of disease. Antibiotics (Basel). **9**(2), 59 (2020)
29. Paharik, A.E., Horswill, A.R.: the staphylococcal biofilm: Adhesins, regulation, and host response. Microbiol. Spectr. **4**(2), 10.1128/microbiolspec.VMBF-0022-2015 (2016)
30. Veerachamy, S., Yarlagadda, T., Manivasagam, G., Yarlagadda, P.K.: Bacterial adherence and biofilm formation on medical implants: a review. Proc. Inst. Mech. Eng. H. **228**(10), 1083–1099 (2014)
31. Bjarnsholt, T.: The role of bacterial biofilms in chronic infections. APMIS Suppl. **136**, 1–51 (2013)
32. Brackman, G., De Meyer, L., Nelis, H.J., Coenye, T.: Biofilm inhibitory and eradicating activity of wound care products against Staphylococcus aureus and Staphylococcus epidermidis biofilms in an in vitro chronic wound model. J. Appl. Microbiol. **114**(6), 1833–1842 (2013)
33. Dowd, S.E., et al.: Survey of bacterial diversity in chronic wounds using pyrosequencing, DGGE, and full ribosome shotgun sequencing. BMC Microbiol. **8**, 43 (2008)
34. Watters, C., et al.: Pseudomonas aeruginosa biofilms perturb wound resolution and antibiotic tolerance in diabetic mice. Med. Microbiol. Immunol. **202**(2), 131–141 (2013)
35. Machado, D., Castro, J., Palmeira-de-Oliveira, A., Martinez-de-Oliveira, J., Cerca, N.: Bacterial vaginosis biofilms: challenges to current therapies and emerging solutions. Front. Microbiol. **6**, 1528 (2016)
36. Gupta, A., et al.: Treatment of H. pylori infection and gastric ulcer: need for novel pharmaceutical formulation. Heliyon. **9**(10), e20406 (2023)
37. Hathroubi, S., Zerebinski, J., Ottemann, K.M.: helicobacter pylori biofilm involves a multigene stress-biased response, including a structural role for flagella. MBio. **9**(5), e01973–18 (2018)
38. Krzyżek, P., Gościniak, G.: A proposed role for diffusible signal factors in the biofilm formation and morphological transformation of helicobacter pylori. Turk J Gastroenterol. **29**(1), 7–13 (2018)
39. Fauzia, K.A., Effendi, W.I., Alfaray, R.I., Malaty, H.M., Yamaoka, Y., Mifthussurur, M.: Molecular mechanisms of biofilm formation in helicobacter pylori. Antibiotics. **13**(10), 976 (2024)
40. Lertsethtakarn, P., Ottemann, K.M., Hendrixson, D.R.: Motility and chemotaxis in campylobacter and helicobacter. Ann. Rev. Microbiol. **65**, 389–410 (2011)
41. Johnson, K.S., et al.: The dCache chemoreceptor TlpA of helicobacter pylori binds multiple attractant and antagonistic ligands via distinct sites. MBio. **12**(4), e0181921 (2021)
42. Hathroubi, S., Hu, S., Ottemann, K.M.: Genetic requirements and transcriptomics of helicobacter pylori biofilm formation on abiotic and biotic surfaces. NPJ Biofilms Microbiomes. **6**, 56 (2020)
43. Olofsson, A., et al.: Biochemical and functional characterization of helicobacter pylori vesicles. Mol. Microbiol. **77**(6), 1539–1555 (2010)

44. Qiao, S., Luo, Q., Zhao, Y., Zhang, X.C., Huang, Y.: Structural basis for lipopolysaccharide insertion in the bacterial outer membrane. Nature. **511**(7507), 108–111 (2014)
45. Egan, A.J.F.: Bacterial outer membrane constriction. Mol. Microbiol. **107**(6), 676–687 (2018)
46. Yonezawa, H., et al.: Outer membrane vesicles of helicobacter pylori TK1402 are involved in biofilm formation. BMC Microbiol. **9**, 197 (2009)
47. Alm, R.A., Bina, J., Andrews, B.M., Doig, P., Hancock, R.E., Trust, T.J.: Comparative genomics of helicobacter pylori: analysis of the outer membrane protein families. Infect. Immun. **68**(7), 4155–4168 (2000)
48. Servetas, S.L., et al.: ArsRS-dependent regulation of homB contributes to helicobacter pylori biofilm formation. Front. Microbiol. **9**, 1497 (2018)
49. Talebi Bezmin Abadi, A., et al.: helicobacter pylori homB, but not cagA, is associated with gastric cancer in Iran. J. Clin. Microbiol. **49**(9), 3191–3197 (2011)
50. Monds, R.D., O'Toole, G.A.: The developmental model of microbial biofilms: ten years of a paradigm up for review. Trends Microbiol. **17**(2), 73–87 (2009)
51. Rani, S.A., et al.: Spatial patterns of DNA replication, protein synthesis, and oxygen concentration within bacterial biofilms reveal diverse physiological states. J. Bacteriol. **189**(11), 4223–4233 (2007)
52. Krzyżek, P., Migdał, P., Grande, R., Gościniak, G.: Biofilm formation of helicobacter pylori in both static and microfluidic conditions is associated with resistance to clarithromycin. Front. Cell. Infect. Microbiol. **12**, 868905 (2022)
53. Yang, F.L., et al.: Proteomannans in biofilm of helicobacter pylori ATCC 43504. Helicobacter. **16**(2), 89–98 (2011)
54. Kulp, A., Kuehn, M.J.: Biological functions and biogenesis of secreted bacterial outer membrane vesicles. Ann. Rev. Microbiol. **64**, 163–184 (2010)
55. Grande, R., et al.: Helicobacter pylori ATCC 43629/NCTC 11639 outer membrane vesicles (OMVs) from biofilm and planktonic phase associated with extracellular DNA (eDNA). Front. Microbiol. **6**, 1369 (2015)
56. Anderson, J.K., et al.: Chemorepulsion from the quorum signal Autoinducer-2 promotes helicobacter pylori biofilm dispersal. MBio. **6**(4), e00379 (2015)
57. Aanniz, T., et al.: Biofilm and Quorum Sensing in Helicobacter pylori. IntechOpen (2022)
58. De la Cruz, M.A., et al.: Gene expression profiling of transcription factors of helicobacter pylori under different environmental conditions. Front. Microbiol. **8**, 615 (2017)
59. Hemmati, F., et al.: Quorum quenching: a potential target for antipseudomonal therapy. Infect. Drug Resist. **13**, 2989–3005 (2020)
60. Azzaya, D., et al.: High antibiotic resistance of helicobacter pylori and its associated novel gene mutations among the Mongolian population. Microorganisms. **8**(7), 1062 (2020)
61. Cimuanga-Mukanya, A., et al.: Synergistic effects of novel penicillin-binding protein 1A amino acid substitutions contribute to high-level amoxicillin resistance of helicobacter pylori. mSphere. **9**(8), e0008924 (2024)
62. Khangai, A., et al.: Exploring alternative treatment choices for multidrug-resistant clinical strains of helicobacter pylori in Mongolia. Microorganisms. **11**(12), 2852 (2023)
63. Balducci, E., Papi, F., Capialbi, D.E., Del Bino, L.: Polysaccharides'structures and functions in biofilm architecture of antimicrobial-resistant (AMR) pathogens. Int. J. Mol. Sci. **24**(4), 4030 (2023)
64. Windham, I.H., Servetas, S.L., Whitmire, J.M., Pletzer, D., Hancock, R.E.W., Merrell, D.S.: Helicobacter pylori biofilm formation is differentially affected by common culture conditions, and proteins play a central role in the biofilm matrix. Appl. Environ. Microbiol. **84**(14), e00391–e00318 (2018)
65. Reshetnyak, V.I., Reshetnyak, T.M.: Significance of dormant forms of helicobacter pylori in ulcerogenesis. World J. Gastroenterol. **23**(27), 4867–4878 (2017)

66. Costa, K., et al.: The morphological transition of helicobacter pylori cells from spiral to coccoid is preceded by a substantial modification of the cell wall. J. Bacteriol. **181**(12), 3710–3715 (1999)
67. Wang, G., Maier, S.E., Lo, L.F., Maier, G., Dosi, S., Maier, R.J.: Peptidoglycan deacetylation in helicobacter pylori contributes to bacterial survival by mitigating host immune responses. Infect. Immun. **78**(11), 4660–4666 (2010)
68. Wong, E.H.J., Ng, C.G., Goh, K.L., Vadivelu, J., Ho, B., Loke, M.F.: Metabolomic analysis of low and high biofilm-forming helicobacter pylori strains. Sci. Rep. **8**(1), 1409 (2018)
69. Njume, C., Afolayan, A.J., Samie, A., Ndip, R.N.: Inhibitory and bactericidal potential of crude acetone extracts of Combretum molle (Combretaceae) on drug-resistant strains of helicobacter pylori. J. Health Popul. Nutr. **29**(5), 438–445 (2011)
70. Raj, D.S., Kesavan, K.D., Muthusamy, N., Umamaheswari, S.: Efflux pumps potential drug targets to circumvent drug resistance – multi drug efflux pumps of helicobacter pylori. Mater Today Proc. **45**(2), 2976–2981 (2021)
71. Ge, X., et al.: Bifunctional enzyme SpoT is involved in biofilm formation of helicobacter pylori with multidrug resistance by upregulating efflux pump Hp1174 (gluP). Antimicrob. Agents Chemother. **62**(11), e00957–e00918 (2018)
72. Wylie, M.R., Windham, I.H., Blum, F.C., Wu, H., Merrell, D.S.: In vitro antibacterial activity of nimbolide against helicobacter pylori. J. Ethnopharmacol. **285**, 114828 (2022)
73. Prasad, A., Devi, A.T., Prasad, M.N.N., Zameer, F., Shruthi, G., Shivamallu, C.: Phyto antibiofilm elicitors as potential inhibitors of Helicobacter pylori. 3 Biotech. **9**(2), 53 (2019)
74. Hałasa, R., Turecka, K., Mizerska, U., Krauze-Baranowska, M.: Anti-helicobacter pylori biofilm extracts from Rubus idaeus and Rubus occidentalis. Pharmaceutics. **16**(4), 501 (2024)
75. Elbestawy, M.K.M., El-Sherbiny, G.M., Moghannem, S.A.: Antibacterial, Antibiofilm and anti-inflammatory activities of eugenol clove essential oil against resistant helicobacter pylori. Molecules. **28**(6), 2448 (2023)
76. Salas-Jara, M.J., Sanhueza, E.A., Retamal-Díaz, A., González, C., Urrutia, H., García, A.: Probiotic lactobacillus fermentum UCO-979C biofilm formation on AGS and Caco-2 cells and helicobacter pylori inhibition. Biofouling. **32**(10), 1245–1257 (2016)
77. Ji, J., Yang, H.: In vitro effects of lactobacillus plantarum LN66 and antibiotics used alone or in combination on helicobacter pylori mature biofilm. Microorganisms. **9**(2), 424 (2021)
78. Zhang, X.L., et al.: The synthetic antimicrobial peptide pexiganan and its nanoparticles (PNPs) exhibit the anti-helicobacter pylori activity in vitro and in vivo. Molecules. **20**(3), 3972–3985 (2015)
79. Grande, R., et al.: Selective inhibition of helicobacter pylori carbonic anhydrases by Carvacrol and Thymol could impair biofilm production and the release of outer membrane vesicles. Int. J. Mol. Sci. **22**(21), 11583 (2021)
80. Grande, R., et al.: Antimicrobial and Antibiofilm activities of new synthesized silver ultra-NanoClusters (SUNCs) against helicobacter pylori. Front. Microbiol. **11**, 1705 (2020)
81. Francesco, V.D., Zullo, A., Hassan, C., Giorgio, F., Rosania, R., Ierardi, E.: Mechanisms of helicobacter pylori antibiotic resistance: an updated appraisal. World J. Gastrointest. Pathophysiol. **2**(3), 35–41 (2011)
82. Zhou, L., Zhang, Y., Ge, Y., Zhu, X., Pan, J.: Regulatory mechanisms and promising applications of quorum sensing-inhibiting agents in control of bacterial biofilm formation. Front. Microbiol. **11**, 589640 (2020)

Docking Based Screening of Potent Natural Components to Combat COVID-19-Associated Brain Fog

Sisir Nandi[1,2] (✉), Rachana Belwal[1], Rishita Dey[1,3], and Asmita Samadder[3] (✉)

[1] Global Institute of Pharmaceutical Education and Research, Kashipur, India
sisir.iicb@gmail.com

[2] Department of Pharmaceutical Chemistry, Global Institute of Pharmaceutical Education and Research, Kashipur, Uttarakhand, India

[3] Cytogenetics and Molecular Biology Lab., Department of Zoology, University of Kalyani, Kalyani, Nadia, India
asmita.samadder@gmail.com; asmitazoo19@klyuniv.ac.in

Abstract. COVID-19 is an infectious disease caused by a novel corona virus SARS-CoV-2 strain (nCoV) and has become a deleterious issue around the globe since December 2019, prominently from Wuhan, China. Millions of lives were destroyed hampering the economy and social security. Even WHO declared it as a dreadful pandemic disease that negatively impacted human unity and pressured the brain to be bound under complete isolation and lock down. SARS-CoV-2 damages the lungs, heart, forced liver, and even neuronal vascular endothelium, resulting in brain fog, lethargy, and tremors. Although COVID-19 can cause a variety of extra-pulmonary symptoms, the heart, brain, and kidneys are especially vulnerable. This vascular component of COVID-19 may contribute to the explanation of the brain fog that some people experience a few months after the virus has cleared. The global COVID-19 pandemic has baffled several scientists and researchers with its versatile consequences and behavior in the etiology of several diseased conditions in post-COVID patients. The indirect and direct negative impact of COVID-19 on memory and other cognitive brain fog in humans is now a major research arena in almost all countries. There hardly exists any allopathic medicine for the treatment of COVID-19-associated brain fog. The current investigation aims to investigate the pathophysiology of the disease and structure-based screening of natural compounds to combat the dreadful post-COVID symptoms of brain fog due to neuro-inflammation followed by degeneration. Thus, a person detected with SARS-CoV-2 infection can be dealt with naturopathy to help in the recovery of memory-loss-function and combat associated neurological complications.

Keywords: SARS-CoV-2 · Brain fog · Memory loss · Luteolin · Quercetin · Curcumin · Resveratrol · Piperine · Linalool · Citronellal · Mahanimbine

A. K. Saxena, A. Saxena (Eds.): GTHTM 2025, pp. 375–395, 2026.
https://doi.org/10.1007/978-3-032-12320-6_22

1 Introduction

The highly contagious SARS-CoV-2 strain, that causes the corona virus disease, was found to be associated with deadly respiratory complications. SARS-CoV-2 was discovered by the WHO to be a new corona virus strain during the COVID-19 outbreak that struck China in December 2019 [1]. The outbreak spread very fast around the world and declared a pandemic that broke down many lakh lives, social security, and the economy of the countries [2]. The world has been facing many COVID-19 waves due to attacks of multiple variants of interest since the outbreak. The first wave was due to the D614G SARS-CoV-2 variant that speeds fast globally. Other variants of interest were Epsilon (GH/452R.V1), Eta (G/484 K.V3; B.1.525), Iota (GH/253G.V1; B.1.526), Kappa (G/452R.V3; B.1.617.1), Zeta (P.2), Theta (P.3, B.1.1.28.3) and Lambda (GR/452Q.V1) whereas Variants of Concern (VOC) such as Alpha (B.1.1.7; 20I/501Y. V1), Beta (B.1.351; GH/501Y.V2), Gamma (P.1; GR/501Y.V3), and Delta (G/478 K. V1; B.1.617.2) were possible reasons for the second wave. Omicron and Delta plus (AY.1 and AY.2) were possible reasons for the third wave attack [3]. The novel corona virus enters through the nose, sinus, and upper throat and reaches into the lungs and intestine shedding in ACE-2 expressing cells [4]. Symptoms in humans are SARS-associated an elevation in body temperature, persistent coughing, dyspnea, pneumonia, multiple organ dysfunction, bronchitis, sore throat, pharyngitis, sinusitis, and other system symptoms include ageusia, anosmia, and frequent diarrhea [5]. The drastic condition was caused by happy hypoxia and circulatory arrests [6]. The novel corona virus created the loss of immunity, inflammatory bowel diseases, and lung dysfunction due to gut-lung microbiota cross-talks [7, 8]. The loss of memory and brain fog-associated issues has been forefront after the outbreak of the second and third waves of COVID-19 infection. Combinatorial issues such as memory loss, unstable mental state, difficulty focusing, and feeling disoriented, headaches, and confusion are symptoms of brain fog. A wide range of neurological symptoms, including ischemic attacks, disorientation, seizures, loss of consciousness, headaches, and ageusia, anosmia, are being experienced by COVID-19 individuals. Stroke patients between the ages of 30 and 40 are experiencing neurological problems that might change their lives. Severe infection, immune system in overdrive, and blood clotting abnormalities also cause the condition of memory loss. In elderly persons, it has lasting consequences in addition to temporary memory loss [9]. The virus is neuro-invasive, meaning that it can enter the brain tissue because the ACE-2 receptor is expressed in brain neuronal cells [10]. Patients who have recovered from COVID-19 have neuropsychological impairments including dizziness, headaches, loss of consciousness, and encephalopathy [11–13]. Maximum death occurs due to severe ischemic stroke, and cerebral thrombosis [14, 15] and Guillain-Barre syndrome (GBS) [16–20]. According to reports, individuals with COVID-19 who spent a significant amount of time in a hospital's critical care facility were at a high risk of developing polyneuropathy and myopathy, other serious neuromuscular difficulties [21, 22]. The lack of oxygen and nutrient supply aggravates the immune response and systemic inflammatory response syndrome (SIRS) that encourage the generation of oxidative stress, and free radicals which initiate neuroinflammation since the central and peripheral neural systems suffer severe neurodegeneration as a result of the release of pro-inflammatory cytokines (PICs). The brain fog linked to long-term COVID-19 is thought to be significantly influenced by neuroinflammation. Cognitive problems in people with chronic symptoms have been linked to elevated levels of pro-inflammatory cytokines, especially the chemokine CCL11. It is thought that this increased inflammatory response impairs neural plasticity and interferes with neurogenesis, which leads to cognitive impairment [23, 24]. During an acute COVID-19 infection, some individuals experience a "cytokine storm," which is defined by an overabundance of immunological activity. Remaining neuroinflammation may continue even after the initial viral assault has been resolved, which might lead to long-term cognitive impairment. It has been proposed that this protracted inflammatory state weakens the blood-brain barrier, allowing immune cells to enter the central nervous system and aggravating neurocognitive impairment [25, 26]. Risk factors associated with extended periods of mechanical ventilation, sepsis, and hypoxia; all are prevalent in severe cases of SARS-CoV-2 [27]. There is no permanent solution in this regard. Therefore, great attention has to be focused on exploring the brain fog issues and solutions to long-term memory loss. The most concerning factor that remains about synthetic drugs like Piracetam and Oxiracetam is that, although these are very popular drugs in use as brain function enhancement as well as neuronal regeneration supplements in several European countries, the FDA has not authorized their application in patients. The U.S. Food and Drug Administration (FDA) has declared Piracetam as "not safe for use" as a dietary supplement in any human with such medical conditions. Since the enforcement of regulations is not that stringent, Piracetam and Oxiracetam remain the most used nootropic drugs [28, 29]. Since the direct relation of these drugs in modulating the neuronal proteins associated with cognition development as well as the exact mechanism of action of these drugs remain a matter of further investigation, the use of such drugs in treating the patients suffering from post-COVID-19 associated human cognition impairment as well as neuronal damage remains a question. Therefore, our focus in this study is to include phyto drugs in the therapeutic management of COVID-19-associated brain fog. Phyto-sourced Luteolin, Quercetin, Curcumin, Resveratrol, Piperine, linalool, citronellal, and Mahanimbine have shown clinical uses against brain fog. However, their exact biochemical mechanisms are yet to be explored. Therefore, the focus of our present study is to investigate the binding affinity and molecular mechanisms of these phyto active components towards acetylcholinesterase (AChE) and butyrylcholinesterase (BChE) which are major brain fog targets associated with memory loss.

2 Materials and Methods

2.1 Activity Data

Research on herbal-based compounds is gaining much importance nowadays on account of their healing power against several diseases and ailments as well as their reduced toxicity, safety, and cost-effective nature. Therefore, these aspects led to the findings of active components that play a preventive role against the current diseased condition "brain fog" associated with COVID-19. Phytocomponents of some herbal

Table 1. Medicinal phytocomponents used in the treatment of COVID-19-associated brain fog impairments.

Sl. No.	Name of herbal plant	Name of bio-active component	Structure of bio-active component	Reference
1.	*Petroselinum crispum, Thymus vulgaris*	Luteolin	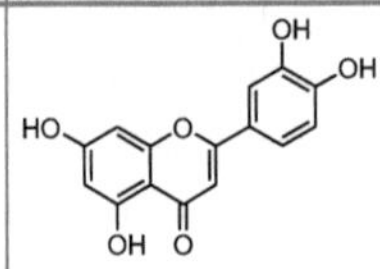	[26–36]
2.	*Allium cepa, Lactuca sativa*	Quercetin	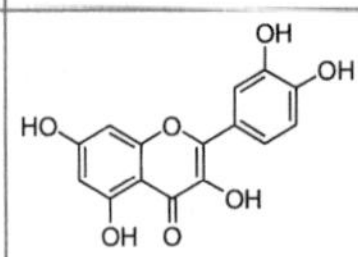	[37–42]
3.	*Curcuma longa*	Curcumin	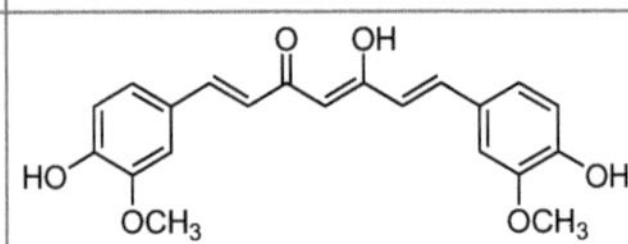	[43–49]
4.	Berries	Resveratrol	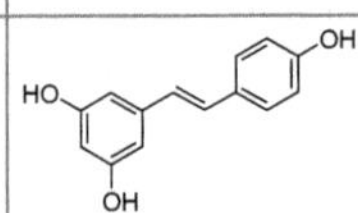	[50–58]
5.	*Piper nigrum*	Piperine	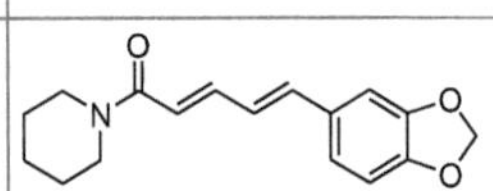	[59–60]
6.	*Lavandulae aetheroleum*	Linalool	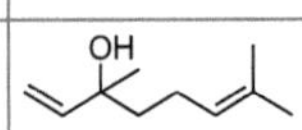	[62]
7.	*Melissae aetheroleum*	Citronellal	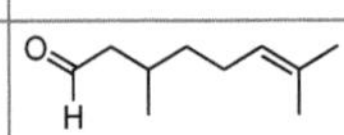	[63]
8.	*Murrayae folium*	Mahanimbine	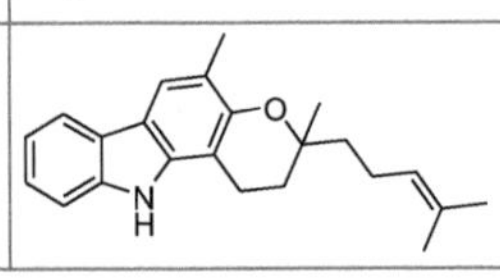	[64]

plants (Table 1) such as *Petroselinum crispum, Thymus vulgaris, Allium cepa, Lactuca sativa, Curcuma longa,* Berries, *Piper nigrum, Lavandulae aetheroleum, Melissae aetheroleum, Murrayae folium* may have affinity for the acetylcholinesterase enzyme (AChE) and butyrylcholinesterase (BChE) enzymes responsible for the cognition and memory. These have been selected for the current structure-based exploration of biochemical mechanisms that might have a therapeutic role in curing memory-related disorders associated with COVID-19.

2.1.1 *Petroselinum crispum, Thymus vulgaris* (Luteolin). Luteolin, also known chemically as tetrahydroxyflavone is a flavonol that has been previously reported to have a scavenging nature against reactive oxygen species, apoptotic activity against tumorigenic cells, and arrest cell cycle [30]. The remarkable function of luteolin in preventing COVID-19-associated brain fog has been revealed by recent research. Mast cells present throughout the body including the brain are part of the immune system mainly responsible for allergic and inflammatory response [31, 32]. Mast cells are also activated by nCoV [33, 34] since there has been no report on SARS-CoV-2 receptor expression [35] but their plastic nature enables it to be induced by viral infections. Histamine, chemokines, tryptase, tumour necrosis factor, and cytokines are released by stimulated mast cells [36, 37]. These mast cells predominantly produce platelet-activating factor, which has been linked to inflammation and microthromboses that cause memory impairment associated with COVID-19 [38, 39]. Luteolin has been considered as the natural, effective compound of mast cell inhibitors [40]. A liposomal formulation of luteolin, derived from olive fruit extract, has been shown to enhance attentional performance in children with autism spectrum disorders and mitigate cognitive dysfunction, commonly referred to as brain fog, in patients with mastocytosis [41].

2.1.2 *Allium Cepa, Lactuca Sativa* (Quercetin). Quercetin is also regarded as effective like luteolin in acting as a mast cell suppressant [42]. Both compounds show anti-viral properties, they are safe to be used and are readily available [43]. Luteolin showed better penetration into the brain [44] than quercetin and exhibited inhibitory action against mast cells as well as microglia, thus reducing neuronal inflammation and cognitive disorders [45, 46]. However, their poor absorption rate in the body is a major drawback but liposomal formulations using olive pomace oil enhanced oral absorption as well as bioavailability of the compounds. Therefore, nano-formulated luteolin and quercetin could be essential therapeutic compounds that may have a role in preventing brain fog associated with COVID-19 [39, 47]. Quercetin has advanced to Phase II of the clinical trial assessing the safety and feasibility of senolytic therapy for Alzheimer's disease. The study is grounded in findings from an experimental Alzheimer's disease mouse model, which demonstrated that the drug combination (quercetin along with dasatinib) utilized in the trial effectively prevents neuronal cell death [48].

2.1.3 *Curcuma Longa* (Curcumin). Curcumin or diferuloylmethane extracted from *Curcuma longa* is a well-known spice used in every household as a colouring agent and preservative [49]. Earlier research has confirmed the free radical scavenging property of curcumin [50] and thus, this ability enables it to be an ideal compound to be used in safeguarding the brain by preserving neuronal architecture, slowing down the aging process, and minimizing brain fog-related signs and symptoms caused due to several disease, viral infections like SARS-CoV and aging therefore preventing cognitive disorders [51–53]. This naturally derived product also improves neuronal function by activating enzymes important for the production of docosahexaenoic acid (DHA) from its precursor molecule alpha-lipoic acid (ALA) [54]. Thus, curcumin might be an effective herbal medicine to be used in curing COVID-19 brain fog due to its ability to protect brain health, brain function, and development, thus reducing memory problems associated with brain fog.

Turmeric extract of concentration 1.5 and 3 mg/mL has been found to maximize the response to methacholine and lessen the trachea's contraction induced by ovalbumin. When taken daily for 30 days, the use of capsule Curcumin 500 mg BD improved the forced expiratory volume (FEV1) for patients with bronchial asthma in comparison to standard therapy [55]. Therefore, it can improve respiratory problems and symptoms following COVID-19. Preclinical and clinical investigations indicate that curcumin may exhibit antidepressant and anxiolytic properties. These findings suggest that curcumin's neurocognitive benefits could be mediated through its modulatory effects on key brain regions, including the amygdala and hypothalamus, which are integral to mood regulation and memory processing [56].

2.1.4 Berries (Resveratrol). Several in vivo and in vitro models showed a different yet remarkable property of resveratrol, a natural phenol present in blueberries, grapes, raspberries, mulberries, etc. [57–60]. Despite possessing anti-oxidative and anti-inflammatory properties, resveratrol also exhibited neuroprotective ability on account of its competence in crossing the blood-brain barrier [53, 61, 62]. The neuroprotective nature of resveratrol is manifested by the prevention of astrocytes and microglial cellular activation which contribute to inflammatory response by ROS generation and pro-inflammatory factors production [61, 63–65]. By its inhibitory action, this phenolic compound reduces cerebral damage and minimizes apoptosis of neuronal cells during adverse conditions such as hypoxia, and trauma [57–60]. These factors made resveratrol an efficacious compound that could be implemented in treating memory loss associated with brain fog. Randomized clinical trials in humans provide evidence that resveratrol enhances cerebral blood flow and vasodilatory responsiveness to hypercapnia. Additionally, it has been shown to improve performance on certain cognitive assessments, subjective cognitive function, and modulate Aβ40 levels in both plasma and cerebrospinal fluid [66].

2.1.5 ***Piper nigrum*** **(Piperine).** The herb *Piper nigrum* is traditionally utilized to treat cough, cold, dyspnea, RTI (respiratory tract infection), recurrent high fever, dysentery, digestive disorders, etc. It also includes proteins, carbs, vitamin C, essential minerals (magnesium, iron, calcium) rich in flavonoids, and carotenes among many other substances. Additionally, it possesses anti-inflammatory, gastroprotective, anti-cancerous, anti-obesity, antidiabetic, antidepressant, antibacterial, and insecticidal role [67]. The three components—*Piper nigrum*, green tea, and *Curcuma longa L*—inhibit both the entry of viruses into host cells and the spread of illness while simultaneously enhancing immunity. Three grams of turmeric paste and five grams of black pepper were combined, split into three parts, and taken three times daily with lukewarm cow milk [68]. 100 mL of green tea was also consumed twice daily. The introduction of piperine with curcumin inhibits the metabolism of curcumin by glucuronidation reaction and therefore increases its bioavailability [69]. Piperine exhibits neuroprotective potential through its influence on locomotor activity, cognitive function, and the biochemical and neurochemical profile of the hippocampus. Research suggests that its effects may be mediated via modulation of neurotransmitter systems, neurotrophic factors, and oxidative stress pathways [70].

2.1.6 ***Lavandulae aetheroleum*** **(Linalool).** *Lavandulae aetheroleum* also known as Lavender can enhance memory function and cognitive function. The effect of aromatherapy using linalool-rich essential oils and their possible interactions with important physiological and biochemical variables implicated in the pathogenesis of depressive disorders has been the subject of several clinical and preclinical studies [71]. Lavender oil was administered as a spray containing 0.29 mg/L to the 144 volunteers and was then analyzed using computerized methods [72]. Lavender oil significantly improves memory and increases concentration time and attention.

2.1.7 *Melissae aetheroleum* **(Citronellal).** Citronellal is present in Melissa oil, which is extracted from the *Melissae aetheroleum.* A number of 72 participants who had been diagnosed with severe dementia received melissa oil. When 10% Melissa lotion was applied to the neck and both arms, a 35% improvement was shown in the patients who were studied using Cohen–Mansfield Scale [73]. Concentration and memory function improvement was seen with 10% cream with alcoholic lemon balm extract was used. Citronella oil inhalation significantly decreased blood pressure, heart rate, and respiration rate, according to participants. Subjective assessments indicated that participants experienced an improvement in mood and a heightened sense of freshness [74].

2.1.8 *Murrayae folium* **(Mahanimbine).** Mahanimbine is an inhibitor of the acetylcholinesterase enzyme, is extracted from the leaves of *Murraya koenigii.* Murai leaf extract containing Mahanimbine lowers cholesterol and brain Cholinesterase activity while also improving cognition. The volunteers received the extract for 30 days, during which time an exteroceptive and interoceptive model demonstrated improved memory in vivo [75].

2.2 2D and 3D Component Structure Treatment

The 2D structures of these 8 components as given in the Table 1 were drawn using ChemDraw 8.0 software. Then these 2D structures were converted into 3D structures, which were further optimized taking the convergence criterion and dielectric constant of 0.001 Kcal/mol and 1.0 respectively using the MM2 force field utilizing the Chem3D Ultra [76]. All the 3D optimized structures were browsed for molecular docking studies taking AChE and BChE as targets.

2.3 Molecular Docking

Molecular docking is an in silico computational technique that aids in the study of the ligand-receptor interaction to determine the target active binding sites and also to obtain the most energetically stable conformation of the ligand-receptor complex so that the energy of interactions, which also represents a score, between ligand and receptor is minimal. It helps to find the important amino acid residues of the target that could bind with the ligand moiety and produce biological activity [77–80].

All 8 components have been selected for molecular docking purposes inside the active binding pocket of the acetylcholinesterase protein, which is the key enzyme responsible for the hydrolysis of the neurotransmitter acetylcholine and is also the target of most of the clinically used drugs for the treatment of cognitive disorders. From the protein data bank, the crystal structure of acetylcholinesterase was downloaded with PDB ID 4EY7. It is co-crystallized with donepezil and was selected as a reference to assign the active cavity for molecular docking studies [81]. Further, these phytocomponents were again docked into the active binding site of butyrylcholinesterase (BChE) having PDB ID: 6SAM. It is found in complexes with 1-(2,3-dihydro-1H-inden2-yl) piperidine-3-yl N-phenyl carbamate as a co-crystal [82]. Butyrylcholinesterase (BChE) is a nonspecific cholinesterase enzyme responsible for choline-based esters. The 3D grid coordinates (X*Y*Z = 15 × 15 × 15) were generated all across the co-crystallized ligand that was attached to the target's active cavity with a grid resolution of 0.4 Å.

The co-crystallized molecule is used as an indicator for determining the position of the ligand's group binding location. Argus Lab 4.0.1 is a highly effective docking simulation program that includes a modular docking module [83]. The optimal position for examining the ligand's mode of binding was determined to be the ligand-receptor complex with the lowest docking score. The BIOVIA Discovery Studio Visualizer has been used to analyze the ligand-receptor complexes. In order to examine how close the docked structure is to the x-ray structure the RMSD difference between the x-ray and the best docked ligand has been calculated and all have been shown below 2.5 Å.

2.4 Challenges and Reliability of Docking

Given the challenges and financial constraints associated with experimental approaches for determining complex structures, computational techniques such as molecular docking are preferred for predicting potential binding modes and affinities. Molecular docking involves evaluating thousands of possible protein-ligand association poses, selecting the one with the lowest energy score as the predicted optimal binding mode. Significant progress has been achieved in docking techniques to improve computational accuracy and efficiency since the groundbreaking work of Kuntz and others [84]. Protein-ligand docking is still one of the most active of these because of its vital function in structure-based drug discovery [85].

3 Results and Discussion

3.1 Interpretation of Predicted Mode of Binding of Ligands' Affinity towards AChE

The eight phytocomponents were successfully docked into the active binding cavity of acetylcholinesterase (AChE). In silico docking predictions of amino acid residues and their binding affinities in terms of dock scores are tabulated in the following Table 2.

3.1.1 Luteolin (2-(3,4-dihydroxyphenyl)-5,7-dihydroxychromen-4-one). The carbonyl group at the 4th position showed hydrogen bonding with PHE338 and TYR337. The hydroxyl group at the 5th and 7th position of chromene showed hydrogen bonding with HIS447 and SER203 respectively. O at 1st position of chromene produced hydrogen bonding with TYR124. The dihydroxy phenyl ring at the 2nd position produced hydrophobic interaction with PHE297 and TRP286 (Fig. 1).

3.1.2 Quercetin (2-(3,4-dihydroxyphenyl)-3,5,7-trihydroxychromen-4-one). The carbonyl group of chromen produced hydrogen bonding with SER203 and GLY121. The hydroxyl group at the 3rd and 4th positions of 2-phenyl produced hydrogen bonding with ASP74 and THR83 respectively. The benzene ring of chromen showed hydrophobic

Table 2. Compounds' predicted mode of interaction with crucial amino acid residues towards AChE.

Sl. No.	Compound's name	Amino acid interaction		Docking score (kcal/mol.)
		H-bonding	Hydrophobic	
1.	Luteolin	TYR124, SER203, TYR337, PHE338, HIS447	TRP286, PHE297	−10.226
2.	Quercetin	ASP74, THR83, GLY 121, SER203, PHE 295	TRP86	−9.469
3.	Curcumin	GLY121, SER203	TRP86, TYR337, HIS447	−12.14
4.	Resveratrol	HIS980	TYR870, PHE871	−12.429
5.	Piperine	GLY121, GLY122, SER203, SER293	TRP86, VAL294, TYR337, TYR341, PHE338, HIS447	−10.671
6.	Linalool	HIS447	VAL294, TYR341, PHE297, PHE338, TRP86, TYR337	−8.426
7.	Citronellal	PHE295, ARG296	TYR341, PHE338, TYR337, HIS447	−10.441
8.	Mahanimbine	HIS447	PHE297, PHE338, TRP86, TRP439, TYR341, TYR337	−13.764

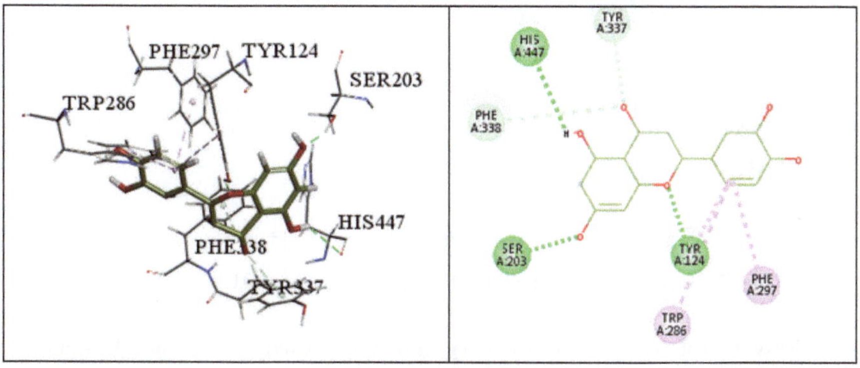

Fig. 1. Best docked pose with the 2D structure of Luteolin-AChE interaction.

attachment with PHE295. The 2-phenyl ring showed hydrophobic bonding with TRP86 (Fig. 2).

3.1.3 Curcumin ((1E,6E)-1,7-bis(4-hydroxy-3-methoxyphenyl) hepta-1,6-diene-3,5-dione). 4-hydroxy group and 3-methoxy group at 1-phenyl ring produced hydrogen bonding with GLY121 and SER203 respectively. The phenyl ring at the 1st position produced hydrophobic interaction with PHE295 while the phenyl ring at the 7th position produced hydrophobic bonding with TYR86, TYR337, HIS447, and PHE338 (Fig. 3).

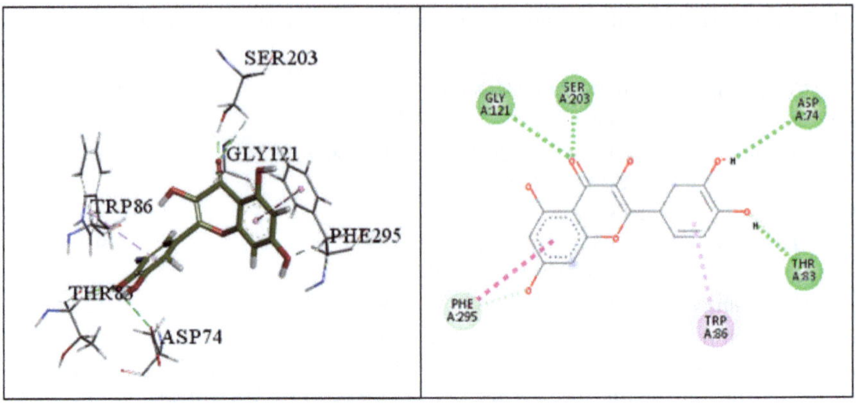

Fig. 2. Best docked pose with the 2D structure of Quercetin-AChE interaction.

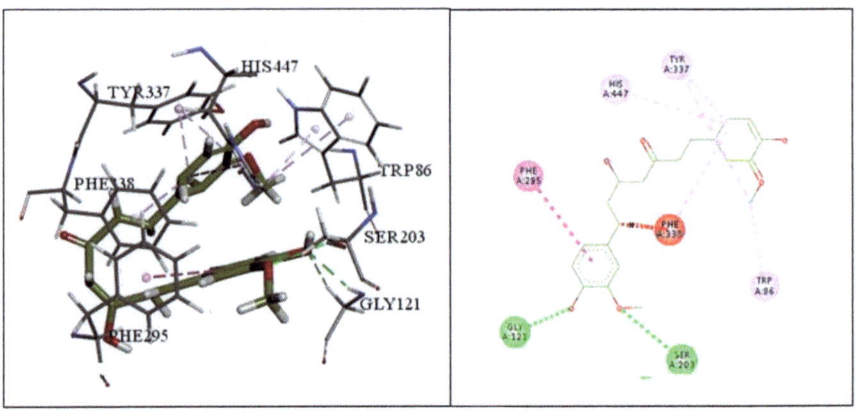

Fig. 3. Best docked pose with 2D image of Curcumin-AChE interaction.

3.1.4 Resveratrol (5-[(E)-2-(4-hydroxy phenyl) ethenyl] benzene-1,3-diol). The 4-hydroxyphenyl ring showed hydrophobic bonding with TYR870, PHE871 and HIS980. The hydroxyl group of 4-hydroxyphenyl produced hydrogen bonding with HIS980 (Fig. 4).

3.1.5 Piperine ((2E,4E)-5-(1,3-benzodioxol-5-yl)-1-piperidin-1-ylpenta-2,4-dien-1-one). The carbonyl group of the penta-2,4-dien chain produced hydrogen bonding with GLY121, GLY122, and SER203. Piperidin ring produced hydrophobic bonding with TYR86, TYR337, and HIS447. The benzene ring of benzodioxol produced hydrophobic interaction with VAL294, TYR341, and PHE338 (Fig. 5).

3.1.6 Linalool (3,7-dimethylocta-1,6-dien-3-ol). The hydroxy group at the 3rd position produced hydrogen bonding with HIS447. The methyl group at 1st position showed hydrophobic interactions with TYR337, TRP86. The methyl group at the 7th position and hydrogen atom at the 4th position showed hydrophobic bonding with PHE338 and the methyl

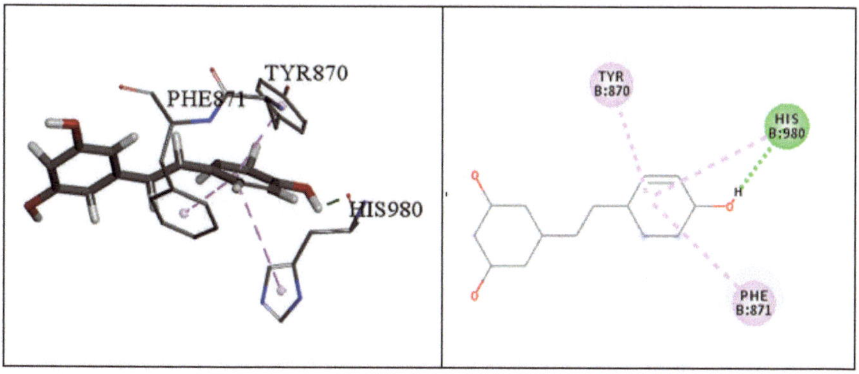

Fig. 4. Best docked pose with the 2D structure of resveratrol-AChE interaction.

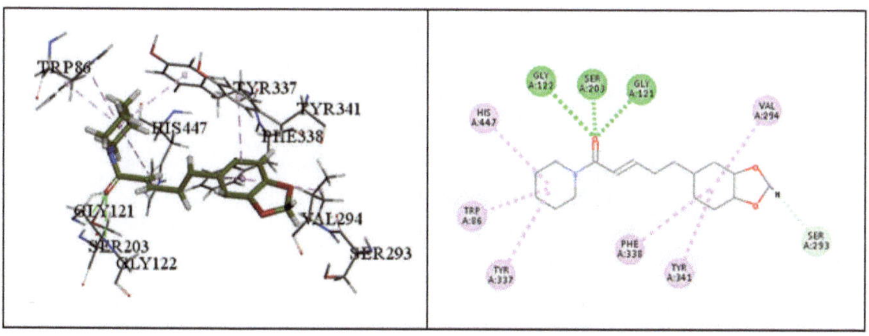

Fig. 5. Best docked pose with the 2D structure of Piperine-AChE interaction.

group at the 8th position produced hydrogen bonding with VAL294, PHE297, PHE338, and TYR341 (Fig. 6).

3.1.7 Citronellal (3,7-dimethyloct-6-enal). The carbonyl group at 1st position produced hydrogen bonding with PHE295 and ARG296. The methyl group at the 3rd position produced hydrophobic bonding with TYR341and the methyl group at the 7th position produced hydrophobic interaction with PHE338 and TYR337. The methyl group at the 8th position produced hydrophobic attachment with TYR337 and HIS447 (Fig. 7).

3.1.8 Mahanimbine (3,5-dimethyl-3-(4-methylpent-3-enyl)-11H-pyrano[3,2-a] Carbazole). Benzene ring of pyranocarbazole showed hydrophobic bonding with PHE338, HIS447 and TYR337. The methyl group at the 3rd position produced hydrophobic bonding with PHE338, TYR337, and HIS447. The oxygen atom showed hydrogen bonding with HIS447. Pyran ring of pyranocarbazole produced hydrophobic interaction with HIS447, and TRP86. The methyl group at the 5th position showed hydrophobic interaction with HIS447 and TRP86. The 4-methyl group of the side chain produced hydrophobic bonding with TYR341, TRP439, TYR337 and TRP86. The methyl group at the 5th position of the side chain produced hydrophobic interaction with TYR341 (Fig. 8).

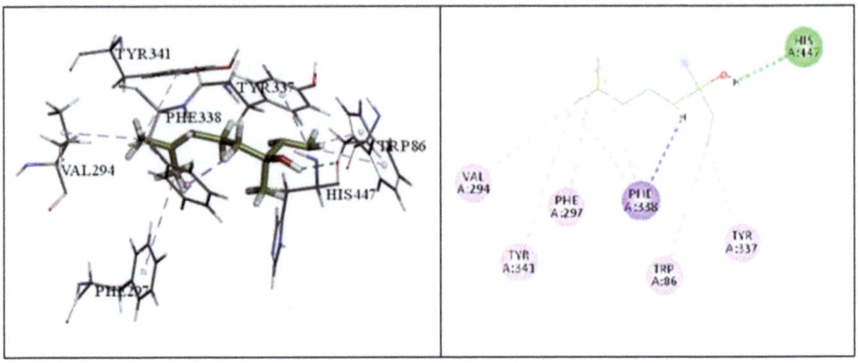

Fig. 6. Best docked pose with the 2D structure of Linalool-AChE interaction.

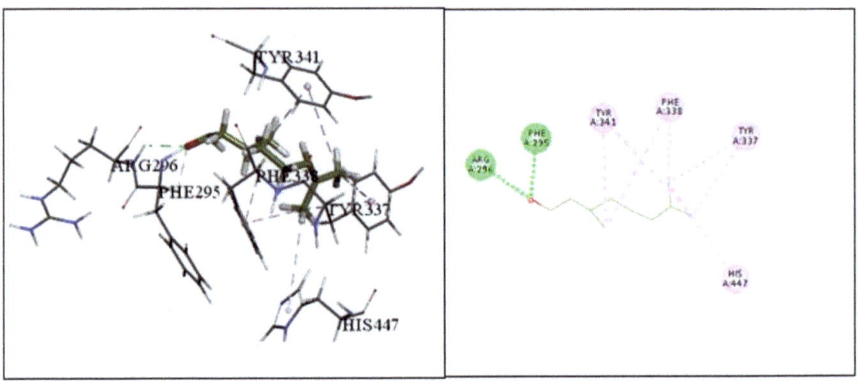

Fig. 7. Best docked pose with the 2D structure of Citronella-AChE interaction.

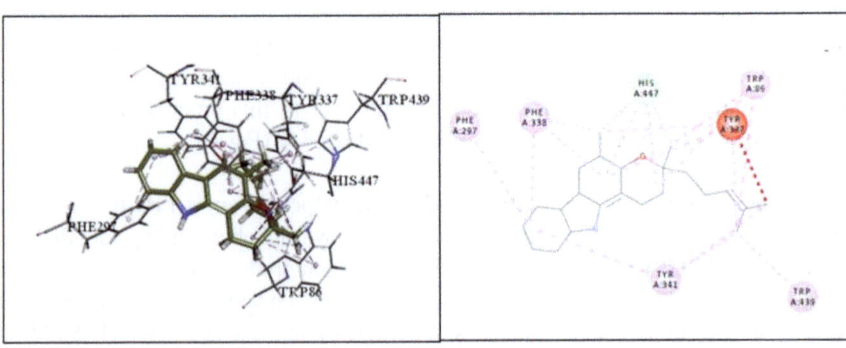

Fig. 8. Best docked pose with the 2D structure of Mahanimbine-AChE interaction.

Table 3. Compounds' predicted mode of interaction with crucial amino acid residues towards BChE.

Sl. No.	Compound's name	Amino acid interaction		Docking score (kcal/mol.)
		H-bonding	Hydrophobic	
1.	Luteolin	HIS438, SER198	PHE329	−8.322
2.	Quercetin	THR120	TRP82, TYR332	−7.621
3.	Curcumin	NA	TYR332, PHE329	−10.157
4.	Resveratrol	LEU286	LEU286	−9.118
5.	Piperine	GLY78, SER198	TRP231, LEU286, VAL288, PHE329, ALA328	−9.81
6.	Linalool	NA	TYR440, MET437, PHE329. ALA328, TYR332	−8.493
7.	Citronellal	NA	ALA328, PHE329, PRO285, TYR332	−8.20
8.	Mahanimbine	NA	PHE329, HIS438, TRP430, MET437, TYR440, ALA328, TYR332	−8.88

3.2 Interpretation of Predicted Mode of Binding of Ligands Affinity Towards BChE

The eight phytocomponents were also effectively docked into the active binding pocket of Butyrylcholinesterase (BChE). In the following Table 3, predicted amino acid residues and their binding affinities are listed with docking score.

3.2.1 Luteolin (2-(3,4-dihydroxyphenyl)-5,7-dihydroxychromen-4-one). The hydroxy group at the 5th position of chromene produced hydrogen bonding with HIS438 and SER198. The pyran ring of chromene showed hydrophobic interaction with PHE329 and HIS438. The phenyl ring at the 2nd position of benzopyran also produced hydrophobic interaction with PHE329 (Fig. 9).

3.2.2 Quercetin (2-(3,4-dihydroxyphenyl)-3,5,7-trihydroxychromen-4-one). The carbonyl group at the 4th position and the hydroxyl group at the 5th position showed hydrogen bonding with THR120. The phenyl ring at the 2nd position produced hydrophobic interaction with TRP82 (Fig. 10).

3.2.3 Curcumin ((1E,6E)-1,7-bis(4-hydroxy-3-methoxyphenyl) hepta-1,6-diene-3,5-dione). The phenyl ring at the 7th position produced hydrophobic interaction with TYR332 and PHE329 (Fig. 11).

3.2.4 Resveratrol (5-[(E)-2-(4-hydroxy phenyl) ethenyl] benzene-1,3-diol). The hydroxy group at the 4th position showed hydrogen bonding with LEU286. Phenyl at the 2nd position of ethenyl produced hydrophobic interaction with LEU286 (Fig. 12).

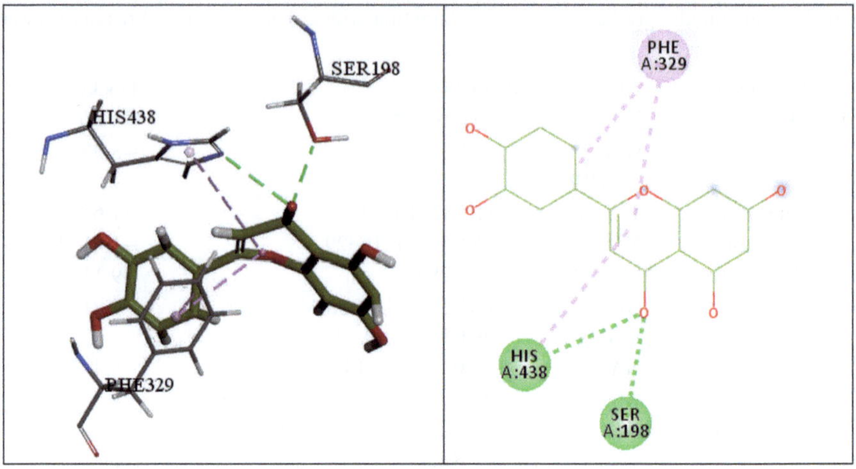

Fig. 9. Best docked pose with the 2D structure of Luteolin-BChE interaction.

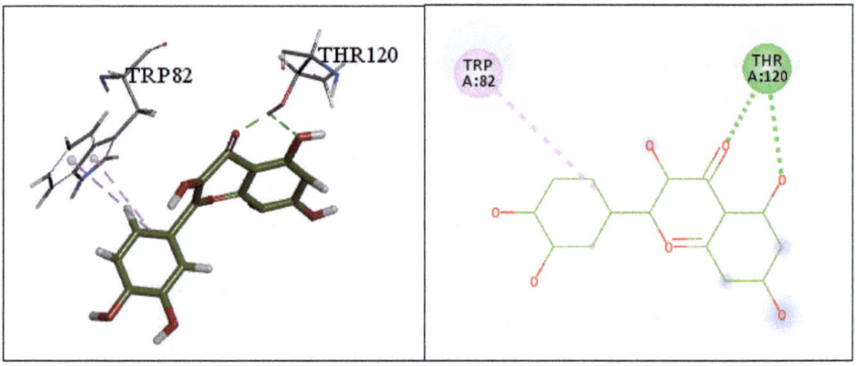

Fig. 10. Best docked pose with the 2D structure of Quercetin-BChE interaction.

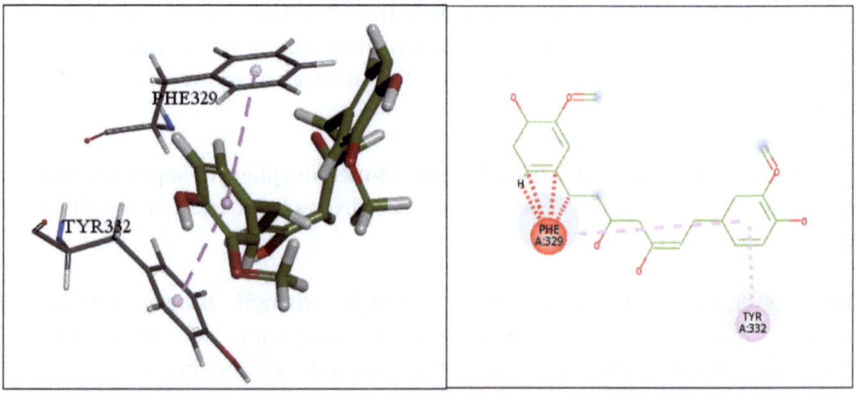

Fig. 11. Best docked pose with 2D image of Curcumin-BChE interaction.

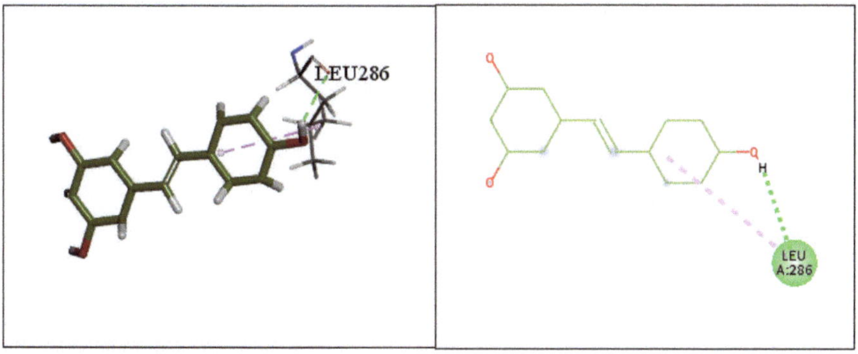

Fig. 12. Best docked pose with the 2D structure of resveratrol-BChE interaction.

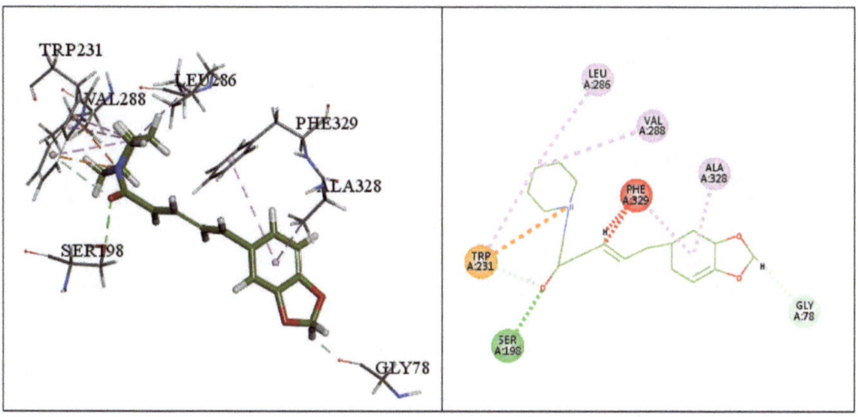

Fig. 13. Best docked pose with the 2D structure of Piperine-BChE interaction.

3.2.5 Piperine ((2E,4E)-5-(1,3-benzodioxol-5-yl)-1-piperidin-1-ylpenta-2,4-dien-1-one). The carbonyl group at the 1st position of pent-2,4-dien produced hydrogen bonding with SER198 and TRP231. The hydrogen atom at the 5th position showed hydrogen bonding with GLY78. Carbon atoms at the 4th position of piperidine produced hydrophobic interaction with LEU286, VAL288, and TRP231. The phenyl ring at the 5th position produced hydrophobic interaction with ALA328 and PHE329 (Fig. 13).

3.2.6 Linalool (3,7-dimethylocta-1,6-dien-3-ol). The methyl group at 1st position produced hydrophobic interaction with ALA328, TYR440, and MET437. The methyl group at the 8th position showed hydrophobic interaction with TYR332, ALA328, and PHE329. Methyl at the 7th position produced hydrophobic interaction with TYR332 (Fig. 14).

3.2.7 Citronellal (3,7-dimethyloct-6-enal). The methyl group at the 3rd position produced hydrophobic attachment with ALA328 and PHE329. Methyl at the 7th position showed hydrophobic interaction with PHE329 and PRO285. The methyl group at the 7th position showed hydrophobic interaction TYR332 (Fig. 15).

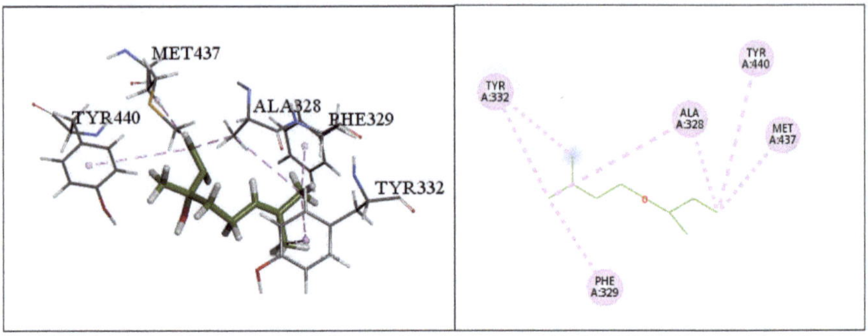

Fig. 14. Best docked pose with the 2D structure of Linalool-BChE interaction.

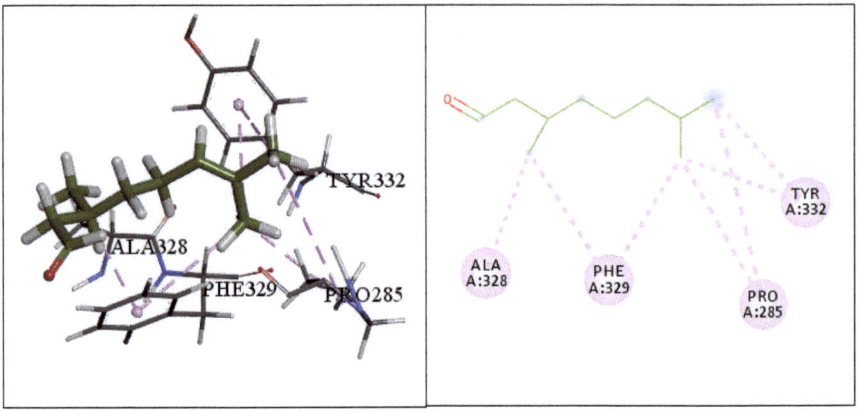

Fig. 15. Best docked pose with the 2D structure of Citronella-BChE interaction.

3.2.8 Mahanimbine (3,5-dimethyl-3-(4-methylpent-3-enyl)-11H-pyrano[3,2-a] carbazole). 1st phenyl ring of carbazole produced hydrophobic interaction with PHE329. The methyl group at the 5th position showed hydrophobic interaction TYR332. The pyran ring and methyl group at the 3rd position produced hydrophobic bonding with PHE329, HIS438, and ALA328. The methyl group at the 4th position of the side chain produced hydrophobic interaction with TRP430, TYR332, and ALA328. C at the 5th position of the side chain produced hydrophobic interaction with MET437, TYR440, TRP430, and ALA328 (Fig. 16).

4 Conclusion

In this study, natural active components were screened and docked within the active cleft of acetylcholinesterase and butyrylcholinesterase receptors to gain insight into the molecular mechanisms of each compound's potential therapeutic binding towards ACHE and BCHE responsible for the memory and neurological cognition. The present

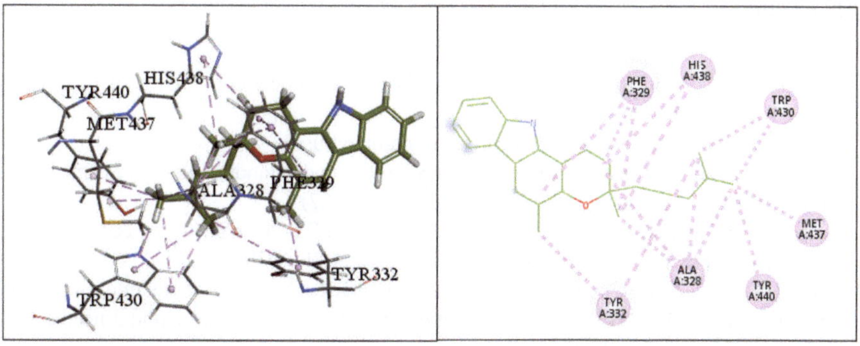

Fig. 16. Best docked pose with the 2D structure of Mahanimbine-BChE interaction.

study showed that luteolin (Parsley, Thyme), quercetin (Onion, Lettuce), curcumin (Curcuma), resveratrol (Berries), piperine (Black pepper), linalool (Lavender oil), citronellal (lemon balm), and mahanimbine (orange jasmine) have a strong affinity to bind the brain acetylcholine as well as butyrylcholinesterase receptors. The docking simulations predicted affinities of these components towards AChE and BChE in terms of docking scores as −13.764 to −8.426 kcal/mol for AchE and −10.157 to −7.621 kcal/mol for BChE. The common amino acid residues such as PHE329, TYR332, and ALA328 interacted with BChE where as His447, TYR337, PHE338, SER203, TRP86, and TYR341 interacted with AChE, which had been predicted as a common mode of binding for the selected phytoactive components to combat COVID-19-associated brain fog.

Despite the adverse respiratory complications experienced by patients during the outbreak of COVID-19, reports have also found that there is alarming concern about brain fog, which is now a prime issue as it directly affects mental health. Reports confirmed that low-grade and chronic inflammation is persistent in both the body and neurons of the patients who have suffered from COVID-19, the consequences of which trail to the experience of brain fog in the patients. Therefore, the purpose of this current article is to scrutinize the etiology, pathways, associated symptoms, and treatments for neuroinflammation associated with it. Although there is a lack of knowledge regarding etiology and specific treatments targeting the post-COVID-19 brain fog, researchers are in search of pharmacological interventions that might improve cognitive functioning by targeting neuroinflammation. There is no evidence of synthetic drugs to combat brain fog; hence, ayurvedic herbal medications might come as a solution in treating problems related to impaired brain function to understand their neurological function for further improving and opening the treatment gateway.

Acknowledgements. Grateful acknowledgements are extended to the respective Institutes for providing the basic infrastructure to carry out the work.

Consent for Publication All the authors have permission to the publication of this research work.

Conflict of Interest None to declare

References

1. Zhou, P., et al.: A pneumonia outbreak associated with a new coronavirus of probable bat origin. Nature. **579**, 270–273 (2020)
2. Worldometer. https://www.worldometers.info/coronavirus/
3. Thakur, V., et al.: Waves and variants of *SARS-CoV-2*: understanding the causes and effect of the COVID-19 catastrophe. Infection. **50**, 309–325 (2022)
4. Belouzard, S., Millet, J.K., Licitra, B.N., Whittaker, G.R.: Mechanisms of coronavirus cell entry mediated by the viral spike protein. Viruses. **4**, 1011–1033 (2012)
5. Chen, P.L., Lee, N.Y., Cia, C.T., Ko, W.C., Hsueh, P.R.: A review of treatment of coronavirus disease 2019 (COVID-19): therapeutic repurposing and unmet clinical needs. Front. Pharmacol. **11**, 584956 (2020)
6. Eastin, C., Eastin, T.: Clinical characteristics of coronavirus disease 2019 in China. J. Emerg. Med. **58**, 711–712 (2020)
7. de Oliveira, G.L.V., Oliveira, C.N.S., Pinzan, C.F., de Salis, L.V.V., Cardoso, C.R.B.: Microbiota modulation of the gut-lung axis in COVID-19. Front. Immunol. **12**, 635471 (2021)
8. Vignesh, R., Swathirajan, C.R., Tun, Z.H., Rameshkumar, M.R., Solomon, S.S., Balakrishnan, P.: Could perturbation of gut microbiota possibly exacerbate the severity of COVID-19 via cytokine storm? Front. Immunol. **11**, 607734 (2020)
9. https://www.ncbi.nlm.nih.gov/pmc/articles/PMC7527190/
10. Whittaker, A., Anson, M., Harky, A.: Neurological manifestations of COVID-19: a systematic review and current update. Acta Neurol. Scand. **142**, 14–22 (2020)
11. Yin, R., et al.: Concomitant neurological symptoms observed in a patient diagnosed with coronavirus disease 2019. J. Med. Virol. **92**, 1782–1784 (2020)
12. Filatov, A., Sharma, P., Hindi, F., Espinosa, P.S.: Neurological complications of coronavirus disease (COVID- 19) encephalopathy. Cureus. **12**, e7352 (2020)
13. Mao, L., et al.: Neurologic manifestations of hospitalized patients with coronavirus disease 2019 in Wuhan, China. JAMA Neurol. **77**, 683–690 (2020)
14. Chen, N., et al.: Epidemiological and clinical characteristics of 99 cases of 2019 novel coronavirus pneumonia in Wuhan, China: a descriptive study. Lancet. **395**, 507–513 (2020)
15. Li, Y., et al.: Acute cerebrovascular disease following COVID-19: a single center, retrospective, observational study. Stroke Vasc. Neurol. **5**, 279–284 (2020)
16. Toscano, G., et al.: Guillain-Barré syndrome associated with SARS-CoV-2. N. Engl. J. Med. **382**, 2574–2576 (2020)
17. Coen, M., et al.: Guillain-Barré syndrome as a complication of SARS-CoV-2 infection. Brain Behav. Immun. **87**, 111–112 (2020)
18. Camdessanche, J., Morel, J., Pozzetto, B., Paul, S., Tholance, Y., Botelho-Nevers, E.: COVID-19 may induce Guillain-Barré syndrome. Rev. Neurol. (Paris). **176**, 516–518 (2020)
19. Sedaghat, Z., Karimi, N.: Guillain Barre syndrome associated with COVID-19 infection: a case report. J. Clin. Neurosci. **76**, 233–235 (2020)
20. Virani, A., Rabold, E., Hanson, T., Haag, A., Elrufay, R., Cheema, T., Balaan, M., Bhanot, N.: Guillain-Barré syndrome associated with SARS-CoV-2 infection. IDCases. **20**, e00771 (2020)
21. Tsai, L.K., et al.: Neuromuscular disorders in severe acute respiratory syndrome. Arch. Neurol. **61**, 1669–1673 (2004)
22. Zhou, C., Wu, L., Ni, F., Ji, W., Wu, J., Zhang, H.: Critical illness polyneuropathy and myopathy: a systematic review. Neural Regen. Res. **9**, 101–110 (2014)
23. Venkataramani, V., Winkler, F.: Cognitive deficits in long Covid-19. N. Engl. J. Med. **387**, 1813–1815 (2022)

24. Gorenshtein, A., Liba, T., Leibovitch, L., Stern, S., Stern, Y.: Intervention modalities for brain fog caused by long-COVID: systematic review of the literature. Neurol. Sci. **45**, 2951–2968 (2024)
25. Nouraeinejad, A.: Brain fog as a long-term sequela of COVID-19. Clin. Med. **5**, 9 (2023)
26. Asadi-Pooya, A.A., et al.: Long COVID syndrome-associated brain fog. J. Med. Virol. **94**, 979–984 (2022)
27. Remsik, J., et al.: Inflammatory leptomeningeal cytokines mediate COVID-19 neurologic symptoms in cancer patients. Cancer Cell. **39**, 276–283e3 (2021)
28. Cohen, P.A., Zakharevich, I., Gerona, R.: Presence of piracetam in cognitive enhancement dietary supplements. JAMA Intern. Med. **180**, 458–459 (2020)
29. Cohen, P.A., Avula, B., Wang, Y.H., Zakharevich, I., Khan, I.: Five unapproved drugs found in cognitive enhancement supplements. Neurol. Clin. Pract. **11**, e303–e307 (2021)
30. Lin, Y., Shi, R., Wang, X., Shen, H.M.: Luteolin, a flavonoid with potential for cancer prevention and therapy. Curr. Cancer Drug Targets. **8**, 634–646 (2008)
31. Olivera, A., Beaven, M.A., Metcalfe, D.D.: Mast cells signal their importance in health and disease. J. Allergy Clin. Immunol. **142**, 381–393 (2018)
32. Galli, S.J., Tsai, M., Piliponsky, A.M.: The development of allergic inflammation. Nature. **454**, 445–454 (2008)
33. Theoharides, T.C.: Potential association of mast cells with COVID-19. Ann. Allergy Asthma Immunol. **126**, 217–218 (2021)
34. Motta Junior, J.D.S., et al.: Mast cells in alveolar septa of COVID-19 patients: a pathogenic pathway that may link interstitial edema to immunothrombosis. Front. Immunol. **11**, 574862 (2020)
35. Xu, H., Zhong, L., Deng, J., Peng, J., Dan, H., Zeng, X., Li, T., Chen, Q.: High expression of ACE2 receptor of 2019-nCoV on the epithelial cells of oral mucosa. Int. J. Oral Sci. **12**, 8 (2020)
36. Bawazeer, M.A., Theoharides, T.C.: IL-33 stimulates human mast cell release of CCL5 and CCL2 via MAPK and NF-kappaB, inhibited by methoxyluteolin. Eur. J. Pharmacol. **865**, 172760 (2019)
37. Kandere-Grzybowska, K., et al.: IL-1 induces vesicular secretion of IL-6 without degranulation from human mast cells. J. Immunol. **171**, 4830–4836 (2003)
38. Demopoulos, C., Antonopoulou, S., Theoharides, T.C.: COVID-19, microthromboses, inflammation and platelet activating factor. Biofactors. **46**, 927–933 (2020)
39. Theoharides, T.C., Antonopoulou, S., Demopoulos, C.A.: Coronavirus 2019, microthromboses, and platelet activating factor. Clin. Ther. **42**, 1850–1852 (2020)
40. Taliou, A., Zintzaras, E., Lykouras, L., Francis, K.: An open-label pilot study of a formulation containing the anti-inflammatory flavonoid luteolin and its effects on behavior in children with autism spectrum disorders. Clin. Ther. **35**, 592–602 (2013)
41. Theoharides, T.C., Stewart, J.M., Hatziagelaki, E., Kolaitis, G.: Brain "fog," inflammation and obesity: key aspects of neuropsychiatric disorders improved by luteolin. Front. Neurosci. **9**, 225 (2015)
42. Okamoto, T.: Safety of quercetin for clinical application (review). Int. J. Mol. Med. **16**, 275–278 (2005)
43. Theoharides, T.C., Cholevas, C., Polyzoidis, K., Politis, A.: Long-COVID syndrome-associated brain fog and chemofog: luteolin to the rescue. Biofactors. **47**, 232–241 (2021)
44. Calis, Z., Mogulkoc, R., Baltaci, A.K.: The roles of flavonols/flavonoids in neurodegeneration and neuroinflammation. Mini Rev. Med. Chem. **20**, 1475–1488 (2020)
45. Devi, S.A., Chamoli, A.: Polyphenols as an effective therapeutic intervention against cognitive decline during normal and pathological brain aging. Adv. Exp. Med. Biol. **1260**, 159–174 (2020)

46. Theoharides, T.C., Stewart, J.M., Hatziagelaki, E., Kolaitis, G.: Brain "fog," inflammation and obesity: key aspects of 2 neuropsychiatric disorders improved by luteolin. Front. Neurosci. **9**, 225 (2015)
47. Fu, X., et al.: Protective role of luteolin against cognitive dysfunction induced by chronic cerebral hypoperfusion in rats. Pharmacol. Biochem. Behav. **126**, 122–130 (2014)
48. Zhu, Y.I., et al.: The Achilles' heel of senescent cells: from transcriptome to senolytic drugs. Aging Cell. **14**, 644–658 (2015)
49. Singh, M., Arseneault, M., Sanderson, T., Murthy, V., Ramassamy, C.: Challenges for research on polyphenols from foods in Alzheimer's disease: bioavailability, metabolism, and cellular and molecular mechanisms. J. Agric. Food Chem. **56**, 4855–4873 (2008)
50. Zhao, B.L., Li, X.J., He, R.G., Cheng, S.J., Xin, W.J.: Scavenging effect of extracts of green tea and natural antioxidants on active oxygen radicals. Cell Biophys. **14**, 175–185 (1989)
51. Sarker, M.R., Franks, S.F.: Efficacy of curcumin for age-associated cognitive decline: a narrative review of preclinical and clinical studies. Geroscience. **40**, 73–95 (2018)
52. Cole, G.M., Teter, B., Frautschy, S.A.: Neuroprotective effects of curcumin. Adv. Exp. Med. Biol. **595**, 197–212 (2007)
53. Kim, J., Lee, H.J., Lee, K.W.: Naturally occurring phytochemicals for the prevention of Alzheimer's disease. J. Neurochem. **112**, 1415–1430 (2010)
54. Wu, A., Noble, E.E., Tyagi, E., Ying, Z., Zhuang, Y., Gomez-Pinilla, F.: Curcumin boosts DHA in the brain: implications for the prevention of anxiety disorders. Biochem. Biophys. Acta. **1852**, 951–961 (2015)
55. Habtemariam, S.: The therapeutic potential of rosemary (Rosmarinus officinalis) diterpenes for Alzheimer's disease. Evid. Based Complement. Alternat. Med. **2016**, 2680409 (2016)
56. Lopresti, A.L., Drummond, P.D.: Efficacy of curcumin, and a saffron/curcumin combination for the treatment of major depression: a randomized, double-blind, placebo-controlled study. J. Affect. Disord. **207**, 188–196 (2017)
57. Virgili, M., Contestabile, A.: Partial neuroprotection of in vivo excitotoxic brain damage by chronic administration of the red wine antioxidant agent, trans-resveratrol in rats. Neurosci. Lett. **281**, 123–126 (2000)
58. Zamin, L.L., et al.: Protective effect of resveratrol against oxygen–glucose deprivation in organotypic hippocampal slice cultures: involvement of PI3-K pathway. Neurobiol. Dis. **24**, 170–182 (2006)
59. Ates, O., et al.: Neuroprotection by resveratrol against traumatic brain injury in rats. Mol. Cell. Biochem. **294**, 137–144 (2007)
60. Della-Morte, D., Dave, K.R., Defazio, R.A., Bao, Y.C., Raval, A.P., Perez-Pinzon, M.A.: Resveratrol pretreatment protects rat brain from cerebral ischemic damage via a sirtuin 1 – uncoupling protein 2 pathway. Neuroscience. **159**, 993–1002 (2009)
61. Wang, Q., et al.: Resveratrol protects against global cerebral ischemic injury in gerbils. Brain Res. **958**, 439–447 (2002)
62. Mokni, M., Elkahoui, S., Limam, F., Amri, M., Aouani, E.: Effect of resveratrol on antioxidant enzyme activities in the brain of healthy rat. Neurochem. Res. **32**, 981–987 (2007)
63. Salehi, B., et al.: Resveratrol: a double-edged sword in health benefits. Biomedicine. **6**, 91 (2018)
64. Zhang, F., Liu, J., Shi, J.S.: Anti-inflammatory activities of resveratrol in the brain: role of resveratrol in microglial activation. Eur. J. Pharmacol. **636**, 1–7 (2010)
65. Candelario-Jalil, E., et al.: Resveratrol potently reduces prostaglandin E2 production and free radical formation in lipopolysaccharide-activated primary rat microglia. J. Neuroinflammation. **4**, 25 (2007)
66. Cicero, A.F.G., Ruscica, M., Banach, M.: Resveratrol and cognitive decline: a clinician perspective. Arch. Med. Sci. **15**, 936–943 (2019)

67. Ashokkumar, K., Murugan, M., Dhanya, M.K., Pandian, A., Warkentin, T.D.: Phytochemistry and therapeutic potential of black pepper [Piper nigrum (L.)] essential oil and piperine: a review. Clin. Phytoscience. **7**, 52 (2021)
68. https://timesofindia.indiatimes.com/life-style/health-fitness/home-remedies/know-the-exact-amount-of-turmeric-you-should-consume-in-a-day/articleshow/66636189.cms
69. Kumar, G., Kumar, D., Singh, N.P.: Therapeutic approach against 2019-nCoV by inhibition of ACE-2 receptor. Drug Res. (Stuttg.). **71**, 213–218 (2021)
70. Assessment Report on Lavandula angustifolia Mill. aetheroleum and Lavandula angustifolia Mill., flos. EMA/HMPC/143183/2010. Committee on Herbal Medicinal Products (HMPC) (2011)
71. Hsieh, T.-Y., Chang, Y., Wang, S.-J.: Piperine provides neuroprotection against kainic acid-induced neurotoxicity via maintaining NGF signalling pathway. Molecules. **27**, 2638 (2022)
72. Dos Santos, É.R.Q., Maia, J.G.S., Fontes-Júnior, E.A., do Socorro Ferraz Maia, C.: Linalool as a therapeutic and medicinal tool in depression treatment: a review. Curr. Neuropharmacol. **20**, 1073–1092 (2022)
73. Buckle, J.: Dementia and Alzheimer's disease. In: Clinical Aromatherapy, 3rd edn. Elsevier, London (2015)
74. Nawrot, J., et al.: Medicinal herbs in the relief of neurological, cardiovascular, and respiratory symptoms after COVID-19 infection a literature review. Cells. **11**, 1897 (2022)
75. Sayowan, W., Siripornpanich, V., Piriyapunyaporn, T., Hongratanaworakit, T., Kotchabhakdi, N., Ruangrungsi, N.: The harmonizing effects of citronella oil on mood states and brain activities. J. Health Res. **26**, 69–75 (2012)
76. Mills, N.: ChemDraw Ultra 10.0. J. Am. Chem. Soc. **128**, 13649–13650 (2006)
77. Briguglio, I., Piras, S., Corona, P., Maria, A.P., Daniela, J., Antonio, C.: SAR and anti-mycobacterial activity of quinolones and triazoloquinolones: an update. Anti-Infect. Agents. **11**, 76–90 (2013)
78. Halgren, T.A.: Merck molecular force field III. Molecular geometries and vibrational frequencies. J. Comput. Chem. **17**, 553–586 (1996)
79. Stahl, M., Rarey, M.: Detailed analysis of scoring functions for virtual screening. J. Med. Chem. **44**, 1035–1042 (2001)
80. Nandi, S., Bagchi, M.C.: 3D-QSAR and molecular docking studies of 4-anilinoquinazoline derivatives: a rational approach to anticancer drug design. Mol. Divers. **14**, 27–38 (2010)
81. Cheung, J., et al.: Structures of human acetylcholinesterase in complex with pharmacologically important ligands. J. Med. Chem. **55**, 10282–10286 (2012)
82. Brazzolotto, X., Kosak, U., Strasek, N., Knez, D., Gobec, S., Nachon, F.: Structure of human butyrylcholinesterase in complex with 1-(2,3-dihydro-1H-inden2-yl)piperidin-3-yl N-phenyl carbamate. Eur. J. Med. Chem. **197**, 112282–112282 (2020)
83. Thompson, M.A.: ArgusLab 4.0.1. Planaria Software LLC, Seattle. Available from: http://www.arguslab.com
84. Kuntz, I.D., Blaney, J.M., Oatley, S.J., Langridge, R., Ferrin, T.E.: A geometric approach to macromolecule-ligand interactions. J. Mol. Biol. **161**, 269–288 (1982)
85. Huang, S.Y., Zou, X.: Advances and challenges in protein-ligand docking. Int. J. Mol. Sci. **11**, 3016–3034 (2010)

Evaluation of Preventive Efficacy of Different Phytocompounds to Combat Hyperglycemia in Mice Model: A Comparative Assessment

Sayoni Dam[1], Sujan Murmu[2], Amar Faruk Saheb[2], Tamanna Khatun[2], Barsha Roy[2], Astha Saha[2], Debapriya Ghosh[2], Ashaporna Biswas[2], Banani Bhattacharjee[1], Arnob Chakrovorty[1], Sisir Nandi[3](✉), and Asmita Samadder[1,2](✉)

[1] Cytogenetics and Molecular Biology Laboratory, Department of Zoology, University of Kalyani, Kalyani, Nadia, India

[2] Cell and Developmental Biology special, Department of Zoology, University of Kalyani, Kalyani, Nadia, India
asmita.samadder@gmail.com

[3] Department of Pharmaceutical Chemistry, Global Institute of Pharmaceutical Education and Research (Affiliated to Veer Madho Singh Bhandari Uttarakhand Technical University), Kashipur, Uttarakhand, India
sisir.iicb@gmail.com

Abstract. Nature and natural components have been in use from time immemorial for healing several ailments of humankind. Presently, pertaining to the side effects incurred by the use of chemicals in synthetic drug materials, people are more inclined towards the usage of natural Phyto-compounds as therapeutic agents against myriads of diseases. In the present study, two important Phyto-compounds, Chlorophyllin (CHL) and Curcumin (CUR), which are our culinary companion and are easily available and affordable, have been used to investigate their efficacy against food additive induced hyperglycemia and its associated complications. The results revealed that both the Phyto-compounds were effective in inhibiting ROS generation in the pancreas and liver and could restrict hyperglycemia induced damage in the morphology of both tissues. Additionally, the architecture of brain and testes were also restored by these phyto-compounds even after the initiation of hyperglycemia in the experimental model. CHL and CUR also effectively inhibited the frequency of formation of comet tailed pancreatic cells when pre-treated prior to hyperglycemic induction. Although both the drugs showed similar results however, the results obtained from the comparative assessment for both the Phyto-compounds showed better results for CHL treatment than that of CUR treated group in terms of inhibition of blood glucose level, ROS generation and other parameters related to hyperglycaemia. Therefore, this study definitely suggests the possibility of using these two Phyto-compounds in the daily diet which would definitely help diabetic patients to survive better with a longer life.

The authors Sayoni Dam, Sujan Murmu, Amar Faruk Saheb, Tamanna Khatun, and Barsha Roy have contributed equally to this work.

A. K. Saxena, A. Saxena (Eds.): GTHTM 2025, pp. 397–409, 2026.
https://doi.org/10.1007/978-3-032-12320-6_23

Keywords: Hyperglycemia · Diabetes · Alloxan · Phytocompounds · Curcumin · Chlorophyllin · Genotoxicity · ROS

1 Introduction

Hyperglycemia, one of the current global burdens for individuals, families, and countries, is characterized by blood glucose levels of 100–125 mg/dL in a fasting condition and greater than 180 mg/dL in phenomenal postprandial [1]. Hyperglycemia contributes to the development of both type 1 and type 2 diabetes. According to the International Diabetes Federation, by 2045, the occurrence of diabetes caused by hyperglycemia is expected to be increased by approximately 46%, which means that 1 in every 8 adults will have diabetes in immediate future. The ailments caused by hyperglycemia include impaired carbohydrate, fat, and protein metabolism, reduced glucose uptake, inactivation of glucose-metabolizing enzymes, liver dysfunction, damage to the tissue architecture of the testes, malformation of sperm, kidney malfunction (such as reduced glomerular filtration rate), podocyte damage, and overall changes in tissue architecture [2]. Hyperglycemia also affects various organs, including the liver, testes, and brain [3].

Now, the most important question is why the prevalence of diabetes is increasing day by day among people of all age groups. Unhealthy lifestyles, overeating of junk food, excessive sugar consumption through sugary drinks and processed foods, which in turn increase cholesterol, blood pressure, and cause excessive physical and mental stress, smoking, etc., contribute to the development of diabetes. However, the most concerning factor is that the agents leading to the development of diabetes are being ingested daily unknowingly through food additives, preservatives, and food colours used in staple foods such as rice, wheat flour, vegetables, fruits and processed food etc. One such compound is Alloxan, a pyrimidine derivative (2,4,5,6-tetraoxypyrimidine 5,6-dioxyuracil), commonly added to wheat flour to enhance its whiteness and shine. It is known to induce hyperglycemia by acting as a toxic glucose analogue. Alloxan enters pancreatic beta cells via the GLUT2 receptor, where it accumulates. Inside the cells, in the presence of reduced glutathione (GSH) and protein-bound -SH groups, Alloxan undergoes reduction, forming dialuric acid. This compound is then oxidised back to Alloxan, initiating a redox cycle that generates reactive oxygen species (ROS) and superoxide radicals. These superoxide radicals are converted to hydrogen peroxide (H_2O_2) with the help of superoxide dismutase. Subsequently, highly reactive hydroxyl radicals are formed through the Fenton reaction. The damage to the DNA of pancreatic beta cells by ROS, causes fragmentation and triggers poly ADP-ribosylation, a DNA repair process [4, 5].

Additionally, Alloxan disrupts calcium homeostasis, raising the cytosolic calcium levels in pancreatic beta cells, which leads to depolarization of the cells and opening of voltage-gated Ca^{2+} channels [6].

Alloxan-induced damage also extends to other organs. In the liver, it causes sinusoidal enlargement, mitochondrial degeneration, disruption of the rough endoplasmic reticulum (RER), nuclear damage, and progressive liver fibrosis [5]. In the brain, Alloxan induces degeneration of cells around the amygdala, hippocampus, pons, and medulla, along with increased activity of acetylcholinesterase [7]. In the testes, Alloxan generates ROS which causes oxidative stress and DNA damage to testicular tissues, thereby deteriorating sperm quality and quantity [8].

Currently, synthetic antihyperglycemic drugs such as Glimepiride, Glipizide, Metformin, Acarbose, and Repaglinide are commonly used to manage blood glucose levels in diabetic patients [9]. However, their associated adverse effects, such as excessive weight loss, decreased appetite, weakness, gastrointestinal upset, renal impairment, and pancreatitis, significantly affect the well-being of diabetic patients [10]. These issues have sparked interest in exploring alternative medications to manage hyperglycemia. Phytocompounds have emerged as promising alternatives due to their safety, efficacy at lower doses, cost-effectiveness, biodegradability, and minimal to no side effects [11–13].

Chlorophyllin (CHL), a compound found in most green leafy vegetables, has demonstrated therapeutic efficacy against hyperglycemia. It has been reported to alleviate food additive-induced genotoxicity, inflammatory responses, mitochondrial dysfunction, and more. Curcumin (CUR), derived from the root of *Curcuma longa*—a commonly used culinary spice—possesses hepatoprotective, anti-allergic, antibacterial, antioxidant, and anti-inflammatory properties. CUR has shown significant reduction in ROS generation, activation of the p53 protein (leading to prevention of genotoxicity), and provides protective effects on the liver, pancreas, and testes. Additionally, CUR can cross the blood-brain barrier, offering protection to brain tissues as well [12].

In the current study, CHL and CUR were selected to investigate their potential anti-hyperglycemic and anti-diabetic effects in a mouse model of Alloxan-induced diabetes. The mice were pre-treated with the phytocompounds before diabetes was induced with Alloxan, and various diabetes-related parameters were assessed. The goals of the study were: i) to evaluate the anti-diabetic potential of CHL pre-treatment in Alloxan-induced diabetic mice, ii) to examine the protective effects of CUR pre-treatment against oxidative stress-induced genotoxicity, and iii) to compare the anti-diabetic efficacy of CHL and CUR pre-treatment in the same model.

2 Materials and Methods

2.1 Reagents and Antibodies

All reagents used in this study were of analytical grade and sourced from Sigma-Aldrich, USA, and other reputable suppliers. The antibodies were obtained from Santa Cruz Biotechnology, USA.

2.2 Animals

In this study, healthy inbred Swiss albino mice (Mus musculus), approximately 6–8 weeks old and weighing around 25 g, were selected. The animals were kept in an environmentally controlled room with 12-h light/dark cycle, temperature of 25 ± 2 °C, and 55 ± 5% humidity and unrestricted access to food and water. Animal handling was carried out in accordance with the guidelines set by the Committee for the Control and Supervision of Experiments on Animals (CPCSEA). All experimental protocols were approved and monitored by the Institutional Animal Ethical Committee, University of Kalyani, West Bengal, India (IAEC-KU) with approval number 892/GO/Re/S/01/CPCSEA; Pr: 20.4.14–28.4.2019.

2.3 Dose Selection of CUR and CHL for Pre-treatment in Mice Model

We adopted the dose selection rationale from Samadder et al., (2021) [11] and Saha et al., (2021) [14] with slight modifications pertaining to the current study for determining the suitable doses of CHL and CUR respectively for oral administration. Four different experimental groups of mice, with three mice in each group, were used:

Group 1: Control – Mice that were not treated.
Group 2: ALX – Mice received 100 mg/kg body weight single intraperitoneal (i.p.) dose of Alloxan.
Group 3: CUR + ALX – Mice were administered a 50 mg/kg body weight dose of CUR orally for seven days, followed by a single 100 mg\kg body weight intraperitoneal injection of Alloxan.
Group 4: CHL + ALX – Mice were administered 50 mg/kg body weight of CHL orally for seven days, followed by a single intraperitoneal injection of Alloxan (100 mg\kg body weight).

2.4 Induction of Diabetes

Diabetes in mice was induced by a single intraperitoneal (i.p.) injection of Alloxan monohydrate (ALX), dissolved in sterile normal saline, at a dose of 100 mg/kg body weight. The animals were provided with a 5% glucose solution to drink overnight to counteract the initial hypoglycemic shock [13].

2.5 Measurement of Serum Glucose

After the conclusion of the treatment period, serum glucose levels were assessed using a standard GOD-POD Glucose kit (Autospan) from Cogent Diagnostics Ltd. [2].

2.6 Comet Assay

The comet assay was performed to assess DNA damage at the single-cell level, following the standard protocol. Fluorescence images were captured using a Carl Zeiss Axio Vert A1-FL-LED microscope, and the frequency of comet-tailed cells was determined from the obtained images. The assay was carried out using single-cell gel electrophoresis according to the established procedure [12].

2.7 DNA Fragmentation Assay

Genomic DNA was extracted from both control and treated mice groups using the phenol-chloroform-isoamyl alcohol extraction method. Briefly, 20 mg of pancreas tissue was homogenised in DNA extraction buffer, incubated at 56 °C, and centrifuged. The aqueous supernatant was mixed with phenol-chloroform-isoamyl alcohol, vortexed, and centrifuged again. The DNA-containing supernatant was collected, spooled with chilled ethanol, and processed further. DNA fragmentation was assessed using agarose gel electrophoresis, and the bands were visualised under a UV transilluminator [5].

2.8 Histopathological Study

After euthanising the mice, tissue samples were collected and fixed in 10% neutral buffered formalin (NBF) for 24 h. The tissues were then embedded in paraffin, sectioned, and mounted onto slides. The sections underwent processing through different alcohol grades, followed by staining with Hematoxylin and Eosin. Finally, the slides were examined under a Carl Zeiss Axio Vert A1-FL-LED microscope at 10x magnification [2, 14].

2.9 Cell Isolation from Pancreatic Tissue and Hepatic Tissue for the Measurement of Reactive Oxygen Species

Hepatic and pancreatic cells were extracted from liver and pancreatic tissues, respectively, of both control and experimental mouse groups in phosphate-buffered saline (PBS). The levels of reactive oxygen species (ROS) in each group were assessed using a spectrofluorimetric technique, with excitation/emission wavelengths of 485 nm/535 nm, according to the protocol outlined by Samadder et al., (2019) [15, 16].

2.10 Assessment of the Expression Level of p53 Protein by Immunofluorescence in Pancreatic Tissue

The expression levels of p53 protein in pancreatic tissue from control and experimental groups were measured using standard immunofluorescence techniques. Primary anti-p53 antibodies (Santa Cruz Biotechnology, Inc., USA) and FITC (Fluorescein-isothiocyanate)-conjugated secondary antibodies (Sigma, USA) were applied. Images were captured using an immunofluorescence microscope (Carl Zeiss LSM 800 microscope) at 10× magnification [16].

2.11 Statistical Analysis

All experiments were performed independently three times. The collected data were used to calculate the mean and standard error. Statistical analysis was conducted using Student's t-test to assess the significance levels. The data presented in graphs are as follows: ##p < 0.01 compared to Control, **p < 0.01, *p < 0.05 compared to ALX, and considered statistically significant based on the Student's t-test.

3 Results and Discussions

Alloxan, a commonly used flour whitener, which after consumption targets a number of important body organs. Alloxan accumulates in pancreatic cells via GLUT-2 receptor, followed by partial destruction of pancreatic islet cells by eliciting genotoxic effects and disrupting the glycemic homeostasis of the body [17]. This glycemic imbalance and generation of ROS lead to the malfunctioning of other body tissues. However, the overall results of the present study reveal that the CUR and CHL pretreatment followed by diabetes induction with alloxan help restore glycemic balance by modulating the parameters related to diabetes and its complications.

3.1 Serum Glucose Level Modulation

Serum glucose level serves as the most important marker of hyperglycemia. The CUR and CHL pre-treatment significantly reduced serum glucose levels in alloxan-induced diabetic mice group. Several literatures claim that alloxan, after entering the body in the presence of reducing agents, oxidises into dialuric acid which reoxidises into alloxan and thereby creates a redox cycle followed by the generation of hydrogen peroxide and superoxide radicals. These superoxide radicals exert deleterious effects on DNA which leads to the apoptosis of pancreatic islet cells [11]. Toxic analogue of glucose, alloxan, also inhibits glucose-metabolising enzymes and reduces the responsiveness of pancreatic beta cells towards insulin formation and release. Oxidative stress induced by ALX damages cellular DNA which in turn partially destroys pancreatic cells, leading to the disruption of glucose homeostasis. The undertaken experiment has exhibited that blood glucose levels were heightened in ALX- treated mice group when compared to the control group. However, the CUR and CHL pre-treatment, followed by the ALX treatment, restricted the elevated blood glucose level. Both the phytocompounds gave good results in this regard, but CHL pre-treatment inhibited blood glucose levels better than CUR pre-treatment (Fig. 1).

3.2 Assessment of DNA Damage by Comet Assay

Comet assay or single cell gel electrophoresis is considered one of the most important experimental techniques for assessing genotoxicity. Frequency and comet tail length increase with increasing DNA damage [11, 18]. In the Comet assay, fragmented DNA migrates towards the anode, forming a structure that resembles the tail of a comet. Comet tail length increased in isolated pancreatic beta cells of the ALX-treated group as compared to the control group. The reason behind this result could be the overproduction of free radicals, inducing membrane leakage. Generation of free radicals is needed for our normal body functioning, it plays an important role in the modulation of

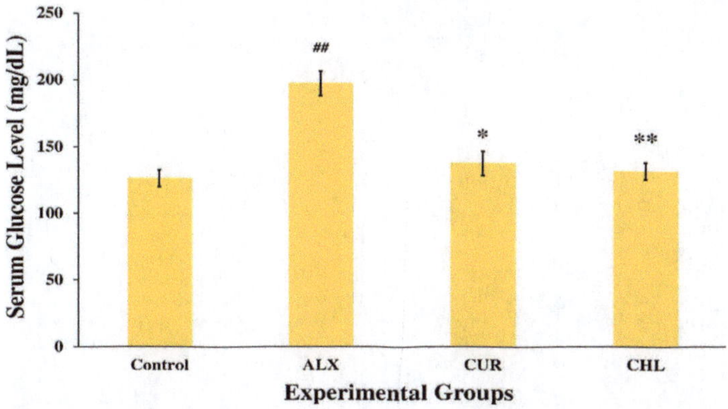

Fig. 1. Graphical representation of serum glucose level in different experimental groups. ##p < 0.01 vs. Control, **p < 0.01, *p < 0.05 vs. ALX, were considered significant for Student's t-test.

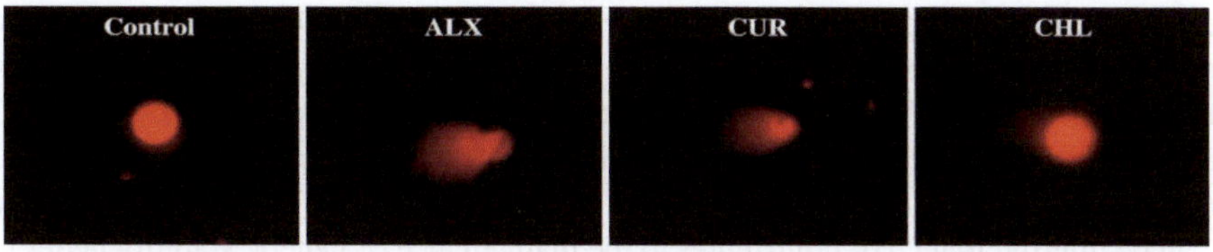

Fig. 2. Assessment of DNA damage of pancreatic cells by Comet assay.

cell survival, cell death, cell signaling and the generation of inflammation-related factors. Overproduction of free radicals exert genotoxic effects. The balance and scavenging of free radicals are required for homeostasis [19, 20]. While comparing ALX treated group with CUR and CHL pretreated group in this regard it was exhibited decreased number of comet tailed cells. Corresponding decrease in length were observed when compared to ALX-treated mice group, indicating potential of these two phytocompounds to protect DNA damage (Fig. 2). Furthermore comet tail length of CHL + ALX treated group was shorter than that of CUR + ALX group, which suggest better efficacy of CHL to provide a bit more protection to DNA.

3.3 Assessment of DNA Damage by DNA Fragmentation Assay

A potent genotoxic compound ALX incur the production of superoxide ion, hydroxyl ion and hydrogen peroxide which after entering the cell causes breakage of DNA strand, exacerbating genomic instability, disruption of genomic integrity and ultimately apoptosis. Pancreatic islet cells are exquisitely sensitive to H_2O_2. The free radicals cause DNA strand breaks and stimulate poly (ADP-ribose) synthetase, which depletes cellular NAD level, contributing to the generation of stress on the cell, followed by apoptosis. Fragmentation of DNA causes smearing in the gel, and from this smearing, the degree of DNA damage can be assessed. Overall DNA fragmentation data of pancreatic

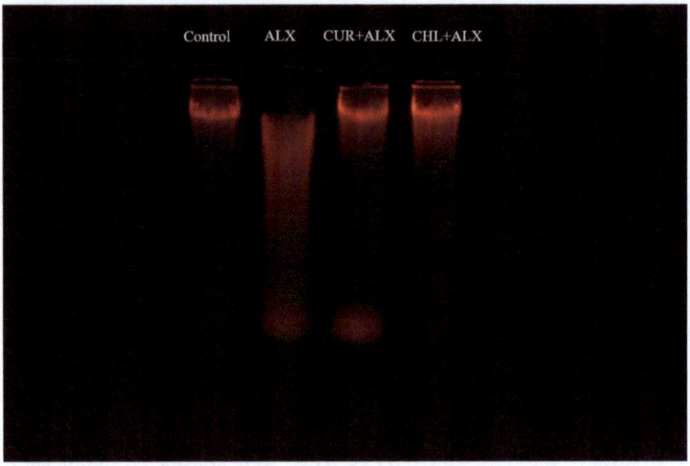

Fig. 3. DNA Fragmentation Assay.

cells found in the current experiment suggest that smearing of DNA in gel electrophoresis significantly increased in ALX-treated group than control, symbolising an increase in DNA damage. DNA smearing gets reduced in the CUR + ALX and CHL + ALX groups. CUR gave satisfactory result, but DNA smearing was reduced more in CHL pre-treated group as compared to CUR pre-treated group (Fig. 3). Thus, from the data, it can be inferred that CHL contributes to the maintenance of genetic stability better [2, 20, 21].

3.4 Histopathological Study

Glucose metabolism impairment and increased oxidative stress exert deleterious effects on different regions of the brain which is considered one of the most sensitive tissues of the body. Disorganisation of brain tissue characterised by tissue necrosis, neuronal disorientation, increased eosinophilia, loss of cellular integrity, and tissue cavitation [7, 21, 22]. Histopathological assessment of the brain tissue of ALX-treated group exhibited increased eosinophilia, tissue disorganisation and tissue shrinkage as compared to the untreated group. In the CHL pretreated group, restoration of brain tissue structure was observed better than CUR pre-treated group. Eosinophilia was reduced more in the CHL pre-treated group. Neuronal cells present in the hippocampal region of the brain back to normalcy as compared to the ALX pre-treated group. All these evidences suggest the protective role of CUR and CHL on brain tissue (Fig. 4a).

Alloxan exerts its effect on liver tissue morphology by disrupting hepatocytes, which leads to the generation of gaps between hepatocytes [5, 14]. In our experiment, it was also exhibited that the appearance of gaps between hepatocytes and sinusoidal size increased after ALX treatment, i.e. overall morphology of hepatic tissue was changed. In the CUR + ALX and CHL + ALX groups, tissue architecture returned to normalcy to some extent. Tissue morphology improved more in the CHL pre-treated group, suggesting the potential of CHL in the improvement of tissue architecture (Fig. 4b).

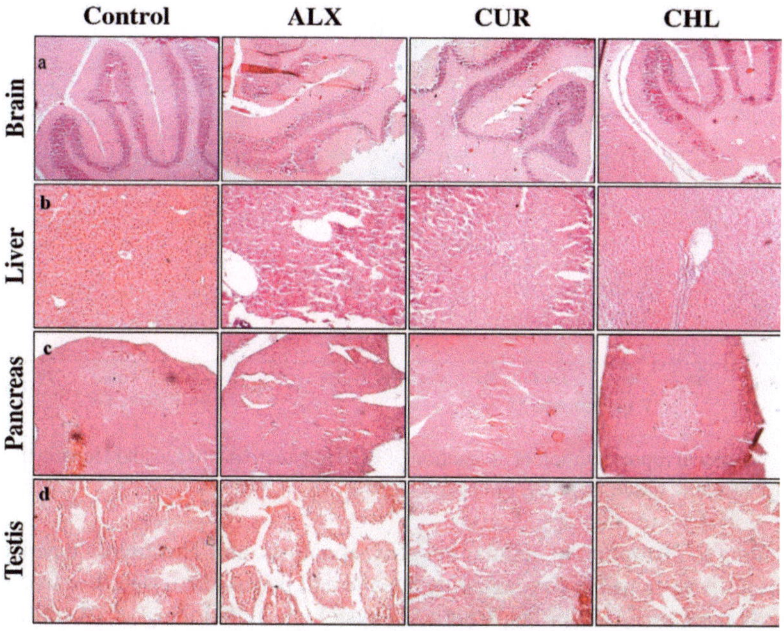

Fig. 4. Histopathological assessment of Hematoxylin and Eosin stained tissue section of (a) Brain, (b) Liver, (c) Pancreas, (d) Testes. Images are 10× magnified.

It was also shown in our study that the ALX severely damages pancreatic islet cells by generating oxidative stress in it. ALX induces its key destructive effect on pancreatic tissue which is characterised by excessive tissue shrinkage and necrosis [20]. In the histological tissue section of the pancreas of ALX treated group, shrinkage of pancreatic tissue, blebbing, loss of cell boundaries and disorganisation of pancreatic acini were evident in a huge amount as compared to the control. In the control group, dense packing of islet cells was visualised, but in the ALX-treated group, this dense cellular packing was reduced. However, in the CUR + ALX and CHL + ALX group, reduction of cell shrinkage and dense packing of islet cells were observed (Fig. 4c).

In the histopathological study of testis reduction in seminiferous tubule size, irregularity in tubule shape, reduction in cellular density, cellular degeneration and increased vacuolization were observed in the ALX-treated group as compared to the untreated group. The CUR and CHL restored tissue architecture. overall cellular shrinkage decreased in pre-treated groups, density of cells increased, and cell vacuole size decreased more in CHL pre-treated group (Fig. 4d).

3.5 Assessment of ROS Generation in Pancreatic and Hepatic Tissue

Alloxan generates ROS by oxidisation in presence of thiol and produce dialuric acid, and this dialuric acid reduce back into alloxan and thereupon creates a redox cycle which in turn produce free radicals like superoxide anion, hydroxyl ion and also produces hydrogen peroxide. This highly reactive hydroxyl ion causes DNA damage,

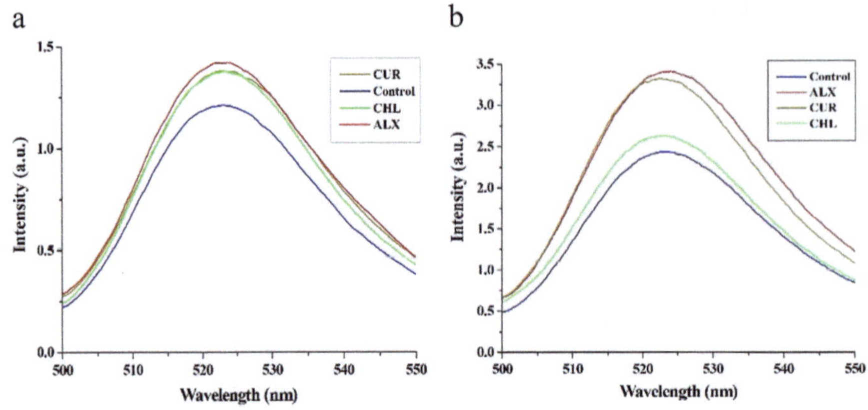

Fig. 5. Graphical representation of estimation of ROS in pancreatic tissue and Hepatic tissue. (a) ROS of Pancreatic tissue; (b) ROS of hepatic tissue.

contributing to disruption of cellular integrity, apoptosis, etc. This highly reactive oxygen species is responsible for damaging hepatic and pancreatic tissue. But when equilibrium is maintained, ROS helps normal physiologic functioning of tissues as well. ALX not only increases ROS production but also disrupts the antioxidant-dependent ROS balancing mechanism [13, 23]. Balance of ROS is very much needed for the proper functioning of mitochondria. For our current study, we isolated cells from pancreatic and hepatic tissue and quantified ROS using the spectrofluorimetric technique. Overall data (Fig. 4) from both tissues indicate the difference in quantity of ROS among experimental groups. Higher ROS intensity was observed in the ALX-treated group, indicating an effect of ALX on the overproduction of ROS. The substantial difference in this regard was seen in the CUR + ALX and CHL + ALX treated groups. Intensity of ROS was lower in the CUR and CHL pretreated group as compared to the ALX treated group, signifying the good effects of these two phytocompounds against ROS generation, thereby providing the protective effect on physiological parameters, however, CHL imposed better efficacy than the CUR group. It was observed that the intensity of ROS decreased more in the CHL pre-treated group in case of both pancreatic (Fig. 5a) and hepatic tissue (Fig. 5b).

3.6 Assessment of Expression of p^{53} Protein in Pancreatic Tissue

The expression pattern of p^{53} is also associated with the ALX treatment. P^{53} plays a protective role in maintaining our genomic integrity by preventing the replication of damaged DNA. Hence, it checks the passage of damaged DNA to the next generation. P^{53}, upon sensing DNA damage, activates the downstream DNA damage repair pathway, or if it is found to be impossible to revert the damage, it upregulates the apoptotic pathway, followed by programmed cell death [2, 24]. To check the expression pattern of p^{53}, the immunofluorescence study of p^{53} protein using FITC-tagged secondary antibody and the nuclear stain PI was undertaken. The results exhibited elevated p^{53} levels in the ALX-treated group than the untreated control group, confirming the genotoxic

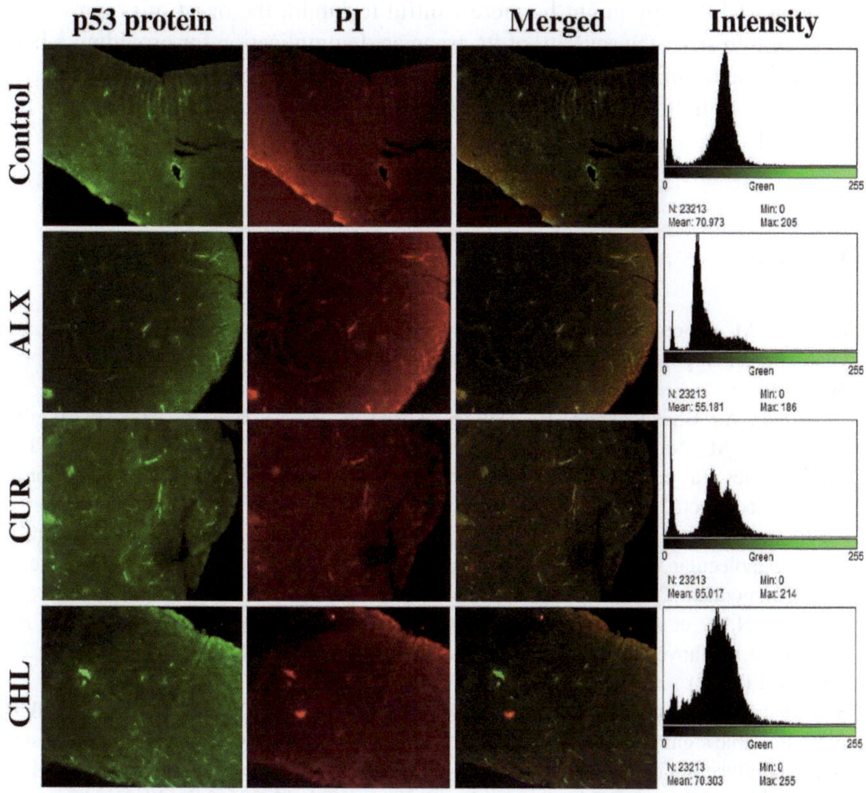

Fig. 6. Immunofluorescence study of expression levels of p^{53} in Pancreatic tissue.

effect of the ALX. The results corroborate with the fact that P^{53} prevents the passage of damaged DNA for maintaining genomic stability. In the mice group treated with the CUR followed by induction with the ALX, the expression of p^{53} was reduced, and in the CHL pre-treated mice group expression of p^{53} was more reduced and found to be more or less similar to the expression pattern of p^{53} in the untreated control mice group [22]. Hence, our result is found to be in line with the concept that oxidative stress exerted by ALX damages cellular DNA, leading to the elevated level of expression of p^{53} protein, which is working hard to protect the cell from DNA damage. The study successfully corroborated with the finding that the CUR and CHL bind to the DNA and protect the same from the deleterious effects of the ALX. In the CHL- administered group, p^{53} expression was more similar to that of the untreated group (Fig. 6).

4 Conclusion

The overall results revealed the fact that to control the occurrence and complications related to diabetes dependency on mainstream medicine would not be sufficient, as it has its own set of side effects and dose dependency issues. Therefore, an alternative or a combinatorial therapy might be more fruitful to inhibit the onset or progression of diabetes or to restrict the severity of its associated complication for providing a better life to the diabetics in future by depending on our regularly used Phyto products derived from the core of mother nature which has been an answer to many diseases from time immemorial.

References

1. Hammer, M., Storey, S., Hershey, D.S., Brady, V.J., Davis, E., Mandolfo, N., Bryant, A.L., Olausson, J.: Hyperglycemia and cancer: a state-of-the-science review. Oncol. Nurs. Forum. **46** (2019)
2. Samadder, A., Bhattacharjee, B., Dey, S., Chakrovorty, A., Dey, R., Sow, P., Tarafdar, D., Biswas, M., Nandi, S.: Enhanced drug carriage efficiency of curcumin-loaded PLGA nanoparticles in combating diabetic nephropathy via mitigation of renal apoptosis. J. Pharmacopunct. **27**, 1–13 (2024)
3. Bhattacharjee, B., Chakrovorty, A., Biswas, M., Samadder, A., Nandi, S.: To explore the putative molecular targets of diabetic nephropathy and their inhibition utilizing potential Phytocompounds. Curr. Med. Chem. **31**, 3752–3790 (2024)
4. Amegan, N.H., et al.: Prevalence and factors associated with hyperglycemia in a rural population of Tanvè and Dékanmey in Benin in 2019. PLOS Global Public Health. **2**(5), e0000471 (2022)
5. Samadder, A., Chakraborty, D., De, A., Bhattacharyya, S.S., Bhadra, K., Khuda-Bukhsh, A.R.: Possible signalling cascades involved in attenuation of alloxan-induced oxidative stress and hyperglycemia in mice by ethanolic extract of Syzygium jambolanum: drug-DNA interaction with calf thymus DNA as target. Eur. J. Pharm. Sci. **44**, 207–217 (2011)
6. Dey, S., Samadder, A., Nandi, S.: Exploring current role of nanotechnology used in food processing industry to control food additives and their biochemical mechanisms. Curr. Drug Targets. **23**, 513–539 (2022)
7. Lakhman, S.S., Kaur, G.: Effect of alloxan-induced diabetes on acetylcholinesterase activity from discrete areas of rat brain. Neurochem. Int. **24**, 159–163 (1994)
8. Kukula, O., Günaydın, C.: Atorvastatin reduces alloxan-induced impairment of aversive stimulus memory in mice. Asian Biomed (Res Rev News). **16**, 71–78 (2022)
9. Samadder, A., Khuda-Bukhsh, A.R.: Nanotechnological approaches in diabetes treatment: a new horizon. World J. Trans. Med. **3**, 84–95 (2014)
10. Chaudhury, A., Duvoor, C., Reddy Dendi, V.S., Kraleti, S., Chada, A., Ravilla, R., Marco, A., Shekhawat, N.S., Montales, M.T., Kuriakose, K., Sasapu, A., Beebe, A., Patil, N., Musham, C.K., Lohani, G.P., Mirza, W.: Clinical review of antidiabetic drugs: implications for type 2 diabetes mellitus management. Front. Endocrinol. (Lausanne). **8** (2017)
11. Samadder, A., Dey, S., Sow, P., Das, R., Nandi, S., Das, J., Bhattacharjee, B., Chakrovorty, A., Biswas, M., Guptaroy, P.: Phyto-chlorophyllin prevents food additive induced Genotoxicity and mitochondrial dysfunction via cytochrome c mediated pathway in mice model. Comb. Chem. High Throughput Screen. **24**, 1618–1627 (2021)
12. Parmar, H.S., Kar, A.: Antidiabetic potential of Citrus sinensis and Punica granatum peel extracts in alloxan treated male mice. Biofactors. **31**, 17–24 (2007)
13. Samadder, A., Das, S., Das, J., Khuda-Bukhsh, A.R.: Relative efficacies of insulin and poly (lactic-co-glycolic) acid encapsulated nano-insulin in modulating certain significant biomarkers in arsenic intoxicated L6 cells. Colloids Surf. B Biointerfaces. **109**, 10–19 (2013)

14. Saha, A., Chakrovorty, A., Bhattacharjee, B., Nandi, S., Samadder, A.: Curcumin mitigates Streptozotocin-induced Genotoxicity in vivo by activating P53 protein and inhibiting oxidative stress and chromosomal aberration. Comb. Chem. High Throughput Screen. (2021)
15. Samadder, A., Das, J., Das, S., Khuda-Bukhsh, A.R.: Dihydroxy-isosteviol-methyl-ester, an active biological component of Pulsatilla nigricans, reduces arsenic induced cellular dysfunction in testis of male mice. Environ. Toxicol. Pharmacol. **34**, 743–752 (2012)
16. Samadder, A., Tarafdar, D., Das, R., Khuda-Bukhsh, A.R., Abraham, S.K.: Efficacy of nano-encapsulated pelargonidin in ameliorating pesticide toxicity in fish and L6 cells: modulation of oxidative stress and signalling cascade. Sci. Total Environ. **671**, 466–473 (2019)
17. Ighodaro, O.M., Adeosun, A.M., Akinloye, O.A.: Alloxan-induced diabetes, a common model for evaluating the glycemic-control potential of therapeutic compounds and plants extracts in experimental studies. Medicina. **53**(6), 365–374 (2017)
18. Cortés-Gutiérrez, E.I., Dávila-Rodríguez, M.I., Fernández, J.L., López-Fernández, C., Gosálbez, A., Gosálvez, J.: New application of the comet assay: chromosome–comet assay. J. Histochem. Cytochem. **59**, 655–660 (2011)
19. Phaniendra, A., Jestadi, D.B., Periyasamy, L.: Free radicals: properties, sources, targets, and their implication in various diseases. Indian J. Clin. Biochem. **30**, 11–26 (2015)
20. Dey, S., Nagpal, I., Sow, P., Dey, R., Chakrovorty, A., Bhattacharjee, B., Saha, S., Majumder, A., Bera, M., Subbarao, N., Nandi, S., Hossen Molla, S., Guptaroy, P., Abraham, S.K., Khuda-Bukhsh, A.R., Samadder, A.: Morroniside interaction with poly (ADP-ribose) polymerase accentuates metabolic mitigation of alloxan-induced genotoxicity and hyperglycaemia: a molecular docking based *in vitro* and *in vivo* experimental therapeutic insight. J. Biomol. Struct. Dyn. **42**, 8541–8558 (2024)
21. Ahmed, R.G.: Evolutionary interactions between diabetes and development. Diabetes Res. Clin. Pract. **92**, 153–167 (2011)
22. Chakrovorty, A., Bhattacharjee, B., Saxena, A., Samadder, A., Nandi, S.: Current naturopathy to Combat Alzheimer's disease. Curr. Neuropharmacol. **21**, 808–841 (2023)
23. Gu, J., Luo, L., Wang, Q., Yan, S., Lin, J., Li, D., Cao, B., Mei, H., Ying, B., Bin, L., Smith, F.G.: Maresin 1 attenuates mitochondrial dysfunction through the ALX/cAMP/ROS pathway in the cecal ligation and puncture mouse model and sepsis patients. Lab. Investig. **98**, 715–733 (2018)
24. Eischen, C.M.: Genome stability requires p53. Cold Spring Harb. Perspect. Med. **6** (2016)

14. Saha, A., Chakravorty, A., [illegible] Gen[illegible] [illegible] high throughput [illegible] Throughput Screen [illegible]
15. [illegible]
16. [illegible]
17. [illegible]
18. [illegible]

Exploring the Antibacterial Potential of Plant-Derived Bioactive Compounds in Food Preservation

Debasish Tripathy[1], Ishika Mandal[2], Deep Talukder[2], Bhabani Shankar Nayak[3(✉)], and Dhirodatta Senapati[3]

[1] Department of Pharmacology, Biju Patnaik University of Technology, Rourkela, Odisha, India

[2] School of Biotechnology, KIIT Deemed to be University, Bhubaneswar, Odisha, India

[3] KIIT School of Pharmacy, KIMS, KIIT Deemed to be University, Bhubaneswar, Odisha, India

bhabanishankar.nayak@kims.ac.in

Abstract. Food preservation is a critical aspect of ensuring food safety and security, and the escalating demand for natural and sustainable antimicrobial agents has led to an increased focus on plant-derived bioactive compounds. A comprehensive analysis of recent studies is presented, encompassing the extraction methods, chemical diversity, and mechanisms of action of these bioactive compounds. The review highlights the rich repertoire of phytochemicals, including Phenolic, Alkaloids, Terpenoids, and essential oils, sourced from various plant species. These compounds exhibit notable antibacterial activities against a spectrum of food-borne pathogens. Emphasis is placed on elucidating the mechanisms underlying the antimicrobial actions, such as disruption of cell membranes, interference with bacterial biofilm formation, and modulation of essential bacterial enzymes. Furthermore, the potential synergistic effects of combining multiple plant-derived compounds or integrating them with conventional preservatives are explored, aiming to enhance the overall antimicrobial efficacy and overcome potential limitations. The impact of extraction methods on the bioactivity of plant-derived compounds is discussed, with a focus on optimizing extraction processes for maximal preservation of bioactive properties. In addition, the review examines the practical applications of these plant-derived compounds in food preservation, considering factors such as dosage, stability, sensory attributes, and regulatory aspects. Current challenges and future directions in the utilization of plant-derived bioactive compounds for food preservation are critically discussed, providing insights into the potential development of novel and effective preservation strategies with minimal impact on food quality and safety. Overall, this review contributes to the growing body of knowledge on sustainable and natural alternatives in food preservation, fostering advancements in the field of antimicrobial research and technology.

Keywords: Antibacterial · Bioactive · Food Preservation · Bacterial biofilm

A. K. Saxena, A. Saxena (Eds.): GTHTM 2025, pp. 411–428, 2026.
https://doi.org/10.1007/978-3-032-12320-6_24

1 Introduction

Food preservation is a critical facet of food science, aiming to enhance the shelf life of products while ensuring their safety and nutritional integrity [1]. Traditional methods often rely on synthetic preservatives, raising concerns about potential health risks and the emergence of antibiotic resistance [2]. As the global demand for safer and more sustainable alternatives intensifies, [3] the exploration of plant-derived bioactive compounds has gained prominence. Plants are known to synthesize a diverse array of secondary metabolites, many of which exhibit antibacterial properties [4]. This review delves into the potential of harnessing plant-derived bioactive compounds for food preservation, acknowledging the urgency driven by the escalating challenges of antibiotic resistance. By elucidating the bioactive components present in various plant sources and evaluating their antibacterial efficacy, this review aims to contribute insights into the development of natural antimicrobial agents, fostering innovation in the field of food science and technology. Through a systematic exploration of the antibacterial potential of plant-derived compounds, this review aspires to inform strategies that align with both health and sustainability goals in the realm of food preservation [5].

The emergence of antibiotic resistance presents a contemporary challenge in healthcare and agriculture [6], urging a re-evaluation of conventional antibiotic usage. Overreliance on antibiotics has led to the development of resistant strains of bacteria, compromising the effectiveness of crucial medical treatments [7]. This crisis necessitates a shift towards alternative solutions, especially in domains like food preservation, where antibiotics have conventionally played a role. As a result, exploring alternative antimicrobial strategies becomes imperative to address the global threat of antibiotic resistance [8].

The rationale for exploring plant-derived bioactive compounds stems from the need for sustainable and natural alternatives to synthetic preservatives, given the concerns surrounding their safety and the rise of antibiotic resistance [9, 10]. Plants are rich sources of diverse secondary metabolites, known for their antimicrobial properties [11]. As consumers increasingly seek healthier and more natural food options, the exploration of plant-derived bioactive compounds aligns with both safety and sustainability goals [12]. This avenue of research not only addresses the need for effective antimicrobial agents in food preservation but also taps into the potential health benefits associated with bioactive compounds derived from plant sources [13].

2 Plant-Derived Bioactive Compounds: Overview and Classification

2.1 Definition and Characteristics of Bioactive Compounds

Bioactive compounds derived from plants represent a diverse group of molecules known for their capacity to exert specific physiological effects in humans and animals. These compounds, often produced as secondary metabolites by plants, possess unique chemical structures and exhibit various biological activities [14]. Bioactive compounds are characterized by their potential to influence health beyond basic nutritional requirements, contributing to the prevention and treatment of diseases.

2.2 Types of Bioactive Compounds Found in Plants

The spectrum of bioactive compounds found in plants is extensive, encompassing a range of secondary metabolites with distinct functionalities. Polyphenols, including flavonoids, phenolic acids, and tannins, are prevalent and are recognized for their antioxidant properties. Alkaloids, such as caffeine and nicotine, contribute to the pharmacological effects of certain plants. Terpenoids, comprising essential oils and carotenoids, also exhibit diverse bioactivities [15]. Additionally, glucosinolates, saponins, and phytosterols are noteworthy bioactive compounds found in various plant species, each with its unique physiological effects.

2.3 Extraction Methods and Challenges

The extraction of bioactive compounds from plant sources involves various techniques, each tailored to the specific characteristics of the compounds of interest. Common methods include solvent extraction, steam distillation, and supercritical fluid extraction [16]. Challenges in extraction processes include optimizing conditions for preserving the integrity and bioactivity of the compounds, selecting suitable solvents, and addressing variations in plant matrices. Additionally, issues related to the standardization of extraction procedures and ensuring sustainability in the sourcing of plant materials pose significant challenges in the extraction of bioactive compounds [17]. Addressing these challenges is crucial for obtaining high-quality extracts with consistent bioactivity for potential applications in food preservation and health-promoting products.

3 Antibacterial Mechanisms of Plant-Derived Bioactive Compounds

3.1 Direct Antibacterial Action

Plant-derived bioactive compounds exhibit potent antibacterial properties (Fig. 1) through various mechanisms, contributing to their efficacy in food preservation and potential therapeutic applications.

3.1.1 Disruption of Cell Membranes. One of the primary mechanisms involves the disruption of bacterial cell membranes. Bioactive compounds, such as certain polyphenols and essential oils, can interact with lipid bilayers, causing structural damage and permeability changes [18]. This disturbance compromises the integrity of bacterial membranes, leading to leakage of cellular contents and eventual cell death.

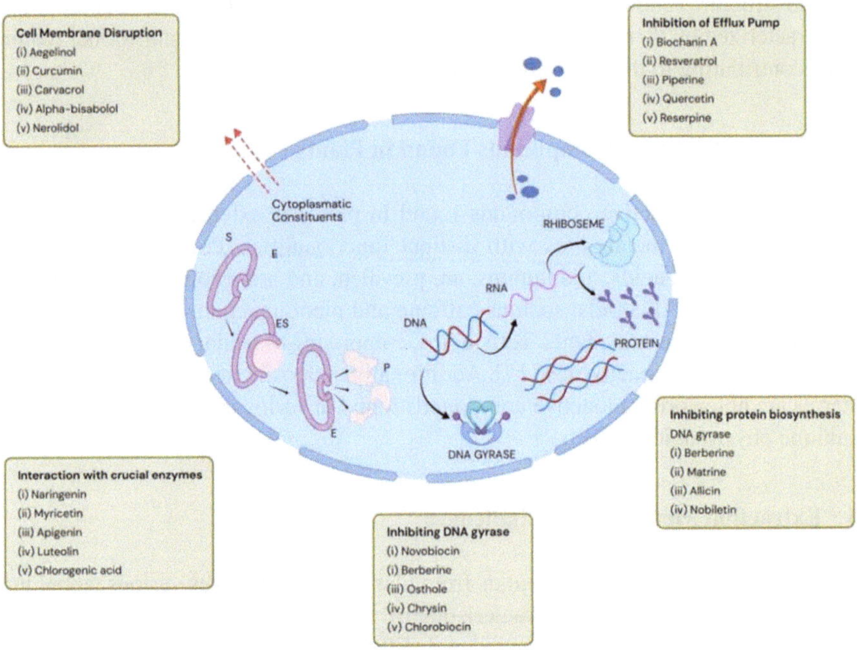

Fig. 1. Direct antimicrobial action.

3.1.2 Inhibition of Cell Wall Synthesis. Another critical antibacterial mechanism involves the inhibition of bacterial cell wall synthesis. Certain bioactive compounds interfere with the enzymes responsible for constructing bacterial cell walls. For instance, phenolic compounds and alkaloids may disrupt the synthesis of peptidoglycan, a crucial component of bacterial cell walls [19]. This interference weakens the structural integrity of the cell, rendering it susceptible to osmotic stress and eventual lysis.

3.1.3 Targeting Essential Enzymes and Proteins. Plant-derived bioactive compounds can selectively target essential enzymes and proteins vital for bacterial survival. These compounds often interfere with key metabolic pathways, such as those involved in DNA replication, protein synthesis, and energy production [20]. By disrupting these processes, bioactive compounds induce cellular dysfunction, preventing bacterial proliferation and promoting antibacterial effects, through which plant-derived bioactive compounds exert direct antibacterial action, provides valuable insights for harnessing their potential in food preservation and combating bacterial infections [2]. This contributes to the development of effective antimicrobial strategies with applications in both the food industry and medical fields (Fig. 1).

3.2 Indirect Antibacterial Action

Plant-derived bioactive compounds demonstrate remarkable antibacterial efficacy through indirect mechanisms, further expanding their potential applications in combating bacterial infections and enhancing food preservation.

3.2.1 Modulation of the Immune Response. Certain bioactive compounds exert indirect antibacterial effects by modulating the host's immune response. Flavonoids and polyphenols, for example, possess immunomodulatory properties that enhance the activity of immune cells [21]. By promoting immune responses, these compounds contribute to the host's ability to fend off bacterial infections, acting as supportive agents in the body's natural defence mechanisms.

3.2.2 Quorum Sensing Inhibition. Quorum sensing (Fig. 2) is a bacterial communication system crucial for coordinating group behaviours, including virulence and biofilm formation. Plant-derived bioactive compounds can inhibit quorum sensing, disrupting the bacterial ability to communicate and coordinate collective actions. This interference impedes the virulence of bacteria [22–24], reducing their pathogenic potential and rendering them more susceptible to immune defences and antimicrobial treatments.

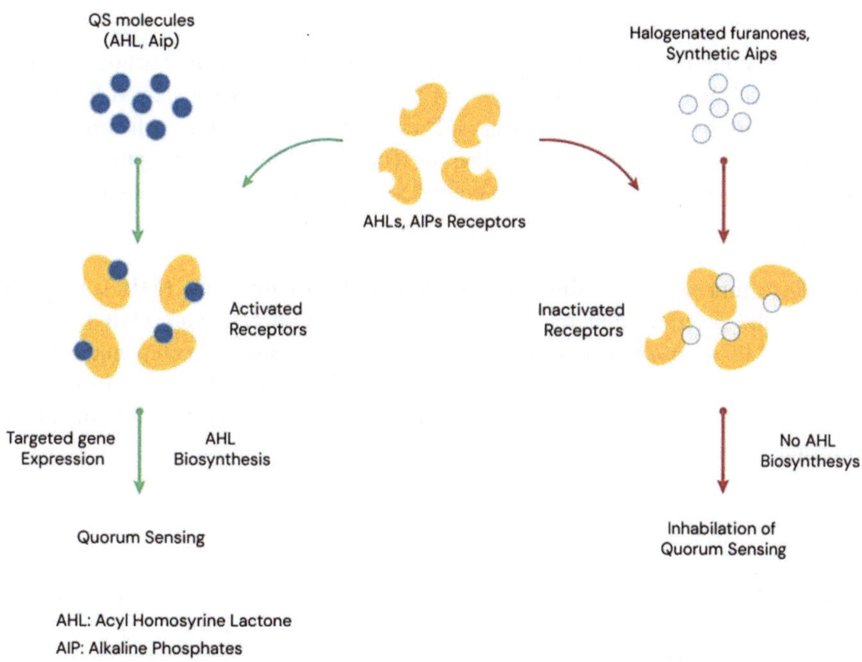

Fig. 2. Quorum Sensing.

3.2.3 Biofilm Disruption. Biofilms, complex bacterial communities encased in a self-produced matrix, pose significant challenges in terms of resistance to antibiotics and immune responses. Bioactive compounds from plants, such as certain essential oils and phenolic compounds, exhibit the ability to disrupt biofilm formation and enhance the disintegration of established biofilms [25]. This action not only prevents the formation of bacterial communities on surfaces but also facilitates the effectiveness of antibacterial agents. These indirect antibacterial mechanisms enhance our grasp of the multifaceted impact of plant-derived bioactive compounds. Exploiting these indirect actions broadens the potential applications of these compounds in both medical and food preservation contexts [26], offering innovative strategies to combat bacterial infections and enhance the safety and shelf life of food products.

4 Evaluation of Plant-Derived Bioactive Compounds for Food Preservation

4.1 In Vitro Studies

In vitro studies play a pivotal role in assessing the potential of plant-derived bioactive compounds for food preservation, providing valuable insights into their antimicrobial properties.

4.1.1 Screening of Bioactive Compounds Against Food-Borne Pathogens. Initial screening involves evaluating the efficacy of bioactive compounds extracted from plants against a spectrum of food-borne pathogens. These studies utilize well-established microbial strains representing common contaminants in food [27]. The screening process aims to identify compounds with robust antimicrobial activity, laying the foundation for further investigation.

4.1.2 Determination of Minimum Inhibitory Concentrations (MIC). Following screening, the determination of Minimum Inhibitory Concentrations (MIC) becomes imperative. MIC values reflect the lowest concentration of a bioactive compound needed to inhibit the growth of a specific microorganism [28]. This quantitative assessment aids in understanding the potency of these compounds and provides critical information for formulating effective preservation strategies.

4.2 In Vivo Studies

In vivo studies transition the evaluation process to more complex settings, providing insights into the safety and efficacy of plant-derived bioactive compounds in practical applications for food preservation.

4.2.1 Animal Models and Safety Assessment. Incorporating animal models allows for a comprehensive safety assessment, evaluating the potential impact of plant-derived compounds on the host organism. These studies consider factors such as toxicity, absorption, distribution, metabolism, and excretion [29]. The results contribute to establishing safe consumption levels and understanding the compounds' physiological effects.

4.2.2 Efficacy of Plant-Derived Compounds in Food Matrices. Evaluating the efficacy of plant-derived compounds within actual food matrices is crucial for practical application. These studies involve incorporating bioactive compounds into various food products and assessing their ability to inhibit microbial growth without compromising sensory attributes. The findings guide the development of effective and consumer-friendly food preservation strategies [30]. This multifaceted evaluation process ensures not only the antimicrobial efficacy of these compounds but also their safety and feasibility in real-world food applications.

5 Challenges and Limitations

Despite the promising potential of plant-derived bioactive compounds for food preservation, several challenges and limitations must be addressed to facilitate their successful integration into commercial applications.

5.1 Synergistic or Antagonistic Interactions with Food Components

The intricate composition of food matrices may give rise to synergistic or antagonistic interactions with plant-derived bioactive compounds. These interactions can influence the overall effectiveness of the compounds in inhibiting microbial growth [31]. Understanding and optimizing these complex interactions are critical to ensure consistent and reliable antimicrobial outcomes across different food products.

5.2 Stability and Bioavailability Issues

The stability of bioactive compounds, influenced by factors such as temperature, light, and oxygen, poses a significant challenge. Maintaining the stability of these compounds during food processing, storage, and preparation is crucial for preserving their efficacy [32]. Additionally, issues related to the bioavailability of these compounds, including their absorption, distribution, metabolism, and excretion within the human body, need careful consideration to ascertain their health benefits.

5.3 Regulatory Considerations and Approval for Commercial Use

The regulatory landscape governing the use of plant-derived bioactive compounds in food preservation is complex. Obtaining approval for commercial use involves navigating stringent regulations and compliance standards. Demonstrating the safety and efficacy of these compounds through rigorous testing and documentation is essential for regulatory approval. Developing standardized methods for assessing the quality and consistency of bioactive compounds is crucial for meeting regulatory requirements and ensuring consumer safety [33]. Overcoming these hurdles will not only enhance the integration of plant-derived bioactive compounds into food preservation practices but also contribute to the development of sustainable and safe solutions for the food industry.

6 Case Studies: Plant-Derived Bioactive Compounds in Food Preservation

6.1 Garlic-Derived Compounds

Garlic, a well-known plant with antimicrobial properties, has been extensively studied for its bioactive compounds, such as allicin (Fig. 3a) and diallyl sulphides. These compounds exhibit broad-spectrum antibacterial effects and have demonstrated efficacy in inhibiting the growth of various food-borne pathogens [34]. Garlic-derived compounds have been explored for their potential as natural preservatives in meat products, dairy, and other perishable foods.

6.2 Oregano and Thyme Essential Oils

Essential oils from oregano and thyme are rich sources of bioactive compounds, including thymol and carvacrol (Fig. 3b). These compounds exhibit strong antimicrobial activity against a wide range of bacteria and fungi. Studies have explored their effectiveness in food preservation, particularly in meat and dairy products [35]. The

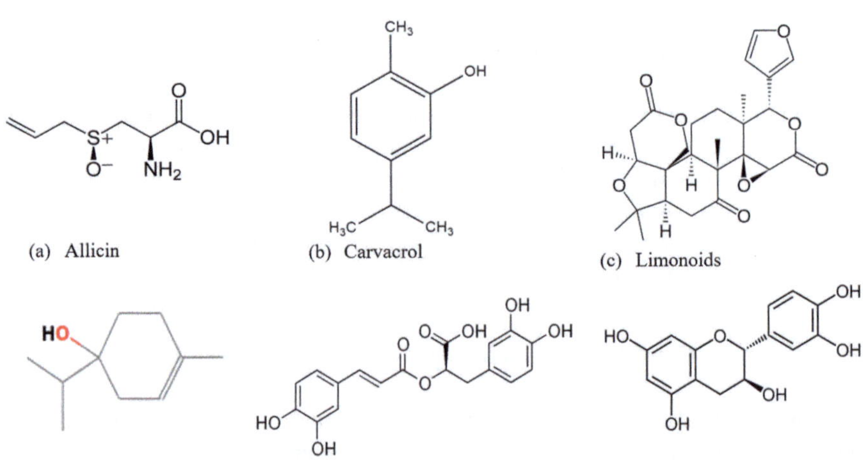

(a) Allicin (b) Carvacrol (c) Limonoids

(d) Terpene-2-ol (e) Rosmarinic acid (f) Catechin

Fig. 3. Various important Plant-Derived Bioactive Compounds used in Food Preservation.

antioxidant properties of these essential oils also contribute to extending the shelf life of food items.

6.3 Citrus Extracts

Citrus extracts, containing compounds such as limonoids (Fig. 3c) and flavonoids, possess notable antimicrobial and antioxidant properties. These extracts have shown promise in inhibiting the growth of spoilage microorganisms and food borne pathogens [36]. The application of citrus extracts in food preservation has been investigated in various products, including beverages and fresh produce.

6.4 Tea Tree Oil

Tea tree oil, derived from the *Melaleuca alternifolia* plant, is recognized for its potent antimicrobial activity. The main bioactive compound, terpinen-4-ol (Fig. 3d), exhibits broad-spectrum antibacterial effects. Research has explored the potential of tea tree oil in preserving perishable foods, demonstrating its efficacy against both Gram-positive and Gram-negative bacteria [37].

6.5 Other Promising Bioactive Compounds

Numerous other plant-derived bioactive compounds have shown promise in food preservation. For example, rosemary extracts with high levels of rosmarinic acid (Fig. 3e) have exhibited antimicrobial effects. Additionally, extracts from green tea, with catechins (Fig. 3f) as the key bioactive compounds, have demonstrated antibacterial properties. These compounds hold potential for application in diverse food products, contributing to both safety and shelf-life extension [38]. The exploration of these compounds in various food matrices provides valuable insights for the development of sustainable and consumer-friendly preservation strategies.

7 Potential Applications and Future Directions

7.1 Incorporation of Plant-Derived Compounds in Food Packaging

The integration of plant-derived bioactive compounds into food packaging materials represents a promising avenue for enhancing food preservation. Antimicrobial compounds, such as essential oils or extracts, can be incorporated into packaging films or coatings, creating an additional barrier against microbial contamination [39]. This approach not only extends the shelf life of packaged foods but also reduces the reliance on synthetic preservatives, aligning with the growing demand for sustainable and eco-friendly packaging solutions.

7.2 Combination Strategies to Enhance Antibacterial Efficacy

Future directions in food preservation involve exploring synergistic combinations of plant-derived bioactive compounds with other natural or conventional antimicrobial agents. Combining compounds with complementary mechanisms of action may enhance antibacterial efficacy, broaden the spectrum of activity, and mitigate the risk of resistance development [40]. Understanding the synergies between different compounds opens new possibilities for developing robust and multifaceted preservation strategies.

7.3 Role of Nanotechnology in Improving Delivery and Stability

Nanotechnology offers innovative solutions for improving the delivery and stability of plant-derived bioactive compounds. Nanoencapsulation techniques enable the controlled release of these compounds, enhancing their efficacy and stability during food processing and storage. Nanoparticles can protect bioactive compounds from degradation, improve solubility, and facilitate targeted delivery [41], presenting exciting opportunities for precision in food preservation applications.

7.4 Biotechnological Approaches for Large-Scale Production

Advancements in biotechnology provide avenues for large-scale production of plant-derived bioactive compounds. Biotechnological methods, including metabolic engineering and synthetic biology, can be employed to optimize the biosynthesis of these compounds in microbial or plant cell cultures [42]. This approach not only ensures a sustainable and reliable supply but also allows for the modification of compound profiles to meet specific application requirements in food preservation.

8 Safety and Consumer Perception

8.1 Toxicity Assessment and Safety Considerations

Ensuring the safety of plant-derived bioactive compounds for use in food preservation requires thorough toxicity assessments. Rigorous studies, including acute and chronic toxicity evaluations, are essential to establish safe consumption levels. Additionally, comprehensive safety considerations should address potential allergic reactions and any interactions with existing food components [43]. Transparent communication of safety assessments is crucial to instil confidence in consumers regarding the use of these compounds in food products.

8.2 Consumer Acceptance and Market Trends

Consumer acceptance plays a pivotal role in the successful integration of plant-derived bioactive compounds into food preservation strategies. Understanding consumer perceptions, preferences, and concerns is essential for market acceptance. Communication strategies that emphasize the natural and sustainable aspects of these compounds can positively influence consumer perception [44]. Monitoring market trends and aligning with the demand for clean label and natural products provides insights into the potential success of these compounds in the food industry.

9 Synergistic Effects and Combinatorial Approaches

9.1 Enhancing Efficacy Through Compound Combinations

Exploring synergistic effects through compound combinations represents a strategic approach to enhance the efficacy of plant-derived bioactive compounds in food preservation. Combinations of compounds with complementary modes of action can result in synergies that improve overall antibacterial effectiveness (Table 1) [45]. This approach not only increases the spectrum of activity but also has the potential to lower the required concentrations of individual compounds, addressing safety concerns.

9.2 Antibacterial Potential of Plant Extract Blends

The antibacterial potential of blends, combining various plant extracts, offers a multifaceted approach to food preservation. Blending extracts with diverse bioactive profiles can lead to a broader range of antimicrobial activity, making it challenging for bacteria to develop resistance. This approach aligns with the trend of developing natural and sustainable food preservation methods, emphasizing the use of plant-derived compounds [46]. Understanding the safety aspects, consumer perceptions, and exploring synergistic effects through combinatorial approaches are integral components of the successful adoption of plant-derived bioactive compounds in food preservation.

10 Scientific Shreds of Evidences

Mohamed M. Yousef, **et al.** ***(2023)*** evaluated potential antibacterial agents from various plant extracts and oils to inhibit the growth of these harmful bacteria and reduce sucrose losses. The screening of bacterial species that grow on beet revealed that a large group of mesophilic bacteria, such as *B. subtilis, L. mesenteroides, P. fluorescens, E. coli, A. baumannii, S. xylosus, E. amnigenus*, and *Aeromonas* species, in addition to a dominant thermophilic species called *Bacillus thermophilus*, were found to be present during the manufacturing of beets. The application of 20 plant extracts and 13 different oils

Table 1. Antibacterial synergistic effect of biomolecules with active molecules [47].

Active molecules combination	Effect	Mechanism of action
Curcumin with Amphotericin B	antibacterial action	A synergistically improved action was observed. In this case, the enzyme inhibitory effect of curcumin was mainly reported as being responsible for enhanced action
Garlic extract with ciprofloxacin	Antibacterial action	The presence of garlic extract improves the inhibitory action of ciprofloxacin
Sulfamethoxazole with proto catechuic acid, ellagic acid, and gallic acid	Synergistic antibacterial action	Increased activity may be due to DNA gyrase and topoisomerase IV enzymes
Kaempferol with norfloxacin and ciprofloxacin	Potent antibacterial action	Combinations show synergism and the reported possible mechanism involved may be DNA gyrase and DNA topoisomerase IV
Silibinin with aminoglycosides	Antibacterial action	Improve the action efficacy of aminoglycosides through a significant inhibitory effect on DNA topoisomerase activity due to the formation of complexes that alter enzyme binding
Carnosic acid with tetracycline	Antimicrobial activity	A possible mechanism reported inhibiting the MDR pumps
Carnosic acid with tetracycline	Antimicrobial activity	A possible mechanism reported inhibiting the MDR pumps
Nisin with thymol	Synergistic antimicrobial activity	Destabilization of bacterial membrane structure resulting in an increased permeability for nisin which leads to bacterial cell lysis
Sulfadiazine with proto catechuic acid and quercetin	Wide range of antimicrobial action	Reported to have synergism, however, the exact mechanisms are unknown
Ceftazidime with quercetin analogs	Potent antimicrobial action	A possible reported mechanism states that cell wall damage due to leakage of potassium. Both compounds act at the same target either at different sites
Sulfamethoxazole with myricetin	Synergistic antimicrobial action	The potency of combination increases due to synergism. Myricetin acts by DNA binding and induces enzymatic DNA breakage
Tetracycline with epigallocatechin gallate	Synergistic antimicrobial action	Enhances the activity of tetracycline against resistant staphylococcal by impairment of tetracycline efflux pump activity and increased intracellular retention of the drug

indicated that the extracts of *Geranium gruinum, Datura stramonium*, and *Mentha spicata* were the best antibacterials to reduce the growth of *B. thermophilus* with inhibition zones equal to 40, 39, and 35 mm, respectively. In contrast, the best active oils for inhibiting the growth of *B. thermophilus* were *Mentha spicata* and *Ocimum bacilicum*, with an inhibitory effect of 50 and 45 mm, respectively. RAPD-PCR with different primers indicated that treating sugar juice with the most effective oils against bacteria resulted in new recombinant microorganisms, confirming their roles as strong antibacterial products. The characterization of *Mentha spicata* and *Ocimum bacilicum* oils using GC/MS analysis identified *cis*-iso pulegone and hexadecanoic acid as the two main bioactive compounds with potential antibacterial activity [48].

Ziyin Li, et al. (2024) investigated the antibacterial properties of *Prunella vulgaris* L (PV) extracts against methicillin-resistant *Staphylococcus aureus* (MRSA) and assessed their mechanistic impact on bacterial cells and cellular functions. UPLC-ESI-MS/MS tentatively identified 28 phytochemical components in the aqueous extract of PV. Exposure to an aqueous extract at ½ MIC and MIC for 5 h resulted in a significant release of intracellular nucleic acid (up to 6-fold) and protein (up to 10-fold) into the extracellular environment. The result showed that the anti-MRSA properties of PV might stem from its ability to disrupt membrane and cell wall integrity, interfere with enzymatic activity, and impede bacterial cell metabolism and the transmission of information and energy that is essential for bacterial growth, ultimately resulting in bacterial apoptosis [49].

Tijana Zeremski, et al. (2024) processed the GC-MS results of four samples of *Helichrysum italicum* essential oil of Serbian origin by chemometric tools, and evaluate the in vitro *and* in silico antimicrobial *activity*. About, 47 compounds were identified, in which the most abundant were γ-curcumene, α-pinene, and *ar*-curcumene, followed by α-ylangene, neryl acetate, *trans*-caryophyllene, italicene, α-selinene, limonene, and italidiones. The essential oil rich in sesquiterpenes (γ-curcumene and *ar*-curcumene chemotype). About 5 strains were sensitive among ten strains of tested microorganisms: *Staphylococcus aureus*, *Listeria monocytogenes*, *Bacillus cereus*, *Saccharomyces cerevisiae,* and *Candida albicans*. Therefore, these microorganism models were used further for in silico molecular docking through the mechanism of ATP-ase inhibitory activity. Results showed that among all compounds from *H. italicum* essential oil, neryl acetate has the highest predicted binding energy. Artificial neural network modeling (ANN) showed that two major compounds γ-curcumene and α-pinene, as well as minor compounds such as *trans-β*-ocimene, terpinolene, terpinene-4-ol, isoitalicene, italicene, *cis-α*-bergamotene, *trans-α*-bergamotene, italidiones, *trans-β*-farnesene, γ-selinene, β-selinene, α-selinene, and guaiol are responsible for the antimicrobial activity of *H. italicum* essential oil. The results of this study indicate that *H. italicum* essential oil samples rich in γ-curcumene, α-pinene, and *ar*-curcumene cultivated in Serbia (Balkan) have antimicrobial potential both in vitro and in silico [50].

Tainá Kreutz, et al. (2024) evaluated the antifungal susceptibility of *Aniba canelilla* (Kunth) Mez (EOAC) and its main compound, 1-nitro-2-phenylethane (NP), against dermatophytes (*Trichophyton rubrum, T. mentagrophytes* and *Microsporum canis*), evidencing antifungal activity with an inhibitory concentration lower than 256 μg/mL. The antifungal activity occurs through a modification of the shape of the conidial structures of the fungus, showing the permeability of the intracellular content due to the visually observed plasmolysis and cytosolic extravasation through an osmotic process [51].

Govindaraj Sriragavi, et al. (2023) isolated actinobacteria from the rhizosphere of bamboo plants collected within the Megamalai forest of the Western Ghats in the Theni zone of Tamil Nadu, India. These actinobacteria were subjected to characterization, and their growth conditions were optimized to enhance the production of bioactive compounds. The strain exhibiting notable antibacterial activity underwent further characterization through 16 s rRNA gene sequencing and subsequent phylogenetic analysis. Employing response surface methodology (RSM), cultural conditions were fine-tuned. Bioactive compounds were extracted from the culture medium using ethyl acetate, and their antibacterial and antioxidant effects were evaluated through disc diffusion and DPPH radical scavenging methods, respectively. Ethyl acetate extracts were analyzed by using FT-IR and GC–MS techniques. About 9 strains of Actinobacteria were isolated from the rhizosphere soil of bamboo. The strain BS-16 displayed remarkable antibacterial activity against three strains: *S. aureus* (19 mm), *B. subtilis* (12 mm), and *S. pyogenes* (10 mm). This strain was identified as *Streptomyces* sp. The ethyl acetate fraction derived from strain BS-16 exhibited dose-dependent antibacterial and antioxidant activities. FT-IR and GC–MS analyses showed the presence of active compounds with antibacterial effects within the extract [52].

11 Key Challenges and Future Prospects for Food Preservation

Despite the considerable promise, key challenges must be addressed for the widespread adoption of plant-derived bioactive compounds in food preservation. Interactions with food components, stability issues, regulatory considerations, and consumer acceptance are critical challenges that require careful attention. Future prospects involve innovative applications, including incorporation into food packaging, synergistic combinations to enhance efficacy, leveraging nanotechnology for improved delivery, and biotechnological approaches for large-scale production. By navigating these challenges and embracing future directions, plant-derived bioactive compounds have the potential to revolutionize food preservation, providing safer, more sustainable, and consumer-friendly alternatives.

12 Recommendations

12.1 Areas for Further Research and Investigation

12.1.1 Mechanistic Insights. Further research is warranted to gain deeper mechanistic insights into the antibacterial actions of specific plant-derived bioactive compounds. Understanding the interactions at the molecular level will enhance the design and optimization of effective preservation strategies.

12.1.2 Combinatorial Approaches. Investigate the synergistic effects of combining different plant-derived compounds and their interactions with conventional preservatives. This exploration can lead to the development of robust and multifaceted preservation methods.

12.1.3 Long-Term Safety Studies. Conduct long-term safety studies to assess the chronic effects of consuming food products preserved with plant-derived bioactive compounds. Continuous monitoring of safety is essential for building consumer confidence and regulatory approval.

12.1.4 Nanotechnology Applications. Explore advanced nanotechnological applications for encapsulating and delivering plant-derived compounds. This avenue could address stability concerns and improve the controlled release of bioactive compounds during food processing and storage.

12.2 Practical Implications for the Food Industry

12.2.1 Integration into Processing. The food industry should consider integrating plant-derived bioactive compounds into food processing methods. This may involve developing standardized extraction processes and determining optimal concentrations for different food matrices.

12.2.2 Consumer Education. Implement educational initiatives to inform consumers about the benefits and safety of plant-derived bioactive compounds. Transparent labeling and clear communication regarding the natural and sustainable aspects can positively influence consumer acceptance.

12.2.3 Collaboration with Regulatory Agencies. Collaborate with regulatory agencies to establish clear guidelines and standards for the use of plant-derived bioactive compounds in food preservation. Proactive engagement with regulators will facilitate the approval process and ensure compliance with safety standards.

12.3 Policy Recommendations and Future Guidelines

12.3.1 Incentives for Research and Development. Governments and industry bodies should provide incentives for research and development in the field of plant-derived bioactive compounds. Grants, subsidies, and support for collaborative projects can accelerate advancements in food preservation technologies.

12.3.2 Regulatory Frameworks. Develop or refine regulatory frameworks to streamline the approval process for incorporating plant-derived bioactive compounds in food products. This involves establishing safety criteria, standardized testing methods, and clear labeling requirements.

12.3.3 Global Collaboration. Encourage international collaboration to establish harmonized guidelines for the use of plant-derived compounds in food preservation. Consistent regulations will facilitate global trade and ensure the safety and quality of preserved food products across borders.

13 Conclusion

In summary, plant-derived bioactive compounds exhibit significant antibacterial potential, making them promising candidates for use in food preservation. These compounds, including those from garlic, oregano, thyme, citrus, and tea tree oil, showcase diverse mechanisms of action such as disrupting cell membranes, inhibiting cell wall synthesis, and targeting essential enzymes. Their direct and indirect antibacterial effects, along with potential applications in food packaging, position them as valuable alternatives to traditional preservatives. Notable case studies highlight the effectiveness of these compounds in various food matrices.

References

1. Teshome, E., Forsido, S.F., Rupasinghe, H.P.V., Olika Keyata, E.: Potentials of natural preservatives to enhance food safety and shelf life: a review. Sci. World J. **2022**, 1–11 (2022)
2. Anand, U., Jacobo-Herrera, N., Altemimi, A., Lakhssassi, N.: A comprehensive review on medicinal plants as antimicrobial therapeutics: potential avenues of biocompatible drug discovery. Metabolites. **9**(11), 258 (2019)-->
3. Kussmann, M., Abe Cunha, D.H., Berciano, S.: Bioactive compounds for human and planetary health. Front. Nutr. **10**, 1193848 (2023)
4. Cowan, M.M.: Plant products as antimicrobial agents. Clin. Microbiol. Rev. **12**(4), 564–582 (1999)
5. Vaou, N., Stavropoulou, E., Voidarou, C., Tsigalou, C., Bezirtzoglou, E.: Towards advances in medicinal plant antimicrobial activity: a review study on challenges and future perspectives. Microorganisms. **9**(10), 2041 (2021)
6. Manyi-Loh, C., Mamphweli, S., Meyer, E., Okoh, A.: Antibiotic use in agriculture and its consequential resistance in environmental sources: potential public health implications. Molecules. **23**(4), 795 (2018)
7. Llor, C., Bjerrum, L.: Antimicrobial resistance: risk associated with antibiotic overuse and initiatives to reduce the problem. Ther. Adv. Drug Saf. **5**(6), 229–241 (2014)
8. Salam, M.A., et al.: Antimicrobial resistance: a growing serious threat for global public health. Healthcare. **11**(3), 1946 (2023)
9. Pinto, L., Tapia-Rodríguez, M.R., Baruzzi, F., Ayala-Zavala, J.F.: Plant antimicrobials for food quality and safety: recent views and future challenges. Foods. **12**(12), 2315 (2023)
10. Dey, S., Samadder, A., Nandi, S.: Exploring current role of nanotechnology used in food processing industry to control food additives and their biochemical mechanisms. Curr. Drug Targets. **23**(5), 513–539 (2022)
11. Keita, K., Darkoh, C., Okafor, F.: Secondary plant metabolites as potent drug candidates against antimicrobial-resistant pathogens. SN Appl. Sci. **4**(8), 209 (2023)
12. Kurek, M., et al.: Antioxidants and bioactive compounds in food: critical review of issues and prospects. Antioxidants. **11**(4), 742 (2022)
13. Hintz, T., Matthews, K.K., Di, R.: The use of plant antimicrobial compounds for food preservation. BioMed Res. Int. **2015**, 246264 (2015)
14. Roy, A., et al.: Flavonoids a bioactive compound from medicinal plants and its therapeutic applications. Biomed. Res. Int. **2022**, 5445291 (2022)

15. Twaij, B.M., Hasan, M.N.: Bioactive secondary metabolites from plant sources: types, synthesis, and their therapeutic uses. Int. J. Plant Biol. **13**(1), 4–14 (2022)
16. Uwineza, P.A., Waśkiewicz, A.: Recent advances in supercritical fluid extraction of natural bioactive compounds from natural plant materials. Molecules. **25**(17), 3847 (2020)
17. Sasidharan, S., Chen, Y., Saravanan, D., Sundram, K.M., Yoga Latha, L.: Extraction, isolation and characterization of bioactive compounds from plants' extracts. Afr. J. Tradit. Complement. Altern. Med. **8**(1), 1–10 (2014)
18. Nazzaro, F., Fratianni, F., De Martino, L., Coppola, R., De Feo, V.: Effect of essential oils on pathogenic bacteria. Pharmaceuticals. **6**(12), 1451–1474 (2013)
19. Khameneh, B., Eskin, N.A.M., Iranshahy, M., Fazly Bazzaz, B.S.: phytochemicals: a promising weapon in the arsenal against antibiotic-resistant bacteria. Antibiotics. **10**(9), 1044 (2021)
20. Divekar, P.A., et al.: Plant secondary metabolites as defense tools against herbivores for sustainable crop protection. Int. J. Mol. Sci. **23**(5), 2690 (2022)
21. Shamsudin, N.F., et al.: Antibacterial effects of flavonoids and their structure-activity relationship study: a comparative interpretation. Molecules. **27**(4), 1149 (2022)
22. Rutherford, S.T., Bassler, B.L.: Bacterial quorum sensing: its role in virulence and possibilities for its control. Cold Spring Harb. Perspect. Med. **2**(11), a012427 (2012)
23. Nandi, S.: Recent advances in ligand and structure based screening of potent quorum sensing inhibitors against antibiotic resistance induced bacterial virulence. Recent Pat. Biotechnol. **10**(2), 195–216 (2016). https://doi.org/10.2174/1872208310666160728104450
24. Kumar, M., Saxena, M., Saxena, A.K., Nandi, S.: Recent breakthroughs in various antimicrobial resistance induced quorum sensing biosynthetic pathway mediated targets and design of their inhibitors. Comb. Chem. High Throughput Screen. **23**(6), 458–476 (2020)
25. Tiwari, V.: Editorial: ESKAPE biofilm: challenges and solutions. Front. Cell. Infect. Microbiol. **13**, 1253439 (2023)
26. Yan, D., Li, Y., Liu, Y., Li, N., Zhang, X., Yan, C.: Antimicrobial properties of chitosan and chitosan derivatives in the treatment of enteric infections. Molecules. **26**(23), 7136 (2021)
27. Thongphichai, W., Pongkittiphan, V., Laorpaksa, A., Wiwatcharakornkul, W., Sukrong, S.: Antimicrobial activity against foodborne pathogens and antioxidant activity of plant leaves traditionally used as food packaging. Foods. **12**(12), 2409 (2023)
28. Andrews, J.M.: Determination of minimum inhibitory concentrations. J. Antimicrob. Chemother. **48**, 5–16 (2001)
29. Eddouks, M., Chattopadhyay, D., Zeggwagh, N.A.: Animal models as tools to investigate antidiabetic and anti-inflammatory plants. J. Evid. Based Complement. Altern. Med. **2012**, 142087 (2012)
30. Samtiya, M., Aluko, R.E., Dhewa, T., Moreno-Rojas, J.M.: Potential health benefits of plant food-derived bioactive components: an overview. Foods. **10**(4), 839 (2021)
31. Donadio, G., et al.: Interactions with microbial proteins driving the antibacterial activity of flavonoids. Pharmaceutics. **13**(5), 660 (2021)
32. Enaru, B., Drețcanu, G., Pop, T.D., Stănilă, A., Diaconeasa, Z.: Anthocyanins: factors affecting their stability and degradation. Antioxidants. **10**(12), 1967 (2021)
33. Dwyer, J.T., Coates, P.M., Smith, M.J.: Dietary supplements: regulatory challenges and research resources. Nutrients. **10**(1), 41 (2018)
34. Shang, A., et al.: Bioactive compounds and biological functions of garlic (*Allium sativum* L.). Foods. **8**(7), 246 (2019)
35. Leyva-López, N., Gutiérrez-Grijalva, E.P., Vazquez-Olivo, G., Heredia, J.B.: Essential oils of oregano: biological activity beyond their antimicrobial properties. Molecules. **22**(6), 989 (2017)
36. Saleem, M., Durani, A.I., Asari, A., Ahmed, M., Ahmad, M., Yousaf, N., Muddassar, M.: Investigation of antioxidant and antibacterial effects of citrus fruits peels extracts using

different extracting agents: phytochemical analysis with *in silico* studies. Heliyon. **9**(4), e15433 (2023)

37. Carson, C.F., Hammer, K.A., Riley, T.V.: *Melaleuca alternifolia* (tea tree) oil: a review of antimicrobial and other medicinal properties. Clin. Microbiol. Rev. **19**(1), 50–62 (2006)
38. Nieto, G., Ros, G., Castillo, J.: Antioxidant and antimicrobial properties of rosemary (*Rosmarinus officinalis,* L.): A Review. Medicines. **5**(3), 98 (2018)
39. Nasution, H., Harahap, H., Julianti, E., Safitri, A., Jaafar, M.: Smart packaging based on polylactic acid: the effects of antibacterial and antioxidant agents from natural extracts on physical-mechanical properties, Colony reduction, perishable food shelf life, and future prospective. Polymers. **15**(20), 4103 (2023)
40. Basavegowda, N., Baek, K.H.: Combination strategies of different antimicrobials: an efficient and alternative tool for pathogen inactivation. Biomedicines. **10**(9), 2219 (2022)
41. Pateiro, M., et al.: Nanoencapsulation of promising bioactive compounds to improve their absorption, stability, functionality and the appearance of the final food products. Molecules. **26**(6), 1547 (2021)
42. Guo, M., et al.: Strategies on biosynthesis and production of bioactive compounds in medicinal plants. Chin. Herb. Med. **16**(1), 13–26 (2023)
43. Lourenço, S.C., Moldão-Martins, M., Alves, V.D.: Antioxidants of natural plant origins: from sources to food industry applications. Molecules. **24**(22), 4132 (2019)
44. Maherani, B., Hossain, F., Criado, P., Ben-Fadhel, Y., Salmieri, S., Lacroix, M.: World market development and consumer acceptance of irradiation technology. Foods. **5**(4), 79 (2019)
45. Basavegowda, N., Baek, K.H.: Synergistic antioxidant and antibacterial advantages of essential oils for food packaging applications. Biomolecules. **11**(9), 1267 (2021)
46. Gutiérrez-Pacheco, M.M., et al.: Mechanisms and applications of citral's antimicrobial properties in food preservation and pharmaceuticals formulations. Antibiotics. **12**(11), 1608 (2023)
47. Choudhury, A.: Potential role of bioactive phytochemicals in combination therapies against antimicrobial activity. J. Pharmacopuncture. **25**(2), 79–87 (2022)
48. Mohamed, M.Y., et al.: Exploring the antibacterial potential of plant extracts and essential oils against *Bacillus thermophilus* in beet sugar for enhanced sucrose retention: a comparative assessment and implications. Front. Microbiol. **14**, 1219823 (2023)
49. Ziyin, L., et al.: Exploring the antibacterial potential and underlying mechanisms of *Prunella vulgaris* L. on methicillin-resistant *Staphylococcus aureus*. Foods. **13**(5), 122 (2024)
50. Tijana, Z., et al.: Combination of chromatographic analysis and Chemometric methods with bioactivity evaluation of the antibacterial properties of *Helichrysum italicum* essential oil. Antibiotics. **13**(6), 499 (2024)
51. Tainá, K., et al.: Antifungal activity of *Aniba canelilla* (Kunth) Mez essential oil and its Main compound 1-Nitro-2-Phenylethane against dermatophytes. Antibiotics. **13**(6), 488 (2024)
52. Govindaraj, S., et al.: Exploring antibacterial properties of bioactive compounds isolated from Streptomyces sp. in Bamboo Rhizosphere Soil. ACS Omega. **8**, 36333–36343 (2023)

In Silico Exploration of the Antidiabetic Capabilities of the Essential Medicinal Plant Euphorbia

Mahvish Jamal[1(✉)] and Mhaveer Singh[1]

[1] School of Pharmaceutical Sciences, IFTM University, Moradabad, India
mahvishjamaal@gmail.com

Abstract. Diabetes mellitus (DM) is a major metabolic syndrome and accounts for about 2.8% of the population worldwide that are affected by any metabolic disease and is projected to rise to 4.4% by 2030. Sulfonylureas, biguanides, α-glucosidase inhibitors, thiazolidinediones, and non-sulfonylureas secretagogues are some of the categories of the drugs available in the market which works via improvement of insulin sensitivity, secretion, complementing insulin, and stimulating glucose uptake. The currently used drugs have been associated with several side effects such as hepatic failure, weight gain, tachycardia and hypothyroidism. Herbal drugs have been a source of cure for several diseases including DM. Herbal drugs have several advantages such as safety, ease in availability, less side effects and more economical in comparison to synthetic drugs. Euphorbia is one of the important medicinal plant of dicotyledons and phytoconstituents (flavanoids, terpenoids and tannins) obtained from this plant find use in the treatment of viral infections, rhinitis, spasms, asthma, and skin diseases, etc. This plant has been reported to have in vivo antidiabetic action but the exact mechanism of action of phytoconstituents of this plant is still unknown and is a matter of investigation. For this purpose docking studies on 3D structures of three proteins involved in type-2 DM viz., protein-tyrosine phosphatase 1B, α-Glucosidase and α-Amylasewere performed using known 25 phytoconstituents of this plant. The conclusion of the study is based on docking energies and availability of important amino acid interactions which sheds light on possible mechanism of phytoconstituents.

Keywords: Diabetes mellitus · Euphorbiacea · Flavanoids · Tannins · Docking

1 Introduction

The Type 2 Diabetes Mellitus (T2DM) is a prevalent metabolic disorder resulting from defective insulin secretion and tissue resistance to insulin [1–3]. Diabetes is a significant global health crisis currently ranked eighth as leading cause of death and disability. In 2022, 830 million people were affected with diabetes which was significantly higher than the cases reported in 1990 (200 million). In 2021, diabetes had a global prevalence of 6.1%, which increased by 90.5% from 1990 when the prevalence was 3.2% and expected to reach 9.8%, affecting 13.1 billion people in 2050 [4–6]. In 2021, the global

A. K. Saxena, A. Saxena (Eds.): GTHTM 2025, pp. 429–446, 2026.
https://doi.org/10.1007/978-3-032-12320-6_25

diabetes prevalence was higher in males than in females with a male-to-female ratio of 1.14 although regional variation was observed [7]. Scientists have explored both pharmacological and non-pharmacological approaches to find out effective treatment for diabetes however no medication has been found completely safe [8–10]. Regular physical exercises are commonly recommended for enhancing insulin sensitivity among non-pharmacological strategies [11, 12]. Among the pharmacological approaches, the main classes of drugs for the treatment diabetes includes sulfonylureas, which increase the release of insulin from pancreatic islets; biguanides, which slows the hepatic production of glucose; peroxisome proliferator-activated receptor-γ (PPARγ) agonists which enhances the action of insulin, and, α-glucosidase inhibitors which delays the absorption of glucose in the intestine [13–15]. These classes of drugs can be used as single preparation or in combination with other hypoglycemic drugs but have drawbacks such weight gain, severe hypoglycemia, side effects from metabolism, and some other issues with target specificity, permeability and solubility [16, 17]. Conventional treatments may cause dependence, side effects, and long-term toxicity [18–20]. Exploring alternative medicines, including natural remedies, can help mitigate these risks. Antioxidants play a crucial role in reducing oxidative stress and neutralizing harmful free radicals [21]. Extensive research has associated oxidative stress with diabetes-related complications [22–23]. Elevated blood sugar levels increase the production of reactive oxygen and nitrogen species while reducing antioxidant enzyme activity. This imbalance disrupts cellular homeostasis, damaging critical biomolecules such as proteins, lipids, and DNA, ultimately leading to harmful byproducts that contribute to disease progression [24, 25]. When the body's antioxidant system fails to counteract this, stress pathways activate, leading to cellular damage and diabetes progression. Research shows that antioxidant therapy protects β cells, preserves their function, and reduces complications. As a result, natural antioxidants are increasingly used due to their therapeutic benefits, safety, affordability, and availability [26]. Natural antioxidants can help prevent or slow type 2 diabetes by reducing oxidative stress, protecting against lipid peroxidation, and supporting antioxidant enzyme function. Natural antioxidant products should also be evaluated in the context of the complex physiological processes that modulate Type 2 diabetes oxidative stress such as glycemic control, postprandial oxidative stress [27], the polyol pathway, high-calorie, high-fat diets, exercise, and sleep. Minimizing processes that induce chronic damaging oxidative stress and maximizing the intake of natural antioxidant products may provide a means of preventing or slowing Type 2 diabetes progression [28, 29].

Euphorbia prostrata belonging to the family, Euphorbiaceae, and generally known as "Hazar dani" is a small annual herb widely distributed all over India particularly in the foothills of Himalayas [30–31]. The herb has been used widely in many conditions such as asthma [32], hemorrhoids [33], diabetes mellitus [34], dysentery [32], hyperlipidemia [35] and has wound healing [36], anti-inflammatory [37], analgesic [38], anthelmintic [39] properties and is also used in diabetic nephropathy [40]. It is also used traditionally to treat many skin infections and snake bite [41]. The phytoconstituents of the *Euphorbia prostrata* have shown both antidiabetic as well as antioxidant properties making the plant indispensable for studies particularly relating to diabetes [42–43]. Based on these evidences, this study aims to evaluate the anti-hyperglycemic activity of antioxidant compounds in the hydroalcoholic extracts of *Euphorbia prostrata* by assessing their binding effectiveness as ligands for three targeted proteins (α-amylase, α-glucosidase, and PTP-1B) using docking methodology. These targets are the validated targets with known antidiabetic compounds as co-crystals. Phytochemical screening of *Euphorbia prostrata* has identified the presence of glycosides, phytosterols, flavonoids, polyphenols, tannins, alkaloids, terpenoids, saponins, and anthraquinones etc. [44]. Through physicochemical and spectroscopic analysis, ten major compounds have been identified: Gallic Acid, Corilagin, 1,2,3-Tri-O-Galloyl-D-Glucose, Geraniin, Tellimagrandin I and II, and Rugosin A, D, E, and G [45, 46]. The dried leaf extracts of *Euphorbia prostrata* contain various hydrolysable ellagitannins, including prostratins A, B, and C, Euphorbin G and H, Flavonoids found in the aerial parts include Kaempferol, Cosmosiin, Rhamnetin 3-Galactoside, Quercetin, Luteolin, and various Quercetin Glycosides. The aerial parts also contain sterols (Beta-Amyrine Acetate, Beta-Sitosterol, Campesterol, Stigmasterol, and Cholesterol), the terpene alcohol Beta-Terpineol and amino acids. From the roots, Myricylic Alcohol and the triterpenes Taraxerol and Tirucallol have been identified [47]. Tannins and flavonoids are potent antioxidants due to their polyphenol structures with multiple hydroxyl groups. They can form hydrogen bonds with reactive oxygen species like singlet oxygen, peroxy nitrite, and hydrogen peroxide, which cause cellular damage. This mechanism suggests their potential role in reducing oxidative stress associated with Type 2 diabetes [48]. Extensive research indicates that phytosterols and diets high in them can support glucose and lipid metabolism and improve insulin sensitivity [49].

The three targeted proteins in this study were previously examined as potential drug targets for Type 2 diabetes mellitus, α-Amylase (2QV4) [50] α-Glucosidase (6C9X) [51], and PTP1B (4I8N) [52].

Alpha (α)-amylase is a calcium-dependent enzyme that helps in digestion of polysaccharides by breaking down of it but it causes postprandial hyperglycemia. α-Amylase inhibitors like acarbose, miglitol, and voglibose have shown as a key target for effectively regulating blood sugar levels. Acarbose, in particular, helps in weight control, reduces cardiovascular risks, delays diabetes onset through beneficially regulating the blood pressure, lipids, platelet aggregation, and vascular health [53].

Alpha-glucosidase inhibitors (AGIs) slowing the absorption of sugars in the gut by temporarily blocking the various alpha-glucosidase enzymes, such as maltase. As AGIs specifically addressed postprandial hyperglycemia which is considered a potential independent risk factor for cardiovascular complications, so they may serve as an appropriate first-line treatment for Type 2 diabetes. Alpha-glucosidase inhibitors (AGIs) do not cause weight gain and hypoglycemia or other severe adverse effects therefor they also considered safe [54].

Protein tyrosine phosphatase 1B (PTP1B) and leukocyte antigen-related tyrosine phosphatase in Protein tyrosine phosphatases, make a key contribution in regulating insulin signal transduction.

For insulin signaling, Tyrosine phosphorylation is essential in the insulin receptor activation loop and PTP1B inhibit this step by dephosphorylating phosphotyrosine residues, which results in negative regulation of insulin signaling [55–58]. Recent studies further associate PTP1B to insulin sensitivity, obesity, and Type 2 diabetes mellitus (T2DM) [59–61]. In addition, PTP1B also work as a negative regulator of leptin signaling, which is essential for metabolic homeostasis. In case obesity, reduced leptin sensitivity results in impaired satiety despite ample energy stores. Evidence suggests that PTP1B influences pancreatic β-cell proliferation, that PTP1B knockout mice exhibit increased β-cell growth and enhanced glucose-induced insulin secretion. These evidences highlight the important role of PTP1B in diabetes, driving interest in the development of PTP1B inhibitors. Extensive reviews on reported PTP1B inhibitors are available in the literature [62, 63].

Computer-aided drug design (CADD) is an effective tool in the field of drug discovery. Structure-based drug design (SBDD) is one of the techniques which helps in identification of interactions important for binding of ligands with the amino acid residues [64–68]. These interactions may help in identifying underlying molecular mechanisms responsible for binding. SBDD mainly involves docking studies where the structure of protein and ligand is known. The docking results are assessed in terms of docking scores, poses of the ligands in the binding pocket as well as some of the novel methodologies which considers number of interactions between the ligands and amino acids present in the binding pocket [69, 70].

The aim of this study was to explore a diverse range of bioactive compounds (phytoconstituents) from all four families of *Euphorbia prostrata* viz. *flavonoids, tannins, phenolic acids, and sterols,* and analyze their interactions with target proteins which may help in elucidating the underlying molecular mechanisms essential for inhibition of the enzymes.

2 Materials and Methods

2.1 Dataset

25 bioactive phytoconstituents used for the docking studies were taken from the literature (Table 1). The molecules belonged to the category of flavonoids (Apigenin, Apigenin 7-Glucoside, Kaempferol, Luteolin, 6-Methoxy Quercetin Glycoside, Quercetin, Quercetin 3-Rhamnoside, Rhamnetin 3-Galactoside), sterols (Beta -Amyrin Acetate, Campesterol, Stigmasterol, Beta -Sitosterol), Phenolic Acids (Ellagic Acid, Gallic Acid), Tannins (Corilagin, Euphorbin G, Euphorbin H, Geraniin, Rugosin A, Rugosin D, Rugosin E, Tellimagradin I, Tellimagradin II), and Terpene Alcohols (1,2,3-Tri-O-Galloyl-D-Glucose, β -Terpineol).

2.2 Preparation of Ligands

The ligands or the molecules taken form the dataset was drawn using the ChemDraw software. The same 2D structures were imported into the Chem3D which converts the 2D structure to 3D. Additional preparation involved the minimization of the structures using the default settings. The structures were saved in the mol format which were then imported in the Autodock 4.2 software where torsions were identified and finally the structures were saved in the pdbqt format for further docking studies.

Table 1. Structure of phytoconstituents of *Euphorbia prostrata.*

S. NO.	COMPONDS	STRUCTURE	REFERENCE
1.	Apigenin	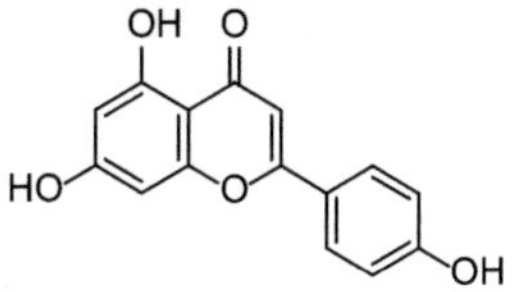	[71, 72]
2.	Apigenin 7-Glucoside	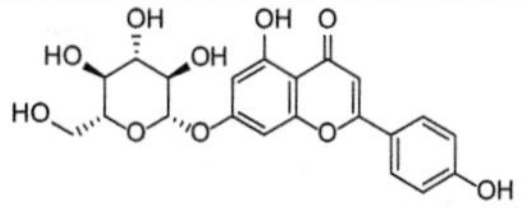	[47]
3.	Kaempferol	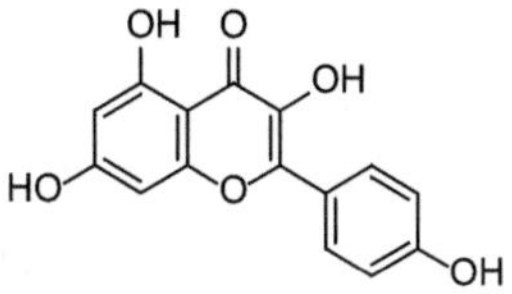	[47]
4.	Luteolin	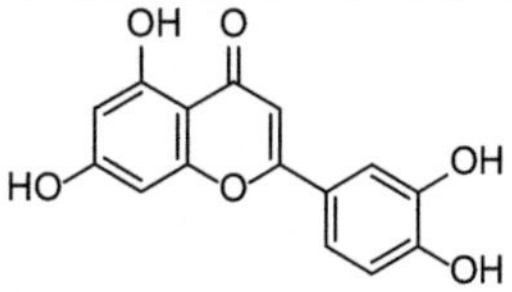	[71, 72]
5.	6-Methoxy Quercetin Glycoside	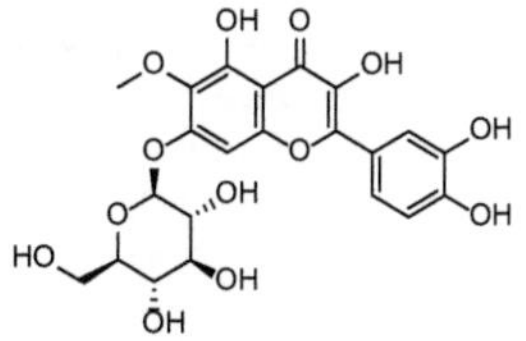	[47]
6.	Quercetin	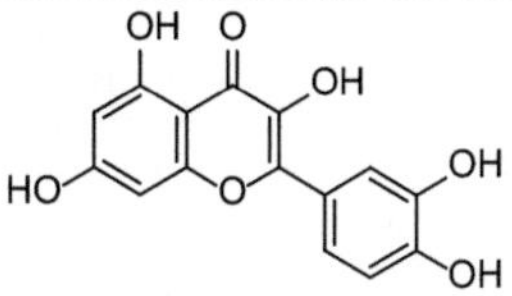	[71, 72]
7.	Quercetin 3-Rhamnoside	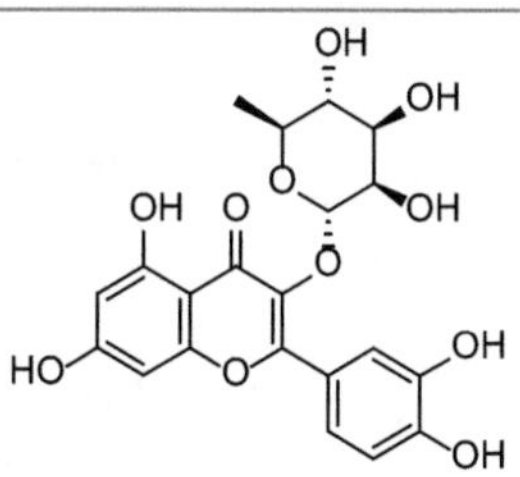	[47]

(continued)

Table 1. (continued)

S. NO.	COMPONDS	STRUCTURE	REFERENCE
8.	Rhamnetin 3-Galactoside	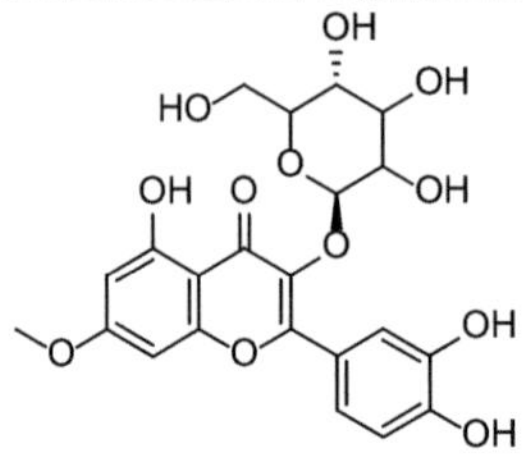	[47]
9.	Beta-AmyrinAcetate	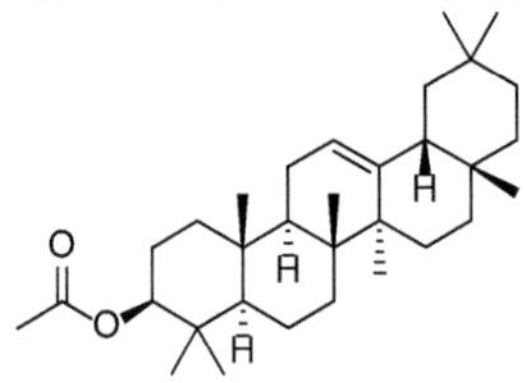	[47]
10.	Campesterol	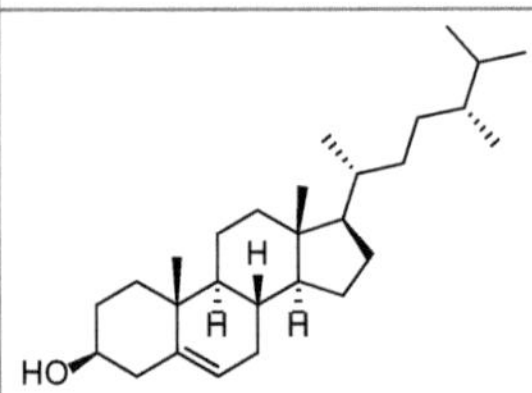	[71]
11.	Stigmasterol	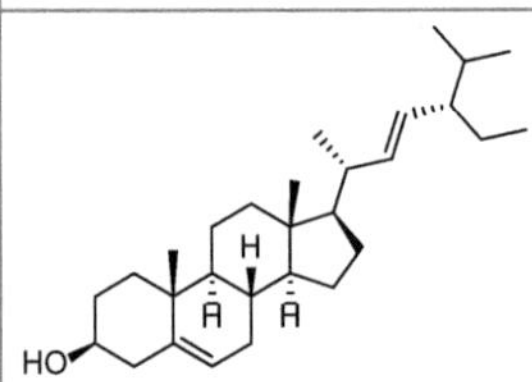	[71]
12.	Beta-Sitosterol	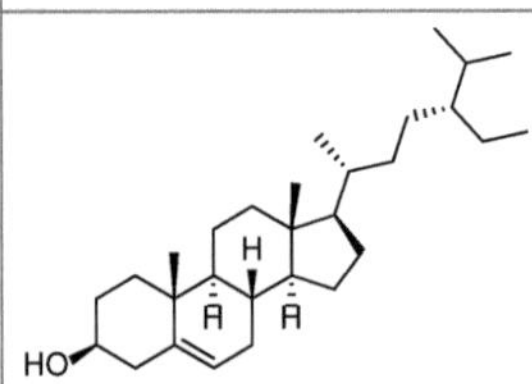	[71]
13.	Ellagic Acid	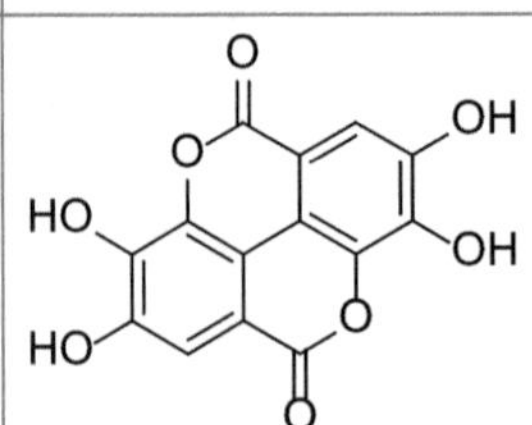	[71]

(continued)

Table 1. (continued)

S. NO.	COMPONDS	STRUCTURE	REFERENCE
14.	Gallic Acid	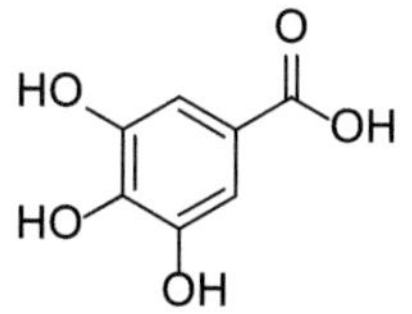	[71–73]
15.	Corilagin	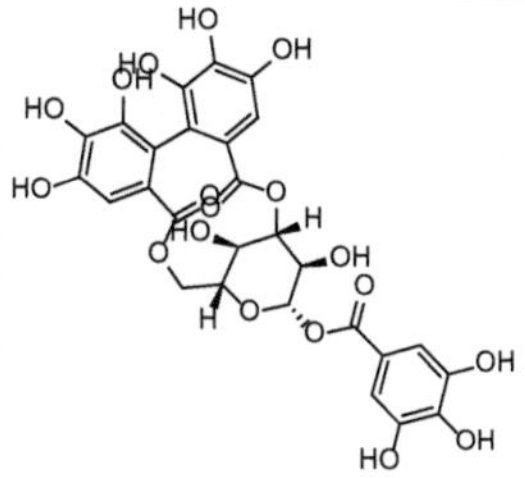	[71]
16.	Euphorbin G	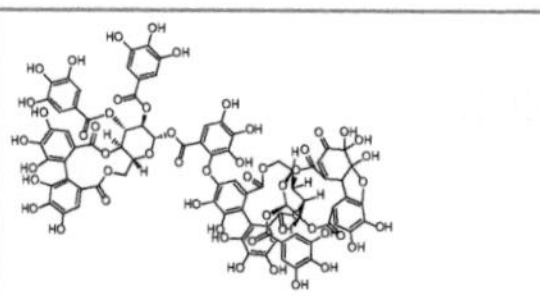	[47]
17.	Euphorbin H	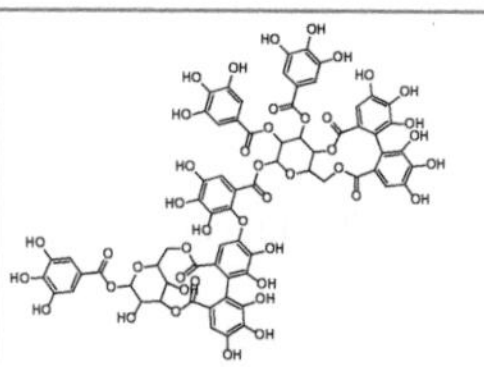	[47]
18.	Geraniin	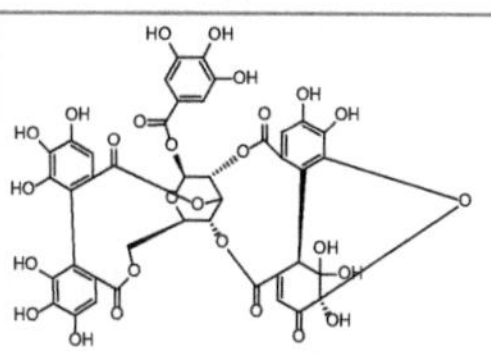	[71]
19.	Rugosin A	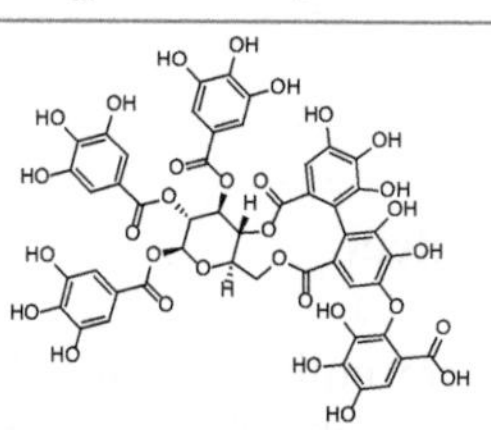	[71]
20.	Rugosin D	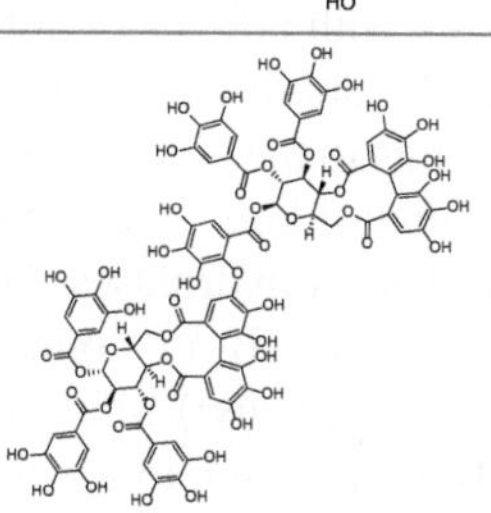	[71]

(continued)

Table 1. (continued)

S. NO.	COMPONDS	STRUCTURE	REFERENCE
21.	Rugosin E	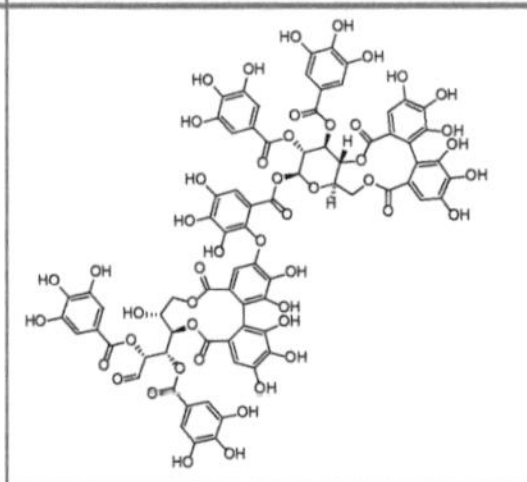	[71]
22.	Tellimagradin I	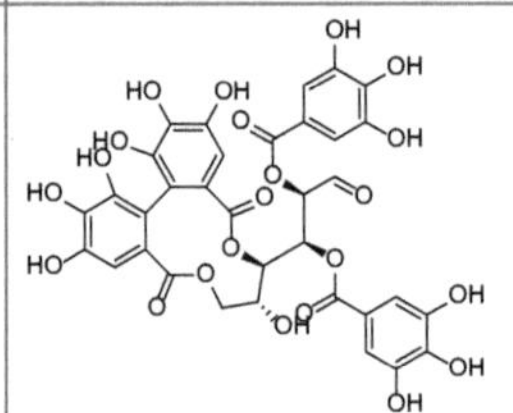	[71]
23.	Tellimagradin II	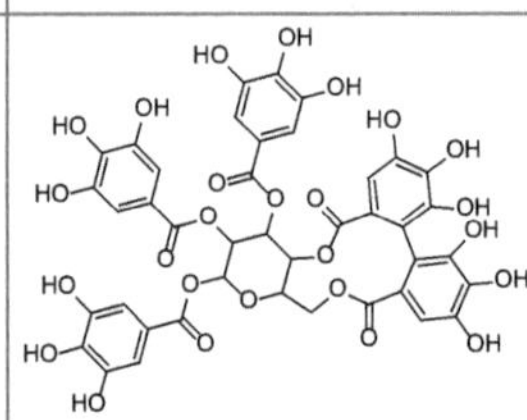	[71]
24.	1,2,3-Tri-O-Galloyl-D-Glucose	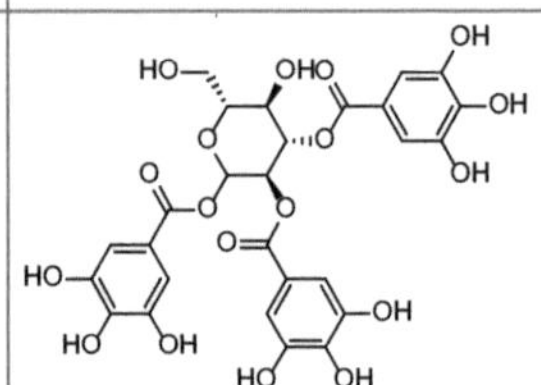	[71]
25.	Beta-Terpineol	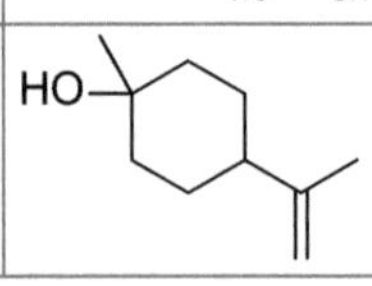	[47]

2.3 Preparation of Protein

Three target proteins were utilized for the docking studies which were downloaded from the protein data bank with pdb ids of 2QV4, 6C9X, and 4I8N. The 2QV4 protein is a human pancreatic alpha-amylase complexed with nitrite and acarbose with a resolution of 1.97 Å and belongs to the class of hydrolases. The 6C9X protein is the crystal structure of the alpha-Glucosidase in complex with voglibose with a resolution of 1.46 Å belonging to the class of hydrolases. The 4I8N protein is the crystal structure of Protein Tyrosine Phosphatase 1B in complex with an inhibitor [(4-{(2S)-2-(1,3-Benzoxazol-2-yl)-2-[(4-fluorophenyl)sulfamoyl]ethyl}phenyl)amino](oxo)acetic acid with a resolution of 2.50 Å belonging to the class of hydrolases. The downloaded target proteins were imported in to the Autodock 4.2 software for minimization. The polar hydrogens were added followed by removal of the non-essential water molecules as referenced from the co-crystallized ligands in each protein. Simultaneously, the missing residues were added followed by addition of the Kollmann charges. The prepared proteins were saved in the pdbqt format for further docking studies.

2.4 Preparation of the Grid Box

The grid box for docking is the specific area in the protein structure where the ligands bind. This is mostly identified by the co-crystallized ligand. The XYZ coordinates of the co-crystallized ligands of the downloaded targets were used to define the grid box. The coordinates for the proteins 2QV4, 6C9X, and 4I8N were (12.384, 48.136, 26.209); (8.190, −6.470, −10.450); and (−11.539, 39.85, 35.54) respectively.

2.5 Docking Studies

The docking of the ligands was carried out using Autodock Vina. This software involves the generation of a configuration files consisting primarily of the protein and ligands in the pdbqt formats along with the XYZ coordinated of the grid box.

3 Results and Discussion

The docking scores of all the 25 phytoconstituents have been described in Table 2. The docking score analysis of the phytoconstituents for each target protein reveal that the binding energies for the pdb id 2QV4 (α-Amylase) range from −5.7 to −10.7 kcal/mol. Similarly, the binding energies for 6C9X (α-Glucosidase) and 4I8N (PTP1B) range from −4.3 to −9.4 and −4.8 to −9.8 kcal/mol respectively. Interestingly, the binding energies for the Beta -Terpineol for each of the studied target protein was lowest among all the phytoconstituents −5.7 for α-Amylase, −4.3 for α-Glucosidase, and −4.8 for PTP1B. This observation indicates a very minor role of Beta-Terpineol in inhibition of either of the three targets. Among the phytoconstituents Euphorbin G showed the highest binding energy (−10.7 kcal/mol) for α-Amylase, Corilagin showed highest binding energy (−9.4 kcal/mol) for α-Glucosidase, and Euphorbin H showed highest binding energy (−9.8 kcal/mol) for PTP1B. It was worth to note that all the phytoconstituents having the highest binding energies belonged to tannins family.

Besides the docking energy or binding energy, the docking studies also consists of interactions of the ligands (phytoconstituents in this study) with the target protein. These interactions shed light on important amino acid residues which are essential for binding with the different atoms of the ligands. In addition, it also reveals the type of interactions required for the binding between the target and the ligand.

The detailed interactions analysis for the three target proteins viz. α-Amylase, α-Glucosidase, and PTP1B have been described in Table 2.

Table 2. Binding energy (kcal/mol) of bio-molecules in *Euphorbia prostrata* to 2QV4, 6C9X, and 4I8N.

Category	Phytoconstituents	Binding Energy (kcal/mol)		
		Alpha Amylase Inhibitors (2QV4)	Alpha Glucosidase Inhibitors (6C9X)	Protein Tyrosine Phosphatase (4I8N)
Flavanoids	Apigenin	−9.5	−7.8	−8.0
	Apigenin 7-Glucoside	−10.3	−9.3	−8.5
	Kaempferol	−9.0	−7.2	−7.7
	Luteolin	−9.6	−8.1	−8.0
	6-Methoxy Quercetin Glycoside	−8.8	−7.8	−7.8
	Quercetin	−9.0	−7.2	−7.7
	Quercetin 3-Rhamnoside	−8.9	−6.9	−7.7
	Rhamnetin 3-Galactoside	−8.7	−7.3	−6.6
Sterols	Beta-Amyrin Acetate	−8.0	−8.7	−7.6
	Campesterol	−9.4	−7.8	−6.3
	Stigmasterol	−9.6	−7.7	−8.6
	Beta-Sitosterol	−8.9	−8.7	−6.3
Phenolic Acids	Ellagic Acid	−9.1	−7.0	−7.2
	Gallic Acid	−6.5	−6.7	−5.8
Tannins	Corilagin	−8.9	−9.4	−8.6
	Euphorbin G	−10.7	−6.9	−7.8
	Euphorbin H	−10.0	−9.0	−9.8
	Geraniin	−8.3	−8.3	−8.2
	Rugosin A	−6.7	−6.4	−5.9
	Rugosin D	−9.7	−8.4	−8.2
	Rugosin E	−8.2	−7.7	−7.5
	Tellimagradin I	−9.2	−8.1	−8.2
	Tellimagradin II	−9.6	−8.8	−9.4
	1,2,3-Tri-O-Galloyl-D-Glucose	−8.9	−8.8	−9.4
Terpene Alcohol	Beta-Terpineol	−9.5	−8.5	−8.1
	Apigenin	−5.7	−4.3	−4.8

3.1 The binding interactions of each protein target has been classified in three categories.

3.1.1 Interaction Analysis of α-Amylase. The co-crystallized ligand of the protein 2QV4 belonging to α-Amylase shows binding with TRP59, THR163, GLY164, ASN105, ALA106, VAL107, GLN63, HIS101, GLU233, ARG195, ASP300, HIS299, and TYR62 residues along with some water molecules. Generally, for understanding activity of newer phytoconstituents it is imperative that the studied phytoconstituents show interaction with some of the important amino acid residues.

The TRP59, GLN63, and TYR62 amino acids were found in the Apigenin, Kaempferol, Luteolin, 6-Methoxy Quercetin Glycoside, Quercetin, and Tellimagradin I phytoconstituents. Interestingly, a closer analysis revealed that TRP59 interaction was present in 23 of the 25 phytoconstituents studied, viz. Apigenin 7-Glucoside, Quercetin 3-Rhamnoside, Rhamnetin 3-Galactoside, Beta-Amyrin Acetate, Campesterol, Stigmasterol, Beta-Sitosterol, Ellagic Acid, Corilagin, Euphorbin G, Euphorbin H, Geraniin, Rugosin A, Rugosin D, Rugosin E, Tellimagradin II, 1,2,3-Tri-O-Galloyl-D--Glucose, Beta-Terpineolin addition to the already mentioned phytoconstituents. The interaction with TYR62 residue was found in Quercetin 3-Rhamnoside, Rhamnetin 3-Galactoside, Beta-Amyrin Acetate, Campesterol, Stigmasterol, Beta-Sitosterol, Corilagin, Tellimagradin II, and Beta-Terpineol. In addition to the above important amino acid residues which are present in most of the phytoconstituents, there are other important residues which are present in lesser number of phytoconstituent for example GLU233, ASP300, and THR163 which were present in the Apigenin 7-Glucoside. In addition, GLU233 and THR163 were also found in Euphorbin G and Rugosin D along with GLU233 and ASP300 which was found in Tellimagradin I. To account for other phytoconstituents, interaction with GLU233 was also found in Euphorbin H, 1,2,3-Tri-O-Galloyl—D-Glucose, and Beta-Terpineol; ASP300 was also found in Luteolin, Quercetin, Ellagic Acid, and Gallic Acid; THR300 was also found in Rugosin E and Tellimagradin II.

3.1.2 Interaction Analysis of α-Glucosidase. The co-crystallized ligand of the protein 6C9Xbelonging to α-glucosidase shows binding with ASP420, ARG404, ASP73, TRP169, ASP197, HIS78 and TYR62 residues along with some water molecules. Among the important residues ASP420 was found in Apigenin, Luteolin, Quercetin 3-Rhamnoside, Ellagic Acid, Corilagin, Geraniin, and Rugosin A, TRP169 was found in Apigenin 7-Glucoside, Quercetin 3-Rhamnoside, Beta-Sitosterol, Gallic Acid, Geraniin, Rugosin A, and Tellimagradin II, and ARG404 was found in Corilagin, Euphorbin H, Euphorbin G, Rugosin D, Rugosin E, and Gallic Acid. In addition to these important amino acid residues, our studies revealed that LYS422, PRO75, and VAL351 were also important to show binding affinity and were present ion 10, 9, and 9 phytoconstituents respectively. Beta-Amyrin Acetate and Campesterol were the two phytoconstituents having all the three residues viz. LYS422, PRO75, and VAL351. The residues LYS422 and PRO75 were common Beta-Amyrin Acetate, Campesterol, Ellagic Acid, and Beta-Terpineol. The residues PRO75 and VAL351 were commonly found in Rhamnetin3-Galctoside, Beta-Amyrin Acetate, and Beta-Sitosterol. In the same way, both LYS422 and VAL351 were found in Kaempferol, Beta-Amyrin Acetate, Campesterol, and Rugosin E. In addition, the other prominent interacting residues found in docking studies were HIS478, TRP271, ALA480, and ASP307 present in 6,7, 8, and 8 phytoconstituents respectively.

3.1.3 Interaction Analysis of PTP-1B. The co-crystal of Protein Tyrosine Phosphatase (PTP-1B) with pdbid:4I8Nshowed interaction with ARG221, SER216, ALA217, PHE182, TYR46, GLY220, MET258, GLN262 along with some important water molecules. The residues ALA217 and TYR46 were found commonly in Luteolin, Quercetin 3-Rhamnoside,

Quercetin, Stigmasterol, Beta-Sitosterol, Tellimagradin I, and Beta -Terpineol. It was interesting to note that individually ALA217 and TYR46 amino acid interactions were found in 19 and 12 phytoconstituents respectively. In addition, the amino acids found in the co-crystallized ligand such as SER216 and PHE182 were also present in our docking studies in the phytoconstituents namely, 6-Methoxy Quercetin Glycoside, Campesterol, Stigmasterol, Beta-Sitosterol, Ellagic Acid, and Euphorbin G showed interactions with PHE182, while Apigenin, Apigenin 7-Glucoside, 6-Methoxy Quercetin Glycoside, 1,2,3-Tri—O-Galloyl-D-Glucose, and Beta-Amyrin Acetate showed interactions with SER216 residue. In addition, 11 interactions for ASP181 and 8 interactions for ASP48 were found prominently in our docking studies.

To summarize the results of the docking in all the three protein targets, it was established that the phytoconstituents belonging to the tannins family showed highest binding energies. The detailed analysis of the interactions found in the phytoconstituents for each protein target has been discussed above. The most important amino acid residue responsible for binding in theα-Amylase enzyme is TRP59 which was found in 23 of the 25 studied phytoconstituents. Among these all the phytoconstituents belonging to the tannins family such as Rugosin D, Rugosin E, Rugosin A, Tellimagradin I, Tellimagradin II, Corilagin, Geraniin, Euphorbin H, Euphorbin G showed interaction with TRP159. Another important residue GLN63 was also found in all the phytoconstituents of the tannins except, Corilagin, Rugosin D. Some of the other important interacting residues such as GLU233 (found in Euphorbin G, Euphorbin H, Rugosin D, and Tellimagradin I), ASP300 (found in Tellimagradin I), and THR163 (found in Euphorbin G, Rugosin D, Rugosin E, and Tellimagradin II) were commonly present in phytoconstituents of tannins family. The phytoconstituents belonging to flavonoid family also showed good binding energies ranging from −8.7 to −10.3. Additionally, all the phytoconstituents belonging to the flavonoid family such as Apigenin, Apigenin 7-Glucoside, Kaempferol, Luteolin, 6-Methoxy Quercetin Glycoside, Quercetin, Quercetin 3-Rhamnoside, Rhamnetin 3-Galactoside showed interaction with the most important residue TRP59. Similarly, the phytoconstituents belonging to tannins family showed interactions with important residues such as ARG404 (Corilagin, Euphorbin G, Euphorbin H, Rugosin D, Rugosin E), and TRP169 (Gernaiin, Rugosin E, and Tellimagradin II), and ASP420 (Corilagin, Gernaiin, and Rugosin A). The phytoconstituents belonging to the flavonoid family did not show prominent interaction with important amino acid residues. The docking studies on PTP1B showed the presence of interacting residue ALA217 in all the phytoconstituents belonging to the tannins family. The same residue was also found in all the phytoconstituents belonging to flavonoids family except Rhamnetin 3-Galactoside.

Based on the above studies, hydrogen bonding and hydrophobic bonds were found as the most important bonding interactions and Apigenin 7-Glucoside, Luteolin among the flavonoids, and Rugosin E, and Tellimagradin I among the tannins were found to be the most important phytoconstituents. The best poses of the most important amino acids is given in Table 3.

Table 3. The best poses of the important phytoconstituents.

S. NO.	Phytoconstituents	3D poses of the docking interactions Alpha Amylase Inhibitors(2QV4)	 Alpha Glucosidase Inhibitors(6C9X)	 Protein Tyrosine Phosphatase(4I8N)
1.	Apigenin 7-Glucoside	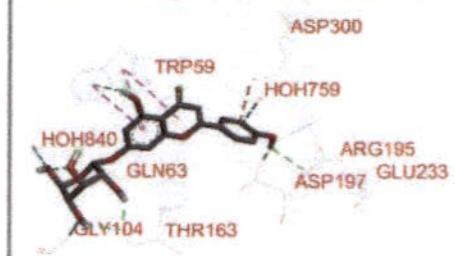	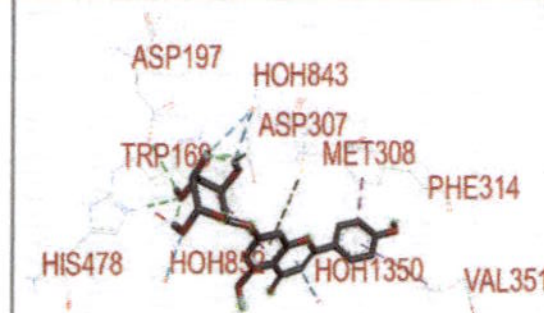	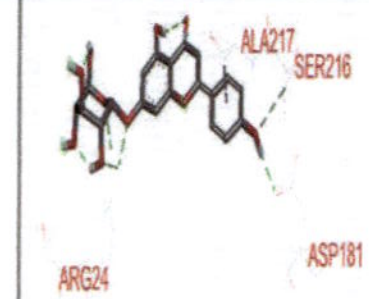
2.	Luteolin	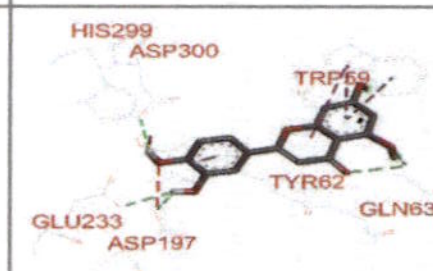	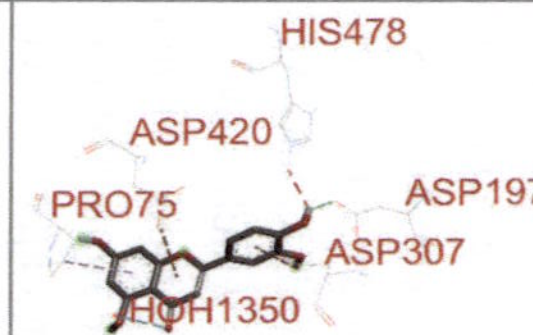	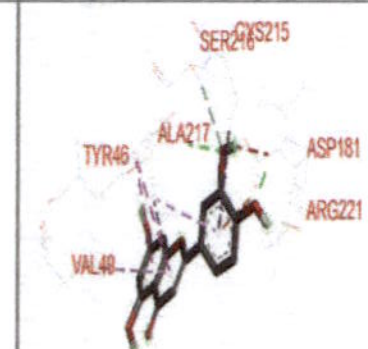
3.	Rugosin A	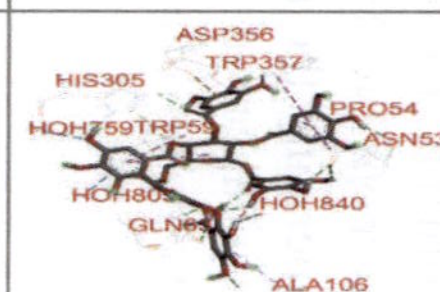	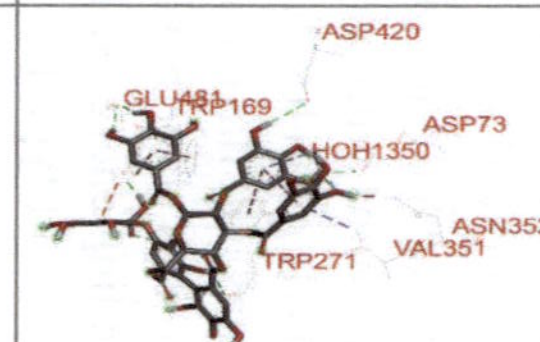	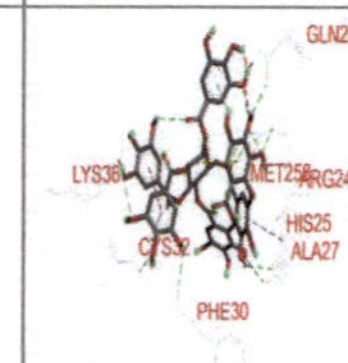
4.	Tellimagradin I	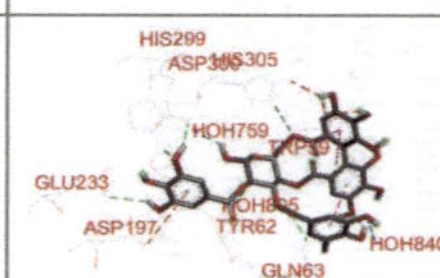	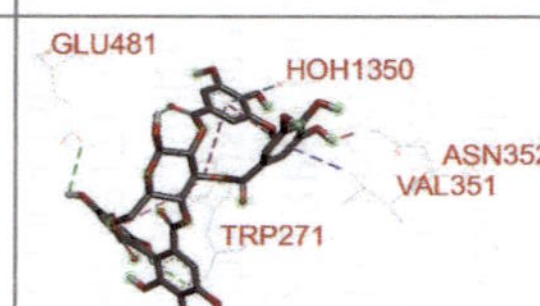	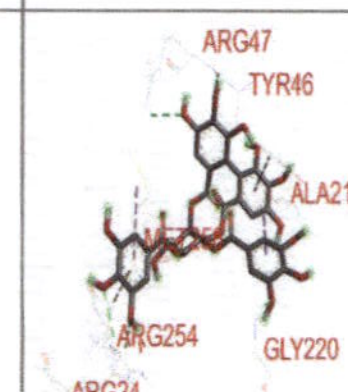

4 Conclusions

Diabetes mellitus is a major issue resulting from insufficient insulin production or the body's ineffective use of insulin, and insulin resistance from lifestyle factors. The drugs available in the market work through different mechanisms but are known to cause serious side effects. In wake of this, it is important to move towards safer alternatives such as herbal drugs. *Euphorbia prostrata* has shown activity towards hyperglycaemia but exact mechanism of its action is still not clear. In this study, we performed the docking studies to elucidate the molecular mechanisms responsible for its action. The structure of the phytoconstituents of the plant were used for docking studies against three important protein targets viz. 2QV4, 6C9X, and 4I8N. The studies showed that phytoconstituents belonging to both tannins (−8.3 to −10.7 kcal/mol) and flavonoids (−8.8 to −10.3 kcal/mol) family showed some of the highest binding energies. The docking interaction studies also showed that phytoconstituents of these two families showed highest number of interactions with important amino acid residues of each protein target, viz. TRP59 for 2QV4, ARG404 for 6C9X, and ALA217 for 4I8N. These studies reveal that it is the tannins and flavonoids which may show highest activity as antidiabetic agents evident through their binding modes (interaction with TRP59, ARG404, AND ALA217), and binding energies. The studies also revealed that hydrogen bonding was the most important bond type for binding followed by hydrophobic bonding. On the basis of number of interactions, as well as interactions with important amino acid residues along with the binding energies, Apigenin 7-Glucoside, Luteolin among the flavonoids, and Rugosin E, and Tellimagradin I among the tannins were found to be the most important phytoconstituents. These studies have helped in the identification of the molecular mechanisms responsible for action of phytoconstituents of the plant *Euphorbia Prostrata* enumerating the important amino acids, interactions as well as specific families of the chemical constituents which may further be tested using in-vitro and in-vivo methods to develop novel antidiabetic agents with minimum side effects.

References

1. Batori, Z., Erdos, L., Somlyay, L.: *Euphorbia prostrata* (Euphorbiaceae), a new alien in the Carpathian Basin. Acta Bot. Hung. **54**, 235–243 (2012)
2. Varsha, A., Tarate, S., Kshirsagar, M., Jagtap, M.: Pharmacognostical and phytochemical investigation of *Euphorbia prostrata*. Int. J. Pharm. Sci. Res. **8**, 980–983 (2023)
3. Sharma, G.D., Tripathi, S.N.: Experimental evaluation of dugdhika (*Euphorbia prostrata w. ait)* for the treatment of 'tamakasvasa' (bronchial asthma). Anc. Sci. Life. **3**, 143 (1984)
4. Gupta, P.: The efficacy of *Euphorbia prostrata* in early grades of symptomatic hemorrhoids—a pilot study. Eur. Rev. Med. Pharmacol. Sci. **15**, 199–203 (2011)
5. Alarcon, A.F.J., Roman, R.R., Perez, G.S., Aguilar, C.A., Contreras, W.C.C., Flores, S.J.L.: Study of the anti-hyperglycemic effect of plants used as antidiabetics. J. Ethnopharmacol. **61**, 101–110 (1998)

6. Larbie, C., Torkornoo, D., Dadson, J.: Anti-diabetic and hypolipidaemic effect of botanicals: a review of medicinal weeds on KNUST campus, Kumasi. J. Appl. Pharm. Sci. **4**, 097–104 (2014)
7. Patil, T.R., Limaye, R.P.: Effect of *Euphorbia prostrata* on the wound healing in excisional wound model in rats. Int. J. Pharmacogn. Phytochem. Res. **9**, 1223–1236 (2017)
8. Hariyadi, D.M., Sahu, V.K.: *Euphorbia prostata* exerts potent anti-inflammatory and anti-arthritic activity in downregulating the increased expression of pro-inflammatory cytokines. Pharm. Sci. **26**, 370–378 (2020)
9. Biwott, T., Kiprop, A., Cherutoi, J., Munyendo, W., Biwott, G.: Analgesic properties of *Euphorbia prostrata* crude extracts. Sci. J. Chem. **3**, 100 (2015)
10. Sharma, S.K., Singh, S., Singh, J.: Anthelmintic effect of *Euphorbia prostrata Ait.* extracts. Indian J. Pharmacol. **43**, 743 (2011)
11. Garg, N., Mannan, A., Mohan, M., Singh, T.G.: Therapeutic efficacy of hydroalcoholic extract of *Euphorbia prostrata Aiton* in NAD-STZ-induced diabetic nephropathy: a multifaceted intervention targeting oxidative stress and inflammation. Obes. Med. **54**, 100579 (2025)
12. Sharma, S.K., Singh, J., Singh, S.: Pharmacognostical and phytochemical investigation of *Euphorbia prostrata ait.* IJPSR. **3**(4), 1043–1048 (2012)
13. Stumvoll, M., Goldstein, B.J., Van Haeften, T.W.: Type 2 diabetes: principles of pathogenesis and therapy. Lancet. **365**, 1333–1346 (2005)
14. Samar, A.A., et al.: Diabetes mellitus: classification, mediators, and complications; a gate to identify potential targets for the development of new effective treatments. Biomed. Pharmacother. **168**, 115734 (2023)
15. Tomic, D., Shaw, J.E., Magliano, D.J.: The burden and risks of emerging complications of diabetes mellitus. Nat. Rev. Endocrinol. **18**, 525–539 (2022)
16. Unai, G.G., et al.: Pathophysiology of type 2 diabetes mellitus. Int. J. Mol. Sci. **21**, 6275 (2020)
17. Ong, K.L.: Global, regional, and national burden of diabetes from 1990 to 2021, with projections of prevalence to 2050: a systematic analysis for the Global Burden of Disease Study 2021. Lancet. **402**, 203–234 (2023)
18. Varghese, J.A., et al.: National Estimates of the adult diabetes care continuum in India, 2019–2021. JAMA Intern. Med. **183**, 963–972 (2023)
19. Wei, J., et al.: The global, regional, and national burden of type 2 diabetes mellitus attributable to low physical activity from 1990 to 2021: a systematic analysis of the global burden of disease study 2021. Int. J. Behav. Nutr. Phys. Act. **22**, 8 (2025)
20. American Diabetes Association: Standards of care in diabetes-2023 abridged for primary care providers. Clin. Diabetes. **41**, 4–31 (2022)
21. Yu, J., Lee, S.-H., Kim, M.K.: Recent updates to clinical practice guidelines for diabetes mellitus. Endocrinol. Metab. (Seoul). **37**, 26–37 (2022)
22. Singh, H., et al.: Role of oxidative stress in diabetes-induced complications and their management with antioxidants. Arch. Physiol. Biochem. **130**, 616–641 (2024)
23. Bhatti, J.S., et al.: Oxidative stress in the pathophysiology of type 2 diabetes and related complications: current therapeutics strategies and future perspectives. Free Radic. Biol. Med. **184**, 114–134 (2022)
24. Reynolds, L., Luo, Z., Singh, K.: Diabetic complications and prospective immunotherapy. Front. Immunol. **14**, 1219598 (2023)
25. Nguyen Vo, T.H., Tran, N., Nguyen, D., Le, L.: An in silico study on antidiabetic activity of bioactive compounds in *Euphorbia thymifolia Linn.* Springerplus. **5**, 1–13 (2016)
26. Huang, D.D., Shi, G., Jiang, Y., Yao, C., Zhu, C.: A review on the potential of Resveratrol in prevention and therapy of diabetes and diabetic complications. Biomed. Pharmacother. **125**, 109767 (2020)

27. Padhi, S., Nayak, A.K., Behera, A.: Type II diabetes mellitus: a review on recent drug based therapeutics. Biomed. Pharmacother. **131**, 110708 (2020)
28. Corathers, S.D., Peavie, S., Salehi, M.: Complications of diabetes therapy. Endocrinol. Metab. Clin. N. Am. **42**, 947 (2013)
29. Barrientos-Avalos, J.R., et al.: Gastrointestinal adverse effects of old and new antidiabetics: how do we deal with them in real life? Rev. Gastroenterol. Méx. (Engl. Ed.). **89**, 521–532 (2024)
30. Feingold, K.R.: Oral and injectable (non-insulin) pharmacological agents for the treatment of type 2 diabetes. In: Endotext. MDText.com, Inc., South Dartmouth (2024)
31. Chong, K., Chang, J.K.J., Chuang, L.M.: Recent advances in the treatment of type 2 diabetes mellitus using new drug therapies. Kaohsiung J. Med. Sci. **40**, 212–220 (2024)
32. Elsayed, N., Aleppo, G., Bannuru, R., Bruemmer, D.: Pharmacologic approaches to glycemic treatment: standards of care in diabetes—2024. Diabetes Care. **47**, S158–S178 (2024)
33. Jain, A.B., Lai, V., Jain, A.B., Lai, V.: Medication-induced hyperglycemia and diabetes mellitus: a review of current literature and practical management strategies. Diabetes Ther. **2024**(15), 2001–2025 (2024)
34. Weinberg Sibony, R., Segev, O., Dor, S., Raz, I.: Drug therapies for diabetes. Int. J. Mol. Sci. **24**, 17147 (2023)
35. Fatima, M.T., Bhat, A.A., Nisar, S., Fakhro, K.A., Al-Shabeeb Akil, A.S.: The role of dietary antioxidants in type 2 diabetes and neurodegenerative disorders: an assessment of the benefit profile. Heliyon. **9**, e12698 (2022)
36. Ghasemi-Dehnoo, M., Amini-Khoei, H., Lorigooini, Z., Rafieian-Kopaei, M.: Oxidative stress and antioxidants in diabetes mellitus. Asian Pac J Trop Med. **13**, 431–438 (2020)
37. Rahimi, R., Nikfar, S., Larijani, B., Abdollahi, M.: A review on the role of antioxidants in the management of diabetes and its complications. Biomed. Pharmacother. **59**, 365–373 (2005)
38. Shafras, M., Sabaragamuwa, R., Suwair, M.: Role of dietary antioxidants in diabetes: an overview. Food Chem. Adv. **4**, 100666 (2024)
39. Tuell, D.S., Los, E.A., Ford, G.A., Stone, W.L.: The role of natural antioxidant products that optimize redox status in the prevention and management of type 2 diabetes. Antioxidant. **12**, 1139 (2023)
40. Montonen, J., Knekt, P., Jarvinen, R., Reunanen, A.: Dietary antioxidant intake and risk of type 2 diabetes. Diabetes Care. **27**, 362–366 (2004)
41. Suresh, V., Reddy, A., Muthukumar, P., Selvam, T.: Antioxidants: pharmacothearapeutic boon for diabetes. In: Antioxidants – Benefits, Sources, Mechanisms of Action. IntechOpen (2021). https://doi.org/10.5772/INTECHOPEN.98587
42. Nawaz, H., et al.: Comparative evaluation of phytochemical composition and antioxidant potential of some medicinally-important euphorbiaceous plants. J. Anim. Plant Sci. **32**, 1703–1712 (2022)
43. Estrada, J.M., Reyes, H.M., Hernández, T.R., Sansininea, A.J.J., Alvarez, A.B.: Contributions from Mexican flora for the treatment of diabetes mellitus: molecules of *Psacalium decompositum* (A. gray) H. rob & brettell. Molecules. **26**, 2892 (2021)
44. Yadav, N.K., Yadav, R.: Medicinal effects, phytochemistry, pharmacology of *Euphorbia prostrata* and promising molecular mechanisms. Chin. J. Integr. Med. **30**, 181–192 (2024)
45. Chen, L., Chen, R., Wei, K.: Constituents of tannins from *Euphorbia prostrata Ait.* Zhongguo Zhong Yao Za Zhi. **17**, 225–255 (1992)
46. Aslam, M.A., et al.: Phytochemical screening, antioxidant, immunomodulatory and analgesic potential of aqueous-ethanolic extracts of *Euphorbia prostrata ait.* and *Crotalaria burhia* Buch–Ham. *Heliyon.* **10**, e27279 (2024)
47. Meghana, B.P., Balasubramanian, T., Suresha, B.S., Rakshitha, S., Soujanya, N.: *Euphorbia prostrata*: a plant review. Asian J. Pharm. Health Sci. **12**, 2713–2717 (2022)

48. Das, K., et al.: In silico studies and evaluation of in vitro antidiabetic activity of berberine from ethanol seed extract of *Coscinium fenestratum (Gaertn.)* Colebr. J. King Saud Univ. Sci. **35**, 102666 (2023)
49. Ahmad, M., et al.: Antioxidant and antibacterial activity of crude methanolic extract of *Euphorbia prostrata* collected from District Bannu (Pakistan). Afr. J. Pharm. Pharmacol. **5**, 1175–1178 (2011)
50. Maurus, R., et al.: Alternative catalytic anions differentially modulate human alpha-amylase activity and specificity. Biochemistry. **47**, 3332–3344 (2008)
51. Tan, K., Tesar, C., Jedrzejczak, R., Joachimiak, A.: The Crystal Structure of the Alpha-Glucosidase (GH 31) from Ruminococcus Obeum ATCC 29174 in Complex with Voglibose. Midwest Center for Structural Genomics (MCSG) (2018). Preprint at https://doi.org/10.2210/pdb6c9x/pdb
52. Reddy, S.M.V.V.V., Rao, K.N., Subramanya, H.: Crystal Structure of Protein Tyrosine Phosphatase 1b in Complex with an Inhibitor [(4-{(2s)-2-(1,3-fluorophenyl) sulfamoyl] ethyl} phenyl) amino] (oxo) Acetic Acid (2012). https://doi.org/10.2210/PDB4I8N/PDB
53. McCarty, M.F., Di Nicolantonio, J.J.: Acarbose, lente carbohydrate, and prebiotics promote metabolic health and longevity by stimulating intestinal production of GLP-1. Open Heart. **2**, e000205 (2015)
54. Van de Laar, F.A.: Alpha-glucosidase inhibitors in the early treatment of type 2 diabetes. Vasc. Health Risk Manag. **4**, 1189–1195 (2008)
55. Eleftheriou, P., Geronikaki, A., Petrou, A.: PTP1b inhibition, a promising approach for the treatment of diabetes type II. Curr. Top. Med. Chem. **19**, 246–263 (2019)
56. Behl, T., et al.: Exploring protein tyrosine phosphatases (PTP) and PTP-1B inhibitors in management of diabetes mellitus. Biomed. Pharmacother. **153**, 113405 (2022)
57. Teimouri, M., et al.: The role of protein tyrosine phosphatase 1B (PTP1B) in the pathogenesis of type 2 diabetes mellitus and its complications. J. Physiol. Biochem. **78**, 307–322 (2022)
58. Liu, Z., Gao, H., Zhao, Z., Huang, M., Wang, S., Zhan, J.: Status of research on natural protein tyrosine phosphatase 1B inhibitors as potential antidiabetic agents: update. Biomed. Pharmacother. **157**, 113990 (2023)
59. Sun, J., Qu, C., Wang, Y., Huang, H., Zhang, M., Zou, W.: Type 2 diabetes mellitus and protein-tyrosine phosphatase 1b. J. Diabetes Metab. Disord. Control. **3**, 180–183 (2016)
60. Hsing, H.Y., Rathnasamy, S., Dianita, R., Wahab, H.A.: Docking-based virtual screening in search for natural PTP1B inhibitors in treating type-2 diabetes mellitus and obesity. Biomed. Res. Ther. **7**, 3579–3592 (2020)
61. Rocha, S., Corvo, M.L., Fernandes, E., Freitas, M.: The emerging target protein tyrosine phosphatase 1B (PTP1B) for type 2 diabetes mellitus management. J. Diabetes Clin. Res. **3**, 99–105 (2021)
62. Kanwal, A., Kanwar, N., Bharati, S., Srivastava, P., Singh, S.P., Amar, S.: Exploring new drug targets for type 2 diabetes: success, challenges and opportunities. Biomedicine. **10**, 331 (2022)
63. Maurus, R., et al.: Alternative catalytic anions differentially modulate human α-amylase activity and specificity. Biochemistry. **47**, 3332–3344 (2008)
64. Ahmed, S., Prabahar, A.E., Saxena, A.K.: Molecular docking-based interactions in QSAR studies on mycobacterium tuberculosis ATP synthase inhibitors. SAR QSAR Environ. Res. **33**, 289–305 (2022)
65. Ahmed, S., Prabahar, A.E., Saxena, A.K.: Molecular docking-based interaction studies on imidazo [1,2-a] pyridine ethers and squaramides as anti-tubercular agents. SAR QSAR Environ. Res. **34**, 435–457 (2023)

66. Kumar, M., Dewangan, H.K., Arya, G.C., Sharmar, R.: Design, development and evaluation of QSAR and molecular modelling of benzothiazole analogues for antibacterial drug discovery. Results Chem. **4**, 100482 (2022)
67. Wang, K., Huang, Y., Wang, Y., You, Q., Wang, L.: Recent advances from computer-aided drug design to artificial intelligence drug design. RSC Med. Chem. **15**, 3978–4000 (2024)
68. Nandi, S., Kumar, M., Saxena, M., Saxena, A.K.: The antiviral and antimalarial drug repurposing in quest of chemotherapeutics to combat COVID-19 utilizing structure-based molecular docking. Comb. Chem. High Throughput Screen. **24**, 1055–1068 (2020)
69. Nandi, S., Kumar, M., Saxena, A.K.: Repurposing of drugs and HTS to combat SARS-CoV-2 Main protease utilizing structure-based molecular docking. Lett. Drug. Des. Discov. **19**, 413–427 (2021)
70. Nandi, S., Kumar, M., Kumari, R., Saxena, A.: Exploring the inhibitory mechanisms of indazole compounds against SAH/MTAN-mediated quorum sensing utilizing QSAR and docking. Drug Target Insights. **16**, 54–68 (2022)
71. Singla, A.K., Pathak, K.: Topical antiinflammatory effects of *Euphorbia prostrata* on carrageenan-induced footpad oedema in mice. J. Ethnopharmacol. **29**, 291–294 (1990)
72. Shahwan, M.J., Al-Qirim, T.M., Daradka, H.: Hypolipidaemic effects of *Euphorbia prostrata* in rabbits. J. Biol. Sci. **9**, 88–91 (2008)
73. Rauf, A., Qaisar, M., Uddin, G., Akhtar, S., Muhammad, N.: Preliminary phytochemical screening and antioxidant profile of *Euphorbia prostrate*. Middle East J. Med. Plants Res. **1**, 09–13 (2012)

Antimicrobial Activity of Aqueous Stem Extract of *Carica papaya* and *Radish leaves*

Vivek Kumar[1], Shahjad[2(✉)], and Arun Kumar[1]

[1] Global Institute of Pharmaceutical Education and Research, Kashipur, India
[2] Department of Pharmacology, Global Institute of Pharmaceutical Education and Research, Kashipur, India
shahzadansari399@gmail.com

Abstract. Background: Traditionally, *Carica papaya L. and Raphanus sativus L.* have been used to cure a variety of illnesses. Their aqueous extracts' antibacterial potential was to be assessed in this investigation. **Methods:** Materials from stems and leaves were macerated, pulverized, and shade-dried. Antimicrobial testing against *E. coli* using the Zone Inhibition Method (Kirby-Bauer method) with ciprofloxacin (10 μg) as positive control, FTIR analysis, and phytochemical screening were carried out. **Results:** phytochemical study identified proteins, carbohydrates, alkaloids, and flavonoids. There were 13 functional groups found by FTIR. At a dosage of 6.25 μg/mL, the combined extract exhibited substantial antibacterial activity with an 8 mm zone of inhibition. **Conclusion:** The extracts show encouraging antibacterial properties, indicating that they may be used as natural antimicrobial agents in medicinal and food preservation applications.

Keywords: Microbial activity · *Carica papaya L* · MIC · Wound healing · Antibiotics · *Raphanus sativus L*

1 Introduction

The most economically significant fruit in the *Caricaceae* family is the pawpaw (*Carica papaya L.*). It is a tree or shrub that grows upright and typically has no branches. Despite being indigenous to Central America, it has spread to many tropical regions. In many regions of the world, people frequently eat the ripe fruit of the pawpaw plant [1–3]. In traditional medicine, *Carica papaya* has been used to cure fever, wounds, and digestive issues. Historically, radish leaves (*Raphanus sativus L.*) have been used as anti-inflammatory and to treat respiratory conditions [4]. On the other hand, the unripe fruit is employed as an abortifacient, galactagogue, diuretic, and mild laxative. Many parts of the plant are used to treat various illnesses; for instance, the roots and seeds are used as abortifacients, and the seeds are used to drive out worms. The leaves, particularly those that have fallen, are used in a variety of ways to cure inflammation, syphilis, gonorrhea, fever, pyrexia, diabetes, and nasty wounds. The abortifacient activity of the seeds, their effects on the germinal epithelium of the seminiferous tubules, the fruit

A. K. Saxena, A. Saxena (Eds.): GTHTM 2025, pp. 447–455, 2026.
https://doi.org/10.1007/978-3-032-12320-6_26

juice's ability to reduce blood pressure, and the leaves' ability to heal wounds are just a few of the scientifically proven uses of *Carica papaya* [4–7]. Many further studies along with Hepatitis, heart disease, ulcers, constipation, stomach disorders, and urinary tract infections are just a few of the conditions that radish extracts have been used to treat historically. *Radish's* antibacterial, anticancer, antioxidant, and anxiety-reducing qualities have also been emphasized in a number of research studies. *Sativus L.* extracts are employed in the cosmetics sector as a skin conditioner, antioxidant, and preservative because of their well-known health benefits. They are specifically recognized as raw ingredients for cosmetics that have been fermented using bacterial strains belonging to the genera Lactobacillus and Leuconostoc [8–10].

2 Material and Methods

2.1 Processing of Plant Sample

Fresh stems and leaves were gathered from Girital Road in Kashipur and thoroughly cleaned with tap water. In order to protect heat-sensitive phytochemicals and guarantee total moisture removal, the stem and leaves were shade-dried at 40 °C. Studies demonstrate that temperatures above 50 °C can degrade flavonoids and alkaloids [9]. The dried roots and leaves were pulverized, using a sterile laboratory mortar and pestle, to obtain a powdered form. Plant materials were first cut into small pieces (2–3 cm) using sterile scissors, then ground using a mortar and pestle. This method is effective for dried plant materials as reported [9]. These were stored in airtight glass containers protected from sunlight until required for analysis.

2.2 Preparation of Extract

Aqueous extraction was performed using the maceration method. The combined extract was prepared in equal proportions (1:1 ratio). Plant powder (50 g) was soaked in distilled water (500 mL) for 24 h with occasional stirring. The mixture was filtered through Whatman No. 1 filter paper and concentrated using a rotary evaporator at 50 °C. Extract yield was calculated and stored at 4 °C until use.

The stem and leaf powder were extracted by the maceration process using the aqueous medium because:

(1) Aqueous extraction at room temperature preserves thermolabile compounds.
(2) Maceration is suitable for water-soluble bioactive compounds
(3) It mimics traditional preparation methods used in ethnomedicine [8, 11].

The stem extract and leaf extract were filtered.

2.3 FTIR

FTIR (Fourier Transform Infrared Spectroscopy) is a method used to obtain the infrared spectrum of absorption or emission of a solid, liquid, or gas. It helps determine the functional groups present in a compound [12].

2.3.1 Role of FTIR. FTIR (Fourier Transform Infrared) spectroscopy was employed in this study to identify and characterize the functional groups present in the aqueous extracts of *Carica papaya* stem and radish leaves. The role of FTIR analysis includes:

- Structural Elucidation: FTIR helped identify specific functional groups (N-H, C=O, C-H, etc.) that may be responsible for antimicrobial activity
- Quality Control: Confirming the presence of bioactive compounds in the extracts
- Fingerprinting: Creating a spectral fingerprint for future standardization of the extracts
- Mechanistic Understanding: Correlating functional groups with observed antimicrobial properties

The FTIR analysis revealed 13 distinct functional groups in the wavelength range of 400–4000 cm^{-1}, including primary amines, isothiocyanates, alkenes, and phenolic compounds, which are known to possess antimicrobial properties [13].

2.4 Phytochemical Screening

The extract was subjected to phytochemical screening using various methods. Phytochemical screening was performed to test for alkaloids, proteins, carbohydrates, and flavonoids. The selected phytochemical tests target the most common bioactive compounds associated with antimicrobial activity in these plant species based on previous literature. The preliminary screening focused on alkaloids, proteins, carbohydrates, and flavonoids as these compounds are well-documented for their antimicrobial properties. Future comprehensive screening will include saponins, sterols, terpenes, and other secondary metabolites to provide a complete phytochemical profile [14].

2.5 Experimental Microorganism

AAKAAR BIOTECH PVT conducts the antimicrobial activity. And they use *E. coli* bacterium to know the antimicrobial activity of the sample. *E. coli* was selected as the test organism because:

(1) it is a representative Gram-negative pathogen,
(2) commonly causes foodborne illnesses,
(3) serves as a standard organism for preliminary antimicrobial screening as per CLSI guidelines [15]. Future studies will include both Gram-positive and Gram-negative bacteria.

2.6 Statistical Analysis

All experiments were performed in triplicate. Data were expressed as mean ± standard deviation. Statistical significance was determined using one-way ANOVA followed by Tukey's post-hoc test. P-values <0.05 were considered statistically significant (See in the Table 4)

3 Determination of Anti-microbial Activity

The Zone Inhibition Method (Kirby-Bauer method) was employed to evaluate antibacterial activity following Clinical and Laboratory Standards Institute (CLSI) guidelines. Mueller-Hinton Agar (MHA) plates were prepared and inoculated with 100 μL of *Escherichia coli* bacterial culture. The inoculum density was standardized to 0.5 McFarland units (1.5×10^8 CFU/mL) using Mueller-Hinton Broth. Sterile paper discs (6 mm diameter) were impregnated with 10 μL of varying extract concentrations (0–100%). Ciprofloxacin (10 μg) served as the positive control, while distilled water was used as the negative control. Plates were incubated at 37 °C for 24 h. The diameter of clear zones around each disc was measured using a digital caliper and recorded in millimeters.

4 Result and Discussion

Papaya is one of the 48 most well-known species in the genus Carica, which is part of the family *Caricaceae*. *Papaya* fruit is grown all over the world and is highly valued by farmers since it requires less space per plant, is simple to grow, and yields higher profits per hectare. The most effective medicinally significant plant material is alkaloids, which were also found in our samples. The biological activities of flavonoids include defense against allergies, inflammations, platelet aggregation, bacteria, ulcers, viruses, and cancers [16]. Plant-derived alkaloids are frequently discovered to possess antibacterial qualities. Our sample also contains protein and carbs, which are necessary for muscular growth and energy production. In India, infectious infections are a leading cause of both morbidity and mortality. Both the prevalence of strains with decreased antibiotic susceptibility and the number of multiple drug-resistant pathogens are steadily rising. In order to determine the presence of bioactive substances that have been reported to confer resistance to plants against bacteria, fungi, and pests, as well as to explain the demonstration of antibacterial activity by the plant extracts used, the antimicrobial activities of *Carica papaya* stem extract and *radish* leaf extract with an aqueous medium were investigated in the current study [17]. The FTIR analysis shows the presence of 13 functional groups in the sample. The results Although the combined extract showed lower activity (8 mm) compared to ciprofloxacin (30 mm), several factors support the significance of this finding: Natural vs. Synthetic Comparison: The extract represents a crude preparation containing multiple compounds, while ciprofloxacin is a pure, synthetic antibiotic optimized for maximum potency. Concentration Considerations: The extract was tested at 6.25 μg/mL, which is significantly lower than many reported plants extract studies, suggesting good potency. Clinical Relevance: According to CLSI guidelines, any zone of inhibition ≥6 mm for crude extracts indicates potential antimicrobial activity worthy of further investigation. Natural Alternative Potential: Natural extracts offer advantages including reduced resistance development, synergistic effects of multiple compounds, and better biocompatibility.

4.1 FTIR Graph of Extract

The FTIR (Fourier Transform Infrared) is done by IIT ROORKEE for the identification of functional groups present in the sample in radish leaves. Both graphs show the presence of different types of compounds. FTIR analysis of plant extracts is a standard approach for identifying predominant functional groups. While extracts contain multiple compounds, FTIR provides a collective fingerprint of major chemical classes present, which helps correlate structure with biological activity. See Figs. 1 and 2 and Table 1. The selected phytochemical tests target the most common bioactive compounds associated with antimicrobial activity. Future comprehensive screening will include saponins, sterols, and terpenes [18].

4.2 Phytochemical Analysis of Extract

Phytochemical analysis of the extract shows the presence of alkaloids, carbohydrates, proteins, and flavonoids (See Table 2)

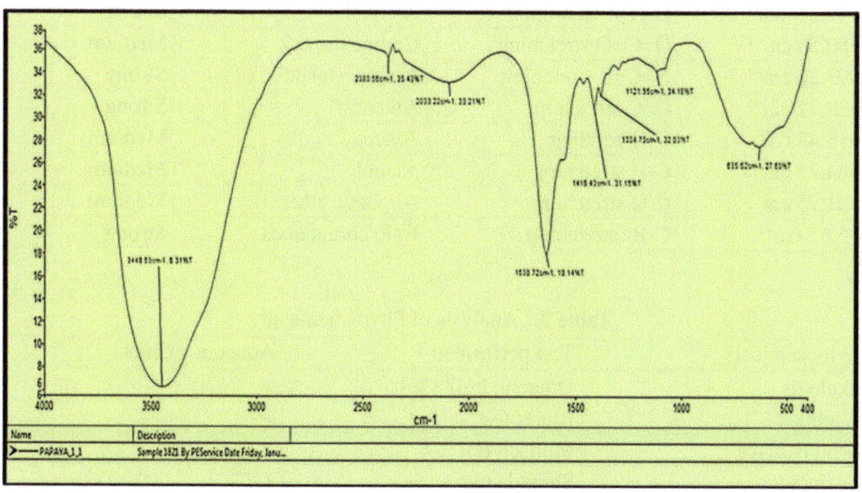

Fig. 1. Graph of Carica papaya.

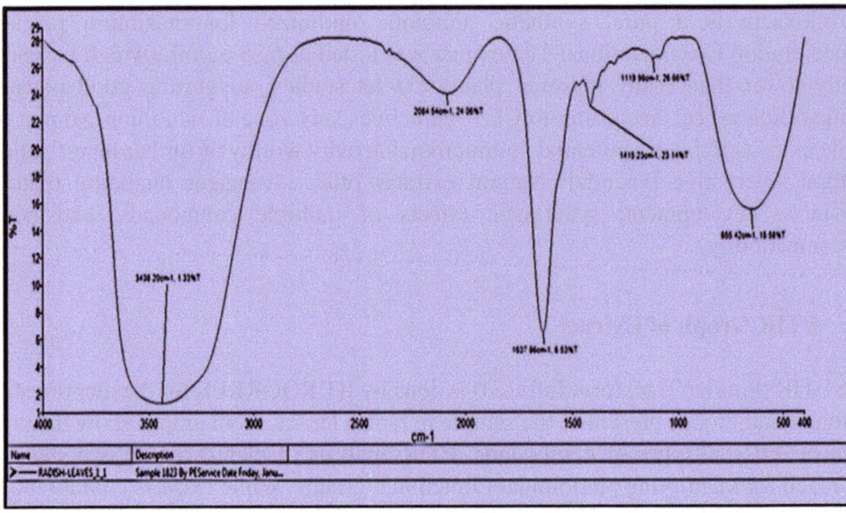

Fig. 2. Graph of Radish leaves.

Table 1. Peak details of Groups.

Peak position	Groups	Class	Peak detail
3438.20 cm^{-1}	N-H stretching	Primary amine	Strong board
2084.54 cm^{-1}	N=C=S stretching	Isothiocyanate	Medium
1637.86 cm^{-1}	C=N stretching	Imine/oxime	Strong
1415 cm^{-1}	C-F stretching	Fluoro compound	Medium
1119.98 cm^{-1}	C-O stretching	Aliphatic ether	Medium
655.42 cm^{-1}		Benzene derivative	Strong
3449.56 cm^{-1}	O-H stretching	Alcohol	Strong
2380.56 cm^{-1}	O=C=O stretching	Carbon dioxide	Medium
2093.22 cm^{-1}	N=C=S stretching	Isothiocyanate	Strong
1638.72 cm^{-1}	C=C stretching	Alkene	Strong
1415.43 cm^{-1}	C-H bending	Alkene	Medium
1384.73 cm^{-1}	C-H stretching	phenol	Medium
1121.95 cm^{-1}	C-O stretching	Aliphatic ether	Medium
635.62 cm^{-1}	C-Br stretching	Halo compounds	Strong

Table 2. Analysis of Phytochemicals.

Phytochemicals	Test performed	Aqueous extract
Alkaloids	Dragenodroff's test	+
Protein	Biuret test	+
Carbohydrates	Molisch test	+
Flavonoids	Shinoda test	+

Table 3. Statistical analysis.

Amount (%/disc)	Plate A	Plate B	Plate C	Average	SD	SEM
PC	30	30	30	30	0	0
0	0	0	0	0	0	0
6.25	8	8	8	8	0	0
12.5	8	7	7	7.333333333	0.57735	0.40825
25	8	6	7	7	1	0.70711
50	0	0	0	0	0	0
100	0	0	0	0	0	0

Table 4. Determination of Concentration.

S. No.	Sample ID	Effective Amount	Zone at Effective Conc. (in mm)
1.	Ciprofloxacin (PC)	10 μg	30
2.	21 (Extract)	6.25	8

4.3 Evaluation of the Antimicrobial Potential of the Extract

The aqueous extract of the stem of *Carica papaya* and radish leaves was screened for its antimicrobial activity against *E. coli*, and the antimicrobial study was done by AAKAAR BIOTECH PRIVATE LTD. LUCKNOW, 21 is our sample code and our extract sample is compared with the standard drug ciprofloxacin 10 μg, and the zone of inhibition is 30. Our sample amount is 6.24 μg at the effective amount and the zone of inhibition is 8 mm. In the present study, the aqueous extracts of *Carica papaya* stem and radish leaf extract showed inhibitory activity against *E. coli* (8 mm); the antibacterial action was shown in the form of the zone of inhibition. The Zone of inhibition assay showing: (A) Positive control - Ciprofloxacin 10 μg (30 mm zone), (B) Test sample - Combined extract at 6.25 μg/mL (8 mm zone), (C) Negative control - Distilled water (no zone). Clear zones indicate bacterial growth inhibition around the discs. (See Tables 3 and 4).

This situation provided the impetus for the search for new antimicrobial substances from various sources like medicinal plants. It is important to investigate scientifically these plants, which have been used in traditional medicines as potential sources of novel antimicrobial compounds. Although the extract showed lower activity (8 mm) compared to ciprofloxacin (30 mm), the activity is statistically significant ($p < 0.05$) and represents a natural alternative with potential for synergistic combinations and reduced resistance development.

5 Conclusion

Based on the results of the present study, it was found that fresh extracts of *Carica papaya* leaves and radish leaf extract showed good antimicrobial activity and were found to have synergistic activity when these two extracts were combined. These extracts can be directly used as natural alternative preventives to control various food poisoning diseases and preserve foodstuff, avoiding the health hazards of chemically antimicrobial agent applications. Given the significant antimicrobial properties observed, the extracts from *Carica papaya* and radish leaves could be explored further for their potential applications in medicine, particularly in treating infections caused by *E. coli*. This could be particularly beneficial in areas where antibiotic resistance is a growing concern in the future the study opens avenues for further research to isolate and characterize the specific compounds responsible for the antimicrobial activity. This could lead to the development of new antimicrobial agents derived from natural sources.

References

1. Nafiu, A.B., Alli-Oluwafuyi, A.M., Haleemat, A., Olalekan, I.S., Rahman, M.T.: Papaya (*Carica papaya* L., Pawpaw). In: Nonvitamin and Nonmineral Nutritional Supplements, pp. 335–359 (2018). https://doi.org/10.1016/B978-0-12-812491-8.00048-5
2. Annegowda, H.V., Bhat, R.: Composition of papaya fruit and papaya cultivars. In: Nutritional Composition of Fruit Cultivars, pp. 497–516 (2015). https://doi.org/10.1016/B978-0-12-408117-8.00021-0
3. Singh, S.P., Sudhakar Rao, D.V.: Papaya (*Carica papaya* L.). In: Postharvest Biology and Technology of Tropical and Subtropical Fruits, p. 4 (2011). https://doi.org/10.1533/9780857092618.86
4. Sharma, A., et al.: *Carica papaya* L. leaves: deciphering its antioxidant bioactives, biological activities, innovative products, and safety aspects. Oxid Med Cell Longev. **2022**, 2451733 (2022). https://doi.org/10.1155/2022/2451733
5. Bhowmik, D.: Traditional and medicinal uses of *Carica papaya*. J. Med. Plants Stud. **1**, 7–15 (2013)
6. Ayodipupo Babalola, B., Ifeolu Akinwande, A., Otunba, A.A., Ebenezer Adebami, G., Babalola, O., Nwuofo, C.: Therapeutic benefits of *Carica papaya*: a review on its pharmacological activities and characterization of papain. Arab. J. Chem. **17** (2024). https://doi.org/10.1016/j.arabjc.2023.105369
7. Kumarasinghe, H.S., et al.: Bioactive constituents from *Carica papaya* fruit: implications for drug discovery and pharmacological applications. Appl. Biol. Chem. **67**, 1–23 (2024). https://doi.org/10.1186/S13765-024-00962-Y/FIGURES/5
8. Noman, O.M., et al.: Comparative study of antioxidant and anticancer activities and HPTLC quantification of rutin in white radish (Raphanus sativus L.) leaves and root extracts grown in Saudi Arabia. Open Chem. **19**, 408–416 (2021). https://doi.org/10.1515/CHEM-2021-0042/ASSET/GRAPHIC/J_CHEM-2021-0042_FIG_006.JPG
9. Goyeneche, R., Rodrigues, C.R., Quispe-Fuentes, I., Pellegrini, M.C., Cumino, A., Di Scala, K.: Radish leaves extracts as functional ingredients: evaluation of bioactive compounds and health-promoting capacities. Waste Biomass Valoriz. **16**, 1003–1014 (2024). https://doi.org/10.1007/S12649-024-02760-5/METRICS
10. Shin, T., Ahn, M., Kim, G.O., Park, S.U.: Biological activity of various radish species. Orient Pharm Exp Med. **15**, 105–111 (2015). https://doi.org/10.1007/S13596-015-0183-9/METRICS
11. Jha, A.K., Sit, N.: Extraction of bioactive compounds from plant materials using combination of various novel methods: a review. Trends Food Sci. Technol. **119**, 579–591 (2022). https://doi.org/10.1016/J.TIFS.2021.11.019

12. Smith, J.D., Li, D.H., Rafferty, M.R.: The implementation research logic model: a method for planning, executing, reporting, and synthesizing implementation projects. Implement. Sci. **15**, 84 (2020). https://doi.org/10.1186/S13012-020-01041-8
13. Palvia: Introduction to spectroscopy [Paperback]. 752 (2008)
14. Silverstein, R.M., Webster, F.X., Kiemle, D.: The Spectrometric Identification of Organic Compounds, p. 512. Wiley, Hoboken (2012)
15. Trease and Evans' Pharmacognosy | ScienceDirect, https://www.sciencedirect.com/book/9780702029332/trease-and-evans-pharmacognosy
16. Clinical and Laboratory Standards Institute (CLSI). Performance Standards for Antimicrobial Susceptibility Testing. 35th ed. CLSI supplement M100 (ISBN 978-1-68440-262-5 [Print]; ISBN 978-1-68440-263-2 [Electronic]). Clinical and Laboratory Standards Institute, USA, 2025
17. AbdelMonein, N.M.: Hepatoprotective activity of chitosan nanocarriers loaded with the ethyl acetate extract of a Stenotrophomonas sp. bacteria associated with the Red Sea sponge amphimedon ochracea in CCl4 induced hepatotoxicty in rats. Adv. Biosci. Biotechnol. **8**, 27–50 (2017). https://doi.org/10.4236/ABB.2017.81003
18. Srinivasan, D., Nathan, S., Suresh, T., Lakshmana Perumalsamy, P.: Antimicrobial activity of certain Indian medicinal plants used in folkloric medicine. J. Ethnopharmacol. **74**, 217–220 (2001). https://doi.org/10.1016/S0378-8741(00)00345-7
19. Deshmukh, O.S.: Phytochemical screening by FTIR spectroscopic analysis of ethanolic root extracts of ethnoveterinary medicinal plants. Int. J. Emerg. Technol. Innov. Res. **7**(9), 636–642 (2020)

12. Smith, J.D., Li, D.H., Rafferty, M.R.: The implementation research logic model: a method for planning, executing, reporting, and synthesizing implementation projects. Implement. Sci. 15, 84 (2020). https://doi.org/10.1186/s13012-020-01041-8
13. Patel: Introduction to cosmetology (Paperback), 753 (2018)
14. Silverstein, R.M., Webster, F.X., Kiemle, D.: The Spectrometric Identification of Organic Compounds, p. 512. Wiley, Hoboken (2005)
15. [illegible] and [illegible]: Pharmacognosy. In: ScienceDirect. https://www.sciencedirect.com/topics/[illegible]/pharmacognosy
16. Clinical and Laboratory Standards Institute (CLSI): Performance Standards for Antimicrobial [illegible] Testing. [illegible] CLSI Supplement M100 [illegible]
[illegible]

QbD Based Extraction of Plumbagin from Plumbago Indica Roots and Its Antioxidant Activity

Gayatri Khosla[1], Neeraj Kumar[2], Disha Dutta[3], and Rama Tyagi[1(✉)]

[1] Galgotias College of Pharmacy, Greater Noida, Uttar Pradesh, India
tyagirama8@gmail.com

[2] Department of Pharmacognosy and Phytochemistry, School of Pharmaceutical Education & Science, New Delhi, India

[3] Devsthali Vidyapeeth College of Pharmacy, Rudrapur, Uttarakhand, India

Abstract. Plumbagin, a naphthoquinone that is found in the root system of the *Plumbago Indica* plant, possesses a variety of therapeutic benefits. Box-Behnken Design (BBD) was utilised to optimise the procedure variables employed in the plumbagin extraction process from *Plumbago Indica* roots. Quality by Design (QbD), and more specifically BBD, was utilized to optimize the extraction parameters, which included extraction duration, the ratio of solvent to drug, and extraction temperature. The quantity of plumbagin found in several different extracts was determined using high-performance, thin-layer chromatography (HPTLC). In conjunction with that, the DPPH method was utilized to assess the antioxidant capabilities of several *Plumbago Indica* extract solutions. The optimal extraction temperature for plumbagin was found to be 52.50 °C, the ratio of the solvent to the drug was found to be 22.50 ml per gram, and the extraction time was found to be 30 min when using BBD. Under these conditions, the plumbagin yield turned out to be 1.981 mg/g. It has been proven that the extract of methanolic solvent possesses DPPH activity, as evidenced by inhibition of 87.11%. It was found that methanol was the most effective solvent for extraction, and the ultrasound-assisted extraction method was the most effective of all the thermal and non-thermal extraction strategies utilized. The current work emphasizes the utilization of QbD in the extraction of potentially active compounds, which helps reduce the amount of time and effort necessary for optimization.

Keywords: Response surface approach · Extraction · Optimization · Plumbagin · *Plumbago indica* · High Performance Thin Layer Chromatography

1 Introduction

Since the beginning of the years, an impressive number of current pharmaceuticals have been derived from organic sources, predominantly plants. Botanical therapeutics possess a significant component in fulfilling the main healthcare demands of humans. Various herbs possess antidiabetic, antimicrobial, antioxidant, and anti-inflammatory properties because of the presence of active compounds. These active constituents have

A. K. Saxena, A. Saxena (Eds.): GTHTM 2025, pp. 457–468, 2026.
https://doi.org/10.1007/978-3-032-12320-6_27

been playing a significant role in preserving human health since prehistoric times. In the pharmaceutical sector, natural products are a crucial part of the drug research programs that are being carried out [1]. Natural treatments are posited to possess the benefit of integrating their potent constituents with numerous other seemingly inert chemicals, which collectively confer an amount of security and effectiveness that surpasses that of their isolated, natural constituents [2].

At the same time, unnatural drugs are currently costly in developing countries, and they are also often adulterated. The Plumbaginaceae family includes the perennial herb known as *Plumbagina indica*, which is widely seen growing in the forests of Uttarakhand, India, and is also cultivated in gardens across the country. The eastern, northern, and southern regions of India are native to wild plants that can be harvested as garden plants. It has also been stated that it is utilised in a range of traditional medicines in Africa and Asia. There is a class of phytoconstituents known as 1,4-naphthoquinones that have been isolated from plants. The phytocompounds possess a variety of therapeutic properties and maintain many pharmacological functions that compel constituents to commit to novel objectives for various diseases [3]. The root system of the traditional therapeutic herb Plumbago indica L. was discovered to be rich in a naphthoquinone called plumbabin. Plumbagin has recently been shown to exhibit a diverse array of physiological actions, including anti-inflammatory, anticancer, antioxidant, antibacterial, antifungal, anti-atherosclerotic, and analgesic characteristics [4].

The optimization of extraction factors to maximize the extraction of bioactive components from plants in just one procedure has gained prominence in this context, which is characterized by the presence of phytoconstituents that exhibit tremendous potential for therapeutic uses. Given this, the RSM approach, a mathematical and statistical strategy that optimizes complex processes by employing multivariate methodologies, is utilized to solve this problem [5]. When compared to other optimisation approaches, this method is labour- and time-saving, and it requires less experiments to evaluate numerous aspects and the interactions between them. For the Optimisation of extraction parameters of various phytoconstituents like flavonoids to function from *Herba Polygoni capitati, Pongamia pinnata, Ficus racemosa, Glycyrrhiza glabra, Atropa balledona,* and *Piper longum*. These are only some of the compounds that have been used [6].

In the current investigation, RSM is utilized to optimize the extraction factors. These parameters include extraction duration, extraction temperature, and the ratio of solvent to drug. Quantification analysis is carried out with high-performance thin-layer chromatography (HPTLC). On the other hand, none of the studies have devised an extraction method that can be used to remove plumbagin from *Plumbago indica* and determine its concentration. As a result, our research makes use of a variety of methods, including the maceration technique, the soxhlet method, the UAE method, and the reflux method, to extract plumbagin. The above techniques have not been published by any other researcher yet.

2 Materials and Methods

2.1 Collection and Authentication of the Plant Raw Material

Khari Baoli in Delhi was the location where the dried roots of P. indica were gathered. Dr. Sunita Garg, who works at NISCAIR in New Delhi, was the one who helped confirm the identification of the roots of P. indica. For future reference, a voucher specimen for the same was deposited at their herbarium with the reference number NISCAIR/ RHMD/ Consult/2020/3696-97.

2.2 Chemicals

Standard plumbagin was acquired from Sigma Aldrich, India. Toluene, acetic acid and glacial acetic acid were purchased from S.D. Fine Chem Limited, India. All other analytical grade chemicals were obtained from S.D. Fine Chem Limited, India

2.3 Preparation of Plant Raw Material

After being thoroughly cleaned to remove cling dust and any other foreign material, the dried roots were rinsed with water to complete the decontamination process. The roots were then pulverized, air-dried, and stored in an air-lock container after being passed through sieves no 14.

2.4 Extraction of Plumbagin

The extraction of plumbagin from *Plumbago indica* was accomplished by the utilisation of four distinct extraction methods, including ultrasound-assisted extraction, soxhelation, reflux, and maceration. These methods were utilised in conjunction with several solvents, including methanol, acetone, dimethylformamide, and dimethyl sulfoxide, which exhibited variable degrees of polarity [7].

2.5 Soxhlet Extraction

10 ml/g of the drug was used in the extraction process, which was conducted with a Soxhlet apparatus at a temperature of 60 °C for 1 h. Following the extraction process, the plant residue was filtered, and the remaining solvent from the filtrate was removed using a rotary evaporator [8].

2.6 Reflux Extraction

The extraction procedure was conducted in a reflux apparatus with a proportion of 10 ml/g of drug and 50 ml of solvent at a temperature of 60^0 C for 1 h. Following the extraction process, the plant residue was filtered, and the remaining solvent from the filtrate was removed using a rotary evaporator under vacuum [9].

2.7 Ultrasound-Assisted Extraction

With a solvent-to-drug ratio of 10 ml/g and a temperature of 60 °C in a sonicator (TOSCHON, SW7), the extraction was performed using the ultrasound-assisted extraction (UAE) method for one hour. Following the extraction process, the plant residue was filtered, and the filtrate was dried using a rotary evaporator under vacuum [10].

2.8 Extraction by Maceration

At room temperature, 2 g of the coarse drug was placed in a container and dipped in 20 ml of solvent (10 g/ml) for a period of 72 h. To obtain a brownish-coloured sticky substance, The filtrate was permitted to evaporate at ambient temperature. Following that, the extract was put aside for further examination [11].

2.9 Evaluation of Various Extraction Approaches for Plumbagin Extraction

The validation of plumbagin in various extracts of Plumbago *indica* was carried out by employing High-Performance Thin Liquid Chromatography (HPTLC) on a Shimadzu HPTLC. The methanol was used to generate a stock solution of standard plumbagin as well as a sample solution of various extracts of *Plumbago indica*. Additionally, dilutions of standard plumbagin ranging from 20 ug/ml to 100 ug/ml were also prepared using the same methanol. A 0.2 μm membrane filter (Axiva) was utilized to filter all the standard and sample solutions before they were applied to the aluminium sheet in the high-performance thin-layer chromatography (HPTLC) apparatus. A mobile phase of Toluene: Ethyl acetate: Glacial acetic acid (7:2:1) was utilized [12]. The detection was carried out at a wavelength of 270 nm. A calibration curve was created by establishing a relationship between the concentration of the standard plumbagin and the peak region. Following that, the linear equation of the calibration curve was utilized to determine the amount of plumbagin present in the various extracts of Plumbago *indica* roots.

2.10 Single Factorial Experiments

After determining the efficient extraction method and the most suitable solvent, single factorial runs were carried out on 3 factors: the extraction time, the solvent-to-drug ratio, and the extraction temperature. The experiment examined the influence of a

Table 1. Ranges of various parameters.

Independent variables	Lower range	Higher range	Coded levels		
			−1	1	+1
Extraction time	20	40	20	30	40
Extraction temperature	35	70	35	52.50	70
Drug-to-solvent ratio	15	30	15	22.5	30

Table 2. Experimental design of dependent and independent variables.

Run	Extraction time (mint.)	Extraction temperature (°C)	Drug-to-solvent ratio (g/ml)	Plumbagin content (mg/g)
1.	30	52.5	22.5	1.98
2.	30	35	15	1
3.	40	70	22.5	1.5
4.	30	52.5	22.5	1.89
5.	30	35	30	1.3
6.	20	52.5	30	1.9
7.	20	52.5	15	1.8
8.	30	70	15	1.2
9.	20	35	22.5	1.6
10.	20	70	22.5	1.2
11.	40	35	22.5	1.6
12.	30	52.5	22.5	1.7
13.	30	70	30	1
14.	30	52.5	22.5	1.8
15.	40	52.5	15	1.79
16.	30	52.5	22.5	1.6
17.	40	52.5	30	1.5

specific factor on plumbagin output by altering one of the parameters while maintaining the other two parameters at their original values. A presentation of the estimated ranges for the various parameters can be seen in Table 1. Using HPTLC, the amount of plumbagin found in each extract was determined.

2.11 Optimization of Extraction Parameters of Plumbagin

Using the Design-expert software (Version 13), which was developed by Stat-Ease, Inc. in the United States, Box-Behnken Design (BBD) was utilized to optimize the extraction factors for plumbagin (Table 2). The experimental design consisted of these seventeen runs. In addition to being designed into 3 levels that were decoded as −1, 0, and + 1 for low, moderate, and maximum levels, accordingly, the three factors that were chosen were given the names Z1, Z2, and Z3 and designed into three levels.

2.12 Quantification of Plumbagin in Various Extracts by HPTLC

Various extracts for BBD trials were examined using HPTLC to quantify plumbagin concentration.

2.13 Phytochemical Evaluations

A preliminary phytochemical evaluation was conducted to examine the existence of various phytochemicals in various extracts of *P. indica*. These phytochemicals include glycosides, alkaloids, flavonoids, phytosterols, phenolics, carbohydrates, triterpenoids, steroids, resins, and tannins.

2.14 Antioxidant Activity Using DPPH

The DPPH technique was utilized in conjunction with a UV spectrophotometer set to 517 nm to evaluate the antioxidant capacity of several extracts of *P. indica* [13]. To establish a standard, aliquots of 20 to 120 ug/ ml were introduced to the various tubes that contained methanol (5 ml) and 1 mM DPPH (0.5 ml). In addition, a blank solution was made by combining 0.5 ml of 1 mM DPPH with 5 ml of methanol. All the dilutions were then kept at room temperature for half an hour in the dark. To determine the antioxidant capacity of various extracts, the following equation was found to be appropriate: The absorbance of the blank solution minus the absorbance of the test sample multiplied by the percentage of scavenging equals the absorbance of the blank solution multiplied by 100. The data are presented as a mean of three.

3 Results and Discussion

3.1 Evaluation of Various Extraction Approaches for Plumbagin Extraction

Preliminary investigations to separate the plumbagin from *plumbago indica* utilising 4 distinct approaches to extraction with various solvents of differing polarity indicated that methanol was the greatest efficacious solvent, while ultrasound-assisted extraction emerged as the superior extraction method for plumbagin. The estimation of plumbagin in all extracts was conducted using HPTLC, with the resulting chromatograms of standard plumbagin and sample plumbagin in the methanolic extract of Plumbago indica displaying R_f value of 6.7 **as** illustrated in Fig. 1. Table 3 R_f value and quantification values of plumbagin in several extracts of Plumbago indica obtained using various extraction procedures.

3.2 Optimization of Extraction Parameters by BBD

After identifying the limits of different variables from single factorial tests, BBD was utilised for additional optimisation. Three factors were combined in 17 distinct ways by BBD. The link between the response variable (plumbagin content) and the tested

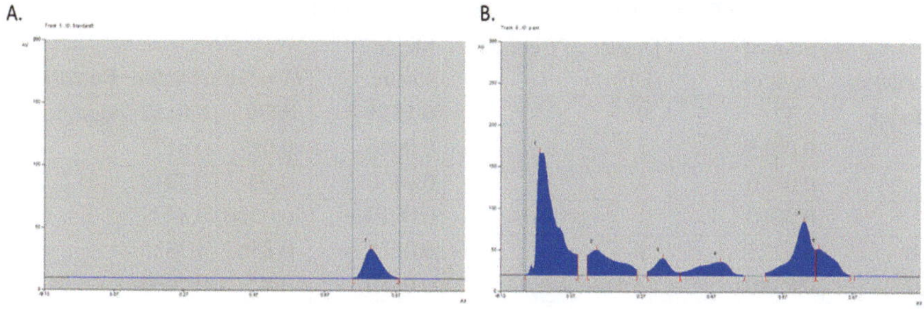

Fig. 1. HPTLC chromatogram of *Plumbago indica* extract and Standard plumbagin.

Table 3. The R_f value and quantification values of plumbagin in different extracts of Plumbagin indica from different extraction methods.

Extraction method	Solvent used	Rf value	Plumbagin content (mg.g)
Soxhlet Extraction	Methanol	0.67	1.54
	Acetone	0.66	0.78
	DMSO	0.63	1.15
	DMF	0.69	0.59
Maceration Extraction	Methanol	0.66	1.16
	Acetone	0.67	0.49
	DMSO	0.69	1.15
	DMF	0.66	0.51
Reflux Extraction	Methanol	0.70	1.32
	Acetone	0.67	0.62
	DMSO	0.68	1.28
	DMF	0.64	0.62
Ultrasound-Assisted Extraction	Methanol	0.68	1.87
	Acetone	0.63	0.67
	DMSO	0.64	1.59
	DMF	0.66	0.83

variables was determined using multiple regression analysis of the test results, yielding the second-order polynomial equation as follows:

Plumbagin Yield = −3.88875–0.085675A + 0.165686B + 0.246100C + 0.000429AB-0.001300 AC-0.000952 BC + 0.001518A2–0.001537B2–0.003524C2

The results of the variance analysis (ANOVA) were provided for the fitted quadratic polynomial model to assess the model's adequacy (Table 4). The regression coefficient (R^2) was 0.8816, indicating a close alignment with the fitted regression model. A disparity of less than 0.2 between anticipated R^2 and adjusted R^2 indicates that the model's fit is exceptional. The noise ratio, referred to as Adequate Precision, was determined to be 7.5124, indicating a favourable ratio beyond four, hence demonstrating the model's efficacy. The absence of fit tests also assesses the model's adequacy. The F-value and p-value were determined to be 1.38 and 0.3698, respectively, indicating a non-significant lack of fit, so rendering the model appropriate.

Table 4. ANOVA for response surface quadratic model.

Variables	Sum of squares	Degree of freedom (DF)	Mean square	F-value	P-value	Remarks
Model	1.37	9	0.1521	5.79	0.0152	significant
A	0.0015	1	0.0015	0.0576	0.8172	
B	0.0450	1	0.0450	1.71	0.2319	
C	0.0010	1	0.0010	0.0386	0.8499	
AB	0.0225	1	0.0225	0.8567	0.3855	
AC	0.0380	1	0.0380	1.45	0.2680	
BC	0.0625	1	0.0625	2.38	0.1668	
A^2	0.0970	1	0.0970	3.69	0.0961	
B^2	0.9331	1	0.9331	35.53	0.0006	
C^2	0.1655	1	0.1655	6.30	0.0404	
Residual	0.1838	7	0.0263			
Lack of Fit	0.0935	3	0.0312	1.38	0.3698	not significant
Pure Error	0.0903	4	0.0226			
Cor Total	1.55	16				

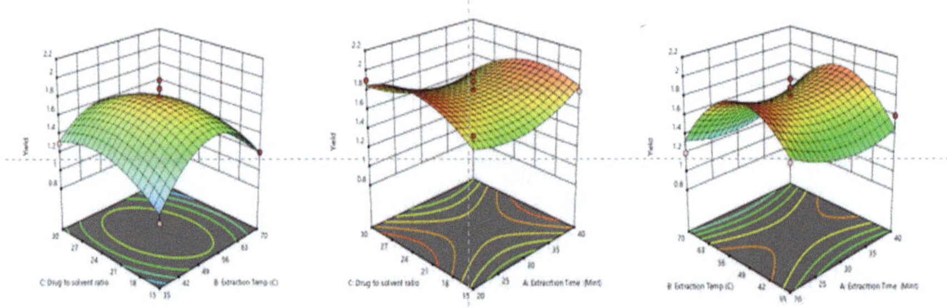

Fig. 2. 3D response surface graphs of extraction time, extraction temperature and drug-to-solvent ratio.

3D response surface plots and contour plots illustrate Figs. 2 and 3 which facilitates a clearer comprehension of the relationships between the responses and variables. The 3D graph indicates that plumbagin yield rises with a spike in the time needed for extraction from 20 min to 30 min and a solvent-to-drug ratio from 15 ml/g to 22.5 ml/g. Nevertheless, the subsequent rise in above mentioned factors indicates a reduction in plumbagin yield. This indicates that both variables are essential for plumbagin extraction. The yield of plumbagin rises with an increase in temperature during the extraction process from 35 °C to 52.5 °C and a solvent-to-drug ratio from 15 ml/g to 22.5 ml/g; however, additional improvements in either factor result in a decline in plumbagin yield. Plumbagin yield improves with rising extraction temperature from 35 °C to 52.5 °C and extraction period from 20 min to 30 min.

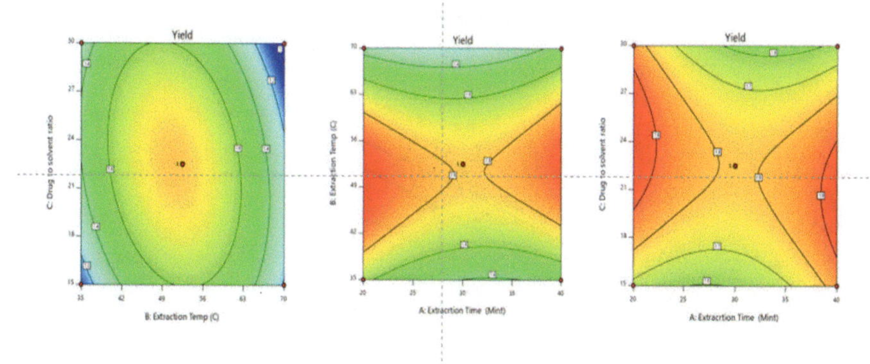

Fig. 3. Contour plot of extraction time, extraction temperature and drug-to-solvent ratio.

The point prediction study indicated that the ideal parameters for extracting plumbagin from Plumbago indica root are an extraction time of 30 min, solvent-to-drug ratio of 22.5 ml/g, and extraction temperature of 52.5 °C. The ideal circumstances yielded an ideal plumbagin concentration of 1.98 mg/g from *Plumbago indica*.

However, no substantial discrepancy was seen when the experimental and anticipated yield were estimated, indicating the response model's excellence for optimisation.

3.3 Phytochemical Screening

Phytochemical analyses of *P. indica* extracts in methanol, acetone, DMSO, and DMF revealed an abundance of quinones, flavonoids and phenolic compounds. The extraction value of plumbagin in various solvents indicates the quantity and nature of phytoconstituents. The methanolic extract of *P. indica* demonstrated the presence of phenolic compounds, quinones, alkaloids, phytosterols, triterpenoids, flavonoids, and glycosides as the principal chemicals, which potentially contributed to its medicinal efficacy.

3.4 Antioxidant Activity

The antioxidant capacity of *P. indica* extracts was assessed by DPPH activity. DPPH free radicals, characterised by a purple hue in the samples being tested, were utilised for the evaluation of the findings. When DPPH solution interacts with samples, it shares a hydrogen atom, yielding a yellow non-radical reduced form known as diphenylpicrylhydrazine. The antioxidant capacity of the different extracts was assessed concerning that of Vitamin C, the standard antioxidant ingredient. The results for assessing the DPPH activity of various P. indica extracts was determined graphically. The methanolic extract of *P. indica* exhibited the maximum antioxidant activity at 87.11%, in comparison to *P. indica* extracts in DMSO, acetone, and DMF which recorded 80.38%, 72.19% and 65.13558% (Fig. 4). This study examined both modern and traditional techniques for extracting plumbagin, a napthaquinone from *Plumbago indica* roots. Additionally, BBD, a contemporary statistical method, was utilised to enhance the extraction

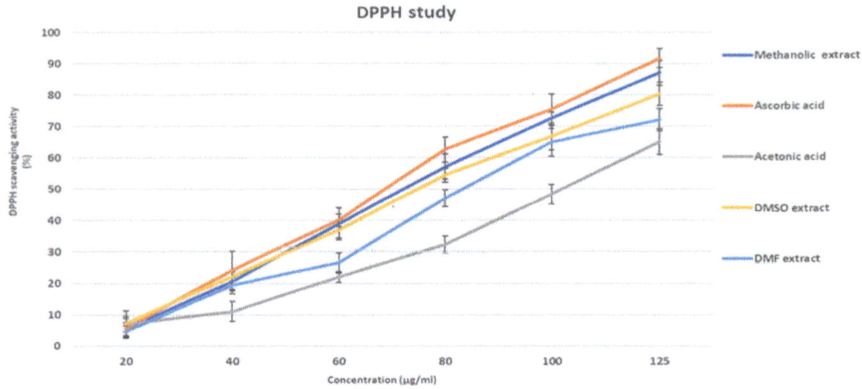

Fig. 4. Scavenging of DPPH radicals of the various extracts of *Plumbago indica* compared with standard drug Ascorbic acid. Each value represents mean ± SD (n = 3).

Table 5. Plumbagin content determined by the HPTLC method in the samples.

Sample No.	Sample	HPTLC Plumbagin (ug/g)
1	Methanolic extract	1.723
2	Acetonic extract	0.847
3	DMFO extract	1.056
4	DMF extract	0.641

characteristics of plumbagin. This optimisation method was chosen due to its cost-effectiveness, efficiency, reduced labour intensity, and several advantages compared to traditional optimisation techniques. It also facilitates the examination of interactions among independent variables.

3.5 HPTLC Study

This study utilised BBD as an effective instrument for optimising the parameters of plumbagin extraction from P. indica roots. The estimation of plumbagin was processed via HPTLC with UV detection at 270 nm. An elution solution comprising toluene, ethyl acetate and glacial acetic acid (7:2:1 v/v) was employed as the mobile phase. Plumbagin content of methanolic, acetone, DMSO and DMF extracts were found to be 1.723 mg/g, 0.847 mg/g, 1.056 mg/g and 0.661 mg/g respectively (Table 5). Nevertheless, no research has established the extraction procedure for isolating and quantifying plumbagin from *Plumbago indica*. Consequently, our study utilises many procedures for plumbagin extraction from *P. indica*, including Soxhlet, reflux, UAE, and maceration, which have not been documented by any other investigator to date. The antioxidant capacity of several extracts of *P. indica* was evaluated using the DPPH technique. The DPPH activity indicates that the methanolic extract demonstrates the strongest antioxidant power, and further characterisation of methanolic extract may prove beneficial for investigators seeking to identify novel medicinal agents.

4 Conclusion

Preliminary single factorial experiments were conducted prior to the use of BBD, and the outcomes obtained from these individual tests were utilised in BBD. Our research determined that methanol is the ideal solvent for extracting plumbagin and that ultrasound-assisted extraction yields a superior amount of plumbagin compared to thermal methods such as Soxhlet and non-thermal methods like maceration. Contemporary technology, specifically ultrasound-assisted extraction, emerged as one of the effective techniques for plumbagin extraction. Employing multiple regression analysis, the outcome variables were aligned with a second-order polynomial equation, and the ideal circumstances for plumbagin extraction from Plumbago indica roots were established using the model equation: extraction time of 30 min, solvent-to-drug ratio of 22.5 ml/g, and extraction temperature of 52.5 °C. In light of these circumstances, the plumbagin content was determined to be 1.98 mg/g, aligning with the anticipated result. The antioxidant capacity of several extracts of *P. indica* was evaluated using the DPPH technique. The DPPH scavenging activity indicates that the methanolic extract demonstrates the strongest antioxidant capability, and the characterisation of this extract may prove beneficial for investigators seeking to identify novel medicinal agents. The results of this investigation will assist future investigators and pharma companies seeking to maximise plumbagin extraction from Plumbago indica roots.

5 Summary

Plumbagin, a napthaquinone found in the roots of Plumbago indica, exhibits diverse pharmacological effects. This study evaluates the accuracy of various contemporary and traditional methods for plumbagin extraction and identifies the key parameters that achieve a high extraction of plumbagin from roots of Plumbago indica in a single run. The current work employed Box-Behnken Design to optimise the process factors for extracting plumbagin from the roots of Plumbago indica. Extraction parameters (extraction duration, solvent-to-drug ratio, and extraction temperature) were optimised using Quality by Design (QbD), through Box-Behnken Design (BBD). The quantification measurement of plumbagin in various extracts was conducted using HPTLC. The experimental results indicated that the Ultrasound-assisted extraction approach was superior to all alternative thermal and non-thermal extraction techniques employed, with methanol identified as the most effective extracting solvent. The best parameters for plumbagin extraction, determined using BBD, were a solvent-to-drug ratio of 22.5 ml/g, an extraction temperature of 52.5 °C, and an extraction duration of 30 min. The plumbagin was produced at a concentration of 2.021 mg/g, which was nearly equivalent to the anticipated value of 1.98 mg/g. Moreover, plumbagin extraction was markedly influenced by all three distinct variables. This study emphasises the application of Quality by Design (QbD), a multivariate statistical technique, in the extraction of functionally active plant-based ingredients, rendering the optimisation process less labour-intensive and more time-efficient than conventional methods. The DPPH scavenging activity indicates that the methanolic extract demonstrates the strongest antioxidant capabilities, so more characterisation of this extract may prove beneficial for scholars seeking to identify new medicinal agents.

References

1. Nasim, N., Sandeep, I.S., Mohanty, S.: Plant-derived natural products for drug discovery: current approaches and prospects. Nucl. **65**, 399–411 (2022). https://doi.org/10.1007/s13237-022-00405-3
2. Ye, L., Zhang, J., Xiao, W., Liu, S.: Efficacy and mechanism of actions of natural antimicrobial drugs. Pharmacol. Ther. **216**, 107671 (2020). https://doi.org/10.1016/j.pharmthera.2020.107671
3. Kumar, A., Gupta, M.K., Singh, R.K., Singh, J.: A review: Plumbago Zeylanica. Int. J. Pharm. Res. Appl. **8**, 1264 (2023). https://doi.org/10.35629/7781-080212641270
4. Khatun, A., Plumbago Indica, L.: A review of its medicinal uses, phytochemistry, pharmacology, and toxicology. Int. J. Herb. Med. **11**, 31–37 (2023). https://doi.org/10.22271/flora.2023.v11.i4a.877
5. Gong, Y., Hou, Z., Gao, Y., Xue, Y., Liu, X., Liu, G.: Optimization of extraction parameters of bioactive components from defatted Marigold (Tagetes Erecta L.) residue using response surface methodology. Food Bioprod. Process. **90**, 9–16 (2012). https://doi.org/10.1016/j.fbp.2010.12.004
6. Gupta, I., Adin, S.N., Aqil, M., Mujeeb, M.: QbD based extraction of Naringin from citrus Sinensis L. Peel and its antioxidant activity. Pharm. Res. **15**, 145–154 (2022). https://doi.org/10.5530/097484900241
7. Calzada-González, F.I., Martínez, J., Vargas-Rodríguez, Y.M., Hernández-Portilla, L.B., Nicolás-Vázquez, M.I., Miranda, R.: Plumbagin extraction: a green study. Nat. Prod. Res. **18**, 1–11 (2023). https://doi.org/10.1177/1934578X231191345
8. Zhang, Q.W., Lin, L.G., Ye, W.C.: Techniques for extraction and isolation of natural products: a comprehensive review. Chinese Med. (United Kingdom). **13**, 1–26 (2018). https://doi.org/10.1186/s13020-018-0177-x
9. Chua, L.S., Latiff, N.A., Mohamad, M.: Reflux extraction and cleanup process by column chromatography for high yield of andrographolide enriched extract. J. Appl. Res. Med. Aromat. Plants. **3**, 64–70 (2016). https://doi.org/10.1016/j.jarmap.2016.01.004
10. Poojar, B., Ommurugan, B., Adiga, S., Thomas, H., Sori, R.K., Poojar, B., Hodlur, N., Tilak, A., Korde, R., Gandigawad, P., et al.: Methodology used in the study. Asian J. Pharm. Clin. Res. **7**, 1–5 (2017). https://doi.org/10.4103/jpbs.JPBS
11. Bitwell, C., Indra, S., et al.: A review of modern and conventional extraction techniques and their applications for extracting phytochemicals from plants. Sci. Afr. **19**, e01585 (2023). https://doi.org/10.1016/j.sciaf.2023.e01585
12. Usmani, Z., Mujeeb, M., Qadir, A., Aqil, M., Mir, S.R., Ahmad, A., Ahmad, V., Warsi, M.H., Usmani, Z.: A validated, rapid and cost-efficient HPTLC method for the quantification of Plumbagin and its antioxidant activity from the different extracts of Plumbago Zeylanica L. J. Planar Chromatogr. Mod. TLC. **33**, 587–597 (2020). https://doi.org/10.1007/s00764-020-00071-8
13. Eldhose, B., Notario, V., Latha, M.S.: Evaluation of phytochemical constituents and in vitro antioxidant activities of Mentha Arvensis. Int. J. Biol. Pharm. Allied Sci. **10**, 157–161 (2021). https://doi.org/10.31032/ijbpas/2021/10.9.1012

Combination of Solid Dispersion and Potential Iontophoresis for Improving Transcorneal In Vitro Permeation of Itraconazole

Swagatika Das[1], Biswaranjan Mohanty[2](✉), Bibaswan Mishra[3], and Bhabani Shankar Nayak[4](✉)

[1] Himalayan Pharmacy Institute, Mjhitar, Rangpo, East Sikkim, India
[2] School of Pharmacy, DRIEMS University, Tangi, Cuttack, Odisha, India
biswaranjan.m@driems.ac.in
[3] Institute of Pharmacy and Technology, Salipur, Cuttack, Odisha, India
[4] KIIT School of Pharmacy, KIMS, KIIT Deemed to be University, Bhubaneswar, Odisha, India
bhabanishankar.nayak@kims.ac.in

Abstract. Background: The delivery of medicament to eye faces difficulties due to the complex structure and presence of different static and dynamic barriers of eye. **Aim:** The present investigates improvement of the aqueous solubility of a poorly soluble itraconazole as solid dispersion with soluplus and to study the effect of different preservatives and application of iontophoresis on transcorneal permeation. **Methods:** Itraconazole solid dispersion was prepared and characterized by FTIR spectroscopy, differential scanning calorimetry and X-ray powder diffraction. Then aqueous isotonic ophthalmic solution of itraconazole (pH 7.2) 0.03% W/V containing different preservatives were developed and in vitro permeation, corneal hydration and antifungal activity were studied. **Results:** FTIR study revealed absence of any significant interactions between Itraconazole and soluplus. The drug was in an amorphous state or as a solid solution in the prepared solid dispersion supported by DSC thermograms and P-XRD study. The aqueous solubility of Itraconazole was found as 386.02 ± 9.4 µg/ml from solid dispersion of drug to soluplus ratio 1:10. In-vitro transcorneal permeation showed maximum percentage permeation of itraconazole from the formulation containing BA (10.93 ± 0.14) followed by formulation containing MP + PP (7.94 ± 0.26) and BKC + EDTA (6.81 ± 0.19) after 2 h. In iontophoretic study, increase in current intensity from 0.5 to 1.5 mA and iontophoresis time duration from 5 min to 15 min resulted in increased in drug permeation. Corneal hydration level was studied to assess the possible corneal damage due to electric treatment. The antifungal activity of the formulations against *Aspergillus flavus* determined by disk diffusion method suggested formulation containing BKC + EDTA showed the maximum zone of inhibition (14.33 ± 0.58 mm) compared to other formulations. **Conclusion:** Hence combinational effect of iontophoresis and BKC + EDTA as preservative has significant role enhancing the corneal permeation and antifungal activity that will potentially enhance the bioavailability of itraconazole to eye.

A. K. Saxena, A. Saxena (Eds.): GTHTM 2025, pp. 469–486, 2026.
https://doi.org/10.1007/978-3-032-12320-6_28

Keywords: Itraconazole · Solid Dispersion · Transcorneal Permeation · Iontophoresis · Corneal Hydration · Antifungal Activity

1 Introduction

Itraconazole is basically a triazoles, derived from the imidazoles. This poorly water soluble drug has received great attention due to its potential antifungal activity [1, 2]. Itraconazole is a poorly soluble drug. Solid dispersion is a technique by which drug candidates can be embedded within a polymeric matrix structure. This system is helpful in improving solubility by reducing the particle size and converting to a somewhat more soluble amorphous form or reducing the crystallinity [3–5].

The ability of topical applied drops to effectively penetrate the natural ocular barriers is limited [6]. Similarly, ophthalmic medicines taken systemically have limited access to the inner eye due to the presence of the blood-aqueous and blood-retinal barriers. Iontophoresis is a technique that involves the application of a direct electrical current to facilitate the transdermal delivery of ionized chemicals onto or through a tissue [7]. Ocular iontophoresis has been found to be an efficacious and noninvasive method for treating ocular tissues, causing minimum discomfort to patients [8, 9]. The efficacy and safety of this method have been demonstrated in successfully and safely delivering the medicine to both anterior and posterior ocular structures, utilizing a low electric current (density) [10–12].

Transcorneal iontophoresis has been found to yield elevated and prolonged drug concentrations in both the cornea and the aqueous humour [7, 13]. The cornea is avascular, allowing for the unimpeded movement of ions without reliance on blood circulation. Both the aqueous and vitreous humors within the eye possess strong electrical conductivity, making the eye highly amenable to iontophoresis [14, 15].

It makes use of carriers that can stabilize supersaturation by inhibiting nucleation/crystal growth in the process of crystalization. Solid dispersions create a supersaturated solution in biological fluids and supersaturation increases the thermodynamic activity of the drug, promoting absorption.

The objective of this study was to enhance the solubility of itraconazole through the utilization of solid dispersion technique employing a polymeric solubilizer known as "soluplus." The primary goal was to develop aqueous formulations of itraconazole [16–18]. Additionally, the study aimed to examine the impact of a preservative on the transcorneal permeation properties of the drug across freshly excised goat corneas, both with and without the application of iontophoresis.

2 Materials and Methods

2.1 Materials

The drug itraconazole was acquired as a gift sample from Jubilant Organosys Ltd., located in New Delhi. Preservatives such as benzyl alcohol (BA), disodium edentate (EDTA), phenylmercuric acetate (PMA), methyl paraben sodium (MP), propyl paraben sodium (PP), benzalkonium chloride (BKC), and thiomersal (THM) were employed in the study. The freeze-dried microbial cultures of *Aspergillus flavus* (MTCC NO 3306) were acquired from the Institute of Microbial Technology in Chandigarh, India. All the remaining compounds used in the experiment were of analytical grade. Fresh goat eyes were procured from a local butcher shop located in Baripada, Odisha, India, within one hour of the animals being slaughtered. The dissection approach for the cornea and the apparatus utilized in the permeation studies were consistent with those previously documented [19].

2.2 Methods

2.2.1 Preparation of Solid Dispersions. Soluplus was solubilized in ethanol, whereas the drug was dissolved in dichloromethane (DCM). The two solutions were combined to obtain a homogeneous solution. Subsequently, the solvent was subjected to evaporation at a temperature of 60 °C for duration of 24 h, following the methodology previously described [20]. The weight ratios between the medication and soluplus were selected as 1:1, 1:2, 1:3, 1:5, and 1:10.

2.2.2 Determination of Solubility. An additional quantity of itraconazole solid dispersions was introduced into water at a temperature of 25 ± 0.5 °C. The suspensions underwent continuous agitation for duration of 48 h. The solution was subsequently subjected to centrifugation for a duration of 30 min at a rotational speed of 5000 rpm (RPM) using a centrifuge manufactured by Remi Equipments Ltd., located in Mumbai, India. The transparent liquid portion was obtained, subjected to filtration, and subsequently examined for the presence of itraconazole using a spectrophotometer (UV-1700, Shimadzu) at a wavelength of 266 nm [21]. Every sample was determined three times.

2.2.3 Characterization of Solid Dispersion of Itraconazole

FT-IR Spectroscopy Studies. The infrared spectrum analysis was conducted utilizing a Fourier transform infrared spectrophotometer (Jasco FT/IR-5300, Japan) within the wavelength range of 4000–400 cm^{-1}, employing KBr pellets [22].

Powder X-Ray Diffraction. The x-ray diffraction patterns of itraconazole, soluplus, a physical mixture of itraconazole, and a solid dispersion of itraconazole were obtained using an x-ray diffractometer (Bruker, Model No: D8 advance, Germany). The diffractometer utilized CuKa radiation (λ = 1.54 A°) and operated at 30 kV and 20 mA. The diffraction angles were measured over the range of $2\theta = 0°–50°$ [23].

Differential Scanning Calorimetry. The thermal properties of itraconazole, soluplus, their physical combination, and solid dispersion of itraconazole were assessed utilizing a differential scanning calorimeter (DSC-60, Shimadzu, Japan). The specimens were crimped within a conventional aluminum pan and subjected to a temperature range of 40 to 250 °C, with a heating rate of 10 °C per min. Throughout the process, a continuous flow of nitrogen at a rate of 30 ml/min was maintained to ensure consistent purging [24].

2.2.4 Preparation of Itraconazole Ophthalmic Solutions (0.03% W/V, pH 7.2) Containing Preservatives. A solid dispersion of itraconazole in distilled water with a concentration of 0.03% (W/V) was formulated. To achieve isotonicity, the dispersion was supplemented with an adequate quantity of sodium chloride. The pH of the solution was modified to a value of 7.2. The solution was supplemented with several substances, including BKC (0.01% w/v), THM (0.005% w/v), PMA (0.002% w/v), EDTA (0.01% w/v), MP (0.02% w/v) and PP (0.01% w/v) in combination, BKC (0.01% w/v) and EDTA (0.01% w/v), or BA (0.5% w/v). The final volume of each solution was adjusted to 100 ml using distilled water.

2.2.5 In Vitro Permeation. The study employed a modified Franz diffusion cell constructed entirely of glass, including an accessible diffusion area of 0.64 cm^2. The cornea that had been recently removed was securely positioned between the donor and receptor compartments, ensuring that its epithelial surface was oriented towards the donor compartment. An aqueous solution of bicarbonate ringer with a pH of 7.4, produced immediately prior to use, was introduced into the receptor compartment, filling it to a volume of 11.5 ml. A volume of one milliliter of the test solution was introduced into the donor compartment positioned above the cornea. To mitigate evaporation from the test fluid, a glass coverslip was employed to close the aperture. The diffusion cell was positioned on a magnetic stirrer, and a Teflon-coated magnetic stir bead was introduced into the receptor compartment, operating at a rotational speed of 50 rpm (rpm). The temperature of the receptor fluid was maintained at 37 ± 0.5 °C. The platinum electrode, serving as the anode, was positioned within the donor compartment, while the second electrode was placed in the receptor compartment for the purpose of iontophoresis. Samples of 1 ml were extracted from the receptor compartment every 5 min, and the concentration of itraconazole was determined using a spectrophotometer (UV-1700, Shimadzu) at a wavelength of 266 nm. The experiment was replicated three times and the average value was computed [25].

2.2.6 Evaluation of Corneal Hydration Levels (HL %). Upon completion of the permeation study, the cornea that was utilized was extracted, and the scleral tissues were meticulously cleansed in accordance with the methodology outlined in the study conducted by Liu Z, et al., 1999 [26]. The epithelium surface was gently swabbed using filter paper and subsequently measured for weight (W_w). Subsequently, the cornea was immersed in a solution of one milliliter of methanol, subjected to an overnight drying process at a temperature of 90 °C, and subsequently reevaluated in terms of weight (W_d). The calculation of corneal hydration level (HL %) was performed using the following formula.

$$\%HL = \left[1 - \left(Wd - Ww\right)\right] \times 100 \tag{1}$$

2.2.7 Antifungal Activity. The antifungal activity of the eye drop formulation against *Aspergillus flavus* was assessed using the paper disc diffusion method, as described in the IP 1985. The medium used for both the antifungal activity and the maintenance of the study was the same. Approximately 0.1 ml of treated cell suspension was utilized to inoculate a volume of 100 ml of molten medium, which had been rendered sterile. Approximately 20 ml of medium that had been inoculated was carefully dispensed into petri dishes made of Borosilicate glass, with an internal diameter of 9 cm. The medium was then left undisturbed until it solidified, as described in reference [27]. Circular paper discs with a diameter of 4 mm were prepared by cutting Whatman No.1 filter paper. The discs were immersed in the solution and three identical discs were then individually placed in the petri plates that contained media infected with microorganisms. The incubation period for each petri dish lasted 48 h at a temperature of 25 °C, following which the dimensions of the inhibitory zones were quantified in millimeters.

3 Result and Discussion

3.1 Solubility Determination

The application of the solid dispersion approach with Soluplus has been found to enhance the water solubility of Itraconazole. The aqueous solubility of Itraconazole was determined to be 62.95 ± 9.97, 135.5 ± 13.59, 171.23 ± 12.94, 264.04 ± 6.78, and 386.02 ± 9.4 μg/ml when the medication was formulated in solid dispersion with a soluplus ratio of 1:1, 1:2, 1:3, 1:5, and 1:10, respectively. The ophthalmic solution was made with a concentration of 0.03% w/v, based on the maximal solubility of Itraconazole from solid dispersion at a ratio of 1:10.

3.2 FT-IR Spectroscopy Studies

The FTIR analysis was conducted on pure itraconazole, soluplus, their physical mixing, and itraconazole solid dispersion in order to assess the chemical compatibility between itraconazole and soluplus. The results of this research are depicted in Fig. 1. The infrared spectra of unmodified Itraconazole exhibited distinct peaks within the range of 400–1700 cm^{-1}. The potential origin of these phenomena can be attributed to the stretching and oscillation of functional groups, such as the -C=C- bonds present in aromatic compounds. The presence of a peak in the spectral range of 1600–1700 cm^{-1} can be ascribed to the stretching and vibration of the carbonyl group (-C=O), whereas peaks corresponding to alkane and amine functional groups were identified in the spectral range of 2900–3000 cm^{-1}. The previous Soluplus sample exhibited distinct peaks at 2935.66, 1692.23, and 1477.21 wavenumbers, which can be attributed to the stretching vibrations of O-H, aromatic C-H, C=O, and C-O-C functional groups, respectively. The spectra of the formulation exhibited identical peaks as well. Furthermore, the prominent spectral peaks detected for Itraconazole prior to and during solid dispersion at the range of 400–2900 cm^{-1} exhibited a high degree of similarity. This observation implies that there are no substantial interactions between Itraconazole and soluplus.

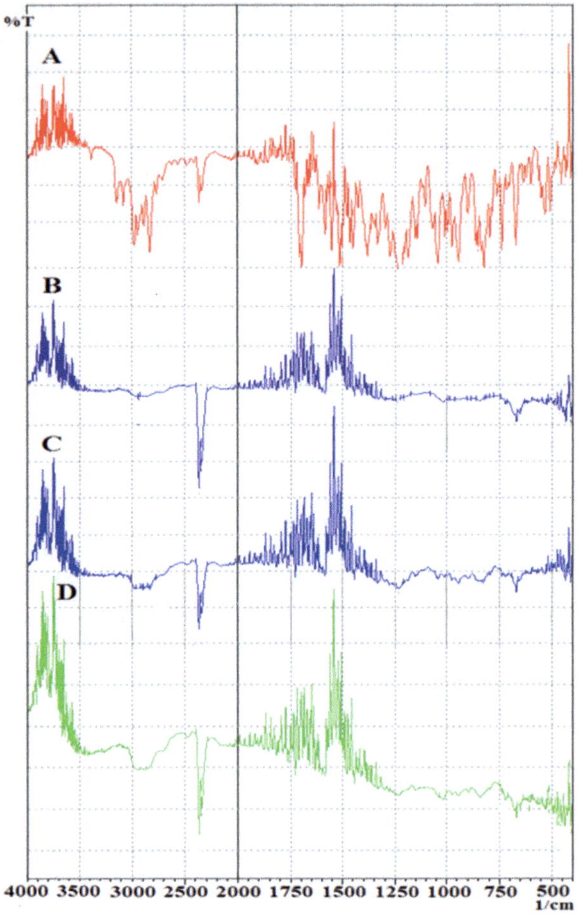

Fig. 1. FT-IR analysis of A) Itraconazole; B) Soluplus; C) Physical mixture and D) Itraconazole solid dispersion.

3.3 Powder X-Ray Diffraction (P-XRD)

Figure 2 illustrates the powder diffraction patterns obtained for itraconazole, soluplus, a physical mixing of itraconazole and soluplus, and itraconazole solid dispersion. The crystalline nature of itraconazole is shown by the existence of distinct peaks in its diffractogram. The amorphous character of soluplus was indicated by the absence of distinct peaks. The diffractogram of itraconazole solid dispersion demonstrates the total attenuation of the typical peaks, suggesting the absence of crystalline properties.

3.4 Thermal Analysis

Figure 3 depicts the differential scanning calorimetry (DSC) thermograms of pure itraconazole, soluplus, the physical mixing of itraconazole and soluplus, and the itraconazole solid dispersion. The thermogram of itraconazole has a distinct endothermic peak

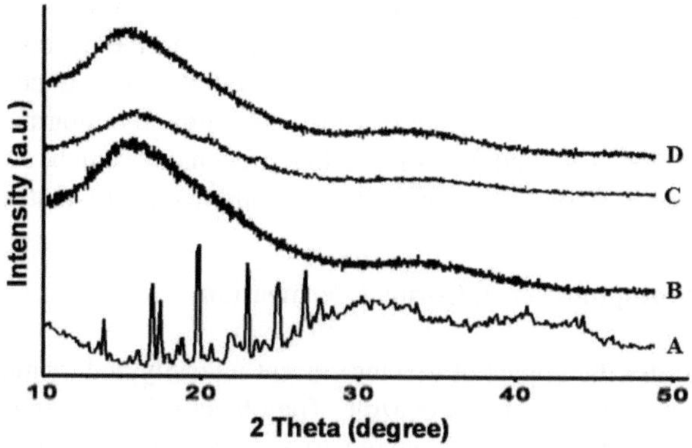

Fig. 2. Powder X-ray diffraction patterns for A) Itraconazole; B) Soluplus; C) Physical mixture and D) Itraconazole solid dispersion.

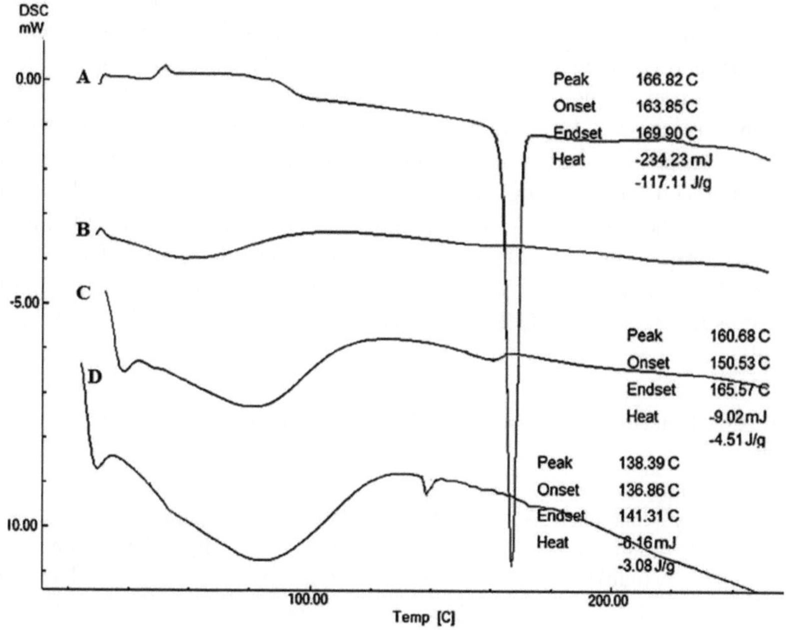

Fig. 3. DSC thermograms of A) Itraconazole; B) Itraconazole solid dispersion; C) Physical mixture and D) Soluplus.

at a temperature of 166.82 °C, indicating a process of heat absorption. This peak is associated with a heat of fusion value of 117.11 J/g, which represents the amount of energy required for the substance to transition from a solid to a liquid state. The thermogram of soluplus exhibited an endothermic peak at a temperature of 138.39 °C, accompanied by a heat of fusion value of −3.08 kJ/g. In the thermogram of the physical combination, an evident endothermic peak was found at a temperature of 160.68 °C. This peak exhibited characteristic of an exothermic reaction, as shown by a heat of fusion value of −4.51 J/g. The observed peak can be attributed to the depressed endothermic behavior of itraconazole. The absence of a melting endotherm of itraconazole was indicated by the disappearance of the endothermic peak in the thermogram of the itraconazole solid dispersion. This finding indicates that itraconazole can be found in either an amorphous state or as a solid solution within the solid dispersion that was created.

3.5 Effect of Iontophoresis on Corneal Permeation

Table 1 demonstrates the impact of varying current intensity and treatment duration on the transcorneal penetration of itraconazole. The objective of this study was to assess the extent of permeation, the percentage of permeation, and potential corneal hydration levels following a 2 h permeation period using varying current intensities and treatment durations. An observed augmentation in drug penetration was observed when either the current strength was raised from 0.5 to 1.5 mA or the length of iontophoresis was extended from 5 to 15 min. When a current of 0.5 mA was administered for durations of 5, 10, or 15 min, the penetration of itraconazole increased by an enhancement factor of 3.17, 5.48, and 10.98, respectively, in comparison to the untreated control. The application of a current intensity of 1.0 mA for exposure durations of 5, 10, or 15 min resulted in an increase in drug permeation by an enhancement factor of 6.54, 10.75, and 15.83 times, respectively, after a total duration of 120 min. In a similar manner, the application of a current of 1.5 mA for durations of 5, 10, or 15 min resulted in a significant increase in the percentage of drug permeation. Specifically, after 120 min of permeation, the drug permeation rates were increased by an enhancement factor of 11.32, 16.91, and 19.92 times higher compared to the untreated control. The hydration level, expressed as a percentage (HL %), of the corneas was evaluated at the conclusion of each trial.

Table 1. Effect of current (intensity) and duration on transcorneal permeation and corneal hydration.

Current intensity (mA)	Current density (mA/cm^2)	Iontophoresis duration (min)	Amount permeation (mcg) (120 min)	Permeation (%) (120 min)	Corneal hydration (%)	Enhancement Factor (EF)
0	0	0	8.37 ± 0.36	2.79 ± 0.12	79.58 ± 1.04	1.00
0.5	0.78	5	26.52 ± 3.41*	8.84 ± 1.14	79.74 ± 0.86	3.17 ± 0.31
0.5	0.78	10	45.91 ± 3.44*	15.30 ± 1.15	79.84 ± 1.21	5.48 ± 0.41
0.5	0.78	15	91.90 ± 3.52*	30.63 ± 1.17	80.36 ± 1.11	10.98 ± 0.42
1.0	1.56	5	54.78 ± 2.17*	18.26 ± 0.72	79.68 ± 0.14	6.54 ± 0.26
1.0	1.56	10	90.05 ± 4.64*	30.02 ± 1.55	81.12 ± 0.78	10.76 ± 0.55
1.0	1.56	15	132.53 ± 3.27*	44.18 ± 1.09	81.64 ± 0.37	15.83 ± 0.39
1.5	2.34	5	94.79 ± 4.06*	31.60 ± 1.35	79.61 ± 1.08	11.32 ± 0.48
1.5	2.34	10	141.61 ± 4.35*	47.20 ± 1.45	80.81 ± 0.87	16.91 ± 0.52
1.5	2.34	15	166.97 ± 7.43*	55.59 ± 2.48	83.93 ± 1.23	19.92 ± 0.89

Values are mean ± SD (n = 3).
Statistically significant ($p < 0.05$*) compared with control without current as determined by 1-way ANOVA followed by Dunnett's test.

To assess potential ocular damage resulting from the applied electric field. The typical hydration range of the cornea is reported to be between 76 and 80%. An increase in corneal hydration level beyond the usual range, namely ranging from 83 to 92%, which corresponds to an elevation of 3 to 7% units or more, signifies potential injury to the corneal epithelium and/or endothelium. The correlation between iontophoretic therapy and enhanced penetration of itraconazole was regularly observed alongside an elevation in corneal moisture levels. The corneal hydration levels following exposure to electrical currents of 0.5 mA for duration of 15 min, and 1.0 or 1.5 mA for duration of 5 min were seen to be within the expected range. However, when the exposure to 1.0 or 1.5 mA exceeded 5 min, the corneal hydration levels were discovered to be outside the normal range, suggesting the occurrence of corneal injury. The moisture level of the cornea was measured to be 80.7%, which suggests the presence of mild corneal injury. Given that the corneal moisture level was found to be less than 83%, it can be inferred that the observed damage has the potential to be reversed.

3.6 Effect of Preservative on Corneal Permeation

The data on corneal permeation of itraconazole from ophthalmic solution stored with various preservatives is presented in Table 2. The findings indicate that the inclusion of BA, BKC, EDTA, THM, a combination of MP and PP, or a combination of EDTA and BKC led to a statistically significant ($p < 0.05$) enhancement in the permeation of itraconazole from the formulations. Conversely, the addition of PMA resulted in a statistically significant ($p < 0.05$) reduction in the permeation of itraconazole from the formulation when compared to the control formulation that did not contain any preservatives.

The formulation comprising BA exhibited the highest % penetration of itraconazole, followed by the formulations containing MP + PP and BKC + EDTA. The study found that the use of BA has been shown to enhance the ability of a positively charged medication such as moxifloxacin to penetrate the cornea [28]. The rise in the penetration and presence of BA can be due to the increased partitioning of itraconazole to the cornea. The ocular transport of insulin [29] and diclofenac [30] has been observed to increase when a combination of microemulsion-based eye drops and penetration enhancers is utilized. The increase in corneal permeability caused by BKC, a type of cationic surfactant, may be attributed to its ability to emulsify and disturb the corneal epithelium [31]. The anionic compound EDTA has been documented to enhance the absorption of several medicines through undamaged corneas [33]. Furthermore, the compound EDTA has the ability to disturb the plasma membrane, resulting in an elevation of intercellular permeability [34, 35]. The addition of EDTA did not result in a statistically significant alteration in the permeability of ofloxacin through the cornea [36]. Nevertheless, the inclusion of EDTA in the formulation resulted in a slight improvement in the penetration rate of itraconazole through goat cornea when compared to the control group. The corneal penetration of fluoroquinolones such as moxifloxacin and gatifloxacin was enhanced by the concurrent use of BAK and EDTA [37, 38].

Table 3 presents the percentage permeation data of itraconazole from various formulations when subjected to iontophoretic treatment at a constant intensity of 1.5 mA for

Table 2. Effect of preservative on cumulative percentage permeation of itraconazole from different formulations.

Preservatives	Cumulative % permeation					Corneal
	15 min	30 min	60 min	90 min	120 min	hydration(%)
CONTROL	0.49 ± 0.15	0.81 ± 0.10	1.11 ± 0.10	1.94 ± 0.09	2.78 ± 0.11	79.77 ± 0.39
BA	3.55 ± 0.20	4.56 ± 0.18	6.85 ± 0.17	8.57 ± 0.08	10.93 ± 0.14	78.78 ± 0.62
BKC	1.88 ± 0.16	2.28 ± 0.07	3.09 ± 0.13	3.51 ± 0.12	4.69 ± 0.09	80.82 ± 0.23
EDTA	0.00 ± 0.00	0.06 ± 0.06	0.54 ± 0.11	0.90 ± 0.11	2.93 ± 0.13	80.15 ± 0.27
EDTA + BKC	2.44 ± 0.13	3.63 ± 0.10	5.30 ± 0.16	5.65 ± 0.13	6.81 ± 0.19	80.24 ± 0.46
MP + PP	3.78 ± 0.21	5.11 ± 0.19	5.99 ± 0.18	6.54 ± 0.21	7.94 ± 0.26	79.89 ± 0.69
PMA	0.32 ± 0.09	0.66 ± 0.11	0.84 ± 0.13	1.31 ± 0.19	1.71 ± 0.12	79.56 ± 0.36
THM	0.32 ± 0.11	0.68 ± 0.13	1.00 ± 0.13	1.73 ± 0.10	2.98 ± 0.16	79.23 ± 0.25

Values are mean ± SD (n = 3); Control was having no preservative.

Table 3. Cumulative % permeation of itraconazole from different formulations with 1.5 mA current for 5 min.

Formulations	Cumulative % permeation					Corneal
	15 min	30 min	60 min	90 min	120 min	hydration(%)
CONTROL	8.44 ± 0.35	16.87 ± 0.52	18.66 ± 0.33	19.12 ± 0.33	19.88 ± 0.11	80.96 ± 0.39
BA	14.79 ± 0.92	26.19 ± 0.71	27.94 ± 0.38	28.62 ± 0.67	31.14 ± 0.46	79.72 ± 0.64
BKC	10.78 ± 0.66	19.41 ± 0.32	22.15 ± 0.26	21.17 ± 0.38	22.47 ± 0.34	82.08 ± 0.34
EDTA	25.94 ± 0.31	32.27 ± 0.48	34.34 ± 0.60	34.69 ± 0.57	35.33 ± 0.25	80.72 ± 0.23
EDTA + BKC	20.04 ± 0.30	21.68 ± 0.37	24.06 ± 0.38	24.38 ± 0.41	24.58 ± 0.37	81.92 ± 0.67
MP + PP	19.79 ± 0.34	31.05 ± 0.40	31.91 ± 0.37	32.45 ± 0.29	35.32 ± 0.47	80.66 ± 0.17
PMA	11.90 ± 1.14	18.58 ± 0.57	17.74 ± 0.36	16.52 ± 0.65	17.16 ± 0.73	80.43 ± 0.70
THM	23.00 ± 0.30	25.92 ± 0.36	27.64 ± 0.53	28.86 ± 0.46	30.82 ± 0.56	79.55 ± 0.65

Values are mean ± SD (n = 3); Control was having no preservative.

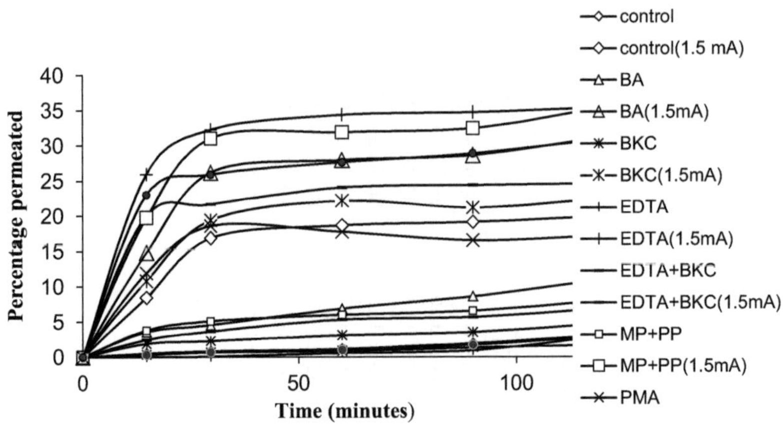

Fig. 4. Cumulative % permeation of itraconazole from different formulation with and without current.

the initial 5 min. The corresponding graphical representation of this data may be observed in Fig. 4. Table 4 displays the relative % permeation data. According to the data presented in the table, it can be observed that the corneas subjected to a consistent intensity of 1.5 mA during the initial 5-min period of the experiment exhibited a notable increase in the percentage permeation of itraconazole from the control formulation. Specifically, the permeation rate was found to be 6.56 times higher compared to the untreated control. In a similar manner, the percentage penetration of itraconazole from formulations that were preserved with various preservatives increased by a factor of 2.85 to 12.10 when a 1.5 mA intensity was applied for the first 5 min, in comparison to the untreated formulations. The application of an intensity of 1.5 mA for the initial 5 min of permeation resulted in a notable increase in corneal hydration when compared to the formulation that was left untreated. Table 5 provides a summary of the apparent permeability and relative permeability of itraconazole, both with and without current intensity. The formulation incorporating EDTA exhibited the highest observed permeability.

3.7 Antifungal Activity

The antifungal study of all the formulations was evaluated by disc diffusion method. Clear zone of inhibition were obtained. Diameter of zone of inhibition were 11.0 ± 1.0, 11.67 ± 0.58, 13.33 ± 1.15, 14.33 ± 0.58, 9.67 ± 0.58, 13.66 ± 0.56, and 13.67 ± 0.58 mm for BA, BKC, EDTA, EDTA+BKC, MP + PP, PMA and THM respectively as presented in Fig. 5.

Table 4. Relative permeation (%) of itraconazole with and without iontophoresis.

Formulations	cum.amt.permeation		cum.% permeation		Relative permeation (%)	corneal hydration (%)	
	No current	1.5 mA	No current	1.5 mA		No current	1.5 mA
CONTROL	8.33 ± 0.30	54.77 ± 0.30†	2.78 ± 0.11	19.88 ± 0.11	6.56	79.77 ± 0.39	80.96 ± 0.39
BA	30.08 ± 0.39*	85.75 ± 1.27†	10.93 ± 0.14	31.14 ± 0.46	2.85	78.78 ± 0.62	79.72 ± 0.64
BKC	12.91 ± 0.23*	61.90 ± 0.94†	4.69 ± 0.09	22.47 ± 0.34	4.81	80.82 ± 0.23	82.08 ± 0.34
EDTA	8.04 ± 0.23	97.36 ± 0.70†	2.93 ± 0.13	35.33 ± 0.25	12.10	80.15 ± 0.27	80.72 ± 0.23
EDTA + BKC	18.74 ± 0.51*	67.73 ± 1.02†	6.81 ± 0.19	24.58 ± 0.37	3.61	80.24 ± 0.46	81.92 ± 0.67
MP + PP	21.86 ± 0.71*	97.28 ± 1.29†	7.94 ± 0.26	35.32 ± 0.47	4.45	79.89 ± 0.69	80.66 ± 0.17
PMA	4.70 ± 0.33*	47.27 ± 2.02†	1.71 ± 0.12	17.16 ± 0.73	10.04	79.56 ± 0.36	80.43 ± 0.70
THM	8.18 ± 0.45	84.90 ± 1.51†	2.98 ± 0.16	30.82 ± 0.56	10.37	79.23 ± 0.25	79.55 ± 0.65

Values are mean ± SD (n = 3).
†Statistically significant ($p < 0.01$) compared with control with current as determined by 1-way ANOVA followed by Dunnett's test
*Statistically significant ($p < 0.01$) compared with control without current as determined by 1-way ANOVA followed by Dunnett's test.

Table 5. Apparent permeability of itraconazole from different formulation with and without current.

Formulations	Papp x 10^5 (without current)	Relative Papp	Papp x 10^5 (with 1.5 mA)	Relative Papp
CONTROL	0.56	1.00	3.37	6.02
BA	1.96	3.52	5.02	2.56
BKC	0.78	1.41	3.67	4.71
EDTA	0.52	0.94	5.04	9.69
EDTA+BKC	1.20	2.17	3.40	2.83
MP + PP	1.27	2.29	5.38	4.24
PMA	0.32	0.58	2.35	7.34
THM	0.55	0.99	4.32	7.85

3.8 Correlation of Corneal Permeation with Antifungal Activity

The observed antifungal activity aligns with the release profile obtained from in vitro permeation studies. These findings confirm that the drug released through the permeation process was sufficient to exert antifungal action. The fungal inhibition zones provide evidence that the diffused drug remained bioactive in the formulation.

4 Conclusion

The aqueous solubility of Itraconazole was increased by solid dispersion technique by using the polymeric solubiliser SOLUPLUS. DSC thermogram and PXRD concludes the presence of amorphous form of Itraconazole in solid dispersion powder. The results show that there was an increasing corneal amount permeation of itraconazole when BA, BKC, EDTA, THM, combination of MP and PP or combination of EDTA & BKC were employed, whereas PMA was found to decrease the amount permeation as compared with control formulation containing no preservatives. Increase in current intensity from 0.5 to 1.5 mA and iontophoresis time duration from 5 min to 15 min resulted in increased in drug permeation. The % permeation of itraconazole from formulations preserved with different preservatives enhanced from 2.85 to 12.10 times due to application of 1.5 mA intensity for initial 5 min as compared with the untreated formulations. Clear zone of inhibition revealed the microbial efficacy of the itraconazole formulations.

Acknowledgements. Authors thanks for providing the research facility by Management of IPT, Salipur, Cuttack, Odisha.

Declarations

Ethical Approval Not applicable

Competing Interests The authors declare that they have no known competing financial interests or personal relationships that could have appeared to influence the work reported in this paper.

Authors' Contributions All authors certify that they have participated sufficiently in the work to take public responsibility for the content, including participation in the concept, design, analysis, writing, or revision of the manuscript. **BSN**: Supervision, Writing- Original draft preparation, Conceptualization, Methodology, and Investigation. **BM**: Visualization, Investigation and data interpretation, **BMi:** Software and Validation, **SD**: Revision, data curation.

Funding This research did not receive any specific grant from funding agencies in the public, commercial, or not-for-profit sectors.

Availability of Data and Materials The data will be available on request.

References

1. Tsai, Y.C., Tsai, T.F.: Itraconazole in the treatment of nonfungal cutaneous diseases: a review. Dermatol. Ther. **9**, 271–280 (2019)
2. El-Sheridy, N.A., Ramadan, A.A., Eid, A.A., El-Khordagui, L.K.: Itraconazole lipid nanocapsules gel for dermatological applications: *in vitro* characteristics and treatment of induced cutaneous candidiasis. Colloids Surf. B: Biointerfaces. **181**, 623–631 (2019)
3. De Mohac, L.M., Raimi-Abraham, B., Caruana, R., Gaetano, G., Licciardi, M.: Multicomponent solid dispersion a new generation of solid dispersion produced by spray-drying. J. Drug Deliv. Sci. Technol. **57**, 101750 (2020)
4. Nair, A.R., Lakshman, Y.D., Anand, V.S., Sree, K.N., Bhat, K., Dengale, S.J.: Overview of extensively employed polymeric carriers in solid dispersion technology. AAPS PharmSciTech. **21**, 1–20 (2020)
5. Pandi, P., Bulusu, R., Kommineni, N., Khan, W., Singh, M.: Amorphous solid dispersions: an update for preparation, characterization, and mechanism on bioavailability, stability, regulatory considerations and marketed products. Int. J. Pharm. **586**, 119560 (2020)
6. Argüeso, P., Woodward, A.M., AbuSamra, D.B.: The epithelial cell glycocalyx in ocular surface infection. Front. Immunol. **12**, 729260 (2021)
7. Wei, D., Pu, N., Li, S.Y., Wang, Y.G., Tao, Y.: ApplicatiKon of iontophoresis in ophthalmic practice: an innovative strategy to deliver drugs into the eye. Drug Deliv. **30**, 2165736 (2023)
8. Helmy, A.M.: Overview of recent advancements in the iontophoretic drug delivery to various tissues and organs. J. Drug Deliv. Sci. Technol. **61**, 102332 (2021)
9. Chen, K., Puri, V., Michniak-Kohn, B.: Iontophoresis to overcome the challenge of nail permeation: considerations and optimizations for successful ungual drug delivery. AAPS J. **23**, 1–5 (2021)
10. Mofidfar, M., Abdi, B., Ahadian, S., Mostafavi, E., Desai, T.A., Abbasi, F., Sun, Y., Manche, E.E., Ta, C.N., Flowers, C.W.: Drug delivery to the anterior segment of the eye: a review of current and future treatment strategies. Int. J. Pharm. **607**, 120924 (2021)
11. Bilmin, K., Synoradzki, K.J., Czarnecka, A.M., Spałek, M.J., Kujawska, T., Solnik, M., Merks, P., Toro, M.D., Rejdak, R., Fiedorowicz, M.: New perspectives for eye-sparing treatment strategies in primary uveal melanoma. Cancer. **14**, 134 (2021)
12. Wu, H., Li, L., Luo, S., Fang, X., Shang, X., Xie, Z., Xiao, X., He, H., Lin, Z., Liu, Z.: Safety and efficacy of repeated crosslinking assisted by transepithelial double-cycle iontophoresis in keratoconus progression after primary corneal crosslinking. Eye. **35**, 3020–3027 (2021)

13. Souza, J.G., Dias, K., Silva, S.A., de Rezende, L.C., Rocha, E.M., Emery, F.S., Lopez, R.F.: Transcorneal iontophoresis of dendrimers: PAMAM corneal penetration and dexamethasone delivery. J. Control. Release. **200**, 115–124 (2015)
14. Dos Santos, G.A., Ferreira-Nunes, R., Dalmolin, L.F., Dos Santos Ré, A.C., Anjos, J.L., Mendanha, S.A., Aires, C.P., Lopez, R.F., Cunha-Filho, M., Gelfuso, G.M., Gratieri, T.: Besifloxacin liposomes with positively charged additives for an improved topical ocular delivery. Sci. Rep. **10**, 19285 (2020)
15. Perez, V.L., Wirostko, B., Korenfeld, M., From, S., Raizman, M.: Ophthalmic drug delivery using iontophoresis: recent clinical applications. J. Ocul. Pharmacol. Ther. **36**, 75–87 (2020)
16. Li, M., Zhang, L., Li, R., Yan, M.: New resveratrol micelle formulation for ocular delivery: characterization and in vitro/in vivo evaluation. Drug Develop. Ind. Pharmacy. **46**, 1960–1970 (2020)
17. Mehra, N., Aqil, M., Sultana, Y.: A grafted copolymer-based nanomicelles for topical ocular delivery of everolimus: formulation, characterization, *ex-vivo* permeation, *in-vitro* ocular toxicity, and stability study. Eur. J. Pharm. Sci. **159**, 105735 (2021)
18. Vigani, B., Rossi, S., Sandri, G., Bonferoni, M.C., Caramella, C.M., Ferrari, F.: Recent advances in the development of *in situ* gelling drug delivery systems for non-parenteral administration routes. Pharmaceutics. **12**, 859 (2020)
19. Melles, G.R., Rietveld, F.J., Beekhuis, W.H., Binder, P.S.: A technique to visualize corneal incision and lamellar dissection depth during surgery. Cornea. **18**, 80–86 (1999)
20. Choi, J.S., Lee, S.E., Jang, W.S., Byeon, J.C., Park, J.S.: Solid dispersion of dutasteride using the solvent evaporation method: approaches to improve dissolution rate and oral bioavailability in rats. Mater. Sci. Eng. C. **90**, 387–396 (2018)
21. Mai, N.N., Nakai, R., Kawano, Y., Hanawa, T.: Enhancing the solubility of curcumin using a solid dispersion system with hydroxypropyl-β-cyclodextrin prepared by grinding, freeze-drying, and common solvent evaporation methods. Pharmacy. **8**, 203 (2020)
22. Maddiboyina, B., Roy, H., Nakkala, R.K., Gandhi, S., Kavisri, M., Moovendhan, M.: Formulation, optimization and characterization of raloxifene hydrochloride loaded PLGA nanoparticles by using Taguchi design for breast cancer application. Chem. Biol. Drug Design. **102**, 457–470 (2023)
23. Roy, H., Nayak, B.S., Maddiboyina, B., Nandi, S.: Chitosan based urapidil microparticle development in approach to improve mechanical strength by cold hyperosmotic dextrose solution technique. J. Drug Deliv. Sci. Technol. **76**, 103745 (2022)
24. Roy, H., Panda, S.P., Panda, S.K., Tripathi, A.K., Srivastava, S.K., Nayak, B.S., Singh, P.K., Singh, G.D.: N-trimethyl chitosan and tripalmitin loaded solid lipid nanoparticles of tofacitinib citrate: characterization and in-vivo anti-inflammatory assessment. J. Drug Deliv. Sci. Technol. **26**, 104789 (2023)
25. Roy, H., Srungarapati, S., Gade, NJ., Gummadi, A., Karunasree, BK., Dakkumalla, M., Maddiboyina, B.: Citicoline loaded nanoemulsion enriched with D-alpha-tocopherol acetate and protein: formulation and in-silico study. J. Drug Deliv. Sci. Technol. 7, 104340 (2023)
26. Liu, Z., Huang, A.J., Pflugfelder, S.C.: Evaluation of corneal thickness and topography in normal eyes using the Orbscan corneal topography system. Br. J. Ophthalmol. **83**, 774–778 (1999)
27. Permana, A.D., Utami, R.N., Layadi, P., Himawan, A., Juniarti, N., Anjani, Q.K., Utomo, E., Mardikasari, S.A., Arjuna, A., Donnelly., R.F.: Thermosensitive and mucoadhesive in situ ocular gel for effective local delivery and antifungal activity of itraconazole nanocrystal in the treatment of fungal keratitis. Int. J. Pharm. **602**, 120623 (2021)
28. Pawar, P.K., Majumdar, D.K.: Effect of formulation factors on *in vitro* permeation of moxifloxacin from aqueous drops through excised goat, sheep, and buffalo corneas. AAPS PharmSciTech. **7**, E 89–94 (2006)

29. Ahuja, M., Singh, G., Majumdar, D.K.: Effect of formulation parameters on corneal permeability of ofloxacin. Sci. Pharm. **76**, 505–514 (2008)
30. Ahuja, M., Dhake, A.S., Majumdar, D.K.: Effect of formulation factors on *in vitro* permeation of diclofenac from experimental and marketed aqueous eye drops through excised goat cornea. Yakugaku Zasshi. **126**, 1369–1375 (2006)
31. Mohanty, B., Mishra, S.K., Majumdar, D.K.: Effect of formulation factors on *in vitro* transcorneal permeation of voriconazole from aqueous drops. J. Adv. Pharm. Technol. Res. **4**, 210 (2013)
32. Kaur, I.P., Smitha, R.: Penetration enhancers and ocular bioadhesives: two new avenues for ophthalmic drug delivery. Drug Dev. Ind. Pharm. **28**, 353–369 (2002)
33. Nakamura, T., Yamada, M., Teshima, M., Nakashima, M., To, H., Ichikawa, N., Sasaki, H.: Electrophysiological characterization of tight junctional pathway of rabbit cornea treated with ophthalmic ingredients. Biol. Pharm. Bull. **30**, 2360–2364 (2007)
34. Rojanasakul, Y., Liaw, J., Robinson, J.R.: Mechanisms of action of some penetration enhancers in the cornea: laser scanning confocal microscopic and electrophysiology studies. Int. J. Pharm. **66**, 131–142 (1990)
35. Thakral, S.: EDTA decreases in vitro transcorneal permeation of fluconazole in phosphate buffer through excised sheep cornea. J. Pharm. Negat. Results. **2**, 24–27 (2011)
36. Rathore, M.S., Majumdar, D.K.: Effect of formulation factors on *in vitro* transcorneal permeation of gatifloxacin from aqueous drops. AAPS PharmSciTech. **7**, E12–7 (2006)
37. Nair, A.B., Aldhubiab, B., Shah, J., Jacob, S., Attimarad, M., Sreeharsha, N., Venugopala, K.N., Joseph, A., Morsy, M.A.: Design, development, and evaluation of constant voltage iontophoresis for the Transungual delivery of Efinaconazole. Pharmaceutics. **15**, 1422 (2023)
38. Ahmed, S., Amin, M.M., Sayed, S.: Ocular drug delivery: a comprehensive review. AAPS PharmSciTech. **24**, 66 (2023)

Development and Assessment of Cubosomal Nanoformulation Loaded with β – Sitosterol for Improving Antioxidant Activity

Ramsha Aslam[1], Vikas Bhatt[2], Priya Pandey[1], and Disha Dutta[1*](✉)

[1] Devsthali Vidyapeeth College of Pharmacy, Affiliated to Veer Madho Singh Bhandari Uttarakhand Technical University, Rudrapur, India
dishadas007@gmail.com

[2] Faculty of Pharmaceutical Sciences, Amrapali University, Haldwani, India

Abstract. The limitations of synthetic drugs, which are commonly characterized by limited bioavailability and poor water solubility, make treating dermatological conditions difficult. According to recent research, β-sitosterol (β-ST), a naturally occurring substance known for its strong anti-inflammatory and antioxidant qualities, may be used as a treatment for conditions affecting the skin. This study explores the encapsulation of β-ST within cubosomal nano-formulations as a solution to the problems of bioavailability and solubility related to hydrophobic bio actives. Glyceryl monooleate (GMO) and P 407 (P407) were used as the lipid phase and Poly vinyl alcohol (PVA) as the aqueous phase in the emulsification process utilized to create CBs. The physicochemical characteristics of the Cubosomes (CBs) loaded with β-ST were thoroughly examined. Particle size, zeta potential, scanning electron microscopy (SEM), encapsulation efficiency, and differential scanning calorimetry (DSC) evaluations were all part of the characterization process. A uniform distribution was suggested by the experimental data, which showed an average particle size of 139.7 nm, a zeta potential of −15.17 mV, and a low polydispersity index (PDI) of 0.225. SEM imaging verified the improved formulation's cubic structure and nanoscale dimensions. The formulation also showed a positive yield and great encapsulation efficiency. Ascorbic acid was used as a comparison standard in the 2,2-Diphenyl-1-picrylhydrazyl (DPPH) experiment to evaluate the antioxidant activity of the β-ST-loaded CBs. According to the study's findings, cubosomal encapsulation offers a viable method to improve antioxidant activity, and facilitating regulated medication release—all of which eventually contribute to long-term storage stability.

Keywords: β- Sitosterol · CBs · Nanoformulation · Skin disease · Topical drug delivery

A. K. Saxena, A. Saxena (Eds.): GTHTM 2025, pp. 487–501, 2026.
https://doi.org/10.1007/978-3-032-12320-6_29

1 Introduction

Delivering drugs effectively to the skin for systemic and local effect without causing irritation and other major problems is a formidable task. The most of the drugs are unable to get through the stratum corneum, a very effective barrier. Few drugs are able to deeply permeate the skin. Drug delivery methods have advanced significantly in recent years, especially with regard to the creation of several nanoformulations [1]. Topical convectional formulations including gels, ointments, lotions and creams have so many drawbacks including partitioning of drugs in the skin surface, greasiness and stickiness of the ointments, insufficient absorption of the drugs through the skin and poor patient compliance [2]. To address these limitations, the concept of novel topical formulations based on nanotechnology, particularly lipid based nanoformulations, is gaining popularity [3]. In recent years, CBs introduced due to their advantageous features (such as safeguarding the drug, enhancing its availability, enhancing solubility, minimizing dosage requirements or prolonging drug release, and delivery of hydrophobic drugs) and are thought to be a viable Lipid-based nanosystem for skin problems. Their unique characteristics and superb affinity with a broad variety of active substances enable their extensive uses, including as drug delivery for chemotherapy and for topical, oral, ophthalmic, and transdermal applications [4, 5].

Therapeutic medications currently available often present challenges, including high costs and limited efficacy, leading to increased interest in new candidates derived from natural resources. One promising compound is β-ST, a phytosterol found in various plant parts such as leaves, fruits, and rhizomes. Numerous studies have highlighted its pharmacological potential, demonstrating capabilities in areas such as anti-inflammatory effects, treatment of alopecia, cancer prevention, angiogenesis, diabetes prevention, antimicrobial properties, and immunomodulatory effects [6–8]. Notably, β-ST has been recognized for its role in treating androgenetic alopecia due to its activity as a 5α-reductase inhibitor [9]. However, despite its beneficial properties, β-ST faces significant limitations, including poor solubility in water and oils, the necessity for high daily dosages (up to 3 g/day), and its crystalline nature at ambient and physiological temperatures, which hinder oral administration and bioavailability. A major challenge in utilizing β-ST therapeutically is the effective incorporation of this compound into drug formulations [10]. To overcome these limitations, the use of CBs as a carrier for topical drug delivery has been suggested, as they offer several advantages over traditional liposomes, potentially enhancing the delivery and efficacy of β-ST in therapeutic applications. This exploration underscores the importance of natural resources in drug development and highlights the potential of advanced drug delivery systems to improve the therapeutic applications of β-ST.

Hence, the aim of this research was to develop a β-ST cubosomal dispersion formulation to improve the solubility of β-ST, get a steady release of drug, enhancing the physical stability and investigating their potential use for improving topical delivery and bioavailability of drugs. The emulsification method was used for the preparation of β-ST CBs by using 2^3 factorial design. GMO and P 407 was used as a lipid phase and PVA was used as a stabilizer. β-ST encapsulated CBs were characterized by, particle size, PDI, zeta potential, DSC, SEM, drug content, and drug entrapment. SEM was used for the surface morphology of CBs and DSC was used to look the interactions between the drugs and lipids in the mixtures [11, 12]. In vitro study was also assessed by using dialysis bag in phosphate buffer solution. The optimised formulation's release dynamics were investigated. In order to figure out how long the optimised formulation will remain stable under various storage conditions, stability studies were conducted last. Over 30 d, the size of the particles and how well they were caught were watched to see if there were any noticeable changes [13]. Antioxidant activity of β-ST was also compared with ascorbic acid. Therefore, the created CBs with the optimized formulation offer a promising method to increase drug solubility, ensure a continuous release of pharmaceuticals, and ensure that β-ST remains stable over an extended period of time.

2 Materials and Methods

2.1 Materials

The phytoconstituent and GMO was procured from Yarrow Chem, Mumbai, India. P 407 was purchased from Yarrow Chem, Mumbai, India and PVA purchased from LOBA Chemie.

2.2 Methods

2.2.1 Preparation of β-ST Loaded CBs by Emulsification Method. As shown in Table 1, CBs dispersions were generated by emulsifying a cube-shaped lipid component made up of GMO and P 407 in PVA-containing water. Firstly, melt Glycerol Monooleate (GMO) and P 407 (P407) in a beaker for the preparation of lipid phase. Use magnetic stirrer and stir the mixture at 70 °C. Dissolve β-ST and PVA in distilled water and heat upto 70 °C for the preparation of aqueous phase. Add heated aqueous phase into the melted lipid mixture then, stir mechanically the mixture at 500 rpm for 2 h. Ensure mixture is fully homogeneous

Table 1. Formulation development chart.

S. No.	Formulation code	Lipids GMO (%w/w)	P (407) (%w/w)	PVA (%w/w)	Drug (mg)	Solvent (ml) Distilled water
1	F1	7	0.35	2	100	240
2	F2	7	0.35	2.5	100	240
3	F3	7	0.35	3	100	240
4	F4	7	5	2	100	240
5	F5	7	5	2.5	100	240
6	F6	7	5	3	100	240
7	F7	7	7.5	2	100	240
8	F8	7	7.5	2.5	100	240
9	F9	7	7.5	3	100	240

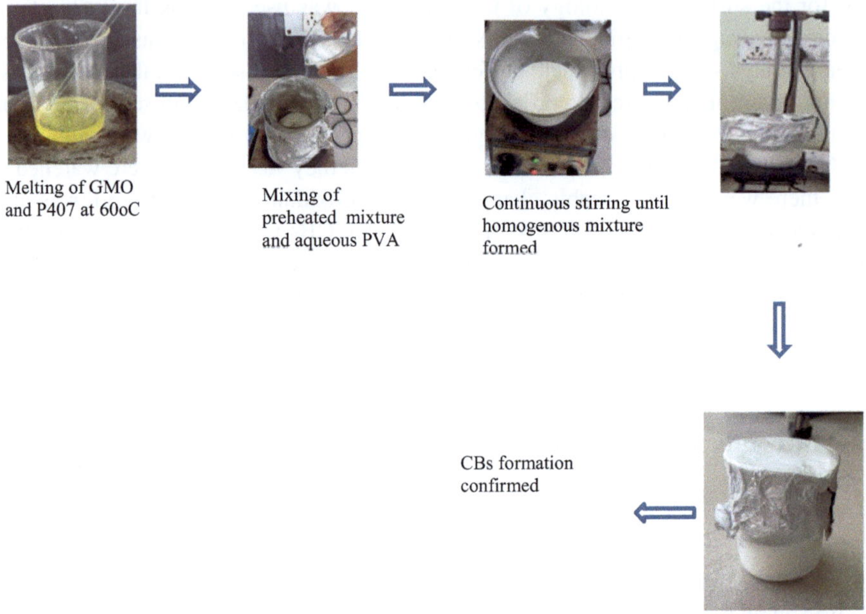

Fig. 1. Illustration of emulsification method.

(cubosomal dispersions). Homogenize the dispersion at 15,000 rpm for 1 min. Cubosomal dispersion will formed. Refrigerate the cubosomal dispersion in amber glass vials for further investigation [14] (Fig. 1).

2.3 Characterization of β-ST Loaded CBs

2.3.1 Determination of Entrapment Efficiency (EE). The EE of the cubosomal formulation was evaluated by centrifuging the sample for one hour at 4 °C and 5000 rpm. This process effectively separated the unencapsulated β-ST from the formulation. The supernatant was subsequently diluted with methanol to enable analysis. After dilution, the supernatant underwent spectrophotometric measurement at a wavelength of 209 nm, with methanol serving as the blank solvent [15, 16]. The EE) was determined using a pre-established calibration curve, calculated according to the following formula:

$$\%EE = \frac{\text{Total amount of drug} - \text{Amount of drug entrapped}}{\text{Total amount of drug}} \times 100$$

This formula provides a quantitative measure of the efficiency with which β-ST was encapsulated within the cubosomal formulation.

2.3.2 Size Distribution and PDI. Dynamic light scattering (DLS) with a Malvern zeta sizer at a set angle was used to measure the particle size and PDI for the CBs formulations. Before measuring the size of a particle, one drop of sample was taken from the selected formulation and then diluted the sample with dispersion medium (distilled water). The test was determined at a temperature of 25 °C [17].

The particle size and PDI of the cubosomal formulations were analyzed using DLS with a Malvern Zetasizer. To prepare the sample for measurement, a drop from the selected formulation was diluted in distilled water. The measurements were performed at a temperature of 25 °C. This approach allows for an evaluation of the size distribution and uniformity of the CBs particles [17].

2.3.3 Zeta Potential Analysis. The zeta potential of the cubosomal formulation was assessed using a Malvern Zetasizer. This measurement evaluates the surface charge of the cubosomal dispersion and provides insights into the stability of the formulation. After dispersing the β-ST loaded CBs in distilled water at 25 °C, the zeta potential was measured in triplicate to ensure accuracy and reliability of the results [18].

2.3.4 Shape and Surface Morphology. The morphology of the optimized β-ST loaded CBs formulation was analyzed using SEM. The morphology of the optimized β-ST-loaded cubosomal formulation was examined using scanning electron microscopy (SEM). This technique provides detailed visual information about the structure and surface characteristics of the CBs, contributing to the understanding of their physical properties and potential functionality [19–22].

2.3.5 Differential Scanning Calorimetry. DSC conducted on the formulated β-ST-loaded CBs to evaluate the solubility of the drug and the effects of thermal changes on the sample's physical state. The thermograms of the pure drug, GMO, PVA, and the optimized CBs formulation were compared over a scanning range of 30–400 °C. The analysis involved a 1-min hold at 30 °C and a heating rate of 10 °C/min, using alumina under a nitrogen atmosphere in sealed aluminum pans [23].

2.3.6 In-Vitro Release Study of Prepared β-ST-Loaded CBs. To conduct the in-vitro release study of β-ST-loaded CBs, a dialysis bag method was employed. The dialysis membranes were immersed overnight in a dissolution medium. After precise weighing, the β-ST cubosomal dispersion was placed inside the dialysis bag, which was then submerged in a 250 mL beaker filled with 5.5 pH phosphate buffer solution. At specified time intervals of 0.5, 1, 2, 3, 4, 5, 6, 7, 8, 15, and 24 h, 5 mL samples were withdrawn and replaced with an equivalent volume of the 5.5 pH phosphate buffer. The absorbance of each sample was measured at 265 nm using a UV spectrophotometer for analysis [24–26].

2.3.7 Drug Release Kinetics. The in-vitro drug release data were analyzed using various mathematical models, including the zero-order model, first-order model, Higuchi diffusion model, Korsmeyer-Peppas model, and Hixon-Crowell model. These models were employed to elucidate the mechanisms underlying the drug release process [27–30].

2.3.8 Stability Studies. A stability study was conducted on the optimized formulation over a period of one month, with storage conditions maintained at temperatures between 2 and 8° Celsius and a relative humidity of 60% [31].

2.3.9 Antioxidant Effect by DPPH Radical Scavenging Activity. It is a widely used method for assessing the antioxidant potential of a compound. This assay tests the ability of a substance to neutralize free radicals. The method involves measuring the decrease in absorbance of the DPPH radical at 517 nm. A reduction in absorbance indicates the scavenging activity of the compound [32].

Prepare β-ST loaded CBs and ascorbic acid at concentrations: 20, 40, 80, 60, 100 μg/ml. Then from these concentration withdrawn 3 ml from each of dilutions and mix 1 ml of 0.1 mM DPPH methanolic solution to each concentration of β-ST loaded CBs and ascorbic acid. Then, incubate the mixture at 37 °C in the dark for 40 min. Use methanol as the sample blank for β-ST loaded CBs and ascorbic acid. Determine the absorbance of sample by utilizing UV spectrophotometer at wavelength of 517 nm [33]. A lower absorbance indicates a stronger scavenging capacity.

The appropriate formula for the estimation of % Radical scavenging activity:

$$\text{Scavenging}(\%) = \frac{(\text{A Control}) - (\text{A Sample})}{(\text{A Control})} \times 100$$

3 Results and Discussion

3.1 Preparation of β-ST CBs

The β-ST loaded CBs were created following the procedure outlined in Table 1. GMO and P 407 were utilized at varying concentrations (% w/w) for CBs formation. PVA served as a stabilizer at a concentration of 2.5% (w/v). PVA contributes to the formation of small particles with a uniform distribution, ensuring they are easily dispersed in a liquid medium. Stabilizers are often incorporated into the cubosomal dispersion to modify its surface properties. The modification of cubosomal nanoparticles with P 407 involved the adsorption of the hydrophobic segments of this polypropylene oxide copolymer onto the outer surface of the CBs.

3.2 Characterization of β-ST Loaded CBs

3.2.1 Entrapment Efficiency. The results indicated the highest exhibited EE of 65.459 ± 0.5%, demonstrating that a significant portion of the drug was successfully encapsulated within the CBs. This high EE can be attributed to the hydrophobic characteristics of the drug, which is soluble in GMO, allowing it to be effectively trapped in the hydrophobic channels of the CBs. These findings suggest that increasing the amount of GMO in the formulation improves EE, while a higher concentration of P407 negatively impacted drug entrapment [34, 35].

Increasing the concentration of PVA resulted in a rise in the viscosity of the dispersion, which contributed to the rigidity of the nanoparticles. This enhanced rigidity led

Results

			Size (d.n...	% Intensity:	St Dev (d.n...
Z-Average (d.nm):	139.7	Peak 1:	185.8	99.6	110.1
PdI:	0.225	Peak 2:	4826	0.4	712.2
Intercept:	0.967	Peak 3:	0.000	0.0	0.000
Result quality	Good				

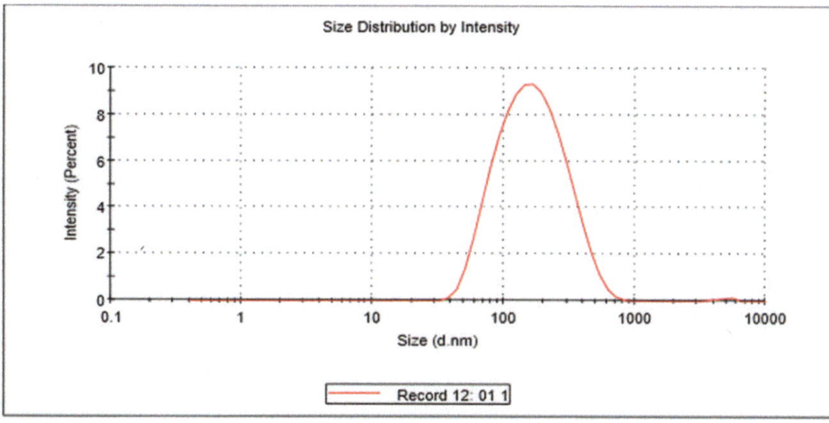

Fig. 2. Particle size of optimized formulation.

to a reduction in the leakage of the encapsulated drug from the dispersions, thereby improving the potential therapeutic efficacy [36, 37].

3.2.2 Particle Size and PDI. Drug penetration through the skin is significantly influenced by the cubosomal dispersion particle size. A particle size of 139.7 nm was measured for the optimized β-ST loaded cubosomal dispersion, demonstrating the successful production of Nano sized particles, which is beneficial for drug delivery applications. Because of their small size, particles in this range usually exhibit better bioavailability, greater surface area, and enhanced drug delivery effectiveness [38, 39]. The uniformity and distribution of particle sizes within the dispersion are reflected by the PDI. A uniform dispersion with a small particle size distribution is suggested by a low PDI value of 0.225. Figure 2 displays the particle size distribution graph for the optimized formulation.

Peak 1: 185.8 nm (Intensity: 99.6%, Std Dev: 110.1 nm), Peak 2: 4826 nm (Intensity: 0.4%, Std Dev: 712.2 nm), Peak 3: Not detected The presence of Peak 1 as the dominant population (99.6%) signifies that the majority of particles are in the nano range, while Peak 2 (0.4%) represents a negligible fraction, possibly aggregates or oversized particles. Additionally, the intercept value of 0.967 and result quality marked as "Good" confirm the reliability and precision of the measurement.

3.2.3 Zeta Potential Analysis. Zeta potential measurements assess the surface charge of the prepared cubosomal dispersion and can predict the long-term stability of the β-ST-loaded CBs [40]. The optimized formulation exhibited a zeta potential of −15.17 mV, indicating that the dispersed particles maintain stability in the medium without aggregation. A negative

System

Temperature (°C):	25.0	Zeta Runs:	12
Count Rate (kcps):	243.2	Measurement Position (mm):	2.00
Cell Description:	Clear disposable zeta c...	Attenuator:	7

Results

			Mean (mV)	Area (%)	St Dev (mV)
Zeta Potential (mV):	-15.17	Peak 1:	-15.05	99.0	4.39
Zeta Deviation (mV):	4.82	Peak 2:	-24.6	1.0	1.72
Conductivity (mS/cm):	0.436	Peak 3:	0.00	0.0	0.00
Result quality	Good				

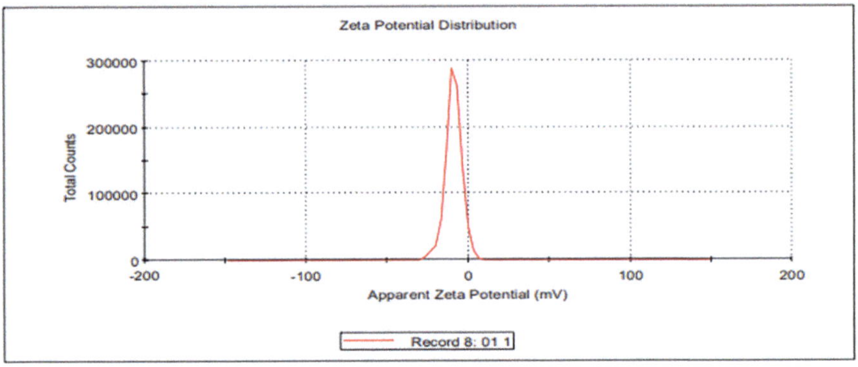

Fig. 3. Zeta potential of the optimized formulation.

zeta potential reflects electrostatic repulsion among the particles, which helps prevent aggregation and ensures the overall stability and quality of the formulation. The zeta potential graph for the optimized formulation is illustrated in the Fig. 3. The mean zeta potential of −15.17 mV reflects a moderately stable colloidal system, suggesting electrostatic repulsion is sufficient to prevent aggregation. Peak 1 with 99% area around −15.05 mV confirms that the majority of particles have a consistent surface charge. The low standard deviation (4.82 mV) indicates tight distribution and consistent measurement. The good result quality ensures that the readings are reliable and accurate.

3.2.4 Shape and Surface Morphology. The shape and surface morphology of optimized β-ST loaded CBs formulation was imaged with SEM shown in Fig. 4. The SEM micrograph clearly shows the morphological characteristics of the cubosomal dispersion at 4500× magnification. The particles appear to be discrete, non-agglomerated, and exhibit a rough surface texture, which is beneficial for increased surface interaction and drug loading. The scale bar of 5 μm suggests the particle sizes are in the nano-to-submicron range, consistent with DLS results (~139.7 nm). The image confirms good dispersion and minimal aggregation, supporting the stability and quality of the cubosomal system.

3.2.5 Differential Scanning Calorimetry. The thermogram of β-ST pure showed a clear, sharp peak at 140 °C, respectively. The DSC thermogram of GMO and P407 showed the peak at Fig. 5 (a) and (b) 27.01 °C and 41.51 °C, respectively. The DSC thermogram of optimized formulation exihibits broadband between 39 °C- 106.92 °C due to the interaction

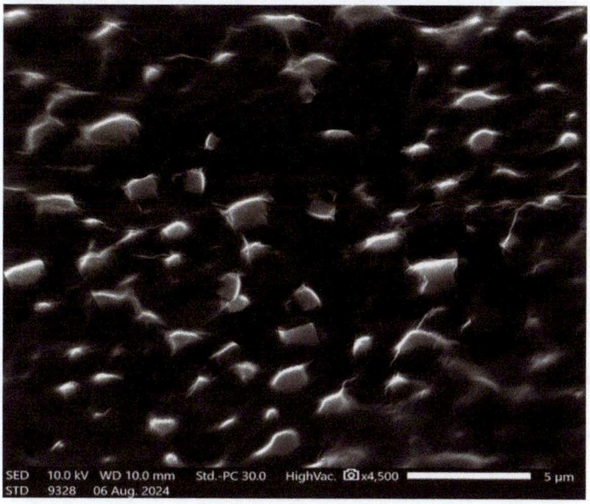

Fig. 4. Morphology of β-ST-LC by Scanning electron microscopy.

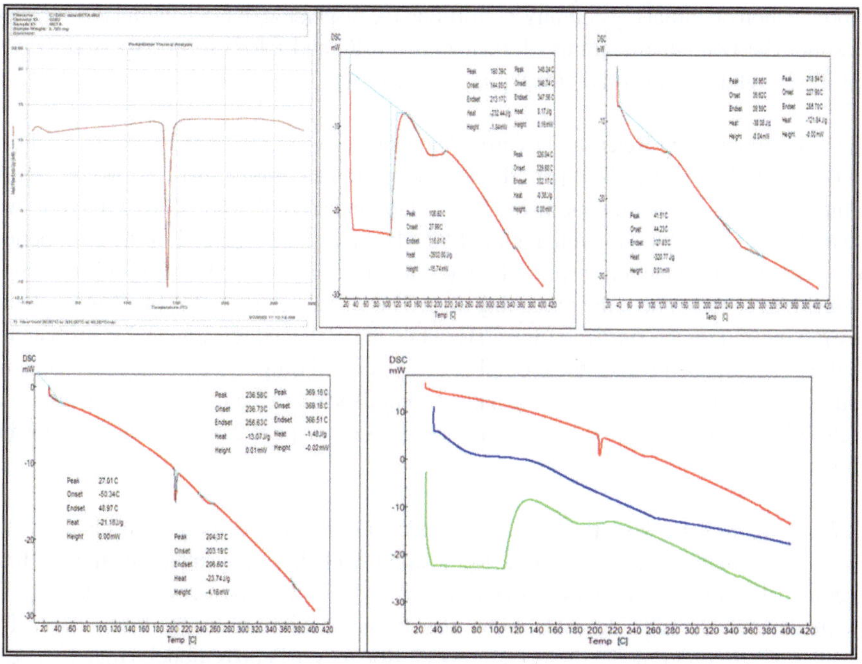

Fig. 5. (a) DSC of pure drug (b) – GMO, -P407, -Beta –Sitosterol CBs.

between the drug and other ingredients of the formulation. This shows the crystalline state of the β-ST loaded cubosomal dispersion. The pure β-ST peak at 140 °C disappears in the formulation, suggesting successful incorporation and interaction with the cubosomal matrix. The broad thermal events observed in the optimized formulation are indicative of

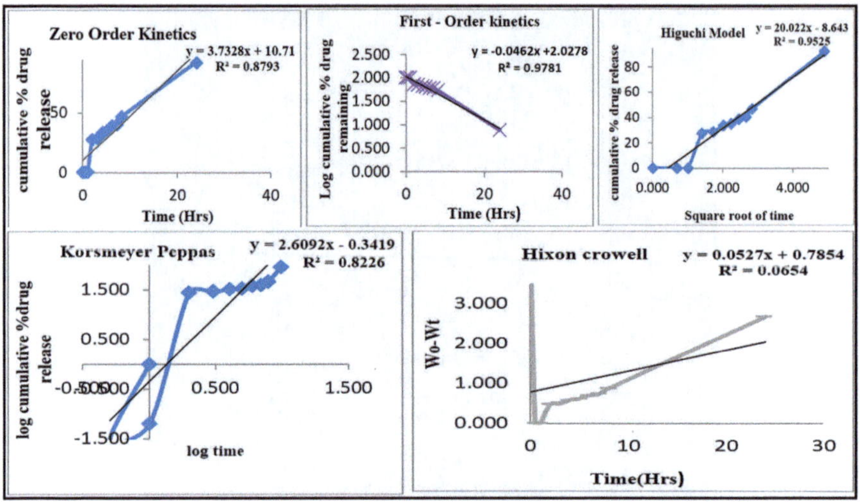

Fig. 6. Drug release kinetics of optimized – formulation.

amorphization or partial crystallinity, confirming the dispersion of β-ST within the excipient network. The DSC thermograms of excipients (GMO and P407) show expected thermal transitions, validating their individual phase behaviors.

3.2.6 In-Vitro Drug Release. In vitro release studies were performed for both β-ST and β-ST loaded CBs over a 24-h period, utilizing phosphate buffer (pH 5.5) as the release medium. The results for the optimized formulation indicated that 27.56% of the drug was released within the first 2 h, with a total of 92.44% released by the end of the 24-h period. The cubosomal formulation displayed a rapid initial release in the first 2 h, followed by a gradual release throughout the remaining time. This initial burst effect is likely attributed to the drug located on the surface of the CBs, while the sustained release can be attributed to the drug encapsulated within the CBs [41, 42]. Sustained drug release is advantageous for reducing dosing frequency, enhancing patient compliance, and maintaining therapeutic drug levels [13, 43]. In contrast, the release characteristics of the pure drug solution were notably different.

3.2.7 Drug Release Kinetics. Several kinetic models were employed to elucidate the release mechanism of the cubosomal formulation, including the zero-order model, first-order model, Korsmeyer-Peppas model, Higuchi model, and Hixon-Crowell model (Fig. 6). The model with R^2 value closest to 1.000 was deemed the best fit for the formulation. The R^2values of zero order model, first-order model, Korsmeyer-Peppas model, Higuchi model, and Hixon-Crowell model were 0.8793, 0.9781, 0.8226, 0.9525 and 0.0654 respectively. Based on the R^2values obtained, both the Higuchi model and zero-order model (Table 2) were identified as the most suitable for describing the release profile. The zero-order model, according to existing literature, indicates that the release process is independent of the concentration of the dissolved drug.

Table 2. R^2 value of different release kinetics model.

Models	Zero-order	First-order	Korsmeyer Peppas	Higuchi	Hixon-crowell
R^2	0.8793	0.9781	0.9525	0.8226	0.064

Table 3. Stability studies.

Storage conditions	Particle size(nm)			EE(%)		
	Initial	30d	90d	Initial	30d	90d
2-8 °C	139.7	141.5	147.6	65 ± 0.553	64.85 ± 0.072	64.18 ± 0.072
25 °C/ 60%RH	139.7	140.2	145.5	65 ± 0.553	64.12 ± 0.193	63.76 ± 0.546

3.2.8 Stability Study. The β-ST-loaded CBs exhibited robust stability over a 90-days period, with no significant changes observed in particle size or EE. A slight increase in particle size and decrease in entrapment efficiency was recorded in 3 months of storage conditions. Initially, the particle size was found to be 139.7 nm. After 3 months of refrigerated storage condition (2-8 °C) and 25 °C/ 60%RH storage condition the particle size was 147.6 nm and 154.5 nm respectively. Similarly, initially %EE was 65 ± 0.553%. After 3 months of storage in refrigerated temperature condition and 25 °C/ 60%RH storage condition the %EE were 64.18 ± 0.072% and 63.76 ± 0.546% respectively. This stability underscores their potential for long-term storage and commercialization [44]. Stability is a critical factor for the successful transition of nanoparticle-based drug delivery systems from the laboratory to practical applications, as summarized in the Table 3.

3.2.9 Antioxidant Activity. In this test, the DPPH radical, originally deep violet in color, becomes colorless or pale yellow upon reaction with a free radical scavenger.

The free radical scavenging activity of various concentration of 20, 40, 60, 80 and 100 μg/ml β-ST loaded CBs was compared with, ascorbic acid as standard at a reaction time of 30 min (Fig. 7). The IC_{50} of β-ST loaded CBs (4.61 μg/ml) was lower in comparison to Ascorbic Acid (17.38 μg/ml) indicating that β-ST loaded CBs has more powerful in antioxidant activity than ascorbic acid as lower the IC50 value higher the antioxidant activity. The results are shown in Table 4.

4 Conclusion

β-ST exhibits a broad spectrum of therapeutic effects, including antidiabetic, antibacterial, anti-inflammatory, and antioxidant properties. Despite its effectiveness in both in vitro and in-vivo studies, its insolubility has hindered its therapeutic use. Recent advancements in drug delivery methods, particularly the development of various nanoformulations, have prompted research aimed at enhancing the efficacy and bioavailability of β-ST. This study focuses on addressing these challenges by incorporating β-ST into a lipid-based nanoformulation using an emulsification technique. The goal is to develop a cost-effective, biodegradable, and stable cubosomal formulation that offers improved solubility, drug entrapment, stability, and a sustained release profile through

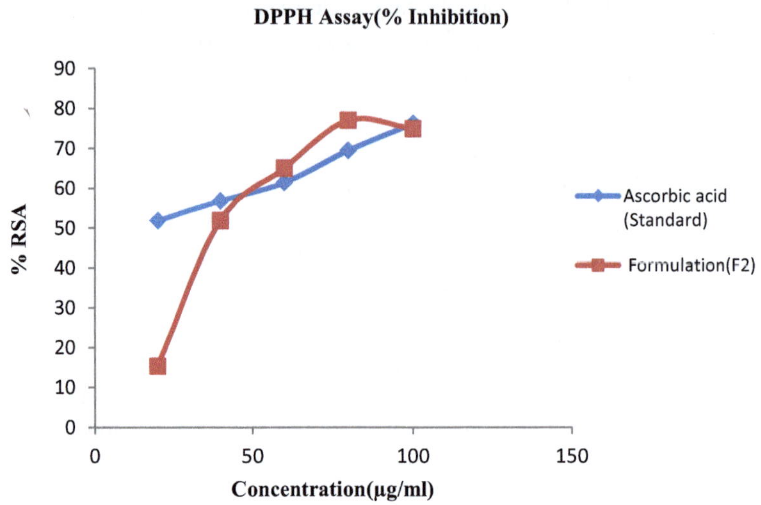

Fig. 7. Showing %RSA of ascorbic acid and β-ST.

Table 4. Antioxidant activity of ascorbic acid (standard) and optimized formulation (F2) at different concentrations.

DPPH Assay (% Inhibition)			
Sr. No.	Concentration(µg/ml)	Ascorbic acid (Standard)	Formulation(F2)
1	20	51.78	15.26946108
2	40	56.76	51.79640719
3	60	61.39	64.97005988
4	80	69.40	76.94610778
5	100	76.16	74.8502994
IC_{50} (µg/ml)		**17.38**	**4.61**

diffusion mechanisms. The optimized cubosomal formulation demonstrated a particle size of 139.7 nm, a PDI of 0.225, a zeta potential of −15.17 mV, and an EE of 65 ± 0.553%.In antioxidant activity assays, the IC_{50} of β-ST loaded CBs was found to be 4.61 µg/ml, significantly lower than that of ascorbic acid, which had an IC_{50} of 17.38 µg/ml. This indicates that β-ST-loaded CBs exhibit stronger antioxidant activity than ascorbic acid. These findings suggest that encapsulating β-ST in a lipid-based nanoformulation can enhance its antioxidant properties, solubility, and release characteristics, while also ensuring long-term storage stability.

Acknowledgement. For providing the required facilities, the authors are grateful to Devsthali Vidyapeeth College of Pharmacy.

Conflict of Interests No conflicts of interest are disclosed by the authors.

Funding Nil

References

1. Goyal, R., Macri, L.K., Kaplan, H.M., Kohn, J.: Nanoparticles and nanofibers for topical drug delivery. J. Control. Release. **240**, 77–92 (2015)
2. Dante, M.D.C.L., Borgheti-Cardoso, L.N., de Abreu Fantini, M.C., Praça, F.S.G., Medina, W.S.G., Pierre, M.B.R., Lara, M.G.: Liquid crystalline systems based on glyceryl monooleate and penetration enhancers for skin delivery of celecoxib: characterization, in vitro drug release, and in vivo studies. J. Pharm. Sci. **107**(3), 870–878 (2018)
3. Sharadha, M., Gowda, D.V., Vishal Gupta, N., Akhila, A.R.: An overview on topical drug delivery system–updated review. Int. J. Res. Pharm. Sci. **11**(1), 368–385 (2020)
4. Zakaria, F., Ashari, S.E., Azmi, I.D.M., Rahman, M.B.A.: Recent advances in encapsulation of drug delivery (active substance) in CBs for skin diseases. J. Drug Deliv. Sci. Technol. **68**, 103097 (2022)
5. Nath, A.G., Dubey, P., Kumar, A., Vaiphei, K.K., Rosenholm, J.M., Bansal, K.K., Gulbake, A.: Recent advances in the use of Cubosomes as drug carriers with special emphasis on topical applications. J. Lipids. **1**, 2683466 (2024)
6. Salen, G., Ahrens, E.H., Grundy, S.M.: Metabolism of β-sitosterol in man. J. Clin. Invest. **49**(5), 952–967 (1970)
7. Prager, N., Bickett, K., French, N., Marcovici, G.: A randomized, double-blind, placebo-controlled trial to determine the effectiveness of botanically derived inhibitors of 5-α-reductase in the treatment of androgenetic alopecia. J. Altern. Complement. Med. **8**(2), 143–152 (2002)
8. Valarmathi, S., Abarna, S., Dharshini, J., Vijai, S.: Cubosomes as advanced nanocarriers for drug delivery. JOPIR. **03**(01), 035–042 (2025)
9. Cabeza, M., Bratoef, E., Heuze, I., Ramírez, E., Sánchez, M., Flores, E.: Effect of β-ST as inhibitor of 5alpha-reductase in hamster prostate. P. W. Pharmcol. Soc. **46**, 153–155 (2003)
10. Prabahar, K., Udhumansha, U., Elsherbiny, N., Qushawy, M.: Microneedle mediated transdermal delivery of β-sitosterol loaded nanostructured lipid nanoparticles for androgenic alopecia. Drug Deliv. **29**(1), 3022–3034 (2022)
11. Chen, L., Liu, R., Liu, C.: Scanning electron microscopy: a potentially useful tool for the diagnosis of soft tissue masses. Exp. Ther. Med. **5**(1), 172–174 (2013)
12. Papadopoulou, V., Kosmidis, K., Vlachou, M.: Comparative study of lipid and surfactant self-assembled nanoparticles using a hydrophilic drug as the active ingredient. Colloids Surf. B Biointerfaces. **113**, 325–332 (2014)
13. Costa, P., Lobo, J.M.S.: Modeling and comparison of dissolution profiles. Eur. J. Pharm. Sci. **13**(2), 123–133 (2001)
14. Salah, S., Mahmoud, A.A., Kamel, A.O.: Etodolac transdermal CBs for the treatment of rheumatoid arthritis: ex vivo permeation and in vivo pharmacokinetic studies. Drug Deliv. **24**(1), 846–856 (2017)
15. Shoman, N.A., Gebreel, R.M., El-Nabarawi, M.A., Attia, A.: Optimization of hyaluronan-enriched CBs for bromfenac delivery enhancing corneal permeation: characterization, ex vivo, and in vivo evaluation. Drug Deliv. **30**(1), 2162162 (2023)
16. Umar, H., Wahab, H.A., Ahmed, N., Fujimura, N.A., Amjad, M.W., Bukhari, S.N.A., Ahmad, W.: Development, optimization and characterization of cisplatin loaded cubosomes for human lung carcinoma. Drug Dev. Ind. Pharm. **20**, 1–14 (2024)
17. Kawish, S.M., Ahmed, S., Gull, A., Aslam, M., Pandit, J., Aqil, M., Sultana, Y.: Development of nabumetone loaded lipid nano-scaffold for the effective oral delivery; optimization, characterization, drug release and pharmacodynamic study. J. Mol. Liq. **231**, 514–522 (2017)

18. Eldeeb, A.E., Salah, S., Ghorab, M.: Formulation and evaluation of CBs drug delivery system for treatment of glaucoma: ex-vivo permeation and in-vivo pharmacodynamic study. J. Drug Deliv. Sci. Tech. **52**, 236–247 (2019)
19. Abdelbary, G., Fahmy, R.H.: Diazepam-loaded solid lipid nanoparticles: design and characterization. AAPS Pharm Sci Tech. **10**, 211–219 (2009)
20. Ghanbarzadeh, S., Hariri, R., Kouhsoltani, M., Shokri, J., Javadzadeh, Y., Hamishehkar, H.: Enhanced stability and dermal delivery of hydroquinone using solid lipid nanoparticles. Colloids Surf. B: Biointerfaces. **136**, 1004–1010 (2015)
21. Malewar, N., Avachat, M., Pokharkar, V., Kulkarni, S.: Controlled release of ropinirole hydrochloride from a multiple barrier layer tablet dosage form: effect of polymer type on pharmacokinetics and IVIVC. AAPS Pharm. Sci. Tech. **14**, 1178–1189 (2013)
22. Senthilkumar, P., Arivuchelvan, A., Jagadeeswaran, A., Subramanian, N., Senthilkumar, C., Mekala, P.: Formulation, optimization and evaluation of enrofloxacin solid lipid nanoparticles for sustained oral delivery. Asian J. Pharm. Clin. Res. **8**, 231–236 (2015)
23. Nasr, M., Ghorab, M.K., Abdelazem, A.: In vitro and in vivo evaluation of CBs containing 5-fluorouracil for liver targeting. Acta Pharm. Sin. B. **5**(1), 79–88 (2015)
24. Peng, X., Zhou, Y., Han, K., Qin, L., Dian, L., Li, G., Pan, X., Wu, C.: Characterization of CBs as a targeted and sustained transdermal delivery system for capsaicin. Drug Des. Dev. Ther. **9**, 4209–4218 (2015)
25. Nithya, R., Jerold, P., Siram, K.: CBs of dapsone enhanced permeation across the skin. J. Drug Deliv Sci Tech. **48**, 75–81 (2018)
26. Rapalli, V.K., Banerjee, S., Khan, S., Jha, P.N., Gupta, G., Dua, K., Hasnain, M.S., Nayak, A.K., Dubey, S.K., Singhvi, G.: QbD-driven formulation development and evaluation of topical hydrogel containing ketoconazole loaded CBs. Mat. Sci. Eng. C. **119**, 111548 (2021)
27. Dash, S., Murthy, P.N., Nath, L., Chowdhury, P.: Kinetic modeling on drug release from controlled drug delivery systems. Acta Pol. Pharm. **67**(3), 217–223 (2010)
28. Setty, C.M., Babubhai, S.R., Pathan, I.B.: Development of Veldecoxib topical gels: effect of formulation variables on the release of valdecoxib. Int. J. Pharm. Pharma. Sci. **2**(1), 70–73 (2010)
29. Kar, N., Chakraborty, S., De, A.K., Ghosh, S., Bera, T.: Development and evaluation of a cedrol-loaded nanostructured lipid carrier system for in vitro and in vivo susceptibilities of wild and drug resistant Leishmania donovani amastigotes. Eur. J. Pharm. Sci. **104**, 196–211 (2017)
30. Hao, J., Fang, X., Zhou, Y., Wang, J., Guo, F., Li, F., Peng, X.: 2011: development and optimization of solid lipid nanoparticle formulation for ophthalmic delivery of chloramphenicol using a box-Behnken design. Int. J. Nanomedicine. **6**, 683–692 (2011)
31. Thakkar, H.P., Desai, J.L., Parmar, M.P.: Application of box-Behnken design for optimization of formulation parameters for nanostructured lipid carriers of candesartan cilexetil. J. Drug Deliv. Sci. Technol. **44**, 431–439 (2018)
32. Prieto, J.M.: Procedure: preparation of DPPH radical, and antioxidant scavenging assay. DPPH Microplate Protoc., 7–9 (2012)
33. Kurechi, T., Kikugawa, K., Kato, T.: Studies on the antioxidants. XIII. Hydrogen donating capability of antioxidants to 2, 2-diphenyl-1-picrylhydrazyl. Chem. Pharm. Bull. **28**, 2089–2093 (1980)
34. Khan, M.M., Madni, A., Torchilin, V., Filipczak, N., Pan, J., Tahir, N., Shah, H.: Lipid-chitosan hybrid nanoparticles for controlled delivery of cisplatin. Drug Deliv. **26**(1), 765–772 (2019)
35. Abd-Elsalam, W.H., Ibrahim, R.R.: Span 80/TPGS modified lipid-coated chitosan nanocomplexes of acyclovir as a topical delivery system for viral skin infections. Int. J. Pharm. **609**, 121214 (2021)

36. Sahoo, S.K., Panyam, J., Prabha, S., Labhasetwar, V.: Residual PVAassociated with poly (D, L-lactide-co-glycolide) nanoparticles affects their physical properties and cellular uptake. J. Control. Release. **82**(1), 105–114 (2002)
37. Nawaz, T., Iqbal, M., Khan, B.A., Nawaz, A., Hussain, T., Hosny, K.M., Abualsunun, W.A., Rizg, W.Y.: Development and optimization of acriflavine-loaded polycaprolactone nanoparticles using Box–Behnken design for burn wound healing applications. Polymers. **14**(1), 101 (2021)
38. Nasr, M., Mokhtar, R., Rania, S.A.: Optimized PEGylated cubosomes: a novel approach for specific delivery of dacomitinib to non-small cell lung cancer cells. Mater. Adv. **6**, 3903–3917 (2025)
39. Naguib, M.J., Hassan, Y.R., Abd-Elsalam, W.H.: 3D printed ocusert laden with ultra-fluidic glycerosomes of ganciclovir for the management of ocular cytomegalovirus retinitis. Int. J. Pharm. **607**, 121010 (2021)
40. Pal, S.L., Jana, U., Manna, P.K., Mohanta, G.P., Manavalan, R.: Nanoparticle: an overview of preparation and characterization. J. Appl. Pharm. Sci. **1**(06), 228–234 (2011)
41. Kurangi, B., Jalalpure, S., Jagwani, S.: Formulation and evaluation of resveratrol loaded Cubosomal Nanoformulation for topical delivery. Curr. Drug Deliv. **18**, 607–619 (2021)
42. Zhang, H., Cui, Y., Zhang, X., Yuan, X., Xu, D., Zhang, L.: Sustained Delivery of Salbutamol from Cubosomal Gel for Management of Paediatric Asthma: In Vitro and in Vivo Evaluation. J. Microencapsul. **39**, 252–260 (2022)
43. Rachmawati, H., Budiputra, D.K., Mauludin, R.: Curcumin nanoemulsion for transdermal application: formulation and evaluation. Drug Dev. Ind. Pharm. **41**(4), 560–566 (2015)
44. Singh, S., Singh, J., Singh, D., Dubey, R.: Stable diosgenin loaded solid lipid nanoparticles: a promising system for delivery of anticancer drugs. Chem. Phys. Lipids. **212**, 46–54 (2018)

26. Sahoo SK [illegible] and [illegible] grinding [illegible]

[illegible]

Formulation and Evaluation of Drug-Phospholipid Complex for Solubility Enhancement

Vishal Kapoor[1], Sumit Durgapal[2], Alka Singh[1], Gauree Kukreti[1](✉), and Archana Dhyani[3](✉)

[1] School of Pharmaceutical Sciences and Technology, Sardar Bhagwan Singh University, Balawala, Dehradun, Uttarakhand, India
gauree25@gmail.com

[2] Department of Pharmaceutics, Uttaranchal Institute of Pharmaceutical Sciences, Uttaranchal University, Premnagar, Dehradun, Uttarakhand, India

[3] School of Pharmacy, Graphic Era Hill University, Clement town, Dehradun, Uttarakhand, India
archana.dhyani89@gmail.com

Abstract. The research work mainly aimed at the formulation, development and in-vitro evaluation of cilnidipine phospholipid complex and its matrix dispersion. Cilnidipine is BCS class II drug that has low solubility and high permeability. The drug is generally used as an antihypertensive agent. The main problem with the drug is its low bioavailability due to its low aqueous solubility. In the present investigation, egg lecithin was used as a phospholipid and a drug-phospholipid complex was prepared using the solvent evaporation method, which was further formulated to matrix dispersion. Cilnidipine and phospholipid were taken in different molar ratios. The solvent was evaporated through Rota evaporator. Due to the stickiness in the prepared drug-phospholipid complex, it was further formulated to prepare matrix dispersion by the addition of PVA as a polymer, which increases the dispersibility of the drug. Formulated drug-phospholipid complex and matrix dispersion were subjected to various characterization studies such as percentage yield, solubility, FTIR, DSC, PDI, Zeta potential and in-vitro drug release studies. The optimized formulation (F1) was selected and was further used for the preparation of matrix dispersion. From the obtained results, it can be concluded that the prepared drug-phospholipid complex and matrix dispersion showed better drug release as compared to pure drug.

Keywords: Bioavailability · Phospholipid complex · Solubility · Dispersion

1 Introduction

Nearly 50% of American adults have hypertension, per the American Heart Association's 2017 recommendations, and the Global Burden of Disease research puts high systolic blood pressure as the main factor in years of quality life loss. Despite improvements in medical treatment, high blood pressure continues to be a significant risk factor for stroke, heart failure, renal failure, atherosclerosis, and dementia, which is increasingly

A. K. Saxena, A. Saxena (Eds.): GTHTM 2025, pp. 503–517, 2026.
https://doi.org/10.1007/978-3-032-12320-6_30

recognized. This is largely because only about half of people with hypertension are aware that they have the condition, and only about half of those who are diagnosed with it receive proper treatment. Two main categories can be used to classify clinical hypertension. Between 85% and 95% of episodes of primary (or essential) hypertension in people have no known cause. In contrast, identified underlying diseases, such as renal artery stenosis, pheochromocytoma, adrenal adenoma, or single-gene alterations, are the source of secondary hypertension [1]. When the blood vessels are under too much pressure (140/90 mmHg or above), the person is diagnosed with hypertension (high blood pressure). Although it is common, if untreated, it can be dangerous [2]. Receiving a blood pressure-lowering drug could be acceptable for individuals who are hypertensive or at significant cardiovascular risk to reduce the risk of stroke and heart conditions. Antihypertensive medications from many kinds have been used in therapeutic settings, including diuretics, as well as beta-blockers, ACE inhibitors, angiotensin II type 1 receptor blockers (ARB), and organic calcium channel blockers (CCBs). Calcium channel blockers are frequently employed as blood pressure-lowering medications. CCBs on the other hand, are said to raise heart rate while lowering blood pressure [3]. A novel and exceptional calcium antagonist known as cilnidipine has a long-lasting, slow-onset vasodilator action. In patients with mild to moderately essential hypertension, cilnidipine is effective as a once-daily antihypertensive medication and has little effect on heart rate and the autonomic nervous system [4].

Poor water solubility and a slow rate of dissolution characterize cilnidipine. Due to its hydrophobic nature, it has good intestinal permeability (log P value 4.7). Due to its poor water solubility and slow dissolving rate, cilnidipine has a reported low oral bioavailability (13%) in humans. Cilnidipine is one of the most often prescribed antihypertensive medications regardless of having low oral absorption qualities because of its effectiveness and safety profile when compared to other calcium channel blockers [5].

Numerous novel strategies have been used to increase the bioavailability of medications, overcoming the gastrointestinal environment barrier. Lipid-based strategies, such as liposome's, solid lipid nanoparticles, phytosomes, etc., have drawn more attention because they can address significant issues with drug delivery. However, there are still some drawbacks to these methods. For instance, liposomes continue to have serious issues with drug leakage, low drug loading capacity, and poor stability. Additionally, solid lipid nanoparticles' variable gelation propensity, high tendency for drug expulsion, and low inherent integration rate limit their overall use [6].

Among these several formulations, drug-phospholipid complexes offer to address some of the key flaws in the current approach because they use a technique that is nearly equivalent to the absorption process that occurs when nutritional components from G.I.T. So, for medications to reach a specific level in the systemic circulation, delivery vehicles are crucial [7].

Even though phospholipid complex can significantly boost bioavailability, its high stickiness causes a poor dissolution due to agglomeration and lump formation, thereby decreasing its oral absorption. Additionally, the semi-solid phospholipid complex's non-flowing properties restrict its use in clinical settings. To disperse phospholipid complex and hence improve the dispersibility of the complex, a matrix dispersion consisting of cilnidipine-phospholipid complex was prepared using PVA [8].

2 Material and Methodology

2.1 Material

Cilnidipine was procured from Yarrow Chemie Products Ltd., Mumbai, and Maharashtra. Egg Lecithin, polyvinylalcohol was purchased from Yarrow Chemie Products Ltd., Mumbai, Maharashtra. Disodium hydrogen phosphate, Potassium dihydrogen phosphate, ethanol was purchased from Loba Chemie Pvt. Ltd., Mumbai, Maharashtra.

2.2 Methodology

2.2.1 Molecular Docking. The molecular docking studies were conducted using Schrodinger software (Maestro 13.1). In molecular docking, the crystal structure was docked insight the active side of Phosphatidylcholine (PDB ID: 1LN1). Molecular docking studies involve selection and preparation of the protein, grid generation, ligand preparation followed by docking. The docking score, binding free energy, hydrogen bonds and pi-pi interaction formed with the amino acids are used to predict their binding affinities and proper alignment of cilnidipine at the active site of the receptor or enzyme [9].

2.3 Preparation of Formulations

2.3.1 Selection of Phospholipid Ratio for Preparation of Complex. Various formulations were prepared by using the solvent-evaporation method. Drug cilnidipine was kept constant and egg lecithin was taken in different weight ratios in a round bottom flask as shown in Table 1. Ethanol was added further to solubilize the drug and phospholipids. The prepared composition was stirred in a magnetic stirrer, at a temperature of 50 °C, keeping the stirring speed at 300 rpm. Subsequently kept in a vacuum evaporator to remove the solvent. The resultant phospholipid drug complex of cilnidipine was collected and placed in desiccators at room temperature [10].

2.3.2 Preparation of Cilnidipine-Phospholipid Complex Based Matrix Dispersion. To the selected quantity of drug and phospholipid, PVA was added. To the ratio of (1:0.5:1) of drug: phospholipid: PVA, ethanol was added to this composition. The mixture was stirred on a magnetic stirrer at 50 °C, keeping the stirring speed constant at 300 rpm. Eventually the solvent was removed by rotary vacuum evaporator. The resultant phospholipid drug complex matrix dispersion of cilnidipine was collected and placed in desiccators at room temperature [10].

Table 1. Formulation table for preparation of cilnidipine-phospholipid complex.

Formulation	Drug: Egg lecithin
F1	1:0.5
F2	1:1
F3	1:1.5
F4	1:2

2.4 Evaluation

2.4.1 Percentage Yield. Different formulations of cilnidipine phospholipid complex and matrix dispersion were evaluated by weighing the practical weight of the formulation after drying [11]. Percentage yield was calculated by:

$$Percentage\ Yield = \frac{Practical\ Yield}{Theoritical\ Yield} \times 100$$

2.4.2 Fourier Transformed Infrared Spectroscopy (FTIR). The FTIR of drug (cilnidipine), phospholipid (egg lecithin),physical mixture of drug and phospholipid, and matrix dispersion were analyzed. FTIR spectrum was recorded between scanning ranges of 400–4000 cm^{-1} and the obtained spectrum was compared with the reference spectrum by observing the peaks of the spectra. The peaks were studied for any noticeable shifts in the representative peaks [12].

2.4.3 Differential Scanning Calorimetry (DSC). DSC is a popular method for spotting changes in a drug's crystallinity. Indicated by the removal of endothermic peaks, the introduction of new peaks, or a modification in shape of the peaks, it can also provide an idea of potential interactions. The thermograms of drug (cilnidipine), phospholipid (egg lecithin), their physical mixture and matrix dispersion were obtained using DSC. The DSC of pure cilnidipine also confirms the melting point of the drug. Each of the samples was sealed in the air-tight aluminum pans and subsequently scanned at the range of 20–200 °C ata speed of 5 °C/min [13].

2.4.4 Determination of Zeta Potential, Particle Size, and Particle Size Distribution. Thezeta potential, particle size and particle size distribution of cilnidipine phospholipid complex and matrix dispersion were determined by adding 20 mg of the formulations in double distilled water separately and further dynamic scattering was conducted usingZeta Sizer (Malvern, M3-Plas) [14].

2.4.5 Determination of Solubility. Briefly, an excess amount of cilnidipine, cilnidipine-phospholipid complex, matrix dispersion was added to 5 ml of double distilled water, pH 1.2 and pH 6.8 in a vial and using water bath shaker (Remi Elektrotech Pvt. Ltd., Mumbai), the temperature was maintained at 37 ± 0.5 °C. The samples were centrifuged to get a clear supernatant and the concentration was determined by using a UV-Spectrophotometer [15].

2.4.6 Determination of Partition Coefficient. The n-octanol and water system was used to determine the partition coefficient of the formulations. The samples of cilnidipine, cilnidipine-phospholipid complex, and matrix dispersion were individually added to the immiscible system of n-octanol and water. Samples were subjected to shaking in a water bath shaker (Remi Elektrotech Pvt. Ltd., Mumbai) at 37 ± 0.5 °C. The vials were kept undisturbed and both layers were separated by centrifugation and the amount of cilnidipine was estimated by using a UV spectrophotometer [16].

2.4.7 In-Vitro Drug Release Studies. In-vitro dissolution was carried out using an IP Type 1 apparatus (Electrolab India Pvt. Ltd).Cilnidipine, cilnidipine-phospholipid complex and matrix dispersion were placed in the dialysis membrane bag 110 with a molecular weight cutoff of 12 to 14 kDa (Hi-media, Thane, India) sealed from both ends and suspended in the buffer under constant stirring and the temperature was maintained at 37 °C. The aliquots of samples were periodically withdrawn at regular time intervals in pH 1.2 HCl and phosphate buffer 6.8, and were followed by replacing the same with an equal volume of the fresh medium. Further, the collected samples were filtered, suitably diluted and analyzed using UV-spectrophotometer to estimate percentage cumulative drug release data at the corresponding time point [16].

2.4.8 In-Vitro Drug Kinetic Studies. To analyze the drug release pattern and estimate the underlying drug release mechanism, data from the diffusion investigation for the optimized formulation were addressed with a model or the curve fitting approach. In-vitro release data was fitted to various kinetic and empirical models, such as zero order, first order Higuchi, and Korsmeyer-Peppas models [17].

3 Results and Discussion

3.1 Molecular Docking

The molecular docking studies were conducted using Schrodinger software (Maestro 13.1). In molecular docking, the crystal structure was docked with the active side of Phosphatidylcholine (PDB ID: 1LN1).

In-silico molecular optimization studies revealed the potential site of interaction between Phosphatidylcholine with, cilnidipine as shown in Fig. 1. Cilnidipine showed the π bond interaction with PHE-199 and ARG-78 and showed the hydrogen bonding interaction with TYR-105 with the docking score − 8.256. The results conclude that cilnidipine can bind to the Phosphatidylcholine (PDB ID: 1LN1) which could further confirm the possible complexation.

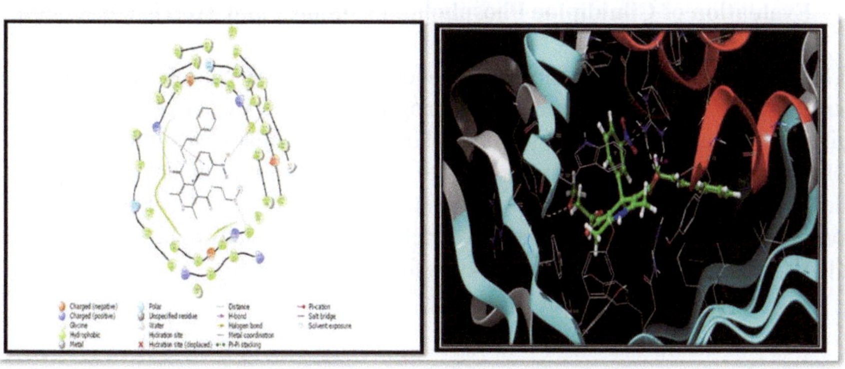

Fig. 1. Molecular Docking of Cilnidipine.

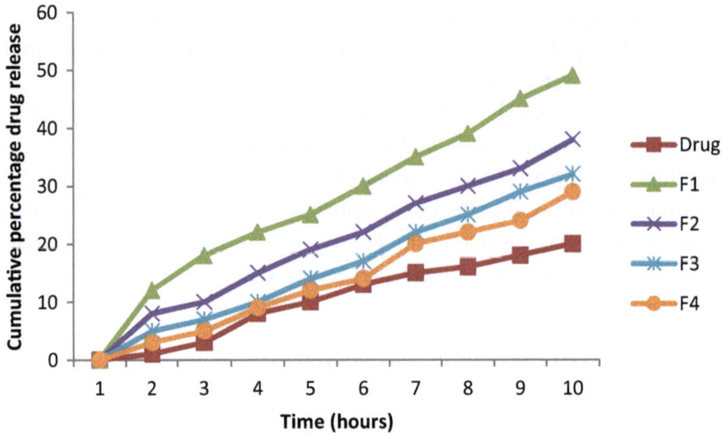

Fig. 2. Cumulative percentage drug release profile of the pure drug and the F1-F4 formulation.

3.2 Preparation of Formulations

3.2.1 Selection of Quantity of Phospholipid. Various ratios of phospholipids were taken along with the constant quantity of drug. The final selection of the best formulation was done based on in-vitro dissolution studies as shown in Fig. 2. The formulation F1 having drug: phospholipid ratio (1:0.5) showed the best drug release as compared to the pure drug. F1 showed approximately 2.5-fold increase in drug release as compared to pure drug. This may have contributed to better interaction between the drug and lecithin at this particular ratio. Further, upon increasing the concentration of phospholipid, the dissolution declined in the order of F1 > F2 > F3 > F4 as shown in Fig. 2. As the quantity of phospholipid increased the viscosity of the formulation also increased which may have resulted in decline in dissolution.

3.2.2 Preparation of Matrix Dispersion. The matrix dispersion comprising of cilnidipine drug phospholipid complex and PVA in a ratio (1:0.5:1) was collected and stored in a desiccator and used for further evaluation.

3.3 Evaluation of Cilnidipine Phospholipid Complex and Matrix Dispersion

3.3.1 Percentage Yield. Different formulations of cilnidipine phospholipid complex and matrix dispersion were evaluated by weighing the practical weight of formulation after drying. The percentage yield of the drug phospholipid complex was found as shown in Table 2. The percentage yield increased with an increase in the amount of phospholipids added. Maximum yield was found with formulation F4 as it has the maximum amount of phospholipid. The results are obtained in correlation with the study done by Nandhini *et al* [18]. Also, the yield of matrix dispersion was found to be 61.91%

Table 2. Percentage Yield of F1-F4 formulations.

Formulation	%Yield
F1	53.87
F2	58.97
F3	61.99
F4	64.23
Cilnidipine-phospholipid based matrix dispersion	61.91

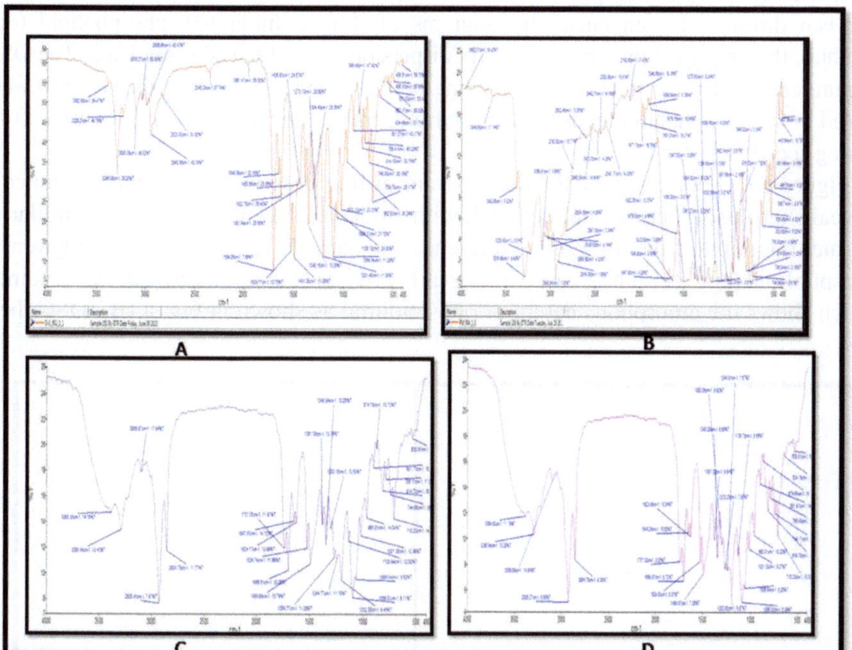

Fig. 3. (A): FTIR spectra of Cilnidipine (B): FTIR spectra of Cilnidipine and excipients physical mixture (C): FTIR spectra of Cilnidipine phospholipid complex (D): FTIR spectra of Cilnidipine phospholipid matrix dispersion.

3.3.2 FTIR Spectroscopy

The drug–excipient compatibility was determined by comparing the spectra of pure drug, drug excipient physical mixture, drug phospholipid complex and drug phospholipid matrix dispersion as shown in Fig. 3 (A), (B), (C) and, (D).The drug showed its characteristic peak at 3289,3093,3030,2942,1700 cm^{-1}due to the presence of –OH group,-NH stretching, -COOH group, CH stretching,-COO group, and-NH bending respectively as shown in Fig. 3 (A).The FTIR spectra of physical mixture (Fig. 3 B) did not show any interaction, thus it can be demonstrated that there was no interaction between cilnidipine and egg lecithin when physically mixed. In the drug phospholipid complex, a new peak appears at 1737 cm^{-1}, corresponding to the ester group formed in the cilnidipine phospholipid complex. This indicates that a bond has formed between cilnidipine and egg lecithin, confirming the complexation between the two, as illustrated in Fig. 3(C). The FTIR spectra of the matrix dispersion, shown in Fig. 3(D), display characteristic peaks of the cilnidipine-phospholipid complex as well as the characteristic peaks of polyvinyl alcohol (PVA). This indicates that PVA has not resulted in the formation of any new bonds, confirming its suitability for creating the matrix dispersion [19].

3.3.3 Differential Scanning Calorimetry (DSC). DSC is commonly used in compatibility analysis that provides abundant information regarding possible interaction between drugs and excipients. Thermograms of drug (cilnidipine), phospholipid (egg lecithin), their physical mixture and matrix dispersion were obtained using DSC. The DSC of pure cilnidipine also confirms the melting point of the drug. Each of the samples were sealed in the air-tight aluminum pans and subsequently scanned at the range of 20–200 °C having speed of 5 °C/min.

Figure 4 (A) depicts the DSC thermogram of free drug cilnidipine that exhibited a typical sharp endothermic melting peak with an onset of 110 °C which was similar to the melting point of the drug in the reference standards [21]. The DSC thermogram of phospholipid (egg lecithin) showed a various broad peak but no sharp peak was obtained which shows the amorphous nature of phospholipid as shown in Fig. 4 (B). DSC ther-

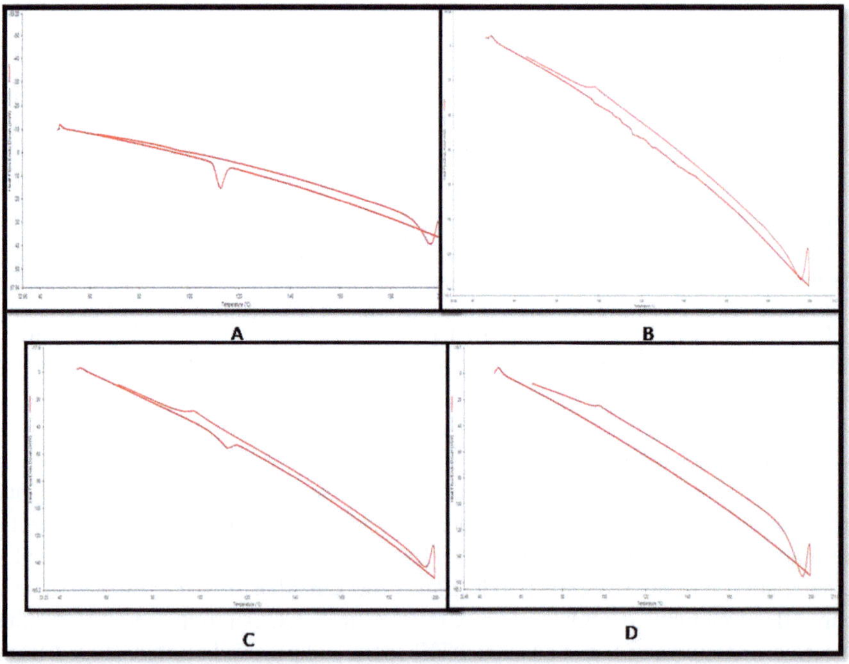

Fig. 4. (A): DSC thermogram of Cilnidipine, (B): DSC thermogram of phospholipid, (C): DSC thermogram of drug cilnidipine-phospholipid complex, (D) DSC curve of cilnidipine-phospholipid complex matrix dispersion.

mogram of drug cilnidipine-phospholipid complex showed a major peak broadening with the onset of 100 °C in contrast to the peak of free cilnidipine as shown in Fig. 4 (C) and thus indicating chemical interaction between cilnidipine and egg lecithin. The DSC curve of cilnidipine-phospholipid complex matrix dispersion as shown in Fig. 4(D) showed no major peaks corresponding to recrystallization were observed, thus it can be concluded that PVA has not caused any significant changes chemically thereby indicating the formation of matrix dispersion [20].

3.3.4 Determination of Particle Size and Particle Size Distribution. The particle size of cilnidipine phospholipid complex and matrix dispersion (F1)was determined by dynamic scattering using Zeta sizer (Malvern, M3-Plas).

The particle size and PDI values of cilnidipine phospholipid complex were found to be 240.4 nm and 0.461respectively.Theaverage particle size and PDI value of cilnidipine phospholipid complex based matrix dispersion were found to be 271.8 nm and 0.482 nm, respectively, as shown in Fig. 5(A). The results showed that the complex and matrix dispersion showed a monodispersed system with smaller particles. They also have the added advantage for the drug-entrapped complex to penetrate, and permeate through the physiological drug barriers [21].

3.3.5 Determination of Zeta Potential. The zeta potential value of cilnidipine phospholipid complex was found to be −3.66 meV and that of matrix dispersion was found to −19.4 meV as shown in Fig. 6 (A), suggesting that matrix dispersion possessed superior

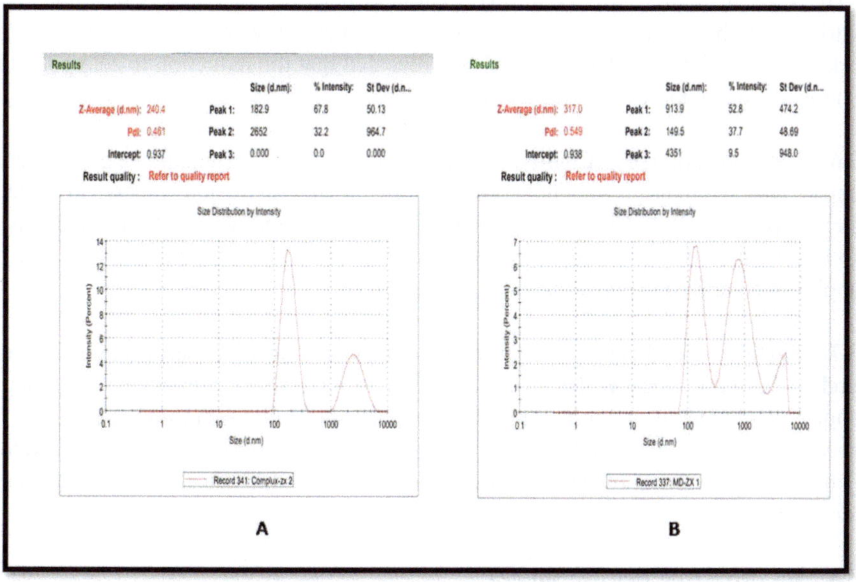

Fig. 5. (A): PDI value of cilnidipine phospholipid complex (F1), 5(B): PDI value of cilnidipine phospholipid complex matrix dispersion (F1).

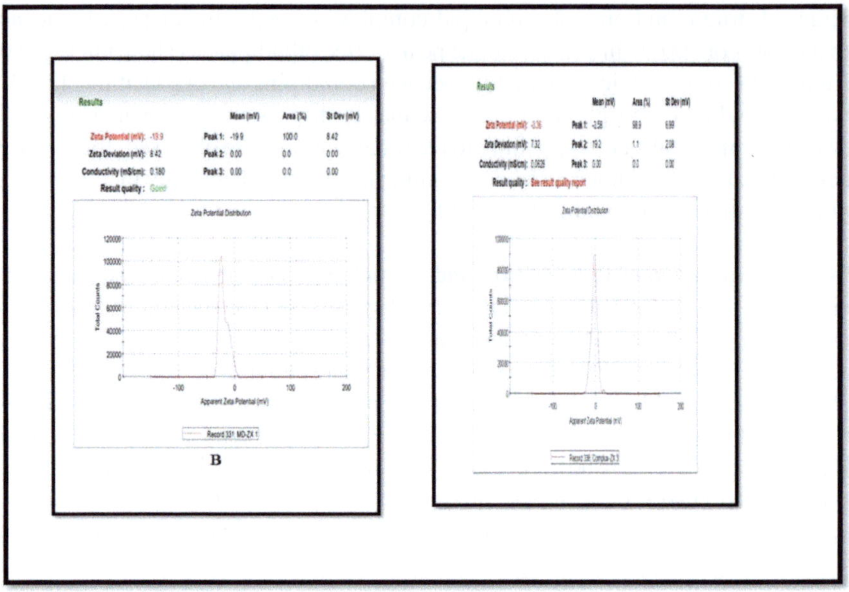

Fig. 6. (A): Zeta potential value of cilnidipine phospholipid complex based matrix dispersion, (B): Zeta potential value of cilnidipine phospholipid complex.

characteristics when it comes to stability whereas cilnidipine phospholipid complex shows zeta value of −4.66 meV which is not electrochemically as stable as matrix dispersion.

The PVA in matrix dispersion also exhibited negative charge which would have contributed to better stability. The results of zeta potential were in co-relation with the work done by Frietas *et al*, their study also revealed that the reduction in zeta potential value leads to instability of complexes [22].

3.3.6 Determination of Solubility. The solubility of cilnidipine, clindipine phospholipid complex and clindipine phospholipid complex based matrix dispersion is shown in Table 3. As cilnidipine is a BCS Class II drug, possessing poor solubility in different mediums, a significant increase in solubility was observed in case of matrix dispersion followed by complex as compared to pure drug.

3.3.7 Partition Coefficient. The partition coefficient (Log P) value of cilnidipine was found to be 4.66, whereas the value decreased to 4.02 when cilnidipine-phospholipid complex was prepared, and further log P value reduced to 3.66 when formulated as matrix dispersion. The study indicated a decline in log P value of cilnidipine-phospholipid and that of matrix dispersion showing enhancement in hydrophilic nature and reduction in hydrophobicity.

3.3.8 In-Vitro Drug Release Studies

The dissolution study revealed that approximately 15% of pure drug was released at 10 h, whereas approximately 48% of drug was released from the drug-phospholipid

Table 3. Solubility determination of cilnidipine, clindipine phospholipids complex and clindipine phospholipids complex based matrix dispersion.

Formulation	Solubility (mg/ml)		
	Water	pH 1.2 HCl	Phosphate buffer pH 6.8
Cilnidipine	0.052 ± 0.576	0.777 ± 2.046	1.254 ± 0.385
Cilnidipine-phospholipid complex	0.965 ± 0.903	1.502 ± 0.691	1.783 ± 0.721
Cilnidipine-phospholipid complex matrix dispersion	2.04 ± 1.04	1.94 ± 0.437	2.146 ± 0.842

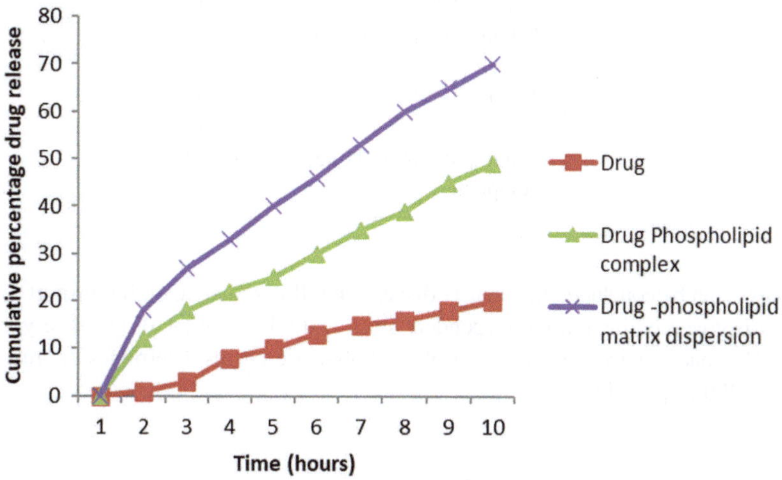

Fig. 7. Cumulative percentage drug release with time.

complex, and about 70% of drug was released from the drug-phospholipid complex matrix dispersion as shown in Fig. 7. The in-vitro dissolution studies suggested that approximately 4.6-fold improvement was obtained in drug release for the matrix dispersion than for pure drug alone. The result suggested that the formulation of clindipine-phospholipid based matrix dispersion can lead to an enhancement of dissolution which further can result in a better therapeutic outcome.

3.3.9 Drug Release Kinetic Studies. The cumulative percentage release data obtained from cilnidipine, optimized cilnidipine-phospholipid complex, and matrix dispersion were fitted into zero order, first order, Higuchi model, Peppas model to understand the mechanism of drug release (Fig. 7). The slopes of regression coefficients of determination (r^2) were listed in Table 4. The results showed that the r^2 values ranged from 0.9364–1.000 for zero order kinetics, 0.9507–0.9364 for first order kinetics, 0.8043–0.8131 for Higuchi model and 0.9439–1.000 for Peppas model. It is known that Peppas model is widely used to confirm whether the release mechanism is Fickian diffusion and non-fickian diffusion. The n (release exponent of Korsmeyer Peppas model) value could be used to characterize different release mechanisms. Since the value of n in case of matrix dispersion is 0.891 which is greater than

Table 4. Kinetic Modeling of Cilnidipine- phospholipid complex and matrix dispersion.

	Complex type	R^2
Zero order(r^2)	Cilnidipine- phospholipid complex	0.978
	Matrix dispersion	0.987
First order(r^2)	Cilnidipine- phospholipid complex	0.971
	Matrix dispersion	0.978
Higuchi model (r^2)	Cilnidipine- phospholipid complex	0.865
	Matrix dispersion	0.8843
Pappas model (r^2)	Cilnidipine- phospholipid complex	0.967
	Matrix dispersion	0.989
		n value
	Cilnidipine- phospholipid complex	0.900
	Matrix dispersion	0.991

0.89, thus it can be concluded that matrix dispersion follows Super case II transport. Case II transport is governed by a time-independent swell and relaxes of the drug delivery system matrix. The macromolecular relaxation of the polymeric chains determines the release in super case II transport (Fig. 8).

4 Conclusion

Cilnidipine phospholipid complex and matrix dispersion were successfully prepared by solvent evaporation method. Cilnidipine phospholipid complex and matrix dispersion showed better solubility profile as compared to the pure drug cilnidipine. The aqueous solubility of cilnidipine increased four-fold when incorporated in the matrix dispersion. The dissolution study revealed that approximately 15% of pure drug was released at 10 h, whereas approximately 48% of drug was released from the drug-phospholipid complex and about 70% of drug was released from drug-phospholipid complex matrix dispersion. The in-vitro dissolution studies suggested that approximately 4.6-fold improvement was obtained in drug release for matrix dispersion than for pure drug alone. Thus, this study holds immense potential to enhance the overall therapeutic outcome of the drug via the preparation of drug-phospholipid complex based matrix dispersion.

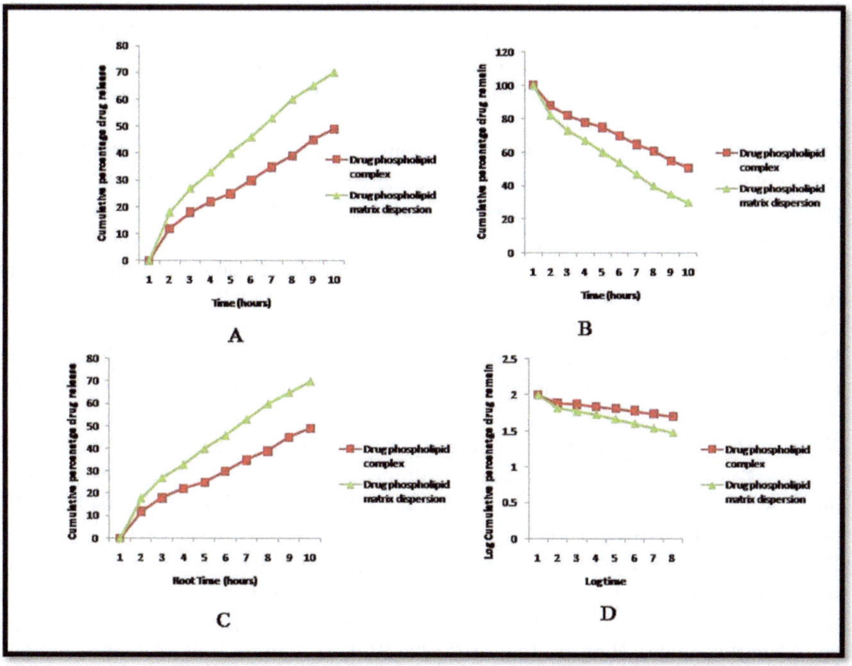

Fig. 8. (A): Zero order release, B: First order release, C: Higuchi's Plot, D: Korsmeyer Peppa's model.

References

1. Harrison, D.G., Coffman, T.M., Wilcox, C.S.: Pathophysiology of hypertension: the mosaic theory and beyond. Circ. Res. **128**, 847–863 (2021). https://doi.org/10.1161/circresaha.121.318082
2. Vallée, A., Gabet, A., Grave, C., Sorbets, E., Blacher, J., Olié, V.: Patterns of hypertension management in France in 2015: the Esteban survey. J. Clin. Hypertens. **22**, 663–672 (2020). https://doi.org/10.1111/jch.13834
3. Kishi, T., Hirooka, Y., Konno, S., Sunagawa, K.: Cilnidipine inhibits the sympathetic nerve activity and improves baroreflex sensitivity in patients with hypertension. Clin. Exp. Hypertens. **31**, 241–249 (2009). https://doi.org/10.1080/10641960902822492
4. Diwan, R., Ravi, P.R., Agarwal, S.I., Aggarwal, V.: Cilnidipine loaded poly (ε-caprolactone) nanoparticles for enhanced oral delivery: optimization using DoE, physical characterization, pharmacokinetic and Pharmacodynamic evaluation. Pharm. Dev. Technol. **26**(3), 278–290 (2021)
5. Zhang, J., Li, J., Ju, Y., Fu, Y., Gong, T., Zhang, Z.: Mechanism of enhanced oral absorption of morin by phospholipid complex based self-nanoemulsifying drug delivery system. Mol. Pharm. **12**(2), 504–513 (2015)
6. Maryana, W., Rachmawati, H., Mudhakir, D.: Formation of phytosome containing silymarin using thin layer-hydration technique aimed for oral delivery. Mater Today Proc. **3**(3), 855–866 (2016)

7. Kuche, K., Bhargavi, N., Dora, C.P., Jain, S.: Drug-phospholipid complex—a go through strategy for enhanced oral bioavailability. AAPS PharmSciTech. **20**(2), 43 (2019)
8. Li, H., Lu, S., Luo, M., Li, X., Liu, S., Zhang, T.: A matrix dispersion based on phospholipid complex system: preparation, lymphatic transport, and pharmacokinetics. Drug Dev. Ind. Pharm. **46**(4), 557–565 (2020)
9. Olusola, O.E., Opeyemi, I., Sunday, S.J., Akeem, O.L., Akinjiyan, M.O.: Molecular docking studies, molecular dynamics and ADME/tox reveal therapeutic potentials of STOCK1N-69160 against papain-like protease of SARS-CoV-2. Mol. Divers. **25**(3), 1761–1773 (2020). https://doi.org/10.1007/s11030-020-10151-w
10. Yeo, S., An, J., Park, C., Kim, D., Lee, J.: Design and characterization of phosphatidylcholine-based solid dispersions of aprepitant for enhanced solubility and dissolution. Pharmaceutics. **12**(5), 407 (2020)
11. Kattamanchi, G., Kalakonda, S.N., Battu, G.R.: Phospholipid complex technique for superior bioavailability of Phytoconstituents. Adv. Pharm. Bull. **7**(1), 35–42 (2017). https://doi.org/10.15171/apb.2017.005
12. Xinyuan, Z., Chunyang, S., Xiya, Z., Tong, L., Yusheng, G., MingxinY, Li, F., Wenqing, W., Jianguo, F.: Preparation of a nanoscale dihydromyricetin-phospholipid complex to improve the bioavailability: *in-vitro* and *in-vivo* evaluations. Eur. J. Pharm. Sci. **5**(2), 104994 (2019). https://doi.org/10.1016/j.ejps.2019.104994
13. Disha, S., Sankha, B., Gupta, G.L., KetanV, Arinjay, J., Laxmi, M., Niraj, P., Putrevu Sreelaya: D-α-tocopheryl polyethylene glycol 1000 succinate surface scaffold polysarcosine based polymeric nanoparticles of enzalutamide for the treatment of colorectal cancer: *In-vitro, in-vivo* characterizations. Heliyon. **10**(3), e25172 (2024)
14. Semalty, A., Semalty, M., Rawat, D.S., Rawat, M.S.M.: Preparation and characterization of phospholipid complexes of naringenin for effective drug delivery. J. Incl. Phenom. Macrocycl. Chem. **67**(3), 253–260 (2024)
15. K. Mulrajani, N. Rajnani AND N. Kurup, Nanocochleates and drug-phospholipid complex: novel approaches for phospholipid based Oral delivery of anti-cancer agents, Indian J. Pharm. Sci. 2023;85(1):13–22.
16. Gurav, S., Usapkar, P., Gurav, N., Nadaf, S., Ayyanar, M., Verekar, R., Bhole, R.: Preparation, characterization, and evaluation (*in-vitro*, *ex-vivo*, and *in-vivo*) of naturosomal nanocarriers for enhanced delivery and therapeutic efficacy of hesperetin. PLoS One. **17**(11), e0274916 (2022). https://doi.org/10.1371/journal.pone.0274916
17. NandhiniS, I.K.: Development and characterization of a nano-drug delivery system containing vasaka phospholipid complex to improve bioavailability using quality by design approach. Res Pharm Sci. **16**(1), 103–117 (2020). https://doi.org/10.4103/1735-5362.305193
18. Dong, W., Ye, J., Wang, W., Yang, Y., Wang, H., Sun, T., Gao, L., Liu, Y.: Self-assembled lecithin/chitosan nanoparticles based on phospholipid complex: a feasible strategy to improve entrapment efficiency and transdermal delivery of poorly lipophilic drug. Int. J. Nanomed. **15**, 5629–5643 (2020)
19. Raghavendra, P.R.M., Kulkarni, S., Sonawane, A., Sugaonkar, S.: Self-nanoemulsifying drug delivery system of Cilnidipine. I J P E R. **55**(3), 1–10 (2021)
20. Rimpy, D., Punna, R.R., Nikita Shantaram, P., Aggarwal, V.: Pharmacodynamic, pharmacokinetic and physical characterization of cilnidipine loaded solid lipid nanoparticles for oral delivery optimized using the principles of design of experiments. Colloids Surf. B Biointerfaces. **193**, 111073 (2020)

21. Suprit, D., Nishikant, A.R., Pradip, W.D., Chandrashekhar, D.B., Michael, P., Vivek, S.D.: Preparation and evaluation of phospholipid-based complex of standardized Centella extract (SCE) for the enhanced delivery of Phytoconstituents. AAPS J. **18**(1), 102–114 (2016)
22. Freitas, C., Müller, R.H.: Effect of light and temperature on zeta potential and physical stability in solid lipid nanoparticle (SLN) dispersions. Int. J. Pharm. **168**(2), 221–229 (1998). https://doi.org/10.1016/S0378-5173(98)00092-1

[illegible]

Invited Lectures (Abstracts)

Machine Learning Techniques for Breast Cancer Diagnosis

Anushree Sinha
Software Engineer, Google LLC, Mountain View, CA
Email: anushreeanushree@gmail.com

Breast cancer remains a leading cause of mortality among women worldwide, emphasizing the critical need for early and accurate diagnosis to improve treatment outcomes and survival rates. This study focuses on developing robust machine learning models for breast cancer prediction using the publicly available Breast Cancer Wisconsin (Diagnostic) dataset. This dataset, comprising features extracted from fine-needle aspirate (FNA) images of breast masses, offers valuable insights into cellular characteristics associated with malignancy.

Our research employs a comparative approach, utilizing a variety of machine learning algorithms, including logistic regression, decision trees, and random forests, to construct predictive models. Each model is trained and evaluated on the dataset, with performance assessed through key metrics such as accuracy, precision, recall, and F1-score. By comparing the effectiveness of these different models, we aim to identify the most suitable approach for accurate breast cancer prediction.

This analysis not only contributes to the expanding field of machine learning in healthcare but also holds the potential to assist clinicians in making more informed diagnostic decisions. The results of this study may pave the way for developing more sophisticated tools to aid in the early detection of breast cancer, ultimately contributing to improved patient care and disease management.

Keywords Breast cancer, machine learning, diagnosis, prediction, FNA images

One Step Forward for the Development of Drug for Non-curable Disease: Clearance of Phase I Human Clinical Trials

Diwan S. Rawat
Senior Professor of Chemistry, University of Delhi and Vice Chancellor, Kumaun University, Nainital, Uttarakhand

A. K. Saxena, A. Saxena (Eds.): GTHTM 2025, pp. 519–525, 2026.
https://doi.org/10.1007/978-3-032-12320-6

The ‘one drug, multiple targets’ concept has gained attention of both the academic and the pharmaceutical industry in recent years. To achieve such goals concept of molecular hybridization was put forward wherein two or more distinct pharmacophores are covalently linked into a single molecule [1, 3]. The development of such molecular frameworks with synthetic selectivity and economic viability is still a challenging task for the pharmaceutical industry. Drugs developed through this approach can be used for the cure of infectious diseases where treatment is limited to few drugs and the known drugs have limitations such as toxicity, pharmacokinetics, pharmacodynamic and drug resistance. The benefit of using molecular hybrid is to activate different or same targets by a single molecule, and increase the therapeutic efficacy and to improve the bioavailability. Molecular hybridization approach has resulted many drug candidates with improved activity profile and some of these compounds are in clinical trials [4]. We have utilized this concept in designing antimalarial molecules and many molecules with aminiquinoline and pyrimidine phamacohpore showed low nano molar activity. Later a massive multi-institutional collaboration was started and over 700 new molecules were studies for Nurr1 activation, a potential target for Parkinson disease model and identified 15 hits out of which 3 compounds have cleared pre-clinical trials and technology has been transferred to NURRON pharmaceuticals for further development [5–8]. These molecules activate the Nurr1 enzyme which is essential for the survival of the dopamine neurons, stops the aggregation of α-synuclein protein in the brain, and promotes autophagy. Systematic studies demonstrated that these compounds can cures the Parkinson induced mice model at 5 mg/kg body weight without any toxicity and recently one of the molecule has successfully cleared phase I human clinical trials.

References

1. Meunier, B.: Acc. Chem. Res. **41** (2008)
2. Shikha, S.S., Sharma, M., Chauhan, P.M.S.: Drug News Perspect. **23**, 632 (2010)
3. de Sena, M.P.P., Franco, L.S., Montagnoli, T.L., Fraga, C.A.M.: Expert Opin. Drug Discov. **19**, 451 (2024)
4. Parkes, A.L., Yule, I.A.: Expert Opin. Drug Discov. **11**, 665 (2016)
5. Manohar, S., Rawat, D.S.: ACS Med. Chem. Lett. **3**, 555 (2012)
6. Tripathi, M., Taylor, D., Khan, S.I., Tekwani, B.L., Ponnan, P., Velpandian, T., Das, U., Rawat, D.S.: ACS Med. Chem. Lett. **10**, 714 (2019)
7. Kim, W., Tripathi, M., Kim, C., Vardhineni, S., Cha, Y., Kandi, S.K., Feitosa, M., Kholiya, R., Sah, E., Thakur, A., Kim, Y., Ko, S., Bhatia, K., Manohar, S., Kong, Y.-B., Sindhu, G., Kim, Y.-S., Cohen, B., Rawat, D.S., Kim, K.-S.: Nat. Commun. **14**, 4283 (2023)
8. Rawat, D.S., Manohar, S., Chinna Rajesh, U., Kumar, D., Thakur, A., Tripathi, M., Linga Reddy, P., Kulangara Kandi, S., Vardhineni, S., Kwang-Soo, Kim, C.-H.: Amino-quinoline based hybrids and uses thereof. Pub no: US 2017/0209441 A1; EP Application No. 13758678; PCT/US2013/28329; WO2013134047 A3, PCT/US2013/028329 (2013). US 11,026,943 B2 (2021)

Protein Function and Structure: Novel Revelation

Indira Ghosh
Former Professor & Dean, School of Computational & Integrative Sciences, Jawaharlal Nehru University, New Delhi 110067

Biological Macromolecules always throw new challenges to inspire biologists to interpret their data, especially mechanistic concepts to build. With the knowledge of too many protein structures in deeper detail, protein structures are well characterized and projected as very compact in nature but functionally highly flexible. "How has this happened?". Many answers are available, like Dynamics of protein structure, different folding with smallest change in sequence, limited secondary structures, thermodynamics of stable protein etc. As the prediction from sequence to structure is reaching a new height, thanks to AI tools, we are intrigued to learn more at the local events in protein structures, like modularity, compactness, interaction space, as these will affect our designing of small molecules to impair the activity.

We discuss here a systematic approach to identify structural modules using component-based approach by considering the arrangement of secondary structural elements (SSEs), strict geometrical sense. For concise representation of contact among SSEs in proteins, "contact string" has been developed, like "fingerprints", which can uniquely represent all different contact patterns, which unravel all possible structures. However, it's interesting to observe that functional proteins are very choosy, only a few patterns occur. Our group has developed ProLego, an application to explore the component aspect of protein structures and provide an intuitive and efficient way to scan the protein topology space. One interesting observation was that the set of topologies that are found to be structurally prevalent are functionally divergent. Identification of the smallest component of protein's modular structural arrangement will provide an insight into its evolution and structural–functional space.

References

1. Khan, T., Ghosh, I.: Modularity in protein structures: study on all-alpha proteins. J. Biomol. Struct. Dyn. **33**, 2667–81 (2015)
2. Khan, T., Panday, S.K., Ghosh, I.: ProLego: tool for extracting and visualizing topological modules in protein structures. BMC Bioinfo. **19**,167 (2018)

Drug Efficacy and Sex Differences in Immune Responses

Lakshmi P. Kotra
Professor, Leslie Dan Faculty of Pharmacy, University of Toronto, and Senior Scientist, Krembil Research Institute, University Health Network

Males and females respond differently to inflammatory insults, and it is also known that sex hormones play a role influencing inflammatory pathways. Endocannabinoid system (ECS), an integral part of inflammatory and immune responses, also behaves differently in males and females. Cannabidiol (CBD), a plant cannabinoid and antagonist to CB2 receptor in the ECS, is a popular molecule and is widely used for its anti-inflammatory properties. CBD is a bicyclic terpenoid derivative with two chiral centers. It is used widely by patients suffering from inflammatory disorders such as arthritis. We studied the anti-inflammatory activities of CBD in male and female mice using an Experimental Autoimmune Encephalomyelitis (EAE) mouse model to understand drug efficacy and sex differences. EAE was induced in C57BL/6J mice by immunization with MOG35-55 peptide, and pertussis toxin. Three cohorts, healthy mice, vehicle-treated and CBD-treated male and female mice were analyzed. CBD-treated cohorts showed noticeable differences between male and female groups, and when compared to the vehicle-treated cohort. These differences are seen in the peripheral inflammatory cells, cytokines as well as infiltration of the peripheral immune cells into the brain and spinal cord. In general, male animals experienced more severe disease onset but responded better than females to the CDB treatment. CBD treated male mice had significantly lower area of inflammation in the spinal cord, fewer brain-invading T cells, and lower count of microglia/macrophages in the spinal cord. These studies suggest that endocannabinoid system and its modulation have an effect on peripheral and central immune system differentially in males and females.

Revolutionizing Healthcare Operations with Sap: Innovation, Integration, and Impact

Prakhar Saraswat
SCM Yuga Technologies Pvt Ltd.

In the rapidly evolving healthcare and life sciences sectors, advanced digital technologies are crucial for ensuring operational excellence, regulatory compliance, and patient-centric care. SAP Advanced Track and Trace for Pharmaceuticals (SAP ATTP) emerges as a transformative solution, enabling pharmaceutical companies to meet global serialization and traceability mandates. This robust platform facilitates end-to-end product tracking, minimizes counterfeit risks, and ensures supply chain transparency, fostering trust among stakeholders. SAP's innovative solutions extend beyond pharmaceuticals, revolutionizing healthcare operations through SAP S/4HANA and SAP Business One ERP. These platforms empower healthcare organizations to streamline processes, enhance data-driven decision-making, and optimize resource utilization. By integrating predictive analytics and real-time insights, they enable superior patient care, compliance with stringent regulatory standards, and cost efficiencies. Further advancements in life sciences and senior care are driven by SAP's commitment to innovation, as highlighted by its strategic focus on digital transformation. Solutions like SAP S/4HANA for Healthcare and SAP Business One cater to the unique needs of hospitals and care facilities, enabling improved patient engagement, operational agility, and staff productivity. The integration of these technologies' underscores SAP's pivotal role in redefining healthcare and life sciences. From secure track-and-trace systems for pharmaceuticals to transformative ERP solutions, SAP drives exceptional outcomes across industries. These innovations enhance patient safety, ensure compliance, and set new benchmarks for operational efficiency, paving the way for a smarter, more connected future in healthcare and life sciences.

Novel Chirality Descriptors for the Prediction Pharmacological and Toxicological Activities of Chiral Molecules

Ramanathan Natarajan[1] and Subhash C. Basak[2]
[1]Saranathan College of Engineering, Tiruchirappalli 620012, Tamil Nadu, India
[2]Retired Adjunct Professor, Department of Chemistry and Biochemistry and Senior Scientist, Natural Resources' Research Institute, University of Minnesota Duluth, MN, USA

The human body contains numerous chiral macromolecules owing to the homochiral nature of the building blocks and therefore is an amazing chiral selector. Thus, the human body will interact with the enantiomers of the drug differently and metabolize each enantiomer by different pathways generating difference in the pharmacological activity. One enantiomer may produce the desired therapeutic activity, while the other may be inactive or, in the worst scenario, produce undesired effects or toxicity to non-target tissues. This revelation had changed the strategy in the pharmaceutical industry for the manufacture of chiral drugs and resulted in the synthesis and "chiral switching" to more and more enantiopure drugs. numerical characterization of chirality was attempted by several others to extend the QSAR approach for predicting pharmacological and toxicological effects of chiral molecules. Most of these approaches in numerical characterization of enantiomers derive only a binary set of descriptors. A single set of chiral indices that is able model the biological response of a SINGLE target may fail to model the biological responses of a different targets that are uncorrelated. Moreover, the neighbourhood of the chiral center (atom) in a molecule plays an important role in the interaction of the chiral molecule with the receptor. Hence, we need a collection of mutually different chirality indices for the QSAR of chiral molecules of different chemical classes or different bioactivities of the same set of molecules. To meet this requirement, a series of relative chirality indices (RCIs) based on the weighted graph models of the neighbourhoods of chiral centers in the molecules is developed. Application of the multidimensional space of the novel family of RCIs in the quantitative chiral structure-activity relationship (QCSAR) studies of different classes of chemicals will be presented.

Keywords Chirality indices; QSAR; molecular descriptors; numerical characterization, enantiomers

Molecular Mechanisms of Allosteric Changes in Drug Targets: Insights from Molecular Dynamics and Machine Learning

Shubhra Ghosh Dastidar
Bose Institute, EN 80, Sector V, Bidhannagar, Kolkata, India

Allosteric changes in the conformations of the biological macromolecules are of tremendous importance for their functions and existence of such mechanism in a system gives a unique opportunity to design drugs with higher precisions. This is because allosteric changes could also be induced by administering small molecules to activate a macromolecule to function or to inactivate it. While state-of-art experiments can suggest the existence of allosteric transitions in a molecular system, the mechanisms and atom-atom interaction network of allostery in a molecule are often non-trivial to understand just from the static models of molecular structures. The insights into allostery are obtainable from the computational modelling and simulations of their dynamics which can show the molecules in their action, revealing a more realistic scenario. This presentation will address this issue with examples of various molecular systems like α, β-Tubulin [1–4], Bcl2 [5], kinases[6–7], all of which are of tremendous importance in the field of drug design to combat cancer. Not only the structural and thermodynamic insights which will be presented, but also our latest initiatives to use machine learning [7] methods to complement the molecular dynamics simulations will be briefly discussed.

Keywords Allostery, Kinase, Tubulin, Molecular Dynamics, Machine Learning

References

1. Basu, D., Ghosh Dastidar, S.: Comput Biol Chem. **108**, 108004, (2024)
2. Majumdar, S., Basu, D., Ghosh Dastidar, S.: J. Chem. Inf. Model. **59**, 2274–2286 (2019)
3. Majumdar, S., Ghosh Dastidar, S.: J. Phys. Chem B. **121**, 118–128 (2017)
4. Majumdar, S., Maiti, S., Ghosh Dastidar, S.: Biochem. **55**, 335–47 (2016)
5. Sinha, S., Maity, A., Ghosh Dastidar, S.: J. Chem. Inf. Model. **58**, 370–382 (2018)
6. Ray Chaudhuri, N., Ghosh Dastidar, S.: J. Chem. Inf. Model. **63**, 224–239 (2023)
7. Ray Chaudhuri, N., Ghosh Dastidar, S.: Biochem. **63**, 1474–1492 (2024)

Machine Learning Application for Detection of the Potential Geroprotectors in the Chemical-Pharmacological Universe

Vladimir Poroikov, *Alexey Lagunin, Leonid Stolbov, Ekaterina Stolbova, Dmitry Druzhilovskiy, Dmitry Filimonov*
Institute of Biomedical Chemistry, Bldg. 10/8, Pogodinskaya Str., Moscow, 119121, Russia E-mail: vladimir.poroikov@ibmc.msk.ru

Aging is defined as a set of age-related changes in the organism, leading to a decrease in its vitality and an increase in the probability of death. Geroprotectors are the substances that slow down the aging process and prevent age-related diseases. The mechanisms of anti-aging action discussed in the literature outline criteria for in silico identification of potential geroprotectors among the pharmacological substances. Intersection of the terms described mechanisms of anti-aging action with the biological activities in PASS (Prediction of Activity Spectra for Substances) revealed the opportunity for development of special computational tool. The PASS GERO program predicts 115 mechanisms of geroprotective action with the average Invariant Accuracy of Prediction in leave-one-out and 20-fold cross-validation equals to 0.9665 and 0.9659, respectively; which demonstrates high accuracy and predictive ability. For validation of this computational tool, we performed the prediction of geroprotective activity profiles for 15 reference compounds. For each reference compound, a certain number of geroprotective action mechanisms are predicted in congruence with known activities. Thus, PASS GERO may be used for selection of new potential geroprotectors by virtual screening. It is reasonable to continue further improving machine learning models, both by replenishing the training sets with new information on potential geroprotectors, and taking into account the evolution of general concepts regarding geroprotection.

This research was supported by the Ministry of Science and Higher Education of the Russian Federation within the framework of Agreement No. 075-15-2024-643.

Keywords Aging, Geroprotectors, In silico evaluation, Machine learning, PASS GERO

Oral Presentation (Abstracts)

Advances in Microsponge Technology for Controlled Drug Delivery: A Comprehensive Review

Aayushman Singh*[1], *and Poonam Gangwar*[2]
*Department of Pharmaceutics, School of pharmaceutical Sciences, Shri Guru Ram Rai University, Patel Nagar, Dehradun

Microsponges are polymeric microparticles with an internal porous structure capable of encapsulating various drugs. This abstract provides an overview of microsponge formulation, encompassing preparation methods, characterization techniques, and applications. The quasi-emulsion solvent diffusion method is a commonly employed technique for microsponge production, involving the formation of an oil-in- water emulsion followed by solvent evaporation. Characterization of microsponges includes particle size analysis, drug loading, encapsulation efficiency, and release kinetics. Microsponges offer advantages such as sustained drug release, improved drug solubility, and enhanced skin penetration. Potential applications span various therapeutic areas, including dermatology, ophthalmology, and oral delivery. Ongoing research focuses on optimizing microsponge formulations for specific drug molecules and therapeutic indications to achieve desired drug release profiles and efficacy. Recent advancements in microsponges have focused on enhancing their loading capacity, stability, and release profiles through the development of novel synthesis techniques, such as solvent evaporation, coacervation, and supercritical fluid processing. Furthermore, the surface modification of microsponges allows for better targeting of specific tissues or skin layers, improving the precision of drug delivery or cosmetic treatments. This review delves into the preparation methods, characterization techniques, and diverse applications of advanced microsponges. It also addresses the challenges associated with scaling up production and regulatory considerations for their widespread adoption. The future of microsponges appears promising, with ongoing research aimed at optimizing their design for personalized medicine, skincare, and more efficient therapeutic outcomes.

Keywords Microsponges, Drug delivery, Sustained release, Quasi-emulsion solvent diffusion, Characterization

A. K. Saxena, A. Saxena (Eds.): GTHTM 2025, pp. 527–531, 2026.
https://doi.org/10.1007/978-3-032-12320-6

Comparative Analysis of Azelnidipine Co-crystal Formation: Solvent Evaporation vs. Wet Grinding for Enhanced Solubility

Aparna Joshi, *and G. Gnana Rajan*
School of Pharmaceutical Sciences SGRR University, Dehradun

Azelnidipine, a calcium channel blocker with poor aqueous solubility, faces challenges in drug formulation and bioavailability. Co-crystallization offers a promising approach to enhance its solubility and dissolution. This study compares two co-crystal formation techniques: wet grinding and solvent evaporation. Notably, the solvent evaporation method yielded well- defined, highly crystalline co-crystals visible to the naked eye, exhibiting superior stability and structural integrity. In contrast, wet grinding produced smaller co-crystals with increased surface area, leading to faster dissolution. Characterization using, DSC, FTIR, and SEM confirmed successful co-crystal formation and molecular interactions. The findings highlight solvent evaporation as a superior method for obtaining high-quality co-crystals, making it a promising approach for improving Azelnidipine pharmaceutical performance.

Keywords Azelnidipine, Co-crystals, Solvent evaporation, Wet grinding, Solubility enhancement, Crystallinity

References

1. Lu, Q., Dun, J., Chen, J.M., Liu, S., Sun, C.C.: Improving solid-state properties of berberine chloride through forming a salt cocrystal with citric acid. Int. J. Pharm. **554**(2019), 14–20 (2018)
2. Cerreia Vioglio, P., Chierotti, M.R., Gobetto, R.: Pharmaceutical aspects of salt and cocrystal forms of APIs and characterization challenges. Adv. Drug Deliv. Rev. **117,** 86–110 (2017)

Formulation and Evaluation of Drug-Phospholipid Complex for Solubility Enhancement

Vishal Kapoor[1], *Gauree Kukreti* [2], ***Archana Dhyani*** *[3], *Sumit Durgapal*[4], *and Alka Singh*[5]

[1,2]School of Pharmaceutical Sciences and Technology, Sardar Bhagwan Singh University, Balawala, Dehradun 248161, Uttarakhand, India.

[3]School of Pharmacy, Graphic Era Hill University, Clement town, Dehradun,248001, Uttarakhand, India.

[4]Department of Pharmaceutics,.Uttaranchal Institute of Pharmaceutical Sciences, Uttaranchal University, Premnagar, Dehradun, 248007, Uttarakhand, India

[5] School of Pharmaceutical Sciences and Technology, Sardar Bhagwan Singh University, Balawala, Dehradun 248161, Uttarakhand, India.

The research work mainly aimed at the formulation development and *in-vitro* evaluation of cilnidipine phospholipid complex and matrix dispersion. Cilnidipine is a BCS class II drug that is having low solubility and high permeability. The drug is generally used as antihypertensive agent. The main problem with the drug is its low bioavailability which is due to the low solubility of the cilnidipine in aqueous medium. In the present investigation, egg lecithin was used as a phospholipid and drug-phospholipid complex was prepared by solvent evaporation method which was further formulated to matrix dispersion. Cilnidipine and phospholipid were taken in different molar ratios by varying the temperature and time keeping the rpm constant. The solvent was evaporated through rota evaporator. Due to the stickiness in the prepared drug- phospholipid complex, it was further formulated to prepare matrix dispersion by the addition of PVA as a polymer which increases the dispersibility of the drug. Formulated drug- phospholipid complex and matrix dispersion were subjected to various characterization that are percentage yield, entrapment efficiency, solubility, FTIR, DSC, PDI, zeta potential and *in - vitro* drug release studies. The optimized formulation 1 was selected and used for the preparation of matrix dispersion. From the above results it can be concluded that the prepared drug-phospholipid complex and matrix dispersion showed a 2-fold increase in the drug release when compared with the pure drug cilnidipine.

Keywords Bioavailability, Phospholipid complex, Solubility, Dispersion

Comparison of 2D and 3D Cancer Cell Cultures: Different Types and Their Characteristics

Asha Gummadi[1], *and Ramaniaha Gangireddy*[2]
Department of Pharmacology, K V S R Siddhartha college of pharmacy, Vijayawada-Andhra Pradesh, India

Cell culture is a widely used in vitro tool for advancing our understanding of cell biology, tissue morphology, disease mechanisms, drug action, protein production, and tissue engineering. Most cancer biology research relies on two-dimensional (2D) cell cultures in vitro. However, 2D cultures have several limitations, including disrupted interactions between cells and their extracellular environment, altered cell morphology, polarity, and division methods. These drawbacks have driven the development of models that better mimic in vivo conditions, such as three-dimensional (3D) cultures. Optimizing culture conditions can enhance our understanding of cancer biology and facilitate biomarker studies and targeted therapies.This review compares 2D and 3D in vitro cultures, along with different types of 3D culture models.

Keywords Co-culture, Cell culture methods, 3D culture, 2D culture, Cancer research

Synergistic Antifungal Activity of Fluconazole Emulgel Containing Various Oils

***Heena Farooqui*[1]*, *Mohd. Vaseem Fateh*[2] *and Sarfaraz Ahmed*[1]
[1]Global Institute of Pharmaceutical Education and Research, Kashipur, Uttarakhand, 244713;
[2]Burman Group of Institutions Roorkee Haridwar Uttarakhand 247667.

This study focuses on the Fuconazole-based solid dispersion incorporated into emulgels with natural oils for antifungal synergistic effect agents like Lemongrass oil, Neem oil, Tulsi oil. Due to fluconazole's low aqueous solubility, solid dispersion formulations were prepared with β-cyclodextrin and PVP as carriers. Among six formulations, F2 was identified as the most effective, demonstrating the highest aqueous solubility (78.41%), drug content (98.68%), and dissolution rate (99.23%). The results indicate that EGF2 has a stronger inhibitory effect on the tested microorganisms compared to Fluconazole gel (1g). Candida albicans: EGF2 exhibited a significantly larger zone of inhibition (27.7 ± 1.7 mm) compared to Fluconazole gel (19.9 ± 3.5 mm), highlighting EGF2's superior antifungal activity against this common pathogen. Candida esophagitis: Similar trends were observed for Candida esophagitis, with EGF2 showing a zone of inhibition of 26.7 ± 2.2 mm, while Fluconazole gel displayed 18.5 ± 0.9 mm. Furthermore, EGF2 showed substantially higher retention on the SC layer (569.62 ± 12.45 μg/cm^2) compared to the marketed formulation (327.15 ± 19.06 μg/cm^2). This higher retention suggests superior adhesive properties or better penetration of EGF2, ensuring more of the active ingredient remains in the skin. The Emulgels also demonstrated reasonable stability in terms of drug content analysis and physicochemical characteristics, making them a promising option for antifungal treatments.

Keywords Gel, Lemongrass, Neem, Emulgel, Antifungal

Formulation and Evaluation of Methyl Methacrylate Grafted Copolymer of Neem Gum By Grafting Technique

Neetu Rajpoot*, *and Dinesh Puri*
School of Pharmacy, Graphic Era Hill University, Dehradun, Uttarakhand

The use of natural polymers as drug delivery and trapping vehicles has become widespread. Natural polymers exhibit greater biocompatibility, biodegradability, and ease of accessibility as compared to synthetic polymers. Additionally, the native natural polymers reactive groups allow for interaction with other functional groups. Such alteration changes the physical and chemical properties of the newly formed polymer and gives it remarkable functions. Neem gum is a transparent, vivid, amber-colored substance that dissolves in cold water and has no harsh taste. It is employed in the production of special purpose foods and as a bulking agent. Neem gum is used in pharmaceuticals as a film coating, thickening agent, slow-release agent, and tablet binder. Additionally, it can be used to improve solubility. In the present study, we aimed to develop grafted copolymer of neem gum (PMMA-g-NG) via grafting neem gum (NG) with methyl methacrylate (MMA) chains by using microwave assisted grafting technique. The free radical induced grafting method was adopted to graft neem gum with methyl methacrylate, using ceric ammonium nitrate (CAN) as a redox initiator. The impact on grafting parameters such as grafting percentage and grafting efficiency (GE), of monomer and initiator concentrations was evaluated. FTIR study proved the grafting phenomenon and PMMA-g-NG showed significant swelling and water retention capacity. The batch with higher grafting efficiency and percentage grafting was selected and characterized by DSC, FT-IR spectroscopy.

Keywords Biocompatibility, Efficiency, Grafting, Microwave, Thickening

Progressive Lipid-Based Nanocarriers to Treat Neurological Disorder

Nirvesh Chaudhri*, *Vaibhav Rastogi*
TMU- Teerthankar Mahaveer University, College of Pharmacy, Moradabad, 244001. India

The recently developed novel nano-techniques have been employed for the sustainable treatment of various diseases like Neurological disorders including Alzheimer's, Parkinson's, Schizophrenia, multiple sclerosis, neuro infections, epilepsy, and malnutrition. The cerebrospinal fluid barrier and other blood-brain barriers defend the Central nervous system from drug particles reaching the brain's intended region. These problems are treated with advanced forms of nanocarriers. The “lipid based Nanocarriers” offer unique benefits compared to other carriers like micelles, liposomes, nanoparticles, etc. Due to their compatibility with the BBB layer, the nano-sized lipid-based carriers system is the ideal solution for medication delivery to the central nervous system. The easy incorporation of drug molecules into neuronal cells, results in enhanced stability and greater concentration. The “lipid-based nanocarriers” are fabricated with various techniques like “high-speed homogenization”, sonication, solvent evaporation, and emulsification methods. In this formulation, solid lipids, liquid lipids, and polymers regulate drug transport to the brain and enhance therapeutic potential in neurological treatment. This approach improves therapeutic outcomes by enabling safe and effective targeting in chronic neurological disease treatment. In this study, we delve into the current advancements in nanotechnology for neurological therapy and elucidate their mechanisms of action.

Keywords Nanotechnology, Lipid Nano carriers, Neurological disease, Drug delivery

Best Poster Award (Abstract)

Harnessing Computational Power – A Synergistic Cadd-AI Approach to Breast Cancer Therapeutics

Vinita Pandey
Global Institute of Pharmaceutical Education & Research (GIPER) Kashipur-244713, Uttarakhand, India

Breast cancer remains a significant global health issue. This challenge underscores the urgent need for novel therapeutic approaches. Computational drug discovery has advanced considerably, and Computer-Aided Drug Design (CADD) and Artificial Intelligence (AI) have emerged as powerful tools that accelerate drug development. This poster will examine the synergistic integration of CADD and AI in breast cancer therapeutics. CADD utilizes computational simulation methods, such as molecular dynamics to predict how prospective therapeutics interact with their biological targets. Additionally, AI tools are being incorporated to enhance prediction accuracy and improve insights into structure–activity relationships. AI- driven machine learning algorithms facilitate biomarker discovery, boost drug candidate optimization and aid target identification. These innovations decrease both time and expenses while enhancing accuracy. This approach leverages computational capabilities and holds significant promise for transforming breast cancer therapy and opens the door for targeted, more effective and personalized treatments.

A. K. Saxena, A. Saxena (Eds.): GTHTM 2025, pp. 533, 2026.
https://doi.org/10.1007/978-3-032-12320-6

Poster Presentations (Abstracts)

Understanding the Medicinal and Therapeutic Potential of Alkaloid-Based Phytocompounds for the treatment of Alzheimer's Disease

Arnob Chakrovorty[1], *Banani Bhattacharjee*[1], *Asmita Samadder*[1,*] *and Sisir Nandi*[2,*]

[1]Cytogenetics and Molecular Biology Lab., Department of Zoology, University of Kalyani, Kalyani, Nadia, 741235, West Bengal, India

[2]Department of Pharmaceutical Chemistry, Global Institute of Pharmaceutical Education and Research (GIPER) (Affiliated to Veer Madho Singh Bhandari Uttarakhand Technical University), Kashipur-244713, Uttarakhand, India.

Alzheimer's disease (AD) is a neurocognitive disorder that affects the central nervous system leading to dementia-like cognitive defects, effecting regular life of the individual. AD is a global problem that demands immediate medical attention. Alkaloid-based phytocompounds (ABP) hold enormous biological activity and therapeutic potential that can be utilized for treatment against AD. ABP administration in streptozotocin-induced AD model was found to improve biochemical parameters of the brain and provide neuroprotection by modulating key proteins. ABP improved brain glucose metabolism, reduced reactive oxygen species in brain tissue and improved the tissue architecture of the cortex and the hippocampus of the brain.

Keywords Alzheimer's disease, Neurodegenerative disease, Phytotherapy, Swiss albino mice

Rationale of Antifunagal and Anti Bacterial Interventions of Imidazole Derivatives

Divya Negi[*1,2],*Akhilesh kumar Mishra*[1],*and Narendra Kumar Singh*[3]

[1]School of Pharmacy, Babu Banarsi Das University, Lucknow, Uttar Pradesh, India-226028

[2]Faculty of Pharmaceutical Sciences, Amrapali University, Lamachaur, Haldwani, Nainital, Uttarakhand, India-263139

[3]Apex Institute of Pharmacy, Chunar, Mirzapur, Uttar Pradesh, India-231304.

A. K. Saxena, A. Saxena (Eds.): GTHTM 2025, pp. 535–552, 2026.
https://doi.org/10.1007/978-3-032-12320-6

Imidazole is a five-member heterocyclic molecule with two nitrogen atoms at positions 1 and 3 in the aromatic ring. It is categorized as a diazole. Clinical trials have utilized a variety of imidazole-based drugs for antifungal, antibacterial, anticancer, antiparasitic, antihypertensive, and anti-inflammatory properties. A remarkable class of chemicals with antimicrobial properties, including potent antifungal and antibacterial agents, are imidazole derivatives, which contain the imidazole ring. The biological, and therapeutic use of compounds containing imidazoles are the main topics of this review. Numerous studies have investigated the antifungal activity of imidazole derivatives, which interfere with the growth of cell membranes and the integrity of the walls of fungal cells. Likewise, their antibacterial properties have been noted in a variety of Gram-positive and Gram-negative bacteria, and they are linked to mechanisms such the suppression of nucleic acid replication, protein synthesis, and cell wall formation. We also investigate the structure-activity relationships of imidazole derivatives, focusing on their pharmacokinetic, toxicological, and biological activities. In addition to increasing the number of effective medications available to treat bacterial and fungal diseases, the discovery of new imidazole-based treatments could contribute to combating the growing threat of drug resistance.

Keywords Imidazole, Antifungal, Antimicrobial

Innovation in Taste-Masking Technologies: Advancing Pediatric Drug Formulations

Divyanshu, *Mohammad Aadil, Raja Babu*
Global Institute of Pharmaceutical Education & Research (GIPER) Kashipur-244713, Uttarakhand, India

Pediatric drug compliance remains a significant challenge in pharmaceutical development, primarily due to the unpleasant taste of active pharmaceutical ingredients (APIs). This study investigates novel taste-masking techniques for developing palatable pediatric formulations. We explored a combination of microencapsulation using Eudragit E100 and cyclodextrin complexation to mask the bitter taste of a model drug, Amoxicillin trihydrate. The taste-masked microspheres were prepared using spray-drying technology and characterized through particle size analysis, encapsulation efficiency, and in vitro dissolution studies. The taste-masking efficiency was evaluated using an electronic tongue and human taste panels. Results demonstrated that the optimized formulation achieved 95% drug encapsulation with a mean particle size of 150 ± 15 µm. The electronic tongue analysis showed significant reduction in bitterness intensity ($p<0.05$), correlating well with human panel evaluations. Drug release studies indicated minimal drug release (< 10%) at salivary pH 6.8, while ensuring complete release (> 90%) at gastric pH 1.2 within 45 minutes. Stability studies confirmed the physical and chemical stability of the formulation over six months. This innovative approach offers a promising solution for improving pediatric medication compliance through effective taste- masking while maintaining therapeutic efficacy.

Wound Healing: An In-depth Review

Fiza, Sristhi Gaur, *Shahjad*, Arun Kumar*
Global Institute of Pharmaceutical Education & Research (GIPER) Kashipur-244713, Uttarakhand, India

Wound healing is a complex and intricate biological process that plays a fundamental role in the maintenance of tissue integrity and the restoration of homeostasis following injuries or surgical procedures. Understanding the mechanisms and factors that govern wound healing is not only crucial for medical professionals but also holds significant implications for patients' overall health and quality of life. This review aims to provide a comprehensive overview of the background and significance of wound healing, shedding light on the various stages of the process and the intricate interplay of cellular and molecular events that underlie it. Wound healing is not a mere localized response to injury; it represents a dynamic and highly regulated cascade of events that involve numerous cell types, signaling pathways, and extracellular matrix components. Furthermore, impaired or delayed wound healing can lead to a range of complications, including chronic wounds, infections, and excessive scarring, all of which can have profound implications for patient outcomes. Exploring the basic biology of wound healing, delving into the stages such as hemostasis, inflammation, proliferation, and remodeling. Each stage examined in detail, the specific cell types and molecular signals that drive the healing process forward. The factors that can influence and potentially hinder wound healing, from systemic conditions like diabetes to local factors such as tissue ischemia and the significance of advanced wound care techniques and therapeutic interventions, including the use of growth factors, tissue engineering, and regenerative medicine, which have the potential to revolutionize wound healing and enhance patient outcomes. This study is essential for healthcare professionals, researchers, and anyone interested in improving the quality of care for patients with wounds.

An Overview of Mood Disorder with a Focus on Depression Related to Vedic Plants

Harsh Sharda, *Deepak Teotia, Arun Kumar, Muneer Alam*
Global Institute of Pharmaceutical Education & Research (GIPER) Kashipur-244713, Uttarakhand, India

Ayurveda places great emphasis on maintaining positive mental well-being. Anything that disrupts the equilibrium of body and mind is recognized to cause a disease. Emotions are fundamental feelings experienced by humans. Each of us has experienced feelings of unhappiness, sadness, or discouragement at various points in our lives. However, when someone experiences feelings of anxiety, hopelessness, helplessness, worthlessness, guilt, irritability, pain, or restlessness, they fall into the classification of depression. Vishada and Avasada are two states resembling depression in Ayurveda. Vishada is characterized by a continual feeling of sadness, a sense of inadequacy stemming from fear of failure, which leads to an inability of both mind and body to operate effectively. Charaka references Vishado Rogavardhananam Agrya, indicating that Vishada is the primary element that exacerbates the disease state. Depression and other mood disorders are particularly severe as they often emerge in early life, are challenging to identify in adolescents, can be mistaken for typical behavioral variations, substance abuse, or other mental health conditions linked with elevated suicide risk. Consequently, treatment is usually commenced quite late and necessitates extended follow-up and prolonged management. Considering the facts presented in this article, we aim to provide a concise overview of the pathophysiology of depression and outline an extensive Ayurvedic strategy to address depression and related mood disorders. It also emphasizes the concealed idea of Ayurvedic psychology that can be applied in medicine to treat depression and enhance mental health.

A Review on Novel Drug Delivery Approach: The Transdermal Drug Patches

Meenakshi[1*]*,and Charu Saxena*[2]

[1*]Research Scholar, Faculty of Pharmacy, Swami Vivekanand Subharti University Meerut

[2]Associate Professor, Faculty of Pharmacy, Department of Pharmaceutics, Swami Vivekanand Subharti University Meerut

Currently, we take about 70% of our prescriptions orally, although their effectiveness is frequently lower than expected. A surface that is easily accessible for administering medications is the human skin. About one-third of the blood in the body circulates through the skin of the average adult, which has a surface area of about two square meters. Developing regulated medication delivery systems has been the pharmaceutical industry's main focus over the last few decades. There are around 200–250 sweat ducts and 10–70 hair follicles per square centimeter of human skin. Because it enables a gradual release of the medicine over time, the transdermal drug delivery system (TDDS) is different from other methods of administering medications to the skin. This decreases the negative effects that are typical with oral drugs and weakens the effect. To transport active substances straight into the bloodstream through the skin, designers create a variety of transdermal patches. Because they are non-invasive, eliminate the need for professional medicine administration, lessen negative effects on the gastrointestinal (GI) system, and encourage patient adherence, transdermal drug delivery (TDDS) systems are considered patient-friendly. Transdermal medication delivery does have many drawbacks, though, including the potential for skin irritation or sensitization, adhesive discomfort, poor skin adherence, high cost, and limited use of certain physicochemical drug features. The several facets of TDDS and new developments in this class of targeted therapeutic formulation in relation to acute and chronic metabolic disorders are included in our current review.

Keywords Drug delivery, Adhesive discomfort, Targeted therapeutic, Regulated medication, and TDDS

The Exosome Expedition: Navigating the Frontiers of Wound Healing with Groundbreaking Studies, Molecular Mysteries, and Therapeutic Innovations

Mehreen Jahan, Nishi Rani, Shahjad, *and Arun Kumar*
Global Institute of Pharmaceutical Education and Research, Kashipur-244713, India

Difficulties in wound healing pose a considerable clinical problem and are a significant source of morbidity in the general population. It is difficult to impossible to perform sequential studies on the same human wound, and often there are no uniform criteria for defining stages of healing or outcome of the repair process. Animal models have provided some valuable information, but the healing process differs in humans and animals, and data obtained from in vitro studies can be difficult to extrapolate to a clinical context. A clearer knowledge of what variables should be measured and what methods are best to measure them would increase opportunities for successful clinical studies. This review aims to address this need, to serve as a roadmap for both current and future researchers in the field of wound healing. Much of our knowledge of the finer details of wound healing has been elucidated through molecular and cell biology, in which the tools of genetic manipulation and cell culture studies have been invaluable. Briefly, this essay intends to synthesize our present knowledge of the processes of inflammation, epithelialization, and tissue repair primarily using evidence obtained from in vitro and in vivo studies. The basic processes involved in healing include inflammation, re-epithelialization, granulation tissue formation, and collagen remodeling. Wound healing aims to reconstruct and preserve the function of the tissue as quickly as possible. However, healing is influenced by many factors including the health and age of the individual, type, size, and site of the wound, and treatment of the wound. As a result, any deviation from the normal sequence and time course of events may result in abnormal healing. This is a vital process that all living organisms undergo at some stage in their lives. Wound healing is often divided into three or four overlapping phases.

Prediction of Squaramides Derivatives as External Data Set Using Validated Qsar for the Design of Potent Anti-tubercular Compounds

Muneer Alam[1,2], *Zeeshan Fatima*[1], *Sisir Nandi*[2] *and Sarfaraz Ahmed*[2]
[1]Amity Institute of Pharmacy, Lucknow, Amity University Uttar Pradesh, Sector 125, Noida- 201313, India

[2]Department of Pharmaceutical Chemistry, Global Institute of Pharmaceutical Education and Research, Kashipur-244713, India.

Tuberculosis (TB) is the most important global infectious killer disease responsible for 1.8 million deaths inclusive of 0.25 million cases of drug resistance TB per year. It has emerged as a world-wide health risk due to drastic death rise in the incidence of multi drug-resistant (MDR) (rifampicin and isoniazid) and extensively drug resistant (XDR) (Fluoroquinolone and at least one of three injectable second line drugs) mycobacterial strains. Among several antitubercular drugs in clinical trials, the bedaquiline (trade name sirturo, code name: TMC207 and R207910) has emerged as a highly promising first drug approved by FDA for the treatment of multi-drug-resistant TB in combination with other drugs. Bedaquiline acts on mycobacterial ATP synthase and is highly effective on replicating as well as on dormant mycobacteria. In view of it several series of squaramides have been synthesized and evaluated for their antitubercular and ATP synthase inhibitory activities. In order to understand the role of physicochemical like hydrophobicity, electronic and steric factors in elisiting the biological response, the Quantitative structure activity relationship (QSAR) developed on Quinolines. It was observed that hydrophobic and steric parameters are more important for explaining the variations in antitubercular activity. Thus validated QSAR was taken as standard model. External data set based on Squaramides compounds have been predicted then it has been shown that some of squaramides have good predicted activity, so some of new chemical entities have been synthesized and biological activity will be carried out against tuberculosis.

Docking Based Qsar Studies on DHFR Inhibitors for the Identification of Anti-malarial Agents

Nishu *and Sarfaraz Ahmed*

Department of Pharmaceutical Chemistry, Global Institute of Pharmaceutical Education and Research, Jaspur Road, Kashipur, Uttarakhand

Malaria is a serious infectious disease caused by Plasmodium parasites, which are transmitted to humans through the bites of infected female Anopheles mosquitoes. According to the World Health Organization's (WHO) World Malaria Report 2024, there were an estimated 263 million cases and 597,000 malaria-related deaths worldwide in 2023. Hence, the continued presence of malaria on the global health agenda makes it imperative to actively seek new antimalarial agents which lead us to Dihydrofolate reductase (DHFR) which is a key enzyme in the folate pathway of Plasmodium species and offers a well-characterized antimalarial enzyme target for drug action. Docking-based Quantitative Structure-Activity Relationship (QSAR) studies provide a powerful avenue for the identification and optimization of DHFR inhibitors. This study introduces integration of molecular docking and QSAR aimed at predicting the binding affinity and biological activity of prospective DHFR inhibitors.

Through the investigation of a dataset of known inhibitors, we derived QSAR models which ca be used in predicting of NCE's.

Review Regarding Human Metapneumovirus (HMPV), A Significant Infection of the Respiratory Tract

Sarfaraj Ahmad, *Harsh Sharda, Kavita Joshi, and Vinita Pandey*
Global Institute of Pharmaceutical Education & Research (GIPER) Kashipur-244713, Uttarakhand, India

The virus known as human metapneumovirus (hMPV), which has been recognized in 2001, primarily affects infections of the upper and lower respiratory tract in youngsters, although it may also impact elderly people along with people with immune systems that are compromised. The primary causative agent for five to ten percent of children who are hospitalizations for severe infections of the respiratory tract is hMPV. Children who suffer from a hMPV infection can get severe bronchiolitis and pneumonia, and the symptoms are identical to those of a human respiratory syncytial virus, or HPV, infection. The first Infection with hMPV typically takes place in early childhood, although re-infections frequently happen at all ages. Because of the virus's sluggish growth in cell culture, molecular techniques (like reverse transcriptase PCR (RT-PCR)) are the favored diagnostic method for identifying hMPV. Several vaccine candidates have demonstrated effectiveness in preventing clinical disease, yet none are currently available for purchase. Our comprehension of hMPV has significantly evolved in recent years, and in this article, we will examine the existing data on the molecular biology and epidemiology of hMPV.

Molecular Docking Analyses of Common Food Additives Having Genotoxicity and Chromosomal Aberration Target Interactions

Sisir Nandi, ***Sarfaraz Ahmed,*** *Sombodhi Bhattacharya, and Anil Kumar Saxena*
Department of Pharmaceutical Chemistry, Global Institute of Pharmaceutical Education and Research (GIPER), Kashipur-244713, India

Food additives have emerged as an essential element of the current food industry. They come from either natural or synthetic sources. Food additives are used specifically to keep food from spoiling, enhance its appearance, flavour, taste, texture, preservation and stabilization of the commercially available food for maintaining the freshness and nutritional value. But the uncontrolled uses of food additives are raising concern about their potential genotoxicity. Very frequently used food additives such as Disodium glycyrrhizate as sweetener, Sunset Yellow as coloring, Tartrazine as coloring, propyl gallate as antioxidant preservation and alloxan as wheat flour whitener may produce chromosomal aberrations and genomic instability. The molecular mechanisms behind these are yet to explore. An attempt has been made to compute the affinity of these food additives towards interacts with the genotoxicity and chromosomal aberration target proteins utilizing structure based docking simulations.

Provide an Overview of PBM, Its Historical Context, and Its Significance in Cellular Healing

Swati Tamta, *and Shahjad*
Global Institute of Pharmaceutical Education and Research College of Pharmacy and Research, Kashipur, Uttarakhand 244713

A growing body of evidence supports the modulation of pain by light exposure. As such, phototherapy is being increasingly utilized for the management of a variety of pain conditions. The modes of delivery, and hence applications of phototherapy, vary by wavelength, intensity, and route of exposure. As such, differing mechanisms of action exist depending upon those parameters. Cutaneous application of red light (660nm) has been shown to reduce pain in neuropathies and complex regional pain syndrome-I, whereas visual application of the same wavelength of red light has been reported to exacerbate migraine headache in patients and lead to the development of functional pain in animal models. Interestingly visual exposure to green light can result in reduction in pain in variety of pain conditions such as migraine and fibromyalgia. Cutaneous application typically requires exposure on the order of minutes, whereas visual application requires exposure on the order of hours. Both routes of exposure elicit changes centrally in the brainstem and spinal cord, and peripherally in the dorsal root ganglia and nociceptors. The mechanisms of photo-biomodulation of pain presented in this review provide a foundation in furtherance of exploration of the utility of phototherapy as a tool in the management of pain.

Harnessing Computational Power – A Synergistic Cadd-AI Approach to Breast Cancer Therapeutics

Vinita Pandey
Global Institute of Pharmaceutical Education & Research (GIPER) Kashipur-244713, Uttarakhand, India

Breast cancer remains a significant global health issue. This challenge underscores the urgent need for novel therapeutic approaches. Computational drug discovery has advanced considerably, and Computer-Aided Drug Design (CADD) and Artificial Intelligence (AI) have emerged as powerful tools that accelerate drug development. This poster will examine the synergistic integration of CADD and AI in breast cancer therapeutics. CADD utilizes computational simulation methods, such as molecular dynamics to predict how prospective therapeutics interact with their biological targets.

Additionally, AI tools are being incorporated to enhance prediction accuracy and improve insights into structure–activity relationships. AI- driven machine learning algorithms facilitate biomarker discovery, boost drug candidate optimization and aid target identification. These innovations decrease both time and expenses while enhancing accuracy. This approach leverages computational capabilities and holds significant promise for transforming breast cancer therapy and opens the door for targeted, more effective and personalized treatments.

Antimicrobial Activity of Aqueous Stem Extract of Carica Papaya and Radish Leaves

Vivek kumar, *Shahjad and Arun Kumar*
Global Institute of Pharmaceutical Education & Research, Kashipur, Uttarakhand, India, 244713

The stem of *Carica papaya* plant and radish leaves were collected and allowed to drying in dark place and ground in mortar pestle. The powered material extracted successively using the maceration process with distilled water all the extract were subjected to systemic phytochemical screening for the presence of phytochemical constituent this indicate the presence of carbohydrate, protein, vitamin C, tannin, alkaloid, flavonoid. To verify the antimicrobial activity of all the extract were determine by cell wall diffusion method in the observation the stem of *Carica papaya* and radish leaves exhibit significant inhibitory activity against all the pathogen extract showed maximum activity. The sample further studied by FTIR it showed 13 functional groups in between the spectra 400 – 4000 nm.

Exploring NaV1.7 Blockers in Itch Management: Insights From Gx201 and Sulfamethoxazole Interactions

Abhishek Kumar Mishra, *and Sailendra Kumar Mahanta*
The Assam Kaziranga University, Koraikhowa, NH-37, Jorhat 785006, Assam, India.

Pruritus, or itching, is a common symptom linked to dermatological and systemic conditions. NaV1.7, a voltage-gated sodium channel subtype, plays a pivotal role in transmitting itch signals through sensory neurons. Pharmacological research shows that inhibiting NaV1.7 can alleviate histamine-dependent itching, often triggered by allergic reactions or inflammation. GX201, a potent NaV1.7 blocker, has shown promise in managing neuropathic pain and reducing itch intensity by decreasing neuronal excitability. Sulfamethoxazole, an antibiotic used to inhibit bacterial growth, paradoxically induces itching through hypersensitivity reactions. These may involve either histamine release via immediate immune responses or cytokine-mediated inflammation in delayed-type reactions. Severe hypersensitivity can lead to conditions like Stevens-

Johnson Syndrome (SJS) or Toxic Epidermal Necrolysis (TEN), further emphasizing the complexity of itching mechanisms. This study explores whether sulfamethoxazole binds to NaV1.7, similarly to GX201, and assesses its potential to alleviate itching through this pathway. Molecular docking analyses revealed that both compounds share the same NaV1.7 binding site but differ in binding energy, with GX201 exhibiting a stronger affinity (-9.29 kcal/mol) compared to sulfamethoxazole (-7.31 kcal/mol). Key interactions, such as hydrogen bonding, aromatic interactions, and hydrophobic effects, contribute to GX201's higher binding stability. Despite sharing a binding site, sulfamethoxazole's histamine-mediated itching underscores the limitations of targeting NaV1.7 alone. Effective itch management may require addressing additional pathways, including histamine receptor activation. These findings highlight the complexity of itch mechanisms and suggest that NaV1.7 inhibition, while promising, may not suffice as a standalone therapeutic strategy for comprehensive itch relief.

Keywords NaV1.7 blockers, Molecular docking, Itch, Neuropathic Pain, Docking energy

Exploring QSAR Modeling for a Series of Pyrazole Derivatives as Acetyl Cholinesterase Inhibitor's for the Management of Alzheimer's Disease

Akshay Kumar, *Jagadish Singh, and Partha Pratim Roy*

Laboratory of Drug Discovery and Ecotoxicology, Department of pharmacy, Guru Ghasidas Vishwavidyalaya (A central University), Bilaspur, 495009, India

Alzheimer's disease is the most prevalent degenerative condition in the elderly, marked by symptoms such as dementia, memory impairment, and changes in behaviour, primarily caused by the degeneration of cholinergic neurons and other contributing factors. Heterocyclic compounds are crucial in the development of new drugs. Among these, pyrazole has emerged as one of the most researched and promising heterocyclic structures for treating Alzheimer's disease over the past decade with therapeutic effects against several key targets. The collected experimental data of 101 compounds targeting AChE was utilized for development of robust QSAR model. The developed models fulfil the good internal fitting (R2 = 0.692; R^2adj = 0.675), internal validation (Q2LOO = 0.63) external validation (Q2F1 = 0.795, Q2F2 = 0.708, CCC ext. = 0.871) MAEext. = 0.246 and RMSE ext. = 0.322). Mechanistic interpretations highlighted the contribution of electropositive and bulky group at 3rd position of pyrazole scaffold. Chlorobenzene group at 2nd nitrogen of pyrazole is very essential for the activity. Electronegative group with fused aromatic ring enhanced the activity. Substitution with aromatic nitrogen fused heterocyclic compound at 3rd position of pyrazole drastically increases the activity.2 H–Pyran attached with –NH2 and CN group at 4rth position of pyrazole responsible for increased activity. Overall, this research underscores the importance of the pyrazole scaffold in the treatment of Alzheimer's, showing potential for inhibitory action.

Keywords QSAR, AChE, Alzheimer and Pyrazole

To Explore Pathogenesis, Current Targets and Mode of Treatment of Trypanosomiasis

Astha Saha[1], *Rukayya*[2], *Debapriya Ghosh*[1], *Ashaporna Biswas*[1], *Sujan Murmu*[1], *Amar Faruk Saheb*[1], *Tamanna Khatun*[1], *Barsha Roy*[1], *Asmita Samadder*[1,3]* *and Sisir Nandi*[2]*
[1]Cell and Developmental Biology Special, Department of Zoology, University of Kalyani, Kalyani, Nadia, 741235, India
[2]Department of Pharmaceutical Chemistry, Global Institute of Pharmaceutical Education and Research (GIPER), Kashipur 244713, India
[3]Cytogenetics and Molecular Biology Lab., Department of Zoology, University of Kalyani, Kalyani, Nadia, 741235, India

Trypanosomiasis is a parasitic disease caused by *Trypanosomabrucei* and *Trypanosomacruzi* responsible for producing Human African Trypanosomiasis (sleeping sickness), and American Trypanosomiasis (Chagas disease). Trypanosomiasis has been the prime concern for global health issues and has been designated as one of the seventeenths neglected tropical diseases (NTD) declared by the World Health Organization (WHO). Trypanosomiasisis most common among low-income countries and communities and claim immediate addressing to save lacks of lives of common human kind. However, there are no specific synthetic drugs known to cure the Trypanosomiasis. Existing chemotherapeutics are unable to kill the protozoa and therefore, the causative microorganisms become modified and transformed to more pathogenic forms. Therefore, this issue calls forprime exploration of the disease pathogenesis, and current targets for *Trypanosomabrucei* and *Trypanosomacruzi*for design and development of potent specific drugs targeting trypanosomalprotozoa inhibition.

Keywords Trypanosomiasis, Drug targets, Pathobiology, Essential proteins

QBD Based Extraction of Plumbagin from Plumbagoindica Roots and Its Antioxidant Activity

Gayatri Khosla[1]*, *Neeraj Kumar*[2], *Disha Dutta*[3], *and Rama Tyagi*[1]
[1]Galgotias College of Pharmacy, Greater Noida, Uttar Pradesh, India, 201310
[2]Department of Pharmacognosy and Phytochemistry, School of Pharmaceutical Education & Science, New Delhi, 110062
[3]DevsthaliVidyapeeth College of Pharmacy, Rudrapur, Pin- 263148, Uttarakhand, India.

Plumbagin, a napthaquinone found in *Plumbago Indica* roots, has various therapeutic characteristics. The research evaluates the effectiveness of several contemporary and traditional plumbagin extraction techniques. To optimize procedure variables for plumbagin extraction from *PlumbagoIndica* roots, Box-Behnken Design (BBD) was used. Quality by Design (QbD), more especially BBD, was used to optimizeextraction parameters (extraction time, solvent-to-drug ratio, and extraction temperature). Using HPTLC, the amount of plumbagin in various extracts was quantified. Moreover, the DPPH method was used to evaluate the antioxidant properties of several *Plumbago Indica* extracts. Methanol is an exceptionally efficient extractive solvent, and the ultrasound-assisted extraction approach yields the maximum amount of plumbagin. The ideal extraction temperature for plumbagin was determined to be 68.489°C, the solvent-to-drug ratio to be 31.249 mL/g, and the extraction duration to be 35.682 minutes using BBD. The yield of plumbagin under these circumstances was 1.981 mg/g, almost identical to the expected value of 1.893 mg/g. With a 79.39% percentage inhibition, the ethanolic extract has demonstrated notable antioxidant activity. Methanol was found to be the most effective extracting solvent, and the ultrasound- assisted extraction method was the most effective of all the thermal and non-thermal extraction techniques employed. The current work emphasizes the application of QbD, in the extraction field of therapeutically active phytoconstituents, which reduces the time and effort required for optimization.

Keywords Response surface approach, Extraction, Optimization, Plumbagin, and *plumbagoindica*

Limonene: A Versatile Ingredient in the Cosmetic Industry

Mohd.Tafseel*, *Disha Dutta, and Ramsha Aslam*
DevsthaliVidyapeeth College of Pharmacy Rudrapur, Uttarakhand

Limonene, a naturally occurring monoterpene hydrocarbon that is mostly derived from citrus fruits, has drawn a lot of interest in the cosmetics business. This substance has several advantages, such as a nice scent, antibacterial properties, and possible skin benefits. Because of its invigorating aroma, limonene is a common ingredient in body care products, colognes, and perfumes. Customers' whole sensory experience is improved by its zesty scent, which is frequently connected to freshness and cleanliness. In addition to its aroma qualities, limonene has antibacterial activity against a variety of microorganisms, which makes it a possible component of skincare products to help stop the growth of fungi and bacteria. Limonene has been researched for its possible advantages for skin health in addition to its scent and antibacterial qualities. According to some research, limonene may have antioxidant qualities that help shield the skin from harm brought on by free radicals. Its ability to increase skin suppleness and lessen wrinkle appearance has also been examined. To completely comprehend the mechanics underlying these possible advantages, more research is necessary. Because of its versatility, limonene is used in a variety of cosmetic items, such as body lotions, soaps, hair care products, colognes, and perfumes. It is anticipated that limonene's use in cosmetics would rise in tandem with consumer demand for sustainable and natural components. However, it is crucial to remember that limonene might irritate certain people, so its application needs to be carefully thought out, particularly in products meant for delicate skin.

Keywords Limonene, Cosmetic industry, Fragrance, Antimicrobial, Skin benefits

Biofilm Formation in Gastric Mucosa by *Helicobacter pylori* Infection and Its Role in Drug Resistance

Pratyusa Biswas, *and Laishram Pradeepkumar Singh*[*]
Laboratory of Translational Biology and Molecular medicine, Department of Zoology, University of Kalyani, Kalyani, Nadia- 741235

*Helicobacter pylori*is the most common, significant pathogenic agent which exhibits global prevalence of approximately 50%. It is associated with several gastrointestinal disorder including peptic ulcer and gastric cancer. Due to the highly increasing rate of infection the *H. pylori* eradication is the most necessary step to decrease the rate of gastric and duodenal ulcer and the gastric cancer prevalence. Despite of having different antibiotic-based medications, the treatment of *H. pylori* infection is hindered due to the emergence of antibiotic-resistance strain throughout the world. One of the most important mechanisms of drug resistance of *H. pylori* include formation of biofilm in gastric mucosa. During infection, some bacteria enter and occupy the gastric epithelium and encase themselves in an extracellular polymeric substance, known as biofilm. The biofilm structure that protects the bacterial colony from any external threat, is responsible for chronic infection of GI tract. Bacterial colony, covered with biofilm is 1000 times more resistant to any antibiotics than planktonic bacteria. Therefore, biofilm can be a novel therapeutic target to reduce the burden of drug-resistance infection of *H. pylori*. This review focuses on the molecular mechanism of biofilm formation of *H. pylori* in gastric mucosa including regulation by quorum sensing. Furthermore, it examines the clinical implications of biofilm-associated drug-resistance. Emerging therapeutic strategies targeting biofilm disruption are also discussed as potential solutions to overcome treatment failure. Understanding the intricate relationship between biofilm formation and drug-resistance offers valuable insights into developing more effective strategies to combat *H. pylori* associated diseases.

Keywords *Helicobacter pylori*, Biofilm, Drug-resistance, Infection, Anti-biofilm Agents

Formulation and Evaluation of Berberine Containing Nano Invasomal Vesicles

Preeti [1*] *Dinesh Puri* [2]
[1] Siddhartha institute of pharmacy
[2]Graphic Era Hill University, Dehradun, Uttrakhand, India

Isoquinoline alkaloid berberine has a variety of medicinal uses but its applications have been hampered by factors such as low bioavailability, minimal absorption, and poor water solubility. Therefore, the goal of this work was to create optimal nanosized berberine invasomes with improved transdermal delivery to overcome the obstacles. The following formulation parameters were investigated by preparing berberine invasomes using a Box-Behnken experimental design: phospholipids %(x1), terpenes %(x2) and terpenes type (x3) on entrapment efficiency (y1) and vesicle size(y2). On the bases of entrapment efficiency inva 21 and inva 27 were found best. The entrapment efficiency was 77 and 78 respectively. Inva 27 was further evaluated for particle size by dls. Particle diameter of inva 27 was 322.0 nm and polydispersity index were 0.2. The study's findings suggest that invasomes have a high potential for transdermal administration of berberine which can increase the topical utility of berberine in several skin diseases.

In Silico Design and Evaluation of TRPV1 Inhibitors Based on Capsaicin for Neuropathic Pain

Priyakshi Chutia, *and Sailendra Kumar Mahanta*
The Assam Kaziranga University, Koraikhowa, NH-37, Jorhat 785006, Assam, India.

Neuropathic pain, a debilitating condition resulting from nerve injury, often involves the over activation of the transient receptor potential vanilloid-1 (TRPV1) receptor, commonly known as the capsaicin receptor. TRPV1 is a non-selective cation channel activated by noxious stimuli such as heat, acidic pH, and capsaicin. It plays a central role in nociceptive signaling, inflammatory pain, and hyperalgesia. Targeting TRPV1 offers a promising therapeutic avenue for managing chronic pain conditions. This study focuses on the in silico design and evaluation of TRPV1 inhibitors derived from capsaicin, aiming to identify potent and selective modulators of this receptor. Computational approaches, including molecular docking, binding site and pocket analysis, and ADMET profiling, were utilized to design and assess the efficacy of novel TRPV1 inhibitors. Existing TRPV1-targeted therapies include both agonists and antagonists. While antagonists like AMG 517 and SB-705498 show efficacy in neuropathic pain management, challenges such as hyperthermia and off-target effects limit their clinical use. Conversely, agonists like capsaicin and resiniferatoxin (RTX) offer prolonged desensitization of nociceptive neurons, providing sustained analgesia. RTX, an ultrapotent capsaicin analog, demonstrates efficacy in preclinical and clinical studies for neuropathic pain, cancer pain, and inflammatory hyperalgesia. Despite promising advancements, translating TRPV1 modulators into clinical practice requires overcoming challenges such as irritancy, thermoregulatory side effects, and selectivity. This study highlights the potential of rational drug design in developing safer, more effective TRPV1 inhibitors for neuropathic pain, emphasizing the importance of computational approaches in accelerating drug discovery.

Keywords TRPV1 receptor, Neuropathic pain, Capsaicin, Resiniferatoxin (RTX), Molecular docking, Rational drug design

Unmasking *Leishmania donovani* Map Kinase12: Comprehensive Study Implies It's Probable Role as a Drug Target

Priyanka H. Mazire*, *Amit Roy*
Department of Biotechnology, Savitribai Phule Pune University (Formerly University of Pune), Ganeshkhind Road, Pune- 411007, Maharashtra, INDIA

Visceral leishmaniasis (VL), a potentially fatal vector-borne disease, which is caused by the intracellular protozoan parasite *Leishmania donovani*, remains a major health problem due to restricted repertoire of drugs, deleterious side effects, high cost and drug resistance. So, there is an instantaneous need not only to search and design new efficacious affordable therapeutic agents with minimal side effects, but also to identify and validate newer drug targets for novel drug development. Mitogen-Activated Protein Kinases (MAPKs) are potential drug targets as they are regulators of diverse cellular processes. In this present study, we report *L. donovani* MAPK12 (LdMAPK12) as a probable virulence factor implying it as a plausible drug target. LdMAPK12 sequence is distinct from human MAPKs, while is highly conserved in different *Leishmania* species. It was observed that LdMAPK12 is expressed in both promastigotes and amastigotes. Higher expression of LdMAPK12 was seen in the virulent and meta cyclic promastigotes in comparison with the avirulent and procyclic promastigotes. Furthermore, pro- inflammatory cytokines reduced, whereas anti-inflammatory cytokines increased LdMAPK12 expression in macrophages. Therefore, this study suggests a probable novel role of LdMAPK12 for parasite virulence and it can be exploited in near future as an important novel target for the development of new therapeutics for VL.

In Silico Studies of 1,2,4-Triazole Derivatives by Targeting DHFR Against *Staphylococcus aureus*: Pharmacophore Generation, Atom-Based 3D-Qsar, Molecular Docking and ADMET

Saket[1], *Ankush*[1], *Arun Kumar*[1], *and Neelam Vashisth*[1]*
Department of pharmaceutical sciences, Gurugram university, Gurugram, Haryana, India- 122018

Antibiotics, which disrupt essential processes for bacterial growth and proliferation, are crucial in combating bacterial infections. However, their misuse and overuse have resulted in increasing antimicrobial resistance, highlighting the urgent need for new antibiotics. Compounds with a 1,2,4-triazole ring exhibit diverse biological activities, with extensive research demonstrating their significant antimicrobial properties. *in silico* approaches are used on previously synthesized ninety-five 1,2,4-triazole derivatives against *staphylococcus aureus* for protein Dihydrofolate reductase. The protein Wild-type *staphylococcus aureus* DHFR in complex with NADPH and trimethoprim (PDB ID 2W9G), were used for the model development against the *S. aureus*. Alongside, the four features Ligand based pharmacophore hypothesis and Atom based 3D-QSAR model was generated with statistical parameters include correlation coefficient for the training set (Q^2=0.7995), regression coefficient for test set (R^2=0.9808) which proves the reliability of model. Contour visualization was done for the best docked and least docked compound. Furthermore, the best docked molecule was taken and through CheMBL database ligand based virtual screening was performed. Later, ADME/T studies were carried out on top 5 compounds from the virtual screened database. Among the five compounds, the compounds CheMBL 208844 and CheMBL 275068 were found to be most potent antimicrobial agents against *S. aureus*. These compounds may be considered as lead for further development of antimicrobial agents.

Keywords 1,2,4-triazole; antimicrobial activity; Atom based 3D-QSAR; Dihydrofolate Reductase; ADME/T

Evaluation of Antifungal Potential of Lansoperazole: A Drug Repurposing Approach

Sarita [1*], *Neha Mathur*[1,] *and Pooja Chawla*[2]

[1*]Amity Institute of Pharmacy, Amity University Uttar Pradesh, Lucknow- 226010, Uttar Pradesh, India

[2]Baba Farid University of Health Sciences Faridkot (Punjab)

The pharmacological options for fungal infections are restricted to azoles, allylamines, echinocandins, and similar categories of medications. The fungal strain exhibits resistance to these treatment classes; hence, we employed a drug repurposing strategy by investigating current medications to validate potential therapeutic advantages through computational methods.

Aim & Objectives: To assess the potential of the new antifungal medicine, we employed antiepileptic medications in this investigation, utilizing in silico methods and the protein of interest for validation. **Method:** Docking was conducted using PRinS3 (Prescience in silico Solution Suite) version V 2.1.0. The X-ESS application conducted molecular dynamics simulations on human lanosterol 14-alpha demethylase (PDB ID: 3LD6) and fungal lanosterol 14-alpha demethylase (PDB ID: 5V5Z). Subsequent *in vitro* investigations were conducted utilizing the fungal strains *Trichophyton rubrum, Microsporum canis and Epidermophyton floccosum* alongside typical pharmacological agents Ketoconazole and Itraconazole for comparative analysis. **Result:** Docking and molecular dynamics simulations demonstrated substantial binding energy against 3LD6 -8.25 kcal /mol & 5V5Z – 8.46 kcal/mol were found & form stable ligand protein complexes for both proteins.The *invitro* activity was carried out for Lansoperazole & MIC value was found to be 200 µg/ml against both dermatophytes *Trichophyton rubrum* , *Epidermophyton floccosum* & 150 µg/ml for *Microsporum canis* while standard drug Ketoconazole against *Trichophyton rubrum* show MIC value as 25 µg/ml Ketoconazole against *Epidermophyton floccosum* show MIC value 25 µg/ml .Ketoconazole and *Microsporum canis* 100 µg/ml. **Summary & Conclusion:** *In silico* and *in vitro* research indicate that Lansoperazole presents as a promising antifungal agent and serves as an alternate treatment for prevalent fungal strains.

Keywords Docking, Drug repurposing, Resistance, Anti-fungal drugs, MIC

Combination of Solid Dispersion and Potential Iontophoresis for Improving Transcorneal *In Vitro* Permeation 'Of Itraconazole

Swagatika Das[1], *Biswaranjan Mohanty*[2*], *Bibaswan Mishra*[3], *and Bhabani Shankar Nayak*[4*]

[1]Himalayan Pharmacy Institute, Mjhitar, Rangpo, East Sikkim, India
[2]School of Pharmacy, DRIEMS University, Tangi, Cuttack, Odisha, India
[3]Institute of Pharmacy and Technology, Salipur, Cuttack, Odisha, India
[4]KIIT School of Pharmacy, KIMS, KIIT Deemed to be University, Bhubaneswar, Odisha, India

The delivery of medicament to eye faces difficulties due to the complex structure and presence of different static and dynamic barriers of eye. **Aim:** The present investigates improvement of the aqueous solubility of a poorly soluble itraconazole as solid dispersion with soluplus and to study the effect of different preservatives and application of iontophoresis on transcorneal permeation. **Methods:** Itraconazole solid dispersion was prepared and characterized by FTIR spectroscopy, differential scanning calorimetry and X-ray powder diffraction. **Results:** FTIR study revealed absence of any significant interactions between Itraconazole and soluplus. The drug was in an amorphous state or as a solid solution in the prepared solid dispersion supported by DSC thermograms and P-XRD study. The aqueous solubility of Itraconazole was found as 386.02± 9.4µg/ml from solid dispersion of drug to soluplus ratio 1:10. In-vitro transcorneal permeation showed maximum percentage permeation of itraconazole from the formulation containing BA (10.93±0.14) followed by formulation containing MP+PP (7.94±0.26) and BKC+EDTA (6.81±0.19) after 2hrs. In iontophoretic study, increase in current intensity from 0.5 to 1.5 mA and iontophoresis time duration from 5 min to 15 min resulted in increased in drug permeation. Corneal hydration level was studied to assess the possible corneal damage due to electric treatment. The antifungal activity of the formulations against *Aspergillus flavus* determined by disk diffusion method suggested formulation containing BKC+EDTA showed the maximum zone of inhibition (14.33±0.58 mm) compared to other formulations. **Conclusion:** Hence combinational effect of iontophoresis and BKC+EDTA as preservative has significant role enhancing the corneal permeation and antifungal activity that will potentially enhance the bioavailability of itraconazole to eye.

Keywords Itraconazole; Solid dispersion; Transcorneal permeation; Iontophoresis; Antifungal activity

Obesity – A Review Article

Swathi. V[1]*, *Dr. Sailendra kumar Mahanta*[2], *Dr.J.N.Suresh Kumar*[3], *M.Harika*[4], *Sk.Nagurbi*[4], *S.Suchitra*[4], *V.Tharuni*[4],*and P.Kethana Kumari*[4]

[1]Associate Professor, Department of Pharmacology, Narasaraopeta Institute of Pharmaceutical Sciences, Narasaraopet, A.P., India.

[2]Department of Pharmacology, School of Pharmacy, The Assam Kaziranga University

[3]Professor and Principal, Department of Pharmaceutics, Narasaraopeta Institute of Pharmaceutical Sciences, Narasaraopet, A.P., India.

[4]Department of Pharmacy, Narasaraopeta Institute of Pharmaceutical Sciences, Narasaraopet, A.P., India.

Obesity is an excessive accumulation of body fat, often induced by a BMI of 30 or higher. It increases the risk of health issues like heart disease and diabetes and is influenced by genetics, environment, and lifestyle choices. Obesity an intricate, multifactorial medical illness that presents substantial dangers to general health and is characterized by an excessive buildup of body fat. Increased risk of many chronic illnesses such as type-2 diabetes, cardiovascular diseases, certain malignancies and musculoskeletal disorders, has been connected to the syndrome. The etiology of obesity beside behavioral, environmental and hereditary factors. Modern lifestyles, marked by increased sedentary behavior and high-calorie, low-nutrient diets, exacerbate the risk of developing obesity. Addressing obesity requires a multifaceted approach including lifestyle modifications, dietary changes, physical activity and in some cases, pharmacological or surgical interventions. Prevention strategies exploit good diet and regular exercise. Effective management and prevention of obesity are crucial for reducing its burden on individuals and healthcare systems globally

Author Index

A. K. Saxena, A. Saxena (Eds.): GTHTM 2025, pp. 553–554, 2026.
https://doi.org/10.1007/978-3-032-12320-6

MIX
Papier aus verantwortungsvollen Quellen
Paper from responsible sources
FSC® C105338
FSC
www.fsc.org

If you have any concerns about our products,
you can contact us on
ProductSafety@springernature.com

In case Publisher is established outside the EU,
the EU authorized representative is:
Springer Nature Customer Service Center GmbH
Europaplatz 3, 69115 Heidelberg, Germany

Printed by Libri Plureos GmbH
in Hamburg, Germany